Lecture Notes in Mathematics

Volume 2313

This series reports on new developments in all areas of mathematics and their applications - quickly, informally and at a high level. Mathematical texts analysing new developments in modelling and numerical simulation are welcome. The type of material considered for publication includes:

1. Research monographs
2. Lectures on a new field or presentations of a new angle in a classical field
3. Summer schools and intensive courses on topics of current research.

Texts which are out of print but still in demand may also be considered if they fall within these categories. The timeliness of a manuscript is sometimes more important than its form, which may be preliminary or tentative.

Titles from this series are indexed by Scopus, Web of Science, Mathematical Reviews, and zbMATH.

Jean-Michel Morel • Bernard Teissier

Editors

Mathematics Going Forward

Collected Mathematical Brushstrokes

 Springer

Editors

Jean-Michel Morel
Ecole Normale Supérieure Paris-Saclay
Paris, France

Bernard Teissier
Université Paris Cité and Sorbonne
Université
CNRS, IMJ-PRG
Paris, France

ISSN 0075-8434   ISSN 1617-9692 (electronic)
Lecture Notes in Mathematics
ISBN 978-3-031-12243-9   ISBN 978-3-031-12244-6 (eBook)
https://doi.org/10.1007/978-3-031-12244-6

Mathematics Subject Classification: 00A05, 00A30, 00A69, 00A71, 00A99, 01A99

This Springer imprint is published by the registered company Springer Nature Switzerland AG
The registered company address is: Gewerbestrasse 11, 6330 Cham, Switzerland

Preface

Upon learning of Catriona Byrne's retirement from Springer's mathematics staff, we needed little convincing to suggest this commemorative volume. As the title suggests, the theme of this volume is the future of mathematics. This is meant in a broad sense: the contributions of the volume range from musings on the future of a particular field to problems for future research to new results closing one door and opening another. And for obvious reasons, this volume does not respect any of the rules of the series for multi-author volumes.

The contributing authors, 44 distinguished mathematicians, have embraced this challenge beautifully. Eighteen have presented a historical perspective of their field, in five cases complemented by problems and conjectures. Eleven have oriented reviews of their subject toward meaningful unsolved problems. There are also five articles on the history of mathematics, four papers presenting conjectures and nine original mathematical contributions.

The very diverse mathematical topics of this volume have been organized alphabetically into 12 parts. Each part is introduced by a teaser page highlighting some features of the articles, inspired by comments of the reviewers and LNM editors.

Most of the contributions begin with heartfelt tributes to Catriona Byrne. Over the past four decades, she has played a very important role in mathematics publishing. We dedicate this volume to her.

On behalf of the editorial board of the Lecture Notes in Mathematics,

Paris, France   Jean-Michel Morel
Paris, France   Bernard Teissier

Contents

Letter from Jean-Pierre Eckmann

Jean-Pierre Eckmann

Dear Catriona,

It is somewhat sad that I should already be writing for your retirement, but I hope we will stay in contact for a long time. Of course, the idea to honor you not just anywhere, but precisely in the Lecture Notes in Mathematics (LNM), marks for me a very special occasion. After all, Springer, the Lecture Notes, the Grundlehren and you, make for a special event for both you and me, as for a family. Indeed, my late father, Beno Eckmann, has had a lifelong contact with Springer. When he and Albrecht Dold founded the LNM (in 1964) you must have just started primary school. But apparently, it must have influenced your professional choice as a mathematician, and after your PhD in 1982 (or actually before it in 1981?) you started to work for Springer. With your well-known drive and energy, you began to interact with my father, who was very fond of you. And thus a "family-tradition" started. Of course, Beno had earlier, important, contacts with the Springer family, with Heinz Götze and Joachim Heinze. Joachim started in 1980, but you joined in 1981. At least, I saw your signature in a special edition of the LNM Volume 1000 (with a contribution by Heinz Hopf (these were unpublished lectures he gave in America after the war)), which my father got as a gift from Springer.

Soon after that, we two must have started to interact. Anyway, we exchanged worries and ideas over the next 40(!) years that you have been with Springer. Of course, as things go these days, Springer has changed owners and policy several

J.-P. Eckmann (✉)
University of Geneva, Geneva, Switzerland
e-mail: Jean-Pierre.Eckmann@unige.ch

J.-M. Morel, B. Teissier (eds.), *Mathematics Going Forward*, Lecture Notes in Mathematics 2313, https://doi.org/10.1007/978-3-031-12244-6_1

1

times, and has now found–hopefully for a long time–some peace in the Nature publishing group. All this was compounded by a profound revolution in the printing and writing business, around the turn of the millenium, about which we had many "philosophical" discussions. But, as they say in German, you, Catriona, stood "wie ein Fels in der Brandung" (Fig. 1).

You must have listened to thousands of talks in your 40 years at Springer, but I found one (https://www.canalc2.tv/video/12658) where you were the speaker! The occasion was

Fig. 1 Video stills from Catriona's 2014 talk

ABES - Agence bibliographique de l'enseignement supérieur
Journées ABES 2014
Du 20-21 mai 2014
Le Corum, MONTPELLIER

and your talk

Les archives numériques :
quelle place dans le processus de recherche d'aujourd'hui?
Dr. Catriona Byrne
Editorial Director, Mathematics, Springer

Naturally, in that talk you described our favorite series, the LNM, and of course, you spoke about Kato's Grundlehren Evergreen *Perturbation Theory for Linear Operators*—this pleased me as the 1966 edition of Kato's book was my "livre de chevet" when I was a student.

You regularly came to Zürich. I add here a photo from 2004 where you visited the Forschungsinstitut für Mathematics ETH (FIM) for its 40th anniversary in 2004 (Fig. 2).

Fig. 2 Catriona and Beno Eckmann, at the 40th anniversary of FIM, 2004

We sometimes talked about real mathematics, and this was always extremely helpful to me: While I was doing my thing at home, you visited many events, and had a clear view of what was going on. I profited enormously from your feeling for new developments in Mathematics. No wonder, since you have handled such a long list of books and journals (Fig. 3):

Dr. Catriona Byrne is responsible for the following book series and journals:

▼ Topics

    all areas of core mathematics; probability theory, quantitative finance, actuarial mathematics, mathematical biosciences, mathematical physics

▼ Book Series (English)

    Stochastic Modelling & Applied Probability
    Encyclopaedia of Math. Sciences, Probability Theory subseries
    Grundlehren der mathematischen Wissenschaften
    Grundlehren Text Editions
    Lecture Notes in Mathematics
    Springer Finance

▼ Other Books

    Collected/Selected Papers

▼ Journals

    Afrika matematika
    Annales mathématiques du Québec
    Annali di Matematica Pura ed Applicata
    Finance and Stochastics
    Inventiones mathematicae
    Jahresbericht der DMV
    Manuscripta mathematica
    Mathematics and Financial Economics
    Mathematische Annalen
    Mathematische Zeitschrift
    Probability Theory and Related Fields
    Publications mathématiques de l'IHES

Fig. 3 Catriona's domain of responsibility (according to Springer), probably there has been much more

Even if you will have to leave these tasks at Springer, I sincerely hope I will have a chance to learn more from you in the future. Thanks for everything, and let's keep in touch.

Jean-Pierre

Part I
Algebraic Geometry

Jean-Louis Colliot-Thélène's article *Une liste de problèmes* is a richly commented list of deep problems close to his heart. They center around rationality questions: finding rational points on algebraic varieties, finding rational curves linking two points, finding zero cycles of degree one on varieties over a global field, determining whether given algebraic varieties are rational or unirational, and much more.

Bernard Teissier's article *Some ideas in need of clarification in resolution of singularities and the geometry of discriminants* presents problems in two different areas of algebraic geometry. The first is resolution of singularities, where Teissier explains problems related to a program for local uniformization of valuations through a relation with toric geometry in possibly infinite dimensions. The second part proposes to clarify the relation between the geometry of the discriminant of miniversal unfoldings and the movements in the space of Morse functions used in cobordism theory.

Pierre Schapira's article *Shifted sheaves for space-time* begins with a condensed history of the evolution from Mikio Sato's microlocal analysis and Kashiwara's index theorem to the microlocal theory of sheaves and its connection with Lagrangian geometry through microsupports. Then Schapira proposes an unexpected connection between the "shift" in derived categories and a simplified model of the "Big Bang". While this text does not strictly belong to algebraic geometry, it is close to it in spirit.

Loring Tu's article *Lefschetz fixed point theorems for correspondences* is a very instructive exploration of the consequences of an inspiring conjecture made by Shimura at the 1964 Woods Hole Conference on algebraic geometry concerning a generalization of the Lefschetz fixed point theorem to smooth and also holomorphic correspondences on varieties.

Une liste de problèmes

Jean-Louis Colliot-Thélène

INTRODUCTION

Dans cette note, je rassemble une liste de problèmes, la plupart bien connus, et restés ouverts depuis de nombreuses années. J'ai réfléchi à la plupart d'entre eux mais n'en revendique pas la propriété. Je mentionne certaines solutions partielles, sans faire un rapport systématique. Je renvoie à [CT87, CT98, Sk01, CT03, CT11, SD11, CT19, W18] pour cela.

Les problèmes portent presque tous sur la généralisation en dimension plus grande que 1 des deux énoncés suivants :

Une conique lisse sur un corps k qui possède un point rationnel sur k est isomorphe, sur k, à la droite projective $\mathbb{P}^1_k$. Ceci donne une paramétrisation biunivoque des points rationnels de la conique par les points rationnels de la droite projective.

Sur un corps de nombres k, si une conique lisse a un point rationnel sur tous les complétés k_v de k, alors elle a un point rationnel sur k. Grâce à Hensel, ce critère est effectif.

Le premier énoncé remonte à l'Antiquité, le second fut établi par Legendre sur les rationnels et par Hilbert sur les corps de nombres, et fut étendu par Minkowski et par Hasse aux quadriques de dimension quelconque.

Beaucoup des problèmes mentionnés ici ont leur source dans mes travaux avec Jean-Jacques Sansuc et avec Peter Swinnerton-Dyer dans les années 1970 et 1980. Certains des problèmes, en particulier ceux sur les intersections de deux quadriques, avaient fait l'objet de rapports non publiés en 1988 et en 2005.

J.-L. Colliot-Thélène

Université Paris-Saclay, CNRS, Laboratoire de mathématiques d'Orsay, 91405, Orsay, France
jean-louis.colliot-thelene@universite-paris-saclay.fr

Date: 10 décembre 2022.

J.-M. Morel, B. Teissier (eds.), *Mathematics Going Forward*, Lecture Notes in Mathematics 2313, https://doi.org/10.1007/978-3-031-12244-6_2

Ich freue mich, an dieser Festschrift zu Ehren von Dr. Catriona Byrne teil-
zunehmen. Für die Einladung, und die Möglichkeit, meine Lieblingsprobleme
zusammenzubringen, bedanke ich mich bei Jean-Michel Morel und Bernard
Teissier, geschäftsführenden Herausgebern von den Springer Lecture Notes in
Mathematics. Mit dem Springer-Verlag habe ich eine spezielle Beziehung. In
den 70er Jahren habe ich oft Bücher am Heidelberger Platz in dem damaligen
West-Berlin gekauft, wie etwa das Buch "Algebraic threefolds with special regards to problems of rationality" von L. Roth (1955). Über die Jahre habe
ich einige Artikel, oft auf Französisch, einmal auf Deutsch, in Springer Zeit-
schriften veröffentlicht. Für die Springer Lecture Notes in Mathematics bin
ich Kunde, Autor, Herausgeber, Gutacher gewesen. Dabei bin ich mehrmals
mit Frau Byrne in Kontakt gewesen.

1. Variétés rationnelles et variétés proches

Soit k un corps algébriquement clos. On dit qu'une variété intègre X sur
k est *rationnelle* si elle est birationnelle à un espace projectif $\mathbb{P}^d_k$, i.e. si son
corps des fonctions $k(X)$ est transcendant pur sur k.

Parmi les exemples classiques de variétés rationnelles, on trouve les variétés
sous-jacentes à un groupe algébrique linéaire connexe, et les variétés projectives qui sont des espaces homogènes de tels groupes. Les quadriques lisses
de dimension au moins 1 rentrent dans ce cadre.

On dit qu'une variété intègre X sur k est *unirationnelle* s'il existe une
application rationnelle dominante d'un espace projectif vers X.

En dimension 1 et en caractéristique zéro en dimension 2, unirationalité implique rationalité. C'est faux dès la dimension 3 (Clemens–Griffiths,
Iskovskikh–Manin, Artin–Mumford).

Parmi les exemples classiques de variétés unirationnelles, on trouve les quotients G/H d'un groupe linéaire connexe G par un sous-groupe fermé H quelconque, non nécessairement connexe. Pour H fini, on connaît des exemples
de tels quotients qui ne sont pas rationnels (Saltman, Bogomolov).

Dans la classification birationnelle des variétés de dimension supérieure
développée vers 1990 (travaux de Kollár, Miyaoka, Mori), ce qui en dimension quelconque joue le rôle des surfaces rationnelles dans la classification des
surfaces, ce sont les variétés rationnellement connexes [K99, AK03]. L'une des
définitions, en caractéristique zéro, est que par deux points (fermés) généraux
d'une telle variété il passe une courbe de genre zéro, et ce sur tout corps
algébriquement clos contenant k. En caractéristique quelconque, la bonne
définition est celle de variété séparablement rationnellement connexe. Les
deux notions coïncident en caractéristique nulle. Dans la suite de ce texte,
par variété rationnellement connexe on entendra variété séparablement rationnellement connexe.

Supposons k de caractéristique zéro. Une variété unirationnelle est rationnellement connexe. La réciproque est un grand problème ouvert. Une
variété lisse, projective, lisse, connexe, à fibré anticanonique ample est appelée
variété de Fano. Un théorème important (Campana, Kollár–Miyaoka–Mori)

dit qu'une variété de Fano est rationnellement connexe. Ainsi toute hypersurface lisse $X \subset \mathbb{P}_k^n$, $n \geqslant 2$, de degré d avec $d \leqslant n$ est rationnellement connexe. Il en est donc ainsi des hypersurfaces cubiques lisses dans $\mathbb{P}_k^n, n \geqslant 3$. Il en est aussi ainsi des intersections complètes lisses de deux quadriques dans $\mathbb{P}_k^n$, $n \geqslant 4$. Ces dernières sont des variétés rationnelles sur le corps algébriquement clos k.

Un autre théorème important (Graber–Harris–Starr, 2003) dit que l'espace total d'une fibration de base rationnellement connexe et de fibres générales rationnellement connexes est rationnellement connexe.

Soient maintenant k un corps quelconque et $\overline{k}$ une clôture algébrique. Nous adoptons ici les conventions suivantes.

On dit qu'une k-variété géométriquement intègre X est rationnelle, resp. rationnellement connexe, si la $\overline{k}$-variété $\overline{X} := X \times_k \overline{k}$ est rationnelle, resp. rationnellement connexe.

On dit qu'une k-variété géométriquement intègre X est k-rationnelle si X est k-birationnelle à $\mathbb{P}_k^d$, i.e. le corps des fonctions $k(X)$ de X est transcendant pur.

On dit qu'une k-variété géométriquement intègre X est stablement k-rationnelle s'il existe des entiers $n \geqslant 0$ et $m \geqslant 0$ tels que $X \times_k \mathbb{P}_k^n$ est k-birationnelle à $\mathbb{P}_k^m$.

On dit qu'une k-variété géométriquement intègre X de dimension d est k-unirationnelle s'il existe une application k-rationnelle dominante d'un espace projectif $\mathbb{P}_k^d$ vers X.

2. Principe de Hasse, approximation faible, obstruction de Brauer–Manin

Étant donnée une variété algébrique X sur un corps k, on note $X(k)$ l'ensemble de ses points rationnels. Pour K/k une extension quelconque de corps, on note $X(K)$ l'ensemble des points rationnels sur K.

On veut donner des critères si possible effectifs permettant de décider si une k-variété donnée X possède un point rationnel.

Pour éviter des répétitions, on va définir un certain nombre de propriétés.

($\mathbf{PR}_X$) L'ensemble $X(k)$ est non vide, i.e. la k-variété possède un point rationnel sur k.

On s'intéresse particulièrement au cas des corps finis, des corps locaux (les corps p-adiques, les corps de séries formelles en une variable sur un corps fini, le corps $\mathbb{R}$ des réels, le corps $\mathbb{C}$ des complexes), et des corps globaux (corps de nombres ou corps de fonctions d'une variable sur un corps fini). Etant donné un corps global k, et une place v de ce corps, on note k_v le corps local complété par rapport à la place v.

Soit désormais k un corps global.

À toute variété algébrique X sur k on associe l'espace $X(\mathbb{A}_k)$ de ses adèles. C'est un sous-ensemble du produit $\prod_v X(k_v)$, non vide si ce produit est non vide. L'espace $X(\mathbb{A}_k)$ est muni d'une topologie naturelle. Si X est projective, alors on a $X(\mathbb{A}_k) = \prod_v X(k_v)$, et la topologie de l'espace des adèles coïncide avec la topologie produit sur $\prod_v X(k_v)$.

On introduit la propriété :

$(\mathbf{PH}_X)$ Soit on a $X(\mathbb{A}_k) = \varnothing$, soit on a $X(k) \neq \varnothing$.

On dit que le principe de Hasse vaut pour une classe de variétés algébriques définies sur k si, pour toute variété X dans cette classe, on a la propriété $\mathbf{PH}_X$.

Pour X/k lisse et géométriquement intègre, on introduit la propriété d'approximation faible :

$(\mathbf{AF}_X)$ L'image de l'application diagonale $X(k) \to \prod_v X(k_v)$ est dense.

Cette propriété implique $\mathbf{PH}_X$. Si $X(k)$ est non vide, elle équivaut au fait que pour tout ensemble fini S de places de k, l'ensemble $X(k)$ est dense dans le produit fini $\prod_{v \in S} X(k_v)$. Il convient de noter que pour certaines classes de variétés, on ne sait pas établir $\mathbf{PH}_X$ pour X dans cette classe, mais que, sous l'hypothèse $X(k) \neq \varnothing$, la propriété $\mathbf{AF}_X$ est facile à établir.

Pour X/k non nécessairement projective, on peut encore considérer une variante de la propriété $\mathbf{AF}_X$. Il s'agit du problème de l'approximation forte. Depuis 2008, il a été étudié du point de vue de l'obstruction de Brauer–Manin, dans plusieurs articles par Fei Xu et moi, Harari, Borovoi, Demarche, Dasheng Wei, Yang Cao, mais nous ne le discuterons pas dans ce texte. Je renvoie à [BD13] et au rapport de Wittenberg [W18, §2.7, §3.2.4, §3.3.4, §3.4.5] pour des résultats et références.

Pour X/k lisse et géométriquement intègre, avec $X(k) \neq \varnothing$, il y a lieu d'introduire la propriété d'approximation "faible faible" [Se92, Chap. 3]) :

$(\mathbf{AFF}_X)$ Il existe un ensemble fini $T = T(X)$ de places de k tel que, pour tout ensemble fini S de places ne rencontrant pas T, l'image de l'application diagonale $X(k) \to \prod_{v \in S} X(k_v)$ est dense.

À tout corps k, à toute variété X sur un corps k, et plus généralement à tout schéma X, on associe son groupe de Brauer–Grothendieck $\mathrm{Br}(X)$ [CTSk21].

Pour k un corps global, la théorie du corps de classes donne des plongements $j_v : \mathrm{Br}(k_v) \hookrightarrow \mathbb{Q}/\mathbb{Z}$, et une suite exacte fondamentale

$$0 \to \mathrm{Br}(k) \to \oplus_v \mathrm{Br}(k_v) \to \mathbb{Q}/\mathbb{Z} \to 0$$

qui généralise la loi de réciprocité quadratique de Gauß.

Etant donnée une variété X sur un corps global k, en utilisant la fonctorialité du groupe de Brauer, les applications $j_v : \mathrm{Br}(k_v) \hookrightarrow \mathbb{Q}/\mathbb{Z}$ induisent un accouplement

$$X(\mathbb{A}_k) \times \mathrm{Br}(X) \to \mathbb{Q}/\mathbb{Z}$$

envoyant un couple $(\{M_v\}, \alpha)$ sur $\sum_v j_v(\alpha(M_v))$. On note $X(\mathbb{A}_k)^{\mathrm{Br}} \subset X(\mathbb{A}_k)$ le noyau à gauche de cet accouplement. Comme remarqué par Manin en 1970, l'application diagonale $X(k) \to X(\mathbb{A}_k)$ induit une inclusion

$$X(k) \subset X(\mathbb{A}_k)^{\mathrm{Br}}.$$

Considérons la propriété :

($\mathbf{BMPH}_X$) Soit on a $X(\mathbb{A}_k)^{\mathrm{Br}} = \varnothing$, soit on a $X(k) \neq \varnothing$.

On dit que l'obstruction de Brauer-Manin au principe de Hasse est la seule pour une classe de variétés algébriques définies sur k si, pour toute variété X dans cette classe, on a la propriété $\mathbf{BMPH}_X$.

Pour X/k projective, on dit que l'obstruction de Brauer-Manin à l'approximation faible est la seule pour X si l'on a :

($\mathbf{BMAF}_X$) L'ensemble $X(k)$ est dense dans $X(\mathbb{A}_k)^{\mathrm{Br}}$.

Cette propriété implique $\mathbf{BMPH}_X$. Il y a une variante où, pour $k_v = \mathbb{R}$ et $k_v = \mathbb{C}$, on remplace $X(k_v)$ par l'ensemble de ses composantes connexes.

Pour les variétés projectives, lisses, géométriquement intègres qui sont rationnellement connexes, en particulier celles qui sont géométriquement unirationnelles, la propriété $\mathbf{BMAF}_X$ implique la propriété d'approximation faible faible $\mathbf{AFF}_X$. Ceci résulte du fait que dans ce cas le quotient $\mathrm{Br}(X)/\mathrm{Br}(k)$ est fini.

Pour les k-variétés projectives et lisses géométriquement intègres, chacune des propriétés définies ci-dessus ne dépend que du corps des fonctions de X : si X et Y sont deux telles k-variétés birationnellement équivalentes, l'une des propriétés vaut pour X si et seulement si elle vaut pour Y.

Si X et Y sont deux k-variétés, et $Z = X \times_k Y$, on a $Z(k) = X(k) \times Y(k)$.
Si $f : X \to Y$ est un k-morphisme, il induit une application $X(\mathbb{A}_k)^{\mathrm{Br}} \to Y(\mathbb{A}_k)^{\mathrm{Br}}$.
Si X, Y sont deux variétés projectives et lisses géométriquement intègres sur un corps de nombres k, et $Z = X \times_k Y$, c'est un résultat de Skorobogatov et Zarhin que l'on a

$$Z(\mathbb{A}_k)^{\mathrm{Br}} = X(\mathbb{A}_k)^{\mathrm{Br}} \times Y(\mathbb{A}_k)^{\mathrm{Br}}.$$

3. POINTS RATIONNELS DES VARIÉTÉS RATIONNELLEMENT CONNEXES SUR UN CORPS GLOBAL

La conjecture suivante fut faite par Sansuc et moi en 1979 pour les surfaces géométriquement rationnelles [CTSa80], et étendue aux variétés rationnellement connexes en toute dimension en 1999 (voir [CT03]).

Conjecture 3.1. *L'obstruction de Brauer-Manin au principe de Hasse et à l'approximation faible pour les points rationnels est la seule obstruction pour les variétés projectives, lisses, rationnellement connexes sur un corps global.*

Avec les notations ci-dessus, ceci dit que pour toute variété X projective, lisse, rationnellement connexe sur un corps global, on a $\mathbf{BMPH}_X$ et $\mathbf{BMAF}_X$.

3.1. Espaces homogènes de groupes algébriques linéaires connexes.
Soit k un corps de nombres.

Pour G un k-groupe semisimple simplement connexe, E un espace principal homogène de G, et X une k-compactification lisse de E, des travaux de Eichler, Kneser, Harder et Tchernousov établirent $\mathbf{PH}_X$. Leurs travaux, et ceux de Platonov, établirent $\mathbf{AF}_X$. Leurs travaux établissent aussi $\mathbf{PH}_X$ et $\mathbf{AF}_X$ pour les variétés projectives X qui sont espaces homogènes d'un groupe algébrique linéaire connexe G.

Pour G un k-groupe algébrique linéaire connexe quelconque, E un espace principal homogène de G, et X une k-compactification lisse de E, on a les propriétés $\mathbf{BMPH}_X$ et $\mathbf{BMAF}_X$ (Voskresenskiĭ pour les tores, Sansuc en général), et donc aussi l'approximation faible faible $\mathbf{AFF}_X$. Ceci vaut aussi si E est un espace homogène de G linéaire connexe lorsque les stabilisateurs géométriques sont connexes (Borovoi).

Par contre, la question suivante est en général ouverte.

Problème 3.1. *Soit G un k-groupe linéaire connexe, $H \subset G$ un k-sous-groupe fini. Soit X une k-compactification lisse du quotient G/H. A-t-on la propriété $\mathbf{BMAF}_X$, ou du moins la propriété $\mathbf{AFF}_X$?*

Comme remarqué par T. Ekedahl et moi en 1988 (voir [Se92, Chap. 3]), une réponse positive pour $\mathbf{AFF}_X$, appliquée à un groupe fini abstrait H plongé dans $GL_{n,k}$ pour n entier convenable, implique que le groupe fini H est le groupe de Galois d'une extension galoisienne finie de corps K/k, propriété qu'on ne sait pas établir pour tous les groupes finis.

Un progrès récent dans cette direction a été accompli par Harpaz et Wittenberg [HW20] pour une classe de groupes finis H comprenant les groupes nilpotents (constants), ce qui leur permet, pour ces groupes, de retrouver et préciser, du point de vue du comportement local, le théorème de Shafarevich que les groupes finis résolubles sont des groupes de Galois sur tout corps de nombres (on trouve la démonstration de ce théorème de Shafarevich dans des ouvrages de Ishkhanov, Lur'e, Faddeev et de Neukirch, Schmidt, Wingberg).

3.2. Surfaces de del Pezzo et variétés de Fano.
Les surfaces de del Pezzo sont les variétés de Fano de dimension 2.

Problème 3.2. *Soit k un corps global. Soit $X \subset \mathbb{P}^n_k$ une intersection complète lisse définie par l'annulation simultanée de formes homogènes $(f_1, \ldots, f_r)$ de degrés respectifs $(d_1, \ldots, d_r)$. Si X est dimension au moins 3 et l'on a $d_1 + \cdots + d_r \leqslant n$, a-t-on $\mathbf{PH}_X$? A-t-on $\mathbf{AF}_X$?*

Soit k un corps de nombres. La méthode du cercle permet d'établir de tels énoncés pour n grand par rapport à la somme des d_i (Birch 1961, Schmidt 1985, Skinner 1997).

C'est une expérience commune que pour les variétés de Fano il est difficile d'exhiber des contre-exemples au principe de Hasse. On consultera [B18] pour un rapport sur cette direction de recherche très active. Sur $k = \mathbb{Q}$, Browning, Le Boudec et Sawin [BLBS20] ont récemment montré que, si l'on ordonne (toutes) les hypersurfaces lisses $X \subset \mathbb{P}^n_{\mathbb{Q}}$ avec $d \leqslant n$ et $n \geqslant 4$ par la hauteur des coefficients, alors 100 % d'entre elles satisfont le principe de Hasse, et une proportion positive a des points rationnels. Pour d'autres résultats "statistiques", on consultera [BBL16, LS16, L18, Bri18, SkSo20].

Considérons maintenant des cas particuliers du problème 3.2.

Problème 3.3. *Soit $n \geqslant 4$ et $X \subset \mathbb{P}^n_k$ une hypersurface cubique lisse sur un corps de nombres. A-t-on* $\mathbf{PH}_X$ *? A-t-on* $\mathbf{AF}_X$ *?*

Si X contient une droite rationnelle $\mathbb{P}^1_k \subset \mathbb{P}^n_k$, alors on a $\mathbf{AF}_X$ [Har94].

Pour $k = \mathbb{Q}$, et $n \geqslant 9$, Heath-Brown utilisa la méthode du cercle pour établir $X(\mathbb{Q}) \neq \varnothing$ pour toute hypersurface cubique lisse, et C. Hooley établit le principe de Hasse pour X lisse dans le cas $n = 8$.

Sur un corps global k de caractéristique $p > 5$, pour $n \geqslant 5$, Zhiyu Tian [T17] a établi $\mathbf{PH}_X$.

Pour $X \subset \mathbb{P}^n_k$ intersection complète lisse sur un corps de caractéristique zéro, on a $\mathrm{Br}(X)/\mathrm{Br}(k) = 0$ si X est de dimension au moins 3. Dans ce cas, sur un corps de nombres, on a donc $X(\mathbb{A}_k)^{\mathrm{Br}} = X(\mathbb{A}_k)$. En dimension 2, par exemple pour les surfaces cubiques et les intersections de deux quadriques dans $\mathbb{P}^4_k$, ce n'est plus nécessairement le cas, il faut tenir compte de l'obstruction de Brauer–Manin.

Problème 3.4. *Soit $X \subset \mathbb{P}^3_k$ une surface cubique lisse sur un corps de nombres. A-t-on* $\mathbf{BMPH}_X$ *? A-t-on* $\mathbf{BMAF}_X$ *?*

Le cas des surfaces diagonales $X \subset \mathbb{P}^3_{\mathbb{Q}}$, d'équation $ax^3 + by^3 + cz^3 + dt^3 = 0$, avec a, b, c, d entiers non nuls, sans facteur cubique, et premiers entre eux dans leur ensemble, a été testé. On sait (Cassels-Guy 1966) que $\mathbf{PH}_X$ ne vaut pas toujours pour ces surfaces, mais dans [CTKaS87] on montra que $\mathbf{BMPH}_X$ vaut lorsque les coefficients sont de valeur absolue plus petite que 100. On a un résultat conditionnel, dû à Swinnerton-Dyer [SD01]. On suppose la finitude des groupes de Tate-Shafarevich des courbes elliptiques sur les corps de nombres. S'il existe un nombre premier $p \neq 3$ qui divise a mais pas bcd, et un nombre premier $q \neq 3$ qui divise b mais pas acd, alors le principe de Hasse vaut pour X, et ce résultat conditionnel implique $\mathbf{PH}_X$ pour toute hypersurface cubique diagonale $X \subset \mathbb{P}^n_{\mathbb{Q}}$ pour $n \geqslant 4$.

Problème 3.5. *Soit $X \subset \mathbb{P}^4_k$ une intersection complète lisse de deux quadriques sur un corps de nombres k. A-t-on* $\mathbf{BMPH}_X$ *?*

On sait que cela vaut si X contient une conique [Sal88, CT90].

Par ailleurs, sous l'hypothèse $X(k) \neq \varnothing$, on a $\mathbf{BMAF}_X$ [SaSk91].

Problème 3.6. *Soit $n \geqslant 5$. Soit $X \subset \mathbb{P}^n_k$ une intersection complète lisse de deux quadriques sur un corps de nombres k. A-t-on $\mathbf{PH}_X$?*

Il est facile de montrer que sous l'hypothèse $X(k) \neq \varnothing$, on a $\mathbf{AF}_X$ [CTSaSD87].

On sait que $\mathbf{PH}_X$ vaut si X contient un ensemble de deux droites conjuguées [CTSaSD87] ou si X contient une conique (Salberger, 1993, non publié).

On sait que $\mathbf{PH}_X$ vaut pour $n \geqslant 8$ [CTSaSD87] et $n = 7$ [HB18].

Sur un corps global de caractéristique $p > 2$, $\mathbf{PH}_X$ été établi pour $n \geqslant 5$ par des méthodes géométriques de déformation par Zhiyu Tian [T17].

Sur tout corps de nombres, Wittenberg [W07] a donné une preuve conditionnelle de $\mathbf{PH}_X$ pour $n \geqslant 5$. Voir la section 3.4 ci-dessous.

3.3. Espaces totaux de fibrations en variétés rationnellement connexes au-dessus de la droite projective.

C'est une classe naturelle de variétés à considérer si l'on veut établir les résultats par récurrence sur la dimension.

Problème 3.7. *Soit X une variété projective et lisse sur un corps de nombres k, munie d'un morphisme $X \to \mathbb{P}^1_k$ dont la fibre générique est rationnellement connexe, et dont les fibres lisses X_m au-dessus des k-points $m \in \mathbb{P}^1(k)$ satisfont $\mathbf{BMHP}_{X_m}$, resp. $\mathbf{BMAF}_{X_m}$. A-t-on $\mathbf{BMHP}_X$, resp. $\mathbf{BMAF}_X$?*

Depuis [CTSaSD87], ce thème a été beaucoup exploré : travaux de Skorobogatov, Harari, Wittenberg, Harpaz, et de nombreux autres auteurs. Je renvoie à [W18] pour des références détaillées.

L'hypothèse sur les fibres est par exemple satisfaite si la fibre générique X_η sur le corps $K = k(\mathbb{P}^1)$ est une compactification lisse d'un espace homogène d'un K-groupe linéaire connexe, à stabilisateurs géométriques connexes (Borovoi).

À une telle fibration on associe une mesure de sa complexité arithmétique : la somme ρ des degrés $[k(m) : k]$ des points fermés $m \in \mathbb{P}^1_k$ dont la fibre $X_m/k(m)$ est non lisse et ne contient pas de composante géométriquement intègre de multiplicité 1.

On a des réponses positives inconditionnelles au problème 3.7 lorsque ρ est (très) petit. Le meilleur résultat général récent est $\rho \leqslant 3$ [HWW21]. Pour ρ quelconque, on a une réponse conditionnelle positive [HW16, HWW21] si l'on accepte une conjecture difficile du type de l'hypothèse de Schinzel. Cette hypothèse, aussi considérée par Bouniakovsky, Dickson, Hardy et Littlewood, Bateman et Horn, affirme que, pour toute famille finie $P_i(t) \in \mathbb{Z}[t]$ de polynômes irréductibles, à coefficients dominants positifs, tels qu'aucun nombre premier ne divise $\prod_i P_i(m)$ pour tout entier m, il existe une infinité d'entiers n tels que chaque $P_i(n)$ soit un nombre premier. L'idée d'utiliser l'hypothèse de Schinzel dans ce cadre remonte à 1979, et a été poursuivie dans divers articles. Elle vient de connaître un rebondissement statistique "inconditionnel" [SkSo20].

Un cas simple est donné par une famille de coniques, d'équation affine

$$y^2 - a(t)z^2 - b(t) = 0,$$

avec $a(t)$ et $b(t)$ polynômes de degrés quelconques. Les fibres de la projection sur l'axe des t satisfont le principe de Hasse. Ici $\rho \leqslant 5$ convient.

Depuis [CTHaSk03] on a aussi beaucoup étudié les équations du type

$$\mathrm{Norm}_{K/k}(\Xi) = P(t)$$

avec Ξ "variable" dans une extension finie K/k et $P(t) \in k[t]$ polynôme non nul. Pour K/k quelconque, les fibres ne satisfont pas en général le principe de Hasse mais elles satisfont la variante avec obstruction de Brauer-Manin.

Sur $k = \mathbb{Q}$, des progrès fondamentaux en combinatoire additive (Green, Tao, Ziegler; Mathiesen) ont permis d'obtenir des résultats inconditionnels avec ρ quelconque. Les résultats de Green, Tao, Ziegler donnent une version de l'hypothèse de Schinzel pour une famille finie de formes linéaires à deux variables sur $\mathbb{Q}$. Pour les exemples de variétés ci-dessus, [BMSk14, HW16] montrent ainsi que lorsque $k = \mathbb{Q}$ et que, dans les équations ci-dessus, le polynôme $a(t)b(t)$, resp. le polynôme $P(t)$, a toutes ses racines dans $\mathbb{Q}$, alors on a $\mathbf{BMAF}_X$ (où X désigne un modèle projectif et lisse des variétés considérées).

3.4. Au-delà des variétés rationnellement connexes. Soit k un corps de nombres. On ne saurait étendre la conjecture 3.1 à toutes les variétés projectives et lisses sur k, comme ce fut montré inconditionnellement par Skorobogatov en 1999. D'autres contre-exemples géométriquement plus simples ont depuis été donnés. Cependant, pour X espace principal homogène d'une variété abélienne A, si l'on ignore la composante connexe de l'élément neutre aux places archimédiennes, $\mathbf{BMAF}_X$ résulte de la finitude conjecturelle du groupe de Tate-Shafarevich de la variété abélienne A.

Skorobogatov (2001) conjecture $\mathbf{BMAF}_X$ pour toute surface X de type $K3$. Si l'on est prêt à utiliser non seulement la finitude des groupes de Tate-Shafarevich mais aussi l'hypothèse de Schinzel, alors une méthode sophistiquée initiée par Swinnerton-Dyer en 1993 permet de prédire un énoncé de type $\mathbf{BMHP}_X$ pour certaines surfaces X fibrées en courbes de genre 1 au-dessus de la droite projective. Parmi ces surfaces, on trouve des surfaces birationnelles à des intersections lisses de deux quadriques dans $\mathbb{P}^4$, mais aussi des surfaces $K3$. La méthode fut développée dans [CTSkSD98, W07]. Sous les dites conjectures, Wittenberg [W07] établit ainsi $\mathbf{PH}_X$ pour toute intersection complète lisse $X \subset \mathbb{P}^n_k$ pour $n \geqslant 5$.

La méthode de [SD01], qui n'utilise "que" l'hypothèse de finitude des groupes de Tate-Shafarevich, a été appliquée par Skorobogatov et Swinnerton-Dyer, et aussi par Harpaz et Skorobogatov [HS16], pour étudier le principe de Hasse pour certaines surfaces de Kummer.

4. Zéro-cycles des variétés sur un corps global

Soit X une variété algébrique sur un corps k. L'indice $I(X)$ de la k-variété X est par définition le pgcd des degrés $[k(P) : k]$ pour tous les points fermés P. C'est aussi le pgcd des degrés des extensions finies K/k telles que $X(K) \neq \varnothing$. Une question plus faible que l'existence d'un point rationnel sur X est celle si l'indice $I(X) = 1$.

Dans le cas des courbes projectives, lisses, géométriquement intègres de genre 0 ou 1, des quadriques de dimension quelconque, et des intersections de deux quadriques, ces deux questions coïncident, mais ce n'est pas le cas en général.

Le groupe $Z_0(X)$ des zéro-cycles sur X est le groupe abélien libre sur les points fermés de X. À un zéro-cycle $z = \sum_P n_P P$ ($n_P \in \mathbb{Z}$) sur la k-variété X on associe son degré $\deg_k(z) := \sum_P n_P[k(P) : k] \in \mathbb{Z}$. L'indice $I(X)$ est donc le générateur positif de l'image de l'application $\deg_k : Z_0(X) \to \mathbb{Z}$.

Sur un corps de nombres k, il est alors naturel de poser la question du principe de Hasse pour la propriété $I(X) = 1$: étant donnée une k-variété projective, lisse, géométriquement intègre X, si on a $I(X_{k_v}) = 1$ pour chaque place v, a-t-on alors $I(X) = 1$? La réponse est non en général (courbes de genre 1, intersections complètes lisses de deux quadriques dans $\mathbb{P}^4$).

Pour une k-variété X, on considère l'accouplement bilinéaire

$$Z_0(X) \times \mathrm{Br}(X) \to \mathrm{Br}(k)$$

$$\left(\sum_P n_P P, \alpha\right) \mapsto \sum_P n_P \mathrm{Cores}_{k(P)/k}(\alpha(P)).$$

Ici $\alpha(P) \in \mathrm{Br}(k(P))$ est l'évaluation de α en P, et on applique ensuite la norme, ou corestriction : $\mathrm{Br}(k(P)) \to \mathrm{Br}(k)$.

On peut dans ce cadre définir une obstruction de Brauer-Manin à l'existence d'un zéro-cycle de degré 1. Comme on verra ci-dessous, on peut aussi définir un analogue de l'obstruction à l'approximation faible.

Pour les zéro-cycles, on a deux conjectures qui, à la différence de la conjecture 3.1, portent sur *toutes* les variétés projectives et lisses, sans restriction sur leur géométrie. Ces conjectures furent faites par Sansuc et moi (1981) dans le cadre des surfaces rationnelles, et étendues au cas général sous la forme ci-dessous dans [CT95, CT99]. Une conjecture proche mais d'aspect assez différent avait été formulée par K. Kato et S. Saito (1983). Voir [W12].

Conjecture 4.1. *Soient k un corps global et X une k-variété projective, lisse, géométriquement intègre sur k. S'il existe une famille de zéro-cycles de degré 1 $z_v \in Z_0(X_{k_v})$ tels que pour tout $\alpha \in \mathrm{Br}(X)$ on ait*

$$\sum_v j_v(z_v, \alpha) = 0 \in \mathbb{Q}/Z,$$

alors il existe un zéro-cycle de degré 1 sur X.

La conjecture est ouverte déjà dans le cas des surfaces cubiques lisses.

Soient X et Y des k-variétés projectives. Soit $\pi : Y \to X$ un k-morphisme. On lui associe un homomorphisme $\pi_* : Z_0(Y) \to Z_0(X)$. On dit qu'un zéro-cycle sur X est rationnellement équivalent à zéro s'il est dans le sous-groupe de $Z_0(X)$ engendré par les $\pi_*(\mathrm{div}_Y(g))$ pour Y variant parmi les courbes normales projectives, la fonction $g \in k(Y)^\times$ variant parmi les fonctions rationnelles non nulles sur une telle courbe Y, et $\pi : Y \to X$ les k-morphismes. Le groupe de Chow $CH_0(X)$ des zéro-cycles de degré zéro sur X est le quotient de $Z_0(X)$ par le sous-groupe des zéro-cycles rationnellement équivalents à zéro. Il est muni d'une flèche degré $CH_0(X) \to \mathbb{Z}$, dont le noyau est noté $A_0(X)$. L'accouplement $Z_0(X) \times \mathrm{Br}(X) \to \mathrm{Br}(k)$ passe au quotient par l'équivalence rationnelle et induit un accouplement bilinéaire $CH_0(X) \times \mathrm{Br}(X) \to \mathrm{Br}(k)$.

La conjecture suivante englobe la conjecture 4.1.

Conjecture 4.2. *Soient k un corps global et X une k-variété projective, lisse, géométriquement intègre sur k. Le complexe*

$$\mathrm{proj}\lim_n CH_0(X)/n \to \prod_v \mathrm{proj}\lim_n CH_0(X_{k_v})^*/n \to \mathrm{Hom}(\mathrm{Br}(X), \mathbb{Q}/\mathbb{Z})$$

induit par la somme des accouplements du groupe de Brauer de X avec les groupes $CH_0(X_{k_v})$, à valeurs dans $\mathrm{Br}(k_v) \subset \mathbb{Q}/\mathbb{Z}$ est une suite exacte.

On note $CH_0(X_{k_v})^* = CH_0(X_{k_v})$ si v est une place non archimédienne, puis $CH_0(X_{k_v})^* = 0$ si v est une place complexe, et pour v réel le quotient de $CH_0(X_\mathbb{R})$ par l'image de la norme $CH_0(X_\mathbb{C}) \to CH_0(X_\mathbb{R})$.

Le théorème suivant [HW16] est l'aboutissement de travaux de Salberger [Sal88], Colliot-Thélène, Swinnerton-Dyer, Skorobogatov, Harari [Har94], Wittenberg [W12]. Un argument relativement élémentaire mais essentiel du travail de Salberger [Sal88] avait été réinterprété par Colliot-Thélène et Swinnerton-Dyer (1994) comme une variante inconditionnelle, adaptée aux zéro-cycles, de l'hypothèse de Schinzel mentionnée au paragraphe 3.

Théorème 4.1. (Harpaz et Wittenberg) *Soient k un corps de nombres, X une k-variété projective et lisse géométriquement intègre, et $\pi : X \to \mathbb{P}^1_k$ un k-morphisme plat à fibre générique une variété rationnellement connexe. Si les fibres lisses X_m au-dessus d'un point fermé m de $\mathbb{P}^1$ satisfont la conjecture 4.1, resp. la conjecture 4.2, alors il en est de même de X.*

Y. Liang [Lia13] avait montré comment on peut établir les conjectures 4.1 et 4.2 pour les variétés projectives et lisses birationnelles à un espace homogène d'un groupe algébrique linéaire connexe à stabilisateurs connexes à partir du résultat pour les points rationnels (connu grâce à Sansuc et Borovoi).

Le théorème ci-dessus s'applique donc à tout $X \to \mathbb{P}^1_k$ comme ci-dessus dont la fibre générique est birationnelle à un espace homogène d'un groupe algébrique linéaire connexe à stabilisateurs connexes.

Si X/k est une courbe projective et lisse de genre quelconque, sous l'hypothèse que le groupe de Tate-Shafarevich de la jacobienne J_X est fini, on a les conjectures 4.1 et 4.2 pour X.

Projet ambitieux. *En admettant la finitude des groupes de Tate-Shafarevich des variétés abéliennes, établir la conjecture 4.1 pour les surfaces diagonales*

$$ax^p + by^p + cz^p + dt^p = 0$$

de degré p premier dans $\mathbb{P}^3_{\mathbb{Q}}$, par une extension de la méthode utilisée pour **PH**$_X$ *et p = 3 par Swinnerton-Dyer [SD01].*

De façon plus générale, pour une variété projective et lisse, on souhaiterait ramener la conjecture 4.1 au cas des courbes. Mais cela semble vraiment hors d'atteinte. Une question plus modeste est : *Suffit-il de connaître la conjecture 4.1 pour toutes les variétés projectives et lisses de dimension 3 pour l'avoir en dimension supérieure ?*

5. Rationalité des variétés et invariants birationnels

Soit X une variété sur un corps k. On dit que deux points $P, Q \in X(k)$ sont R-liés s'il existe un ouvert $U \subset \mathbb{P}^1_k$ et un k-morphisme $U \to X$ avec $P, Q \in f(U(k))$. La R-équivalence sur $X(k)$ est la relation d'équivalence engendrée par cette relation.

Étant donnés un corps k de caractéristique zéro et une k-variété projective, lisse, géométriquement connexe, l'ensemble $X(k)/R$ et le sous-groupe $A_0(X) \subset CH_0(X)$ formé des classes de zéro-cycles de degré zéro, sont des invariants k-birationnels des k-variétés projectives et lisses, et ils sont réduits à un élément si la k-variété X est stablement k-rationnelle, ou plus généralement facteur direct birationnel d'un espace projectif.

Pour toute k-variété X projective, lisse, géométriquement intègre sur un corps k disons de caractéristique zéro, $i \geqslant 1$ et $j \in \mathbb{Z}$, on dispose des groupes de cohomologie non ramifiée $H^i_{nr}(k(X)/k, \mathbb{Q}/\mathbb{Z}(j))$, à coefficients dans les racines de l'unité tordues j fois. Ces groupes sont des invariants k-birationnels, réduits à $H^i(k, \mathbb{Q}/\mathbb{Z}(j))$ si X est stablement k-birationnelle à un espace projectif. On consultera [CT19] pour un rapport récent sur ces invariants. On a $H^2_{nr}(k(X)/k, \mathbb{Q}/\mathbb{Z}(1)) = \mathrm{Br}(X)$.

5.1. Unirationalité.

Problème 5.1. *Soit k un corps infini. Soit X une k-variété projective et lisse, rationnellement connexe. Supposons $X(k)$ non vide.*

(a) L'ensemble $X(k)$ des points rationnels est-il Zariski dense dans X ?

(b) La k-variété X est-elle k-unirationnelle ?

C'est connu pour les surfaces cubiques lisses, mais ces questions sont ouvertes pour les surfaces rationnelles quelconques. Pour ces surfaces, une réponse affirmative découlerait d'une réponse affirmative à la question suivante [CTSa80, §V] :

Problème 5.2. *Les torseurs universels [CTSa80, §II.C] sur les k-surfaces rationnelles projectives et lisses sont-ils des k-variétés (stablement) k-rationnelles dès qu'ils possèdent un point rationnel ?*

C'est connu pour les surfaces de Châtelet [CTSaSD87], et plus généralement les sufaces fibrées en coniques sur $\mathbb{P}^1_k$ avec au plus 4 fibres géométriques singulières, mais déjà le cas des surfaces de del Pezzo X de degré 4 avec groupe de Picard $\mathrm{Pic}(X)$ de rang un est ouvert.

Une réponse affirmative à ce problème dans le cas $k = \mathbb{C}(\mathbb{P}^1)$ impliquerait l'unirationalité sur $\mathbb{C}$ des variétés complexes de dimension 3 fibrées en coniques au-dessus du plan projectif $\mathbb{P}^2_{\mathbb{C}}$, ce qui est une question ouverte bien connue.

5.2. **R-équivalence.** Soit k un corps p-adique. Soit $f : X \to Y$ un k-morphisme projectif et lisse de k-variétés lisses géométriquement intègres, à fibres des variétés rationnellement connexes. C'est un théorème de Kollár (1999) que pour $m \in Y(k)$ l'ensemble $X_m(k)/R$ associé à la fibre X_m est fini, et que son cardinal est semi-continu supérieurement pour la topologie p-adique sur $Y(k)$: pour $m \in Y(k)$, en tout point n d'un voisinage ouvert convenable de m, l'ordre de $X_n(k)/R$ est au plus celui de $X_m(k)/R$ [K04].

Problème 5.3. (Kollár) *Sous les conditions ci-dessus, l'ordre de $X_m(k)/R$ est-il localement constant pour la topologie p-adique sur $Y(k)$?*

Problème 5.4. *Soient k un corps parfait de dimension cohomologique 1 et X une k-variété projective, lisse, (séparablement) rationnellement connexe. Supposons $X(k) \neq \varnothing$.*
(a) L'ensemble $X(k)/R$ est-il réduit à un point ?
(b) A-t-on $A_0(X) = 0$?
(c) Ces propriétés valent-elles au moins si k est un corps C_1 ?

Je renvoie à [CT11, §10] pour une discussion de divers cas concrets, tant de corps que de types de variétés. La question (b) a une réponse affirmative pour les surfaces rationnelles. La question (c) a une réponse affirmative pour les intersections complètes lisses de deux quadriques dans $\mathbb{P}^n_k$ pour $n \geqslant 4$. Ici encore, une réponse affirmative à la question (a) dans le cas $k = \mathbb{C}(\mathbb{P}^1)$, et déjà la finitude de $X(k)/R$ dans ce cas, impliquerait l'unirationalité sur $\mathbb{C}$ des variétés complexes de dimension 3 fibrées en coniques au-dessus du plan projectif $\mathbb{P}^2_{\mathbb{C}}$.

5.3. **Rationalité des intersections de deux quadriques.** Soit $X \subset \mathbb{P}^n_k$ une intersection complète lisse de deux quadriques $f = g = 0$ sur un corps k. Une telle variété est k-rationnelle si elle possède une droite $\mathbb{P}^1_k$. Un théorème d'Amer assure que X contient une droite $\mathbb{P}^1_k$ si et seulement si la quadrique d'équation $f + tg = 0$ sur le corps $k(t)$ (où t est une variable) contient un $\mathbb{P}^1_{k(t)}$, i.e. si et seulement si la forme quadratique $f + tg$ sur le corps $k(t)$ contient deux hyperboliques.

Pour k algébriquement clos, on retrouve le fait que $X \subset \mathbb{P}_k^n$ contient une droite si $n \geqslant 4$, et est rationnelle.

Pour k un corps C_1, le corps $k(t)$ est C_2. Dans ce cas $X \subset \mathbb{P}_k^n$ contient une droite si $n \geqslant 6$, et est donc k-rationnelle. C'est le meilleur résultat possible : pour $k = \mathbb{C}(z)$ corps des fonctions rationnelles en une variable, Hassett et Tschinkel [HT21] donnent un exemple de $X \subset \mathbb{P}_k^5$ qui n'est pas stablement k-rationnelle.

Pour $n = 5$, sur un corps quelconque, un théorème récent [BW19], valable sur tout corps, dit que la k-variété X est k-rationnelle si et seulement si elle contient une droite $\mathbb{P}_k^1$. On n'a pas par contre de critère pour la k-rationalité stable.

Soit k un corps p-adique. Pour $n \geqslant 8$ on a $X(k) \neq \varnothing$. Commençons par raffiner certains des résultats de [CTSaSD87, Chap. 3]. C'est un théorème [PS10, HB10, HHK09, L13, PS14] que toute forme quadratique en au moins 9 variables sur un corps de fonctions d'une variable sur un corps p-adique est isotrope. Ainsi toute forme quadratique en au moins 11 variables sur $k(t)$ s'annule sur un vectoriel de dimension 2 sur $k(t)$. Via le théorème d'Amer, ceci implique que pour $n \geqslant 10$ toute intersection de deux quadriques $X \subset \mathbb{P}_k^n$ contient une droite $\mathbb{P}_k^1$. Donc pour $n \geqslant 10$, si X est une intersection complète lisse, elle est k-rationnelle.

Problème 5.5. *Soit $X \subset \mathbb{P}_k^n$ une intersection lisse de deux quadriques sur un corps p-adique k. Supposons $X(k) \neq \varnothing$.*

(a) Que peut-on dire sur la k-rationalité (stable) de $X \subset \mathbb{P}_k^n$ pour $6 \leqslant n \leqslant 9$?

(b) On a $X(k)/R = \{*\}$ pour $n \geqslant 7$. Que peut-on dire pour $n = 5, 6$?

(c) Résultats et questions analogues pour le groupe $A_0(X) \subset CH_0(X)$ des classes de zéro-cycles de degré zéro.

Si $p \neq 2$ et $n = 6$, on a $A_0(X) = 0$ [PS95]. Par la méthode de spécialisation [V15, CTP16] on devrait pouvoir donner des exemples d'intersections lisses X de deux quadriques dans $\mathbb{P}_k^5$, avec $X(k) \neq \varnothing$, qui ne sont pas stablement k-rationnelles. Voir [CTP16, Thm. 1.21] et [HT21, §9, §10].

On peut aussi se poser la question de la rationalité sur le corps $\mathbb{R}$ des réels.

Problème 5.6. *Soit $X \subset \mathbb{P}_\mathbb{R}^n$, $n \geqslant 4$, une intersection complète lisse de deux quadriques. Supposons que $X(\mathbb{R})$ est non vide et connexe (pour la topologie réelle). Ceci implique-t-il que X est (stablement) $\mathbb{R}$-rationnelle ?*

Pour $n = 4$, l'hypothèse implique que X est $\mathbb{R}$-rationnelle. Pour $n = 5$, Hassett et Tschinkel [HT21] ont montré que X est $\mathbb{R}$-rationnelle si et seulement si X contient une droite $\mathbb{P}_\mathbb{R}^1$. Ainsi pour $n = 5$ on peut avoir $X(\mathbb{R})$ connexe non vide et X non $\mathbb{R}$-rationnelle. La question de la $\mathbb{R}$-rationalité stable est ouverte. Pour $n = 6$, Hassett, Kollár et Tschinkel (2020) ont montré que $X(\mathbb{R})$ connexe non vide équivaut à X $\mathbb{R}$-rationnelle. En dimensions supérieures le problème est ouvert.

5.4. **Cohomologie non ramifiée.** La question classique de la rationalité (stable) des hypersurfaces cubiques lisses dans $\mathbb{P}^4_{\mathbb{C}}$ amène à considérer le problème suivant, qui est lié à l'étude des cycles de codimension deux (voir [V15, Thm. 1.10, 3.1, 3.3] et [CT15, Thm. 5.4, 5.6, 5.8]).

Problème 5.7. *Soit $X \subset \mathbb{P}^n_{\mathbb{C}}$, $n \geqslant 4$ une hypersurface lisse de degré $d \leqslant n$. Soit K un corps contenant $\mathbb{C}$. Pour $n = 4, 5$, l'application*

$$H^3(K, \mathbb{Q}/\mathbb{Z}(2)) \to H^3_{nr}(K(X)/K, \mathbb{Q}/\mathbb{Z}(2))$$

est-elle un isomorphisme pour tout corps K contenant $\mathbb{C}$?

Pour $n \geqslant 6$ et tout $d \leqslant n$, c'est connu [CT15, Thm. 5.6]. Pour $n = 5$ et $d = 3$, la réponse est affirmative, cela résulte [CT15, Thm. 5.8] d'un théorème de C. Voisin (2006). Pour $n = 4$ et $d = 3$, c'est un problème en général ouvert [V15].

Soient $\mathbb{F}$ un corps fini et ℓ un nombre premier différent de la caractéristique de $\mathbb{F}$. Soit $X/\mathbb{F}$ une variété projective et lisse géométriquement connexe de dimension d. Le groupe $H^3_{nr}(\mathbb{F}(X)/\mathbb{F}, \mathbb{Q}_\ell/\mathbb{Z}_\ell(2))$ est un analogue supérieur de la partie ℓ-primaire du groupe de Brauer $\mathrm{Br}(X)$ d'une variété $X/\mathbb{F}$. C'est une extension d'un groupe fini par un groupe divisible. Pour $d = 2$, i.e. X une surface, on a $H^3_{nr}(\mathbb{F}(X)/\mathbb{F}, \mathbb{Q}_\ell/\mathbb{Z}_\ell(2)) = 0$ (corps de classes supérieur). A. Pirutka a donné des exemples de variétés géométriquement rationnelles de dimension 5 avec $H^3_{nr}(\mathbb{F}(X)/\mathbb{F}, \mathbb{Q}_2/\mathbb{Z}_2(2)) \neq 0$. F. Scavia et F. Suzuki viennent de donner un exemple de variété de dimension 4 pour laquelle ce groupe est non nul.

Problème 5.8. *Pour toute variété projective et lisse intègre X de dimension 3, le groupe $H^3_{nr}(\mathbb{F}(X)/\mathbb{F}, \mathbb{Q}_\ell/\mathbb{Z}_\ell(2))$ est-il divisible ? Est-il nul ? Est-ce déjà le cas pour les variétés rationnellement connexes ?*

On sait l'établir pour quelques classes intéressantes de variétés : les variétés fibrées en coniques au-dessus d'une surface [PS16], et les hypersurfaces cubiques lisses dans $\mathbb{P}^4_{\mathbb{F}}$. La question est liée à une forme forte de la conjecture de Tate entière pour les 1-cycles sur les variétés de dimension 3 sur un corps fini, et à la validité de la conjecture 4.2 ci-dessus pour les surfaces sur un corps global de caractéristique positive [CT99, CTK13]. Le lien entre la conjecture de Tate entière pour les 1-cycles sur les variétés sur un corps fini et la conjecture 4.1 sur un corps global de caractéristique positive avait été fait par S. Saito en 1989. On sait établir $H^3_{nr}(\mathbb{F}(X)/\mathbb{F}, \mathbb{Z}/2) = 0$ pour $X \subset \mathbb{P}^4_{\mathbb{F}}$ une hypersurface cubique lisse (car.($\mathbb{F}) \neq 2$).

Problème 5.9. *Soit $\mathbb{F}$ un corps fini, $\mathrm{car}(\mathbb{F}) \neq 2$, et soit $X \subset \mathbb{P}^5_{\mathbb{F}}$ une hypersurface cubique lisse. A-t-on $H^3_{nr}(\mathbb{F}(X)/\mathbb{F}, \mathbb{Z}/2) = 0$?*

6. Points rationnels et indice des variétés algébriques.

Problème 6.1. *Soit $X \subset \mathbb{P}^n_k$, $n \geqslant 4$, une intersection complète lisse de deux quadriques sur un corps k de dimension cohomologique 1. Pour $n = 5$, a-t-on $X(k) \neq \varnothing$?*

Pour $n \geqslant 6$, c'est vrai et facile. Pour $n = 4$, la réponse est négative. La démonstration repose sur la construction de très grands corps.

Soit C une courbe géométriquement intègre sur le corps des réels avec $C(\mathbb{R}) = \varnothing$, par exemple la conique d'équation homogène $x^2 + y^2 + t^2 = 0$. On sait que le corps $\mathbb{R}(C)$ est de dimension cohomologique 1. C'est une question ouverte si c'est un corps C_1. Plus généralement on demande s'il y a un analogue du théorème de Graber, Harris et Starr :

Problème 6.2. *Toute variété rationnellement connexe X sur le corps $K = \mathbb{R}(C)$ a-t-elle un point rationnel ?*

On ne sait déjà pas si pour toute variété projective et lisse rationnellement connexe X sur $\mathbb{R}$ avec $X(\mathbb{R}) = \varnothing$ il existe un $\mathbb{R}$-morphisme de la conique sans point vers X.

Problème 6.3. *Existe-t-il un entier $n \geqslant 4$ tel que toute hypersurface cubique lisse $X \subset \mathbb{P}^n_k$ sur un corps k de dimension cohomologique 1 possède un point rationnel, ou du moins satisfasse $I(X) = 1$?*

Les corps p-adiques ne sont pas des corps C_2. On a cependant la question :

Problème 6.4. (Kato et Kuzumaki) *Pour toute hypersurface $X \subset \mathbb{P}^n_k$ de degré d sur un corps p-adique, si l'on a $n \geq d^2$, a-t-on $I(X) = 1$?*

Ceci a été établi par Kato et Kuzumaki (1985) lorsque le degré d est un nombre premier. C'est ouvert déjà pour $d = 4$.

Problème 6.5. (Cassels et Swinnerton-Dyer) *Soient k un corps et $X \subset \mathbb{P}^n_k$ une hypersurface cubique. Si l'on a $I(X) = 1$, a-t-on $X(k) \neq \varnothing$?*

D. Coray (1976) montra qu'il en est ainsi sur un corps p-adique. Pour $X \subset \mathbb{P}^3_k$ une surface cubique lisse sur un corps quelconque, il montra que l'hypothèse $I(X) = 1$ entraîne l'existence sur X d'un point fermé de degré 1, 4 ou 10. La question si on peut éliminer 10 et 4 est restée ouverte. L'analogue de la question pour les surfaces de del Pezzo de degré 2 a une réponse négative (Kollár et Mella).

Problème 6.6. (Serre) *Soient k un corps, G un groupe algébrique linéaire connexe sur k, et E un espace principal homogène sous G. Si l'indice $I(E)$ est égal à 1, a-t-on $E(k) \neq \varnothing$?*

Pour les espaces homogènes non principaux, la propriété ne vaut pas, des contre-exemples ont été construits par Florence et par Parimala.

Problème 6.7. *Soient k un corps, $\mathrm{car}(k) = 0$, et $X \subset \mathbb{P}^4_k$ une hypersurface cubique lisse sans point rationnel, i.e. d'indice $I(X) = 3$. Soit Y/k une k-variété projective lisse géométriquement connexe de dimension au plus 2. S'il existe une k-application rationnelle de X vers Y, a-t-on $I(Y) = 1$?*

Par la classification k-birationnelle des surfaces géométriquement rationnelles, le problème se ramène au cas où Y est une surface cubique lisse k-minimale. Ce cas semble résister aux formules de degré à la Rost [M03, Z10] qui avaient permis d'étendre le théorème d'Hoffmann [H95] restreignant les dimensions possibles pour les couples de quadriques anisotropes admettant une application rationnelle entre elles.

7. Groupes algébriques linéaires

Soit G un groupe algébrique linéaire réductif connexe sur un corps k. L'ensemble $G(k)/R$ est naturellement muni d'une structure de groupe. Tout élément de ce groupe est d'ordre fini. Si K/k est une extension transcendante pure, l'homomorphisme $G(k)/R \to G(K)/R$ est bijectif.

Soit D une algèbre centrale simple (de rang fini) sur un corps k. Soit $G = SL_{1,D} \subset GL_{1,D}$ le groupe algébrique des éléments de norme réduite 1. C'est un k-groupe algébrique semisimple simplement connexe. Un théorème de Voskresenskiĭ utilisant un théorème de Platonov identifie dans ce cas $G(k)/R$ au groupe $SK_1(D)$ quotient du groupe des éléments de $D^\times$ de norme réduite 1 par le groupe $[D^\times, D^\times]$ engendré par les commutateurs.

Des travaux de Platonov, Yanchevskiĭ, Merkurjev, Chernousov ont identifié le quotient $G(k)/R$ pour beaucoup de groupes classiques, tant simplement connexes qu'adjoints, et ont au passage établi sa commutativité. Le problème général suivant reste cependant ouvert.

Problème 7.1. *Soient k un corps et G un k-groupe algébrique réductif connexe. Le groupe quotient $G(k)/R$ est-il commutatif?*

Voskresenskiĭ (1977) avait posé la question pour G linéaire connexe sur un corps quelconque. Pour un groupe linéaire connexe non réductif, sur un corps non parfait, F. Scavia (2021) a donné une réponse négative.

Pour les deux problèmes suivants, on consultera le rapport de P. Gille [Gi07].

Problème 7.2. *Soient k un corps et G un k-groupe réductif connexe. Si pour tout corps K contenant k, le groupe $G(K)/R$ est trivial, ceci implique-t-il que G est facteur direct birationnel d'une k-variété k-rationnelle?*

Problème 7.3. *Soit k un corps parfait de dimension cohomologique $\leqslant 3$. Si G est un k-groupe semisimple simplement connexe, a-t-on $G(k)/R = 1$?*

Ce problème est motivé par les travaux de Suslin sur le groupe $SK_1(D)$ d'une algèbre simple centrale. On a un certain nombre de résultats lorsque la dimension cohomologique est $\leqslant 2$.

Problème 7.4. *Soit G un groupe algébrique linéaire connexe sur le corps $\mathbb{R}$ des réels. La $\mathbb{R}$-variété G est-elle $\mathbb{R}$-rationnelle, i.e. le corps des fonctions de G est-il transcendant pur sur $\mathbb{R}$?*

Problème 7.5. *Soit k un corps de type fini sur $\mathbb{Q}$. Si G est un k-groupe linéaire connexe, le quotient $G(k)/R$ est-il fini?*

C'est connu pour G un k-tore (Colliot-Thélène et Sansuc 1977) et pour k un corps de nombres (P. Gille 1997). Pour $G = SL_{1,D}$ le k-groupe algébrique des éléments de norme 1 dans une algèbre centrale simple D sur k, le problème se traduit ainsi :

Problème 7.6. *Si D est une algèbre centrale simple sur un corps k de type fini sur $\mathbb{Q}$, le groupe $SK_1(D)$ est-il fini ?*

Problème 7.7. *Soit $H/\mathbb{C}$ un groupe linéaire connexe et $G \subset H$ un sous-groupe fermé connexe. Le quotient H/G est-il une variété rationnelle ?*

C'est une question célèbre, déjà pour $H = GL_{n,\mathbb{C}}$ et $G = PGL_{m,\mathbb{C}}$. Pour traiter cette question, on peut essayer d'utiliser la cohomologie non ramifiée.

Soit $k = \mathbb{C}$, et soit X une compactification lisse de $GL_{n,\mathbb{C}}/G$, avec $G \subset GL_{n,\mathbb{C}}$ sous-groupe algébrique fermé connexe. Pour tout corps K contenant $\mathbb{C}$, et $i = 1, 2$ on sait que

$$H^i(K, \mathbb{Q}/\mathbb{Z}(i-1)) = H^i_{nr}(K(X)/K, \mathbb{Q}/\mathbb{Z}(i-1)).$$

Pour $i = 2$, ceci dit que le groupe de Brauer de $X \times_\mathbb{C} K$ est réduit à l'image de $\mathrm{Br}(K)$, énoncé essentiellement dû à Bogomolov. Dans une série d'articles, Merkurjev [M17] et Sanghoon Baek [B21] ont établi $H^3_{nr}(\mathbb{C}(X)/\mathbb{C}, \mathbb{Q}/\mathbb{Z}(2)) = 0$ pour de nombreuses classes de groupes réductifs G.

Problème 7.8. *Dans chacun de ces cas, pour tout corps K contenant $\mathbb{C}$, la flèche $H^3(K, \mathbb{Q}/\mathbb{Z}(2)) \to H^3_{nr}(K(X)/K, \mathbb{Q}/\mathbb{Z}(2))$ est-elle un isomorphisme ?*

Avec les méthodes décrites dans [CT15, §5], on pourrait essayer de résoudre ce problème via l'étude des cycles de codimension deux d'une bonne compactification lisse de $GL_{n,\mathbb{C}}/G$.

Références

[AK03] C. Araujo et J. Kollár, Rational curves on varieties, in *Higher dimensional varieties and rational points* (Budapest, 2001), 13–68, Bolyai Soc. Math. Stud., **12**, Springer, Berlin, 2003.

[B21] S. Baek, Degree 3 unramified cohomology of classifying spaces for exceptional groups, J. Pure Appl. Algebra **225** (2021).

[BW19] O. Benoist et O. Wittenberg, Intermediate Jacobians and rationality over arbitrary fields, à paraître aux Annales scientifiques de l'École Normale Supérieure.

[BD13] M. Borovoi et C. Demarche, Manin obstruction to strong approximation for homogeneous spaces, Comment. Math. Helv. **88**, No. 1, 1–54 (2013).

[Bri18] M. Bright, Obstructions to the Hasse principle in families, manuscripta math. **157** (2018), no. 3-4, 529–550.

[BBL16] M. Bright, T. D. Browning et D. Loughran, Failures of weak approximation in families, Compos. Math. **152** (2016), no. 7, 1435–1475.

[B18]   T. D. Browning, How often does the Hasse principle hold ? in *Algebraic geometry : Salt Lake City 2015*, 89–02, Proc. Sympos. Pure Math., **97.2**, Amer. Math. Soc., Providence, RI, 2018.

[BLBS20]   T. D. Browning, P. Le Boudec et W. Sawin, The Hasse principle for random Fano hypersurfaces, prépublication, math arXiv :2006.02356v1

[BMSk14]   T. Browning, L. Matthiesen et A N Skorobogatov, Rational points on pencils of conics and quadrics with many degenerate fibres, Ann Math, **180** (2014) 381–402

[CT87]   J.-L. Colliot-Thélène, Arithmétique des variétés rationnelles et problèmes birationnels, in *Proceedings of the International Congress of Mathematicians*, Berkeley, California 1986, (1987), Tome I, 641–653.

[CT90]   J.-L. Colliot-Thélène, Surfaces rationnelles fibrées en coniques de degré 4, in *Séminaire de théorie des nombres de Paris 88-89*, Progress in Mathematics, Birkhäuser (1990), p. 43–55.

[CT95]   J.-L. Colliot-Thélène, L'arithmétique du groupe de Chow des zéro-cycles (exposé aux Journées arithmétiques de Bordeaux, Septembre 93), Journal de théorie des nombres de Bordeaux **7**(1995) 51–73.

[CT98]   J.-L. Colliot-Thélène, The Hasse principle in a pencil of algebraic varieties, in *Proceedings of the Tiruchirapalli conference* (India, January 1996), ed. M. Waldschmidt and K. Murty, Contemporary Mathematics **210**(1998) 19–39.

[CT99]   J.-L. Colliot-Thélène, Conjectures de type local-global sur l'image de l'application cycle en cohomologie étale, in *Algebraic K-Theory* (1997), W. Raskind and C. Weibel ed., Proceedings of Symposia in Pure Mathematics **67**, Amer. Math.Soc. (1999) 1–12.

[CT03]   J.-L. Colliot-Thélène, Points rationnels sur les fibrations (notes d'un cours donné à Budapest en septembre 2001), in *Higher Dimensional Varieties and Rational Points*, Bolyai Society Mathematical Series, **12**, Springer-Verlag, 2003, edited by K. J. Böröczky, J. Kollár and T. Szamuely, 171–221.

[CT11]   J.-L. Colliot-Thélène, Variétés presque rationnelles, leurs points rationnels et leurs dégénérescences, in *Arithmetic Geometry* (CIME 2007), Springer LNM **2009** (2011), p. 1–44.

[CT15]   J.-L. Colliot-Thélène, Descente galoisienne sur le second groupe de Chow : mise au point et applications Documenta Mathematica, Extra Volume : Alexander S. Merkurjev's Sixtieth Birthday (2015) 195–220.

[CT19]   J.-L. Colliot-Thélène, Non rationalité stable sur les corps quelconques, in *Birational Geometry of Hypersurfaces* (Gargnano del Garda, 2018), A. Hochenegger, M. Lehn et P. Stellari ed., Lecture Notes of tjhe Unione Matematica Italiana, Springer (2019) p. 73–110.

[CTHaSk03] J.-L. Colliot-Thélène, D. Harari et A.N. Skorobogatov, Valeurs d'un polynôme à une variable représentées par une norme, in *Number theory and algebraic geometry*, M. Reid and A. Skorobogatov eds., London Math. Soc. Lecture Note Series **303**, Cambridge University Press, 2003, pp. 69–89.

[CTK13] J.-L. Colliot-Thélène et B. Kahn, Cycles de codimension 2 et H^3 non ramifié pour les variétés sur les corps finis, J. K-Theory **11** (2013) 1–53.

[CTSa80] J.-L. Colliot-Thélène et J.-J. Sansuc, La descente sur les variétés rationnelles, in *Journées de géométrie algébrique d'Angers* (Juillet 1979), éd. A. Beauville, Sijthoff & Noordhoff (1980), 223–237.

[CTKaS87] J.-L. Colliot-Thélène, D. Kanevsky et J.-J. Sansuc, Arithmétique des surfaces cubiques diagonales, in *Diophantine Approximation and Transcendence Theory*, Springer L.N.M. **1290** (ed. G. Wüstholz) (1987) 1–108.

[CTP16] J.-L. Colliot-Thélène et A. Pirutka, Hypersurfaces quartiques de dimension 3 : non-rationalité stable, Ann. Sci. Éc. Norm. Sup. 4e série, t. **49** (2) (2016) 371–397.

[CTSaSD87] J.-L. Colliot-Thélène, J.-J. Sansuc et Sir Peter Swinnerton-Dyer, Intersections of two quadrics and Châtelet surfaces, I, J. für die reine und angew. Math. **373** (1987) 37–107 ; II, ibid. **374** (1987) 72–168.

[CTSk21] J.-L. Colliot-Thélène et A. N. Skorobogatov, *The Brauer–Grothendieck group*, Ergebnisse der Mathematik und ihrer Grenzgebiete. 3. Folge, **71**. Springer, Cham, 2021.

[CTSkSD98] J.-L. Colliot-Thélène, A. N. Skorobogatov et Sir Peter Swinnerton-Dyer, Hasse principle for pencils of curves of genus one whose Jacobians have rational 2-division points, Inventiones math. **134** (1998) 579–650.

[Gi07] P. Gille, Le problème de Kneser–Tits, in *Séminaire Bourbaki*, Volume 2007/2008. Astérisque **326**, 39–82, Exp. no. 983 (2009).

[Har94] D. Harari. Méthode des fibrations et obstruction de Manin, Duke Math. J. **75** (1994) 221–260.

[HHK09] D. Harbater, J. Hartmann et D. Krashen, Applications of patching to quadratic forms and central simple algebras, Invent. math. **178** (2009), no. 2, 231–263.

[HS16] Y. Harpaz et A.N. Skorobogatov, Hasse principle for Kummer varieties, Algebra & Number Theory **10** (2016) 813–841.

[HW16] Y. Harpaz et O. Wittenberg, On the fibration method for zero-cycles and rational points, Annals of Mathematics **183** (2016), no. 1, 229–295.

[HW20] Y. Harpaz et O. Wittenberg, Zéro-cycles sur les espaces homogènes et problème de Galois inverse, Journal of the American Mathematical Society **33** (2020), no. 3, 775–805.

[HWW21] Y. Harpaz, D. Wei et O. Wittenberg, Rational points on fibrations with few non-split fibres, prépublication, septembre 2021.

[HT21] B. Hassett et Yu. I. Tschinkel, Rationality of complete intersections of two quadrics over nonclosed fields, with an appendix by Jean-Louis Colliot-Thélène, L'Enseign. Math. **67** (2021), no. 1-2, 1–44.

[HB10] D.R. Heath-Brown, Zeros of systems of p-adic quadratic forms, Compos. Math. **146** (2010), no. 2, 271–287.

[HB18] D.R. Heath-Brown, Zeros of pairs of quadratic forms, J. reine angew. Math. **739** (2018) 41–80.

[H95] D. Hoffmann, Isotropy of quadratic forms over the function field of a quadric, Math. Z. 220 (1995), no. 3, 461–476.

[K99] J. Kollár. *Rational curves on algebraic varieties.* Ergebnisse der Mathematik und ihrer Grenzgebiete, 3. Folge, **32**, 2nd ed., Springer-Verlag, 1999.

[K04] J. Kollár, Specialization of zero-cycles, Publ. Res. Inst. Math. Sci. **40** (2004), no. 3, 689–708.

[L13] D. B. Leep, The u-invariant of p-adic function fields, J. reine angew. Math. **679** (2013), 65–73.

[Lia13] Y. Liang., Arithmetic of 0-cycles on varieties defined over number fields, Ann. Scient. École Norm. Sup. 4e série **46** (2013) 35–56.

[L18] D. Loughran, The number of varieties in a family which contain a rational point, J. Eur. Math. Soc. (JEMS) **20** (2018), no. 10, 2539–2588.

[LS16] D. Loughran et A. Smeets, Fibrations with few rational points, Geom. Funct. Anal. **26** (2016), no. 5, 1449–1482.

[M03] A. Merkurjev, Steenrod operations and degree formulas, J. reine angew. Math. **565** (2003), 13–26.

[M17] A. Merkurjev, Invariants of algebraic groups and retract rationality of classifying spaces, in *Algebraic Groups : Structure and Actions*, Proceedings of Symposia in Pure Mathematics, **94**, AMS (2017), 277–294.

[PS95] R. Parimala et V.Suresh, Zero-cycles on quadric fibrations : finiteness theorems and the cycle map, Invent. math. **122** (1995), no. 1, 83–117.

[PS10] R. Parimala et V. Suresh, The u-invariant of the function fields of p-adic curves, Ann. of Math. **172** (2010) 1391–1405.

[PS14] R. Parimala et V. Suresh, Period-index and u-invariant questions for function fields over complete discretely valued fields, Invent. math. **197** (1) (2014) 215–235.

[PS16] R. Parimala et V. Suresh, Degree 3 cohomology of function fields of surfaces, IMRN **2016** :14 (2016) 4341–4374.

[Sal88] P. Salberger, Zero-cycles on rational surfaces over number fields, Invent. math. **91** (1988), no. 3, 505–524.

[SaSk91] P. Salberger, A.N. Skorobogatov, Weak approximation for surfaces defined by two quadratic forms, Duke Math. J. **63** (1991), no. 2, 517–536.

[Se92] J-P. Serre, Topics in Galois Theory, Research Notes in Mathematics **1** (1992) Jones and Bartlett Publishers.

[Sk01] A. N. Skorobogatov, *Torsors and rational points*, Cambridge tracts in Mathematics **144** (2001).

[SkSo20] A. N. Skorobogatov et E. Sofos, Schinzel Hypothesis with probability 1 and rational points, prépublication, 2020, arXiv :2005.02998 [math.NT]

[SD01] Sir Peter Swinnerton-Dyer, The solubility of diagonal cubic surfaces, Ann. Sci. École Norm. Sup. 4e série **34** (2001) 891–912.

[SD11] Sir Peter Swinnerton-Dyer, Topics in Diophantine Geometry, in *Arithmetic Geometry* (CIME 2007), Springer LNM **2009** (2011).

[T17] Zhiyu Tian, Hasse principle for three classes of varieties over global function fields, Duke Math. J. **166** (2017), no. 17, 3349–3424.

[V15] C. Voisin, Unirational threefolds with no universal codimension 2 cycle, Invent. math. 201 (2015), 207–237.

[W07] O. Wittenberg, *Intersections de deux quadriques et pinceaux de courbes de genre 1*, Lecture Notes in Mathematics **1901**, Springer-Verlag, Berlin, 2007.

[W12] O. Wittenberg, Zéro-cycles sur les fibrations au-dessus d'une courbe de genre quelconque, Duke Mathematical Journal **161** (2012), no. 11, 2113–2166.

[W18] O. Wittenberg, Rational points and zero-cycles on rationally connected varieties over number fields, in *Algebraic Geometry : Salt Lake City 2015*, Part 2, p. 597–635, Proceedings of Symposia in Pure Mathematics **97**, American Mathematical Society, Providence, RI, 2018.

[Z10] K. Zainoulline, Degree formula for connective K-Theory, Invent. math. **179** (2010) 507–533.

Some Ideas in Need of Clarification in Resolution of Singularities and the Geometry of Discriminants

Bernard Teissier

> *Mathematics only exists in a living community of mathematicians that spreads understanding and breathes life into ideas both old and new. The real satisfaction from mathematics is in learning from others and sharing with others. All of us have clear understanding of a few things and murky concepts of many more. There is no way to run out of ideas in need of clarification.*
>
> *(W.P. Thurston, quoted in [20])*

Personal Note

Among the different ways of sharing ideas are discussions, letters, videos, platforms such as MathOverflow or Images des Mathématiques, and publications. Publications are subject to more precise rules, and require more prolonged effort because they are not only a means of communication but also the main repository of ideas and results.

That is why the activity of Catriona over four decades has been so useful for the mathematical community. She has a unique way of imagining possible publications, encouraging without pushing, showing great patience and understanding adapted to each author (or editor). She possesses an amazingly rich perception of the mathematical community, knowing of so many mathematicians not only what they do, but also what they are. Catriona really cares about authors (or editors) as people as well as about the quality of the texts. Adding to this an inexhaustible energy, Catriona plays a unique and very important role in the spreading of understanding, as Thurston writes, and thus for the progress of our science. As an expression of gratitude and friendship, I wish to dedicate to her an exposition of some of the problems I have come across and so illustrate the last sentence of Thurston's quote.

B. Teissier (✉)
Université Paris Cité and Sorbonne Université, CNRS, IMJ-PRG, Paris, France
e-mail: bernard.teissier@imj-prg.fr

© The Author(s), under exclusive license to Springer Nature Switzerland AG 2023
J.-M. Morel, B. Teissier (eds.), *Mathematics Going Forward*, Lecture Notes in Mathematics 2313, https://doi.org/10.1007/978-3-031-12244-6_3

1   Problems Related to Resolution of Singularities

In his 1964 paper Hironaka introduced the general concept of *embedded resolution* of a singular space X embedded in a non-singular variety Z. That is a birational morphism $b: Z' \to Z$ with Z' non-singular, such that the strict transform of X is non-singular and transversal in Z' to the exceptional divisor of b, which is mapped to the singular locus of X. Indeed, Hironaka's proof builds Z' as the result of a sequence of blowing-ups with non-singular centers. If X is a toric variety equivariantly embedded in a non-singular toric variety Z, it is more natural to seek a birational toric morphism $Z' \to Z$ of non-singular toric varieties such that the strict transform of X is non-singular and transversal to the toric boundary of Z'. For toric varieties over an algebraically closed field, this was proved to exist in [33, §6] and [16]. The process is purely combinatorial and therefore blind to the characteristic of the field.

If $\mathcal{O}$ is the local ring of a formal branch C over an algebraically closed field k, its normalization is $k[[t]]$ and the set of values which the t-adic valuation v takes on elements of $\mathcal{O}$ is a numerical semigroup $\Gamma \subset \mathbf{N}$. It is finitely generated. For tradition's sake, we denote by $g + 1$ its minimal number of generators. The associated graded ring $\mathrm{gr}_v \mathcal{O}$ of the valuation v restricted to $\mathcal{O}$ (see [29, 33, §2]) is isomorphic to the semigroup algebra $k[t^\Gamma]$ of Γ with coefficients in k and thus corresponds to an affine toric variety $C^\Gamma \subset \mathbf{A}^{g+1}(k)$. If $\xi_1, \ldots, \xi_g$ are elements of $\mathcal{O}$ whose v-values generate Γ, the image of the formal embedding of C in the affine space $\mathbf{A}^{g+1}(k)$ determined by the ξ_i can be degenerated to C^Γ inside $\mathbf{A}^{g+1}(k)$ in such a way that some toric embedded resolutions of $C^\Gamma \subset \mathbf{A}^{g+1}(k)$ also give embedded resolutions of $C \subset \mathbf{A}^{g+1}(k)$. All this is blind to the characteristic of the field k. An instance of this, in the complex analytic world, first appeared in [15] and recently the case of reduced plane curve singularities has been settled (and more) in [12] and also in [10, Corollary 7.11].

An attempt to generalize this leads to the following, where k is an algebraically closed field.

Problem A   *Let $X \subset \mathbf{A}^n(k)$ be a reduced affine algebraic variety over k. Do there exist algebraic embeddings $\mathbf{A}^n(k) \subset \mathbf{A}^N(k)$ such that:*

(1) *The intersection of the image of X (resp. $\mathbf{A}^n(k)$) with the torus of $\mathbf{A}^N(k)$ is dense in X (resp. $\mathbf{A}^n(k)$).*
(2) *There exist birational equivariant maps $\pi : Z \to \mathbf{A}^N(k)$ of non-singular toric varieties such that the strict transform X^π of X (resp. the strict transform of $\mathbf{A}^n(k)$), which exists by (1), is non-singular and transversal to the toric boundary of Z.*
(3) *The ideal in the ring $\mathcal{O}_X(X)$ of the singular subspace of X is generated, up to integral closure, by monomials in the coordinates of $\mathbf{A}^N(k)$.*

If the embedding $X \subset \mathbf{A}^n(k) \subset \mathbf{A}^N(k)$ satisfies the first two conditions, we call it a *torific embedding* for X. For example, an isolated hypersurface singularity which is non-degenerate with respect to its Newton polyhedron is torifically embedded in its ambient space. See [37] and [3] for generalizations. Tevelev has shown in [36] that any embedded resolution diagram of irreducible projective varieties

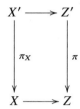

can be embedded in a diagram:

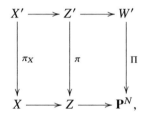

where $Z \to \mathbf{P}^N$ is an embedding, the map Π is a birational toric map of non-singular varieties for a toric structure on $\mathbf{P}^N$, the images of Z and X have dense intersections with the torus of $\mathbf{P}^N$, their strict transforms are the non-singular varieties Z' and X', and they are transversal to the toric boundary of W'.

In this sense, in characteristic zero where we have Hironaka's theorem, and more generally whenever embedded resolution can be proved, toric embedded resolutions are cofinal among embedded resolutions of a given irreducible projective variety X.

Coming back to affine or local torific embeddings, the problem of course is to prove the existence of torific embeddings without assuming embedded resolution, in a way which hopefully would also work in positive characteristic. As a bonus, torific embeddings should, as in the case of curves, contain important geometric information on the singularities of X, which do not seem, in dimension ≥ 3, to be legible in the resolution by blowing ups. To my knowledge there are two approaches to this problem:

– Mourtada's approach (see [24, 26]) is based on a deep vision of the relationship between components of the exceptional divisor (divisorial valuations centered in Z) of an embedded resolution of X and contact subvarieties of the jet schemes on Z associated to the embedding $X \subset Z$. Suitable irreducible components of the contact varieties mentioned above correspond to divisorial valuations centered in $\mathbf{A}^n(k)$ and the equations of each one give an embedding $\mathbf{A}^n(k) \subset \mathbf{A}^N(k)$ such that the divisorial valuation is the trace on $\mathbf{A}^n(k)$ of a *monomial* divisorial valuation on $\mathbf{A}^N(k)$. Then a tropical/toroidal argument explains how to produce a torification. Anyway, that is the idea, and it proves extremely fruitful in spite of the complexity of the computation of the equations of irreducible components.

Mourtada realizes this program in a number of important cases, which I shall not detail here.
– My approach is more directly inspired by the case of curves presented above. As in Zariski's approach, it begins with local uniformization of valuations. The reason is that if $\mathcal{O}_{X,x}$ is the local algebra of a singularity and v is a rational[1] valuation centered in $\mathcal{O}_{X,x}$, the associated graded algebra $\mathrm{gr}_v \mathcal{O}_{X,x}$ is again isomorphic to the semigroup algebra $k[t^\Gamma]$ of the semigroup Γ of values taken by v on $\mathcal{O}_{X,x}$.

If Γ is finitely generated, then we have again an affine toric variety and we can show that toric resolutions of this toric variety, which are blind to the characteristic, provide local uniformization of the valuation after a suitable re-embedding of (X, x) (see [29]).

However, the semigroup Γ is not at all finitely generated in general and we have to think of $\mathrm{Spec}k[t^\Gamma]$ as being of infinite embedding dimension, this embedding dimension being in fact an ordinal, see [29, corollary 3.10]. Such a toric variety is defined by an infinite collection of binomials and does not have a resolution so we have to show that a *"finite partial"* embedded resolution extends to a local uniformization of v. In order to do that we need equations for the degeneration of $\mathcal{O}_{X,x}$ to its graded algebra. This is something we can do for *complete* equicharacteristic noetherian local domains.

Indeed, if the noetherian equicharacteristic local domain R is complete, there exists for any rational valuation of R an embedding of the formal space corresponding to R into the space where the generalized toric variety corresponding to $\mathrm{gr}_v R$ resides; it is given by the Valuative Cohen Theorem of [29].

Since R is noetherian, the semigroup Γ is well ordered and combinatorially finite in the sense that there are finitely many distinct expressions of an element of Γ as a sum of other elements. As a consequence of being well ordered it has a unique minimal system of generators $(\gamma_i)_{i \in I}$, the index set I being an ordinal $\leq \omega^{h(v)}$ where $h(v)$ is the height, or rank, of the valuation, which is $\leq \dim R$. Taking variables $(u_i)_{i \in I}$, one can consider the k-vector space of all formal sums $\Sigma_{e \in E} d_e u^e$ where E is any set of monomials in the u_i and $d_e \in k$. Since the values semigroup Γ is well ordered and combinatorially finite this vector space is in fact, with the usual multiplication rule, a k-algebra $k[\widehat{(u_i)_{i \in I}}]$ which we endow with a weight w by giving u_i the weight γ_i. Combinatorial finiteness means that there are only finitely many monomials with a given weight, and we can enumerate them according to the lexicographic order of exponents (see [29, §4]). Thus, we can embed the set of monomials $u * m$ in the well ordered lexicographic product $\Gamma \times \mathbf{N}$. Combinatorial finiteness also implies that the initial form of every series with respect to the weight filtration is a polynomial so that the corresponding graded algebra of $k[\widehat{(u_i)_{i \in I}}]$ is

---

[1] This means that the inclusion $R \subset R_v$ of R in the valuation ring of the valuation not only satisfies $m_v \cap R = m_R$ for maximal ideals, but also there is no residual extension: $k = R/m_R = k_v = R_v/m_v$.

the polynomial algebra $k[(U_i)_{i \in I}]$ with $U_i = \mathrm{in}_w u_i$, graded by giving U_i the degree γ_i.

The k-algebra $k\widehat{[(u_i)_{i \in I}]}$ is endowed with a monomial valuation given by the weight: $w(\Sigma_{e \in E} d_e u^e) = \min_{d_e \neq 0} w(u^e)$. This valuation is rational since all the γ_i are > 0. Note that 0 is the only element with value ∞ because here ∞ is an element larger than any element of Γ. With respect to the "w ultrametric" given by $u(x, y) = w(y - x)$, the algebra $k\widehat{[(u_i)_{i \in I}]}$ is spherically complete (see [31], theorem 4.2) and has most of the properties of power series algebras, except for noetherianity unless the set I is finite, in which case it is isomorphic to the usual power series ring, with weights on the variables.

The γ_i are the degrees of a minimal set of homogeneous generators $(\overline{\xi}_i)_{i \in I}$ of the Γ-graded k-algebra $\mathrm{gr}_\nu R$. The first part of the valuative Cohen theorem asserts that one can choose representatives $(\xi_i)_{i \in I}$ in R of the $(\overline{\xi}_i)_{i \in I}$ in such a way that $u_i \mapsto \xi_i$ determines a surjective continuous (with respect to the valuations) map of k-algebras

$$\pi : k\widehat{[(u_i)_{i \in I}]} \to R$$

whose associated graded map with respect to the filtrations associated to the valuations is the surjective graded map of k-algebras

$$\mathrm{gr}\pi : k[(U_i)_{i \in I}] \to \mathrm{gr}_\nu R, \quad U_i \mapsto \overline{\xi}_i.$$

If the valuation ν is of rank one or the set I is finite, any set of representatives $(\xi_i)_{i \in I}$ of the $(\overline{\xi}_i)_{i \in I}$ is eligible.

Since even when the set I is infinite the non-zero homogeneous components of $\mathrm{gr}_\nu R$ are one-dimensional k-vector spaces, and in fact $\mathrm{gr}_\nu R$ is isomorphic to the semigroup algebra $k[t^\Gamma]$ (see [33, Proposition 4.7]), the kernel of the map $\mathrm{gr}\pi$ is a prime ideal generated by binomials $(U^m - \lambda_{mn} U^n)_{(m,n) \in M}$ with $\lambda_{mn} \in k^*$, and the second part of the valuative Cohen theorem states that the kernel of π is generated, up to closure in the w ultrametric, by *overweight deformations* of those binomials: series whose initial forms with respect to the weight are those binomials.

Geometrically this corresponds to equations defining the image of an embedding, via the series ξ_i, of the formal germ corresponding to R in an infinite-dimensional weighted affine space in such a way that the original valuation is the trace on R of a monomial valuation on the ambient space.

In this infinite-dimensional space, our singularity can be degenerated in a faithfully flat way to the "toric variety" defined by the binomials $U^m - \lambda_{mn} U^n$ of $k[(U_i)_{i \in I}]$ (see [33, Proposition 2.3]). However, if the number of variables is infinite, there is no resolution of singularities for such a generalized toric variety. It is truly "infinitely singular" and in fact for valuations centered in $k[[x, y]]$ this corresponds exactly to the "infinitely singular" case where the valuation is the order of vanishing of a series on a very transcendental (non-Puiseux) curve in the plane whose strict transforms remains singular in infinitely many point blowing ups (see [33, Examples 4.20 and 4.22]).

On the other hand, we know that for rational Abhyankar valuations (= of rational rank equal to the dimension of R), one can prove that after a birational modification of R to an R' still dominated by the valuation ring of v, we can obtain that the semigroup of values of the valuation v on R' is finitely generated. For this reason, rational Abhyankar valuations on an equicharacteristic excellent local domain with an algebraically closed residue field can be uniformized (see [29], and [19] and [6] for different approaches for algebraic function fields). This leads to the following conjecture for non-Abhyankar valuations, whose semigroup cannot be finitely generated (see also [34]):

Let R be a complete equicharacteristic local domain with algebraically closed residue field and v a rational valuation centered in R and of rational rank $r <$ dim R. Let $(\gamma_i)_{i \in I}$ be the minimal system of generators for the semigroup Γ of v on R. There exists a nested system of finite subsets $B_\alpha \subset I$ with $\bigcup_\alpha B_\alpha = I$ and for each B_α a prime ideal K_α of R such that R/K_α is of dimension r and endowed with an Abhyankar valuation v_α whose semigroup is generated by the $(\gamma_i)_{i \in B_\alpha}$. We have $\bigcap_\alpha K_\alpha = (0)$ and for each $x \in R$ we have $v(x) = v_\alpha(x \mod.K_\alpha)$ for large enough B_α. Finally each R/K_α is an overweight deformation of an affine toric variety and for large enough B_α toric embedded uniformizations of the valuation v_α also uniformize the valuation v on R.

It is a convenient way to express that the valuation can be uniformized by "finite partial" embedded toric resolutions of Spec$k[t^\Gamma]$, an adapted form of torific embedding for the valuation. The slogan is: *Approximating a rational non-Abhyankar valuation v by rational Abhyankar semivaluations[2] should provide torific embeddings for v.*

In order to apply this to our algebraic situation we have to deduce a torific embedding for an algebraic local ring from a torific embedding of a complete local ring to which we can apply the valuative Cohen theorem. For that it suffices to solve the following problem (see [33, *Proposition 5.19*] and [18]):

Problem B *Given an excellent equicharacteristic local domain R and a valuation v centered in R, show that there exists a prime ideal H of the m_R-adic completion $\hat{R}$ such that $H \cap R = (0)$ and v extends to a valuation $\hat{v}$ of $\hat{R}/H$ with the same value group.*

This means that the graded inclusion $\mathrm{gr}_v R \subset \mathrm{gr}_{\hat{v}} \hat{R}/H$ is birational.

If the valuation is of rank one, the proof is in [18, Theorem 2.1]. For Abhyankar valuations, the proof is in [29, 7.2].

Conjecture 9.1 in [18] to the effect that after a birational v-modification R' of R one can even have that the semigroup of $\hat{R'}/H'$ is the same as that of R' has been

---

[2] A semivaluation of R is a valuation of a quotient of R by a prime ideal.

disproved by Cutkosky in [5, Theorem 1.5, Theorem 1.6] even when completion is replaced by henselization.

Since we assume that the residue field k is algebraically closed, local uniformization of rational valuations entails local uniformization for all valuations (see [33, Proposition 3.20]). By (quasi-)compactness of the Riemann–Zariski manifold, the space of valuations centered in R is (quasi-)compact and therefore there are finitely many valuations such that the collections of morphisms uniformizing them by toric embedded uniformizations uniformizes all valuations centered in R. Now the problem is to:

Problem C *Prove that those torific embeddings can be combined into one embedding for* $\operatorname{Spec} R$ *where a toric birational map will simultaneously uniformize all valuations and thus provide a local embedded resolution of singularities for* R.

One can find some inspiration in [12, §3] as well as in the local tropicalization methods of [28] and [10].

2 Problems Related to the Geometry of Discriminants of Miniversal Unfoldings

Let $f(z_1, \ldots, z_n) \in \mathbf{R}\{z_1, \ldots, z_n\}$ be a series without constant term and such that it has an algebraically isolated critical point at the origin, which means that $\dim_{\mathbf{R}} \mathbf{R}\{z_1, \ldots, z_n\}/(\frac{\partial f}{\partial z_1}, \ldots, \frac{\partial f}{\partial z_n}) < \infty$. This dimension is the Milnor number μ of the isolated critical point associated to the complexification of the series $f(z_1, \ldots, z_n)$ (see [23]). A function with an algebraically isolated critical point is finitely determined, so we may assume that f is a polynomial. Let us consider an unfolding of the function f, say

$$F(z, t) = f(z_1, \ldots, z_n) + \sum_{k=1}^{\mu-1} t_k g_k(z_1, \ldots, z_n),$$

which is miniversal (see [2, Chap. 8]) if the images of the functions $1, g_1, \ldots, g_{\mu-1}$, which again we may take to be polynomials, even monomials, form a basis of the real vector space $\mathbf{R}\{z_1, \ldots, z_n\}/(\frac{\partial f}{\partial z_1}, \ldots, \frac{\partial f}{\partial z_n})$, which we shall henceforth assume.

This unfolding defines a germ of a stable map (see [13, Chap. III, Theorem 3.4] or [25, Chap 5])

$$\mathbf{F} = (F, t) \colon (\mathbf{R}^n \times \mathbf{R}^{\mu-1}, 0) \to (\mathbf{R} \times \mathbf{R}^{\mu-1}, 0)$$

expressed in the natural coordinates $(\lambda, t) = (\lambda, t_1, \ldots, t_{\mu-1})$ on $\mathbf{R} \times \mathbf{R}^{\mu-1}$ by:

$$\lambda \circ \mathbf{F} = f(z_1, \ldots, z_n) + \sum_{k=1}^{\mu-1} t_k g_k(z_1, \ldots, z_n),$$

$$t_k \circ \mathbf{F} = t_k \quad \text{for } k = 1, \ldots, \mu - 1.$$

Because $\mathbf{F}$ is a stable map, it can be Thom-stratified (see [13, 21]) and there exist "polycylinders" $U = \mathbf{B}_n \times \mathbf{B}_{\mu-1} \subset \mathbf{R}^n \times \mathbf{R}^{\mu-1}$ and $V = \mathbf{B}_1 \times \mathbf{B}_{\mu-1} \subset \mathbf{R} \times \mathbf{R}^{\mu-1}$ such that $F^{-1}(V) \cap U$ is a neighborhood of 0 in which the critical locus C is non-singular, and $C \cap (\partial \mathbf{B}_n \times \mathbf{B}_{\mu-1}) = \emptyset$. The only critical points which appear in that neighborhood are those which tend to 0 as $t \to 0$, and each fiber $\mathbf{F}^{-1}(\lambda, t)$ for $(\lambda, t) \in V$ is transversal to $\partial \mathbf{B}_n \times \{t\}$. We shall freely assume that the closed balls $\mathbf{B}_e \subset \mathbf{R}^e$ are "small enough".

Now recall that, assuming of course that 0 is a critical point, up to a change of variables our function can be written as

$$f(z_1, \ldots, z_n) = \sum_{j=1}^{q^+} z_i^2 - \sum_{j=q^++1}^{q^++q^-} z_j^2 + \tilde{f}(z_{q^++q^-+1}, \ldots, z_n),$$

where $\tilde{f}$ is of order ≥ 3 and has the same Milnor number as f.

Then the algebra $\mathbf{R}\{z_1, \ldots, z_n\}/(\frac{\partial f}{\partial z_1}, \ldots, \frac{\partial f}{\partial z_n})$ is naturally isomorphic to $\mathbf{R}\{z_{q^++q^-+1}, \ldots, z_n\}/(\frac{\partial \tilde{f}}{\partial z_{q^++q^-+1}}, \ldots, \frac{\partial \tilde{f}}{\partial z_n})$. A miniversal unfolding $\tilde{F}$ of $\tilde{f}$ is miniversal for f, the only difference between F and $\tilde{F}$ being a fixed difference of indices between the Morse singularities appearing in the unfoldings.

From now on we shall assume that the order of $f(z_1, \ldots, z_n)$ is ≥ 3. Then we may choose $g_i = z_i$ for $i = 1, \ldots, n$ and g_k of order ≥ 2 for $k > n$. The equations for the critical locus $C \subset \mathbf{B}_n \times \mathbf{B}_{\mu-1}$ of the unfolding F are

$$\frac{\partial F}{\partial z_i} = \frac{\partial f}{\partial z_i} + t_i + \sum_{k=n+1}^{\mu-1} t_k \frac{\partial g_k}{\partial z_i} = 0 \text{ for } i = 1, \ldots, n,$$

showing that C is non-singular and of dimension $\mu - 1$.

Shrinking the balls $\mathbf{B}_n$, $\mathbf{B}_{\mu-1}$ if necessary, we assume that the map $\nu: C \to \mathbf{B} \times \mathbf{B}_{\mu-1}$ induced by $\mathbf{F}$ is finite by the Weierstrass preparation theorem (an analytic map with a finite fiber is locally finite). Its image is the real part D of a complex hypersurface, the discriminant $D(\mathbf{C})$ of the complexification of the morphism $\mathbf{F}$ (see [30, §5]). We have seen that C is non-singular, and the map $\nu: C \to D$ is finite. As image of C, and because all maps in sight are algebraic, the discriminant D is a semialgebraic hypersurface in $\mathbf{B}_1 \times \mathbf{B}_{\mu-1}$.

A miniversal unfolding of an algebraically isolated singularity of hypersurface is a versal deformation so that we can lift to D the properties of discriminants of miniversal deformations, and in particular the product decomposition theorem of [35, Chap. III, Théorème 2.1] and [30, Theorem 4.8.1, Cor. 4.8.2] which remains true in real geometry and implies that non-singular points of D are the images of Morse singularities in C. It also implies that at a general point of the codimension one components of the singular locus of the complexification $D(\mathbf{C})$, the singular locus is locally isomorphic either to a cusp ($y^2 - x^3 = 0$) times $\mathbf{C}^{\mu-2}$ (cusp type) or to a node ($y^2 \pm x^2 = 0$) times $\mathbf{C}^{\mu-2}$ (node type).

Proposition 2.1 *The zero set in the critical locus C of the hessian $h_z(F)$ of F with respect to the variables $z_1, \ldots, z_n$ is of codimension one.*

Proof In the real space or in the complexification the image of the zero set of the hessian is the part of the singular locus which is the closure of the set of points of cusp type. The real part of a complex point of cusp type is a real point of cusp type, locally isomorphic to a cusp times $\mathbf{R}^{\mu-2}$.   □

The singular locus of D is of codimension one and its image Δ in $\mathbf{B}_{\mu-1}$ is a semialgebraic hypersurface containing the bifurcation locus Σ and the conflict strata in the sense of bifurcation theory. Indeed, a point $t = (t_1, \ldots, t_{\mu-1})$ is in $\mathbf{B}_{\mu-1} \setminus \Delta$ if and only if the corresponding function $F_t = F(z, t): \mathbf{R}^n \to \mathbf{R}$ is an excellent[3] Morse function in $\mathbf{B}_n$, all of whose Morse singularities tend to 0 as $t \to 0$. In particular the Maxwell set, which corresponds to functions F_t attaining at least twice their absolute minimum, is contained in Δ because it is the image of a singular stratum of D (see [30, 5.4.1] and Michel Coste's examples in [9]).

We know that the geometry of the complex discriminant hypersurface $D(\mathbf{C})$ contains important information on the geometry of the hypersurface of $\mathbf{C}^n$ defined by $f(z_1, \ldots, z_n) = 0$, its deformations and in particular its Milnor fiber (see [30]). The geometry of the discriminant hypersurface is very special. For example, tangent hyperplanes to the discriminant hypersurface at non-singular points tending to the origin all have as limit the hyperplane $\lambda = 0$ (see [30, §5, Remark 3]) and as we have seen a general plane section of $D(\mathbf{C})$ has only cusps and nodes as singularities (see [30, 4.8.2]).

The geometry of the discriminant D in the real case also contains important information. I would like to state two problems concerning this geometry:

Given $f(z_1, \ldots, z_n) \in \mathbf{R}[z_1, \ldots, z_n]$ as above, Michel Herman asked, in the early 1990s, the following question:

If in a neighborhood of 0 the family of hypersurfaces $f(z_1, \ldots, z_n) = \lambda$ is topologically trivial for $|\lambda|$ small enough, do there exist a neighborhood U of 0 and an unfolding $f(z_1, \ldots, z_n) + sg(s, z_1, \ldots, z_n)$ such that for $s \neq 0$ and small enough, the function $f(z_1, \ldots, z_n) + sg(s, z_1, \ldots, z_n)$ has no critical point in U?

---

[3] Meaning Morse function with distinct critical values.

The geometric translation of this statement is that under the hypothesis of topological triviality, for a suitable representative of the germ $\mathbf{F} \colon (\mathbf{R}^n \times \mathbf{R}^{\mu-1}, 0) \to (\mathbf{R} \times \mathbf{R}^{\mu-1}, 0)$, the map $p \circ v \colon C \to \mathbf{B}_{\mu-1}$ which we have seen above is not surjective. Indeed, if that is the case the complement of the image of C being semialgebraic we can find (see [4, Theorem 2.2.5]) in that complement a germ of a semialgebraic arc $t_1(s), \ldots, t_{\mu-1}(s)$ with $t_j(0) = 0$, which will give the unfolding we seek. The converse follows from the versality of the unfolding.

From the equations of the critical locus, we see that it can be endowed with coordinates $z_1, \ldots, z_n, t_{n+1}, \ldots, t_{\mu-1}$ and then the map $p \circ v \colon C \to \mathbf{R}^{\mu-1}$ can be written as follows:

$$t_j \circ (p \circ v) = -\left(\frac{\partial f}{\partial z_j}(z_1, \ldots, z_n) + \sum_{k=n+1}^{\mu-1} t_k \frac{\partial g_k}{\partial z_j}(z_1, \ldots, z_n) \right) \text{ for } 1 \le j \le n,$$

$$t_j \circ (p \circ v) = t_j \text{ for } n+1 \le j \le \mu - 1.$$

The Jacobian matrix of the map $p \circ v$ is therefore related to the hessian matrix $H_z(F)$ of F with respect to the variables $z_1, \ldots, z_n$ as follows:

$$\mathrm{Jac}(p \circ v) = \begin{pmatrix} -H_z(F) & (-\frac{\partial g_k}{\partial z_j}) \\ 0 & \mathrm{Id}_{\mu-1-n} \end{pmatrix}$$

Taking determinants gives :

$$\mathrm{jac}(p \circ v) = (-1)^n h_z(F),$$

and considering signs gives, at each point of C where $\mathrm{jac}(p \circ v) \ne 0$,

$$\mathrm{sign}(\mathrm{jac}(p \circ v)) = (-1)^n (-1)^{\mathrm{index} H_z(F)}.$$

For $t \in \mathbf{R}^{\mu-1} \setminus \Delta$, let $N_i(t)$ be the number of critical points of index i of the Morse function F_t on $\mathbf{B}_n$. Then by definition of the local topological degree (see [11]), we have the equality

$$\deg(p \circ v) = (-1)^n \sum_{i=0}^{n} (-1)^i N_i(t),$$

which is independent of $t \in \mathbf{B}_{\mu-1} \setminus \Delta$.

As t approaches the origin, the discriminant D flattens towards the hyperplane $\lambda = 0$ (see [30, 5.5]). We shall only use the fact that for $t \in p(D)$ the line (λ, t), $\lambda \in \mathbf{R}$, has a maximum intersection point $(\lambda_{\max}(t), t)$ and a minimum intersection point $(\lambda_{\min}(t), t)$ with D, which both tend to 0 with t, and D does not contain the λ axis. If we denote by $X_{\lambda,t}$ the fiber $\mathbf{F}^{-1}(\lambda, t) \subset \mathbf{B}_n \times \{t\}$, we note that since we may assume the $X_{\lambda,t}$ meet $\partial \mathbf{B}_n \times \{t\}$ transversally, all the fibers $X_{\lambda,t}$ for $\lambda > \lambda_{\max}(t)$ (resp $\lambda < \lambda_{\min}(t)$) are diffeomorphic to $X_{\lambda,0}$ with $\lambda > 0$ (resp $X_{\lambda,0}$ with $\lambda < 0$).

By a direct application of Morse theory (see [14, Chapitre 13, exerc. 2.12] and [1, Lemma]), we have for small enough $\epsilon > 0$ the following relations between Euler–Poincaré characteristics:

$$\chi(X_{\lambda_{\max}+\epsilon,t}) - \chi(X_{\lambda_{\min}-\epsilon,t}) = 2\sum_{i=0}^{n}(-1)^i N_i(t) \text{ if } n \text{ is odd}$$
$$\chi(X_{\lambda_{\max}+\epsilon,t}) - \chi(X_{\lambda_{\min}-\epsilon,t}) = 0 \qquad\qquad \text{ if } n \text{ is even.}$$

One can verify that if the family $f(z_1, \ldots, z_n) = \lambda$ is topologically trivial for $|\lambda|$ small enough, so is the family $f(z_1, \ldots, z_n) + w^2 = \lambda$ and as we saw, adding squares of new variables does not change the geometry of the miniversal unfolding.

Topological triviality of the $X_{\lambda,0}$ implies $\chi(X_{\lambda_{\max}+\epsilon,t}) - \chi(X_{\lambda_{\min}-\epsilon,t}) = 0$, so that we have:

Proposition 2.2 *For all n the hypothesis of local topological triviality of the family* $f(z_1, \ldots, z_n) = \lambda$ *implies that the local topological degree of the map* $p \circ v: C \to \mathbf{B}_{\mu-1}$ *is zero.*

And what we want to prove is that this map is not surjective.

When $n = 2$ the result was proved by Gusein–Zade in [17] using an ingenious argument to construct explicit unfoldings without critical points using induction on the Milnor number and resolution of singularities of curves.

In [32] it was suggested to use elimination of critical points as in the proof of the h-cobordism theorem (see [8, 22]). In other words, is the condition $\sum_{i=0}^{n}(-1)^i N_i(t) = 0$ sufficient to make it possible to eliminate all the critical points of a Morse function F_t by moving t in $\mathbf{B}_{\mu-1}$? More generally, we wish to ask the question:

Problem D *Does the geometry of the discriminant D reflect the various configurations of critical points of Morse functions which can appear in differential geometry: For example, can one find values of $t \in \mathbf{B}_{\mu-1}$ such that all the critical points of the same index of F_t are at the same level (have the same critical value)? Can one describe the obstruction to performing elimination of critical points of the functions F_t by movements of $t \in \mathbf{B}_{\mu-1}$?*

For example, it is explained in [31] that if one can find values of $t \in \mathbf{B}_{\mu-1}$ such that all non-degenerate local minima of F_t (stable attractors) are at the same level, one obtains a proof of Thom's catastrophe-theoretic version of the Gibbs phase rule. It states that the maximum number of local minima which a Morse function F_t can have in $\mathbf{B}_n$ (the number of coexisting phases of the system) is at most the codimension in $\mathbf{R}^{\mu-1}$ of the Thom stratum T_0 of the origin, plus one. Since along the Thom stratum the morphology does not vary, the coordinates of a space transversal to T_0 and of complementary dimension in $\mathbf{B}_{\mu-1}$ are the essential parameters of the system.

There are some indications towards a possible proof in [32]. Since the hessian matrix of F is a well defined matrix valued and very special function on the critical locus C, it may be that the study of its image in the Grothendieck–Witt ring, as in [27], is useful. Geometrically, the question is whether the closures of all the open sets of the discriminant hypersurface D which are the images by ν of the sets of points of C where the index of $H_z(F)$ has a given value, have to intersect and whether the closures of some other sets with different indices have to meet in codimension one cusp-type components of the singular locus of D. Two-dimensional slices of the discriminant transversal to such components correspond to plane configurations studied by Jean Cerf (Cerf diagrams, see [7]).

From this viewpoint, we are interested in a dynamical version of Problem **D**: *the problem is to understand which among the deformations of functions that are used in differential geometry, for example those used in the h-cobordism theorem (see [22] and [8]), can be realized by the "small" movements of t in* $\mathbf{B}_{\mu-1}$.

References

1. V.I. Arnol'd. Index of a singular point of a vector field, the Petrovski–Oleinik inequality and mixed Hodge structures. *Funkt. Anal. i. Priloj.* **12**(1), 1–11 (1978).
2. V.I. Arnol'd, S.M. Gusein-Zade and A.N. Varchenko. *Singularities of differentiable maps, Vol. I.* Monographs in Math., Birkhäuser, Basel (1985).
3. F. Aroca, M. Gómez-Morales and H. Mourtada. Groebner Fan and embedded resolution of ideals on toric varieties. *arXiv:2202.10874* (2022).
4. J. Bochnak, M. Coste and M.F. Roy. *Real Algebraic Geometry.* Ergebnisse der Math., Folge 3, Vol. **36**, Springer (1998).
5. S.D. Cutkosky. Extensions of valuations to the Henselization and Completion. *Acta Mathematica Vietnamica* **44**, 159–172 (2019).
6. S.D. Cutkosky. Local uniformization of Abhyankar valuations. To appear in *Mich. Math. J.* Available at *Michigan Math. J.* Advance Publication, 1–33, DOI: https://doi.org/10.1307/mmj/20205888 (2021)
7. J. Cerf. La stratification naturelle des espaces de fonctions différentiables réelles et le théorème de la pseudo-isotopie. *Publications Mathématiques de l'IHES* **39**, 5–173 (1970).
8. J. Cerf et A. Gramain. *Le théorème du h-cobordisme (Smale),* Cours professé au printemps 1966 à la Faculté des Sciences d'Orsay. Secrétariat Mathématique de l'Ecole Normale Supérieure (1968). Available at https://www.maths.ed.ac.uk/~v1ranick/surgery/cerfgram.pdf
9. M. Coste. Une sorcière, trois parapluies, un poisson. In: *Images des Mathématiques* (2010). https://images.math.cnrs.fr/Une-sorciere-trois-parapluies-un-poisson.html
10. M.A. Cueto, P. Popescu Pampu and D. Stepanov. Local tropicalization of Splice Type Surface Singularities. *Preprint,* https://arxiv.org/pdf/2108.05912.pdf
11. D. Eisenbud and H.I. Levine. An algebraic formula for the degree of a C^∞ map germ. *Ann. Math. (2)* **106**(1), 19–44 (1977).
12. A. de Felipe, P. González Pérez and H. Mourtada. Resolving singularities of curves with one toric morphism. *arXiv:2110.11276v1* (2021).
13. C.G. Gibson, K. Wirthmüller, A.A. du Plessis and E.J.N. Looienga. *Topological stability of smooth mappings.* Lecture Notes in Mathematics **552**, Springer (1976).
14. C. Godbillon. *Eléments de topologie algébrique.* Hermann, Collection Méthodes, Paris (1971).
15. R. Goldin and B. Teissier. Resolving singularities of plane analytic branches with one toric morphism. In: *Resolution of singularities (Obergurgl, 1997),* pp. 315–340, Progr. Math. **181**, Birkhäuser, Basel (2000).

16. P.D. González Pérez and B. Teissier. Embedded resolutions of non necessarily normal affine toric varieties. *C. R. Math. Acad. Sci. Paris* **334**(5), 379–382 (2002).
17. S.M. Gusein-Zad. On the existence of deformations without critical points (the Teissier problem for functions of two variables). *Functional Analysis and Its Applications* **31**(I), 58–60 (1997).
18. F.J. Herrera Govantes, M.A. Olalla Acosta, M. Spivakovsky and B. Teissier. Extending a valuation centered in a local domain to the formal completion. *Proc. London Math. Soc.* **105**(3), 571–621 (2012). doi: https://doi.org/10.1112/plms/pds002
19. H. Knaf, F.-V. Kuhlmann. Abhyankar places admit local uniformization in any characteristic. *Ann. Sci. École Norm. Sup.* **38**(4), 833–846 (2005).
20. F. Laudenbach and A. Papadopoulos. W.P. Thurston and French Mathematics. *EMS Surveys in Mathematical Sciences* **6**(1/2), 33–81 (2019). DOI https://doi.org/10.4171/EMSS/32
21. J. Mather. How to stratify mappings and jet spaces. In: *Plans sur Bex 1975*, Springer Lecture Notes in Mathematics **535**, pp. 128–176 (1976).
22. J. Milnor. *Lectures on the h-cobordism Theorem*, Princeton University Press (1965).
23. J. Milnor. *Singular points of complex hypersurfaces*. Annals of Math. Studies **61**, Princeton U.P. (1968).
24. H. Mourtada. Jet schemes and generating sequences of divisorial valuations in dimension two. *Michigan Math. J.* **66**(1), 155–174 (2017).
25. D. Mond and J. Nuño-Ballesteros. *Singularities of Mappings*. Grundlehren der mathematischen Wissenschaften **357**. Springer International Publishing (2020).
26. H. Mourtada and C. Plénat. Jet schemes and minimal toric embedded resolutions of rational double point singularities. *Comm. Algebra* **46**(3), 1314–1332 (2018).
27. S. Pauli and K. Wickelgren. Applications to $\mathbf{A}^1$-enumerative geometry of the $\mathbf{A}^1$-degree. *Research in the Mathematical Sciences* **8**(2), 24 (2021).
28. P. Popescu Pampu and D. Stepanov. Local tropicalization. In: *Algebraic and Combinatorial aspects of Tropical Geometry, Proceedings Castro Urdiales 2011*, E. Brugallé, M.A. Cueto, A. Dickenstein, E.M. Feichtner and I. Itenberg (editors), pp. 253–316, Contemporary Mathematics **589**, AMS (2013).
29. B. Teissier. Overweight deformations of affine toric varieties and local uniformization. In: *Valuation theory in interaction, Proceedings of the second international conference on valuation theory, Segovia-El Escorial, 2011*. Edited by A. Campillo, F-V. Kuhlmann and B. Teissier, pp. 474–565, European Math. Soc. Publishing House, Congress Reports Series, Sept. (2014). Available at https://webusers.imj-prg.fr/~bernard.teissier/documents/Overweightfinal.pdf
30. B. Teissier. The hunting of invariants in the geometry of discriminants. In: *Real and complex singularities (Proc. Ninth Nordic Summer School NAVF Sympos. Math., Oslo, 1976)*, pp. 565–678. Sijthoff and Noordhoff, Alphen aan den Rijn (1977).
31. B. Teissier. Appendix 3: Bifurcations and Gibbs phase rule. In: *Oeuvres Mathématiques de René Thom*, Vol. III, Documents Mathématiques **19**, pp. 167–170, Société Mathématique de France (2022).
32. B. Teissier. Autour d'une question de Michel Herman. Preprint, available at https://webusers. imj-prg.fr/~bernard.teissier/old-papers.html
33. B. Teissier. Valuations, deformations, and toric geometry. In: *Valuation Theory and its applications*, Vol. II, pp. 361–459, Fields Inst. Commun. **33**, AMS., Providence, RI (2003). Available at https://webusers.imj-prg.fr/~bernard.teissier/documents/NewValfinal.pdf
34. B. Teissier. Approximating rational valuations by Abhyankar semivaluations. In: *Workshop on Asymptotic methods in commutative algebra*, Oberwolfach Reports **13**, Issue 4, pp. 3214–3217 (2016).
35. B. Teissier. Cycles évanescents, sections planes, et conditions de Whitney. In: *Singularités à Cargèse*, Astérisque **7–8**, SMF (1973).
36. J. Tevelev. On a question of B. Teissier. *Collectanea Math.* **65**(1), 61–66 (2014). (Published online February 2013. DOI https://doi.org/10.1007/s13348-013-0080-9)
37. J. Tevelev. Compactifications of subvarieties of tori. *American journal of mathematics* **129**(4), 1087–1104 (2007).

Shifted Sheaves for Space-Time

Pierre Schapira

> *On the occasion of the retirement of Catriona Byrne, with my*
> *sincere friendship*

Personal Note

Books are of fundamental importance in Science, for the future, for the readers of course and also for the authors: what better than a book to make your ideas spread? I have a long history of publishing books with Springer, starting with LNM 126 in 1970! But my most significant experience was the publication of the book 292, with Masaki Kashiwara, published in 1990. Catriona Byrne played a prominent role in the process, making suggestions and corrections, offering criticism and, more importantly, providing psychological support. Thank you Catriona!

The book mentioned above is about the microlocal theory of sheaves. Here, three names have to be quoted: Jean Leray, who invented sheaf theory in the 40s, when he was a prisoner of war; Alexander Grothendieck, who gave to this theory its full strength by developing it in the framework of categories[1] and emphasizing the functorial point of view through "the six operations"; and Mikio Sato, who introduced the microlocal point of view, showing that what we think of as "local" on a manifold is in some sense global, it is the projection on the manifold of phenomena occurring in the cotangent bundle. The book with Kashiwara mentioned above is a formulation of this fundamental idea in the language of sheaves.

But what is sheaf theory? It is the mathematical treatment of the dichotomy local/global. Some objects, or properties, are completely different when viewed locally or globally. Some have no local existence, although they globally exist, and

---

[1] We refer to [5] for a philosophical look at this theory.

P. Schapira (✉)
Sorbonne Université, CNRS IMJ-PRG, Paris, France
e-mail: pierre.schapira@imj-prg.fr
http://webusers.imj-prg.fr/~pierre.schapira

© The Author(s), under exclusive license to Springer Nature Switzerland AG 2023   43
J.-M. Morel, B. Teissier (eds.), *Mathematics Going Forward*, Lecture Notes
in Mathematics 2313, https://doi.org/10.1007/978-3-031-12244-6_4

others only exist locally. The Möbius strip is a popular mathematical illustration of this fact: this strip may be locally oriented but traversing around it once, the orientation is reversed. This dichotomy is present in many everyday phenomena, especially in politics, where it is the source of strong conflicts.

And what does microlocal[2] mean? If you are a point x on a real manifold M, to be local means to observe everything in a small ball around you. But the manifold admits a tangent bundle $\tau_M : TM \to M$ and its dual, the cotangent bundle $\pi_M : T^*M \to M$. At x, roughly speaking, the tangent bundle $T_x M$ is the vector space of all light rays passing through x and its dual $T_x^* M$ is the space of all walls passing through x that block the light. The tangent space is more intuitive but the cotangent space, a symplectic manifold, is more important. It is the "phase space" of the physicists. In a word, to be microlocal does not mean to replace M with T^*M but to work on M with T^*M in mind. For example, the micro-support of a sheaf F on M is the set of co-directions in which the sheaf F does not propagate. The micro-support is co-isotropic for the symplectic structure of T^*M, similarly as the characteristic variety of a coherent D-module in the complex case.[3] Indeed, the micro-support of the sheaf of holomorphic solutions of a coherent D-module is nothing but the characteristic variety of the D-module.

The smallest non-empty co-isotropic sets (assuming some weak regularity condition such as, for example, being subanalytic) are the Lagrangian subvarieties. The sheaves whose micro-support are Lagrangian are exactly the constructible sheaves. A sheaf is constructible if there exists a stratification of M along which it is locally constant. Constructible sheaves are of fundamental importance in various branches of mathematics and also in physics. Generically, when it is smooth and the rank of the projection π_M is constant, a Lagrangian submanifold Λ is the conormal bundle to its projection, a submanifold of M. But when the rank of the projection is no longer constant, the projection becomes singular and caustics appear. In order to calculate asymptotic expansions in a neighborhood of a caustic, Viktor Maslov [8] introduced the index of a closed curve in a Lagrangian submanifold. The so-called Maslov index was studied and reformulated by several authors, including [1, 6], until Masaki Kashiwara gave a very simple and elegant description of it (see [4, Appendix]).

When a "pure sheaf" F (a notion that we shall not explain here, referring to [4, § 7.5]) is microlocally supported by a smooth Lagrangian manifold Λ, at each point p of Λ one can attach to F a half integer, called its shift at p, and this shift jumps when the rank of the projection does so. This should be related to what is called "phase transition" in physics.

Let us illustrate this point with space-time and the expansion of the universe (the Big Bang). Of course, what will be written now is purely mathematical, is very rough and does not have the pretension of corresponding to any physical reality. Let

---

[2] See [11] for a detailed survey.

[3] This is an important and difficult theorem of [10] later reformulated and proved purely algebraically in [2].

us represent the universe as a ball of dimension n (of course, for us $n = 3$) whose radius R grows linearly with the time t so that we represent space-time as a closed cone in $\mathbb{R}^4$ with vertex at $t = 0$, similarly as a light cone in a Minkowski space. One asks: what happens for $t < 0$? If one replaces the space-time with the constant sheaf supported by it, the sheaf $\mathbf{k}_{\{|x|\leq t\}}$ (for a given field $\mathbf{k}$), defined on $t \geq 0$, we need to extend it naturally for $t < 0$. The micro-support of this sheaf at the boundary is the interior conormal. If we extend it naturally for $t < 0$ we get the exterior conormal which is the micro-support of the constant sheaf on the open cone. In [3, Exa. 3.10, 3.11] we construct a "distinguished triangle"

$$\mathbf{k}_{\{|x|<-t\}}[n] \rightarrow K \rightarrow \mathbf{k}_{\{|x|\leq t\}} \xrightarrow[\psi]{+1}$$

and the micro-support of K outside the zero-section is the smooth Lagrangian manifold associated with the Hamiltonian isotopy $(x; \xi) \mapsto (x - t\xi/|\xi|; \xi)$. Hence, we get a sheaf K which corresponds to our intuition for $t \geq 0$, and which is the open cone *shifted by the dimension* for $t < 0$, the space at time t_0 being more or less the dual of the space at time $-t_0$ (Fig. 1).

Fig. 1 Before the Big Bang

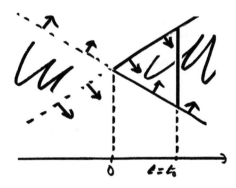

One can modify the Lorentzian case encountered above and consider a similar situation on the n-dimensional unit sphere $M = \mathbb{S}^n$ ($n \geq 2$) endowed with the canonical Riemannian metric. In this case, the sheaf obtained has a shift which jumps by the dimension minus one when $t \in \pi\mathbb{Z}$.

What does it mean to be shifted? For sheaves this is a very common notion. What is called "a sheaf" is indeed an object of the derived category, represented by a complex of sheaves, and the shift of a complex is something familiar and elementary. However, one would like the space at each time t to have a Riemannian structure and the notion of shifted Riemannian structure, or that of shifted Lorentzian structure, has not appeared in the literature, in contrast to what happens with symplectic geometry. Shifted symplectic geometry is a part of what is called derived geometry, based on the new language of ∞-categories.

Conclusion

These two examples of a shift appearing in sheaf theory could suggest another point of view on the Big Bang, a topic which should be treated with lot of care since it attracts non-scientists and is the occasion of much nonsense. Nevertheless we can mention the paper [7], and there is a vast physics literature on this subject, notably under the impulse of Roger Penrose (see for example [9]).

Acknowledgements It is a pleasure to thank Stéphane Guillermou for his advice.

References

1. V.I. Arnold. On a characteristic class entering into conditions of quantization. (In Russian) *Funkcional. Anal. i Prilozen* **1**, 1–14 (1967).
2. O. Gabber. The integrability of the characteristic variety. *Amer. Journ. Math.* **103**, 445–468 (1981).
3. S. Guillermou, M. Kashiwara and P. Schapira. Sheaf quantization of Hamiltonian isotopies and applications to nondisplaceability problems. *Duke Math Journal* **161**, 201–245 (2012).
4. M. Kashiwara and P. Schapira. *Sheaves on manifolds*. Grundlehren der Mathematischen Wissenschaften **292**, Springer-Verlag, Berlin (1990).
5. R. Krömer. *Tool and object*. Science Networks, Historical Studies **32**, Birkhäuser Verlag, Basel (2007).
6. J. Leray. *Analyse Lagrangienne et mécanique quantique*. Collège de France (1976).
7. Y. Manin and M. Marcolli. Big Bang, Blowup, and Modular Curves: Algebraic Geometry in Cosmology. *SIGMA* **10** (2014). arXiv:1402.2158.
8. V.P. Maslov. *Theory of perturbations and asymptotic methods*. (In Russian) Moskow Gos. Univ. (1965).
9. R. Penrose. *Cycles of Time: an extraordinary view of the universe*. Bodley Head (2012).
10. M. Sato, T. Kawai and M. Kashiwara. Microfunctions and pseudo-differential equations. In: *Hyperfunctions and pseudo-differential equations (Proc. Conf., Katata, 1971; dedicated to the memory of André Martineau)*, pp. 265–529, Lecture Notes in Math. **287**, Springer, Berlin (1973).
11. P. Schapira. Microlocal analysis and beyond. In: *New spaces in Mathematics*, edited by M. Anel and G. Catren, pp. 117–152, Cambridge University Press (2021). arXiv:1701.08955.

Lefschetz Fixed Point Theorems for Correspondences

Loring W. Tu

Dedicated to Catriona Byrne on the occasion of her retirement from Springer

2000 Mathematics Subject Classification Primary 58C30; Secondary 32Hxx

The classical Lefschetz fixed point theorem states that the number of fixed points, counted with multiplicity ± 1, of a smooth map f from a compact oriented manifold M to itself can be calculated as the alternating sum $\sum (-1)^k \operatorname{Tr} f^*|_{H^k(M)}$ of the trace of the induced homomorphism in cohomology.[1] This alternating sum is called the *Lefschetz number* $L(f)$ of the map f. As a corollary, if the Lefschetz number $L(f)$ is nonzero, then f has at least one fixed point.

In 1964, at the AMS Woods Hole Conference in Algebraic Geometry, Shimura conjectured an analogue for a holomorphic map of the Lefschetz fixed point theorem. Shimura's conjecture got the people at the conference all excited, and there was a workshop to prove it. At the end of the conference, there were two proofs—an algebraic proof by Verdier, Mumford, Hartshorne, and others, along more or less classical lines from the Grothendieck version of Serre duality, and an

---

[1] Throughout this article $H^*(M)$ denotes de Rham cohomology [4] and the fixed points are assumed to be nondegenerate.

L. W. Tu (✉)
Department of Mathematics, Tufts University, Medford, MA, USA
e-mail: loring.tu@tufts.edu

© The Author(s), under exclusive license to Springer Nature Switzerland AG 2023
J.-M. Morel, B. Teissier (eds.), *Mathematics Going Forward*, Lecture Notes
in Mathematics 2313, https://doi.org/10.1007/978-3-031-12244-6_5

analytic proof by Atiyah and Bott. Grothendieck generalized the algebraic proof in [9, Cor. 6.12, p. 131] and Atiyah and Bott generalized the analytic proof to the Atiyah–Bott fixed point theorem for an elliptic complex in [1, Th. 1, p. 246] and [2, Th. A, p 377].

There was a bit of controversy about this, because afterwards, Shimura's name disappeared from this theorem. It is now called the holomorphic Lefschetz fixed point theorem and the more general version is the Atiyah–Bott fixed point theorem. Shimura was quite upset about this. The principals in this story have all passed away, Atiyah and Shimura in the last 2 years. Fortunately, while they were still living, I was able to interview Michael Atiyah, Raoul Bott, Goro Shimura, and John Tate about the holomorphic Lefschetz fixed point theorem and in 2015 I published an article [13] in the hope of setting the history straight.

In Shimura's recollection, he had conjectured more than the holomorphic Lefschetz fixed point theorem. He said he had made a conjecture for an algebraic correspondence, which for a complex projective variety is the same as a holomorphic correspondence, but he could not remember the statement nor did he keep any notes. He believed that his conjecture for a holomorphic correspondence should have number-theoretic consequences for a Hecke correspondence and higher-dimensional automorphic forms. This article is an exploration of Shimura's forgotten conjecture, first for a smooth correspondence, then for a holomorphic correspondence in the form of two conjectures, and finally an open problem involving an extension to holomorphic vector bundles over two varieties and the calculation of the trace of a Hecke correspondence.

The coincidence locus of two set maps f, $g \colon N \to M$ is the subset of N on which they agree. A coincidence locus is sometimes the fixed-point set of a correspondence and vice versa, but the two types of sets are not the same. In Lefschetz's original paper [11] he obtained a coincidence locus formula for two continuous maps of manifolds. The fixed-point formula for a smooth correspondence (Theorem 5.1) in this article agrees with Lefschetz's coincidence formula when the coincidence is a correspondence. Thus, Theorem 5.1 is essentially already in Lefschetz [11]. It is also a special case of [5] for the trivial group action and of [6, Theorem 4.7, p. 15] for the trivial sheaf. Since Lefschetz's time, there have been many generalizations and variants of his coincidence and fixed-point formulas [5–7, 10, 12]. I offer this article in the hope that a simple-minded proof of a simple-minded statement in the smooth case may spur some interest in the holomorphic case.

At the end of the article, I include as historical documents some emails concerning the conjecture from Shimura to Atiyah and me in 2013. I would like to thank Jeffrey D. Carlson, Mark Goresky, Jacob Sturm, and the anonymous referee for many helpful comments and suggestions.

Fig. 1 A correspondence Γ
on X

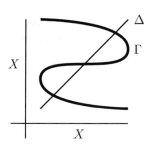

1 Correspondences

Definition 1.1 Let X be a topological space. A ***correspondence*** on X is a subspace $\Gamma \subset X \times X$ such that the two projections $\pi_i : \Gamma \subset X \to X, i = 1, 2$, are covering maps of finite degree (Fig. 1).

A correspondence Γ on X may be viewed as the graph of a multivalued function from X to X whose value at $p \in X$ is the set $\pi_2 \pi_1^{-1}(p)$. By symmetry, it can also be the multivalued function $\pi_1 \pi_2^{-1}$.

We have defined a correspondence in the continuous category. Clearly, it can also be defined in the categories of smooth manifolds and smooth maps, complex manifolds and holomorphic maps, and algebraic varieties and regular maps.

2 Lefschetz Number of a Smooth Correspondence

Suppose $\pi : N \to M$ is a C^∞ covering map of degree r. Denote by $\mathcal{A}^k(N)$ the vector space of smooth k-forms on N. For $\omega \in \mathcal{A}^k(N)$ and $p \in M$, define a k-covector $(\pi_* \omega)_p$ at p on M by

$$(\pi_* \omega)_p (v_1, \ldots, v_k) = \sum_{q_i \in \pi^{-1}(p)} \omega_{q_i}(v_1^i, \ldots, v_k^i),$$

where $v_1, \ldots, v_k \in T_p M$ and $v_1^i, \ldots, v_k^i$ are the unique tangent vectors in $T_{q_i}(N)$ such that $\pi_* v_j^i = v_j$. As p varies over M, the k-covector $(\pi_* \omega)_p$ becomes a k-form $\pi_* \omega$ on M. This defines a pushforward map $\pi_* : \mathcal{A}^k(N) \to \mathcal{A}^k(M)$ of smooth k-forms on N. Since $\pi_* d = d\pi_*$, the pushforward induces a linear map $H^k(N) \to H^k(M)$ in cohomology, also denoted by π_*.

A smooth correspondence induces a linear map on the cohomology of the manifold M by

$$\pi_{1*} \pi_2^* : H^*(M) \to H^*(M).$$

Definition 2.1 The *Lefschetz number* $L(\Gamma)$ of a smooth correspondence Γ is defined to be the alternating sum of the traces of the linear map $\pi_{1*}\pi_2^*$ on $H^k(M)$:

$$L(\Gamma) = \sum_{k=0}^{n} (-1)^k \operatorname{Tr} \pi_{1*}\pi_2^* : H^k(M) \to H^k(M).$$

3 Fixed Points of a Smooth Correspondence

A *fixed point* of a smooth correspondence Γ on a manifold M is a point p in M such that $(p, p) \in \Gamma \cap \Delta$ in $M \times M$, where Δ is the diagonal. The correspondence is called *transversal* if Γ intersects Δ transversally in $M \times M$. In this case, the fixed points are said to be *nondegenerate*. Nondegenerate fixed points are isolated.

When the manifold M is oriented and the correspondence is transversal, we can assign a *multiplicity* or *index* to each fixed point p in the usual way: $\iota_\Gamma(p) = \pm 1$ depending on whether the orientation on the tangent space $T_{(p,p)}(M \times M)$ agrees or disagrees with the orientation on the direct sum $T_{(p,p)}\Gamma \oplus T_{(p,p)}\Delta$. The intersection number $\#(\Gamma, \Delta)$ is then the sum $\sum \iota_\Gamma(p)$, where the sum runs over all fixed points p of the correspondence Γ. When the manifold M is compact, the number of nondegenerate fixed points is finite and the intersection number is defined.

4 The Trace of a Smooth Correspondence

We show how to calculate the trace of a correspondence in terms of differential forms.

Proposition 4.1 *Let $\Gamma \subset M \times M$ be a smooth correspondence on a compact oriented smooth manifold M, $\psi_1, \ldots, \psi_m$ closed $(n - k)$-forms on M representing a basis for $H^{n-k}(M)$, and $\psi_1^*, \ldots, \psi_m^*$ closed k-forms representing the dual basis for $H^k(M)$. Then on $H^k(M)$,*

$$\operatorname{Tr} \pi_{1*}\pi_2^* = \sum_{i=1}^{m} \int_\Gamma \pi_1^* \psi_i \wedge \pi_2^* \psi_i^*.$$

Proof Let $[a_j^i]$ be the matrix of the linear operator $\pi_{1*}\pi_2^*$ on $H^k(M)$:

$$\pi_{1*}\pi_2^*(\psi_j^*) = \sum a_j^i \psi_i^*.$$

Then

$$a_j^i = \int_M \psi_i \wedge \pi_{1*} \pi_2^* \psi_j^*$$

$$= \frac{1}{r} \int_\Gamma \pi_1^* \psi_i \wedge \pi_1^* \pi_{1*} \pi_2^* \psi_j^* \quad \left(\text{because } \int_M \tau = \frac{1}{r} \int_\Gamma \pi_1^* \tau\right)$$

$$= \int_\Gamma \pi_1^* \psi_i \wedge \pi_2^* \psi_j^* \quad \left(\text{because } \omega \wedge \pi_1^* \pi_{1*} \tau = r\omega \wedge \tau\right).$$

Therefore,

$$\operatorname{Tr} \pi_{1*} \pi_2^* = \sum_i a_i^i = \sum_i \int_\Gamma \pi_1^* \psi_i \wedge \pi_2^* \psi_i^*. \qquad \Box$$

5   The Lefschetz Fixed Point Theorem for a Smooth Correspondence

Theorem 5.1 (Lefschetz Fixed Point Theorem for a Smooth Correspondence)
Suppose Γ is a transversal smooth correspondence on a compact, oriented smooth n-manifold M. Then the Lefschetz number of Γ is

$$L(f) = \sum_{\text{fixed points } p} \iota_\Gamma(p).$$

Our proof largely emulates the approach of Griffiths and Harris in their account of the Lefschetz fixed point formula for a smooth self-map [8, Chap. 3, Sec. 4, pp. 419–422], but generalized to a smooth correspondence. The main idea is quite simple. By Poincaré duality, the intersection number #(Γ, Δ) of the correspondence Γ with the diagonal Δ can be calculated as the integral of the wedge product of the differential forms representing their Poincaré duals. On the other hand, with the trace formula of Proposition 4.1, the Lefschetz number of the correspondence Γ can also be calculated in terms of differential forms. The two expressions in differential forms turn out to be equal.
Proof Let $\psi_1, \ldots, \psi_s$ be closed forms on M representing a basis for $H^*(M)$, and $\psi_1^*, \ldots, \psi_s^*$ closed forms representing the dual basis for $H^*(M)$. Note that the forms ψ_i, ψ_j^* run over all degrees, but ψ_i and ψ_i^* have complementary degrees in n. By the Künneth formula, $\pi_1^* \psi_i \wedge \pi_2^* \psi_j$ represent a basis for the cohomology $H^*(M \times M)$. It is proven in [8, p. 420] that the Poincaré dual of the diagonal Δ is given by

$$\eta_\Delta = \sum_i (-1)^{\deg \psi_i^*} \pi_1^* \psi_i \wedge \pi_2^* \psi_i^*.$$

Then

$$L(\Gamma) = \sum_k (-1)^k \operatorname{Tr} \pi_{1*} \pi_2^* |_{H^k(M)}$$

$$= \sum_k (-1)^k \sum_{\deg \psi_i = n-k} \int_\Gamma \pi_1^* \psi_i \wedge \pi_2^* \psi_i^* \qquad \text{(Proposition 4.1)}$$

$$= \int_\Gamma \sum_i (-1)^{\deg \psi_i^*} \pi_1^* \psi_i \wedge \pi_2^* \psi_i^* \qquad (\psi_i \text{ runs over all degrees)}$$

$$= \int_\Gamma \eta_\Delta \qquad \text{(by the formula for } \eta_\Delta)$$

$$= \int_M \eta_\Gamma \wedge \eta_\Delta \qquad \text{(def. of the Poincaré dual } \eta_\Gamma)$$

$$= \#(\Gamma \cdot \Delta) = \sum_{\text{fixed points } p} \iota_\Gamma(p). \qquad \qquad \qquad \square$$

6 A Conjecture for a Holomorphic Correspondence

Let Γ be a *holomorphic correspondence* on a complex manifold M of complex dimension n, that is, a complex submanifold of $M \times M$ such that the two projections $\pi_i : \Gamma \to M$ are holomorphic covering maps. As for a smooth correspondence, a fixed point of the holomorphic correspondence Γ is a point $p \in M$ such that (p, p) is in the intersection $\Gamma \cap \Delta$ in $M \times M$, where Δ is the diagonal in $M \times M$. The correspondence Γ is said to be *transversal* if Γ intersects the diagonal Δ transversally in $M \times M$.

Denote by $\mathcal{O}$ the sheaf of holomorphic functions and $\mathcal{A}^{p,q}$ the sheaf of C^∞ (p, q)-forms on M. Let $\Gamma(M, \mathcal{A}^{p,q})$ be the space of global sections of $\mathcal{A}^{p,q}$; these are simply the C^∞ (p, q)-forms on M. The sheaf $\mathcal{O}$ has an acyclic resolution

$$0 \to \mathcal{O} \to \mathcal{A}^{0,0} \xrightarrow{\bar\partial} \mathcal{A}^{0,1} \xrightarrow{\bar\partial} \mathcal{A}^{0,2} \xrightarrow{\bar\partial} \cdots$$

and the cohomology $H^k(M, \mathcal{O})$ is the cohomology of the differential complex of global sections

$$\Gamma(M, \mathcal{A}^{0,0}) \xrightarrow{\bar\partial} \Gamma(M, \mathcal{A}^{0,1}) \xrightarrow{\bar\partial} \Gamma(M, \mathcal{A}^{0,2}) \xrightarrow{\bar\partial} \cdots .$$

(For background on sheaf cohomology, see [14].)

For a holomorphic covering map $f : N \to M$, both the pullback f^* and the pushforward f_* of C^∞ $(0, k)$-forms are cochain maps of the complexes $\Gamma(N, \mathcal{A}^{0,\bullet})$

and $\Gamma(M, \mathcal{A}^{0,\bullet})$. Since the projection maps $\pi : \Gamma \to M$ are holomorphic covering maps, both the pullback $\pi_2^* : H^*(M, \mathcal{O}) \to H^*(\Gamma, \mathcal{O})$ and the pushforward $\pi_1^* H^*(\Gamma, \mathcal{O}) \to H^*(M, \mathcal{O})$ in cohomology are well-defined. Thus, the holomorphic correspondence Γ induces linear maps of cohomology groups

$$\pi_{1*}\pi_2^* : H^k(M, \mathcal{O}) \to H^k(M, \mathcal{O}), \quad k = 0, \ldots, n.$$

The **holomorphic Lefschetz number** $L(\Gamma, \mathcal{O})$ of Γ is defined to be an alternating sum of traces as before:

$$L(\Gamma, \mathcal{O}) = \sum_{k=0}^{n} (-1)^k \operatorname{Tr} \pi_{1*}\pi_2^* : H^k(M, \mathcal{O}) \to H^k(M, \mathcal{O}).$$

The holomorphic Lefschetz number is a global invariant. Next we define the local contribution at each fixed point. Since a correspondence is a holomorphic covering map of M via π_1, locally it is the graph of a holomorphic function f. At a fixed point p, let $J(\Gamma)$ be the Jacobian matrix of the holomorphic function f with respect to any holomorphic coordinate system.

Conjecture 6.1 *If Γ is a transversal holomorphic correspondence on a compact complex manifold M, then the holomorphic Lefschetz number of Γ is given by*

$$L(\Gamma, \mathcal{O}) = \sum_{\text{fixed points } p} \frac{1}{1 - \det J(\Gamma)_p}.$$

I do not have any evidence for this conjecture other than that it specializes to the correct formula when the correspondence Γ is the graph of a holomorphic map $f : M \to M$. Of course, the simplicity of the statement plays in its favor.

7 Extension to Holomorphic Vector Bundles

In their seminal paper on the fixed point theorem for elliptic complexes [3], Atiyah and Bott extended, as a corollary of their general theorem, the Lefschetz fixed point theorem to a holomorphic vector bundle for a self-map of a compact complex manifold.

To get an idea of what needs to be generalized for a holomorphic correspondence, we give here a brief summary of the Atiyah–Bott result for a holomorphic vector bundle. For more details, consult [3, Section 4, pp. 455–459]. Let E be a holomorphic vector bundle over a compact complex manifold M and $f : M \to M$ a holomorphic map. Denote by $\Gamma(E)$ the vector space of C^∞ sections of E over M and by $\Lambda^{p,q}$ the C^∞ vector bundle of (p, q)-covectors on M. The smooth sections of $E \otimes \Lambda^{p,q}$ are the E-valued (p, q)-forms on M. The $\bar{\partial}$-operator on (p, q)-forms extends to E-valued (p, q)-forms by acting as the identity on E and as $\bar{\partial}$ on the forms. There is then a differential complex

$$\Gamma(E) \xrightarrow{\bar{\partial}} \Gamma(E \otimes \Lambda^{0,1}) \xrightarrow{\bar{\partial}} \Gamma(E \otimes \Lambda^{0,2}) \xrightarrow{\bar{\partial}} \cdots .$$

The cohomology $H^*(M, \mathcal{O}(E))$ of M with coefficients in E is defined to be the cohomology of this complex of E-valued (p, q)-forms.

Now let F be a holomorphic vector bundle over the complex manifold M and let f^*F be its pullback under the holomorphic map $f : M \to M$. The map $f : M \to M$ induces a linear map of C^∞ sections $f^* : \Gamma(F) \to \Gamma(f^*F)$ by sending a section $s \in \Gamma(F)$ to

$$(f^*s)(x) = (s \circ f)(x) = s(f(x)) \in F_{f(x)} = (f^*F)_x, \quad x \in M$$

where $F_{f(x)}$ is the fiber of F at $f(x)$. In order to obtain an endomorphism of $\Gamma(F)$, Atiyah and Bott introduced the notion of a **lifting** of the map f to the bundle F. It is a holomorphic bundle map $\varphi : f^*F \to F$ over M. A lifting φ induces a linear map $\varphi_* : \Gamma(f^*F) \to \Gamma(F)$ by composition: $\varphi_*(s) = \varphi \circ s$. The holomorphic map $f : M \to M$ and a lifting $\varphi : f^*F \to F$ together define an endomorphism of $\Gamma(F)$:

$$\Gamma(F) \xrightarrow{f^*} \Gamma(f^*F) \xrightarrow{\varphi_*} \Gamma(F).$$

Applied to $F = E \otimes \Lambda^{0,k}$, this will then induce an endomorphism

$$(f, \varphi)^* : H^*(M, \mathcal{O}(E)) \to H^*(M, \mathcal{O}(E))$$

and the Lefschetz number of the triple (f, φ, E) is defined to be

$$L(f, \varphi, E) := \sum_{k=1}^{n} (-1)^k \operatorname{Tr} (f, \varphi)^* \big|_{H^k(M, \mathcal{O}(E))}, \quad n = \dim_{\mathbb{C}} M. \tag{1}$$

Theorem 7.1 (Atiyah and Bott [3, Theorem 4.12, p. 458]) *Let E be a holomorphic vector bundle over a compact complex manifold M, $f : M \to M$ a transversal holomorphic self-map, and $\varphi : f^*E \to E$ a holomorphic bundle map. Then*

$$L(f, \varphi, E) = \sum_{f(p)=p} \frac{\operatorname{Tr} \varphi_p}{\det(1 - f_{*,p})}.$$

In this theorem, a **transversal** map is one whose graph intersects the diagonal transversally in $M \times M$, $\varphi_p : E_{f(p)} = E_p \to E_p$ is a complex linear map, and $f_{*,p}$ is the differential of f on the holomorphic tangent space of M at p.

For a holomorphic correspondence Γ and a holomorphic vector bundle E over M, the lifting of a self-map needs to be replaced by some notion of a lifting of the correspondence Γ to the bundle E, which should be a holomorphic bundle map over Γ. Then a plausible conjecture should have the same form as Theorem 7.1.

Conjecture 7.1 *Let E be a holomorphic vector bundle over a compact complex manifold M, $\Gamma \subset M \times M$ a transversal holomorphic correspondence, φ a suitably defined lifting of Γ to E, and $L(\Gamma, \varphi, E)$ a suitably defined Lefschetz number. Then the Lefschetz number $L(\Gamma, \varphi, E)$ satisfies*

$$L(\Gamma, \varphi, E) = \sum_{f(p)=p} \frac{\mathrm{Tr}\, \varphi_p}{\det\left(1 - J(\Gamma)_p\right)},$$

where $J(\Gamma)_p$ is the Jacobian matrix of Γ at (p, p).

In Shimura's emails to Michael Atiyah and Loring Tu in June 2013 (see Appendix), he actually claimed more. He said he had conjectured at Woods Hole in 1965 a Lefschetz fixed point formula for an algebraic correspondence between two holomorphic vector bundles on two algebraic varieties of the same dimension. The statement of this forgotten conjecture remains a mystery.

Stated more generally, Shimura's intention might have been the following (as formulated by Mark Goresky in a recent private communication):

Find and prove a holomorphic Lefschetz fixed point theorem that can be used to calculate the trace of a Hecke correspondence on the holomorphic cohomology, coherent cohomology, or $\bar{\partial}$-cohomology, of a Hermitian locally symmetric space.

Appendix

Email from Goro Shimura to Loring Tu, June 13, 2013

Dear Loring,

It is nice to hear from you. I remember that you sent me your book in collaboration with Bott. Here is my belated thanks for the book!

As for that fixed point formula I can say the following.

In the case of Riemann surfaces, Eichler's result is quite general, and so it was definitely meaningless to conjecture something only for Riemann surfaces.

What I conjectured was a formula for an algebraic correspondence, not just for a map, between two algebraic varieties of the same dimension, so that it generalizes Eichler's formula. (Naturally, we have to (I had to) formulate it in terms of holomorphic bundles.) I thought it might be applicable to automorphic forms on the higher-dimensional spaces.

$\vdots$

As I understand it, the Atiyah–Bott formula deals with only a map, not a correspondence, and so it does not include Eichler's formula, nor does it prove my conjecture. Therefore I think it is an open problem to prove it for a correspondence. Am I wrong?

$$\vdots$$

With best regards,
Goro Shimura

Email from Goro Shimura to Michael Atiyah, June 19, 2013

Dear Michael,

$$\vdots$$

Frankly I am incapable of telling you what exactly my conjecture was. Probably I made notes, but I don't think I can find them.

I can tell you that it concerned an algebraic correspondence between two holomorphic bundles on two base algebraic varieties of the same dimension, consistent with an algebraic correspondence on the base varieties. I formulated it so that it becomes Eichler's formula in the one-dimensional case, and also it becomes a special case of the Lefschetz fixed point formula when the bundles are trivial. I was not considering real analyticity.

$$\vdots$$

With very best regards,
Goro

Note Added in Proofs
Mark Stern proves both Conjectures 6.1 and 7.1 for compact Kähler manifolds in his paper [15, Th. 3.4 and 3.11]. His lifting in Conjecture 7.1 Is a holomorphic bundle map $\varphi \colon \pi_2^* E \to \pi_1^* E$ over the correspondence Γ, where $\pi_i \colon \Gamma \to M$ are the two projections. Shimura's conjecture on two holomorphic bundles on two varieties and its number-theoretic applications remain open.

References

1. M. Atiyah and R. Bott, A Lefschetz fixed point formula for elliptic differential operators, Bull. Amer. Math. Soc. 72 (1966), 245–250.

2. M. Atiyah and R. Bott, A Lefschetz fixed point formula for elliptic complexes: I, Ann. of Math. 86 (1967), 374–407.
3. M. Atiyah and R. Bott, A Lefschetz fixed point formula for elliptic complexes: II, Applications, Ann. of Math. 88 (1968), 531–545.
4. R. Bott and L. W. Tu, *Differential Forms in Algebraic Topology*, third corrected printing, Springer, New York, 1995.
5. I. Dell'Ambrogio, H. Emerson, R. Meyer, An equivariant Lefschetz fixed-point formula for correspondences, Doc. Math. 19 (2014), 141âĂŞ-194.
6. M. Goresky, R. MacPherson, Local contribution to the Lefschetz fixed point formula, Invent. Math. 111 (1993), no. 1, 1–33.
7. M. Goresky, R. MacPherson, The topological trace formula, J. Reine Angew. Math. 560 (2003), 77–150.
8. P. Griffiths and J. Harris, *Principles of Algebraic Geometry*, John Wiley, New York, 1978.
9. A. Grothendieck and L. Illusie, Formule de Lefschetz, Exp. III, in SGA 5 (1966–67), Lecture Notes in Math. 589, Springer, 1977, 73–137.
10. M. Kuga and J. H. Sampson, A coincidence formula for locally symmetric spaces, Amer. J. Math. 94 (1972), 486–500.
11. L. Lefschetz, Intersections and transformations of complexes and manifolds, Transactions of A.M.S. 28 (1926), 1–49.
12. L. Taelman, *Sheaves and functions modulo p*, *Lectures on the Woods Hole trace formula*, London Math. Soc. Lecture Note Series vol. 429, Cambridge University Press, Cambridge, 2016.
13. L. W. Tu, On the genesis of the Woods Hole fixed point theorem, Notices of the AMS 62 (2015), 1200–1205.
14. L. W. Tu, Introduction to sheaf cohomology, ArXiv Math.AT preprint, 26 pages.
15. M. Stern, Fixed point theorems from a de Rham perspective, Asian J. Math. 13 (2009), pp. 65–88.

Part II
Dynamical Systems

The article *Vous avez dit qualitatif ?* of Alain Chenciner is a fascinating mathematical, historical, epistemological and philosophical presentation of a large part of the theory of dynamical systems.

The article *A tale o' pi by pelota* of Ariel Amir and Tadashi Tokieda offers a physical explanation of a result of Gregory Galperin stating that counting the number of collisions between a large mass and a small mass sliding freely on a floor, and a wall, also computes the decimals of pi.

Chère Catriona

Je ne remercierai jamais assez Marcel Berger de m'avoir proposé de lui succéder dans le comité éditorial des Grundlehren et ainsi de m'avoir fait faire votre connaissance. J'avais hésité mais un mot de vous m'avait convaincu : les Grundlehren, c'est une collection pour le long terme, on a le temps ! Et en effet, jamais de délais comminatoires, des rapporteurs à qui l'on laisse le temps de lire, de peser, de décider s'ils pensent que le texte qu'ils ont en main est bien destiné à cette collection qui se veut de référence. Et c'est toujours avec gentillesse et précision que vous prenez acte, orientez, ... Un jour, sachant que je lisais le russe, vous avez proposé de me donner un lot de livres russes qui s'étaient accumulés chez Springer au cours de années ; j'y découvris un bijou : Основные понятия теории вероятностей, première version russe, en 1974, du livre d'Andreï Kolmogorov sur les fondements de la théorie des probabilités que Springer avait publié en allemand en 1933 (Grundbegriffe der Wahrscheinlichkeitsrechnung). Vous rencontrant quelque temps après je vous avais dit mon émerveillement devant ce livre qui, au moment même où je préparais un cours sur la théorie de l'information de Shannon m'avait fait comprendre ce que signifie le choix d'une tribu. Et, émouvante surprise, il portait une dédicace de Kolmogorov lui-même. Devant votre étonnement je vous avais naturellement proposé de vous rendre ce précieux exemplaire mais avec un sourire vous m'aviez dit de le garder.

Kolmogorov est présent dans ce qui suit mais ce sont Poincaré et Thom qui en sont les personnages principaux. Deux français, n'était-ce pas là un bon prétexte pour laisser dans cette langue qui se fait rare chez les scientifiques ce texte issu d'une conférence en 2016 au séminaire de Philosophie et Mathématique de l'ENS ?

Alain Chenciner

Un petit dessin en hommage à votre père (vous reconnaitrez les personnages)

Vous avez dit "qualitatif" ?

Alain Chenciner

Pour Catriona

Les topologues sont les enfants de la nuit [38]

1 Introduction

Dans [22], Youri Iliashenko décrit joliment l'évolution de l'étude mathématique des équations différentielles en distinguant trois périodes :

– celle de Newton : une équation différentielle est donnée. Résolvez-la !
– celle de Poincaré : une équation différentielle est donnée. Décrivez le comportement qualitatif des solutions, sans la résoudre !
– celle d'Andronov : aucune équation différentielle n'est donnée. Décrivez les propriétés qualitatives des solutions !

Dès la deuxième période, caractérisée par la considération de la figure que forment les solutions d'une équation dans l'*espace des phases*, topologie et théorie de la mesure jouent un rôle prépondérant. Dans la troisième, c'est l'espace formé par un ensemble d'équations différentielles, et les structures algébrico – géométriques engendrées par une équation "générique" (ou une famille générique d'équations), qui devient la question, ce point de vue prenant toute sa force chez Thom qui en fait une source de modèles.

Bien entendu, les choses ne sont jamais aussi tranchées : études de stabilité des mécaniciens, théorie des perturbations des astronomes, espace des solutions de Lagrange, théorème d'oscillation de Sturm, ... mais l'irruption de la topologie dans l'étude des équations différentielles (espace des phases, espaces fonctionnels) est bien un phénomène majeur. Et je suis tenté de renverser la phrase de Thom mise en

A. Chenciner (✉)
Département de mathématique, Université Paris VII, Observatoire de Paris, IMCCE (UMR 8028), Paris, France
e-mail: alain.chenciner@obspm.fr
http://www.imcce.fr/Equipes/ASD/person/chenciner/

© The Author(s), under exclusive license to Springer Nature Switzerland AG 2023
J.-M. Morel, B. Teissier (eds.), *Mathematics Going Forward*, Lecture Notes in Mathematics 2313, https://doi.org/10.1007/978-3-031-12244-6_6

exergue et d'affirmer que, même si l'algèbre présente une transparence formelle et une efficacité impressionnante, le sens, et donc la lumière, est plutôt du côté de la topologie. La compréhension d'un phénomène peut-elle être autre que qualitative ?

N.B. *Une première version de ce texte fut écrite à l'occasion d'un exposé au séminaire de philosophie et mathématique de l'E.N.S. le 15 février 2016. Je me suis inspiré librement de passages de [14, 16–18], recopiant même littéralement certains d'entre eux.*

2   En guise de hors-d'œuvre

Les trois périodes qu'Iliashenko identifie dans l'étude mathématique des équations différentielles peuvent être déjà reconnues dans l'étude des équations polynomiales à une variable :

– 1) un polynôme $P(x) = \sum_{i=0}^{n} a_i x^i$ est donné. Trouver les racines réelles, i.e. les $x \in \mathbb{R}$ tels que $P(x) = 0$;

– 2) un polynôme $P(x) = \sum_{i=0}^{n} a_i x^i$ est donné. Décrire qualitativement l'application $P : P_1(\mathbb{C}) \to P_1(\mathbb{C})$ et en particulier l'ensemble des racines ;

– 3) aucun polynôme n'est donné. Décrire qualitativement les racines et leurs bifurcations lorsque les coefficients du polynôme varient.

La première période culmine dans la résolution explicite des équations de degré inférieur ou égal à 4 (Cardan, Tartaglia), la deuxième dans la description de l'application P comme revêtement ramifié de la sphère de Riemann sur elle-même (Riemann...), la dernière dans l'étude de la géométrie des discriminants (Thom...) et la description des bifurcations qui en découle (figure 1).

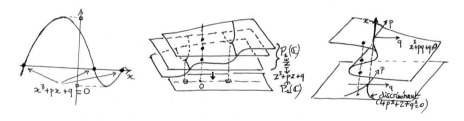

Fig. 1 Les trois périodes de l'étude des racines de polynômes

3   Intégrer une équation différentielle ?

Dès la découverte du calcul différentiel par Leibniz et Newton, on a ramené divers problèmes de géométrie ou de dynamique à la recherche des fonctions $x(t) = \big(x_1(t), x_2(t), \cdots, x_n(t)\big)$ d'une variable t à valeurs dans $\mathbb{R}^n$ qui satisfont à une relation $F(t, x, x', x'', \cdots) \equiv 0$, où $x', x'', \cdots$ désignent les dérivées successives de x. On dit que F est une équation différentielle (ou encore un système d'équations différentielles).

A priori, rien que du calcul et en effet, si l'équation est par exemple *linéaire* et *autonome* (i.e. F est indépendante de t), on sait "calculer" ses solutions : par exemple, dans le cas *scalaire* ($n = 1$), les solutions des équations $x' - 2x = 0$, $x'' = 0$, $x'' + \omega^2 x = 0$ sont respectivement ke^{2t}, $at + b$, $c \sin \omega t + d \cos \omega t$, où k, a, b, c, d sont des constantes. Mais la solution de l'équation du pendule simple $x'' + \omega^2 \sin x = 0$ fait déjà intervenir des fonctions elliptiques ... En l'absence de formule explicite, ce qui est le cas général, on peut chercher, comme le faisait Euler, une solution sous la forme $x(t) = a_0 + a_1 t + a_2 t^2 + \cdots$ d'une série dont les coefficients a_i sont obtenus par identification terme à terme mais se pose alors la question de la convergence : par exemple, la série divergente d'Euler $\sum_{i=0}^{+\infty} (-1)^i i! t^{i+1}$, est une solution formelle de l'équation différentielle $t^2 x' + x = t$.

Lorsque l'équation peut être résolue par rapport à la dérivée d'ordre le plus élevé, par exemple, dans le cas le plus important en physique et en mécanique des équations d'ordre 2, si elle s'écrit $x'' = f(t, x, x')$, on la ramène à une équation du premier ordre en prenant comme inconnues les dérivées d'ordre non maximal, ce qui donne dans l'exemple

$$(x', y') = \big(y, f(t, x, y)\big).$$

Paradigme du *déterminisme*, le théorème général d'existence et d'unicité locale des solutions est dû dans ce cas au baron Cauchy (qu'il faut accompagner des noms de Lipschitz, Peano, Arzela). Il permet en retour de définir de nouvelles fonctions (par exemple l'exponentielle) comme unique solution d'une équation différentielle qui possède telle ou telle propriété. Mais il y a loin d'un théorème d'existence des solutions ou même de leur explicitation sous la forme de développements en séries à la compréhension de leur comportement. Un nouvel objet doit être introduit qui sera le lieu privilégié d'une analyse *qualitative* des solutions.

4   L'espace des phases et ses habitants

C'est en transformant une équation différentielle en *champ de vecteurs* dans l'*espace des phases*, c'est-à-dire en un (système d') équation(s) du premier ordre, que l'on commence à l'appréhender qualitativement. Dans l'exemple ci-dessus, supposant que f ne dépende pas de t, il s'agit du champ sur $\mathbb{R}^n \times \mathbb{R}^n = \mathbb{R}^{2n}$ défini par

$$X(x, y) = \big(y, f(x, y)\big),$$

qu'on interprète comme la donnée en chaque point $(x, y) \in \mathbb{R}^{2n}$ du vecteur de coordonnées $\big(y, f(x, y)\big)$. Résoudre l'équation revient à tracer les *courbes intégrales* $\big(x(t), y(t)\big)$ ayant pour vecteur vitesse en chaque point (x, y) le vecteur

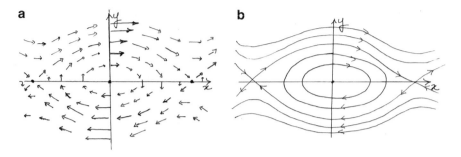

Fig. 2 (**a**) L'équation du pendule ; (**b**) ses courbes intégrales

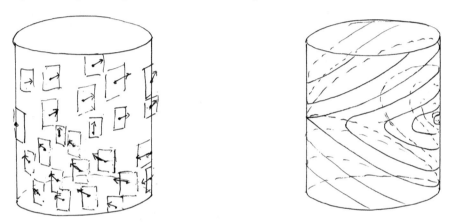

Fig. 3 Le véritable espace des phases du pendule (voir [10, 11])

$X(x, y)$ (figure 2 pour l'équation du pendule, voir également [10]). Nul besoin de calculer pour se faire une idée, certes encore grossière, des solutions.

La notion de champ de vecteurs se généralise immédiatement aux variétés différentiables. Un exemple très simple est encore donné par le pendule, la nature angulaire de la coordonnée x impliquant que le véritable espace des phases est un cylindre (figure 3).

Remarque En représentant simultanément l'ensemble des solutions comme un *feuilletage* de l'espace des phases, on rend en particulier visible le théorème d'existence et d'unicité des solutions à ceci près que, l'équation considérée étant autonome, chaque courbe intégrale correspond à une infinité de solutions ne différant l'une de l'autre que par une translation correspondant aux différents choix du point origine $x(0)$. Pour des équations scalaires ($n = 1$), on peut rajouter le temps et représenter les graphes des solutions qui sont alors complètement séparées comme dans un cristal liquide cholestérique. La figure 4 illustre ceci sur l'équation du second ordre autonome la plus simple $\ddot{x} = 0$.

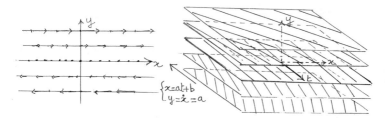

Fig. 4 Séparation des graphes des solutions de $x'' = 0$

Lecture Le début du livre [1] pour comprendre comment la simple représentation géométrique de l'ensemble des configurations que peut présenter un système permet de résoudre par la topologie un problème difficilement accessible au calcul.

5 Intégrer ou décrire ? Le problème des trois corps

On sait résoudre le problème des deux corps, qui se ramène à l'attraction par un centre fixe ; par contre, la tension entre la "résolution" du problème des trois corps par Sundman et la preuve de "non-intégrabilité" de ce problème par Poincaré illustre de façon exemplaire la difficulté inhérente à la notion même de résolution d'une équation différentielle et la nécessité d'une description qualitative des solutions.

5.1 Sundman

Le théorème de Sundman (articles de 1907 et 1909 reproduits dans [34] énonce la possibilité d'écrire des développements convergents pour les solutions du problème des trois corps dont le moment cinétique n'est pas nul. Sundman montre que cette hypothèse proscrit toute collision triple, puis que les collisions doubles se régularisent comme points de branchement. Une transformation algébrique et un changement de temps lui permettent alors de construire la solution sous la forme d'une série qui converge pour toutes les valeurs du nouveau temps. On peut certes présenter ce résultat comme une "résolution du problème des trois corps", mais d'un point de vue pratique ces séries n'apportent rien, d'abord parce qu'elles convergent très lentement, ensuite parce qu'aucun renseignement qualitatif sur la nature de la solution n'est lisible sur la série qui la représente. Quant aux solutions voisines, elles sont tout simplement absentes de la représentation.

Or ce sont justement ces solutions voisines qui sont importantes, non seulement parce que les conditions initiales ne sont jamais connues exactement mais également parce que c'est sur les *équations aux variations* que se lisent les exposants des solutions périodiques et que c'est de la non nullité de ces derniers que Poincaré fait découler la non intégrabilité.

5.2 Poincaré

Bruns [4] avait montré[1] la non-existence d'intégrales premières du problème
newtonien des trois corps qui soient algébriques en les vitesses, autres que celles qui
sont conséquences des symétries du problème, à savoir énergie et moment cinétique.
Par une méthode complètement différente intimement reliée au comportement des
solutions périodiques, Poincaré montre dans les chapitres V et VI des *Méthodes
nouvelles* la non-existence dans le problème planétaire des trois corps de nouvelles
intégrales premières qui soient analytiques sur (une partie de) l'espace des phases,
mais également en les masses planétaires supposées suffisamment petites.
    Comparant son résultat à celui de Bruns, il écrit :

> Le théorème qui précède est plus général en un sens que celui de M. Bruns, ... Mais, en un
> autre sens, le théorème de M. Bruns est plus général que le mien ; j'établis seulement, en
> effet, qu'il ne peut pas exister d'intégrale algébrique pour toutes les valeurs suffisamment
> petites des masses ; et M. Bruns démontre qu'il n'en existe pour aucun système de valeurs
> des masses. ([30] , tome I, section 85)

Si les "solutions périodiques génériques" (i.e., les solutions dont les exposants
sont non nuls) étaient denses dans l'espace des phases, ou simplement si elles
formaient un "ensemble d'unicité",[2] cela impliquerait immédiatement la non-
intégrabilité. En effet, l'existence d'un ensemble complet d'intégrales premières
commutant deux à deux et presque partout indépendantes impliquerait que chaque
orbite périodique sur laquelle les intégrales premières sont indépendantes a tous
ses exposants nuls. Malheureusement, cette propriété de densité, bien que vrai-
semblable, n'est toujours pas prouvée, mais elle a manifestement guidé l'intuition
de Poincaré. La preuve donnée dans les *Méthodes nouvelles* repose sur une
délicate analyse de l'abondance des coefficients non nuls dans le développement
de Fourier de la fonction perturbatrice (qui s'annule avec les masses des planètes),
ce qui revient essentiellement à montrer la non nullité des exposants des solutions
périodiques et interdit donc à ces dernières de former des "tores résonants" (i.e. des
continua remplissant des tores invariants périodiques) comme elles le font dans le
cas "non perturbé" où les masses planétaires s'annulent (figure 5), en rapport étroit
avec le passage de la figure 4 (gauche) à la figure 2 (gauche) obtenu en ajoutant
à l'équation $x'' = 0$ une petite perturbation $\omega^2 \sin x$: du continuum d'équilibres
$y = 0$ ne subsistent que les points $x = 0 \mod \pi, y = 0$).

---

[1] En fait, l'article de Bruns contenait une faute qui fut relevée et corrigée par Poincaré.

[2] i.e., un ensemble tel qu'une fonction analytique s'annulant dessus est identiquement nulle.

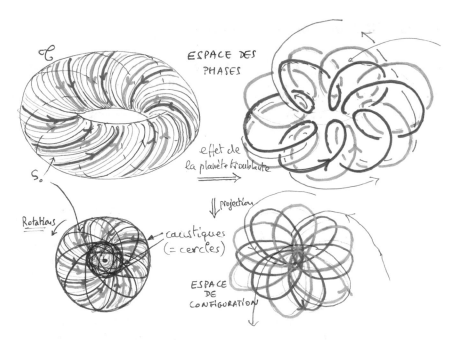

Fig. 5 Destruction de tores résonants (d'après [16])

5.3 Poser les bonnes questions

L'approche qualitative des problèmes est parfaitement exprimée par Poincaré dès 1881 dans l'introduction à la première partie de son *Mémoire sur les courbes définies par une équation différentielle* ([28], voir aussi [14]) : le problème des trois corps y apparaît déjà comme la motivation de son étude qualitative globale des équations différentielles :

> Prenons, par exemple, le problème des trois corps : ne peut-on pas se demander si l'un des corps restera toujours dans une certaine région du ciel ou bien s'il pourra s'éloigner indéfiniment ; si la distance de deux corps augmentera, ou diminuera à l'infini, ou bien si elle restera comprise entre certaines limites ? Ne peut-on pas se poser mille questions de ce genre, qui seront toutes résolues quand on saura construire qualitativement les trajectoires des trois corps ? Et, si l'on considère un nombre plus grand de corps, qu'est-ce que la question de l'invariabilité des éléments des planètes, sinon une véritable question de géométrie qualitative, puisque, faire voir que le grand axe n'a pas de variations séculaires, c'est montrer qu'il oscille constamment entre certaines limites. Tel est le vaste champ de découvertes qui s'ouvre devant les géomètres.

6 Stabilité dans le problème restreint

Transition vers la troisième période dans la description d'Iliashenko, le problème de la stabilité, bien que concernant ici une équation précise (une lune de masse nulle tournant autour d'une terre dont l'orbite serait plane et circulaire), ne fait appel qu'à certaines propriétés qualitatives de celle-ci et non à sa forme exacte.

6.1 Hill

La figure 6 représente les *régions de Hill* dans le problème restreint circulaire plan des trois corps.

Introduites par Hill [20] en 1878, un an avant la thèse de Poincaré, dans son étude du mouvement de la lune, ces régions fournissent une preuve purement topologique, indépendante de toute résolution des équations, d'une forme de stabilité dans le problème ; si la *constante de Jacobi* – i.e. l'énergie du problème dans un repère tournant qui fixe le soleil et la terre – est suffisamment négative, la lune reste confinée à un disque centré sur la terre dont le bord est le contour apparent d'une des composantes connexes de l'hypersurface d'énergie correspondante dans l'espace des phases (de dimensions 4).

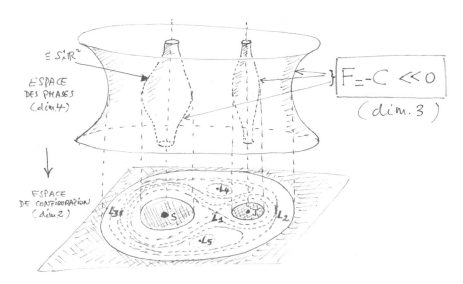

Fig. 6 Régions de Hill (d'après [16])

6.2   Poincaré

Le résultat de Hill, bel exemple de cette *théorie qualitative des équations différentielles* que Poincaré développera dans une série de travaux, n'exclut pas la possibilité de collisions de la lune avec la terre. On sait (voir [16]) que Poincaré croira avoir démontré l'impossibilité de telles collisions dans son Mémoire de 1889 *Sur le problème des trois corps et les équations de la dynamique* et, sa faute une fois reconnue, ne prouvera dans le chapitre XXVI des *Méthodes nouvelles de la mécanique céleste* qu'un résultat de stabilité en moyenne, qu'il appellera *Stabilité à la Poisson*. Il faudra attendre la théorie KAM (acronyme de Kolmogorov, Arnold, Moser), initiée en 1954 par Kolmogorov [23], pour disposer d'un résultat rigoureux d'existence de solutions quasi-périodiques impliquant la stabilité dans le problème considéré, mais entre temps, le résultat de Poincaré aura engendré la *théorie ergodique*, sorte de compromis entre les approches purement topologique et purement analytique des équations, dans laquelle ce ne sont plus les solutions individuelles que l'on cherche à déterminer, mais des ensembles de solutions supports de mesures de probabilité invariantes. Quant à l'erreur du mémoire de 1889, sa correction est à l'origine de la compréhension des *intersections homoclines et hétéroclines de variétés invariantes*[3] (voir [16] figures 18 et 21) et a donné naissance à la bien mal nommée "théorie du chaos". L'une des figures les plus connues attachées à ces phénomènes est celle du *fer à cheval* de Smale (voir [10]) ; il est intéressant de rappeler que c'est en étudiant un article de Cartwright et Littlewood montrant, contrairement à ce qu'il avait conjecturé, l'existence d'équations dissipatives qui possèdent de façon robuste une infinité de solutions périodiques, que Smale découvrit ce dernier. Bel exemple, s'il en faut, de l'intrication du quantitatif et du qualitatif.

6.3   KAM

La vision géométrique de l'espace des phases joue un rôle particulièrement important dans la démonstration par Kolmogorov de la persistance sous l'effet d'une perturbation suffisamment petite de solutions quasi-périodiques "suffisamment non résonantes". Le premier geste géométrique est de considérer non pas simplement une solution quasi-périodique et le problème de Cauchy associé, mais son adhérence dans l'espace des phases. On obtient ainsi un *ensemble invariant dynamiquement significatif,* un *tore invariant lagrangien* dont on peut chercher une *fonction génératrice* comme solution de l'équation de Hamilton-Jacobi ; c'est ce qui avait permis à Poincaré de construire les *séries de Lindstedt* dont Lindstedt lui-même n'avait pu construire que les premiers termes (voir [16]). On assiste bien ici au passage de

---

[3] Celles attachées aux solutions périodiques *hyperboliques* ayant survécu à la perturbation.

l'espace des phases comme pur lieu de calcul à l'espace de phases comme lieu à la fois de topologie et de théorie de la mesure.[4]

7 Généricité

Hassler Whitney ayant démontré qu'étant donné un fermé quelconque F dans $\mathbb{R}^n$, il existe une fonction $f : F \to \mathbb{R}^n$ de classe C^∞ telle que $f^{-1}(0) = F$, parler de la forme d'une hypersurface de niveau d'une telle fonction est désespéré : la différentiabilité n'est pas une restriction plus forte que la simple continuité. Par contre, comme Poincaré, Thom sait bien que, si l'on accepte d'écarter les situations trop singulières, une approche géométrique de tels objets redevient possible. Pour le premier, c'est l'étude des singularités d'équations différentielles définies par des polynomes "les plus généraux de leur degré", l'introduction des probabilités dans l'étude des systèmes dynamiques ou l'affirmation de l'imparité du nombre de géodésiques fermées sur une surface convexe ; pour le second ce seront les stratifications d'espaces fonctionnels et la *théorie des singularités*.

7.1 Le général et le particulier : mesure et catégorie

7.1.1 Poincaré

Dans sa thèse [31], Anne Robadey étudie l'usage des notions de "cas général" et "cas particulier", voire "exceptionnel", dans la formulation et la démonstration de certains "théorèmes" de Poincaré. Un exemple, vivement critiqué par Morse, est l'affirmation que sur une surface convexe, le nombre de géodésiques fermées plongées est impair. Cette affirmation, évidemment fausse pour la sphère ronde, est cependant vraie "en général". D'un autre côté, Poincaré a parfaitement compris que dès qu'un système dynamique atteint un certain niveau de complexité, chercher à décrire précisément toutes les solutions perd son sens et que l'on ne peut qu'essayer de donner des propriétés de la plupart d'entre elles. La notion de probabilité semble bien adaptée à ce but et Poincaré, nullement arrêté par les probabilités continues qui effrayaient tant Joseph Bertrand[5] développe des considérations profondes qui donneront naissance à la *théorie ergodique*. Le *théorème de récurrence*, qu'il démontre afin d'obtenir la forme de stabilité dans le problème restreint des trois corps évoquée dans la section 6.2) en est l'exemple inaugural.

---

[4] Partant d'un système complètement intégrable, on obtient en effet un ensemble de mesure positive de tores invariants.

[5] Poincaré comprend que c'est la notion d'ensemble de mesure nulle qui est bien définie, indépendemment du choix d'une densité de probabilité pourvu que celle-ci ait une régularité suffisante (voir [16] section 8.2, [31] section 3.5.3).

7.1.2 Borel, Baire

De même qu'il est illusoire de chercher à décrire toutes les solutions d'une équation différentielle donnée, il est naturel de chercher à décrire qualitativement non pas toutes les équations mais un sous-ensemble assez "gros" et si possible "naturel" de celles-ci. La notion de "cas général" ou, comme on dit aujourd'hui, de "propriété générique" est multiforme, les points de vue topologique (ensemble maigre au sens de Baire) et probabiliste (ensemble de mesure nulle) s'opposant le plus souvent (voir [7, 8]). On peut voir là l'opposition entre qualitatif et quantitatif mais, comme le montre le beau livre d'Oxtoby [26], la réalité est plus subtile : bien que définissant des univers étrangers l'un à l' autre, les deux principales notions de cas général – reposent sur des théories (théorie de la mesure et théorie de la catégorie de Baire) ayant beaucoup de similarités.

7.2 De la stabilité structurelle aux figures universelles de bifurcations

7.2.1 Andronov, Pontryagin

Introduite en 1937 sous le nom de *systèmes grossiers* par Andronov et Pontryagyn [2], la stabilité structurelle d'une équation, autrement dit le maintien des propriétés qualitatives de ses solutions lorsqu'elle est légèrement perturbée (d'une manière qu'il faut bien entendu préciser techniquement) semble être une condition nécessaire à la pertinence d'une telle équation dans la description d'un phénomène. Si pour des équations dans le plan cette propriété se révèle être générique, il n'en est plus du tout de mème en dimension supérieure (voir [32]).

7.2.2 Thom

Chez Thom, cette exclusion des cas trop particuliers prend une forme très aboutie dans l'étude des espaces de fonctions indéfiniment dérivables (C^∞) sur une variété. Si l'on accepte d'ignorer un sous-ensemble de codimension infinie (i.e. un sous-ensemble que n'importe quelle famille de fonctions à un nombre fini de paramètres pourra éviter au prix d'une éventuelle petite perturbation), on peut affirmer que l'ensemble des zéros possède une structure donnant prise à une étude géométrique. C'est la *théorie des singularités*[6] (voir [9]) qu'à la suite de Morse et Whitney, Thom développe à l'aide des deux outils techniques majeurs que sont le *lemme de transversalité dans les espaces de jets* et la notion de *stratification*. La classification des singularités de germes de fonctions C^∞ de petite codimension est à l'origine de

---

[6] et en particulier la notion de *déploiement versel*.

la *Théorie des catastrophes* (voir [27]) dans laquelle ce sont les figures universelles de bifurcation et non plus les équations particulières qui servent de modèles ; mais l'échec d'une telle classification dans le cas des systèmes dynamiques, où aucune relation d'équivalence n'est satisfaisante, fait qu'une description probabiliste qui oublie les trajectoires exceptionnelles s'impose dans la plupart des cas.

8 Symétries

La notion de généricité que nous venons d'évoquer est très générale puisque qu'elle porte sur l'espace de toutes les équations différentielles ou celui de tous les difféomorphismes sur une variété.

Mais les équations provenant de la physique ou de la mécanique telles le problème des n corps ont des symétries dont le rôle dans la nature qualitative de leurs solutions est déterminant et ce sont donc les sous-espaces d'équations différentielles ou de difféomorphismes déterminés par ces symétries qui, *bien que leur codimension soit infinie*, sont seuls pertinents.

8.1 Symétries et intégrales premières : Noether

Ces équations sont en général les *équations aux variations* (ou *équations d'Euler-Lagrange*) exprimant la stationnarité de l' *intégrale d'action* $\int_a^b L \mathrm{d}t$ associée à un Lagrangien L qui, dans l'exemple du problème des n corps, est une fonction des positions et des vitesses de chacun des corps. Un théorème justement célèbre d'Emmy Noether (voir [24]) montre qu'à chaque champ de vecteurs sur l'espace des configurations[7] dont le flot local laisse invariant le Lagrangien est associée une *intégrale première*, c'est-à-dire une fonction des positions et des vitesses qui reste constante au cours d'un mouvement. Ainsi, l'invariance par les rotations implique la *conservation du moment cinétique*, le caractère autonome, c'est-à-dire le fait que le Lagrangien ne dépende pas du temps, implique la *conservation de l'énergie*.

8.2 Invariants intégraux : Poincaré, Cartan

Déjà présente chez Lagrange (voir [21]), la *structure symplectique* de l'espace des phases[8] prend également son origine dans la nature variationnelle des équations

---

[7] i.e. l'espace des positions dans le cas du problème des n corps.

[8] Techniquement, cette structure apparaît de façon naturelle du côté cotangent c'est-à- dire dans l'espace des positions - impulsions.

d'Euler-Lagrange. Impliquant en particulier la conservation de la mesure (théorème dit de Liouville), cadre dans lequel prend sens le théorème de récurrence évoqué dans la section 7.1.1, c'est l'exemple typique d'un *invariant intégral* : il faut lire au début du troisième tome des *Méthodes nouvelles* ([30] chapitre XXII) la manière superbe qu'a Poincaré d'introduire la notion d'invariant intégral comme intégrale infinitésimale des *équations aux variations* le long d'une solution (voir [16] paragraphe 8.1). Dans son premier livre [5] Elie Cartan généralisera l'invariant intégral de Poincaré en y incluant le temps.

8.3   *Minimiser l'action sous contrainte de symétrie*

Un exemple de raisonnement de type topologique dont l'idée revient encore à Poincaré (voir [15]) est la recherche de solutions du problème des *n* corps ayant un comportement qualitatif prescrit. En particulier, la recherche des solutions "les plus simples" ayant certaines propriétés de symétrie[9] dans le sens où elles minimisent l'action lagrangienne parmi les chemins ayant ces symétries, a conduit à la découverte de nouvelles classes de solutions, les *chorégraphies* et les *Hip-Hops* (voir [25]). Là encore, comme pour les solutions quasi-périodiques de Kolmogorov évoquées dans la section 6.3, la résolution du problème de Cauchy serait inopérante, la théorie étant incapable de déterminer a priori les conditions initiales de telles solutions.

9   Images, formes, noms

9.1   *La faune de la dynamique qualitative : Smale*

Toute une zoologie (une botanique ?) s'est constituée, simultanément pour les équations différentielles et pour leur version discrète, les difféomorphismes : singularités, solutions périodiques, solutions quasi-périodiques, ensembles invariants, leurs variétés stables, instables ou centrales, points homoclines ou hétéroclines, dynamique symbolique, attracteurs et leurs bassins, transitoires, mesures invariantes, entropies métriques ou topologiques, ... L'article-programme [32] que Smale publie en 1967 insiste en particulier sur l'itération des difféomorphismes, version à temps discret (stroboscopie) des équations différentielles. Il met en place les notions (en particulier l'*ensemble non errant* ou Ω-*set*, la *no cycle condition*, la Ω-stabilité) lui permettant d'établir la classification d'un sous-ensemble de difféomorphismes qui, s'il est générique en dimension deux, ne l'est plus du tout en dimension supérieure où une classification raisonnable semble utopique. Je renvoie pour un panorama

---

[9] il s'agit ici de symétries discrètes.

du domaine en 1985 à mon article [10] de l'*Encyclopædia Universalis*. Depuis, la théorie a certes évolué mais les bases conceptuelles sont restées dans une large mesure celles établies par Poincaré. Les images, par contre, issues d'ordinateurs de plus en plus puissants, sont devenues incomparablement plus riches et précises mais il arrive souvent qu'une image plus grossière et délibérément déformée parle mieux à l'imagination . . . et qu'une image trop fidèle nous trompe (le théorème de Cauchy semble faux sur les portraits de phase de certaines équations comportant un très petit paramètre).[10]

9.2   L'importance des images

Les premiers dessins de solutions d'une équation différentielle sous la forme de courbes intégrales feuilletant un espace des phases semblent être ceux de Joukowski et de Poincaré, tous deux dans le cas d'un espace des phases de dimension deux. C'est là que sont répertoriés les divers types de points singuliers génériques, nœuds, cols, foyers et que des théorèmes (Euler-Poincaré, Poincaré Bendixson) régissent l'organisation globale de ces éléments. Bien entendu, dans leurs études de stabilité linéaire de mouvements séculaires, Lagrange et Laplace calculaient déjà des valeurs propres d'équations différentielles linéaires (avant même que la théorie spectrale des matrices soit née) mais la représentation géométrique était absente.

Pour Thom, le choix radical est entre magie et géométrie :

> (. . .) je ne suis pas sûr que dans un univers où tous les phénomènes seraient régis par un schéma mathématiquement cohérent, mais dépourvu de contenu imagé, l'esprit humain serait pleinement satisfait. Ne serait-on pas alors, en pleine magie ? Dépourvu de toute possibilité d'intellection, c'est-à-dire d'interpréter géométriquement le schéma donné, où l'homme cherchera à se créer malgré tout par des images appropriées une justification intuitive au schéma donné, ou sombrera dans une incompréhension résignée que l'habitude transformera en indifférence. En ce qui concerne la gravitation, il n'est pas douteux que la seconde attitude a prévalu ; car nous n'avons, en 1968, pas moins de raisons de nous étonner de la chute d'une pomme que Newton. Magie ou géométrie, tel est le dilemme que pose toute tentative d'explication scientifique. De ce point de vue, les esprits soucieux de compréhension n'auront jamais, à l'égard des théories qualitatives et descriptives, des présocratiques à Descartes, l'attitude méprisante du scientisme quantitatif. [37]

Mais image ne s'oppose pas à schéma mathématique cohérent et souvent elle guide ce dernier : le *fer à cheval* de Smale en témoigne superbement. Un exemple tiré de [12] (voir un résumé dans [13]) concerne le comportement de certaines familles à deux paramètres génériques de difféomorphismes de $\mathbb{R}^2$: pour des valeurs des paramètres voisines d'une certaine courbe Γ, les sous-ensembles

---

[10] Rappelons qu'il est effectivement faux si la régularité de l'équation est plus faible que lipschitzienne. Qu'alors l'unicité ne soit plus assurée est mis en avant par Joseph Boussinesq dans [3] pour justifier de la nécessité de choix (divins ? une autre forme de qualitatif . . . voir l'analyse de [40]) dans l'évolution des phénomènes vitaux.

invariants, courbe, orbite périodique ou ensemble de Cantor, que possèdent les difféomorphismes correspondants s'organisent au voisinage d'un paraboloïde dans l'espace produit du plan des paramètres par le plan de phase. On peut, en première approximation, considérer ce paraboloïde comme le déploiement dans la direction de la courbe Γ d'un plan sur lequel agirait un difféomorphisme générique du disque conservant les aires (figure 7).

L'image – imitant celle de la classique *bifurcation de Hopf* – est bien entendu heuristique et ne participe aucunement à la démonstration qui nécessite d'envisager un à un les divers sous-ensembles invariants, mais elle guide qualitativement la recherche de ceux-ci : un difféomorphisme conservatif appartenant à un sous-ensemble de codimension infinie de l'ensemble des difféomorphismes, on comprend qu'il puisse rassembler les traits d'une infinité de difféomorphismes non conservatifs ; par exemple, il existe au voisinage de Γ un ensemble de Cantor de valeurs des paramètres pour lesquelles le difféomorphisme correspondant possède comme seul ensemble invariant autre que le point fixe une courbe fermée invariante non normalement hyperbolique sur laquelle il est conjugué à une rotation diophantienne, alors que, le difféomorphisme conservatif générique possède un ensemble de Cantor de telles courbes invariantes.[11] De même, toute la complexité de la dynamique conservative dans les *zônes d'instabilité* entre les courbes invariantes se retrouve mais les différents éléments (orbites périodiques elliptiques ou hyperboliques, orbites homoclines, orbites dont l'adhérence est un ensemble de Cantor, ...) apparaissent isolément pour différentes valeurs des paramètres appartenant à une infinité de *bulles* (figure 10 de [13]).

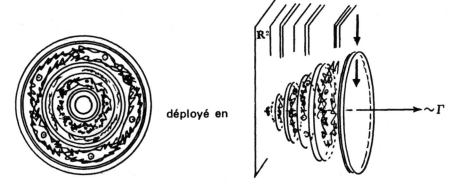

déployé en

Fig. 7 (d'après [12])

---

[11] C'est le théorème de la courbe invariante de Moser, qui fait partie de la galaxie K.A.M..

9.3   La convention du Nom

Finalement, ce qui est en jeu, c'est le langage, le problème des nominations. Nommer exige l'oubli de détails mais la signification est à ce prix. Poincaré ne disait pas autre chose dans [29] en montrant le caractère "conventionnel" de la considération du groupe des déplacements dans la description des mouvements d'un "solide" :

> Quand l'expérience nous apprend qu'un certain phénomène ne correspond pas du tout aux lois indiquées, nous l'effaçons de la liste des déplacements. Quand elle nous apprend qu'un certain changement ne leur obéit qu'approximativement, nous considérons ce changement, par une convention artificielle, comme la résultante de deux autres changements composants. Le premier composant est regardé comme un déplacement satisfaisant rigoureusement aux lois dont je viens de parler, tandis que le second composant, qui est petit, est regardé comme une altération qualitative. Ainsi nous disons que les solides naturels ne subissent pas seulement de grands changements de position, mais aussi de petites flexions et de petites dilatations thermiques.

et en conclusion :

> Tout comme la catégorie de l'espace représentatif, le concept général de groupe est une forme de notre entendement et le groupe des déplacements relève d'une suite de décisions conventionnelles qui adaptent, dans un équilibre réfléchi, notre expérience à la catégorie : En résumé, les lois en question ne nous sont pas imposées par la nature, mais sont imposées par nous à la nature. Mais si nous les imposons à la nature, c'est parce qu'elle nous permet de le faire. Si elle offrait trop de résistance, nous chercherions dans notre arsenal une autre forme qui serait pour elle plus acceptable.

De même, dans le domaine des équations différentielles, le fait qu'une forme reconnaissable et descriptible n'apparaisse qu'au prix de l'exclusion d'équations trop dégénérées était considéré par Thom comme le signe du caractère "naturel" de la théorie :

> Chez les Modernes, imbus de description mathématique, cette même distinction réapparaît avec la distinction classique : "Signal-Bruit". Il n'existe, c'est bien connu, aucun critère intrinsèque permettant, dans un ensemble de données expérimentales, de séparer ce qui va constituer le "Signal" (considéré comme objet scientifiquement recevable), du résidu numérique, qu'on rejettera dans un "Bruit" rebelle à l'analyse. Le signal provient toujours d'une nomologie préexistante, c'est-à-dire d'un ensemble de règles mathématiques censées être valables pour la description des faits considérés. Déjà Aristote avait bien vu que la science ne devait s'occuper que de phénomènes "naturels", c'est-à-dire de phénomènes qui se présentent "le plus souvent" ($\omega\sigma\ \varepsilon\pi\iota\ \tau o\ \pi o\lambda\upsilon$) ; les autres phénomènes, relevant de l'accident, en seront en principe exclus . . . [39]

Difficile enfin d'illustrer l'importance du Nom sans évoquer Alexandre Grothendieck dont la démarche est opposée : il ne s'agit plus d'oublier des détails ou des cas trop particuliers mais au contraire de rechercher dans une généralité maximale, incluant des cas rejetés jusqu'alors comme pathologiques, le naturel[12] d'une théorie. Comme l'écrit Pierre Cartier, le nom précède ici la découverte :

---

[12] à prendre ici au sens catégorique, c'est-à-dire au sens de morphisme de foncteurs.

C'est un maître de la dénomination, il en use comme d'une de ses stratégies intellectuelles majeures. Il a un talent particulier pour nommer les choses avant de se les approprier et de les conquérir, et beaucoup de ses choix terminologiques sont remarquables. [6]

9.4   La variété infinie et joyeuse des formes

(...) il y a une certaine opposition entre géométrie et algèbre. Le matériau fondamental de la géométrie, de la topologie, c'est le continu géométrique ; étendue pure, instructurée, c'est une notion mystique par excellence. L'algèbre, au contraire, témoigne d'une attitude opératoire fondamentalement "diaïrétique". Les topologues sont les enfants de la nuit ; les algébristes, eux, manient le couteau de la rigueur dans une parfaite clarté. ([38])

Pour Thom, qui a souvent affirmé que c'est dans les parties floues et mal formalisées – mais génératrices de formes – des mathématiques qu'il se passe quelque chose qui l'intéresse, le sens est "clairement" du côté de la nuit : s'accordant mal avec la transparence, il ne se déploie que dans une certaine opacité riche de formes rêvées, y compris dans des espaces de dimension supérieure à trois (voir [17]), et ce bien que tout l'édifice des mathématiques semble reposer sur la construction "évidente" des entiers. N'est-ce pas ce que suggère la phrase suivante, qui clôt l'extrait dans lequel Thom affirme péremptoirement que "Tout ce qui est rigoureux est insignifiant", phrase que j'avais essayé de commenter dans [18] ?

Depuis la rupture galiléenne, le savant a toujours recherché le point faible de la nature ; il a toujours essayé d'exploiter les automatismes, la "stupidité" de la nature : la physique est tout entière fondée sur ce manque d'imagination têtu des forces naturelles. Mais de la répétition indéfinie du même acte, l'addition de un, naissent les entiers naturels, l'arithmétique, d'où émerge, en grande partie, la grandiose construction des mathématiques. Ceci nous montre comment, d'un fond d'événements indistinguables, peut sortir la variété infinie et joyeuse des formes. [35]

10   Conclusion

La difficulté qu'il y a à extraire de l'expression explicite d'une solution d'une équation différentielle un renseignement utilisable était exprimée on ne peut plus clairement par Sturm dès 1836 dans l'introduction de ([33]) :

[...] On ne sait les intégrer que dans un très petit nombre de cas particuliers hors desquels on ne peut pas même en obtenir une intégrale première ; et lors même qu'on possède l'expression de la fonction qui vérifie une telle équation, soit sous forme finie, soit en série, soit en intégrales définies ou indéfinies, il est le plus souvent difficile de reconnaître dans cette expression la marche et les propriétés caractéristiques de cette fonction. Ainsi par exemple, on ne voit pas si dans un intervalle donné elle devient nulle ou infinie, si elle change de signe, et si elle a des valeurs *maxima* ou *minima*. Cependant la connaissance de ces propriétés renferme celle des circonstances les plus remarquables que peuvent offrir les nombreux phénomènes physiques et dynamiques auxquels se rapportent les équations différentielles dont il s'agit. S'il importe de pouvoir déterminer la valeur de la fonction inconnue pour une valeur isolée quelconque de la variable dont elle dépend, il n'est pas moins nécessaire de discuter la marche de cette fonction, ou en d'autres termes, d'examiner la forme et la sinuosité de la courbe dont cette fonction serait l'ordonnée variable, en prenant pour abscisse la variable indépendante. Or on peut arriver à ce but par la seule considération des équations différentielles en elles-mêmes, sans qu'on ait besoin de leur intégration.

Le surgissement d'une *théorie qualitative* des équations différentielles (ou plus généralement des *sytèmes dynamiques*, i.e. des actions de groupe générales) était donc inévitable mais on peut se demander si l'acception de cet adjectif a évolué de Poincaré à Thom, Smale ou Arnold. Ne s'agit-il pas chez chacun d'une tentative de description d'un espace des phases "qui fasse sens", avec des outils topologiques, géométriques, probabilistes, symboliques, mais également algébriques et analytiques ? Ceci implique en général que, comme le rappelle Thom, on ne décrive pas une équation isolée mais ce qu'ont en commun un ensemble d'équations et que l'on peut alors nommer. Bien sûr, Poincaré, en particulier dans ses *Méthodes nouvelles*, reste plus près des équations originales et des structures supplémentaires (intégrales premières, invariants intégraux, symétries) qu'elles recèlent, mais il a déjà très clairement la compréhension de la nécessité d'oublier les cas "trop particuliers" si l'on veut tendre à des classifications. Quant à Thom, ce sont surtout les figures universelles de bifurcations qui l'intéressent mais les *catastrophes généralisées* qu'elles engendrent résistent souvent à la description.

Opposer qualitatif à quantitatif n'est ici guère pertinent car on calcule aussi sur des classes d'homologie ou des lois de probabilité et identifier une forme normale n'est après tout que chercher un "bon" changement de coordonnées, version non linéaire de la diagonalisation d'une matrice. Et puis, quoi de plus résolument analytique que les problèmes de petits dénominateurs liés aux mouvements quasi-périodiques, rencontrés par les astronomes dès l'origine de la théorie des perturbations ? Or ces questions sont intimement liées au problème de la stabilité et plus précisément à la généricité de la diffusion dans les perturbations de systèmes hamiltoniens complètement intégrables à au moins trois degrés de liberté, toutes questions dans lesquelles quantitatif et qualitatif se mêlent étroitement.

Finalement, tout sens n'est-il pas de nature qualitative et la théorie qualitative des systèmes dynamiques, qui cherche à faire apparaître des structures identifiables, n'est-elle pas simplement, comme le souhaitait Sturm, la théorie des équations différentielles en ce qu'elles nous disent quelque chose sur les phénomènes qu'elles sont censées représenter ? La discussion ci-dessous, sur laquelle Jean Petitot a attiré mon attention, clôt l'article [36] et me servira de conclusion :

– Dr. Bodmer : What do you mean by a non-quantitative model ? You are still describing a system of equations.
– Dr. Thom : No, I mean a geometric-algebraic structure.
– Dr. Bodmer : How is that defined except by a set of equations ?
– Dr. Thom : By a set of equations defined only up to a homeomorphism. It is a topological configuration. Its study requires qualitative thinking instead of quantitative thinking. I am sorry, but I don't think that quantitative thinking is the answer for all things in nature. In linguistics, it is certainly not the case.

– Dr. Bodmer : I think the distinction between qualitative and quantitative is merely a matter of quantity !

– Dr. Thom : No, No, No.

   (Editor's note : On this hopeful note of common agreement, the conference was ended.)

L'auteur remercie Jean-Michel Morel pour ses relectures constructives et le referee anonyme pour son impitoyable chasse aux coquilles, maladresses et références manquantes.

Références

1. V. Arnold. *Equations différentielles ordinaires*. Ed. Mir-Moscou (1974).
2. A. Andronov & L. Pontryagin. Systèmes grossiers. *Dokl. acad. sci. URSS*, Vol. XIV, n° 5 (1937).
3. J. Boussinesq. *Conciliation du véritable déterminisme avec l'existence de la vie et de la liberté morale*. Paris (1878).
4. H. Bruns. Über die Integrale des Vielkörperproblems. *Acta Mathematica*, Vol. 11, 25–96 (1887–1888).
5. E. Cartan. *Leçons sur les invariants intégraux*. Hermann (1922).
6. P. Cartier. Un pays dont on ne connaîtrait que le nom (Grothendieck et les "motifs"). In : *Le réel en mathématiques : psychanalyse et mathématiques, sous la direction de Pierre Cartier et Nathalie Charraud*, éditions Agalma (2004).
7. A. Chenciner. Vous avez dit "générique" ? *Séminaire de philosophie et mathématique, E.N.S.* 25 avril (1983).
8. A. Chenciner. Stabilité structurelle et ergodicité. *Journal de Physique*, Supplément au n° 8, tome 39 (1978).
9. A. Chenciner. Singularités des fonctions différentiables : la théorie mathématique et ses applications. *Encyclopædia Universalis* (1981).
10. A. Chenciner. Systèmes dynamiques différentiables. *Encyclopædia Universalis* (1985). Cet article ayant été supprimé de la version électronique de l'E.U. sous le prétexte technique d'une difficulté à reproduire les formules (sic !, voir [19]), une copie de la version papier peut être consultée à l'année 1985 dans https://perso.imcce.fr/alain-chenciner/preprint.html.
11. A. Chenciner. Connaissez-vous le pendule ? *Gazette des mathématiciens* 86, octobre (2000).
12. A. Chenciner. Bifurcations de points fixes elliptiques.
    I. *Publications de l'I.H.E.S.* n° 61, 67–127 (1985)
    II. *Inventiones Mathematicæ* 80, 81–106 (1985)
    III. *Publications de l'I.H.E.S.* n° 66, 5–91 (1988).
13. A. Chenciner. Perturbing a planar rotation : normal hyperbolicity and angular twist. In : *Geometry in History* (ed. S.G. Dani and A. Papadopoulos), Springer (2019).
14. A. Chenciner. De la mécanique céleste à la théorie des systèmes dynamiques, aller et retour : Poincaré et la géométrisation de l'espace des phases. *Actes de la conférence "Epistémologie des systèmes dynamiques", Paris Décembre 1999*, Hermann (2001).
15. A. Chenciner. *A note by Poincaré. Regular and Chaotic Dynamics*, V. 10, N° 2 (2005).
16. A. Chenciner. *Poincaré and the three-body problem*. Bourbaphy 2012, Birkhäuser (2014).
17. A. Chenciner. Ombres et traces. Texte écrit à l'occasion d'un exposé au séminaire de l'Association des amis de Jean Cavaillés (ENS, 29 novembre 2014). In : *René Thom : portrait mathématique et philosophique* sous la direction d'Athanase Papadopoulos, Editions du CNRS (2018).
18. A. Chenciner. Le vrai, le faux, l'insignifiant. Conférence dans le colloque Phénomath, 22 mai (2015).

19. A. Chenciner. Le niveau baisse. *Gazette des mathématiciens* 168, 66–67, avril (2021).
20. G.W. Hill. Researches in Lunar Theory. *American Journal of Mathematics*, Vol. 1 n° 1, 5–26 (1878).
21. P. Iglesias. Les origines du calcul symplectique chez Lagrange. *L'Enseignement Mathématique*, t. 44, 257–277 (1998).
22. Y. Iliashenko. Attracteurs des systèmes dynamiques et généricité. *Image des maths* (2005).
23. A.N. Kolmogorov. On the Conservation of Conditionally Periodic Motions under Small Perturbation of the Hamiltonian. *Dokl. akad. nauk SSSR*, vol. 98, 527–530 (1954).
24. Y. Kosmann-Schwarzbach. *Les théorèmes de Noether : invariance et lois de conservation au XXe siècle*. Éditions de l'École Polytechnique, 2e édition (2006).
25. R. Montgomery. *N-body choreographies*. *Scholarpedia* http://www.scholarpedia.org/article/N-body_choreographies.
26. J.C. Oxtoby. *Measure and category*. Springer. First edition (1971). Second edition (1980).
27. J. Petitot. La théorie des catastrophes. *Encyclopædia Universalis*.
28. H. Poincaré. Mémoire sur les courbes définies par une équation différentielle, partie 1 (1881). In : *Œuvres*, tome I.
29. H. Poincaré. On the foundations of geometry. *The Monist* (1898). http://www.jstor.org/stable/27899007?seq=1#page_scan_tab_contents On trouve le texte original de Poincaré sur le site http://www.mathkang.org/cite/confA01.html.
30. H. Poincaré. *Les méthodes nouvelles de la mécanique céleste*. Gauthier-Villars et fils, 3 volumes (1892), (1893), (1899).
31. A. Robadey. *Différentes modalités du travail sur le général dans les recherches de Poincaré sur les systèmes dynamiques*. Thèse Paris 7 et Observatoire de Paris, Janvier (2006). https://halshs.archives-ouvertes.fr/tel-00011380/.
32. S. Smale. Differentiable dynamical systems. *Bull. Amer. Mat. Soc.* 73, 747–817 (1967).
33. C.F. Sturm. Mémoire Sur le Équations différentielles linéaires du second ordre. *Journal de Liouville* I, 106–186 (1836).
34. K. Sundman. Mémoire sur le problème des trois corps. *Acta Mathematica*, Vol. 36, 105–179 (1913).
35. R. Thom. La science malgré tout. *Encyclopædia Universalis*, Organum (1973) (copyright 1968).
36. R. Thom. A Mathematical Approach to Morphogenesis : Archetypal Morphologies. In : *Heterospecific Genome Interaction*, Wistar Institute Symposium Monograph 9, Wistar Institute Press, Tel Aviv (1969).
37. R. Thom. *Stabilité structurelle et morphogénèse* (1968) (publié par Benjamin en 1972).
38. R. Thom. Les racines biologiques du symbolique. (1976), édité dans *Circé*, pp. 40–51 (1978).
39. R. Thom. *Entre la fécondité du Faux et l'insignifiance du Vrai : la voie étroite de la Science...* Academia dei Lincei, Roma, oct. (1989).
40. M. Zerner. Origine et réception des articles de Boussinesq sur le déterminisme. In : *Contra los titanes de la rutina*, Comunidad de Madrid, Publicaciones Oficiales (1994).

Tale o' pi by pilota

Ariel Amir and Tadashi Tokieda

Take masses M, m, and slide M toward m along a frictionless floor ended by a wall; all collisions, between mass m and mass M as well as between wall and mass m, are elastic.

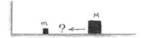

Denote by π_d the total number of inter-mass and wall-mass collisions that occur when $M/m = 100^d$, for $d = 0, 1, 2, \ldots$ *Gedankenexperiment* when $d = 0$, $M/m = 1$: we would then witness collisions 1) m-M, then 2) wall-m, then 3) m-M, after which M escapes to infinity and no more collisions occur. Thus $\pi_0 = 3$.

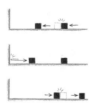

A. Amir
School of Engineering and Applied Sciences, Harvard University, Cambridge, MA, USA
e-mail: arielamir@seas.harvard.edu

T. Tokieda (✉)
Department of Mathematics, Stanford University, Stanford, CA, USA
e-mail: tokieda@stanford.edu

© The Author(s), under exclusive license to Springer Nature Switzerland AG 2023
J.-M. Morel, B. Teissier (eds.), *Mathematics Going Forward*, Lecture Notes
in Mathematics 2313, https://doi.org/10.1007/978-3-031-12244-6_7

If d is large so that $M/m \gg 1$, then m busily rebounds back and forth in the narrowing gap between the wall and the approaching M. After being hit by m many times M begins to recede, the gap widens, and eventually the experiment finishes.

Data reveal

π_0	π_1	π_2	π_3	π_4	π_5	$\cdots$
3	31	314	3141	31415	314159	$\cdots$

Yes, $\pi_d/10^d$ turns out to replicate exactly the leading $d+1$ digits of π.

This result was discovered by Gregory Galperin [1]. In our little note we offer a derivation which, though merely approximate, has the virtue of being physically natural.

Per cycle

m slides toward wall, wall-m collide, m slides toward M, m-M collide

how do the velocities V, v of M, m change? It is intuitive that

$$V_2 \approx V_0 \quad \text{and} \quad v_2 - v_0 \approx 2V_0 \quad \text{if} \quad M/m \gg 1$$

i.e. V hardly changes but v changes by $2V$ per 2 collisions: $\mathrm{d}v/\mathrm{d}\left(\dfrac{\pi_d}{2}\right) \approx 2V$ or

$$\mathrm{d}\pi_d \approx \frac{\mathrm{d}v}{V}.$$

The prudent among us may check the intuition by calculation. The conservations of momentum and energy give for the wall-m collision

$$\begin{bmatrix} V_1 \\ v_1 \end{bmatrix} = \begin{bmatrix} 1 & 0 \\ 0 & -1 \end{bmatrix} \begin{bmatrix} V_0 \\ v_0 \end{bmatrix}$$

and for the m-M collision

$$\begin{bmatrix} V_2 \\ v_2 \end{bmatrix} = \frac{1}{M+m} \begin{bmatrix} M-m & 2m \\ 2M & -(M-m) \end{bmatrix} \begin{bmatrix} V_1 \\ v_1 \end{bmatrix} \approx \begin{bmatrix} 1 & 0 \\ 2 & -1 \end{bmatrix} \begin{bmatrix} V_1 \\ v_1 \end{bmatrix} \quad \text{if} \quad M/m \gg 1.$$

Unsurprisingly these matrices represent reflections, hence have eigenvalues ± 1. They compose as

$$\begin{bmatrix} V_2 \\ v_2 \end{bmatrix} \approx \begin{bmatrix} 1 & 0 \\ 2 & -1 \end{bmatrix} \begin{bmatrix} 1 & 0 \\ 0 & -1 \end{bmatrix} \begin{bmatrix} V_0 \\ v_0 \end{bmatrix} = \begin{bmatrix} 1 & 0 \\ 2 & 1 \end{bmatrix} \begin{bmatrix} V_0 \\ v_0 \end{bmatrix},$$

in particular $v_2 - v_0 \approx 2 V_0$.

Now on the Vv-plane, the phase point $\begin{bmatrix} V \\ v \end{bmatrix}$ hops alternately between the lower plane $\{v < 0 \,|\, \text{wall} \overset{m}{\leftarrow} M\}$ and the upper plane $\{v > 0 \,|\, \text{wall} \overset{m}{\rightarrow} M\}$, while being constrained on the ellipse

$$\frac{1}{2} M V^2 + \frac{1}{2} m v^2 = \text{constant energy } E.$$

Once $\begin{bmatrix} V \\ v \end{bmatrix}$ enters the 'escape wedge' $\{V > 0 \text{ and } |v| < |V|\}$, however, no more collisions occur. On account of hopping, $\begin{bmatrix} V \\ v \end{bmatrix}$ traces the ellipse with density $\frac{1}{2}$. Therefore

$$\pi_d \approx \frac{1}{2} \int_{\text{outside wedge}} \frac{dv}{V}.$$

Since $M/m \gg 1$, the ellipse becomes very tall and very thin, making the integral outside the wedge practically equal to the integral along the whole ellipse. Restricting it to the first quadrant and compensating with a factor of 4,

$$\pi_d \approx \frac{1}{2} \cdot 4 \int_0^{v_{\max} = \sqrt{2E/m}} \frac{dv}{\sqrt{\dfrac{2E - mv^2}{M}}} = 2 \arcsin 1 \cdot \frac{\sqrt{2E/m}}{\sqrt{2E/M}} = \pi \cdot \sqrt{M/m}.$$

The relation $d\pi_d \approx dv/V$, key to the derivation above, is deducible from adiabatic invariance, too. Let w be the width of the gap between the wall and M. We have $V = -\dot{w}$. The time it takes m to slide over w is $w/v \approx 1/\dot{\pi}_d$, inversely proportional to the rate of collision. As long as $|V| \ll |v|$, which holds if $M/m \gg 1$,

$$\text{adiabatic invariant } \ vw \approx \text{constant} \quad \Longrightarrow \quad \frac{\dot{\pi}_d}{\dot{v}} \approx -\frac{w\dot{\pi}_d}{v\dot{w}} \approx \frac{1}{V}.$$

Pilota in the title is a ball game, a sort of squash played in its rustic form with palms of hands, popular throughout southern Europe; children hit the ball against a wall, hit the rebounding ball again, etc. The word is Basque, a loan from neighboring Romance languages (French *pelote*, Spanish *pelota*). And Hebrew is read from right to left, whereas Japanese palindromes are reversed syllable by syllable.

Reference

1. G. Galperin. Playing pool with π. *Regul. Chaotic Dyn.* **8**, 375–394 (2003).

Part III
Finance

Chris Rogers' paper *What next?* is an excellent introduction to this part. In it, he stresses the new fields opening to mathematics because "computational tools make it so much easier to explore than before; new applications demand it; our capacity to explore theoretical mathematical questions has not expanded to the same extent; applications is where we will find new theoretical questions." Even if, he says, "a lot of this could as well be done by physicists, engineers, statisticians, or computer scientists – and is", the point is that the mathematical upbringing is irreplaceable. The author ends up expanding on one class of problems on controlled Markov processes where "it may possibly be that recent computational techniques may allow progress".

In *Some remarks on enlargement of filtration and finance*, Monique Jeanblanc gives a short overview of enlargement (or expansion) of filtration, providing a survey of recent important results, some open questions, and applications in mathematical finance.

In *Modern extreme value theory at the interface of risk management, Bayesian networks and heavy-tailed time series*, Paul Embrechts, Claudia Klüppelberg and Thomas Mikosch present three very different examples of recent research from the realm of modern extreme value theory, at the interface of risk management, Bayesian networks and heavy-tailed time series.

The strong contribution *Limits of limit-order books* of Christopher Almost, John Lehoczky, Steven Shreve and Xiaofeng Yu shows that the diffusion scaling methodology developed to study heavy traffic limits of queueing systems can be adapted to Poisson limit-order book models. The authors show that while the Poisson model is infinite-dimensional with state recording the number of orders at each tick on a doubly infinite price-tick grid, the diffusion limit is low dimensional.

What Next?

L. C. G. Rogers

When I was invited to contribute to this collection, the suggestion from Springer was to try to address the question, "Where is mathematics going?" Well, prediction is notoriously hard to do, especially when it concerns the future, so maybe some reflections on how mathematics has developed in the past could inform where the subject might be going.

When we start our mathematical education, it is easy to think that the lectures we are attending present what mathematics is; it is only later that we realise that what we learned was only what mathematics was at the time. For example, one hundred years ago saw the publication of Watson's definitive account of Bessel functions [9], a work of great scholarship for sure, but a topic that features little in contemporary research. Areas of mathematics open up, are worked for some years, and in time pass into history, only the highlights becoming part of what all students are taught. Most of us feel the necessity to work on subjects that interest our colleagues, otherwise we will not be published or invited to conferences. To stay relevant, we have to be constantly aware of how the subject is changing round us.

New mathematics can come from pure curiosity; or from the needs of applications; or from changes in technology; but in my view it is the last of these which has been most important in the past fifty years. The availability of powerful laptop computing and easy-to-use packages have completely transformed how we do mathematics. Where now is the art form of integration, which those of my generation spent long hours mastering and practising? Coded up in symbolic math packages. How about Watson's Bessel function identities? The same. Why do we need the painstaking drudgery of power-series expansions for special functions when our packages can evaluate them rapidly and accurately? Even those who

L. C. G. Rogers (✉)
Statistical Laboratory, University of Cambridge, Cambridge, UK
e-mail: L.C.G.Rogers@statslab.cam.ac.uk

© The Author(s), under exclusive license to Springer Nature Switzerland AG 2023
J.-M. Morel, B. Teissier (eds.), *Mathematics Going Forward*, Lecture Notes
in Mathematics 2313, https://doi.org/10.1007/978-3-031-12244-6_8

consider themselves pure mathematicians have to be familiar with computational techniques, either to manage laborious symbolic calculations, or more simply to generate random instances of mathematical structures to check whether a conjecture about them might possibly be true.

The insistence on precise statements and correct proof is still the essence of mathematics. This delivers us truths as absolute as humanly possible, as durable as stone, and which can be put together to build structures of amazing beauty and great strength, like great castles. If we stay inside our castles (as many of our colleagues do) we can find endless interest, but we have limited influence on the world outside. If we go outside, our focus shifts from the creation of mathematical results to the use of mathematical methods to answer questions that matter. So for example, in various stochastic algorithms, such as simulated annealing, Markov Chain Monte Carlo, particle filtering, machine learning,..., the ideas of the algorithm can be very simply explained—the mathematics here is not particularly deep. A basic program can easily be written to implement the algorithm; however, your first attempt will at best do moderately well on very simple cases, while whole communities of academics spend their lives adapting and improving the algorithms to cope with more and more challenging examples. It is my belief that mathematics in the next fifty years will deal much more with applications than in the last fifty, because

- computational tools make it so much easier to explore than before;
- new applications demand it;
- our capacity to explore theoretical mathematical questions has not expanded to the same extent;
- applications is where we will find new theoretical questions.

A lot of this could as well be done by physicists, engineers, statisticians, or computer scientists—and is. Can the mathematical training bring something different of value? I believe it can, particularly when things don't work out with the computations, for example. Here the insistence on accuracy pays off. Once coding bugs are weeded out,[1] a common reason the code is hitting problems is because some of the numerical values being computed are either far too big, or far too small. At this point, an understanding of asymptotics is often key to sorting out the issues, and this is something that mathematicians are typically quite good at.[2]

And now let me offer you just one class of problems where I think it may possibly be that recent computational techniques may allow progress ...

---

[1] As an aside, I remember vividly a talk given by a colleague whose code was still not working by the time he had to give the talk; his immortal line was, "There's just one bug left".

[2] So while we may not much need Watson [9] for Bessel functions identities, we may still need it for asymptotic results, and these are beyond the scope of computation—one up for old mathematics!

Controlled Markov Processes

Preliminaries

Informally, a *Markov process*[3] is just a random process $(X_0, X_1, \ldots)$ with the defining property that where it goes next depends only on where it is now, not on how it got there. So in symbols

$$\mathbb{P}(X_{t+1} \in dy \mid X_0 = x_0, \ldots, X_t = x_t) = P(dy \mid x_t) \tag{1}$$

for some *transition function P*. A simple example is random walk $(X_t)_{t \geq 0}$ in d dimensions

$$X_{t+1} = X_t + \varepsilon_{t+1}, \qquad t \geq 0, \tag{2}$$

where the ε_t are independent identically-distributed (IID) $\mathbb{R}^d$-valued random variables, and $X_0 \in \mathbb{R}^d$ is some initial position.

A *controlled Markov process* is like a Markov process, but at each time t you get to choose a control a_t and where the process goes next depends on X_t and a_t:

$$\mathbb{P}(X_{t+1} \in dy \mid X_0 = x_0, a_0 = \alpha_0, \ldots, X_t = x_t, a_t = \alpha_t) = P(dy \mid x_t, \alpha_t). \tag{3}$$

Of course, the control a_t chosen is random, because it will in general depend on the history so far, but it cannot be *anticipating*, that is, it can only depend on information known at time t, not on information to be revealed in the future.

But there is no point introducing controls if we do not at the same time introduce some *objective* which is to be optimized by choice of those controls. A wide range of examples fit into the framework of an *additive objective*, where the aim of the controller is to achieve[4]

$$\max_{a_0, \ldots, a_{T-1}} \mathbb{E}\left[\sum_{t=0}^{T-1} R_t(X_t, a_t) + R_T(X_T) \right], \tag{4}$$

where $T > 0$ is some given time horizon, possibly infinite. The interpretation is clear; you gain a reward $R_t(x, a)$ at time t if $X_t = x$ and you choose action $a_t = a$, and at the final time you gain a reward which is a function of where you reached at time T. Let's suppose from now on that the objective we are given is of the form (4).

---

[3] We just discuss discrete-time Markov processes, which is all that is needed here.

[4] Of course, we should have sup instead of max in (4), but you understand we assume for ease of exposition that the supremum is achieved.

So we could turn the random walk (2) into a controlled Markov process by allowing you to choose a control a_t so that

$$X_{t+1} = X_t + a_t + \varepsilon_{t+1}, \qquad t \geq 0. \tag{5}$$

Given the dynamics (3) and the objective (4) for the problem, what we want to do is to determine optimal controls $a_t^* = u_t^*(X_t)$ which achieve the maximum objective (4). It has been known [3] for around seventy years how to answer this, by solving the *Bellman equations* for the *value function* $(V_t)_{t=0}^T$:

$$V_t(x) = \max_a \mathbb{E}\left[R_t(x, a) + V_{t+1}(X_{t+1}) \,\middle|\, X_t = x \right] \quad (t < T) \tag{6}$$

$$V_T(x) = R_T(x). \tag{7}$$

So you start at the end, and work back one time-step at a time. The optimal control a_t^* is the value of a achieving the maximum in (6), which will of course in general depend on x.

While the Bellman equations (6)–(7) solve the optimal control problem *in principle*, in practice there are very few examples where the value function can be found in closed form; so the issues are

(1) How can we determine numerically an approximation to the optimal policy?
(2) How can we establish bounds on the performance of an approximately optimal policy?

It is impossible and unnecessary to survey here the enormous range of approaches offered to answer these questions, but a few observations are in order:

- if we have found what we think is an approximation $(u_t)_{t=0}^{T-1}$ to the optimal control $(u_t^*)_{t=0}^{T-1}$, we could always simulate what happens when we use this control, and thereby obtain an lower bound[5] for the objective;
- the value function (V_t) is of secondary interest, but the Bellman equations require us to calculate it in order to access the optimal policy;
- in high dimensions, characterizing the value function is numerically difficult;
- the Bellman equations do not give us an upper bound for the value;
- even for the very simplest class of problems, optimal stopping problems, these issues persist.

---

[5] ... subject to simulation error ....

Optimal Stopping Problems

In an *optimal stopping* problem, you are allowed to stop a Markov process at a time of your choosing, and you get a reward at that time which depends on the time and the place when you stop. So your actions have no influence on the dynamics of X, you simply collect a reward $g(\tau, X_\tau)$ at the time τ you choose to stop. For an optimal stopping problem, the Bellman equations become

$$V_t(x) = \max \left\{ g(t, x), \ \mathbb{E}\left[V_{t+1}(X_{t+1}) \mid X_t = x\right] \right\} \quad (t < T) \tag{8}$$

$$V_T(x) = g(T, x), \tag{9}$$

and the value function can be expressed equivalently as

$$V_t(x) = \max_{\tau \in \mathcal{T}_{[t,T]}} \mathbb{E}\left[g(\tau, X_\tau) \mid X_t = x \right], \tag{10}$$

where $\mathcal{T}_{[t,T]}$ denotes the class of stopping times with values in $\{t, \ldots, T\}$.

About fifty years after the start of dynamic programming, there came a remarkable advance in the understanding of optimal stopping problems, discovered independently and contemporaneously by Haugh and Kogan [6] and by me [7]. The result[6] was that the characterization (10) of the value function for an optimal stopping problem has an alternative dual characterization

$$V_t(x) = \max_{\tau \in \mathcal{T}_{[t,T]}} \mathbb{E}\left[g(\tau, X_\tau) \mid X_t = x \right]$$

$$= \inf_{M \in \mathcal{M}} \mathbb{E}\left[\sup_{t \le s \le T} \left\{ g(s, X_s) - M_s + M_t \right\} \mid X_t = x \right], \tag{11}$$

where $\mathcal{M}$ is the space of all martingales. Moreover, the infimum is achieved by taking M to be the martingale M^* of the Snell envelope of the supermartingale $Y_t = V_t(X_t)$:

$$\Delta M_t^* \equiv M_t^* - M_{t-1}^* = V_t(X_t) - \mathbb{E}\left[V_t(X_t) \mid \mathcal{F}_{t-1} \right], \tag{12}$$

where $\mathcal{F}_t$ is the σ-field generated by the process X up to time t. This was an important step for several reasons:

- the characterization (11) provides an *upper* bound for any martingale M;
- having chosen a martingale M, the expression (11) can be evaluated approximately by simulation;
- since it is not necessary to find the value function, the approach is less susceptible to the curse of dimensionality than the standard primal approach (10).

---

[6] ... foreshadowed by work of Davis and Karatzas [4] ....

Andersen and Broadie [1] quickly realised that a 'good' stopping time for an optimal stopping problem could be used to generate a 'good' martingale to use in the dual characterization (11), and the dual approach to optimal stopping problems has been extensively studied in the past twenty years.

Unfinished Business

If there is a dual characterization of optimality for an optimal stopping problem, it is natural then to ask whether there is a dual characterization of optimality for a more general dynamic programming problem. And there is—see [8]. Though most of that paper discusses the problem in the 'weak' formulation (which handles the effect of controls on the evolution by considering the induced change of measure), in Sect. 4.3 of [8] the 'strong' formulation is given, which for the sake of completeness is reproduced here now.

Suppose that the evolution of the controlled process can be specified in terms of the control sequence (a_t) by

$$X_{t+1} = \xi(t, X_t, a_t, \varepsilon_{t+1}) \tag{13}$$

for some function ξ, where the ε_t are IID $U(0, 1)$ random variables; this is exactly what happens when we simulate the evolution of the controlled process. If we are given a sequence of functions (h_t), we define

$$Ph_{t+1}(x, a) = \mathbb{E}\, h_{t+1}(\xi(t, x, a, \varepsilon_{t+1})). \tag{14}$$

The result Theorem 5 of [8] then says the following.

Theorem 1

$$V_0(X_0) = \min_{(h_t)} \mathbb{E}\left[\sup_{(a_t)} \sum_{t=0}^{T-1} \left\{ R_t(X_t, a_t) - h_{t+1}(X_{t+1}) + Ph_{t+1}(X_t, a_t) \right\} + R_T(X_T) \right]. \tag{15}$$

The minimum is achieved by taking $h_t = V_t$.

How could we practically approximate the value function V in such a problem, either using the Bellman equation characterization of V or using the characterization of Theorem 1?

Of course, the theory takes us only a little way, what matters is how quickly and accurately we can compute numerical values in examples. In recent years, advances in machine learning have opened up new avenues for computation, and this has been applied quite successfully to optimal stopping problems—see, for example, the paper [2] of Becker *et al*. Machine learning learns from a large set of training examples a numerical approximation to some arbitrary function which maps inputs to outputs, as in the classic 'Hello World' MNIST dataset[7] of hand-written digits. In that instance, the inputs are the bitmaps of the scanned hand-written digits, and the outputs are the true values of the digits written.

A dynamic programming problem fits into this framework, as first observed by Han [5]. If we think of an additive objective where we try to minimize the expected sum of non-negative costs, then each training example would consist of an $N_{sim} \times T$ array of values for ε, the function to be discovered is the map which takes state and time to action, the output from the calculation would be the expectation of total cost, and the target value would be zero in all cases.

Could this be used to discover approximately the value function using Theorem 1? At the time of [8], I could see no computationally-feasible approach to (15). The main obstacle was that if we were to try to evaluate the right-hand side of (15) for a particular sequence (h_t) we would need to compute the supremum inside the expectation. Now of course this can be done sample path by sample path, but for each sample path we have in effect to solve a (deterministic) dynamic-programming problem, which in general will be quite slow. And this calculation is having to be done N_{sim} times for each sequence (h_t); and the search over sequences (h_t) will require a large number of such calculations. So a slow calculation sits inside a loop inside a loop ...

Could machine learning be a way to a solution?[8] The issues which make evaluating (15) difficult persist, but there are at least well-developed infrastructures for machine learning which gets one into an application quite quickly. On the downside, generic methods rarely work well out of the box on a real example, so we can expect that extensive modification will be needed. A further disadvantage of machine learning in dynamic programming is that the number of timesteps typically cannot be too large; in effect, the time required to solve for T timesteps will be $O(T^2)$, because evaluating the value function at time t requires us to call the values of the value function at times $s > t$. Maybe Theorem 1 will remain a curiosity.

Our Catriona

And all these years as we have been going about our research, and attending meetings around the globe, our paths have been intersecting with that unique phenomenon of the mathematical world, Catriona Byrne. Whether the meeting is in Oberwolfach, Exeter, Milan, Osaka, or Toronto, up she pops with a carry-on bag of recent Springer titles and an order book, a briefcase of manuscripts under consideration, darting here and there in the coffee breaks, or just chatting with the

---

[7] https://deepai.org/dataset/mnist.

[8] Remarkably, the numerical approach used by Haugh and Kogan [6] twenty years ago was based on machine learning.

delegates sitting out an uninteresting talk, sometimes earnest, sometimes laughing, sometimes just talking about this and that, but always elegant and charming. I have spent many a happy hour in her company, whether skipping a dull talk over coffee, or elsewhere, and value her friendship. She is one of the very few mathematical editors who really gets out into the community, and knows first-hand what is going on in the subject at large; she has actually done mathematical research; her ability to stretch Springer's travel budget is almost relativistic—they will not find a comparable replacement. The mathematical community will certainly miss her as she begins her well-earned retirement, but hopefully we will keep in touch. I've suggested some possible topics for my academic colleagues, but just in case Catriona is short of ideas for the years ahead, maybe compile a good travel guide based on her extensive experience? Or perhaps a novel about what mathematicians *really* get up to!? The profession could do with some sympathetic promotion!

But whatever it is to be, dear Catriona, we all wish you a long and happy retirement—though we will likely not meet up so often in the years ahead, the pleasure will be greater when we do!

References

1. L. Andersen and M. Broadie. Primal-dual simulation algorithm for pricing multidimensional American options. *Management Science* **50**(9), 1222–1234 (2004).
2. S. Becker, P. Cheridito and A. Jentzen. Deep optimal stopping. *Journal of Machine Learning Research* **20**, 74 (2019).
3. R.E. Bellman. The theory of dynamic programming. *Bulletin of the American Mathematical Society* **60**(6), 503–515 (1954).
4. M.H.A. Davis and I. Karatzas. A deterministic approach to optimal stopping. In: *Probability, Statistics and Optimisation* (ed. F. P. Kelly), pp. 455–466, New York Chichester: John Wiley & Sons Ltd (1994).
5. J. Han and W. E. Deep learning approximation for stochastic control problems. arXiv preprint arXiv:1611.07422 (2016).
6. M.B. Haugh and L. Kogan. Pricing American options: a duality ap- proach. *Operations Research* **52**(2), 258–270 (2004).
7. L.C.G. Rogers. Monte Carlo valuation of American options. *Mathematical Finance* **12**(3), 271–286 (2002).
8. L.C.G. Rogers. Pathwise stochastic optimal control. *SIAM Journal on Control and Optimization* **46**(3), 1116–1132 (2007).
9. G.N. Watson. *A treatise on the theory of Bessel functions.* Cambridge University Press (1995).

Some Remarks on Enlargement of Filtration and Finance

Monique Jeanblanc

Personal Note

It was a great pleasure to write this paper in honour of my friend Catriona, who so efficiently manages her duties for Springer, providing useful editorial advice, improving the quality of the first versions of submitted books, promoting Springer volumes at many conferences, and contacting authors both old and new. Like many of us, I met Catriona quite often in numerous mathematical workshops. The time we spent together (too short) was always a pleasure. I will miss her during the forthcoming meetings and wish her a pleasant retirement.

1 Introduction

The information about the world is different for all of us. Some of us are specialists in history, others in philosophy and so on. The same is true of financial markets: some of the agents have information, say, in US market, others in European market. We restrict this general framework to the case where two groups of agents on the market may have different information about the dynamics of the same traded asset. If one of them has more information than the remaining ones, and if this new information is useful, she can make profit or even arbitrages.

M. Jeanblanc (✉)
Laboratoire de Mathématiques et Modélisation d'Evry, Université Paris-Saclay, CNRS, Univ Evry, Evry, France
e-mail: monique.jeanblanc@univ-evry.fr

Examples

- The Monty Hall problem is a well-known situation where you can make profit from extra information:

  Suppose you are on a game show, and you are given the choice of three doors: Behind one door is a car; behind the others, goats. You pick a door, say No. 1, and the host, who knows what's behind the doors, opens another door, say No. 3, which has a goat. He then says to you, "Do you want to pick door No. 2?" To increase your probability of winning, you have to change your choice.
- William Duer is widely considered the first to have used his privileged knowledge in a scheme that involved speculating on bank stocks in 1789. Six months later, he resigned from his position after it was discovered that he was taking advantage of his access to confidential information in order to speculate on stocks and bonds (see investopedia).
- Another example took place on 19th June 1815, the day after the battle of Waterloo.

  Nathan Rothschild, who knew about Napoleon's defeat beforehand (thanks to a spy or a carrier pigeon) went to the London Stock Exchange and proceeded to sell his English stocks, causing others to do the same. The resulting Stock Market crash (market impact) enabled Nathan Rothschild's agents to then buy up these assets.

The American Securities and Exchange Commission (the SEC) defines insider trading as follows: An Insider is an officer, director, 10% stockholder or anyone who possesses inside information because of his or her relationship with the Company or with an officer, director or principal stockholder of the Company. Rule 10b-5's application goes considerably beyond just officers, directors and principal stockholders. This rule also covers any employee who has obtained material non-public corporate information, as well as any person who has received a tip from an Insider of the Company concerning information about the Company that is material and nonpublic, and trades (i.e., purchases or sells) the Companys stock or other securities. This is illegal (see the SEC web page for recent cases and more information).

One of the IFRS (International Financial Reporting Standard) rules is that an entity need not undertake an exhaustive search of all possible markets to identify the principal market or, in the absence of a principal market, the most advantageous market.

In both cases, the notion of different information is advanced, and a goal is to try to model the new information and its impact on the market. The new information mainly concerns the behaviour of prices in the future, for example that the firm will close a part of its activity next month. We assume that a group of agents has access to the information described by a filtration $\mathbb{F}$ and we shall model some kinds of new information. Then the insider can use the knowledge of this information to construct a portfolio and have a better terminal wealth.

Here, we do not distinguish "illegal" insider trading from trading with new information. We are only trying to see the impact of new information on the dynamics of prices.

2 Mathematical Facts

2.1 Problem of Enlargement of Filtration

At the beginning this problem was purely a mathematical problem. In the 70s, Itô [24] underlined that, in the case of a Brownian motion B with natural filtration $\mathbb{F}$, in order to give a meaning to $\int_0^t B_1 dB_s$, or more generally to $\int_0^t \theta_s B_1 dB_s$ where θ is an $\mathbb{F}$-adapted process (note that the process $\Psi = (\Psi_t = B_1, \forall t \geq 0)$ is not $\mathbb{F}$-adapted) it is natural to enlarge the filtration $\mathbb{F}$ with the random variable B_1 and to obtain the decomposition of B as a semimartingale[1] in this enlarged filtration (this is known as the Brownian bridge). Recall that the set of semimartingales is the larger space of processes which makes it possible to define a "good" stochastic integration (Bitcheler–Dellacherie–Mokobodzki Theorem [3, Section 1.2.1]). At the same time, independently of each other, P-A. Meyer and D. Williams asked the question: what can be said about $\mathbb{F}$-martingales when one introduces the smallest filtration containing $\mathbb{F}$ and turns a given random time into a stopping time?

A first fact is that the martingale property is not stable under enlargement of filtration. More precisely, if $(\Omega, \mathcal{G}, \mathbb{P})$ is a probability space endowed with two filtrations $\mathbb{F}$ and $\mathbb{K}$ with $\mathbb{F} \subset \mathbb{K}$ (i.e., $\mathcal{F}_t \subset \mathcal{K}_t, \forall t \geq 0$), and if X is a $(\mathbb{P}, \mathbb{F})$-martingale, X can fail to be a $(\mathbb{P}, \mathbb{K})$-martingale. The general problem of enlargement of filtration is the following one. Let $\mathbb{K}$ be a filtration larger than $\mathbb{F}$: under which conditions are all $\mathbb{F}$-martingales $\mathbb{K}$-semimartingales and obtain the $\mathbb{K}$-semimartingale decomposition of any $\mathbb{F}$-martingale. The condition is called the $(\mathcal{H}')$-hypothesis by Jacod [25] and many other authors. As usual, for a filtration $\mathbb{K}$, we denote by $\mathcal{P}(\mathbb{K})$ the predictable σ-algebra and by $\mathcal{O}(\mathbb{K})$ the optional σ-algebra on $\Omega \times \mathbb{R}_+$.

Example 2.1 Let X be an $\mathbb{F}$-martingale of the form $X_t = \mathbb{E}[X_\infty | \mathcal{F}_t]$ where $X_\infty \in \mathcal{F}_\infty$ is integrable, and $\mathcal{K}_t = \mathcal{F}_\infty, \forall t \geq 0$. Then $\mathbb{E}[X_\infty | \mathcal{K}_t] = X_\infty \neq X_t$, and X is not a $\mathbb{K}$-martingale.

Example 2.2 Let $\mathbb{F}$ be the filtration generated by a Brownian motion B and $\mathcal{K}_t = \mathcal{F}_{t+\delta}, \forall t \geq 0$, where $\delta > 0$. In that case, B is not a $\mathbb{K}$-semimartingale (see, e.g., [3, Example 1.19]).

---

[1] A semimartingale is the sum of a local martingale and a process with finite variation. When the finite variation part can be chosen as a predictable process, the semimartingale is said to be special and the decomposition with predictable part is unique.

For financial purposes, semimartingales play an important role: Let S be the $\mathbb{F}$-adapted price process, (eventually d-dimensional) locally bounded and assume that the riskless asset is constant (equal to 1). The fundamental theorem of asset pricing claims that No Free Lunch with Vanishing Risk (NFLVR) holds in $\mathbb{F}$ if and only if there exists a strictly positive $\mathbb{F}$-martingale L such that SL is an $\mathbb{F}$-local martingale, or equivalently if there are no arbitrages (see, e.g., Björk [8, Chapter 10]). Another weaker result states that there is No Unbounded Profits with Bounded Risk (NUPBR) if and only if there exists an $\mathbb{F}$-strictly positive *local* martingale L such that SL is an $\mathbb{F}$-local martingale. These two characterisations require that asset prices are semimartingales under historical probability. The (local) martingale L is called a (local) deflator.

The financial definition of NFLVR and NUPBR is too long to be given in this note, and would not be particularly useful. Instead we refer the reader to Delbaen and Schachermayer [15] and Björk [8, Chapter 10] for NVLVR and Kabanov, Kardaras and Song [31] for NUPBR.

Despite an extensive literature, very few cases are solved and very few concrete examples are known (see a list of examples in [3] and [35]). Studies are mainly concerned by

- Initial Enlargement: A filtration $\mathbb{F}$ being given and ζ being a random variable, one sets $\mathbb{F}^{(\zeta)} = \mathbb{F} \vee \sigma(\zeta)$ (this is the case in Itô, where $\zeta = B_1$). This problem was solved in a quite general setting by Jacod [26]. We shall give a proof under a restrictive condition and recall the general result, without proof.
- Progressive Enlargement: A nonnegative random variable τ is given. We denote by A the indicator process $A_t = \mathbb{1}_{\{\tau \leq t\}}$ and by $\mathbb{A} = (\mathcal{A}_t, t \geq 0)$ its natural filtration. A filtration $\mathbb{F}$ being given, one sets $\mathcal{G}_t = \mathcal{F}_t \vee \mathcal{A}_t, \forall t \geq 0$ (up to a regularization, so that $\mathbb{G}$ is continuous on right). In other words, $\mathbb{G}$ is the smallest filtration containing $\mathbb{F}$ and turning τ into a stopping time. This corresponds to the question of Meyer and Williams. The first mathematical study was done by Barlow [7] for a specific class of random times, called honest times.
- Others: The new information can be the knowledge of a random variable at some random time (see Corcuera and Valkeika [14]) or more generally, two filtrations $\mathbb{F}$ and $\widetilde{\mathbb{F}}$ being given, one studies the enlargement of $\mathbb{F}$ with $\widetilde{\mathbb{F}}$, i.e., $\mathbb{F} \vee \widetilde{\mathbb{F}}$. This general problem was intensively studied by Protter and various coauthors [32, 36].

Up until now, four lecture notes have been dedicated to enlargement of filtration: Aksamit and Jeanblanc [3], Jeulin [29], Jeulin and Yor [30] and Mansuy and Yor [35]. Chapter 20 of Dellacherie et al. [16] contains a very general presentation of enlargement of filtration theory, based on fundamental results of the theory of stochastic processes, developed in the previous chapters and books by the same authors. Chapter X in Jacod [25] presents deep results in a general setting. Chapter 12 in Yor [40] and the book of Mansuy and Yor [35] focus on the case where all martingales in the reference filtration $\mathbb{F}$ are continuous (hypothesis **(C)**). The paper of Nikeghbali [37] also assume hypothesis **(C)** and the fact that τ avoids all $\mathbb{F}$-stopping times (hypothesis **(A)**). A survey containing many exercises can be found in Mallein and Yor [34, Chapter 10]. Protter [38] and Jeanblanc et al. [28] have devoted a chapter of their books to the subject. The lecture by Song [39] contains a general study of the subject. The book of Hillairet and Jiao [23] contains applications to portfolio optimization.

Quite surprisingly, applications of enlargement of filtration theory to finance started only at the end of the 90s with the thesis of Amendinger [4] and independently in the paper by Grorud and Pontier [21].

A basic example is that of point processes with bounded variation, hence the compensated martingale exists in any filtration larger than its natural one. Of course, the compensator depends on the filtration. See [1] and [17].

We shall need the notion of projections, that we recall now. If $\mathbb{H}$ is a filtration satisfying $\mathbb{H} \subset \mathbb{K}$, and Y is a $\mathbb{K}$-adapted process such that $Y_\vartheta \, \mathbb{1}_{\{\vartheta < \infty\}}$ is integrable for any $\mathbb{H}$-stopping time ϑ, the $\mathbb{H}$-*optional projection* of Y is the $\mathbb{H}$-optional process $^{o,\mathbb{H}}Y$ such that $\mathbb{E}[Y_\vartheta \, \mathbb{1}_{\{\vartheta < \infty\}} | \mathcal{H}_\vartheta] = \,^{o,\mathbb{H}}Y_\vartheta \, \mathbb{1}_{\{\vartheta < \infty\}}$, for any $\mathbb{H}$-stopping time ϑ. This optional projection satisfies $\mathbb{E}[Y_t | \mathcal{H}_t] = \,^{o,\mathbb{H}}Y_t$, for all $t \geq 0$. If Y is a càdlàg $\mathbb{K}$-martingale, then $^{o,\mathbb{H}}Y$ is an $\mathbb{H}$-martingale. See, e.g., [3, Section 1.3.1].

2.2 Particular Cases

2.2.1 Discrete Time

In discrete time, any integrable $\mathbb{H}$-adapted process X is an $\mathbb{H}$-special semimartingale. Indeed $X_n = M_n + V_n, \forall n \geq 0$, where

$$M_n = M_{n-1} + X_n - \mathbb{E}[X_n | \mathcal{H}_{n-1}], \quad V_n = V_{n-1} + \mathbb{E}[X_n - X_{n-1} | \mathcal{H}_{n-1}]$$

and $M_0 = X_0, V_0 = 0$. The process M is a martingale, and V is predictable with finite variation. Therefore, if X is an $\mathbb{F}$-martingale and $\mathbb{K}$ a larger filtration, X is a $\mathbb{K}$-special semimartingale and its semimartingale decomposition reduces to computation of the conditional expectations $\mathbb{E}[X_n | \mathcal{H}_{n-1}]$. See Choulli and Deng [12] or Blanchet and Jeanblanc [9] for examples, as well as for initial enlargement and progressive enlargement.

These authors also present the study of arbitrages due to the new information. This is a difficult problem, and the proofs are similar to those in continuous time.

2.2.2 Immersion

Let $\mathbb{F} \subset \mathbb{K}$. Immersion holds between $\mathbb{F}$ and $\mathbb{K}$ if any $\mathbb{F}$-local martingale is a $\mathbb{K}$-local martingale: this is equivalent to, for any $t \geq 0$, the σ-fields $\mathcal{F}_\infty$ and $\mathcal{K}_t$ being conditionally independent given $\mathcal{F}_t$, i.e., $\forall t \geq 0$, $\forall K_t \in \mathcal{K}_t, \forall F_\infty \in \mathcal{F}_\infty$, both being square integrable

$$\mathbb{E}[K_t\, F_\infty | \mathcal{F}_t] = \mathbb{E}[K_t | \mathcal{F}_t]\, \mathbb{E}[F_\infty | \mathcal{F}_t]\,.$$

This case (also called the $(\mathcal{H})$-hypothesis) was presented in Brémaud and Yor [10] and this hypothesis is assumed in many studies, in particular for progressive enlargement in a credit risk framework. Roughly speaking, it means that the new information contained in $\mathcal{K}_t$ has no influence on the past information $\mathcal{F}_t$. Note that if $\mathbb{F} \subset \mathbb{K} \subset \mathbb{G}$ and $\mathbb{F}$ is immersed in $\mathbb{G}$, then $\mathbb{F}$ is immersed in $\mathbb{K}$ (but $\mathbb{K}$ can fail to be immersed in $\mathbb{G}$). A nice property is that under immersion, NFLVR is preserved. Indeed, if L is an $\mathbb{F}$-deflator, it is a $\mathbb{G}$-positive martingale as well and a $\mathbb{G}$-deflator, the process SL being an $\mathbb{F}$ and a $\mathbb{G}$-local martingale.

Immersion is not stable under change of probability (see [13]).

Example An important example is the one introduced by Lando [33]. Given a filtered probability space $(\Omega, \mathcal{G}, \mathbb{F}, \mathbb{P})$ and a non-negative $\mathbb{F}$ adapted process λ, as well as a random variable Θ, independent of $\mathbb{F}$ with unit exponential law, one defines

$$\tau = \inf\left\{t\,:\,\int_0^t \lambda_s\, ds \geq \Theta\right\}\,.$$

Then, since obviously $\mathbb{F}$ is immersed in $\mathbb{F} \vee \sigma(\Theta)$ (by independence) and $\mathbb{F} \subset \mathbb{G} \subset \mathbb{F} \vee \sigma(\Theta)$, where $\mathbb{G}$ is the progressive enlargement of $\mathbb{F}$ by τ, the filtration $\mathbb{F}$ is immersed in $\mathbb{G}$.

3 Initial Enlargement

A filtered probability space $(\Omega, \mathcal{G}, \mathbb{F}, \mathbb{P})$ and a $\mathcal{G}$-measurable random variable ζ being given, one sets $\mathbb{F}^{(\zeta)} = \mathbb{F} \vee \sigma(\zeta)$. We assume that $\mathcal{F}_0$ is trivial, and, if necessary, we take the smallest right-continuous filtration containing $\mathbb{F}^{(\zeta)}$. Note that $\mathcal{F}_0^{(\zeta)} = \sigma(\zeta)$.

This is a generalisation of the problem studied by Itô, for which $\zeta = B_1$.

We now present two important results on measurability:

For any fixed $t > 0$, every $\mathcal{F}_t^{(\zeta)}$-measurable random variable $Y_t^{(\zeta)}$ is of the form $Y_t^{(\zeta)} = y_t(\omega, \zeta(\omega))$ where $y_t(\cdot, u)$ is, for any u, an $\mathcal{F}_t$-measurable random variable.

Every $\mathbb{F}^{(\zeta)}$-predictable process $Y^{(\zeta)}$ is of the form $Y_t^{(\zeta)} = y_t(\omega, \zeta(\omega))$ where $(t, \omega, u) \mapsto y_t(\omega, u)$ is a $\mathcal{P}(\mathbb{F}) \otimes \mathcal{B}(\mathbb{R})$-measurable function [29, Lemma 3.13].

We shall now simply write $y_t(\zeta)$ for $y_t(\omega, \zeta(\omega))$. The result on predictable processes cannot be extended in full generality to optional processes, but no counterexample is known.

3.1 Jacod's Conditions

We start with a particular case, for which the proof is easy, found simultaneously by Grorud and Pontier [21] and Amendinger [4]. Then, we shall state (without proof) the general result of Jacod.

Assumption ($\mathcal{E}$) The $\mathbb{F}$-conditional law of ζ is equivalent to η, the law of ζ. More precisely there exists a non-negative $\mathcal{O}(\mathbb{F}) \otimes \mathcal{B}(\mathbb{R}_+)$-measurable map $(\omega, t, u) \to p_t(\omega, u)$ càdlàg in t such that

- for every u, the process $(p_t(u))_{t \geq 0}$ is a strictly positive $\mathbb{F}$-martingale,
- for every $t \geq 0$, the measure $p_t(u)\eta(du)$ equals $\mathbb{P}(\tau \in du \mid \mathcal{F}_t)$, in other words, for any Borel bounded function h, for any $t \geq 0$

$$\mathbb{E}[h(\zeta)|\mathcal{F}_t] = \int_{\mathbb{R}} h(u) p_t(u)\eta(du).$$

Assumption ($\mathcal{E}$) is also called Jacod's equivalence assumption.

Lemma 3.1 *Under Assumption ($\mathcal{E}$), the process L defined as $L_t = \dfrac{1}{p_t(\zeta)}, t \geq 0$ is a $(\mathbb{P}, \mathbb{F}^{(\zeta)})$-martingale. Let $\mathbb{P}^*$ be the probability measure defined on $\mathbb{F}^{(\zeta)}$ as*

$$d\mathbb{P}^*_{|\mathcal{F}_t^{(\zeta)}} = L_t \, d\mathbb{P}_{|\mathcal{F}_t^{(\zeta)}}. \tag{1}$$

Under $\mathbb{P}^*$, the random variable ζ is independent of $\mathcal{F}_t$ for any $t \geq 0$ and, moreover

$$\mathbb{P}^*_{|\mathcal{F}_t} = \mathbb{P}_{|\mathcal{F}_t} \text{ for any } t \geq 0, \qquad \mathbb{P}^*_{|\sigma(\zeta)} = \mathbb{P}_{|\sigma(\zeta)}.$$

Proof Obviously, one has $L_0 := \frac{1}{p_0(\zeta)} = 1$. Setting $L_t(u) := \dfrac{1}{p_t(u)}, \forall t \geq 0$, for any bounded Borel function h and any $\mathcal{F}_s$-measurable bounded random variable K_s and $s \leq t$, one has

$$\mathbb{E}[L_t h(\zeta) K_s] = \mathbb{E}\left[K_s \int_{\mathbb{R}} L_t(u)h(u)p_t(u)\eta(du) \right] = \mathbb{E}\left[K_s \int_{\mathbb{R}} h(u)\eta(du) \right]$$

$$= \mathbb{E}[K_s] \int_{\mathbb{R}} h(u)\eta(du) = \mathbb{E}[K_s]\mathbb{E}[h(\zeta)].$$

For $t = s$, we obtain $\mathbb{E}[L_s h(\zeta) K_s] = \mathbb{E}[K_s]\mathbb{E}[h(\zeta)]$, hence $\mathbb{E}[L_t h(\zeta) K_s] = \mathbb{E}[L_s h(\zeta) K_s]$. Since h and K_s are arbitrary and generate $\mathcal{F}_s^{(\zeta)}$, it follows that L is an $\mathbb{F}^{(\zeta)}$-martingale. Thus, for each $t \geq 0$, we can define the probability measure $\mathbb{P}^*$ on $\mathcal{F}_t^{(\zeta)}$ by $d\mathbb{P}^*|_{\mathcal{F}_t^{(\zeta)}} = L_t \, d\mathbb{P}|_{\mathcal{F}_t^{(\zeta)}}$. The equivalence of $\mathbb{P}^*$ and $\mathbb{P}$ on $\mathcal{F}_t^{(\zeta)}$ for each $t \in [0, \infty)$ follows from the strict positivity of L_t. For any bounded Borel function h, any $\mathcal{F}_t$-measurable bounded random variable K_t, and denoting by $\mathbb{E}^*$

the expectation under $\mathbb{P}^*$, the above computations yield

$$\mathbb{E}^*\left[h(\zeta)K_t\right] = \mathbb{E}[h(\zeta)]\,\mathbb{E}[K_t]. \tag{2}$$

For $h = 1$ (resp. $K_t = 1$), one obtains $\mathbb{E}^*[K_t] = \mathbb{E}[K_t]$ (resp. $\mathbb{E}^*[h(\zeta)] = \mathbb{E}[h(\zeta)]$) and the assertions $\mathbb{P}^*|_{\mathcal{F}_t} = \mathbb{P}|_{\mathcal{F}_t}$ and $\mathbb{P}^*|_{\sigma(\zeta)} = \mathbb{P}|_{\sigma(\zeta)}$ are proven. Thus the identity (2) can be rewritten as $\mathbb{E}^*\left[h(\zeta)K_t\right] = \mathbb{E}^*[h(\zeta)]\,\mathbb{E}^*[K_t]$, which shows that the random variable ζ and the σ-field $\mathcal{F}_t$ are independent under $\mathbb{P}^*$. □

Corollary 3.1 *Under the probability measure* $\mathbb{P}^*$, $\mathbb{F}$ *is immersed in* $\mathbb{F}^{(\zeta)}$.

Proof Since under $\mathbb{P}^*$, the random variable ζ and the σ-field $\mathcal{F}_\infty$ are independent, the assertion follows. □

Under Assumption $(\mathcal{E})$, we study a financial market $(\Omega, \mathcal{G}, \mathbb{F}, \mathbb{P}, S)$ with null interest rate. If the prices S are $(\mathbb{P}, \mathbb{F})$ martingales, then $\mathbb{P}^*$ defined above is an equivalent martingale measure for the market $(\Omega, \mathcal{G}, \mathbb{F}^{(\zeta)}, S)$.

Note that if ζ satisfies $(\mathcal{E})$ under $\mathbb{P}$, it satisfies $(\mathcal{E})$ under any probability measure equivalent to $\mathbb{P}$. If the financial market is such that there exists an $\mathbb{F}$-equivalent martingale measure $\mathbb{Q}$, then, denoting by $p^{\mathbb{Q}}(\zeta)$ the density of ζ under $\mathbb{Q}$, it follows that $\mathbb{Q}^*$ is an equivalent probability measure where

$$\mathbb{Q}^*|_{\mathcal{F}_t^{(\zeta)}} = \frac{1}{p_t^{\mathbb{Q}}(\zeta)}\mathbb{Q}|_{\mathcal{F}_t^{(\zeta)}}. \tag{3}$$

Proposition 3.1 *Under Assumption* $(\mathcal{E})$, *any* $(\mathbb{P}, \mathbb{F})$-*local martingale* X *is a* $(\mathbb{P}, \mathbb{F}^{(\zeta)})$-*special semimartingale with decomposition*

$$X_t = X_t^{(\zeta)} + \int_0^t \frac{\mathrm{d}\langle X, p_\cdot(u)\rangle_s|_{u=\zeta}}{p_{s-}(\zeta)},$$

where $X^{(\zeta)}$ *is a* $(\mathbb{P}, \mathbb{F}^{(\zeta)})$-*local martingale*.

Proof If X is a $(\mathbb{P}, \mathbb{F})$-martingale, it is a $(\mathbb{P}^*, \mathbb{F}^{(\zeta)})$-martingale. Indeed, since $\mathbb{P}$ and $\mathbb{P}^*$ are equal on $\mathbb{F}$, X is a $(\mathbb{P}^*, \mathbb{F})$ martingale, hence, using the fact that ζ is $\mathbb{P}^*$ independent of $\mathbb{F}$, it is a $(\mathbb{P}^*, \mathbb{F}^{(\zeta)})$ martingale. Noting that $\mathrm{d}\mathbb{P} = p_t(\zeta)\mathrm{d}\mathbb{P}^*$ on $\mathcal{F}_t^{(\zeta)}$, Girsanov's theorem yields that the process $X^{(\zeta)}$, defined by $X_t^{(\zeta)} = X_t - \int_0^t \frac{\mathrm{d}\langle X, p_\cdot(u)\rangle_s|_{u=\zeta}}{p_{s-}(\zeta)}$, is a $(\mathbb{P}, \mathbb{F}^{(\zeta)})$-martingale. □

The general result of Jacod [26] (proved 20 years before the equivalence result) is the following

Theorem 3.1 *Assume that there exists a non-negative* $\mathcal{O}(\mathbb{F}) \otimes \mathcal{B}(\mathbb{R})$-*measurable map* $(\omega, t, u) \to p_t(\omega, u)$ *càdlàg in* t *such that*

- *for every* u, *the process* $(p_t(u))_{t \geq 0}$ *is a non-negative* $\mathbb{F}$-*martingale*,

- *denoting by η the law of ζ, for every $t \geq 0$, the measure $p_t(u)\eta(du)$ equals $\mathbb{P}(\tau \in du \mid \mathcal{F}_t)$, in other words, for any Borel bounded function h, for any $t \geq 0$*

$$\mathbb{E}[h(\zeta)|\mathcal{F}_t] = \int_{\mathbb{R}} h(u) p_t(u) \eta(du) .$$

Then, any $(\mathbb{P}, \mathbb{F})$-local martingale X is a $(\mathbb{P}, \mathbb{F}^{(\zeta)})$-special semimartingale with canonical decomposition

$$X_t = X_t^{(\zeta)} + \int_0^t \frac{d\langle X, p.(u)\rangle_s|_{u=\zeta}}{p_{s-}(\zeta)}, \tag{4}$$

where $X^{(\zeta)}$ is a $(\mathbb{P}, \mathbb{F}^{(\zeta)})$-local martingale.

The proof is more delicate, the process $1/p(\zeta)$ is well defined (since, from Jacod [26, Corollaire 1.11], $p(\zeta) > 0$) but is no longer a martingale. Jacod [26] mentioned (page 25) that it would be possible to use Girsanov-type results in the absolute continuity condition (page 15), but that in any case, the difficulties are due to measurability conditions in both approaches. The assumption of absolute continuity is also called the $(\mathcal{J})$-assumption.

Under the $(\mathcal{J})$ hypothesis

(a) every $\mathbb{F}^{(\zeta)}$-optional process $Y^{(\zeta)}$ is of the form $Y_t^{(\zeta)}(\omega) = y_t(\omega, \zeta(\omega))$ for some $\mathbb{F} \otimes \mathcal{B}(\mathbb{R}^d)$-optional process $(y_t(\omega, u), t \geq 0)$ (see Fontana [19]).

(b) Let $Y_T^{(\zeta)}$ be an $\mathcal{F}_T^{(\zeta)}$-measurable integrable random variable. Then, for $s < T$:

$$\mathbb{E}\big(Y_T^{(\zeta)}|\mathcal{F}_s^{(\zeta)}\big) = \frac{1}{p_s(\zeta)} \mathbb{E}\big(y_T(u) p_T(u)|\mathcal{F}_s\big)\Big|_{u=\zeta} .$$

(c) Characterization of $(\mathbb{P}, \mathbb{F}^{(\zeta)})$-martingales in terms of $(\mathbb{P}, \mathbb{F})$-martingales: The process $Y^{(\zeta)}$ is a $(\mathbb{P}, \mathbb{F}(\zeta))$-martingale if and only if, for any u, the process $y(u)p(u)$ is a $(\mathbb{P}, \mathbb{F})$ martingale.

We now present the propagation of the predictable representation property.

We assume that there exists a $(\mathbb{P}, \mathbb{F})$-local martingale X such that any $(\mathbb{P}, \mathbb{F})$-local martingale Y can be represented as

$$Y_t = Y_0 + \int_0^t \varphi_s dX_s \tag{5}$$

for some $\varphi \in \mathcal{P}(\mathbb{F})$.

Then (see Fontana [19]), under the $(\mathcal{J})$-hypothesis, every $(\mathbb{P}, \mathbb{F}^{(\zeta)})$-martingale $Y^{(\zeta)}$ admits a representation

$$Y_t^{(\zeta)} = Y_0^{(\zeta)} + \int_0^t \Phi_s dX_s^{(\zeta)}, \tag{6}$$

where $\Phi \in \mathcal{P}(\mathbb{F}^{(\zeta)})$ and $Y_0 \in \mathcal{F}_0^{(\zeta)}$. Here $X^{(\zeta)}$ is the $(\mathbb{P}, \mathbb{F}^{(\zeta)})$-martingale part (given in (4)) of the $(\mathbb{P}, \mathbb{F}^{(\zeta)})$-semimartingale X introduced in (5).

3.2 Brownian Bridge

The Brownian Bridge is obtained when studying the initial enlargement of a Brownian filtration $\mathbb{F}$ (generated by the Brownian motion B) with the random variable B_1. Note that Jacod's absolute condition is not satisfied at time 1. Nevertheless, it is not difficult to prove (see Jeulin [29, Th. 3.23] or [3, Proposition 4.1]) that

$$B_t^{(B_1)} := B_t - \int_0^{t \wedge 1} \frac{B_1 - B_s}{1 - s} ds, \ 0 \le t \le 1$$

is an $\mathbb{F}^{(B_1)}$-martingale, a first step being to prove the existence of the integral. Then, using elementary computations, it is easy to prove that $B^{(B_1)}$ is a martingale, and by Lévy's Theorem, it is a Brownian motion. This main example presents another point of interest: even if B is a semimartingale, not all $\mathbb{F}$-martingales are $\mathbb{G}$-semimartingales (see [10]).

Theorem 3.2 *Let X be an $\mathbb{F}$-local martingale with representation $X_t = X_0 + \int_0^t \varphi_s dB_s$ for an $\mathbb{F}$-predictable process φ satisfying $\int_0^1 \varphi_s^2 ds < \infty$ a.s. Then, the following conditions are equivalent:*

(a) *the process X is an $\mathbb{F}^{(B_1)}$-semimartingale;*
(b) $\int_0^1 |\varphi_s| \frac{|B_1 - B_s|}{1 - s} ds < \infty$ $\mathbb{P}$-a.s.;
(c) $\int_0^1 \frac{|\varphi_s|}{\sqrt{1 - s}} ds < \infty$ $\mathbb{P}$-a.s.

If these conditions are satisfied, the $\mathbb{F}^{(B_1)}$-semimartingale decomposition of X is, for $t \le 1$,

$$X_t = X_0 + \int_0^{t \wedge 1} \varphi_s dB_s^{(B_1)} + \int_0^{t \wedge 1} \varphi_s \frac{B_1 - B_s}{1 - s} ds. \tag{7}$$

This is an example where some $\mathbb{F}$-martingales are $\mathbb{F}^{(B_1)}$-semimartingales, but not all of them.

Note that, in a Brownian filtration, Yor's criterion [3, Section 4.3] is more general than Jacod's condition.

Application Consider a financial market with null interest rate and risky asset $dS_t = S_t(b dt + \sigma dB_t)$, $S_0 = x$, driven by a Brownian motion B and $\zeta = S_T$. The arbitrage is obvious (no need for mathematics) and the conditional density does not exist on $[0, T]$. If one takes $\zeta = S_T + \epsilon$, where ϵ is a discrete random variable, independent of $\mathbb{F}$, Jacod's absolute continuity assumption is satisfied (see Amendinger et al. [5]).

4   Progressive Enlargement

Here $\mathbb{F}$ is a given filtration, τ a *finite* random time and $\mathbb{G}$ the progressive enlargement: roughly speaking, $\mathcal{G}_t = \mathcal{F}_t$ on $t < \tau$ and $\mathcal{G}_t = \mathcal{F}_t \vee \sigma(\tau)$ after τ.

There are typically two cases: before τ and after τ. Before τ, there is no new information, except τ has not yet occurred. After τ, the time when τ has occurred is known. This is easy to illustrate with a "financial" example. Let S be the price of a risky asset (e.g., a Black and Scholes dynamic), and assume zero interest rate. Let $\tau = \inf\{t : S_t = \sup_{u \leq T} S_u\}$. If an agent has access to the progressive enlargement: before τ she will buy the stock, say at time 0 at price S_0, and wait till τ, when she will sell the stock making arbitrage. She can also realize an arbitrage after τ: at time τ, she short sells the stock at price S_τ and delivers it at price $S_t < S_\tau$ at any time t after τ. This kind of random time is called an honest time (see below).

The $\mathbb{F}$-dual optional projection A^o of A is the optional process such that for any non-negative bounded $\mathbb{F}$-optional process Y such that Y_τ is integrable

$$\mathbb{E}[Y_\tau] = \mathbb{E}\left[\int_{[0,\infty)} Y_s dA_s^o\right].$$

The $\mathbb{F}$-dual predictable projection A^p of A is the $\mathbb{F}$-predictable process such that for any non-negative bounded $\mathbb{F}$-predictable process Y, such that Y_τ is integrable,

$$\mathbb{E}[Y_\tau] = \mathbb{E}\left[\int_{[0,\infty)} Y_s dA_s^p\right]. \tag{8}$$

Two processes are important: the optional projection of $1 - A$, denoted Z, and the optional projection of $1 - A_-$, denoted $\widetilde{Z}$, i.e., $Z = {}^o(1 - A)$, and $\widetilde{Z} = {}^o(1 - A_-)$.

Comment 4.1 One can prove (see, e.g., [3, Proposition 1.4]) that $Z = m - A^o = M - A^p$, where m and M are $\mathbb{F}$-martingales and that $\widetilde{Z} = m - A_-^o$. The decomposition $Z = M - A^p$ is the Doob–Meyer decomposition of Z.

Note that

$$Z_t = \mathbb{P}(\tau > t | \mathcal{F}_t), \; \widetilde{Z}_t = \mathbb{P}(\tau \geq t | \mathcal{F}_t). \tag{9}$$

Defining Z as in (9) can create some difficulties. Indeed the equality is valid a.s. for any t, and (except if Z is continuous) prevents us for defining the process Z (the union of negligible sets can fail to be negligible)

4.1 Before τ

Lemma 4.1 *For any $\mathcal{G}_t$-measurable random variable $Y_t^{\mathbb{G}}$, there exists an $\mathcal{F}_t$-measurable random variable Y such that $Y_t^{\mathbb{G}} 1\!\!1_{\{t<\tau\}} = Y_t 1\!\!1_{\{t<\tau\}}$. If X_T is integrable and $\mathcal{F}_T$-measurable, one has*

$$\mathbb{E}[X_T 1\!\!1_{\{T<\tau\}} | \mathcal{G}_t] = 1\!\!1_{\{t<\tau\}} \frac{\mathbb{E}[X_T Z_T | \mathcal{F}_t]}{Z_t}.$$

Proof By definition of $\mathbb{G}$, the existence of Y is obvious. The uniqueness is not granted. The second assertion follows from the first, taking conditional expectation with respect to $\mathcal{F}_t$ the equality $\mathbb{E}[X_T 1\!\!1_{\{T<\tau\}} | \mathcal{G}_t) = 1\!\!1_{\{t<\tau\}} Y_t$. Note that $Z > 0$ on $\{t < \tau\}$. See Elliott et al. [18, Section 3.1]. □

The $\mathbb{G}$-predictable processes can be described in terms of a family of $\mathbb{F}$-predictable processes:

Lemma 4.2 *For any $\mathbb{G}$-predictable bounded process $Y^{\mathbb{G}}$, there exists a bounded $\mathbb{F}$-predictable process Y and a map $y : \mathbb{R}^+ \times \mathbb{R}^+ \times \Omega \to \mathbb{R}$, which is $\mathcal{B}(\mathbb{R}^+) \times \mathcal{P}(\mathbb{F})$-measurable and bounded such that $Y_t^{\mathbb{G}} = Y_t 1\!\!1_{\{t\leq\tau\}} + y(t,\tau) 1\!\!1_{\{\tau<t\}}$. (See Jeulin [29, Lemma 4.4].)*

The $\mathbb{G}$-*compensator* of A is the $\mathbb{G}$-predictable increasing process $\Lambda^{\mathbb{G}}$ such that

$$\widetilde{M} := A - \Lambda^{\mathbb{G}} \tag{10}$$

is a $\mathbb{G}$-martingale; this process is flat after τ (i.e., $\Lambda_{t\wedge\tau}^{\mathbb{G}} = \Lambda_t^{\mathbb{G}}$). From Lemma 4.2, there exists an $\mathbb{F}$-predictable increasing process Λ such that $\Lambda_t^{\mathbb{G}} = \Lambda_{t\wedge\tau}, \forall t \geq 0$. Furthermore, $\Lambda_t^{\mathbb{G}} 1\!\!1_{\{t\leq\tau\}} = 1\!\!1_{\{t\leq\tau\}} \int_0^t \frac{dA_s^p}{Z_{s-}}$ (see, e.g., Proposition 2.15, page 37 in [3]). The process Λ is not uniquely defined after τ (except if $Z_- > 0$) and, hereafter, we choose

$$d\Lambda_t = \frac{dA_t^p}{Z_{t-}} 1\!\!1_{\{Z_{t-}>0\}}, \forall t \geq 0, \Lambda_0 = 0. \tag{11}$$

As an application of the above and by definition of dual projections, we obtain the following result (see, e.g., Jeanblanc and Li [27]), which is useful for pricing defaultable claims:

For any bounded $\mathbb{F}$-predictable process K,

$$\mathbb{E}[K_\tau 1\!\!1_{\{\tau\leq T\}} | \mathcal{G}_t] = K_\tau 1\!\!1_{\{\tau<t\}} + 1\!\!1_{\{\tau\geq t\}} \frac{\mathbb{E}[\int_t^T K_u dA_u^p | \mathcal{F}_t]}{Z_t}.$$

For any bounded $\mathbb{F}$-optional process K,

$$\mathbb{E}[K_\tau \mathbb{1}_{\{\tau \leq T\}}|\mathcal{G}_t] = K_\tau \mathbb{1}_{\{\tau \leq t\}} + \mathbb{1}_{\{\tau > t\}} \frac{\mathbb{E}[\int_t^T K_u dA_u^o |\mathcal{F}_t]}{Z_t}.$$

Lemma 4.3 *Under the two assumptions* **(A)** *and* **(C)**, *any (càdlàg)* $(\mathbb{P}, \mathbb{F})$-*local martingale* X *stopped at time* τ *is a* $(\mathbb{P}, \mathbb{G})$-*semimartingale with decomposition*

$$X_{t \wedge \tau} = X_t^{\mathbb{G}} + \int_0^{t \wedge \tau} \frac{d\langle X, M \rangle_s}{Z_{s-}},$$

where $X^{\mathbb{G}}$ *is a* $(\mathbb{P}, \mathbb{G})$-*local martingale. Here* M *is the martingale part in the Doob–Meyer decomposition of* Z.

Proof Let $Y_s^{\mathbb{G}}$ be a $\mathcal{G}_s$-measurable random variable. There exists an $\mathcal{F}_s$-measurable random variable y_s such that $Y_s^{\mathbb{G}} \mathbb{1}_{\{s < \tau\}} = y_s \mathbb{1}_{\{s < \tau\}}$, hence, if X is an $\mathbb{F}$-martingale, for $s < t$,

$$\mathbb{E}(Y_s^{\mathbb{G}}(X_{t \wedge \tau} - X_{s \wedge \tau})) = \mathbb{E}(Y_s^{\mathbb{G}} \mathbb{1}_{\{s < \tau\}}(X_{t \wedge \tau} - X_{s \wedge \tau}))$$

$$= \mathbb{E}(y_s \mathbb{1}_{\{s < \tau\}}(X_{t \wedge \tau} - X_{s \wedge \tau}))$$

$$= \mathbb{E}\left(y_s(\mathbb{1}_{\{s < \tau \leq t\}}(X_\tau - X_s) + \mathbb{1}_{\{t < \tau\}}(X_t - X_s))\right).$$

From the definition of Z and (8),

$$\mathbb{E}\left(y_s \mathbb{1}_{\{s < \tau \leq t\}} X_\tau\right) = -\mathbb{E}\left(y_s \int_s^t X_u dZ_u\right).$$

From the integration by parts formula (taking into account the continuity of Z and X)

$$\int_s^t X_u dZ_u = -X_s Z_s + Z_t X_t - \int_s^t Z_u dX_u - \langle X, Z \rangle_t + \langle X, Z \rangle_s.$$

We have also

$$\mathbb{E}\left(y_s \mathbb{1}_{\{s < \tau \leq t\}} X_s\right) = \mathbb{E}(y_s X_s (Z_s - Z_t))$$

$$\mathbb{E}\left(y_s \mathbb{1}_{\{t < \tau\}}(X_t - X_s)\right) = \mathbb{E}(y_s Z_t (X_t - X_s))$$

hence, from the martingale property of X,

$$\mathbb{E}(Y_s^{\mathbb{G}}(X_{t\wedge\tau} - X_{s\wedge\tau})) = \mathbb{E}(y_s(\langle X, M\rangle_t - \langle X, M\rangle_s))$$

$$= \mathbb{E}\left(y_s \int_s^t \frac{d\langle X, M\rangle_u}{Z_u} Z_u\right) = \mathbb{E}\left(y_s \int_s^t \frac{d\langle X, M\rangle_u}{Z_u} \mathbb{E}(\mathbb{1}_{\{u<\tau\}}|\mathcal{F}_u)\right)$$

$$= \mathbb{E}\left(y_s \int_s^t \frac{d\langle X, M\rangle_u}{Z_u^\tau} \mathbb{1}_{\{u<\tau\}}\right) = \mathbb{E}\left(y_s \int_{s\wedge\tau}^{t\wedge\tau} \frac{d\langle X, M\rangle_u}{Z_u}\right).$$

The result follows.   □

The general case is more delicate. See, e.g., [16, Section 76] or [3, Theorem 5.1].

Theorem 4.1 *Every càdlàg* $\mathbb{F}$*-local martingale* X *stopped at time* τ *is a special* $\mathbb{G}$*-semimartingale with the canonical decomposition*

$$X_t^\tau = X_t^{\mathbb{G}} + \int_0^{t\wedge\tau} \frac{d\langle X, m\rangle_s}{Z_{s-}}, \tag{12}$$

where $X^{\mathbb{G}}$ *is a* $\mathbb{G}$*-local martingale and m is as defined in Comments 4.1.*

Arbitrages Before τ Introduce $\widetilde{R} := R\mathbb{1}_{\{\widetilde{Z}_R=0<Z_{R-}\}} + \infty\mathbb{1}_{\{\widetilde{Z}_R=0<Z_{R-}\}^c}$, where $R := \inf\{t : Z_t = 0\}$. The following conditions are equivalent.

(1) The $\mathbb{F}$-stopping time $\widetilde{R}$ is infinite.
(2) For any $\mathbb{F}$-local martingale X, there exists a non-negative $\mathbb{G}$-local martingale ζ such that $X^\tau\zeta$ is a $\mathbb{G}$-local martingale, where X^τ is the stopped process (non-arbitrage of the first kind).

See [3, Theorem 5.46] for a proof.

Arbitrages Under the $(\mathcal{E})$ **Hypothesis**   Under the $(\mathcal{E})$ hypothesis, if discounted prices are $(\mathbb{P}, \mathbb{F})$-martingales, $\mathbb{P}^*$ (defined in (1)) is an equivalent martingale measure on $\mathbb{G}$. Otherwise, if there exists an equivalent martingale measure $\mathbb{Q}$ for $\mathbb{F}$-adapted discounted prices, $\mathbb{Q}^*$ defined in (3) is an equivalent martingale measure in $\mathbb{G}$.

4.2   Some Facts on the Predictable Representation Property

We assume that the predictable representation property holds in the filtration $\mathbb{F}$, i.e., there exists an $\mathbb{F}$-local martingale X such that every $\mathbb{F}$-local martingale Y can be represented as $Y_t = Y_0 + \int_0^t \varphi_s dX_s$ for some $\varphi \in \mathcal{P}(\mathbb{F})$.

Under some conditions, the predictable representation property propagates to $\mathbb{G}$. For example, (see Fontana [19]) under the $(\mathcal{J})$-hypothesis $(X^{\mathbb{G}}, \widetilde{M})$ has the predictable representation property in $\mathbb{G}$ where $X^{\mathbb{G}}$ is the $\mathbb{G}$ martingale part of the $\mathbb{G}$-semimartingale X and $\widetilde{M}$ is the $\mathbb{G}$-martingale defined in (10).

We refer to [3, Section 5.6] for more information.

We now study the relationship between $\mathbb{G}$-martingales and $\mathbb{F}$-martingales obtained in [11, proposition 2.2].

Proposition 4.1 *Under the* $(\mathcal{J})$*-hypothesis, a* $\mathbb{G}$*-optional process of the form* $Y^{\mathbb{G}} := \widetilde{y}\,\mathbb{1}_{[\![0,\tau[\![} + \widehat{y}(\tau)\,\mathbb{1}_{[\![\tau,\infty[\![}}$*, where* $\widetilde{y}$ *and* $\widehat{y}(u)$ *are* $\mathbb{F}$*-optional processes, is a* $\mathbb{G}$*-martingale if and only if the following two conditions are satisfied*

(a) *for* η*-a.e* u, $\left(\widehat{y}_t(u)\,p_t(u), t \geq u\right)$ *is an* $\mathbb{F}$*-martingale;*
(b) *the process* y *is an* $\mathbb{F}$*-martingale, where*

$$y_t := \mathbb{E}(Y_t|\mathcal{F}_t) = \widetilde{y}_t Z_t + \int_0^t \widehat{y}_t(u)\,p_t(u)\eta(\mathrm{d}u) . \tag{13}$$

Under the $(\mathcal{J})$-hypothesis, if the function $(\omega, u) \rightarrow X(\omega, u)$ is $\mathcal{F}_T \otimes \mathcal{B}(\mathbb{R}_+)$-measurable and bounded, then

$$\mathbb{E}[X(\tau)|\mathcal{G}_t] =$$

$$\mathbb{1}_{\{t<\tau\}}\frac{1}{Z_t}\mathbb{E}[\int_{]t,\infty]} X(u)p_T(u)\eta(\mathrm{d}u)|\mathcal{F}_t] + \mathbb{1}_{\{\tau\leq t\}}\frac{1}{p_t(\tau)}\mathbb{E}[X(u)p_T(u)|\mathcal{F}_t],$$

for $t \leq T$ and

$$\mathbb{E}[X(\tau)|\mathcal{G}_t] = \mathbb{1}_{\{t<\tau\}}\frac{1}{Z_t}\int_{]t,\infty]} X(u)p_t(u)\eta(\mathrm{d}u) + \mathbb{1}_{\{\tau\leq t\}}X(\tau),$$

for $T < t$ (see e.g., [3, lemma 5.24]).

4.3 Immersion

Immersion is easily characterized in a progressive enlargement setting: $\mathbb{F}$ is immersed in $\mathbb{G}$ if and only if, $\forall t \geq 0$

$$\mathbb{P}(\tau > t|\mathcal{F}_t) = \mathbb{P}(\tau > t|\mathcal{F}_\infty)$$

(see, e.g., [3, lemma 3.8]). This implies that Z is decreasing and $Z = 1 - A^o$ (see [3, Proposition 3.9]).

Many models of "default risk" are constructed as follows (see Gueye and Jeanblanc [22]). Let $\mathbb{F}$ be a given filtration and K a càdlàg increasing $\mathbb{F}$-adapted process, and define

$$\tau = \inf\{t : K_t \geq \Theta\},$$

where Θ is a random variable independent of $\mathbb{F}$ with unit exponential law. Then $\mathbb{P}(\tau > t | \mathcal{F}_t) = e^{-K_t}$, and immersion holds between $\mathbb{F}$ and $\mathbb{G}$. If K is continuous, τ avoids $\mathbb{F}$ stopping times. If not, the jump times of K are the $\mathbb{F}$-stopping times not avoided by τ.

Comment 4.2 Let us point out a "technical" difficulty. Assume that K has no jumps at constant time, which implies $\mathbb{P}(\tau = t) = 0, \forall t > 0$. This does not imply that $Z = \widetilde{Z}$ (the equality meaning that the two processes are indistinguishable). Indeed, $Z = 1 - A^o$ whereas $\widetilde{Z} = 1 - A^o_-$.

We have underlined that immersion is not stable under change of probability. However, let us point out that if a price process S is given on $(\Omega, \mathcal{G}, \mathbb{F}, \mathbb{P})$ and the interest rate is null, if the corresponding financial market satisfies NFLVR, and if the $(\Omega, \mathcal{G}, \mathbb{G}, \mathbb{P}, S)$ financial market satisfies NFLVR, then, under $\mathcal{I}$-hypothesis, one can choose a $\mathbb{G}$-equivalent martingale measure such that immersion holds. If the market $(\Omega, \mathcal{G}, \mathbb{F}, \mathbb{P})$ is complete and discounted prices are $(\mathbb{P}, \mathbb{F})$-martingales, under any $\mathbb{G}$-equivalent martingale measure immersion holds.

4.4 Honest Times

Honest times were introduced by Barlow [7].

Definition 4.1 A random time τ is an $\mathbb{F}$-honest time if, for every $t > 0$, there exists an $\mathcal{F}_t$-measurable random variable τ_t such that $\tau = \tau_t$ on $\{\tau < t\}$.

Fontana et al. [20] assume that τ is honest, and the following conditions (a), (b) and (c)

(a) The restricted financial market $(\Omega, \mathbb{F}, \mathbb{P}, S)$ satisfies NFLVR.
(b) The random time τ avoids all $\mathbb{F}$-stopping times (Condition **(A)**).
(c) The martingale part of the semimartingale S is a continuous $\mathbb{F}$-local martingale $M^S = \left(M^S_t\right)_{t \geq 0}$ which has the $\mathbb{F}$-predictable representation property in the filtration $\mathbb{F}$.

Then, they prove that the $(\mathbb{F}, \mathbb{P}, S)$ market is complete and

(1) NUPBR holds in the enlarged market on the time horizon $[0, \tau]$,
(2) there exists an explicit arbitrage opportunity in the enlarged market on the time horizon $[0, \tau]$ and on the interval $[\tau, \vartheta]$ for an explicit $\mathbb{G}$ stopping time ϑ (see [2, Theorem 3]),
(3) NFLVR fails to hold in the enlarged market on the time horizon $[0, \tau]$,
(4) NUPBR fails to hold in the enlarged financial market on the global time horizon $[0, \infty]$.

In the case of honest time, for a $(\mathbb{P}, \mathbb{F})$-local martingale X

$$X_t = X_t^{\mathbb{G}} + \int_0^{t \wedge \tau} \frac{1}{Z_{s-}} d\langle X, m \rangle_s - \int_\tau^{\tau \vee t} \frac{1}{1 - Z_{s-}} d\langle X, m \rangle_s,$$

where $X^{\mathbb{G}}$ is a $(\mathbb{P}, \mathbb{G})$-local martingale.

Any $\mathbb{G}$-optional process can be written

$$Y = L\, \mathbb{1}_{[\![0,\tau]\!]} + J\, \mathbb{1}_{[\![\tau,\infty]\!]} + K\, \mathbb{1}_{[\![\tau,\infty[\![},$$

where L and K are $\mathbb{F}$-optional processes and J is an $\mathbb{F}$-progressively measurable process.

Example 4.1 We recall Barlow's counterexample given in [7, p. 319] to show that a $\mathbb{G}$-optional process cannot always be decomposed as $L\, \mathbb{1}_{[\![0,\tau[\![} + K\, \mathbb{1}_{[\![\tau,\infty[\![}$, where L and K are $\mathbb{F}$-optional processes. Let B be a Brownian motion, $\mathbb{F}$ its natural filtration $\vartheta = \inf\{t : |B_t| = 1\}, \tau = \sup\{t \leq \vartheta : B_t = 0\}$ and $\mathbb{G}$ the progressive enlargement of $\mathbb{F}$ with τ. The process X defined as $X_t = \mathbb{1}_{\{t \geq \tau\}} \mathrm{sgn}(B_\vartheta)$ is right-continuous and $\mathbb{G}$-adapted, hence $\mathbb{G}$-optional. Moreover X is a $\mathbb{G}$-martingale. Obviously, if the pair (L, K) exists, then $L = 0$ and one can choose K to be $\mathbb{F}$-predictable, since $\mathcal{O}(\mathbb{F}) = \mathcal{P}(\mathbb{F})$. Then $\Delta X_\tau = K_\tau$ would be $\mathcal{G}_{\tau-}$-measurable, which contradicts the $\mathbb{G}$-martingale property of X.

Lemma 4.4 *Assume that $(\Omega, \mathbb{F}, \mathbb{P}, S)$ is a complete market satisfying NFLVR on the time horizon [0,T]. If τ is a finite honest time which satisfies (A), there are classical arbitrages before τ for $(\Omega, \mathbb{G}, \mathbb{P}, S)$ and classical arbitrages after τ for $(\Omega, \mathbb{G}, \mathbb{P}, S)$.*

Proof See [3, Section 5.8.1] and the examples in [2]. ☐

5   Information Drift

Assume that B is an $\mathbb{F}$-Brownian motion and $B_t^{\mathbb{K}} = B_t + \int_0^t k_s ds$ a $\mathbb{K}$-Brownian motion where $\mathbb{F} \subset \mathbb{K}$. When S is the $\mathbb{F}$-adapted price of an asset, one has (in the Brownian case)

$$
\begin{aligned}
dS_t &= S_t(b_t dt + \sigma_t dB_t) \\
&= S_t((b_t \sigma_t + k_t)dt + \sigma_t dB_t^{\mathbb{K}}).
\end{aligned}
$$

The quantity k is called the information drift. See [6] or [5] for more information.

In the case of portfolio optimisation, when the interest rate is null, denoting by X the wealth associated to a self-financing portfolio, i.e., $dX_t = \pi_t dS_t$, $X_0 = x$, one computes easily

$$\sup_{\pi \in \mathbb{F}} \mathbb{E}[\ln(X_T)] = \ln x, \quad \sup_{\pi \in \mathbb{F}^{\mathbb{K}}} \mathbb{E}[\ln(X_T)] = \ln x + \mathbb{E}\left[\int_0^T k_t^2 dt\right].$$

6 Conclusion and Open Problems

We hope to have given a presentation of enlargement problems. As we mentioned at the beginning, many problems remain to be solved. For example, solve an optimal stopping problem in an enlarged filtration, compare the solution of a BSDE in two filtrations, give the $\mathbb{G}$-decomposition of any martingale when τ is a random time in a Poisson filtration (see [2] for some examples). It would be interesting to provide some tests to detect insider trading (as in [21]). The reverse problem of shrinkage is to give the $\mathbb{F}$-decomposition of the optional projection of a $\mathbb{K}$-semimartingale and has no general solution.

Acknowledgements The author thanks warmly the language editor for improving the English a lot and the two referees for providing some help to improve the paper.

References

1. A. Aksamit, M. Jeanblanc and M. Rutkowski. Integral representations of martingales for progressive enlargements of filtrations. *Stochastic Process. Appl.* **129**(4), 1229–1258 (2019).
2. A. Aksamit, T. Choulli, J. Deng and M. Jeanblanc. Arbitrages in a progressive enlargement setting. In: *Arbitrage, Credit and Informational Risks*, C. Hillairet, M. Jeanblanc and Y. Jiao (eds.). pp. 53–86, World Scientific, Singapore (2014).
3. A. Aksamit and M. Jeanblanc. *Enlargement of filtration with finance in view*. Springer Brief (2017).
4. J. Amendinger. *Initial enlargement of filtrations and additional information in financial markets*. PhD thesis, Technischen Universität Berlin (1999).
5. J. Amendinger, D. Becherer and M. Schweizer. A monetary value for initial information in portfolio optimization. *Finance and Stochastics* **7**(1), 29–46 (2003).
6. S. Ankirchner. *Information and Semimartingales*. PhD thesis, Humboldt Universität Berlin (2005).
7. M.T. Barlow. Study of filtration expanded to include an honest time. *Zeitschrift für Wahrscheinlichkeitstheorie und verwandte Gebiete* **44**, 307–323 (1978).
8. T. Björk, T. *Arbitrage Theory in Continuous Time*. Second edition, Oxford University Press (2004).
9. Ch. Blanchet-Scalliet and M. Jeanblanc. Enlargement of filtration in discrete time. In: *From Probability to Finance, Lecture Notes of BICMR Summer School on Financial Mathematics*, pp. 71–144, Springer (2020).

10. P. Brémaud and M. Yor. Changes of filtration and of probability measures. *Zeitschrift für Wahrscheinlichkeitstheorie und verwandte Gebiete* **17**, 550–566 (2013).
11. G. Callegaro, M. Jeanblanc and B. Zargari. Carthagian enlargement of filtrations. *ESAIM: Probability and Statistics* **17**, 550–566 (2013).
12. T. Choulli and J. Deng. Non-arbitrage for Informational Discrete Time Market. *Stochastics* **89**(3–4), 1–26 (2017).
13. D. Coculescu, M. Jeanblanc and A. Nikeghbali. Default times, non arbitrage conditions and change of probability measures. *Finance and Stochastics* **16**(3), 513–535 (2012).
14. J.M. Corcuera and V. Valdivia. Enlargements of Filtrations and Applications. Preprint, arXiv:1201.5870 (2012).
15. F. Delbaen and W. Schachermayer. *The Mathematics of Arbitrage*. Springer, Berlin (2005).
16. C. Dellacherie, B. Maisonneuve and P.-A. Meyer. *Probabilités et Potentiel*, chapitres XVII–XXIV, Processus de Markov (fin). Compléments de calcul stochastique. Hermann, Paris (1992).
17. P. Di Tella and M. Jeanblanc. Martingale representation in the enlargement of the filtration generated by a point process. *Stochastic Process. Appl.* **131**, 103–121 (2021).
18. R.J. Elliott, M. Jeanblanc and M. Yor. On models of default risk. *Math. Finance* **10**, 179–196 (2000).
19. C. Fontana The strong predictable representation property in initially enlarged filtrations under the density hypothesis. *Stochastic Process. Appl.* **128**(3), 1007–1033 (2018).
20. C. Fontana, M. Jeanblanc and S. Song. On arbitrages arising with honest times. *Finance Stoch.* **18**, 515–543 (2014).
21. A. Grorud and M. Pontier. Insider trading in a continuous time market model. *Int. J. Theor. Appl. Finance* **1**, 331–347 (1998).
22. D. Gueye and M. Jeanblanc. Generalized Cox model for default times. Forthcoming in *Mathematics Going Forward,* Lecture Notes in Mathematics 2313, ch. 9, Springer.
23. C. Hillairet and Y. Jiao. *Portfolio Optimization with Different Information Flow*. ISTE Press, Elsevier (2017).
24. K. Itô. Extension of stochastic integrals. In: *Proceedings of the International Sympsoium in Stochastic Differential Equations (RIMS, Kyoto Univ., Kyoto, 1976)*, pp. 95–109, Wiley (1976).
25. J. Jacod. *Calcul stochastique et Problèmes de martingales*. Lecture Notes in Mathematics **714**, Springer-Verlag, Berlin (1979).
26. J. Jacod. Grossissement initial, hypothèse H' et théorème de Girsanov. In: *Grossissements de filtrations: exemples et applications*, Lecture Notes in Mathematics, Séminaire de Calcul Stochastique 1982–83, vol. 1118, Springer-Verlag (1987).
27. M. Jeanblanc and L. Li. Characteristics and constructions of default times. *SIAM J. Financial Math.* 11(3), 720–749 (2020).
28. M. Jeanblanc, M. Yor and M. Chesney. *Martingale Methods for Financial Markets*. Springer-Verlag, Berlin (2007).
29. Th. Jeulin. *Semi-martingales et grossissement de filtration*. Lecture Notes in Mathematics **833**, Springer-Verlag (1980).
30. Th. Jeulin and M. Yor. *Grossissements de filtrations: exemples et applications*. Lecture Notes in Mathematics **1118**, Springer-Verlag (1985).
31. Y. Kabanov, C. Kardaras and S. Song. On Local Martingale Deflators and Market Portfolios. *Finance and Stochastics* **20**(4), 1097–1108 (2016).
32. Y. Kchia and P. Protter. Progressive Filtration Expansions via a Process, with Applications to Insider Trading. *Int. J. Theor. Appl. Finan.* **18**, 1550027 (2014).
33. D. Lando. On Cox processes and credit risky securities. *Review of Derivatives Research* **2**, 99–120 (1998).
34. B. Mallein and M. Yor. Exercices sur les temps locaux de semi-martingales continues et les excursions browniennes. Preprint, arXiv:1606.07118 (2016).
35. R. Mansuy and M. Yor. *Random Times and Enlargements of Filtrations in a Brownian Setting*. Lectures Notes in Mathematics **1873**, Springer (2006).

36. L. Neufcourt and Ph. Protter. Expansion of a filtration with a stochastic process: the information drift. Preprint arXiv:1902.06780 (2019).
37. A. Nikeghbali. An essay on the general theory of stochastic processes. *Probability Surveys* **3**, 345–412 (2006).
38. P.E. Protter. *Stochastic Integration and Differential Equations*. Second edition, Springer, Berlin (2005).
39. S. Song. Local solution method for the problem of enlargement of filtration. Preprint, arXiv:1302.2862 (2013).
40. M. Yor. *Some Aspects of Brownian Motion, Part II: Some Recent Martingale Problems*. Lectures in Mathematics, ETH Zürich, Birkhäuser, Basel (1997).

Modern Extreme Value Theory at the Interface of Risk Management, Bayesian Networks and Heavy-Tailed Time Series

Paul Embrechts, Claudia Klüppelberg, and Thomas Mikosch

1 Introduction

In the summer of 2022, Dr. Catriona Byrne, Catriona to mathematicians all over the world, will retire as Springer's Editorial Director for Mathematics. We are delighted to contribute to this written bouquet of thanks. Perhaps it sounds a bit easy to say "life as a mathematician without Catriona will never be the same". Perhaps "easy" but "true"! We all have many lovely memories of meeting and discussing with Catriona at numerous conferences, farewell lectures, academic festive events, departmental visits,.... Over lunch or dinner, she always provided us with valuable advice when it came to publishing a book or editing a journal under the umbrella of Springer. However, what we will also sorely miss are the exchanges on societal issues well beyond mathematical academia. Only on February 24, 2022, Paul Embrechts, for instance, received such a lovely lunch-follow-up-e-mail from her. We copy it here, in part, as it so much highlights Catriona's wonderful personality:

"Dear Paul, Thank-you again for making time for Richard and myself yesterday and for your invitation to a delicious lunch at the Dozentenfoyer (always a delight). ... Here is Matheus Grasselli's [fiction] book with Izaias Almada: "The Venetian Files: The Secret of Financial Crises". ... And today I received my copy of the

P. Embrechts
Department of Mathematics, ETH Zürich, Zurich, Switzerland
e-mail: paul.embrechts@math.ethz.ch

C. Klüppelberg (✉)
Department of Mathematics, Technical University of Munich, Garching, Germany
e-mail: cklu@ma.tum.de

T. Mikosch
Department of Mathematics, University of Copenhagen, Copenhagen, Denmark
e-mail: mikosch@math.ku.dk

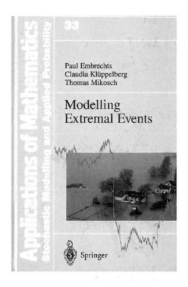 

Fig. 1 The different covers of [7]; the original first edition of 1997 (left) and a later printing (right). The first edition appeared as volume 33 in the Springer Series "Applications of Mathematics: Stochastic Modelling and Applied Probability". For its first edition, the book was planned to appear in the traditional yellow jacket. Claudia and Thomas visited the Springer-Verlag in Heidelberg and discussed the possibility of adding a figure on the cover. We still appreciate Catriona's support for this, at the time, novel idea

book "Scotland and the Flemish people" (photos att.): it is interesting. I am looking forward to reading it. Very best wishes, also to Gerda, Catriona."

The personal attentions in that e-mail are the "Flemish" (as indeed Paul is) and the "also to Gerda", Paul's wife.

Claudia and Thomas fondly recall their visit to the Springer-Verlag in Heidelberg during which Catriona showed enthusiastic support for our project. When you look at the two covers in Fig. 1, you surely notice that the subtitle for Insurance and Finance appeared on the cover of the reprinted later edition(s) (right) but not on the original first 1997 edition (left). This somehow reflects the fact that in the first instance we indeed concentrated more on methodology, though very much with applied interpretations and reader guidelines in mind. Before we decided to change the cover by including the subtitle, we had an intensive discussion about whether to write for or in. It became for as we clearly and academically honestly wanted to convey the message that here is a theory, EVT, that has considerable promise for applications in both insurance and finance but that the field still had to gain maturity before it really could become an in in the title. In our negotiations with Springer-Verlag, we also clearly stated that the appearance of finance in the title should not lead to an increase in the sale-price. Not all publishers adhere to this principle. Catriona very much understood and supported our concerns. On the occasion of

Catriona's 25 years anniversary at Springer-Verlag we wrote a thank-you poem. It seems fit to include it here so it may reach a wider audience.

1.1 An Extremal ode to Catriona Byrne

Three from Zurich said "let's write a book"
You cannot imagine how long that adventure took
As topic they choose to work on extremes
Not a bad choice among possible themes
For applications they considered insurance and finance
And discussed on and off with actuaries for guidance
One went to Groningen, the other to Mainz
The third to London? . . . No, he declines
Being away from Rösti and FIM
They ran into difficulties finishing the thing
Sitting in a boat merrily rowing about
One of them threatened to throw the others out
The project needs completion without delay
In comes an angel, if we may say
Dressed in yellow, with the crest of a horse
Catriona took charge and said "I endorse"
Under her wings extremes smoothed out
And brought the project in the hands of the crowd
Now it stands proud on so many a shelf
For the quality it can happily vow for itself
The authors went to places far away
To Munich, Copenhagen, but one said "I stay"
So did Catriona twenty-five years on
"I like it in yellow and do not want to be gone"
Give her your manuscript, let her bundle your thought
Into a book everyone wished they had bought

Dear Catriona, ad multos annos!
Paul, Claudia, Thomas
Zurich, Munich, Copenhagen

Fig. 2 The three co-authors of [7], Paul, Thomas and Claudia, hard at work on the book during a longer stay at the Mathematical Research Institute of Oberwolfach (MFO). This important stay formed part of MFO's Research in Pairs programme

1.2 New Aspects of Extreme Value Theory

In [7] we presented the state of the art of one-dimensional EVT for independent identically distributed random variables and stationary time series in discrete-time with applications to insurance and finance. Data were analysed and figures designed with SPLUS, a licenced early version of the modern R-software. No comprehensive extreme value package was available at that time. By now, a multitude of R-packages for extreme value statistics exist; see https://cran.r-project.org.

In the mean time, EVT has grown in various directions, this both theoretically as well as through its applications. Concerning the former, models for continuous-time stochastic processes and random fields have been developed. Further, multivariate EVT has reached maturity. More R-packages for extreme value statistics have been written, and new EVT related programming tools, like for instance in MATLAB or Python are available or on the way. We have witnessed a true explosion of applications, and this in all areas of science. This growth in applications is no doubt due to an increase risk awareness worldwide. An obvious example relates to environmental concern and climate change. Further, the availability of large amounts of data in the context of data science poses interesting challenges. These developments are well documented in numerous textbooks, journal articles, and in particular in the journal *Extremes*, also published by Springer; see https://www.springer.com/journal/10687.

In this paper we present three very different areas of EVT, where each of us introduces their current research interests. Paul Embrechts presents some fundamental theorems of Quantitative Risk Management, where heavy-tailed models like the multivariate t-distribution play an important role. Claudia Klüppelberg presents some recent research on max-linear Bayesian networks allowing for a causal analysis of risk events, and Thomas Mikosch browses through recent developments on extremes for serially dependent sequences with marginal power-law tails.

2 The Fundamental Theorems of Quantitative Risk Management (QRM) *(by Paul Embrechts)*

When [7] was published, EVT had already enjoyed numerous applications in insurance, the world of finance however still very much believed in the bell curve, the normal or Gaussian distribution. Events like the 1987 crash or the 1998 downfall of LTCM, a famous hedge fund, opened the eyes of many in the finance profession that there indeed was life beyond the normal tail. Especially the regulators for banking and insurance pressed for more realistic modelling assumptions underlying the calculation of regulatory capital. In the mid-nineties the estimation of high quantiles, far in the tail of a distribution became crucial. The key concept became Value-at-Risk (VaR), a regulatory risk measure that became legally mandatory to report on for the financial industry. The fact that quantiles well beyond the 90% quantile had to be estimated called for EVT-based technology. Though we were aware of these developments, [7] contains only one line on this topic (p. 370): "In the context of risk management, RiskMetrics [543] forms an interesting software environment in which various of the techniques discussed so far, especially concerning quantile (VaR) estimation, are to be found." The "RiskMetrics [543]" refers to a document of JP Morgan for the calculation of regulatory (risk) capital underlying the trading book of a bank. This one line meanwhile has grown into thousands of publications, with a book like [13] solely on the topic of VaR comprising 624 pages! Below we discuss some results from this world of Quantitative Risk Management. They concern the First and Second Fundamental Theorems of Quantitative Risk Management (1st FTQRM and 2nd FTQRM).

The 1st FTQRM goes back to Section 3.4 in the publication [8]. These results were first announced at the ETH Risk Day of 1998 in a talk by Daniel Straumann with the title "Tempting fallacies in the use of correlation"; see https://www2.math.ethz.ch/finance/ETH_Risk_Day.html. For further details, see Section 8.3 in [17].

2.1   The 1st FTQRM

The ingredients of the 1st FTQRM are:

- The multivariate normal distribution
- Multivariate normal variance mixtures
- Spherical and elliptical distributions
- Two important risk measures: VaR and ES
- Stress tests and stress scenario sets

Definition 2.1 (Multivariate Normal Distribution) $\mathbf{X} = (X_1, \ldots, X_d)^\top$ has a *multivariate normal distribution* if $\mathbf{X} \overset{\mathrm{d}}{=} \boldsymbol{\mu} + A\mathbf{Z}$, where $\mathbf{Z} = (Z_1, \ldots, Z_k)^\top$, $Z_i \sim N(0, 1)$ independent, $A \in \mathbb{R}^{d \times k}$, and $\boldsymbol{\mu} \in \mathbb{R}^d$. Here $\overset{\mathrm{d}}{=}$ means equality in distribution. We shall write

$$\mathbf{X} \sim N_d(\boldsymbol{\mu}, \Sigma), \quad \Sigma = AA^\top.$$

$\square$

The multivariate normal distribution has a very special property:

$$\mathbf{X} \sim N_d(\boldsymbol{\mu}, \Sigma) \quad \Longleftrightarrow \quad \forall \mathbf{a} \in \mathbb{R}^d : \quad \mathbf{a}^\top \mathbf{X} \sim N(\mathbf{a}^\top \boldsymbol{\mu}, \mathbf{a}^\top \Sigma \mathbf{a});$$

it yields a characterisation of multivariate Gaussianity through Gaussianity of all one-dimensional projections.

In the following definition we go slightly beyond the multivariate normal by allowing a common, multiplicative, independent stress factor on the components of the matrix A and hence on those of the variance-covariance matrix Σ.

Definition 2.2 (Multivariate Normal Variance Mixtures) A random vector $\mathbf{X}$ has a *(multivariate) normal variance mixture distribution* if

$$\mathbf{X} \overset{\mathrm{d}}{=} \boldsymbol{\mu} + \sqrt{W} A\mathbf{Z},$$

where $\mathbf{Z} \sim N_k(\mathbf{0}, I_k)$ with I_k the $k \times k$ identity matrix, $W \geq 0$ is a random variable independent of $\mathbf{Z}$, $A \in \mathbb{R}^{d \times k}$, and $\boldsymbol{\mu} \in \mathbb{R}^d$. The vector $\boldsymbol{\mu}$ is called the *location vector* and $\Sigma = AA^\top$ the *scale* (or *dispersion*) *matrix*. $\square$

An important example is the following.

Example 2.1 Let $\mathbf{X} \sim t_d(\nu, \boldsymbol{\mu}, \Sigma)$ have a multivariate t-distribution with $\nu > 0$ degrees of freedom and density

$$f_{\mathbf{X}}(\mathbf{x}) = \frac{\Gamma(\frac{1}{2}(\nu + d))}{\Gamma(\nu/2)(\pi \nu)^{d/2} \Sigma^{1/2}} \left(1 + \frac{(\mathbf{x} - \boldsymbol{\mu})^\top \Sigma^{-1}(\mathbf{x} - \boldsymbol{\mu})}{\nu} \right)^{-(\nu+d)/2}, \quad \mathbf{x} \in \mathbb{R}^d,$$

where Γ is Euler's Gamma function. Then $\mathbf{X}$ is a multivariate normal variance mixture as in Definition 2.2, where W has an inverse gamma distribution with mean $\mathbb{E}[W] = \nu/(\nu - 1)$ (if $\nu > 1$) and covariance matrix $\operatorname{cov}(\mathbf{X}) = \nu/(\nu - 2)\Sigma$ (if $\nu > 2$). We shall write

$$\mathbf{X} \sim M_d(\boldsymbol{\mu}, \Sigma, \widehat{F}_W), \quad \widehat{F}_W(s) = \mathbb{E}[e^{-sW}], \quad s \geq 0.$$

It is not difficult to see that the contour surfaces of multivariate normal mixture distributions are ellipsoids. In order to derive properties of these distributions, it helps to formulate a slightly more general class of distributions. For this we base the construction on the geometric definition of ellipticity.

Definition 2.3 (Spherical Distribution) A random vector $\mathbf{Y}$ has a *spherical distribution* if for every orthogonal matrix $U \in \mathbb{R}^{d \times d}$ (i.e. $UU^\top = U^\top U = I_d$) we have $\mathbf{Y} \stackrel{d}{=} U\mathbf{Y}$ (hence, $\mathbf{Y}$ is distributionally invariant under rotations and reflections). $\quad\square$

Theorem 2.4 (Characterization of Spherical Distributions) *Let* $\|\mathbf{t}\| = (t_1^2 + \cdots + t_d^2)^{1/2}, \mathbf{t} \in \mathbb{R}^d$. *The following are equivalent:*

(1) $\mathbf{Y}$ *is spherical.*
(2) *There exists a function* $\Psi : [0, \infty) \to \mathbb{R}$, *called the* characteristic generator, *such that* $\mathbb{E}[e^{i\mathbf{t}^\top \mathbf{Y}}] = \Psi(\|\mathbf{t}\|^2), \mathbf{t} \in \mathbb{R}^d$.
(3) *For all* $\mathbf{a} \in \mathbb{R}^d$, $\mathbf{a}^\top \mathbf{Y} \stackrel{d}{=} \|\mathbf{a}\| Y_1$ *(linear combinations are of the same type).*

We shall write $\mathbf{Y} \sim S_d(\Psi)$. $\quad\square$

Theorem 2.5 (Stochastic Representation) *The following representation holds:*

$$\mathbf{Y} \sim S_d(\Psi) \quad \Longleftrightarrow \quad \mathbf{Y} \stackrel{d}{=} R\mathbf{S},$$

where the radial part $R \geq 0$ *is independent of the angular part*

$$\mathbf{S} \sim U(\{\mathbf{x} \in \mathbb{R}^d : \|\mathbf{x}\| = 1\}).$$

Here U stands for the uniform distribution. $\quad\square$

Definition 2.6 (Elliptical Distribution) A random vector $\mathbf{X} = (X_1, \ldots, X_d)^\top$ has an *elliptical distribution* if for $\mathbf{Y} \sim S_k(\Psi)$, $A \in \mathbb{R}^{d \times k}$, $\mathbf{X} \stackrel{d}{=} \boldsymbol{\mu} + A\mathbf{Y}$ with location vector $\boldsymbol{\mu} \in \mathbb{R}^d$ and scale matrix $\Sigma = AA^\top$. $\quad\square$

Remark 2.1 By Theorem 2.5, an elliptical random vector admits the stochastic representation

$$\mathbf{X} \stackrel{d}{=} \boldsymbol{\mu} + RA\mathbf{S}, \quad R, \mathbf{S} \text{ as in Theorem 2.5.}$$

From Theorem 2.4 (2) it follows that the characteristic function of an elliptical random vector $\mathbf{X}$ is

$$\mathbb{E}[e^{it^\top \mathbf{X}}] = e^{it^\top \mu} \Psi(t^\top \Sigma t), \quad t \in \mathbb{R}^d.$$

We shall write $\mathbf{X} \sim E_d(\mu, \Sigma, \Psi)$.

Recall from finance and insurance the two following risk measures, *Value-at-Risk* (VaR) and *Expected Shortfall* (ES).

Definition 2.7 For $0 < \alpha < 1$ and X a random variable with distribution function F_X,

$$\text{VaR}_\alpha(X) = F_X^{-1}(\alpha) = \inf\{x \in \mathbb{R} : F_X(x) \geq \alpha\}.$$

If $\mathbb{E}[X] < \infty$,

$$\text{ES}_\alpha(X) = \frac{1}{1-\alpha} \int_\alpha^1 \text{VaR}_\delta(X) d\delta$$

$$= \mathbb{E}[X \mid X > \text{VaR}_\alpha(X)] \quad \text{if } F_X \text{ is continuous.}$$

$\square$

Remark 2.2 Note that ES_α is always subadditive, whereas in general, VaR_α is not; see Examples 2.25 and 2.26 in [17].

The importance of Theorems 2.4 and 2.5 becomes clear from the following result.

Theorem 2.8 (Subadditivity of VaR for Elliptical Models) Let $L_i = \lambda_i^\top \mathbf{X}$, for $\lambda_i \in \mathbb{R}^d$, $i = 1, \ldots, n$ with $\mathbf{X} \sim E_d(\mu, \Sigma, \Psi)$, then for $1/2 < \alpha < 1$:

$$\text{VaR}_\alpha\left(\sum_{i=1}^n L_i\right) \leq \sum_{i=1}^n \text{VaR}_\alpha(L_i).$$

Proof Consider a generic $L = \lambda^\top \mathbf{X} \overset{d}{=} \lambda^\top \mu + \lambda^\top A\mathbf{Y}$ for $\mathbf{Y} \sim S_k(\Psi)$. By Theorem 2.4 (3),

$$\lambda^\top A\mathbf{Y} \overset{d}{=} \|\lambda^\top A\| Y_1,$$

so $L \overset{d}{=} \lambda^\top \mu + \|\lambda^\top A\| Y_1$. By translation invariance and positive homogeneity of VaR_α we obtain

$$\text{VaR}_\alpha(L) = \lambda^\top \mu + \|\lambda^\top A\| \text{VaR}_\alpha(Y_1).$$

Applying this result to $L = \sum_{i=1}^{n} L_i = (\sum_{i=1}^{n} \lambda_i)^\top \mathbf{X}$ and to each $L_i = \lambda_i^\top \mathbf{X}$, $i = 1, \ldots, n$, we obtain

$$
\begin{aligned}
\mathrm{VaR}_\alpha \left(\sum_{i=1}^{n} L_i \right) &= \sum_{i=1}^{n} \lambda_i^\top \boldsymbol{\mu} + \| \sum_{i=1}^{n} \lambda_i^\top A \| \mathrm{VaR}_\alpha(Y_1) \\
&\leq \sum_{i=1}^{n} \lambda_i^\top \boldsymbol{\mu} + \left(\sum_{i=1}^{n} \| \lambda_i^\top A \| \right) \mathrm{VaR}_\alpha(Y_1) \\
&= \sum_{i=1}^{n} (\lambda_i^\top \boldsymbol{\mu} + \| \lambda_i^\top A \| \mathrm{VaR}_\alpha(Y_1)) \\
&= \sum_{i=1}^{n} \mathrm{VaR}_\alpha(L_i),
\end{aligned}
$$

where we used the Cauchy–Schwarz inequality and the fact that for $1/2 < \alpha < 1$, $\mathrm{VaR}_\alpha(Y_1) \geq 0$. Taking $\lambda_i = \mathbf{e}_i$ (i-th unit vector), we obtain the subadditivity

$$
\mathrm{VaR}_\alpha (\sum_{i=1}^{n} X_i) \leq \sum_{i=1}^{n} \mathrm{VaR}_\alpha(X_i).
$$

$\square$

For the definition of a coherent risk measure ρ, see Definition 8.1 in [17]; in particular, ES_α is always a coherent risk measure. It follows from Theorem 2.8 that VaR_α is coherent for the class of elliptical random vectors (or distribution functions).

There is a link between the notion of coherence and stress testing. For this we consider the set $\mathcal{M}$ of linear portfolios based on a fixed d-dimensional random vector $\mathbf{X}$:

$$
\mathcal{M} = \{ L : L = m + \lambda^\top \mathbf{X}, m \in \mathbb{R}, \lambda \in \mathbb{R}^d \}.
$$

For a positive-homogeneous risk measure $\rho : \mathcal{M} \to \mathbb{R}$, we define a risk-measure function $r_\rho(\lambda) = \rho(\lambda^\top \mathbf{X})$, $\lambda \in \mathbb{R}^d$. There is a one-to-one relationship between ρ and r_ρ given by

$$
\rho(m + \lambda^\top \mathbf{X}) = m + r_\rho(\lambda).
$$

Define the scenario set

$$
S(\rho) = \{ \mathbf{x} \in \mathbb{R}^d : \mathbf{u}^\top \mathbf{x} \leq r_\rho(\mathbf{u}) \, \forall \mathbf{u} \in \mathbb{R}^d \},
$$

hence S_ρ is an intersection of half-spaces, so that S_ρ is a closed convex set. The precise form of S_ρ depends on the distribution of $\mathbf{X}$ and on ρ. If $\rho = \text{VaR}_\alpha$, S_ρ has the interpretation of a depth set in robust statistics. Also recall that a risk measure ρ is law-invariant if for every random variable X, $\rho(X)$ only depends on the distribution function F_X of X. Typical examples are VaR and ES.

We are now in the position to formulate the 1st FTQRM.

Theorem 2.9 (1st FTQRM) *Let $\mathbf{X} \sim E_d(\boldsymbol{\mu}, \Sigma, \Psi)$ and ρ be any positive-homogeneous, translation-invariant and law-invariant risk measure on $\mathcal{M}$. Then the following hold:*

(1) For all $L = m + \boldsymbol{\lambda}^\top \mathbf{X} \in \mathcal{M}$,

$$\rho(L) = m + \boldsymbol{\lambda}^\top \boldsymbol{\mu} + \sqrt{\boldsymbol{\lambda}^\top \Sigma \boldsymbol{\lambda}} \rho(Y_1), \quad Y_1 \in S_1(\Psi).$$

(2) If $\rho(Y_1) \geq 0$, then ρ is subadditive on $\mathcal{M}$. In particular, we obtain that in the case of elliptical portfolios, VaR is coherent.

(3) If $\mathbb{E}[X]$ exists, then for all $L = m + \boldsymbol{\lambda}^\top \mathbf{X} \in \mathcal{M}$ and $(\rho_{ij} = \mathcal{P}(\Sigma)_{ij}$ with $\mathcal{P}(\Sigma)$ the "correlation matrix" associated to Σ, see [17], p. 176),

$$\rho(L - \mathbb{E}[L]) = \sqrt{\sum_{i=1}^{d} \sum_{j=1}^{d} \rho_{ij} \lambda_i \lambda_j \rho(X_i - \mathbb{E}[X_i]) \rho(X_j - \mathbb{E}[X_j])}.$$

(4) If $\text{cov}(\mathbf{X})$ exists and $\rho(Y_1) > 0$, then for all $L \in \mathcal{M}$,

$$\rho(L) = \mathbb{E}[L] + k_\rho \sqrt{\text{var}(L)}$$

for some $k_\rho > 0$ depending on ρ.
(5) If Σ^{-1} exists and $\rho(Y_1) > 0$, then

$$S_\rho = \{\mathbf{x} \in \mathbb{R}^d : (\mathbf{x} - \boldsymbol{\mu})^\top \Sigma^{-1} (\mathbf{x} - \boldsymbol{\mu}) \leq \rho(Y_1)^2\}.$$

$\square$

2.2 Discussion

(1) For a proof, see [17], Theorem 8.28.
(2) Theorem 2.9 (1) yields an explicit form of $\rho(L)$ as a function of the portfolio structure and the defining factors of the underlying elliptical distributions.
(3) Theorem 2.9 (2) implies that, under fairly general conditions on the risk measure ρ, we obtain subadditivity of ρ on $\mathcal{M}$.

(4) Theorem 2.9 (3) computes the overall portfolio stress loss $\rho(L - \mathbb{E}[L])$ as an aggregation of the individual stress factor contributions $\rho(X_i - \mathbb{E}[X_i])$, $i = 1, \ldots, d$ and forms the basis of the standard formula under the regulatory guidelines Solvency 2 for insurance companies.
(5) Theorem 2.9 (4) has the important implication that in this case, the optimal portfolio under ρ coincides with the Markowitz variance-minimizing portfolio.
(6) Theorem 2.9 (5) represents the scenario (stress) sets S_ρ defined above as ellipsoids.

Important Comment The 1st FTQRM holds true under the important condition of ellipticity of the underlying risk factors. A special case corresponds to the multivariate normal and the multivariate t. In such cases, QRM becomes a fairly standard exercise. This brings us to the (much more) important "2nd FTQRM". If the conditions underlying the 1st FTQRM, in particular $\mathbf{X} \sim E_d(\boldsymbol{\mu}, \Sigma, \Psi)$, do not hold, then the results (1)–(5) from Theorem 2.9 typically do not hold. Moreover, in practice they may fail in a rather dramatic way!

Remark 2.3 We wrote "2nd FTQRM" above in quotation marks, indeed, as stated, it is not a theorem, but more a summary statement on numerous results existing in the literature. For an example, see e.g. Section 8.4.4 in [17].

3 Max-Linear Bayesian Networks *(by Claudia Klüppelberg)*

Graphical models can represent multivariate distributions in an intuitive way and, hence, facilitate the statistical analysis of high-dimensional data. Such models are usually modular so that high-dimensional distributions can be described and handled by a careful combination of lower-dimensional factors. Furthermore, graphs are natural data structures for algorithmic treatment. Conditional independence and Markov properties are essential features for graphical models. The book by Lauritzen [16] masterly describes the fundamental mathematical and statistical theory of graphical models.

Moreover, graphical models can allow for causal interpretation, often provided through a recursive system on a directed acyclic graph (DAG), and the max-linear model (1) below is a specific example. The book by Pearl [18] provides a rich source for the study of causality for statistical data.

We present below some conditional independence properties of max-linear Bayesian networks, which emphasize their difference to linear networks.

3.1   An Extremal Graphical Model

Max-linear Bayesian networks as introduced in [10] model causal dependence between extreme events. It is specified by a random vector $X = (X_1, \ldots, X_d)$, a directed acyclic graph $\mathcal{D} = (V, E)$ with nodes $V = \{1, \ldots, d\}$, non-negative *coefficients* $c_{ij} \geq 0$ for $i, j \in V$ (summarized in the *coefficient matrix* $C = (c_{ij})$), and independent random variables $Z_1, \ldots, Z_d$. These, known as *innovations*, have support $\mathbb{R}_{>} := (0, \infty)$ and have atom-free distributions. Then X is defined by the recursive system

$$X_i = \bigvee_{j \in \mathrm{pa}(i)} c_{ij} X_j \vee Z_i, \quad i = 1, \ldots, d, \tag{1}$$

where $\mathrm{pa}(i)$ (parents of i) denotes the set of nodes j where there is a directed edge from j to i. The equation system (1) has solution

$$X_i = \bigvee_{j \in \mathrm{an}(i) \cup \{i\}} b_{ij} Z_j, \quad i = 1, \ldots, d, \tag{2}$$

where $\mathrm{an}(i)$ (ancestors of i) denotes the set of nodes j where there is a directed path from j to i, and b_{ij} is a maximum taken over all the products along such paths (see [10], Theorem 2.2). Any such path that realizes this maximum is called *max-weighted* under C.

Above we have used the following standard notation. A *path* in a DAG $\mathcal{D}$ is a sequence of nodes $i_0, i_1, \ldots, i_k$ such that $i_\ell \to i_{\ell+1}$ or $i_{\ell+1} \to i_\ell$ is an edge in $\mathcal{D}$ for each $\ell = 0, \ldots, k$. A *directed path* has edges $i_\ell \to i_{\ell+1}$ for all ℓ. A *collider* on a path is a node i_ℓ in a path such that $i_{\ell-1} \to i_\ell \leftarrow i_{\ell+1}$.

Whereas linear Bayesian networks are based on *classical linear algebra*, max-linear Bayesian networks are based on *tropical linear algebra* in the max-times semiring $(\mathbb{R}_{\geq}, \odot, \cdot)$, defined by

$$a \odot b := a \vee b = \max(a, b), \quad a \cdot b := ab \quad \text{for } a, b \in \mathbb{R}_{\geq}.$$

These operations extend to $\mathbb{R}_{\geq}^d$ coordinate-wise and to corresponding matrix multiplication for $R \in \mathbb{R}_{\geq}^{m \times n}$ and $S \in \mathbb{R}_{\geq}^{n \times p}$ as

$$(R \odot S)_{ij} = \bigvee_{k=1}^{n} r_{ik} s_{kj}.$$

In tropical linear algebra, $B = (b_{ij})$ is called the *Kleene star* matrix and a max-weighted path is called a *critical* path.

For max-linear Bayesian networks, X is Markov with respect to its DAG. However, tropical linear algebra has various consequences concerning conditional independence properties and statistical analysis of the model.

Fig. 3 Diamond graph (left) and with the node $K = \{2\}$ being observed, as indicated in red. If $c_{42}c_{21} \geq c_{43}c_{31}$, it holds that $X_1 \perp\!\!\!\perp X_4 \mid X_2$

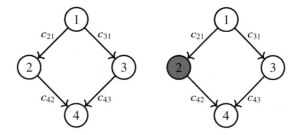

3.2 Conditional Independence

Graphical models identify conditional independence relations through *separation criteria* applied to a graph. The standard separation criteria is given by the following definition.

Definition 3.1 Two nodes $i, j \in V$ are *d-connected* given a set $K \in V \setminus \{i, j\}$ if there is a path π from j to i such that all colliders on π are in $K \cup an(K)$ and no non-collider on π is in K. For three disjoint subsets I, J, K of the node set V, the node set K *d-separates* I and J if no pair of nodes $i \in I$ and $j \in J$ is d-connected relative to K.

It was observed already in [14] and investigated in detail in [1] that the conditional independence properties for max-linear Bayesian networks are very different from those in linear Bayesian networks. In particular, they are often not faithful to their underlying DAG $\mathcal{D}$. This means that the above d-separation criterion on the DAG typically will not identify all valid conditional independence relations, in contrast to the situation for most Bayesian networks based on discrete random variables or linear structural equations.

We present three examples to explain some of the relevant issues and refer to [1] for details.

Example 3.1 (Diamond) Consider the DAG in Fig. 3.
The path $1 \to 2 \to 4$ is max-weighted if and only if $c_{42}c_{21} \geq c_{43}c_{31}$. If this is the case, the joint distribution of (X_1, X_2, X_4) has the representation

$$X_1 = Z_1, \quad X_2 = c_{21}X_1 \vee Z_2,$$

and

$$X_4 = c_{42}X_2 \vee Z_4 \vee c_{43}X_3$$
$$= c_{42}(Z_2 \vee c_{21}Z_1) \vee Z_4 \vee c_{43}(Z_3 \vee c_{31}Z_1)$$
$$= c_{42}Z_2 \vee c_{42}c_{21}Z_1 \vee Z_4 \vee c_{43}Z_3 \vee c_{43}c_{31}Z_1$$
$$= c_{42}Z_2 \vee c_{42}c_{21}Z_1 \vee Z_4 \vee c_{43}Z_3 \quad \text{since } c_{42}c_{21} \geq c_{43}c_{31}$$
$$= c_{42}X_2 \vee Z_4 \vee c_{43}Z_3$$

and hence we have $X_1 \perp\!\!\!\perp X_4 \mid X_2$, which does *not* follow from the d-separation criterion. Here, the fact that $1 \to 2 \to 4$ is *max-weighted* renders the path $1 \to 3 \to 4$ unimportant for the conditional independence $X_1 \perp\!\!\!\perp X_4 \mid X_2$, even if $1 \to 3 \to 4$ were also max-weighted (that is, even if $c_{42}c_{21} = c_{43}c_{31}$).

In Example 3.1, the complicating issue was associated with paths being max-weighted or not. However, this is not the only way standard d-separation fails. In Example 3.2, the complications are associated with double colliders along a path.

Example 3.2 (Cassiopeia) A max-linear Bayesian network on the graph in Fig. 4 will satisfy $X_1 \perp\!\!\!\perp X_3 \mid X_{\{4,5\}}$ for all coefficient matrices C. This can be seen from the following calculations, where we assume that $c_{ji} = b_{ji} = 1$ for all edges in this DAG and let $x_K = (x_4, x_5)$. Then $X_i = Z_i$ for $i = 1, 2, 3$ and

$$\begin{bmatrix} x_4 \\ x_5 \end{bmatrix} = \begin{bmatrix} Z_1 \vee Z_2 \vee Z_4 \\ Z_2 \vee Z_3 \vee Z_5 \end{bmatrix} \geq \begin{bmatrix} Z_4 \\ Z_5 \end{bmatrix} \quad \text{and} \quad \begin{bmatrix} x_4 \\ x_5 \end{bmatrix} \geq \begin{bmatrix} Z_1 \vee Z_2 \\ Z_2 \vee Z_3 \end{bmatrix}.$$

We have three situations for (x_4, x_5) corresponding to

$$x_4 < x_5, \qquad\qquad x_4 > x_5, \qquad\qquad x_4 = x_5$$

$$\begin{bmatrix} x_4 \\ x_5 \end{bmatrix} \geq \begin{bmatrix} Z_1 \vee Z_2 \\ Z_3 \end{bmatrix}, \quad \begin{bmatrix} x_4 \\ x_5 \end{bmatrix} \geq \begin{bmatrix} Z_1 \\ Z_2 \vee Z_3 \end{bmatrix}, \quad \begin{bmatrix} x_4 \\ x_5 \end{bmatrix} \geq \begin{bmatrix} Z_1 \\ Z_3 \end{bmatrix} \text{ and } Z_2 = x_4 = x_5.$$

Hence, all Z_i are bounded in all three cases. Moreover, Z_1 and Z_3 never occur together in any inequality, rendering $X_1 \perp\!\!\!\perp X_3 \mid X_{\{4,5\}}$. However, this conditional independence statement does *not* follow from the d-separation criterion since the path from 1 to 3 is d-connecting relative to $\{4, 5\}$.

Example 3.2 shows that max-linear Bayesian networks are often not faithful to d-separation, but d-separation is also not *complete* for conditional independence in these networks. That is, there are conditional independence statements which are valid for any choice of coefficients C, but cannot be derived from d-separation.

Also, in contrast to standard results for linear Bayesian networks, some conditional independence relations are highly *context-specific*, i.e. depend drastically on the particular values of the conditioning variables, as in Example 3.3.

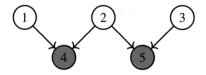

Fig. 4 The Cassiopeia graph with observed nodes $K = \{4, 5\}$. Here it holds that $X_1 \perp\!\!\!\perp X_3 \mid X_{\{4,5\}}$

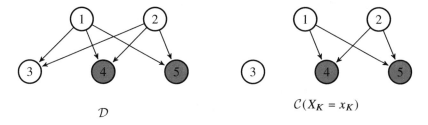

$\mathcal{D}$ $\mathcal{C}(X_K = x_K)$

Fig. 5 The left-hand figure displays what we call the tent DAG $\mathcal{D}$. For all coefficients equal to 1, the source DAG $\mathcal{C}(X_K = x_K)$ when the observed nodes are $K = \{4, 5\}$ with observed values $x_4 = x_5 = 2$, is obtained from the left-hand figure by removing the edges $1 \to 3$ and $2 \to 3$, which become redundant in the context $\{X_4 = X_5 = 2\}$

Example 3.3 (Tent) Consider the DAG $\mathcal{D}$ on the left in Fig. 5 with all coefficients $c_{ji} = 1$. Let $K = \{4, 5\}$ be the set of observed nodes; we seek all independence relations conditionally valid in the context $X_4 = X_5 = 2$. Writing out the model (1) we find

$$X_1 = Z_1, \quad X_2 = Z_2, \quad X_3 = Z_3 \vee X_1 \vee X_2,$$
$$X_4 = Z_4 \vee X_1 \vee X_2 = 2,$$
$$X_5 = Z_5 \vee X_1 \vee X_2 = 2.$$

Since $Z_1, \ldots, Z_5$ are a.s. different when the innovations have atom-free distributions, it holds apart from a null-set that $X_1 \vee X_2 = Z_1 \vee Z_2 = 2$. This introduces bounds on the innovations; we must have $Z_1, Z_2, Z_4, Z_5 \leq 2$ and it also holds that $X_3 \geq 2$. Further, we then have

$$X_1 = Z_1, \quad X_2 = Z_2, \quad X_1 \vee X_2 = 2, \quad X_3 = Z_3 \vee 2,$$
$$X_4 = Z_4 \vee 2 = 2,$$
$$X_5 = Z_5 \vee 2 = 2,$$

whence we conclude that $X_3 \perp\!\!\!\perp (X_1, X_2) \mid X_4 = X_5 = 2$, since now the dependence of X_3 on X_1, X_2 has disappeared. This independence statement is reflected in the lack of edges $1 \to 3$ and $2 \to 3$ in the source DAG $\mathcal{C}(X_4 = X_5 = 2)$, shown to the right in Fig. 5. $\qquad\square$

The paper [1] provides a complete description of valid conditional independence statements for a given coefficient matrix C, conditional independence statements that hold for all C supported on a given DAG $\mathcal{D}$, as well as those that depend on the specific values of the conditioning variables. This is achieved by introducing three separation criteria. These are less restrictive than d-separation, as they focus on paths that are *max-weighted* (as in Example 3.1), do not have *multiple colliders* (as in Example 3.2), and, for a given context, refer to the source DAG, obtained by removing edges that are *redundant* in the context (as in Example 3.3).

Definition 3.2 says that a $*$-connecting path is d-connecting with at most one collider.

Definition 3.2 A path between nodes j and i in a DAG is $*$-*connecting* relative to a given node set K if and only if it is one of the paths in Fig. 6. Conversely, K $*$-*separates* j and i if they are not $*$-connecting given K. For three disjoint subsets I, J and K of the node set V, the node set K $*$-separates I and J if no pair of nodes $i \in I$ and $j \in J$ is $*$-connected relative to K. □

The characterization of conditional independence relations in a max-linear model is based on $*$-separation in different graphs, corresponding to three situations. The different graphs are derived from the so-called *reachability graph* $\mathcal{D}^*$ of $\mathcal{D}$, which has an edge if and only if there is a path from j to i, or if $j = i$. Indeed, $B = (b_{ji})$ as in (2) is a weighted reachability matrix.

In [1] we formulate three different theorems, exemplified by Examples 3.1, 3.2 and 3.3, to clarify conditional independence for max-linear Bayesian networks. All three have the following structure, using $*$-separation ($\perp_*$) in appropriate derived DAGs.

Theorem 3.3 *Let X be a max-linear Bayesian network over a directed acyclic graph $\mathcal{D} = (V, E)$. Then for all $I, J, K \subseteq V$,*

$$I \perp_* J \mid K \text{ in } \widetilde{\mathcal{D}} \implies X_I \perp\!\!\!\perp X_J \mid X_K.$$

The DAG $\widetilde{\mathcal{D}}$—derived from $\mathcal{D}$, C, and the specific *context* $\{X_K = x_K\}$—depends on the situation. Their characterization is the main focus in [1], and we distinguish the following three:

(1) $\widetilde{\mathcal{D}} = \mathcal{C}(X_K = x_K)$: The coefficient matrix C is fixed and the context $\{X_K = x_K\}$ is specific, thus yielding conditional independence relations that are valid for the particular values x_K (as in Example 3.3).

(2) $\widetilde{\mathcal{D}} = \mathcal{D}_K^*(C)$: The coefficient matrix C is fixed, which yields independence relations that may depend on C which are valid for all possible contexts (as in Example 3.1).

(3) $\widetilde{\mathcal{D}} = \mathcal{D}_K^*$: The coefficient matrix C is arbitrary with support included in $\mathcal{D}$ and this yields conditional independence relations that are universally valid under these conditions (as in Example 3.2).

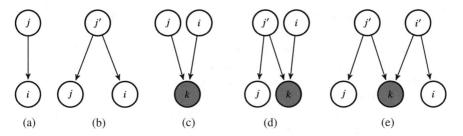

Fig. 6 Types of $*$-connecting paths between i and j. Nodes that are colored red are in K

3.3 Discussion

(1) In max-linear Bayesian networks there are more conditional independence properties than in a linear Bayesian network. This can be seen e.g. from Example 3.1, where the path $1 \to 3 \to 4$ is d-connecting, but $1 \perp\!\!\!\perp 4 \mid 2$.

(2) The above separation criteria follow the form of the moralization procedure for establishing d-separation, where also separation is checked in a derived graph, the moral graph.

(3) In Section 6 of [1] we give conditions for the above separation criteria to be *complete* in the sense that they yield all conditional independence statements that are valid under the specified conditions.

4 A Light History of Heavy Tails *(by Thomas Mikosch)*

There is no clear-cut definition of the notion of heavy-tailed distribution. In our 1997 book [7] we made an attempt to define different classes of heavy-tailed distributions via certain properties they have in common. A large part of the monograph was devoted to the subexponential class which we understood as *the* natural family of univariate distributions for modeling risks in finance and insurance: a distribution F on $[0, \infty)$ is *subexponential* if for iid random variables $X_1, \ldots, X_n$ with distribution F and any $n \geq 2$ the tails of their sum $S_n = X_1 + \cdots + X_n$ and maximum $M_n = \max_{i=1,\ldots,n} X_i$ are equivalent:

$$\frac{\mathbb{P}(S_n > x)}{\mathbb{P}(M_n > x)} \to 1, \qquad x \to \infty, \tag{3}$$

we also write $\mathbb{P}(S_n > x) \sim \mathbb{P}(M_n > x)$ for this equivalence. For example, if we interpret S_n as the aggregated claim amount in an insurance portfolio and M_n as the largest claim size in it, then the probability for S_n being extremely large is due to M_n. In catastrophe insurance, M_n is something like a claim size caused by a 9/11 or Fukushima event: a single event may lead to bankruptcy of the whole insurance company.

Subexponential distributions are defined via the simple relation (3). However, (3) does not tell us much about the structure of the distribution F of the X_i. The tail structure of subexponential F was studied starting in the 1960s. It was well understood already in the 1980s. It turned out that subexponential distributions cover power-law tails (these distributions have certain infinite power-moments, i.e., they are *very* heavy-tailed), but also distributions like the lognormal, or Weibull with tail $\overline{F}(x) = 1 - F(x) = \exp(-x^\tau)$, $x > 0$, for $\tau \in (0, 1)$, are subexponential. The two latter distributions have all power-moments but no exponential ones, i.e., they have *medium* heavy tails. All subexponential distributions are popular in insurance for modeling claim size distributions, and also in finance. For example, the lognormal is the marginal distribution of the celebrated Black–Scholes model for prices of speculative assets.

Major parts of our book [7] were devoted to distributions with power-law tails. In mathematical terms, we dealt with the regularly varying distributions on $[0, \infty)$ with tail index $\alpha > 0$,

$$\mathbb{P}(X > x) = x^{-\alpha} L(x), \qquad x > 0, \tag{4}$$

where L is a *slowly varying* (flat) function compared to the power law $x^{-\alpha}$, i.e., for arbitrarily small $\varepsilon > 0$ and sufficiently large x, $L(x)$ can be sandwiched between $x^{-\varepsilon}$ and x^ε. This function is an infinite-dimensional nuisance parameter which makes the statistical analyses of the tails in (4) a nightmare, for example, the estimation of α is a hard problem which gave rise to dozens of scientific articles and book chapters.

The regularly varying distributions can easily be extended to the real line by introducing a *tail-balance condition*: for some p_+, $p_- \geq 0$ such that $p_+ + p_- = 1$,

$$\mathbb{P}(|X| > x) = x^{-\alpha} L(x) \text{ and } \mathbb{P}(\pm X > x) \sim p_\pm \mathbb{P}(|X| > x), \qquad x \to \infty. \tag{5}$$

This means that X has a regularly varying *radial* part $|X|$ and an *angular* part $X/|X|$ which determines the probability of the sign $+1$ or -1 of X given that $|X|$ is large. The extension (5) of a regularly varying random variable enabled us in [7] to conduct an extreme value analysis of iid random variables (X_i) with such a marginal distribution, but we were also able to derive extreme value theory (EVT) for *linear processes*:

$$X_t = \sum_{j=0}^{\infty} \psi_j Z_{t-j}, \qquad t \in \mathbb{Z}. \tag{6}$$

This class contains the backbone of classical time series analysis: the causal ARMA processes; see the textbook treatment by Brockwell and Davis [3]. Under summability conditions on the real sequence (ψ_j) and regular variation of the marginal distribution of the iid noise (Z_t) (in the sense of (5)) we could present a theory for the extremes of (X_t) which completely parallels the theory for an iid sequence with regularly varying marginals. This could be achieved by employing the so-called *single big jump principle*: given we observe a large value of X_t satisfying (6), this value is caused by a single big value of the Z_{t-j}. Keeping this principle in mind, it is rather easy to understand the asymptotic behavior of the extremes of (X_t). A related EVT can also be derived for linear processes with more general subexponential iid noise (Z_t). The single big jump principle is the heuristic guide to success in this case too.

With the advancement of financial econometrics, starting in 1982 with the introduction of Robert Engle's [9] ARCH model for returns of speculative assets, new interesting time series models entered the field. These models exhibit zero autocorrelations, i.e., they are white noise—in agreement with real-life time series of financial returns. EVT for a special case of these models (the ARCH(1), *autoregressive conditionally heteroscedatic process of order 1*) was provided rather early on: in the 1989 paper by de Haan, Resnick, Rootzén and de Vries [12]. Fortunately, some very qualified people had met and written this paper: Holger Rootzén and Sid Resnick—two pioneers on extremes for time series, Laurens de Haan—a classic of EVT, Casper de Vries—a financial econometrician who was aware of Engle's work, and all of them were aware of a forthcoming paper by Charles Goldie [11] who had derived the marginal tails of the ARCH(1) which turned out to be of power-law type (5) with a constant $L(x)$.

The three authors of this paper had the pleasure to meet the aforementioned researchers many times over the last 30 years, to discuss scientific matters with them. In particular, Charles Goldie took the effort to read (and significantly improve upon) our book [7] but also the 2016 Springer monograph [4], jointly written with Ewa Damek and Darek Buraczewski. A main driving force for the latter book was the goal to better understand the structure of ARCH-type models. These are highly non-linear models, in their simplest form, the ARCH(1) is given by

$$X_t = \sigma_t Z_t, \qquad \sigma_t^2 = \alpha_0 + \alpha_1 X_{t-1}^2, \qquad t \in \mathbb{Z},$$

for positive constants α_i and an iid sequence (Z_t) with mean zero and unit variance. In contrast to linear processes (6), the multiplicative structure of the recursion $\sigma_t^2 = \alpha_0 + \alpha_1 \sigma_{t-1}^2 Z_{t-1}^2$ makes the analysis of the marginal tails of (X_t) a very difficult problem. In this case, no single big jump principle is available, i.e., it is not possible to identify a single value Z_t which is responsible for the largest values in the sample $X_1, \ldots, X_n$. Still, the Markovian structure of ARCH(1) allowed the authors of [12] to derive the EVT for this model.

When dealing with the extremes of time series, one needs to understand the extremal dependence structure of this series. With the exception of Gaussian time series, the extremes cannot be dealt with by covariances or correlations as in classical time series analysis. Extremes happen in the very far out tails of the distribution of the series. Due to dependence it is not enough to study the marginal tail behavior: typically, extremes of time series appear in *clusters,* in groups of large positive or negative values roughly at the same time. We see these clumps in financial return series in times of crises; see Fig. 7 for an illustration.

Therefore one needs to understand the simultaneous tails of a time series. This means that we are seeking for the joint tail behavior of the lagged vectors $\mathbf{X}_h = (X_0, \ldots, X_h)$, $h \geq 0$ (here we only consider real-valued X_i, but the same problem arises for vector-valued time series).

But how can one define a tail of a random vector $\mathbf{X}$?

A guide to answering this question can be found in Resnick's classical 1987 Springer book [19]. He dealt with multivariate EVT when the marginal distributions are regularly varying in the sense of (5). In addition to the regular variation of the radial part $|\mathbf{X}|$ of a random vector $\mathbf{X} \in \mathbb{R}^d$ we also need to have control of its angular part, i.e., of the likelihood of the directions $\mathbf{X}/|\mathbf{X}|$. A possible definition of *multivariate regular variation of* $\mathbf{X}$ *with index* $\alpha > 0$ is the following one: $|\mathbf{X}|$ is regularly varying with index $\alpha > 0$ and there exists a random vector Θ with values on the unit sphere $\mathbb{S}^{d-1} = \{\mathbf{x} : |\mathbf{x}| = 1\}$ of $\mathbb{R}^d$ such that for every $r > 1$ and "nice" subsets of $\mathbb{S}^{d-1}$,

$$\mathbb{P}\left(\left(\frac{|\mathbf{X}|}{x}, \frac{\mathbf{X}}{|\mathbf{X}|}\right) \in (r, \infty) \times S \,\Big|\, |\mathbf{X}| > x\right) \to r^{-\alpha}\, \mathbb{P}(\Theta \in S), \qquad x \to \infty. \tag{7}$$

We observe that $\mathbb{P}(Y > r) = r^{-\alpha}$, $r > 1$, defines a Pareto(α) distribution. Thus, the limit relation (7) can be interpreted as follows: given that $|\mathbf{X}|$ is large the scaled radial and angular parts $|\mathbf{X}|/x$ and $\mathbf{X}/|\mathbf{X}|$ are asymptotically independent, i.e., the limits Y and Θ are independent. The univariate case is already covered by the tail-balance condition (5) with $\mathbb{P}(\Theta = \pm 1) = p_\pm$: in this case we have only two directions.

Now we can define a *regularly varying time series*: take a strictly stationary sequence (X_t) and require that (7) holds for every lagged vector $\mathbf{X} = \mathbf{X}_h$, $h \geq 0$. This idea can really be made to work. The first paper where this idea was exploited was written by Davis and Hsing [5] in 1995. They paved the way for the extreme value analysis of regularly varying time series in the aforementioned sense.

An elegant reformulation of the regular variation property of a strictly stationary time series was provided by Basrak and Seghers [2] in 2009: the strictly stationary sequence (X_t) is regularly varying with index $\alpha > 0$ if for every lagged vector $\mathbf{X}_h$, $h \geq 0$, the following distributional convergence relation holds

$$\mathbb{P}\left(x^{-1}\mathbf{X}_h \in \cdot \,\big|\, |X_0| > x\right) \xrightarrow{w} \mathbb{P}\left(Y\,(\Theta_0, \ldots, \Theta_h) \in \cdot\right), \qquad x \to \infty. \tag{8}$$

Here Y is a Pareto(α)-distributed random variable independent of the sequence of random variables $(\Theta_t)_{t \geq 0}$, constituting the so-called *spectral tail process.* This process describes how a large value $|X_0|$ at time zero propagates into the extreme future of the time series. For example, the solution to the AR(1) equation $X_t =$

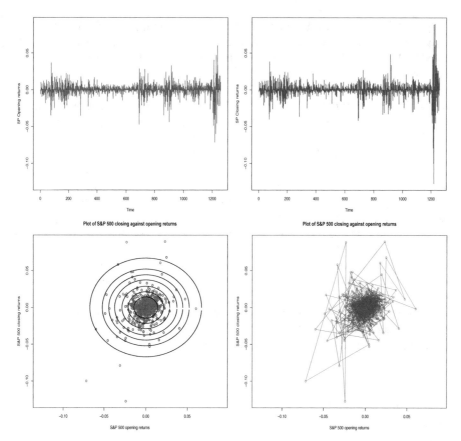

Fig. 7 *Top:* the 1258 log-returns $X_t = \log P_t - \log P_{t-1}$ of the daily S&P 500 stock index's opening (left) and closing (right) prices P_t from the period 11 May, 2015, to 8 May, 2020. The influence of the corona virus pandemic is clearly visible in the last 60 values of these time series. *Bottom:* scatterplot of the closing against the opening log-returns of the same stock index time series. The circles in the left graph have radii corresponding to the 80, 90, 95, 97, 98, 99, 99.5% quantiles of the distances of the scatter points from the origin. The red dots outside these circles may be interpreted as "extremes of different degrees". In the right plot each point is connected with two points, corresponding to the log-return time series on the previous and next days. It is striking that a point with a large norm is typically connected with two points with large norms: this is an indication of extremal clusters. This is particularly evident between 11 February, 2020, and 8 May, 2020: this part of the time series is colored by red points and lines

$\varphi X_{t-1} + Z_t, t \in \mathbb{Z}$, for some $\varphi \in (-1, 1)$ with regularly varying iid noise (Z_t) is regularly varying with spectral tail process $\Theta_t = \Theta_0 \varphi^t$ where $\mathbb{P}(\Theta_0 = \pm 1) = \lim_{x \to \infty} \mathbb{P}(\pm X_0 > x)/\mathbb{P}(|X_0| > x)$, i.e., a shock to the AR(1) process at time zero vanishes exponentially fast while the spectral tail process carries the sign Θ_0 of this shock into the future.

Over the last 10–15 years a group of researchers in EVT has focused one the probabilistic and statistical analyses of regularly varying processes. These do not only include time series, but also graphical models, random fields, random sets, max-stable structures, random recurrence equations, branching processes,... We refer to the recent Springer book by Kulik and Soulier [15] for a state-of-the-art overview.

One of the reasons for the popularity of regularly varying structures is that univariate regular variation is extended to the multivariate case in a "very natural way". Efforts have been made to extend the entire univariate subexponential class (including distributions which have all power-moments) to the multivariate case. The paper by Samorodnitsky and Sun [20] gives an overview of the results achieved. The authors indicate that the existing definitions of multivariate subexponentiality are either trivial (requiring only subexponentiality of the marginal distributions without any dependence assumptions between the components of the vector) or too complicated to be of general practical use (i.e., these definitions depend on particular applications like ruin probabilities, and correspond to some special sets in $\mathbb{R}^d$). A possible reason for these observations may be that the classical definition of a subexponental distribution is based on convolutions (the denominator in (3) can be replaced by $n \overline{F}(x)$—it is simply a suitable normalization of the tail $\mathbb{P}(x^{-1} S_n \in (1, \infty))$ corresponding to the set $(1, \infty)$). In contrast, regular variation is a natural domain of attraction condition both for partial sums and maxima in the uni- and multivariate cases (a glance at (8) shows that this limit relation does not depend on the structure of some underlying sets).

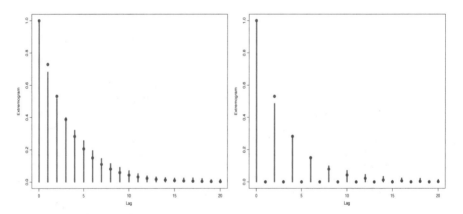

Fig. 8 *Top:* Sample extremogram $\widehat{\rho}_+(h)$, $h = 1, \ldots, 20$, of the extremogram $\rho_+(h)$ in (9) for a stationary AR(1) process $X_t = \varphi X_{t-1} + Z_t$ with iid student-t noise with 3 degrees of freedom. *Left:* $\varphi = 0.9$ *Right:* $\varphi = -0.9$. The estimation is based on the sample size 100 000. The blue dots and red vertical lines represent the theoretical values $\rho_+(h)$ and their estimators $\widehat{\rho}_+(h)$, respectively

We conclude with an application of the regular variation property of a univariate time series (X_t) to some analysis of extremal dependence clusters: the *extremogram*. This name was coined by Davis and Mikosch [6] for the following limiting function for $h \geq 1$:

$$\rho_{\pm}(h) = \lim_{x \to \infty} \mathbb{P}\big(\pm X_h > x \mid X_0 > x \big)$$

$$= \lim_{x \to \infty} \mathbb{P}\big(\pm x^{-1} X_h \in (1, \infty) \mid X_0 > x \big)$$

$$= \mathbb{P}(\pm Y \Theta_h > 1, \Theta_0 = 1). \qquad (9)$$

The latter relation is a direct consequence of the defining property (8) of the spectral tail process (Θ_t) and the continuous mapping theorem . The extremogram sequence $(\rho_{\pm}(h))_{h \geq 0}$ constitutes an autocorrelation function for the extreme events that, given $|X_0|$ is large, so are X_h or $-X_h$ after h lags. We illustrate the extremogram and its estimation for an AR(1) process in Fig. 8.

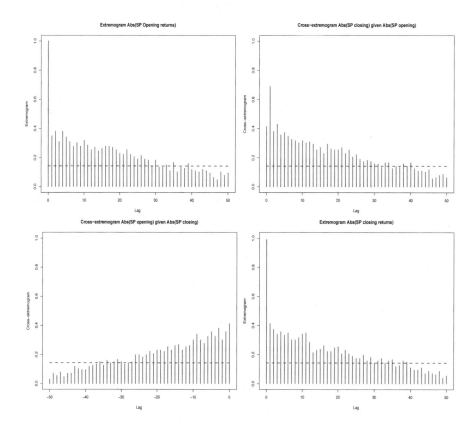

Fig. 9 Sample extremograms of the absolute values of the opening (X_t) (*top left*) and closing (Y_t) (*bottom right*) log-returns of the S&P 500 series; see Fig. 7 for a visualization of this series. Sample cross-extremograms for $\{|Y_h| > x\}$ given $\{|X_0| > x\}$ (*top right*) and for $\{|X_h| > x\}$ given $\{|Y_0| > x\}$ (*bottom left*) are also shown. The thresholds x are the corresponding 90% empirical quantiles of the absolute values of the data. The blue line indicates a 95% confidence band corresponding to the iid case: if a vertical red line at lag h is outside these bands this is an indication of extremal dependence between $|X_0|$ and $|X_h|$

The extremogram can be defined in numerous other ways. We illustrate this in Fig. 9. There we show the sample extremograms $\widehat{\rho}_+(h)$ (i.e., estimators of the extremograms $\rho_+(h)$) of the absolute values of the daily opening (X_t) and closing (Y_t) log-returns of the S&P 500 series from Fig. 7 but also the estimators of the cross-extremograms

$$\lim_{x \to \infty} \mathbb{P}\big(|Y_h| > x \mid |X_0| > x\big) \qquad \text{and} \qquad \lim_{x \to \infty} \mathbb{P}\big(|X_h| > x \mid |Y_0| > x\big).$$

References

1. C. Améndola, C. Klüppelberg, S. Lauritzen and N.M. Tran. Conditional independence in max-linear Bayesian networks. *Ann. Appl. Probab.* **32**(1), 1-45 (2022).
2. B. Basrak and J. Segers. Regularly varying multivariate time series. *Stoch. Proc. Appl.* **119**, 1055–1080 (2009).
3. P.J. Brockwell and R.A. Davis. *Time Series: Theory and Methods.* Second edition. Springer Series in Statistics. Springer, New York (1991).
4. D. Buraczewski, E. Damek and T. Mikosch. *Stochastic Models with Power-Laws. The Equation $X = AX + B$.* Springer, New York (2016).
5. R.A. Davis and T. Hsing. Point process and partial sum convergence for weakly dependent random variable s with infinite variance. *Ann. Probab.* **23**, 879–917 (1995).
6. R.A. Davis and T. Mikosch. The extremogram: a correlogram for extreme events. *Bernoulli* **15**, 977–1009 (2009).
7. P. Embrechts, C. Klüppelberg and T. Mikosch. *Modelling Extremal Events for Insurance and Finance.* Springer, Heidelberg (1997).
8. P. Embrechts, A. McNeil and D. Straumann. Correlation and dependence in risk management: properties and pitfalls. In: Dempster, M.A.H. (Ed.): *Risk Management: Value at Risk and Beyond,* pp. 176–223, Cambridge University Press, Cambridge (2002).
9. R.F. Engle. Autoregressive conditional heteroscedasticity with estimates of the variance of United Kingdom inflation. *Econometrica* **50**, 987–1007 (1982).
10. N. Gissibl and C. Klüppelberg. Max-linear models on directed acyclic graphs. *Bernoulli,* P**24**(4A), 2693–2720 (2018).
11. C.M. Goldie. Implicit renewal theory and tails of solutions of random equations. *Ann. Appl. Probab.* **1**, 126–166 (1991).
12. L. de Haan, S.I. Resnick, H. Rootzén and C.G. de Vries. Extremal behaviour of solutions to a stochastic difference equation with applications to ARCH processes. *Stoch. Proc. Appl.* **32**, 213–224 (1989).
13. P. Jorion. *Value at Risk: The New Benchmark for Managing Financial Risk* (3rd ed.) McGraw-Hill, New York (2006).
14. C. Klüppelberg and S. Lauritzen. Bayesian networks for max-linear models. In: F. Biagini, G. Kauermann and T. Meyer-Brandis (Eds.) *Network Science – An Aerial View,* pp. 79–97, Springer International Publishing (2019).
15. R. Kulik and P. Soulier. *Heavy-Tailed Time Series.* Springer, New York (2020).
16. S.L. Lauritzen. *Graphical Models.* Oxford University Press, Oxford (1996).
17. A.J. McNeil, R. Frey and P. Embrechts. *Quantitative Risk Management: Concepts, Techniques and Tools.* Princeton University Press, Princeton (2015).
18. J. Pearl. *Causality: Models, Reasoning, and Inference* 2nd Ed. Cambridge University Press, Cambridge (2009).

19. S.I. Resnick. *Extreme Values, Regular Variation, and Point Processes.* Reprint 2008. Springer, New York (1987).
20. G. Samorodnitsky and J. Sun. Multivariate subexponential distributions and their applications. *Extremes* **19**, 171–196 (2016).

Limits of Limit-Order Books

Christopher Almost, John Lehoczky, Steven Shreve, and Xiaofeng Yu

1 Introduction

1.1 Background

On electronic exchanges buyers and sellers of financial assets are matched by a continuous double auction, whose operation is described below. Although some agents submitting orders may be acting as *market makers*, posting both buy and sell orders, these agents no longer have the favored position of floor specialists who received orders and could either match those with other received orders or execute

Christopher Almost was partially supported by the Natural Sciences and Engineering Research Council of Canada PGS D3 under Grant No. 358495. This paper includes personal views and opinions of Dr. Almost, and they have not been reviewed or endorsed by Waterfront International Ltd.

Steven Shreve was partially supported by the National Science Foundation under Grant No. DMS-0903475.

Xiaofeng Yu's views represented here are those of the author and not those of Millennium International Management LP.

C. Almost
Waterfront International Ltd., Toronto, ON, Canada

J. Lehoczky
Department of Statistics and Data Science, Carnegie Mellon University, Pittsburgh, PA, USA
e-mail: jpl@stat.cmu.edu

S. Shreve (✉)
Department of Mathematical Sciences, Carnegie Mellon University, Pittsburgh, PA, USA
e-mail: shreve@andrew.cmu.edu

X. Yu
Millennium International Management LP, New York, NY, USA

© The Author(s), under exclusive license to Springer Nature Switzerland AG 2023
J.-M. Morel, B. Teissier (eds.), *Mathematics Going Forward*, Lecture Notes in Mathematics 2313, https://doi.org/10.1007/978-3-031-12244-6_11

the orders for their own account. Instead, all agents place orders of essentially four types,[1] *market buy orders*, *market sell orders*, *limit buy orders*, and *limit sell orders*.

A limit buy/sell order specifies the number of shares to be bought/sold and the price at which the transaction is to take place. Allowable prices are on a discrete grid. The grid points are called *ticks*. If there is no limit sell/buy order in the book that matches the price associated with an arriving limit buy/sell order, the arriving limit order is queued at that price for later execution or cancellation. If there is a matching limit order queued in the book, the arriving limit order is partially or fully executed against the existing order, depending on the size of the existing order. Any part of the arriving limit order not executed is queued.

In contrast, market buy/sell orders accept the best price available in the book, and if the arriving order exhausts the limit orders queued at a particular price, it moves on to the limit orders queued at the next best price. The "best price" at which a market buy order executes is the lowest price at which a limit sell order is queued, and this is called the *best ask price*, or simply the *ask price*. The "best price" at which a market sell order executes is the highest price at which a limit buy order is queued, and this is called the *best bid price*, or simply the *bid price*.

On most exchanges, when an order arrives that can execute against limit orders queued at a particular price, the queued limit orders are executed in order of arrival to the queue, the oldest being executed first. For this reason, limit orders are often submitted to establish time priority in case the agent wants to later execute. Exchanges permit submitting agents to cancel limit orders before they are executed, and indeed most limit orders are canceled rather than being executed.

A *zero-intelligence Poisson model* of the limit-order book dynamics assumes the market and limit buy and sell orders arrive at different prices according to independent Poisson processes, where the intensity of the Poisson processes may depend on the state of the limit-order book. "Zero-intelligence" denotes the fact that there is no attempt to model the motivation of the individual agents who are submitting the orders. In these models, orders are cancelled according to independent exponential random variables. These limit-order book models are akin to queueing models of telephone or computer communication traffic in which the statistics of the traffic are modeled but not the reasons for the traffic. In Sect. 1.2, we present a brief history of and the evidence for the efficacy of these types of models.

---

[1] Certain agents are permitted to place other types of orders, e.g., *iceberg orders*, which become visible to other agents only gradually as they are executed. These are important if one is to study strategic play among agents. That is not our goal, and thus to avoid unnecessary complications, we restrict our attention to the four principal types of orders introduced here.

1.2 Survey of the Literature

For general background on limit-order books, both empirical studies and models, one may consult Biais et al. [9], Cont [13, 14], Gould et al. [27], Hautsch and Huang [28], Parlour and Seppi [43], and Szabolcs and Farmer [46].

Poisson models of order arrivals have a long history, predating electronic exchanges. Early models by Garman [23], Amihud and Mendelson [4] and Mendelson [42] posit Poisson arrivals of buy and sell orders to a market that clears periodically. In [23] and [4], this clearing is facilitated by a market maker.

In a step toward zero-intelligence models, Roşu [44] and Kruk [37] build models with Poisson arrivals of buy and sell orders but with prices posted strategically by the agents submitting the orders.

Cohen et al. [12] build a zero-intelligence model for a continuous double auction with order queues only at the bid and ask prices. Domowitz and Wang [18], who include a nice discussion of the operation of limit-order books, extend this model to allow order queues at finitely many prices and compute the stationary distribution for their model. Kelly and Yudovina [34] and Luckock [40] permit Poisson arrivals to a continuum of prices and also compute stationary distributions. One of the few papers in which arriving orders are not assumed to have constant size is Kruk [36], in which arrivals are modeled by renewal processes and order sizes are randomly distributed. Under a fluid scaling of time/volume, but not prices, a limiting evolution for the order book is obtained.

Smith et al. [45] and Daniels, Farmer, Gillemot et al. [17] build a zero-intelligence Poisson model and conduct extensive simulations together with dimensional analysis and mean field approximations to obtain predictions about price volatility, market depth, the size and variability of the bid-ask spread, the price impact of submitting a market order, and the probability that a limit order is filled. The orders are of constant size, but [45] reports that when comparing their simulations to those with random order size, as long as the distribution of order size has a thin tail, the same qualitative results are obtained. Farmer et al. [19] report good agreement between the predictions in [45] and [17] and data from the London stock exchange.

For an interesting experiment on the ability of zero-intelligence trading to achieve price discovery, see [24].

By scaling price ticks as well as time and order volume, one can obtain limiting models that governed by partial differential equations in the case of fluid scaling (see Gao and Deng [22], Horst and Kreher [29], and Horst and Paulsen [31]) and by measure-valued stochastic differential equations or stochastic partial differential equations in the case of diffusion scaling or multiple time scales (see Bayer et al. [7], Horst and Kreher [30], Lakner et al. [39], and Lakner et al. [38]). Kirilenko et al. [35] develop a model with three time scales and both fluid and diffusion scalings. Maglaras et al. [41] build a fluid model from the outset, i.e., not obtained as the limit of a stochastic model. Horst and Xu [32] introduce Hawkes random measures, an extension of Hawkes processes that have been used to capture the clustering of arrivals observed in data. See the references in [32] for entry into the literature on this use of Hawkes processes.

The literature on Markov models of limit-order books is too extensive to survey here. We mention only Huang et al. [33], who build a two-time-scale model and fit it to data, Gonzalez [25, 26], and the references therein.

Finally, a number of authors have used microstructure models based on Poisson processes or more general processes to obtain macro models for price movement. Among them, in addition to several papers already mentioned, are Abergel and Jedidi [1], Bak et al. [6], Bayraktar et al. [8], Blanchet and Chen [10], Föllmer and Schweizer [20] and Föllmer [21].

1.3   A Representative Model

In the remainder of this paper, we introduce diffusion modeling of limit-order books by describing a particular model in some detail. This model is inspired by Cont et al. [16], who construct a zero-intelligence Poisson model and use Laplace transform analysis to obtain analytical conclusions about the stationary distribution. The work of Cont, et. al. [16] raises the question of whether another queueing theory methodology, determining heavy-traffic (diffusion) limits of zero-intelligence Poisson models, is possible and useful. Steps in that direction have been taken by Cont and de Larrard [15], Avellaneda et al. [5] and Chávez-Casillas and Figueroa-Lopez [11], all of whom consider queues only at the bid and ask. When one of these is depleted in [15], the system is reinitialized. In [5] the model of [15] is extended to include "hidden liquidity" at the best prices. In [11] the depletion of a best price leads to orders arriving between the new best prices to replenish the system. In contrast to these works, in the model presented below, orders queue outside the bid-ask spread and rules governing order arrivals and departures continue to apply even when a best price queue is depleted.

2   Description of the Representative Model

In this section we present the diffusion limit of a sequence of zero-intelligence Poisson models. We describe the model and principal results here, and refer to Almost et al. [3] for proofs. This work is based on the PhD dissertations of Almost [2] and Yu [47].

To demonstrate the viability of this approach we choose the simplest zero-intelligence Poisson model in which the determination of this limit is nontrivial. In our model, all orders are the same size. Market buy and sell orders arrive as Poisson processes with intensity $\lambda_0 > 0$ and $\mu_0 > 0$, respectively. Limit buy orders arrive at prices one and two ticks below the ask price as Poisson processes with intensities $\lambda_1 > 0$ and $\lambda_2 > 0$, respectively. Limit sells arrive at prices one and two ticks above the bid price as Poisson processes with intensities $\mu_1 > 0$ and $\mu_2 > 0$, respectively. We build a sequence of models indexed by the positive integers $n = 1, 2, \ldots$. The six parameters $\lambda_0, \lambda_1, \lambda_2, \mu_0, \mu_1, \mu_2$ are common to all these models and may be chosen arbitrarily subject to the following condition:

Assumption 2.1 There are two numbers $a > 1$ and $b > 1$ satisfying $a + b > ab$ such that

$$a\lambda_0 = b\mu_0,$$

$$\lambda_1 = (a - 1)\lambda_0,$$

$$\lambda_2 = (a + b - ab)\lambda_0,$$

$$\mu_1 = (b - 1)\mu_0,$$

$$\mu_2 = (a + b - ab)\mu_0.$$

In many cases in practice, the bid and ask prices are thousands of ticks away from price zero. This permits us to avoid boundary conditions by assuming that the price ticks are on a doubly infinite grid.

Buy orders that are two or more ticks below the bid price are subject to cancellation at rate $\theta_b/\sqrt{n} > 0$ (per order). Analogously, sell orders that are two or more ticks above the ask price are subject to cancellation at rate $\theta_s/\sqrt{n} > 0$ (per order).

Because there are two types of queued orders, limit buys and limit sells, we remove ambiguity by creating at each price tick a process that is the number of limit buys queued at that tick if there are any and is the *negative* of the number of limit sells queued at that tick if there are any. Obviously, there cannot be both limit buys and limit sells queued at the same price. If at a generic price tick we denote this process in the n-th model at time t by $Q^n(t)$, the sign of $Q^n(t)$ carries the order type information. We illustrate this in Fig. 1, where the positive bars correspond to queued buy orders and the negative bars correspond to queued sell orders at the prices marked on the horizontal axis. Up arrows labeled λ_0, λ_1 and λ_2 indicate the price locations of the arriving buy orders at the rates indicated by the λ-labels, which will build positive bars or, in the case of the market orders arriving at the ask price, will shorten the negative bar at that price. Similarly, down arrows labeled μ_0, μ_1 and μ_2 show the price locations of the arriving sell orders at the rates indicated by the μ-labels, which will build negative bars or, in the case of market orders arriving at the bid price, will shorten the positive bar at that price.

It is apparent in Fig. 1 that a large bid-ask spread cannot persist. In fact, the typical limit-order book configuration in the n-th system has a one-tick spread, as shown in Fig. 2, or a two-tick spread. In Fig. 2, we have labeled queue length processes U^n, V^n, W^n, X^n, Y^n and Z^n, where the latter three are negative in the

Fig. 1 Limit-order book

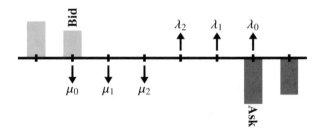

figure. We denote by p_u, p_v, p_w, p_x, p_y and p_z the respective prices at which the queues of orders U^n, V^n, W^n, X^n, Y^n and Z^n are posted. We define $Q^n :=$ $(U^n, V^n, W^n, X^n, Y^n, Z^n)$.

Fig. 2 Typical limit-order book

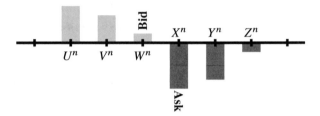

It turns out that under Assumption 2.1, in the n-th model the components of $Q^n(nt)$ are either of size $O(\sqrt{n})$ or size $o(\sqrt{n})$. Thus we accelerate time by a factor of n, divide the queue length by $\sqrt{n}$, and seek the limit of the scaled system as $n \to \infty$. In other words, we define the scaled queued order process vector by

$$\widehat{Q}^{(n)}(t) = \frac{1}{\sqrt{n}} Q^n(nt), \quad t \geq 0,$$

and its individual components are denoted $\widehat{U}^{(n)}$, $\widehat{V}^{(n)}$, etc. We study $\lim_{n \to \infty} \widehat{Q}^{(n)}$. Note that we do not scale the prices. We study the limiting process not just at a single price tick, but rather consider the system of processes at all price ticks simultaneously. This is potentially a countably-infinite-dimensional process, but because of cancellations most components of this process are zero, and we can restrict our attention at any time to the queues at only six price ticks.

To describe the limiting system, let us denote by

$$U^*(t), V^*(t), W^*(t), X^*(t), Y^*(t), Z^*(t)$$

the limit of the scaled number of orders at six adjacent price ticks at scaled time t, the ticks chosen so that

$$V^*(t) > 0, \quad Y^*(t) < 0. \tag{1}$$

Condition (1) will remain in force until Theorem 7.2 of [3] below. Recall that p_u, p_v, p_w, p_x, p_y and p_z are the respective prices at these six adjacent ticks. One can show that there are always six adjacent price ticks with this property, although this set of six ticks is not always uniquely determined. The bid price in the limiting system is p_x if $X^*(t)$ is positive, is p_w if $X^*(t) \leq 0$ and $W^*(t) > 0$, or is p_v if both $X^*(t) \leq 0$ and $W^*(t) \leq 0$. We call the bid price in the limiting system the *essential bid price* in recognition of the fact that it may not be the limit of the bid prices in the sequence of pre-limit models, as we discuss below. We adopt the same terminology for the ask, calling the ask price in the limiting system the *essential ask price*.

Let us consider the case $W^*(t) = 0$ and $X^*(t) < 0$, in which the essential bid is p_v and the essential ask is p_x (see Fig. 3). This means in the pre-limit sequence of models, at the same six price ticks, there are queue length processes U^n, V^n, W^n, X^n, Y^n, Z^n with

$$V^n(nt) = O(\sqrt{n}) > 0, \quad W^n(nt) = o(\sqrt{n}), \quad X^n(nt) = O(\sqrt{n}) < 0. \qquad (2)$$

Let us assume further that

$$U^n(nt) = O(\sqrt{n}) > 0, \quad Y^n(nt) = O(\sqrt{n}) < 0. \qquad (3)$$

In the n-th pre-limit model, $W^n(nt)$ can be either positive, negative or zero. If $W^n(nt)$ is positive, p_w is the bid, but if $W^n(nt)$ is negative, p_w is the ask. It is thus possible that the bid in every pre-limit model is p_w, even though the essential bid in the limiting system is p_v. Similarly, the ask in every pre-limit model could be p_w, even though the essential ask in the limiting system is p_x

Fig. 3 Limiting system for (2) and (3)

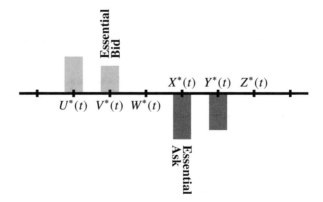

In the limiting system, at Lebesgue almost every time, the essential ask and essential bid differ by two ticks as in Fig. 3. On the zero-measure set of remaining times, the essential ask and bid differ by three ticks. This limiting system, however, is an approximation to pre-limit models in which the spread between ask and bid prices is tighter. In fact, the limit of the fraction of time in which the spread is one tick in the pre-limit models is

$$2 - \frac{a+b}{ab} \tag{4}$$

(see [3, Remark 4.11]), and these parameters can be chosen to correspond to empirical observation. Indeed, Assumption 2.1 permits us to choose $a = b = 1 + \varepsilon$ for $0 < \varepsilon < 1$, and as ε ranges over $(0, 1)$, the expression in (4) also ranges over $(0, 1)$.

We return to the six limiting processes $U^*, V^*, W^*, X^*, Y^*, Z^*$ at adjacent price ticks for an interval of time in which (1) prevails. During this time interval, we designate V^* and Y^* the pair of *bracketing processes* and call the pair W^* and X^* the *interior processes*.

Definition 2.1 Let B be a standard Brownian motion, and define

$$P_\pm^B(t) := \int_0^t \mathbb{I}_{\{\pm B(s)>0\}} ds, \quad t \geq 0,$$

to be the occupation times of the positive and negative half-lines. Let $\sigma_\pm$ be positive numbers, and define

$$\Theta := \frac{1}{\sigma_+^2} P_+^B + \frac{1}{\sigma_-^2} P_-^B. \tag{5}$$

We call the process

$$Z = B \circ \Theta^{-1} \tag{6}$$

a *two-speed Brownian motion* with speed σ_+^2 when positive and speed σ_-^2 when negative.

Properties of two-speed Brownian motion and its relation to skew Brownian motion are developed in [3, Appendix A]. According to [3, Theorem 4.19], the pair of interior processes (W^*, X^*) behaves like

$$\left(\max\{B_{w,x}, 0\}, \min\{B_{w,x}, 0\} \right), \tag{7}$$

where $B_{w,x}$ is a two-speed Brownian motion. This two-speed Brownian motion has speed σ_+^2 when it is positive and speed σ_-^2 when it is negative, where

$$\sigma_+ = \sqrt{2(\lambda_0 + b\lambda_1)}, \tag{8}$$

$$\sigma_- = \sqrt{2(\mu_0 + a\mu_1)}. \tag{9}$$

The subscripts of $B_{w,x}$ indicate that it arises when we take (W^*, X^*) to be the pair of interior processes. In particular, W^* is a Brownian motion with variance σ_+^2 when $B_{w,x}$ is positive, and is zero when $B_{w,x}$ is negative. On the other hand, X^* is a Brownian motion with variance σ_-^2 when $B_{w,x}$ is negative, and is zero when $B_{w,x}$ is positive. During the interval of time in which (1) holds and we observe this behavior by (W^*, X^*), we refer to (W^*, X^*) as a *split Brownian motion*.

When $B_{w,x}$ is positive, the limiting system has a two-tick spread between the essential ask at p_y and the essential bid at p_w, as shown in Fig. 4.

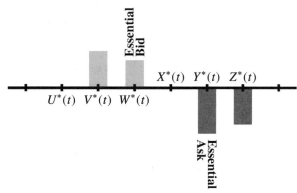

Fig. 4 Limiting system when $B_{w,x}$ is positive

When $B_{w,x}$ is negative, the limiting system again has a two-tick spread, this time between the essential ask at p_x and the essential bid at p_v, as shown in Fig. 3. When $B_{w,x}$ is zero, the limiting system has a three-tick spread between the essential ask at p_y and the essential bid at p_v, as shown in Fig. 5.

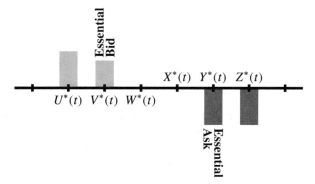

Fig. 5 Renewal state

We call this third configuration a *renewal state* for the limiting system.

During an interval of time in which (1) holds and X^* is negative (see Fig. 3), and hence by (7), W^* is zero, the combination of limit buy arrivals and cancellations will cause U^* to be frozen at the level

$$\kappa_L := \frac{\lambda_2 \mu_1}{\theta_b \lambda_1}. \tag{10}$$

In general, the queue one tick below the essential bid is frozen at κ_L. In particular, when W^* is positive, V^* is frozen at κ_L. Analogously, when X^* is negative, Y^* is frozen at

$$\kappa_R := -\frac{\mu_2 \lambda_1}{\theta_s \mu_1}. \tag{11}$$

Because after reaching zero, $B_{w,x}$ immediately crosses zero infinitely many times, immediately after each renewal state, there are infinitely many renewal states with $V^* = \kappa_L$, $W^* = 0$, $X^* = 0$, $Y^* = \kappa_R$. The queues U^* and Z^* are frozen at κ_L and κ_R, respectively. All other queues are zero.

During an interval of time in which (1) holds and $X^*(t)$ is negative, U^* is frozen at κ_L. Therefore, we could regard U^* and X^* as the bracketing queues and regard V^* and W^* as the interior queues. Analogously to (7), during this time interval V^* and W^* behave like

$$\big(\max\{B_{v,w}, 0\}, \min\{B_{v,w}, 0\} \big), \tag{12}$$

where $B_{v,w}$ is a two-speed Brownian motion different from $B_{w,x}$ in (7). But when (1) holds and X^* is negative, W^* is zero, so $B_{v,w}$ is positive and V^* behaves like a Brownian motion with variance σ_+^2. On the other hand, according to (7), X^* behaves like a Brownian motion with variance σ_-^2. These Brownian motions have constant correlation [3, Remark 5.16]

$$-\rho = \frac{\lambda_1 + \mu_1}{\sqrt{(\lambda_0 + b\lambda_1)(\mu_0 + a\mu_1)}} = \frac{2(\lambda_1 + \mu_1)}{\sigma_+ \sigma_-}, \tag{13}$$

where $-1 < \rho < 0$. Because of our convention that sell orders are queued with negative sign, the positivity of $-\rho$ implies that the correlation ρ between the actual number of orders at the essential bid and essential ask is negative.

We began this discussion under the assumption (1), but eventually either V^* or Y^* will reach zero. The following theorem addresses the possibility that V^* reaches zero, which it can do only when X^* is negative.

Theorem 2.1 of [3] *If at some time t_0, $V^*(t_0) = v_1 > 0$ and $X^*(t_0) = x_1 < 0$, then the probability that V^* reaches zero before X^* is θ_0/α, where*

$$\alpha := -\arctan\left(-\frac{1-\rho^2}{\rho}\right), \quad \theta_0 := \arctan\left(\frac{\sigma_+\sqrt{1-\rho^2}\,|x_1|}{\sigma_- v_1 + \sigma_+ \rho x_1}\right) \in (0, \alpha). \tag{14}$$

If the system is in the renewal state in which W^* and X^* are both zero, V^* is positive, and Y^* is negative, then as discussed above, V^* will be forced to the value κ_L and Y^* will be forced to the value κ_R. When $B_{w,x}$ goes on a negative excursion, so that X^* goes on a negative excursion, V^* has a chance to reach zero. If it fails to do so before the negative excursion of $B_{w,x}$ ends, then V^* is "snapped back" to κ_L. Similarly, when $B_{w,x}$ goes on a positive excursion, so that W^* goes on a positive excursion, then Y^* has a chance to reach zero. During this time, W^* behaves like a Brownian motion with variance σ_+^2, Y^* behaves like a Brownian motion with variance σ_-^2, and the correlation between W^* and Y^* is $-\rho$. If Y^* fails to reach zero before the positive excursion of $B_{w,x}$ ends, then it is "snapped back" to κ_R.

Now consider an excursion of length ℓ of X^* away from zero. Just before the beginning of this excursion, $B_{w,x}$ was positive, so p_v was one tick below the best bid price and V^* was frozen at κ_L. During this excursion of X^*, V^* has a chance to reach zero. Let $p_{V^*}(s, \ell)$ to be the density for the first passage time of V^* from κ_L to 0 during an excursion of X^* of length ℓ. (The length of the excursion is relevant because V^* and X^* have correlation $-\rho \neq 0$.) In [3, Lemma 7.4] there is the explicit formula

$$
p_{V^*}(s, \ell)ds
$$

$$
:= \mathbb{P}\{\tau_{V^*}^E \in ds \,|\, \lambda(E) = \ell\}
$$

$$
= \frac{\sqrt{2\pi(1-\rho^2)\ell^3}\,\pi^2\sigma_+ \sin\alpha}{2\kappa_L\alpha^3(\ell-s)\sqrt{s(\ell-s\cos^2\alpha)}} \exp\left(-\frac{\kappa_L^2}{2\sigma_+^2(1-\rho^2)s} \cdot \frac{\ell - s\cos 2\alpha}{(\ell-s)+(\ell-s\cos 2\alpha)}\right)
$$

$$
\times \sum_{n=1}^{\infty}(-1)^{n-1}n^2 I_{n\pi/(2\alpha)}\left(\frac{\kappa_L^2}{2\sigma_+^2(1-\rho^2)s} \cdot \frac{\ell - s}{(\ell-s)+\ell-s\cos 2\alpha}\right)ds, \, 0 < s < \ell.
$$

Define

$$
p_{V^*}(\ell) = \int_0^\ell p(s, \ell)ds
$$

to be the probability V^* transitions from κ_L to 0 during the excursion of X^* of length ℓ. Similarly, define

$$
p_{y^*}(s, \ell)
$$

$$
:= \mathbb{P}\{\tau_{y^*}^E \in ds \,|\, \lambda(E) = \ell\}
$$

$$
= \frac{\sqrt{2\pi(1-\rho^2)\ell^3}\,\pi^2\sigma_- \sin\alpha}{2|\kappa_R|\alpha^3(\ell-s)\sqrt{s(\ell-s\cos^2\alpha)}} \exp\left(-\frac{\kappa_R^2}{2\sigma_-^2(1-\rho^2)s} \cdot \frac{\ell - s\cos 2\alpha}{(\ell-s)+(\ell-s\cos 2\alpha)}\right)
$$

$$
\times \sum_{n=1}^{\infty}(-1)^{n-1}n^2 I_{n\pi/(2\alpha)}\left(\frac{\kappa_R^2}{2\sigma_-^2(1-\rho^2)s} \cdot \frac{\ell - s}{(\ell-s)+\ell-s\cos 2\alpha}\right), \quad 0 < s < \ell,
$$

which is the density of the passage time of Y^* from κ_R to zero during an excursion of W^* of length ℓ, and further define

$$p_{Y^*}(\ell) = \int_0^\ell p_{V^*}(s, \ell)\mathrm{d}s.$$

Finally, set

$$\lambda_- := \frac{1}{\sigma_-} \int_0^\infty \frac{p_{V^*}(\ell)\mathrm{d}\ell}{\sqrt{2\pi\ell^3}}, \qquad \lambda_+ := \frac{1}{\sigma_+} \int_0^\infty \frac{p_{Y^*}(\ell)\mathrm{d}\ell}{\sqrt{2\pi\ell^3}}.$$

Theorem 2.2 of [3] *Starting from the renewal state of Fig. 5, the probability that V^* vanishes before Y^* (Fig. 6), and thus there is a downward price change is $\lambda_-/(\lambda_+ + \lambda_-)$. The probability of an upward price change is $\lambda_+/(\lambda_+ + \lambda_-)$.*

The characteristic function of the distribution of the time between renewal state transitions conditioned on the direction of the transition is provided by [3, Theorem 7.9].

Fig. 6 New renewal state

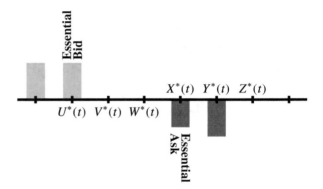

Finally, we observe that although the limiting processes in our model behave like Brownian motions for periods of time, they are actually more complicated than Brownian motions. In fact, they are not semimartingales. To understand why this is the case, consider the process V^*. We have just seen that when $B_{w,x}$ in (7) is negative, so that W^* is zero and X^* is negative, then V^* behaves like a Brownian motion. However, when $B_{w,x}$ becomes positive, then W^* is positive and p_w is the essential bid. In this case, V^*, being one tick below the essential bid, is frozen at κ_L. But immediately after it reaches zero, the Brownian motion $B_{w,x}$ has infinitely many changes of sign, which causes V^* to repeatedly diffuse and then jump back to κ_L. The jumps in V^* are not absolutely summable, and this creates an unbounded variation non-martingale component to V^*. Fortunately, the squares of the jumps are summable, which permits us to study V^* through the tool of Poisson random measures.

3 Conclusion

This work shows that diffusion scaling methodology developed to study heavy traffic limits of queueing systems can be adapted to Poisson limit-order book models. While the Poisson model is infinite dimensional with state recording the number of orders at each tick on a doubly infinite price-tick grid, the diffusion limit is low dimensional. In this work, the state of the diffusion limit is the volume at two price ticks, the essential bid and essential ask, and the location of these price ticks.

While not addressed in this work, it is natural to consider optimal execution in the limiting model. The arrival of a buy order of size $O(\sqrt{n})$ in the n-th model at the essential bid is queued behind $O(\sqrt{n})$ other orders and will execute in $O(\sqrt{n})$ time. The diffusion scaling accelerates times by the factor n, and hence in the limiting model this order will execute instantaneously. On the other hand, a buy order placed one tick below the essential bid must wait until this tick becomes the essential bid, and then will execute instantaneously. These instantaneous executions eliminate the need to keep track of an order's priority at its price tick, and consequently the optimal execution problem in the limiting model is low dimensional.

There are many ways the model of this paper can be generalized. To move toward a more realistic model, the most important generalization is to allow cancellations at and near the best bid and ask price. Other generalizations are to allow arrivals at more price ticks, time-dependent or volume-dependent arrival rates, random order sizes, and renewal rather than Poisson arrival processes.

References

1. F. Abergel and A. Jedidi. A mathematical approach to order book modeling. *International J. Theoretical Applied Finance* **16**(5), 1350025 (2013).
2. C. Almost. *Diffusion scaling of a limit-order book model: The symmetric case.* Ph.D. dissertation, Department of Mathematical Sciences, Carnegie Mellon University, https://kilthub.cmu.edu/articles/thesis/Diffusion_Scaling_of_a_Limit-Order_Book_Model_The_Symmetric_Case/8160173 (2018).
3. C. Almost, J. Lehoczky, S. Shreve and X. Yu. Diffusion Limit of Poisson Limit-Order Book Models. https://arxiv.org/abs/2008.01155 (2022).
4. Y. Amihud and H. Mendelson. Dealership market. *J. Financial Economics* **8**, 31–53 (1980).
5. M. Avellaneda, J. Reed and S. Stoikov. Forecasting prices in the presence of hidden liquidity. *Algorithmic Finance* **1**, 35–43 (2011).
6. P. Bak, M. Paczuski and M. Shubik. Price variations in a stock market with many agents, *Physica A* **246**(3-4), 430–453 (1997).
7. C. Bayer, U. Horst and J. Qiu. A functional limit theorem for limit order books with state dependent price dynamics. *Ann. Appl. Probab.* **27**, 2753–2806 (2017).
8. E. Bayraktar, U. Horst and R. Sircar. A limit theorem for financial markets with inert investors. *Math. Operations Research* **31**, 789–810 (2006).
9. B. Biais, P. Hillion and C. Spatt. An empirical analysis of the limit order book and the order flow in the Paris bourse. *J. Finance* **50**, 1655–1689 (1995).
10. J. Blanchet and X. Chen. Continuous-time modeling of bid-ask spread and price dynamics in limit order books. *arXiv:1310.1103* (2013).

11. J. Chávez-Casillas and J. Figueroa-López. A one-level limit order book model with memory and variable spread. *Stochastic Processes Applications* **127**, 2447–2481 (2017).
12. K.J. Cohen, R.M. Conroy and S.F. Maier. Order flow and the quality of the market. In: *Market Making and the Changing Structure of the Securities Industry*, ed., Y. Amihud, T. Ho and R. Schwartz, Lexington Books, Lexington, MA (1985).
13. R. Cont. Empirical properties of asset returns: stylized facts and statistical issues. *Quantitative Finance* **1**, 223–236 (2001).
14. R. Cont. Statistical modeling of high frequency financial data: facts, models and challenges. *IEEE Signal Processing Magazine* **28**, 16–25 (2011).
15. R. Cont and A. de Larrard. Markovian limit order market. *SIAM J. Financial Mathematics* **4**, 1–25 (2012).
16. R. Cont, S. Stoikov and R. Talreja. A stochastic model for order book dynamics. *Operations Research* **58**, 549–563 (2010).
17. M.G. Daniels, J.D. Farmer, L. Gillemot, G. Iori and E. Smith. Quantitative model of price diffusion and market friction based on trading as a mechanistic random process. *Physical Review Letters* **90**(10), 108102 (2003).
18. I. Domowitz and J. Wang. Auctions as algorithms. *J. Econ. Dynamics and Control* **18**, 29 (1994).
19. J.D. Farmer, P. Patelli and I. Zovko. The predictive power of zero intelligence in financial markets. *Proc. Nat. Acad. Sci. USA* **102**, 2254–2259 (2005).
20. H. Föllmer and M. Schweizer. A microeconomic approach to diffusion models for stock prices. *Math. Finance* **3**, 1–23 (1993).
21. H. Föllmer. Stock price fluctuation as a diffusion in a random environment. *Phil. Trans. Royal Society of London, Series A* **347**, 471–483 (1994).
22. X. Gao and S.J. Deng. Hydrodynamic limit of order-book dynamics. *Probability in the Engineering and Informational Sciences* **32**, 96–125 (2018).
23. M. Garman. Market microstructure. *J. Financial Economics* **3**, 257–275 (1976).
24. D. Gode and S. Sunder. Allocative efficiency of markets with zero-intelligence traders: Market as a partial substitute for individual rationality. *J. Political Economy* **101**, 119–137 (1993).
25. F. Gonzalez. *Limit order book models and applications*. Ph.D. thesis, Department of Statistics and Data Science, Carnegie Mellon University (2018).
26. F. Gonzalez and M. Schervish. Instantaneous order impact and high-frequency strategy optimization in limit order books. *Market Microstructure and Liquidity* **3**(2), 1850001 (2017). https://doi.org/10.1142/S2382626618500016
27. M. Gould, M. Porter, S. Williams, M. McDonald, D. Fenn and S. Howison. Limit order books. *Quantitative Finance* **13**, 1709–1742 (2013).
28. N. Hautsch and R. Huang. Limit order flow, market impact and optimal order sizes: Evidence from NASDAQ TotalView-ITCH data. http://ssrn.com/abstract=1914293 (2011).
29. U. Horst and D. Kreher. A weak law of large numbers for a limit order book model with fully state dependent order dynamics. *SIAM J. Financial Mathematics* **8**, 314–343 (2017).
30. U. Horst and D. Kreher. Second order approximation for limit order books. *Finance and Stochastics* **22**, 827–877 (2018).
31. U. Horst and M. Paulsen. A law of large numbers for limit order books. *Math. Operations Research* **42**, 1280–1312 (2017).
32. U. Horst and W. Xu. A scaling limit for limit order books driven by Hawkes processes. *SIAM J. Financial Mathematics* **10**, 350–393 (2019).
33. W. Huang, C. Lehalle and M. Rosenbaum. Simulating and analyzing order book data: the queue reactive model. *J. American Statistical Assoc.* **110**, 107–122 (2014).
34. F. Kelly and E. Yudovina. A Markov model of a limit order book: Thresholds, recurrence, and trading strategies. *Math. Operations Research* **43**, 181–203 (2018).
35. A. Kirilenko, R. Sowers and X. Meng. A multiscale model of high-frequency trading. *Algorithmic Finance* **2**, 59–98 (2013).
36. L. Kruk. Functional limit theorems for a simple auction. *Math. Operations Research* **28**, 716–751 (2003).

37. L. Kruk. Limiting distribution for a simple model of order book dynamics. *Central European Journal of Mathematics* **10**, 2283–2295 (2012).
38. P. Lakner, J. Reed and F. Simatos. Scaling limits for a limit order book model using the regenerative characterization of Lévy trees. *Stochastic Systems* **7**, 342–373 (2017).
39. P. Lakner, J. Reed and S. Stoikov. High frequency asymptotics for the limit order book. *Market Microstructure and Liquidity* **2**(1), 1650004 (2016).
40. H. Luckock. A steady-state model of a continuous double auction. *Quantitative Finance* **3**, 385–404 (2003).
41. C. Maglaras, C. Moallemi and H. Zheng. Optimal execution in a limit order book and an associated microstructure market impact model. *Working paper* (2015).
42. H. Mendelson. Market behavior in a clearing house. *Econometrica* **50**, 1505–1524 (1982).
43. C. Parlour and D. Seppi. Limit order markets. *Handbook of Financial Intermediation and Banking* **5**, 63–95 (2008).
44. I. Roşu. A dynamic model of the limit order book. *Rev. Financial Studies* **22**, 4601–4641 (2009).
45. E. Smith, J.D. Farmer, L. Gillemot and S. Krishnamurthy. Statistical theory of the continuous double auction. *Quantitative Finance* **3**, 481–514 (2003).
46. M. Szabolcs and J.D. Farmer. An empirical behavioral model of liquidity and volatility. *J. Economic Dynamics Control* **32**, 200–234 (2008).
47. X. Yu. *Diffusion scaling of a limit-order book: The asymmetric case.* Ph.D. dissertation, Department of Mathematical Sciences, Carnegie Mellon University (2018). https://kilthub. cmu.edu/articles/thesis/Diffusion_Scaling_of_a_Limit-Order_Book_The_Asymmetric_Case/ 9730988

Part IV
Geometry

Jean-Pierre Bourguignon's article *Spinors in 2022* skillfully presents a short history of spinors, an exposition of their use in Physics, the differential operators connected to them, their use in differential geometry and their connection to fundamental theorems such as the Atiyah–Singer index theorem. The paper ends with a well supported advocacy for "a true spinorial geometry".

Carles Casacuberta's article *Cohomological localizations and set-theoretical reflection* studies the problem of cohomological localization of spaces and spectra. While homological localizations are a fundamental tool in algebraic topology, the existence of cohomological localizations is connected to problems of set theory, in particular the existence of large cardinals.

Parvadeh Joharinad and Jürgen Jost's article *Geometry of data* explores a viewpoint on metric geometry, in particular curvature and convexity, motivated by data analysis which of course deals with finite sets of points. It offers very interesting perspectives on the simplicial complexes associated with a covering of a topological space, such as the Čech complex.

Vitali Milman's article *A chapter about Asymptotic Geometric Analysis: Isomorphic position of centrally symmetric convex bodies* concerns inequalities on the volumes of various convex bodies in n-dimensional Euclidean space associated in various ways, such as slicing, with a given one and its images (called positions) by non-degenerate endomorphisms of the ambient space. The goal is to find inequalities involving only constants that are independent of the dimension.

Ieke Moerdijk's article *A mysterious tensor product in topology* proposes an alternative to the tensor product of operads defined by Boardman and Vogt, which does not have all the desirable invariance features. The properties of this alternative definition and its relation to that of Boardman and Vogt are sources of open problems.

Spinors in 2022

Jean-Pierre Bourguignon

> *Dedicated to Catriona BYRNE on the occasion of her retirement, in recognition of the great support provided over the years and of opportunities offered.*

In the first two decades of the twenty-first century, spinors have continued to be omnipresent in Physics, and they attracted a lot of attention in Mathematics too. Some aspects that were discussed in the 1995 *Postface* (cf. [5]) have further developed but new ones have appeared. Between 1998 and 2022, arXiv lists 1879 articles containing the word *spinor* in their title. Even the very restrictive expression *Spinorial Geometry* appears 16 times in arXiv in the same period and several comprehensive surveys or books have appeared dealing with spinors (cf. [6], [19] and [8]).

Here is an outline of the topics that are covered in the article:

1. The Initial and Persistent Strangeness of Spinors;
2. The Unabated Importance of Spinors in Physics;
3. The Role of Natural Differential Operators on Spinor Fields;
4. Towards a True Spinorial Geometry?

1 The Initial and Persistent Strangeness of Spinors

Spinors appeared in 1913 as an exotic representation of the orthogonal group in the classification work of Élie Cartan (cf. [10] and [11] for the real case that appeared in 1914). The construction was completed in 1935 by Richard Brauer and Hermann Weyl (cf. [9]).

J.-P. Bourguignon (✉)
CNRS-Institut des Hautes Études Scientifiques, Bures-sur-Yvette, France
e-mail: JPB@ihes.fr

© The Author(s), under exclusive license to Springer Nature Switzerland AG 2023
J.-M. Morel, B. Teissier (eds.), *Mathematics Going Forward*, Lecture Notes in Mathematics 2313, https://doi.org/10.1007/978-3-031-12244-6_12

As we will discuss in the next section, it is the relevance of spinors in the context of Quantum Physics that made them fundamental objects in this science.

Still, the fact that, to make sense of them, a bilinear form, most of the time a non-degenerate one, has to be chosen has made their use in Differential Geometry somewhat problematic if one insists on using general coordinates. This is precisely the last statement one finds in Élie Cartan's book *La théorie des spineurs* (cf. [12]) published in 1935. It is very clear for him though that the difficulty can be easily overcome by using *moving frames*, one of his favourite tools. Of course the development of Bundle Theory in the middle of the twentieth century allowed the situation to be clarified. The book [15] by Claude Chevalley has been an important reference from the algebraic point of view.

This is why, in their book *Spin Geometry* (cf. [30]), that has played an important role in the development of the subject, H. Blaine Lawson and Marie-Louise Michelsohn consider that, in some sense, Michael F. Atiyah and Isadore M. Singer had to reinvent spinors for mathematical purposes in connection with their work on the Index Theorem.

It is a fact that, in spite of their evidence and importance, some fundamental questions related to spinors remained open for a long time. This was the case for the dependence of spinors on the metric for which erroneous statements appeared in the Physics literature. After an initial contribution by Ernst Binz and Regina Pferschy (cf. [4]), Paul Gauduchon and I provided a full solution to this question in 1992 (cf. [7]). A simpler approach to deal with this issue has been developed by Christian Bär, Paul Gauduchon and Andrei Moroianu in 2003 (cf. [3]) that also allows one to deal with the case of Lorentzian metrics, an important extension.

More fundamentally, the fact that spinors are difficult to comprehend continues to be perceptible. Lecturing in 2013 at the conference at IHÉS on the occasion of my leaving the directorship, Sir Michael Atiyah entitled his talk *"What is a spinor?"* and claimed at the beginning of his presentation that *"I spent most of my life working with spinors in one form or another and I do not know what a spinor is"*. Therefore, it appeared natural to me to give the same title to the *Atiyah lecture* I gave in January 2021, the first in the series. Indeed, in this lecture, I discussed some of the peculiar features of spinors that make them difficult to comprehend.

2   The Unabated Importance of Spinors in Physics

The introduction of spinors in Physics goes back to the early development of Quantum Physics. Two articles are usually quoted in this respect: one by Wolfgang Pauli (cf. [38]) and one by Paul Adrien Maurice Dirac (cf. [16]). The second one introduces the *Dirac equation*, that is fundamental in Quantum Mechanics. It is first order in the field and invariant under the Lorentz group, while the Schrödinger equation cannot be since it involves taking one derivative in the time variable and two derivatives in the space variables. In Minkowski space, the solution comes from taking a *square root of the D'Alembertian*. In doing that, there is one price one has to pay, namely the initial equation on ordinary functions has to be replaced by one on spinor fields, on which the Dirac operator acts.

Further developments involved taking this approach systematically, with the introduction of two fundamentally different types of particles: *fermions*, particles with half-integral spins represented by spinor-type fields subject to the Pauli exclusion principle, and *bosons*, having integral spins represented by tensor fields. In [39], Roger Penrose also showed how to formulate General Relativity using spinors.

In the 1990s the idea of having physical theories that would admit *supersymmetries*, i.e. transformations mixing the two types of particles became very hot. It had in particular a considerable impact in the research in relation with the Standard Model of Elementary Particles. Underpinning these developments was of course the possibility of having a *Supergeometry*, a point of view that I considered from a general mathematical point of view in the postface *Spinors in 1995* (cf. [5]).

Of course today spinors are more than ever at the heart of the developments in Physics, even though the results of the experiments obtained in 2012 at the Light Hadron Collider at CERN in Geneva, which concluded the search for the Higgs Boson, did not lead to a confirmation of a role for supersymmetric particles. Nevertheless, publications dealing with supersymmetry continue to flourish with more than 500 articles in arXiv containing the word *supersymmetric* in their title since January 2021.

Although a number of results on this topic were already available in 1995, in the Postface I did not address *Killing Spinors*, which are special spinor fields representing infinitesimal supersymmetries. As they involve another fundamental operator on spinor fields, the Penrose operator, I leave the discussion on them to the next section dedicated to natural differential operators on spinor fields. They also have a major role from a purely mathematical point of view in relation to holonomy.

3   The Role of Natural Differential Operators on Spinor Fields

As mentioned in the previous section, the Dirac operator played a key role in turning spinors into central objects in Mathematics and in Physics.

Actually, there are only two universally defined differential operators on spinor fields: the *Dirac operator*, a first order self-adjoint operator mapping spinor fields into themselves, and the *Penrose operator*, also a first order operator mapping spinor fields to differential 1-forms tensored with spinor fields. Full definitions can be found on pages 61–70 of the book *A Spinorial Approach to Riemannian and Conformal Geometry* (cf. [8]).

When the reference metric used to define spinors is Riemannian, the Dirac operator is a square root of the Laplace-Beltrami operator, hence elliptic. Its key role in the Index Theorem comes from the fact that, in a specific sense, its symbol generates all symbols from a topological point of view. This was Atiyah and Singer's main motivation to consider it. To prove the Index Theorem for it is sufficient to establish the general theorem provided one uses homotopy arguments.

Spinor fields annihilated by the Penrose operator (this equation is sometimes called the *twistor equation*) and eigenspinors for the Dirac operator are called *Killing spinors* (their equation can be found on page 131 of [8]). They satisfy an over-determined system with the consequence that metrics admitting a Killing spinor have to be rather special: their Ricci curvature is a multiple of the metric (the metric is called an *Einstein metric*) and their curvature tensor has a special form. The vector field dual to the 1-form component of their square is a Killing vector field, and the fact that they are often viewed as *square roots of Killing fields*, hence their name.

It is worth noting that the concept was introduced by Roger Penrose with a completely different objective. He was indeed looking into ways to generate first integrals for the geodesic flow of the Kerr metric (cf. [40]).

For some time, a description of metrics admitting Killing spinors eluded. It was Christian Bär (cf. [2]) who, in 1993, came up with the appropriate construction, namely a cone endowed with a cone metric over the manifold admits a parallel spinor, connecting the situation with special holonomy. This situation mimics what happens for the round sphere embedded in Euclidean space: on the sphere the spinor bundle is trivialised by Killing spinors induced by constant (hence parallel) spinor fields in the flat Euclidean space.

Killing spinors also appear in relation with the limiting case of the *Friedrich inequality* (cf. [17]) on eigenvalues of the Dirac operator (cf. page 131 of [8]). The connection with conformal geometry is also underpinning a refined version of this inequality, called the *Hijazi inequality* (cf. [24]).

4   Towards a True Spinorial Geometry?

Let us start by recalling the status of a question whose connection with spinors does not appear at first sight, namely whether a given manifold admits a metric with positive scalar curvature. In the end, the answer, not yet fully complete as I will explain, is rather subtle, going quite deeply into the differential topology of the manifold and, to establish it, spinorial data play a key role.

In 1975, Jerry L. Kazdan and Frank Warner proved that, for $n \geq 3$, all n-dimensional manifolds admit metrics with negative scalar curvature (cf. [28]). The analogous statement for metrics with positive scalar curvature is not true, in spite of the weakness of the scalar curvature as a Riemannian invariant. In the last 60 years, understanding this phenomenon mobilised efforts by many people. We will concentrate here on closed manifolds (i.e. compact without boundary), although the question for non compact manifolds or manifolds with boundary has recently attracted a lot of attention.

The initial key remark was made in 1963 by André Lichnerowicz in [31] to the effect that, on a spin manifold endowed with a metric with positive scalar curvature, non-vanishing harmonic spinors (i.e spinors annihilated by the Dirac operator) cannot exist. This follows directly from the formula, due to Erwin Schrödinger and rediscovered by André Lichnerowicz, establishing that the difference between the square of the Dirac operator and the connection Laplacian applied to spinor fields is the scalar curvature up the positive factor $\frac{1}{4}$. One just has to integrate the formula against the harmonic spinor. The Index Theorem implies that some spin manifolds must have non-trivial harmonic spinors, e.g. $4k$-dimensional manifolds with non-vanishing Â-genus. The result has been refined by Nigel Hitchin in 1974 (cf. [25]) using a more involved Dirac operator and more sophisticated KO-invariants introduced by Atiyah whose non-vanishing obstructs the existence of metrics with positive scalar curvature.

Since these important initial contributions going back to the 1970s, there has been a large number of articles dealing with many aspects of the problem.

First of all, sticking to spin manifolds, Misha Gromov and H. Blaine Lawson showed in 1980 (cf. [22]) that, for simply connected manifolds of dimension $n \geq 5$, the answer depends only on the Spin-bordism class of the manifold. In the same article, they prove that all simply connected non-spin manifolds of dimension at least 5 admit metrics with positive scalar curvature. The fact that the Lichnerowicz-Hitchin obstructions form a complete set for simply connected spin manifolds was established by Stephan Stolz in 1992 (cf. [43]).

For non simply connected manifolds, Gromov and Lawson introduced in [21] original constructions of new obstructions to the existence of a metric with positive scalar curvature.

Later, many refinements to deal with the presence of a fundamental group have been obtained so that many names should be listed. Here, I will limit myself mentioning Jonathan Rosenberg and Stephan Stolz (cf. [41] et [42]), who have substantially and repeatedly contributed to the subject. Stolz just produced a comprehensive report (cf. [44]) stating in great detail where the problem stands with still many new results continuing to be obtained.

Let me sum up what this special question reveals. On a manifold, for a spin structure to exist a topological constraint has to be satisfied, namely the vanishing of the second Stiefel-Whitney class. This allows us to consider globally defined spinor fields and, among them, harmonic spinors play a special role. In some sense, they are analogous to harmonic forms which, as we know from the De Rham theory, detect the cohomology. In some sense, what is for the moment missing in our situation is how a local invariant such as the scalar curvature affects the way harmonic spinors can develop over the manifold with the consequence that their mere existence forces the scalar curvature to be negative somewhere.[1]

---

[1] To lift any ambiguity, let us recall that the global existence of a harmonic spinor on a closed manifold ensures that there is a point where its norm achieves its maximum; at this point, its Hessian is non-positive, hence its Laplacian non-negative. If, at such a point, one considers the

Let me now address some other issues and explain further why, in my opinion, the role of spinors to deal with Riemannian questions has not yet reached its full potential. As just alluded to, the first key remark is that we may be missing an appropriate geometric picture for spinors. Hence, we are back to the question *"What is a spinor?"*!

Before discussing new developments since 1995, let us come back to topological constraints related to spin structures. The broader class of manifolds admitting a $spin_c$-structure, that includes all complex manifolds, has been considered; for a $spin_c$ structure to exist, the topological condition is that the second Stiefel-Whitney class is the mod 2 reduction of an integral class, the first Chern class. Some geometric results on them have been obtained by Andrei Moroianu (cf. [35, 36]) and Roger Nakad (cf. [37]).

Let us review some recent results concerning various objects connected to spinors since their study has continued:

- Several generalisations of Killing spinors have been considered, e.g. by Friedrich and E.C. Kim (cf. [20] for example) and by Marc Herzlich and Andrei Moroianu (cf. [23]) for $spin_c$ manifolds; another variation on the notion of Killing spinor has been used in [1] by Ilke Agricola, Simon G. Chiossi, Thomas Friedrich, and Jos Höll to investigate SU_3- and G_2-manifolds; this involves using results from hypersurface theory and the building of conical manifolds;
- Spinors have been used to study hypersurfaces by several authors, in particular for the representations of surfaces by Robert Kusner and Nick Schmitt (cf. [29]) and Friedrich (cf. [18]), but also for more general hypersurfaces notably by Andrzej Trautmann (cf. [45]), Sebastian Montiel (cf. [32]) and Bertrand Morel (cf. [33]);
- A new (and interesting) functional combining spinor fields and maps from surfaces to Riemannian manifolds has been introduced by Qun Chen, Jürgen Jost, JiaYu Liu and GuoFang Wang in 2004 (cf. [13]). One of its important features in two dimensions is its conformal invariance. Its critical points are called *Dirac-harmonic maps*. It is naturally inspired by supersymmetric considerations coming from Theoretical Physics. Some regularity results concerning Dirac-harmonic maps have been obtained. including a removability of singularities theorem (more recent results can be found in [14]);
- Other considerations inspired by the energy-momentum right hand side of the Einstein equations in General Relativity have also led to several articles, often in relation with generalisations of Killing spinors; here we can quote [34] by Bertrand Morel;

---

scalar product of the two sides of the Schrödinger-Lichnerowicz formula applied to the harmonic spinor and performs the usual manipulation making the Laplacian of its square norm appear, one gets three terms whose sum must vanish: the Laplacian of the square norm, the square norm of its covariant derivative and the product of the scalar curvature by its square norm. This is only possible if the scalar curvature is non-positive.

- So far, mathematicians focused their attention on spinor fields of spin $\frac{1}{2}$; spinors with spin $\frac{3}{2}$ should be investigated with the analog of the Dirac operation, *the Rarita-Schwinger operator*, that acts on them; a link with infinitesimal deformations of Einstein metrics has been suggested by Bernard Julia (cf. [27]) (for a recent contribution by Yasushi Homma and Uwe Semmelmann, one can consult [26]).

References

1. I. Agricola, S.G. Chiossi, T. Friedrich and J. Höll. Spinorial Description of SU_3- and G_2-Manifolds. *J. Geom. Phys.* **98**, 535–555 (2015).
2. C. Bär. Real Killing Spinors and Holonomy. *Commun. Math. Phys.* **154**, 509–521 (1993).
3. C. Bär, P. Gauduchon and A. Moroianu. Generalized Cylinders in Semi-Riemannian and Spin Geometry. *Math. Z.* **249**(3), 545–580 (2005).
4. E. Binz and R. Pferschy. The Dirac Operator and the Change of Metric. *C.R. Math. Rep. Acad. Sci. Canada*, **V**, 269–274 (1983).
5. J.-P. Bourguignon. Postface: Spinors in 1995, In: C. Chevalley *The Algebraic Theory of Spinors* (second edition), 199–210, Springer (1997).
6. J.-P. Bourguignon, Th. Branson, A. Chamseddine, O. Hijazi and R.J. Stanton (eds). *Dirac Operators: Yesterday and Today*. Proc. Summer School-Workshop held in Beirut in 2001, Int. Press. (2005).
7. J.-P. Bourguignon and P. Gauduchon. Spineurs, opérateurs de Dirac et variations de métriques. *Commun. Math. Phys.* **144**, 581–599 (1992).
8. J.-P. Bourguignon, O. Hijazi, J.-L. Milhorat, A. Moroianu and S. Moroianu. *A Spinorial Approach to Riemannian and Conformal Geometry*. Monographs Math., European Math. Soc. (2015).
9. R. Brauer and H. Weyl. Spinors in n Dimensions. *Amer. J. Math.* **57**, 425–449 (1935).
10. É. Cartan. Les groupes projectifs qui ne laissent invariante aucune multiplicité plane. *Bull. Soc. Math. France* **41**, 53–96 (1913).
11. É. Cartan. Les groupes projectifs continus réels qui ne laissent invariante aucune multiplicité plane. *J. Math. Pures Appl.* (6) **10**, 149–186 (1914).
12. É. Cartan. *La théorie des spineurs*, Gauthier-Villars, Paris (1937); second edition, *The Theory of Spinors*, Hermann, Paris (1966).
13. Q. Chen, J. Jost, J.Y. Liu and G.F. Wang. Dirac-Harmonic Maps. *Math. Z.* **254**, 409–432 (2006).
14. Q. Chen, J. Jost, G. Wang and M. Zhu. The Boundary Value Problem for Dirac-Harmonic Maps. *J. Eur. Math. Soc.* **15**, 997–1031 (2013).
15. C. Chevalley. *The Algebraic Theory of Spinors*. Columbia Univ. Press (1954); second edition, Springer (1997).
16. P.A.M. Dirac. The Quantum Theory of the Electron. *Proc. Roy. Soc.* **A117**, 610–624 (1928).
17. Th. Friedrich. Der erste Eigenwert des Dirac-Operators einer kompakten Riemannschen Mannigfaltigkeit nichtnegativer Skalarkrümmung. *Math. Nachr.* **97** (1980), 117–146.
18. Th. Friedrich. On the Spinor Representation of Surfaces in Euclidean 3-space. *J. Geom. Phys.* **28**, 143–157 (1998).
19. Th. Friedrich, *Dirac-Operatoren in der Riemannschen Geometrie*. Adv. Lectures in Math., Fr. Viehweg & Sohn (1997); English edition: *Dirac Operators in Riemannian Geometry*, Grad. Stud, Math. **25**, Amer. Math. Soc. (2000).
20. Th. Friedrich and E.C. Kim, Some Remarks on the Hijazi Inequality and Generalizations of the Killing Equation for Spinors. *J. Geom. Phys.* **37**, 1–14 (2001).

21. M. Gromov and H.B. Lawson Jr. Spin and Scalar Curvature in the Presence of a Fundamental Group. *Ann. of Math.* **111**(2), 209–230 (1980).

22. M. Gromov and H.B. Lawson Jr. The Classification of Simply Connected Manifolds of Positive Scalar Curvature. *Ann. of Math.* **111**(3), 423–434 (1980).

23. M. Herzlich and A. Moroianu. Generalized Killing Spinors and Conformal Eigenvalue Estimates for spin$_c$ Manifolds. *Ann. Global Anal. Geom.* **17**, 341–370 (1999).

24. O. Hijazi. A Conformal Lower Bound for the Smallest Eigenvalue of the Dirac Operator and Killing Spinors. *Commun. Math. Phys.* **104**, 151–162 (1986).

25. N. Hitchin. Harmonic Spinors, *Adv. Math.* **14**, 1–55 (1974).

26. Y. Homma and U. Semmelmann. The Kernel of the Rarita-Schwinger Operator on Riemannian Spin Manifolds, *Commun. Math. Phys.* **370**, 853–871 (2019).

27. B. Julia. Système linéaire associé aux équations d'Einstein. *C. R. Acad. Sc. Paris* **295**, 113–116 (1982).

28. J.L. Kazdan and F.W. Warner. Existence and Conformal Deformation of Metrics with Prescribed Gaussian and Scalar Curvatures. *Ann. of Math.* **101**(2), 317–331 (1975).

29. R. Kusner and N. Schmitt. The Spinor Representation of Surfaces in Space. arXiv: dg-ga/961005, 1–52 (1996).

30. H.B. Lawzon Jr. and M.L. Michelsohn. *Spin Geometry*, Princeton Math. Series **38**, Princeton Univ. Press (1989).

31. A. Lichnerowicz. Spineurs harmoniques. *C.R. Acad. Sci. Paris* **A257**, 7–9 (1963).

32. S. Montiel. Dirac Operators and Hypersurfaces. *Proc. 9th Int. Workshop Diff. Geom.* **9**, 1–15 (2005).

33. B. Morel. Surfaces in S^3 and H^3, *Actes Sém. théorie spectrale et géométrie Institut Fourier* **23**, 131–144 (2004-2005).

34. B. Morel. The Energy-Momentum Tensor as a Second Fundamental Form. arXiv: math/0302205, 1–13 (2003).

35. A. Moroianu. Parallel and Killing Spinors on spin$_c$ Manifolds. *Commun. Math. Phys.* **187**, 417–428 (1997).

36. A. Moroianu. Spin$_c$ Manifolds and Complex Contact Structures, *Commun. Math. Phys.* **193**, 661–673 (1998).

37. R. Nakad. Lower Bounds for the Eigenvalues of the Dirac Operator on Spin$_c$ Manifolds. *J. Geom. Phys.* **60**, 1634–1642 (2010).

38. W. Pauli. Zur Quantenmechanik des magnetischen Elektrons. *Z. Physik* **43**, 601–623 (1927).

39. R. Penrose. A Spinor Approach to General Relativity. *Ann. Phys.* **10**(2), 171–201 (1960).

40. R. Penrose and R. Rindler. *Spinors and Space-Time*, Cambridge Univ. Press, Cambridge (1984).

41. J. Rosenberg. C*-Algebras, Positive Scalar Curvature, and the Novikov Conjecture. *Publ. Math. Inst. Hautes Études Sci.* **58**, 197–212 (1983).

42. J. Rosenberg and S. Stolz. A "Stable" Version of the Gromov-Lawson Conjecture. In: *The Czech Centennial*, 405-418, Contemp. Math. **181**, American Mathematical Society (1995).

43. S. Stolz. Simply Connected Manifolds of Positive Scalar Curvature. *Ann. of Math.* **136**(3), 511–540 (1992).

44. S. Stolz. Positive Scalar Curvature – Constructions and Obstructions. arXiv: math-DG/2202.05904, 1–34 (2022).

45. A. Trautmann. The Dirac Operator on Hypersurfaces. *Acta Phys. Polon. B* **26**, 1283–1310 (1995).

Cohomological Localizations and Set-Theoretical Reflection

Carles Casacuberta

Personal Note
This article is dedicated to Dr. Catriona Byrne on her retirement, with warm thanks for three decades of friendly collaboration and, above all, for her lifelong support of the mathematical community.

Introduction

The technique of computing homotopy groups of spaces one prime at a time was pioneered by Serre [30]. A remarkable result derived from Serre's work states that for every prime p the homotopy groups $\pi_k(S^n)$ of the n-sphere with $n \geq 2$ contain nonzero p-torsion elements for infinitely many values of k.

In 1961 Adams discovered that spheres can be embedded into CW-complexes with countably many cells in such a way that homology groups and homotopy groups are transformed into their p-local versions, and furthermore he proved that odd-dimensional spheres localized in this sense at primes $p \geq 5$ become homotopy associative H-spaces [2]. Subsequently, localization of 1-connected spaces at primes was thoroughly developed by several authors, including Bousfield–Kan [10], Hilton–Mislin–Roitberg [17], Mimura–Nishida–Toda [24], Sullivan [32], etc. Among many achievements, this technique opened the way into rational homotopy theory [28, 33].

Supported by MCIN/AEI/10.13039/501100011033 under grant PID2020-117971GB-C22.

C. Casacuberta (✉)
Departament de Matemàtiques i Informàtica, Universitat de Barcelona (UB), Barcelona, Spain
e-mail: carles.casacuberta@ub.edu

© The Author(s), under exclusive license to Springer Nature Switzerland AG 2023
J.-M. Morel, B. Teissier (eds.), *Mathematics Going Forward*, Lecture Notes in Mathematics 2313, https://doi.org/10.1007/978-3-031-12244-6_13

From a category-theoretical point of view, localizing spaces at a prime p is equivalent to inverting up to homotopy the collection of all maps $X \to Y$ that induce isomorphisms $H_n(X; \mathbb{Z}_{(p)}) \cong H_n(Y; \mathbb{Z}_{(p)})$ for all n, where H_n denotes singular homology and $\mathbb{Z}_{(p)}$ are the integers localized at p. Adams designed a convenient machinery for this purpose, involving idempotent monads, categories of fractions, and Brown representability, and showed that his construction of homological localizations was feasible for arbitrary representable homology theories. However, his presentation of results in [3, 4] contained a set-theoretical inaccuracy which was later repaired by Bousfield in [7]. We explain this in more detail in Sect. 1.

Bousfield extended his approach to spectra [8], and it was in the realm of stable homotopy where homological localizations were best understood and most useful, especially towards the study of chromatic phenomena [29]. Every finite p-local spectrum X is the homotopy inverse limit of its *chromatic tower* of localizations $L_{E(n)}X$, where $L_{E(0)}$ is rationalization and $L_{E(1)}$ is localization with respect to p-local complex K-theory. The homology theories $E(n)_*$ were defined by Johnson and Wilson [20] after earlier work of Brown–Peterson [12] and Morava. This result, known as the *chromatic convergence theorem*, opened the way to impressive advances in the calculation of homotopy groups of spheres.

Bousfield also showed that the Kan–Quillen model structure on the category of simplicial sets [27] can be modified by incorporating E_*-equivalences into the collection of weak equivalences for some spectrum E, and the fibrant spaces in the resulting model structure are the E_*-local Kan complexes. This idea was broadly generalized in what is nowadays called *Bousfield localizations* of model categories, and has found applications in various mathematical disciplines. By an E_*-equivalence we mean a map $X \to Y$ inducing isomorphisms $E_n(X) \cong E_n(Y)$ for all n.

In an unpublished paper [9], Bousfield considered localizations with respect to *cohomology* theories, in which case one seeks to invert up to homotopy the E^*-equivalences for a spectrum E, i.e., the maps $X \to Y$ inducing isomorphisms $E^n(Y) \cong E^n(X)$ for all n. He never supplied a proof of the existence of arbitrary cohomological localizations, although he showed that in many examples the class of E^*-equivalences coincides with the class of F_*-equivalences for some homology theory F_*. This was worked out further by Hovey in [18], where he conjectured that every cohomological Bousfield class is indeed a homological Bousfield class. While this is still an open problem in the category of spectra, a counterexample was found by Stevenson in the derived category of a non-Noetherian ring [31].

Although the lack of examples of cohomological Bousfield classes that are not homological Bousfield classes has diminished the practical interest of constructing cohomological localizations in homotopy theory, the problem of whether the existence of cohomological localizations can be proved or not using the ZFC axioms of set theory has remained as a challenging logical problem.

A first step was made in [13] by showing that *Vopěnka's principle* from set theory implies the existence of localizations with respect to arbitrary cohomology theories on simplicial sets. This result was based on previous knowledge of locally presentable and accessible categories, where it had been shown that Vopěnka's principle implies the existence of localizations onto limit-closed subcategories [1].

Using other methods, Przeździezcki proved in [26] that an E^*-localization can be constructed in ZFC if each of the spaces constituting E is a homotopy retract of a compact space. Another step came in [6] by showing that if arbitrarily large supercompact cardinals exist, then E^*-localization exists for all spectra E.

The existence of supercompact cardinals cannot be proved in ZFC, since they are inaccessible. They have an important place in the large-cardinal hierarchy [21], where Vopěnka's principle also belongs (much higher up). Large cardinals appear naturally in several areas of mathematics. For example, the existence of Grothendieck universes—an assumption very often made to justify the use of "small" sets—is equivalent to the existence of inaccessible cardinals.

The general form of the *reflection principle* in set theory says informally that every property of the universe of all sets is shared by some set. This principle can be formalized in different ways, some of which are related to large-cardinal axioms.

The concept of *structural reflection* was introduced by Bagaria and discussed in detail in [5]. It states that for a class $\mathcal{C}$ of structures of the same type there is a cardinal κ such that every $X \in \mathcal{C}$ has a logically equivalent substructure of cardinality smaller than κ and isomorphic to some $Y \in \mathcal{C}$. This assertion is implied by the Löwenheim–Skolem theorem if the class $\mathcal{C}$ is defined by an upward absolute formula—that is, a formula whose truth in a transitive model implies its truth in every larger model. For classes defined by formulas of higher complexity, structural reflection requires the existence of large cardinals.

In our case, the class of E_*-equivalences for a spectrum E can be defined by an upward absolute formula with E as a parameter. Consequently, the existence of arbitrary E_*-localizations can be proved in ZFC. Although this was of course known since [7], we emphasize that a proof can be given by means of a basic set-theoretical argument; see Theorem 2.1 below.

However, the complexity of defining E^*-equivalences seems to be higher in the Lévy hierarchy, since no upward absolute formula has been found for this purpose. The difficulty is that, in order to formalize the statement that a space X is E^*-acyclic, the collection of all functions from X to E has to be considered in some way, and sets of functions (for example, $2^{\mathbb{N}}$) are not upward absolute in general.

While it is conceivable that a proof of the existence of cohomological localizations can be given in ZFC, it is unreasonable to expect an explicit counterexample in ZFC, since such a counterexample would invalidate most of the large-cardinal hierarchy. In fact, the statement that arbitrary cohomological localizations exist is perhaps equivalent to some large-cardinal principle.

1 Homology Theories and Cohomology Theories

By a *homology theory* we mean a generalized homology in the sense of the Eilenberg–Steenrod axioms [15], and similarly for a *cohomology theory*. We only consider homology and cohomology theories that are *reduced* (i.e., vanishing on a point) and *representable*, that is, defined by a spectrum as follows. A *spectrum* E is a collection of pointed spaces E_k for $k \geq 0$ together with structure maps $S^1 \wedge E_k \to E_{k+1}$, where S^n denotes the n-sphere and $\wedge$ is the smash product, i.e., $A \wedge B$ is obtained by collapsing the one-point union $A \vee B$ within the product $A \times B$ for pointed spaces A and B. Every spectrum E yields a homology theory E_* by defining

$$E_n(X) = \operatorname{colim}_k [S^{n+k}, E_k \wedge X]$$

for $n \in \mathbb{Z}$ and every pointed space X, where $[-, -]$ denotes pointed homotopy classes of maps, and a cohomology theory E^* as

$$E^n(X) = \operatorname{colim}_k [S^k \wedge X, E_{n+k}]$$

for $n \in \mathbb{Z}$ as well. For convenience we assume that E is an Ω-*spectrum*, that is, the adjoint maps $E_k \to \Omega E_{k+1}$ of the structure maps of E are weak homotopy equivalences, where Ω denotes loops. Then $E^n(X) \cong [X, E_n]$ for $n \geq 0$.

Note that X appears in the target of maps $S^{n+k} \to E_k \wedge X$ for homology while it appears in the source of maps $S^k \wedge X \to E_{n+k}$ for cohomology. This fact implies covariance of E_* but contravariance of E^*, a fundamental difference.

A map $X \to Y$ of spaces is called an E_*-*equivalence* if it induces isomorphisms $E_n(X) \cong E_n(Y)$ for all n, and it is called an E^*-*equivalence* if it induces isomorphisms $E^n(Y) \cong E^n(X)$ for all n. An E_*-*localization* of a space X is a terminal E_*-equivalence going out of X, that is, an E_*-equivalence $\eta_X \colon X \to L_E X$ such that for every E_*-equivalence $f \colon X \to Y$ there is a map $g \colon Y \to L_E X$ such that the composite $g \circ f$ is homotopic to η_X, and g is unique up to homotopy with this property. An E^*-*localization* is defined analogously.

1.1 Categories of Fractions

The approach undertaken by Adams in [3, 4] to construct E_*-localizations for every homology theory E_* is summarized in this section. He worked in the category $\mathcal{H}$ whose objects are CW-complexes with basepoint and whose morphisms are pointed homotopy classes of maps. We write $f \simeq g$ to denote that f and g are homotopic.

If S denotes either the class of E_*-equivalences or the class of E^*-equivalences for a spectrum E, then S admits a *calculus of left fractions* as defined by Gabriel–Zisman in [16] as follows:

(i) S is closed under compositions.

(ii) For every pair of maps $s: W \to X$ and $f: W \to Y$ where $s \in S$ there are maps $g: X \to Z$ and $t: Y \to Z$ with $t \in S$ such that $g \circ s \simeq t \circ f$.

(iii) For every map $s: W \to X$ in S and every pair of maps $f, g: X \to Y$ with $f \circ s \simeq g \circ s$ there is a map $t: Y \to Z$ in S such that $t \circ f \simeq t \circ g$.

Condition (ii) is satisfied by choosing a homotopy pushout of f and s, and condition (iii) holds using a homotopy coequalizer of f and g, with a Mayer–Vietoris argument in both cases; cf. [7, Lemma 3.6]. In addition to (i), for composable maps s and t, if two of s, t and $t \circ s$ are in S then the third is also in S.

The *category of fractions* $S^{-1}\mathcal{H}$ has the same objects as $\mathcal{H}$ and morphisms from X to Y are equivalence classes of zig-zags

$$X \xrightarrow{f} Z \xleftarrow{s} Y$$

where $s \in S$, and two zig-zags $(f, s): X \to Z \leftarrow Y$ and $(f', s'): X \to Z' \leftarrow Y$ are defined to be equivalent if there is a space Z'' equipped with maps $g: Z \to Z''$ and $g': Z' \to Z''$ in S such that $g' \circ f' \simeq g \circ f$ and $g \circ s \simeq g' \circ s'$. Composition is defined using (ii), and it is well defined thanks to (iii).

In the category $S^{-1}\mathcal{H}$ each map $s: X \to Y$ in S has an inverse, namely the zig-zag $(\mathrm{id}, s): Y \to Y \leftarrow X$. Moreover, there is a canonical functor $Q: \mathcal{H} \to S^{-1}\mathcal{H}$ sending every map $f: X \to Y$ to $(f, \mathrm{id}): X \to Y \leftarrow Y$, and Q is universal among functors from $\mathcal{H}$ sending all maps in S to isomorphisms; see [16, Proposition 2.4].

However, there is a famous difficulty with the category $S^{-1}\mathcal{H}$, namely we need to prove that it is *locally small*, i.e., it has only a set of morphisms between any two objects (not a proper class). As explained in the next subsection, this is feasible if S is the class of E_*-equivalences for a spectrum E, yet it is still an open problem (in ZFC) when S is the class of E^*-equivalences.

Once this difficulty is solved, Brown's representability theorem [11] ensures the existence of a right adjoint $R: S^{-1}\mathcal{H} \to \mathcal{H}$ to $Q: \mathcal{H} \to S^{-1}\mathcal{H}$, i.e., for all spaces X and Y there is a natural bijective correspondence

$$S^{-1}\mathcal{H}(QX, Y) \cong \mathcal{H}(X, RY).$$

In order to use Brown representability, it is necessary that, for a fixed space Y, the functor sending each space X to $S^{-1}\mathcal{H}(QX, Y)$ be set-valued rather than class-valued. This is the reason why we need that the category $S^{-1}\mathcal{H}$ be locally small.

The adjoint pair Q, R yields an idempotent functor $L_E: \mathcal{H} \to \mathcal{H}$, namely the composite RQ equipped with the unit η of the adjunction. This idempotent functor is an E_*-localization on $\mathcal{H}$. Indeed, for every X the map $\eta_X: X \to L_E X$ is in S, and it is a terminal map in S going out of X, as desired.

The properties of E_*-localization of spaces are analogous to those of the passage from abelian groups to $\mathbb{Q}$-vector spaces by formally inverting nonzero integers. In fact, if we choose $E_* = H_*(-; \mathbb{Q})$ as our homology theory (hence $\mathcal{S}$ is the class of singular homology equivalences with rational coefficients), then the resulting idempotent functor on the homotopy category $\mathcal{H}$ of pointed CW-complexes extends Sullivan's rationalization of 1-connected spaces [33]. This was one of the motivations of Adams' work, although the behaviour of $H_*(-; \mathbb{Q})$-localization on arbitrary spaces is much more difficult to describe than in the case of 1-connected spaces.

We next address the problem of proving that $\mathcal{S}^{-1}\mathcal{H}$ is locally small.

1.2 Solution-Set Conditions

A standard way to prove that a category of fractions is locally small is to impose the existence of a cofinal subset of the class $\{s\colon Y \to Z \mid s \in \mathcal{S}\}$ for every fixed Y. This cofinality condition was stated as Axiom 3.4 in [4] and also considered by Deleanu in [14], and reads as follows:

(A) For every space Y there is a subset $A = \{s_\alpha\colon Y \to Z_\alpha\}$ of $\mathcal{S}$ such that for every map $s\colon Y \to Z$ in $\mathcal{S}$ there is a map $s_\alpha\colon Y \to Z_\alpha$ in A and a map $g\colon Z \to Z_\alpha$ such that $g \circ s \simeq s_\alpha$.

This condition ensures that each zig-zag $(f, s)\colon X \to Z \leftarrow Y$ represents the same morphism in $\mathcal{S}^{-1}\mathcal{H}$ as $(g \circ f, s_\alpha)\colon X \to Z_\alpha \leftarrow Y$ for some α, and there is only a set of those. Consequently, $\mathcal{S}^{-1}\mathcal{H}(X, Y)$ is a set for all X and Y, as wanted.

Unfortunately, there seems to be no way to check a priori that condition (A) holds for E_*-equivalences nor for E^*-equivalences. This is the reason why Adams' approach was not considered to be conclusive at that moment.

However, as observed by Fiedorowicz in [4, §8], the fact that $\mathcal{S}^{-1}\mathcal{H}$ is locally small can also be inferred from the following solution-set condition, which is much more useful than (A):

(B) For all spaces X and Y there is a set of zig-zags $B = \{(f_\alpha, s_\alpha)\colon X \to Z_\alpha \leftarrow Y\}$ with $s_\alpha \in \mathcal{S}$ such that for every $(f, s)\colon X \to Z \leftarrow Y$ with $s \in \mathcal{S}$ there exists $(f_\alpha, s_\alpha) \in B$ and a map $g\colon Z_\alpha \to Z$ such that $g \circ f_\alpha \simeq f$ and $g \circ s_\alpha \simeq s$.

In other words, condition (B) imposes that the category of zig-zags from X to Y where the backward arrow is in $\mathcal{S}$ has a weakly initial small subcategory. This ensures that each (f, s) represents the same morphism as (f_α, s_α) for some α, and hence it follows again that $\mathcal{S}^{-1}\mathcal{H}(X, Y)$ is a set for all X and Y.

Condition (B) holds for the class $\mathcal{S}$ of E_*-equivalences if E is any spectrum. The following argument is a rewriting of [4, Lemma 8.3] or [7, Lemma 11.3].

Theorem 1.1 (Existence of Homological Localizations: Topological Proof)
E_*-localization exists for every spectrum E.

Proof In order to prove that E_*-localization of CW-complexes exists for every spectrum E it is sufficient to prove that condition (B) does hold for the class of E_*-equivalences.

Given X and Y, let κ be an infinite cardinal bigger than the cardinality of the sets of cells of X and Y and bigger than the cardinality of the abelian group $E_*(S^0)$. It then follows by means of the Atiyah–Hirzebruch spectral sequence that the cardinality of $E_*(X)$ and $E_*(Y)$ is smaller than κ, and therefore the cardinality of $E_*(X \vee Y)$ is also smaller than κ, since $E_*(X \vee Y) \cong E_*(X) \oplus E_*(Y)$.

Let B be a set of representatives of all homeomorphism classes of zig-zags $(f_\alpha, s_\alpha) \colon X \to Z_\alpha \leftarrow Y$ where Z_α has less than κ cells and s_α is an E_*-equivalence. Suppose given $(f, s) \colon X \to Z \leftarrow Y$ where s is an E_*-equivalence. If W_0 denotes the image of the map $X \vee Y \to Z$ induced by (f, s), then the homomorphism $\varphi_0 \colon E_*(W_0) \to E_*(Z)$ induced by the inclusion $W_0 \subset Z$ is an epimorphism. Since E_* commutes with filtered colimits, every homology class in $\ker \varphi_0$ vanishes on some subcomplex of Z obtained by adding finitely many cells to W_0. Hence we can choose a subcomplex W_1 of Z with less than κ cells such that $W_0 \subset W_1 \subset Z$ and the homomorphism $E_*(W_0) \to E_*(W_1)$ sends all the elements of $\ker \varphi_0$ to zero. The inclusion $W_1 \subset Z$ induces again an epimorphism on E_*-homology and we can iterate the same construction in order to obtain a nested sequence of subcomplexes W_n such that if $W = \cup_{n=1}^\infty W_n$ then W still has less than κ cells and the inclusion $W \subset Z$ is now an E_*-equivalence. Moreover, the composite map $X \vee Y \to W$ yields maps $g \colon X \to W$ and $t \colon Y \to W$ whose composites with the inclusion $W \subset Z$ are equal to f and s respectively. Hence (g, t) is homeomorphic to an element of B, and condition (B) is fulfilled.   □

In this proof, the fact that E_* is a covariant functor that commutes with filtered colimits is essential. Thus, while this approach works well for homology theories, there seems to be no way to check in ZFC that condition (B) holds in the case of cohomology theories. Nevertheless, as we explain in Sect. 2, condition (B) does hold for the class of E^*-equivalences for any spectrum E if we assume the existence of sufficiently large cardinals—indeed, too large to be available in ZFC.

2 Set-Theoretical Reflection

2.1 Cardinality and Rank

In ZFC set theory, no set can be an element of itself and no descending sequence for the membership relation can be infinite. The *rank* of a set X is defined inductively as the smallest ordinal greater than the ranks of all the elements of X. In particular, the rank of every ordinal is equal to itself.

Cardinality and rank are different concepts. For example, the set $\mathbb{R}$ of real numbers has rank $\omega + 1$ (where ω is the first infinite ordinal) but uncountable cardinality, and the set $\{\mathbb{R}\}$ has cardinality 1 but rank $\omega + 2$. In what follows, the cardinality of a set X will be denoted by $|X|$.

A set X is called *transitive* if every element of X is also a subset of X, that is, if X has the property that whenever $a \in X$ and $b \in a$ then $b \in X$. The *cumulative hierarchy* of sets is defined by transfinite recursion as $V_0 = \emptyset$, $V_{\alpha+1} = \mathcal{P}(V_\alpha)$ where $\mathcal{P}$ denotes power set, and $V_\lambda = \cup_{\alpha < \lambda} V_\alpha$ if λ is a limit ordinal. It follows by induction that each V_α is transitive and $V_\alpha \subset V_\beta$ if $\alpha < \beta$. Every set X is a member of some set in this hierarchy [19, Lemma 6.3], and the rank of X is the smallest ordinal α such that $X \in V_{\alpha+1}$. Hence V_α is the set of all sets of rank smaller than α. The union $V = \cup_\alpha V_\alpha$ is the set-theoretical universe or *von Neumann universe*. If κ is an inaccessible cardinal then V_κ is a model for ZFC set theory, and $|V_\kappa| = \text{rank}(V_\kappa) = \kappa$.

2.2 Structures

A summary of terminology and basic facts about languages, structures and theories can be found in [1, Ch. 5], [6, § 1] or [19, Ch. 12], among many other places.

For a regular cardinal λ and a set S, a λ-*ary S-sorted signature* consists of a set of *operation symbols*, a set of *relation symbols* and an *arity* function that assigns to each operation symbol an ordinal $\alpha < \lambda$, a sequence of *input sorts* $\langle s_i \mid i < \alpha \rangle$ and an *output sort* $s \in S$ (then we denote the corresponding operation symbol by $\prod_{i<\alpha} s_i \to s$), and to each relation symbol an ordinal $\beta < \lambda$ and a sequence of sorts $\langle s_j \mid j < \beta \rangle$. An operation symbol of arity 0 is called a *constant*.

A σ-*structure* for a signature σ is an S-sorted set $X = \{X_s \mid s \in S\}$ equipped with an *interpretation* of σ, that is, a function $\prod_{i<\alpha} X_{s_i} \to X_s$ for each operation symbol $\prod_{i<\alpha} s_i \to s$ (including a distinguished element in X_s for each constant of sort s) and a relation on X for each relation symbol. A *homomorphism* of σ-structures is an S-sorted function that preserves operations and relations.

The *language* of a λ-ary signature σ is made of a set of *variables* of cardinality λ and *formulas* involving variables, operations and relations in σ, equality, negation, implication, conjunctions and disjunctions of cardinality smaller than λ, and finitely many quantifiers over sets of variables of cardinality smaller than λ. Languages with $\lambda = \aleph_0$ are called *finitary*. For example, the language of ZFC set theory is one-sorted and finitary with a binary relation symbol $\in$, which is interpreted as membership.

Variables that appear unquantified in a formula are called *free*. Formulas without free variables are called *sentences*, and a set of sentences is called a *theory*.

For each language, a *satisfaction relation* is defined inductively for formulas with an assignment for their free variables. Thus, for a formula $\varphi(x_i)_{i\in I}$ with free variables x_i, we say that the sentence $\varphi(a_i)_{i\in I}$ *holds* in a σ-structure X if $a_i \in X$ for all $i \in I$ and φ is satisfied in X under the variable assignment $x_i \mapsto a_i$. A σ-structure X is called a *model* for a set of formulas if each of these formulas holds in X under any variable assignment.

For example, the signature of the theory of pointed simplicial sets is ω-sorted with unary operations d_i^n of sorts $n \rightarrow n - 1$ (faces) and s_i^n of sorts $n \rightarrow n + 1$ (degeneracies) for $0 \leq i \leq n$, and a constant of sort 0 (the basepoint), and the axioms of this theory are the simplicial identities [23]. Homomorphisms between models are basepoint-preserving simplicial maps.

A *parameter* in a formula is a set which is fixed in every variable assignment. Every formula φ of the language of ZFC set theory with a parameter p defines a class $C = \{X \mid \varphi(X, p)\}$, meaning that C consists of all sets X for which $\varphi(X, p)$ holds in V. A formula φ with a parameter p is called *absolute* for a set or a proper class M with $p \in M$ if φ holds in V if and only if it holds with its quantifiers relativized to M; then one also says that M is *elementary* for φ.

An *elementary embedding* between σ-structures is a function $f: X \rightarrow Y$ that preserves and reflects truth, that is, for every formula $\varphi(x_i)_{i \in I}$ of the language of σ with free variables x_i, and for all $a_i \in X$, the sentence $\varphi(a_i)_{i \in I}$ holds in X if and only if $\varphi(f(a_i))_{i \in I}$ holds in Y. Thus, every elementary embedding $X \rightarrow Y$ is, in particular, an injective homomorphism of σ-structures.

2.3 Reflection Principles

The Löwenheim–Skolem theorem is a central result in first-order logic. Its simplest form states that every infinite model for a countable language has a countable elementary submodel [19, Theorem 12.1]. More generally, for every infinite σ-structure X and every infinite cardinal $\kappa \geq |\sigma|$, if $\kappa < |X|$ then there exists an elementary substructure $Y \subset X$ of cardinality κ (downward version) and if $\kappa > |X|$ then there exists an elementary extension $X \subset Y$ of cardinality κ (upward version).

Another version specializes to a finite set of formulas (or equivalently one formula) and reads as follows. Given any formula φ of ZFC set theory and given an infinite cardinal κ, there is a set M of cardinality κ which is elementary for φ. Moreover, M can be chosen as an extension of any given set of cardinality κ, and it can be chosen transitive if we remove the restriction on its cardinality [19, Theorem 12.14]. In this situation, one says that the formula φ is *reflected* by M. This is called the *reflection principle* and it is usually referred to by saying that every formula that holds in V already holds in V_α for some α; see [5] for more details and a historical perspective.

The following variant, called *structural reflection*, was used in [5, 6]. A class C of σ-structures closed under isomorphic images is *reflected* by a cardinal κ if every $X \in C$ has an elementary substructure $Y \in C$ of cardinality smaller than κ. The Löwenheim–Skolem theorem implies that every isomorphism-closed class of σ-structures defined by a formula that is absolute for transitive classes is reflected by any uncountable cardinal larger than $|\sigma|$. In fact it is sufficient that the formula be *upward* absolute, in a sense that we next discuss.

2.4 The Lévy Hierarchy

The existence of a cardinal reflecting a class C depends on the *Lévy complexity* [22] of a formula defining C. A formula φ is called Σ_0 or Π_0 if it does not contain unbounded existential quantifiers $\exists$ nor unbounded universal quantifiers $\forall$, that is, all quantifiers in φ are of the form $\exists x \in a$ or $\forall x \in a$ where a is some set. For $n \geq 1$, a formula is called Σ_n if it has the form $\exists x \, \varphi(x)$ where φ is Π_{n-1}, and it is called Π_n if it has the form $\forall x \, \varphi(x)$ where φ is Σ_{n-1}.

One of the consequences of the reflection principle is that for every n there exist arbitrarily large cardinals α such that V_α is a Σ_n-elementary substructure of V, that is, a Σ_n formula with parameters in V_α holds in V_α if and only if it holds in V.

Every Σ_0 formula is absolute for transitive classes; see [19, Lemma 12.9]. Likewise, Σ_1 formulas are upward absolute while Π_1 formulas are downward absolute. The reason is that if in a transitive model M there exists a set x with a property expressed by a Σ_0 formula, then every model containing M also has a set x with that property (in fact, the same x), and if every x in M has a property expressed by a Σ_0 formula then the same holds in each transitive submodel of M.

As an example, the clause $a \subseteq b$ is formalized by the Σ_0 formula $\forall x \in a \, (x \in b)$, which is absolute between two models $M \subset N$ if M is transitive and $a, b \in M$ (we need transitivity to ensure that $x \in a$ implies $x \in M$). The claim "a is the set of all subsets of b" can be formalized with the Π_1 formula $\forall x \, (x \in a \leftrightarrow x \subseteq b)$, and its truth is not preserved upwards, since the set $\mathcal{P}(\mathbb{N})$ of all subsets of $\mathbb{N}$ is countable in any countable transitive model of ZFC but uncountable in V.

As another example, the claim "x is finite" can be formalized with a Σ_1 formula stating that there is a bijection between x and some finite ordinal, and the assertion that "a is the set of finite subsets of b" can be expressed as follows:

$$x \in a \leftrightarrow \exists n < \omega \, \exists f (f \text{ is a function from } n \text{ to } b \text{ with image } x). \qquad (1)$$

Moreover, (1) can be rewritten by stating that there is a transitive model of a sufficiently large finite fragment of ZFC containing a and b in which at least the pairing and union axioms hold and in which (1) is true, as in [6, Example 2.3]. This is a Σ_1 statement. Consequently, quantifying over finite subsets of some given set b can be done by means of Σ_1 formulas with b as a parameter.

2.5 Existence of Localizations

For fixed simplicial sets X and Y, a simplicial set Z equipped with pointed maps $X \to Z$ and $Y \to Z$ can be viewed as a model of a theory over an ω-sorted signature σ_{XY} consisting of unary operations $n \to n - 1$ for $n \geq 1$ and $n \to n + 1$ for $n \geq 0$ plus a constant of sort 0 (to be interpreted as faces, degeneracies, and basepoint in Z) and, in addition, for all n, a constant of sort n for each element of X_n and another constant of sort n for each element of Y_n, to be interpreted as images in Z of the simplices of X and Y. This signature σ_{XY} need no longer be countable but $|\sigma_{XY}| = |X| + |Y|$. The axioms are the simplicial identities for Z together with a statement that the functions $f \colon X \to Z$ and $s \colon Y \to Z$ determined by the constants are simplicial maps, and we need to add as another axiom that the Kan condition holds for the simplicial set Z. Our choices guarantee that homomorphisms $g \colon Z \to Z'$ of σ_{XY}-structures satisfy $g \circ f = f'$ and $g \circ s = s'$, since constants are preserved by homomorphisms.

Theorem 2.1 has the same statement as Theorem 1.1, but a very different proof. It has been written using and adapting results from [6].

Theorem 2.1 (Existence of Homological Localizations: Set-Theoretical Proof)
E_*-localization exists for every spectrum E.

Proof Our aim is to infer that condition (B) from Sect. 1 holds for the class $\mathcal{E}_{XY}$ of Kan simplicial sets Z equipped with pairs of pointed maps $f \colon X \to Z$ and $s \colon Y \to Z$, where X, Y are fixed simplicial sets and s is an E_*-equivalence. For this, we view each such Z as a σ_{XY}-structure for the signature σ_{XY} defined above.

All the terms in a definition of $\mathcal{E}_{XY}$ are absolute, except for a formula stating the fact that s is an E_*-equivalence, which we next analyze following [6, Theorem 9.2]. A map is an E_*-equivalence if and only if its cofibre is E_*-acyclic, and a space A is E_*-acyclic if and only if the spectrum $E \wedge A$ is weakly contractible, i.e., all its homotopy groups vanish. As detailed in the proof of [6, Proposition 9.1], this fact can be expressed with formulas that contain quantifiers involving finite sets of simplices of the spaces $E_n \wedge A$, where $\{E_n\}$ is the set of constituents of E and fibrant replacements are used when needed. Hence we can write a Σ_1 formula φ with $p = \{X, Y, E\}$ as a set of parameters such that $\mathcal{E}_{XY} = \{Z \mid \varphi(Z, p)\}$.

Pick an uncountable cardinal κ larger than the ranks of X, Y and E and such that $|V_\kappa| = \kappa$. Given any $Z \in \mathcal{E}_{XY}$, the reflection principle ensures that there is a regular cardinal $\lambda > \kappa$ such that $Z \in V_\lambda$ and $\varphi(Z, p)$ holds in V_λ. By the Löwenheim–Skolem theorem, there is an elementary submodel $N \subset V_\lambda$ with $|N| < \kappa$ such that $Z \in N$ and the transitive closure of $\{X, Y, E\}$ is also in N. By elementarity, $\varphi(Z, p)$ holds in N. However, Z need not be a *subset* of N, since N is not transitive.

Now let $\pi \colon N \to M$ be the unique isomorphism where M is transitive—this uses the Mostowski collapse; see [19, Theorem 6.15] for details—and let $j \colon M \to N$ be the inverse isomorphism. If we pick $z = \pi(Z)$, then z is also a σ_{XY}-structure since j is an isomorphism and $j(\sigma_{XY}) = \sigma_{XY}$ because the transitive closure of $\{X, Y\}$ is in N, and $z \in V_\kappa$ since $z \subset M$ and $|M| < \kappa$. Moreover, $\varphi(z, p)$ holds in M because $\varphi(j(z), j(p))$ holds in N, as $j(p) = p$ since the transitive closure of p is in N. Using the fact that Σ_1 formulas are upward absolute, we infer that $\varphi(z, p)$ holds in V, which means that $z \in \mathcal{E}_{XY}$. Moreover, the restriction $j|z \colon z \to Z$ is an elementary embedding of σ_{XY}-structures. This means that there is an injective map $z \to Z$ with $z \in \mathcal{E}_{XY} \cap V_\kappa$, so condition (B) holds. $\qquad\square$

In the case of formulas of higher Lévy complexity, the Löwenheim–Skolem theorem is not sufficient to ensure reflectivity. Instead, elementary embeddings from the universe V into convenient models are needed [5].

The *critical point* of an elementary embedding $j: V \to M$ is the smallest cardinal κ such that $j(\kappa) \neq \kappa$. Then all sets of rank smaller than κ are fixed by j, and $j(\kappa) > \kappa$. The existence of nontrivial elementary embeddings of V cannot be proved in ZFC, since if there is one then its critical point is a measurable cardinal. Indeed, many kinds of large cardinals are defined as critical points of elementary embeddings $V \to M$ with suitable conditions on M.

A cardinal κ is called *supercompact* if for every ordinal λ there exists an elementary embedding $j: V \to M$ with M transitive and with critical point κ such that $j(\kappa) > \lambda$ and M is closed under sequences of length λ; see [19, 21]. As evidenced by the following result, which is based on [6, Theorem 5.2], supercompact cardinals yield structural reflection for Σ_2 formulas.

Theorem 2.2 (Existence of Cohomological Localizations) E^*-*localization exists for every E if arbitrarily large supercompact cardinals exist.*

Proof Now we aim to prove that condition (B) from Sect. 1 holds for the class $\mathcal{E}^{XY}$ of Kan simplicial sets Z with pairs of pointed maps $f: X \to Z$ and $s: Y \to Z$, where X, Y are simplicial sets and s is an E^*-equivalence.

For this, we need to prove that the class of E^*-equivalences for a spectrum E can be defined by means of a Σ_2 formula. This was done in [6, Theorem 9.3] and it is summarized as follows. A map of pointed simplicial sets is an E^*-equivalence if and only if its cofibre A is E^*-acyclic, and this means that the function spaces $\text{map}(A, E_n)$ are weakly contractible for all n, where $E = \{E_n\}$ and we assume that E is an Ω-spectrum without loss of generality. In order to formalize the fact that $\text{map}(A, E_n)$ is weakly contractible for all n, we write the following Σ_2 statement: "A is a simplicial set and there exists a function f with domain $\mathbb{N}$ such that, for all $n \in \mathbb{N}$, $f(n)$ is a simplicial set and, for every x and every $k \in \mathbb{N}$, x is a k-simplex of $f(n)$ if and only if x is a simplicial map $A \wedge \Delta[k]_+ \to E_n$, and $f(n)$ is weakly contractible". Here $\Delta[k]_+$ denotes a standard k-simplex with a disjoint basepoint.

Thus, let φ denote a Σ_2 formula defining the class $\mathcal{E}^{XY}$ with $p = \{X, Y, E\}$ as a set of parameters, and let κ be a supercompact cardinal larger than the rank of p. Given $Z \in \mathcal{E}^{XY}$, pick a regular cardinal λ bigger than κ such that $Z \in V_\lambda$ and V_λ is Σ_2-elementary in V (here we use again the reflection principle). Let $j: V \to M$ be an elementary embedding with M transitive and critical point κ such that $j(\kappa) > \lambda$ and M is closed under sequences of length λ. This implies that M contains V_λ.

Next, observe that V_λ is Σ_1-elementary in M, since every Σ_1 formula that holds in M also holds in V, and V_λ is Σ_2-elementary in V. Consequently, Σ_2 formulas are upward absolute between V_λ and M. Since φ is a Σ_2 formula, $\varphi(Z, p)$ holds in V_λ and hence it holds in M. Since Z and $j(Z)$ are in M and $Z \in V_\lambda$ and M is closed under λ-sequences, the restriction $j|Z: Z \to j(Z)$ is in M. Furthermore, Z is a σ_{XY}-structure in M since $j(\sigma_{XY}) = \sigma_{XY}$, and $j|Z$ is an elementary embedding of σ_{XY}-structures.

We have that $\text{rank}(Z) < \lambda < j(\kappa)$ in V and also in M, since M contains V_λ. Therefore, as witnessed by Z, in M there exists a σ_{XY}-structure z of rank smaller than $j(\kappa)$ for which $\varphi(z, j(p))$ holds, and there is an elementary embedding $z \to Z$. By elementarity of j, the corresponding statement is true in V; that is, there exists a σ_{XY}-structure z of rank smaller than κ and $\varphi(z, p)$ holds in V, and there is an elementary embedding $z \to Z$. In other words, every $Z \in \mathcal{E}^{XY}$ has a substructure in $\mathcal{E}^{XY} \cap V_\kappa$. Hence the elements of $\mathcal{E}^{XY} \cap V_\kappa$ form a solution set, and condition (B) holds, as needed. $\qquad\qquad\square$

If no bound is imposed on the complexity of a formula defining a class $\mathcal{C}$, then the existence of a cardinal reflecting $\mathcal{C}$ follows from the Vopěnka principle. We omit the details of this claim and refer to [6].

3   Conclusions

In practice, in order to construct localizations with respect to proper classes of maps, sometimes one assumes the existence of *Grothendieck universes*, and moves to a higher universe whenever the construction of a category of fractions requires it. However, the assumption that for each set X there exists a Grothendieck universe $\mathcal{U}$ with $X \in \mathcal{U}$ is equivalent to the assumption that for each cardinal κ there exists an inaccessible cardinal $\lambda > \kappa$.

What we have proved in Theorem 2.2 is that, if there is a supercompact cardinal larger than a given spectrum E and larger than two given simplicial sets X, Y, then $S^{-1}\mathcal{H}(X, Y)$ is a set (not a proper class), where $\mathcal{H}$ is the homotopy category of pointed simplicial sets and $S^{-1}\mathcal{H}$ denotes the category of fractions for the class S of E^*-equivalences. In consequence, we obtain an E^*-localization functor without having to "pass to a higher universe", since the category $S^{-1}\mathcal{H}$ happens to be locally small, which was the only pending requirement for Adams' argument in [4] to work. It is also remarkable that Adams' construction of E_*-localizations could have been made precise at that time by just using the reflection principle and the Löwenheim–Skolem theorem to infer that $S^{-1}\mathcal{H}(X, Y)$ is a set for all X and Y (Theorem 2.1).

In conclusion, what is the situation now? We know that the existence of a proper class of supercompact cardinals ensures the existence of arbitrary cohomological localizations, but we do not know the precise logical strength of the claim that cohomological localizations exist. There are two possibilities:

(1) It can be proved in ZFC that cohomological localizations exist.
(2) The claim that cohomological localizations exist is itself a large-cardinal principle.

We do not consider the possibility that somebody may find a counterexample in ZFC, since this would imply that the existence of a proper class of supercompact cardinals is inconsistent with ZFC. This would make inconsistent an enormous segment of the large-cardinal hierarchy. Although this is not impossible, it is extremely unlikely.

Possibility (1) cannot be discarded, although this has remained a challenge for almost fifty years. Hence, possibility (2) seems the most plausible one.

One approach to try to prove that possibility (1) is the winning one would be to verify Hovey's conjecture [18], according to which for every spectrum E there is another spectrum F such that the class of E^*-equivalences is equal to the class of F_*-equivalences. If this were true, then the existence of cohomological localizations would be provable in ZFC by Theorem 2.1 above.

However, a solution of Hovey's conjecture seems still out of reach, in spite of the fact that it has been shown to be false in derived categories of rings [31]. It is not even known whether the collection of cohomological Bousfield classes of spectra is a set or a proper class, while it has been known since Ohkawa's work in [25] that there is only a set of distinct homological Bousfield classes.

Acknowledgements I thank Joan Bagaria for revising and correcting the set-theoretical content of the manuscript.

References

1. J. Adámek and J. Rosický. *Locally Presentable and Accessible Categories*. London Math. Soc. Lecture Note Ser., vol. 189, Cambridge University Press, Cambridge (1994).
2. J.F. Adams. The sphere, considered as an H-space mod p. *Quart. J. Math.* **12**, 52–60 (1961).
3. J.F. Adams. Idempotent functors in homotopy theory. In: *Manifolds – Tokyo 1973*, pp. 247–253, Univ. of Tokyo Press, Tokyo (1975).
4. J.F. Adams. *Localisation and completion, with an addendum on the use of Brown–Peterson homology in stable homotopy*. Lecture notes by Z. Fiedorowicz on a course given at The University of Chicago in Spring 1973. Revised and supplemented by Z. Fiedorowicz, arXiv:1012.5020 (2010).
5. J. Bagaria. Large cardinals as principles of structural reflection. arXiv:2107.01580 (2021).
6. J. Bagaria, C. Casacuberta, A.R.D. Mathias and J. Rosický. Definable orthogonality classes in accessible categories are small. *J. Eur. Math. Soc.* **17**, 549–589 (2015).
7. A.K. Bousfield. The localization of spaces with respect to homology. *Topology* **14**, 133–150 (1975).
8. A.K. Bousfield. The localization of spectra with respect to homology. *Topology* **18**, 257–281 (1979).
9. A.K. Bousfield. Cohomological localizations of spaces and spectra. Unpublished (1979).
10. A.K. Bousfield and D.M. Kan. *Homotopy Limits, Completions and Localizations*. Lecture Notes in Math., vol. 304, Springer-Verlag, Berlin-Heidelberg (1972).
11. E.H. Brown. Abstract homotopy theory. *Trans. Amer. Math. Soc.* **119**, 79–85 (1965).
12. E.H. Brown and F.P. Peterson. A spectrum whose $\mathbb{Z}_p$-cohomology is the algebra of reduced p-th powers. *Topology* **5**, 149–154 (1966).
13. C. Casacuberta, D. Scevenels, and J.H. Smith. Implications of large-cardinal principles in homotopical localization. *Adv. Math.* **197**, 120–139 (2005).
14. A. Deleanu. Existence of the Adams completion for CW-complexes. *J. Pure Appl. Algebra* **4**, 299–308 (1974).
15. S. Eilenberg and N.E. Steenrod. Axiomatic approach to homology theory. *Proc. Natl. Acad. Sci. USA* **31**, 117–120 (1945).

16. P. Gabriel and M. Zisman. *Calculus of Fractions and Homotopy Theory*. Ergeb. Math. Grenzgeb., vol. 35, Springer-Verlag, Berlin-Heidelberg (1967).
17. P. Hilton, G. Mislin and J. Roitberg. *Localization of Nilpotent Groups and Spaces*. North-Holland Math. Studies, vol. 15, North-Holland, Amsterdam (1975).
18. M. Hovey. Cohomological Bousfield classes. *J. Pure Appl. Algebra* **103**, 45–59 (1995).
19. T. Jech. *Set Theory. The Third Millenium Edition, Revised and Expanded*. Springer Monographs in Math., Springer-Verlag, Berlin-Heidelberg (2003).
20. D.C. Johnson and W.S. Wilson. BP operations and Morava's extraordinary K-theories. *Math. Z.* **144**, 55–75 (1975).
21. A. Kanamori. *The Higher Infinite: Large Cardinals in Set Theory from Their Beginnings*. Perspectives in Mathematical Logic, Springer-Verlag, Berlin-Heidelberg (1994).
22. A. Lévy. *A Hierarchy of Formulas in Set Theory*. Mem. Amer. Math. Soc., vol. 57, Amer. Math. Soc., Providence (1965).
23. J.P. May. *Simplicial Objects in Algebraic Topology*. The Univ. of Chicago Press, Chicago (1967).
24. M. Mimura, G. Nishida and H. Toda. Localization of CW-complexes. *J. Math. Soc. Japan* **23**, 593–624 (1971).
25. T. Ohkawa. The injective hull of homotopy types with respect to generalized homology functors. *Hiroshima Math. J.* **19**, 631–639 (1989).
26. A.J. Przeździecki. Homotopical localizations at a space. *Topology Appl.* **126**, 131–143 (2002).
27. D.G. Quillen. *Homotopical Algebra*. Lecture Notes in Math., vol. 43, Springer-Verlag, Berlin-Heidelberg (1967).
28. D.G. Quillen. Rational homotopy theory. *Ann. of Math. (2)* **90**, 205–295 (1969).
29. D. Ravenel. Localizations with respect to certain periodic homology theories. *Amer. J. Math.* **106**, 351–414 (1984).
30. J.-P. Serre. Groupes d'homotopie et classes de groupes abéliens. *Ann. of Math. (2)* **58**, 258–294 (1953).
31. G. Stevenson. Derived categories of absolutely flat rings. *Homol. Homotop. Appl.* **16**, 45–64 (2014).
32. D. Sullivan. Genetics of homotopy theory and the Adams conjecture. *Ann. of Math. (2)* **100**, 1–79 (1974).
33. D. Sullivan. Infinitesimal computations in topology. *Publ. Math. IHÉS* **47**, 269–331 (1977).

Geometry of Data

Parvaneh Joharinad and Jürgen Jost

Personal Note of Jürgen Jost
I first met Catriona Byrne at the Arbeitstagung *in Bonn in June 1982, that is, 40 years ago. She had just started to work at Springer a few months before. I remember that encounter quite well, as it led to my first book project, a volume in the* Lecture Notes in Mathematics, *with the title* Harmonic Maps between Surfaces. *Thus, it is very fitting that this tribute to her work will appear in the* Lecture Notes in Mathematics *series.*

Over those 40 years, several more book projects were successfully guided by her and her colleagues. More than that, mutual understanding and trust on publishing books and editing journals developed, and a frequent exchange of advice was both helpful and enjoyable. In particular, she successfully implemented the inauguration and the start of the Journal of the European Mathematical Society, *of which I became the founding editor-in-chief.*

Her competence, her energy, and her dedication to high quality mathematical publishing will be profoundly missed after her retirement.

1 Introduction

Many data sets come with a basic geometric structure, distances between data points. It is therefore natural to use geometric methods to analyze such data. The deepest geometric concepts, however, were developed in the nineteenth century for

P. Joharinad · J. Jost (✉)
Max Planck Institute for Mathematics in the Sciences, Leipzig, Germany
e-mail: jjost@mis.mpg.de

© The Author(s), under exclusive license to Springer Nature Switzerland AG 2023   183
J.-M. Morel, B. Teissier (eds.), *Mathematics Going Forward*, Lecture Notes
in Mathematics 2313, https://doi.org/10.1007/978-3-031-12244-6_14

smooth manifolds, more precisely Riemannian manifolds, and the most fundamental concept there is curvature. In the twentieth century, notions of curvature were successfully generalized to more general classes of spaces. Still, those spaces, like geodesic length spaces, are typically not discrete, in contrast to data sets. Thus, we have found it desirable to rethink fundamental geometric concepts from a more abstract perspective that also naturally includes discrete spaces. Of course, there are ideas and approaches that we can build upon, most importantly those pioneered by Gromov [21, 22]. From such a perspective, the distinction between discrete and connected spaces is partly one of scale. From a large scale perspective, spaces from those two classes may look alike.

Such a large scale perspective is still quantitative, hence geometric, and is therefore different from a qualitative topological approach. Nevertheless, as we shall see, there are important links between the two. In particular, we can look at the successful topological data analysis method of persistent homology from a geometric perspective.

Topological data analysis asks when balls in a metric space (X, d) intersect. This is a qualitative concept, but the data analysis method of persistent homology makes this quantitative through the dependence on the radii of the balls. Geometric data analysis, as we conceive it in this contribution, asks how much balls have to be enlarged to intersect. And as we shall see, this is captured by a suitable concept of curvature. And curvature, from a general perspective as adopted here, quantifies convexity. Therefore, convexity and its strengthening as hyperconvexity will be our basic concepts.

We thank the referee for useful comments.

2 Preliminaries from Metric Geometry

Let (X, d) be a metric space. $x, y, \ldots$ will be points in X, and they thus have a distance $d(x, y)$. A continuous path $c : [0, 1] \longrightarrow X$ with $x = c(0), y = c(1)$ has length

$$l(c) := \sup \sum_{i=1}^{i=n} d(c(t_i), c(t_{i-1})).$$

The supremum here is taken over all partitions of $[0, 1]$, with $t_0 = 0, t_n = 1$. (X, d) is called a *length space* if for all x, y,

$$d(x, y) = \inf\{l(c) : \text{c is a path between } x \text{ and } y\}.$$

A length space (X, d) is called *geodesic* if this infimum is always realized, that is, any $x, y \in X$ can be connected by a *shortest path* $c : [0, 1] \longrightarrow X$, i.e.

$$d(x, y) = l(c).$$

Thus, the distance between x and y is realized by some curve, a shortest geodesic.

Every complete locally compact length space is a geodesic space. However, there is another way to determine whether a complete metric space is a geodesic (resp. length) space by checking the existence of mid-points (resp. approximate midpoints).

Definition 2.1 $m \in X$ is a *midpoint* between x, y if

$$d(x, m) = d(m, y) = \frac{1}{2}d(x, y).$$

We may also say that a pair of points $x, y \in X$ has *approximate midpoints* if for every $\epsilon > 0$ there exists an $m_\epsilon \in X$ with

$$\max\{d(m_\epsilon, x), d(m_\epsilon, y)\} \leq \frac{1}{2}d(x, y) + \epsilon.$$

We observe

Lemma 2.1 *Every pair of points in a geodesic space (resp. length space) has at least one midpoint (resp. approximate midpoints).*

The converse is true provided that the metric space is complete. □

In the sequel,

$$B(x, r) := \{y \in X : d(x, y) \leq r\}$$

will always be the *closed* ball centered at x with radius $r \geq 0$.

Definition 2.2 (X, d) is *totally convex* if for any $x_1, x_2 \in X, r_1, r_2 > 0$ with

$$r_1 + r_2 \geq d(x_1, x_2),$$

we have

$$B(x_1, r_1) \cap B(x_2, r_2) \neq \emptyset.$$

Any radii r_i will be > 0 in the sequel.

Again, an easy lemma

Lemma 2.2 *Geodesic spaces are totally convex.* □

Length spaces are not necessarily totally convex, as they need not be complete. An example is $\mathbb{R}^2 \setminus 0$ with the length structure induced by the Euclidean distance. But even when a length space is complete, it need not be totally convex. For instance, we can take the metric space consisting of two points and countably many intervals connecting them, the nth interval having length $1 + \frac{1}{n}$.

Let us formulate Definition 2.2 as

Principle 2.1 Two balls that can intersect do intersect.

We shall now introduce a fundamental quantity. For

$$r_1 + r_2 \geq d(x_1, x_2)$$

we put

$$\rho((x_1, x_2), (r_1, r_2)) := \inf_{x \in X} \max_{i=1,2} \frac{d(x_i, x)}{r_i}, \tag{1}$$

$$\rho(x_1, x_2) := \sup_{r_1, r_2} \rho((x_1, x_2), (r_1, r_2)). \tag{2}$$

If $\rho(x_1, x_2) = 1$ for each pair of points $x_1, x_2 \in X$, then the existence of approximate midpoints is guaranteed, and X is a length space provided that it is a complete metric space. If, moreover, the infimum is attained for each pair by some $x_0 \in X$, then X is a geodesic space provided that it is complete.

Another obvious

Lemma 2.3 *When X is complete the supremum in (2) is realized by $r_1 = r_2 = \frac{1}{2}d(x_1, x_2)$, that is*

$$\rho(x_1, x_2) = \inf_{x \in X} \max_{i=1,2} \frac{2d(x_i, x)}{d(x_1, x_2)}. \tag{3}$$

Moreover, $\rho(x_1, x_2) = 1$ is achieved for some x when

$$d(x_1, x) + d(x_2, x) = d(x_1, x_2),$$

that is, when x is a midpoint of x_1, x_2.   □

Thus, we want to find points between two points x_1 and x_2, and quantify to what extent that can fail.

Therefore, in the realm of complete metric spaces, the more (2) deviates from 1 the less is the chance to approximate distances by lengths of connecting paths.

A key idea now is to extend this to three points.

3   Tripod Spaces

Definition 3.1 A geodesic length space (X, d) is a *tripod space* if for any three points $x_1, x_2, x_3 \in X$, there exists a *median*, that is, a point $m \in X$ with

$$d(x_i, m) + d(x_j, m) = d(x_i, x_j), \text{ for } 1 \leq i < j \leq 3.$$

We note that for a median, we have

$$d(x_1, x_2) + d(x_2, x_3) + d(x_3, x_1) = 2(d(x_1, m) + d(x_2, m) + d(x_3, m)).$$

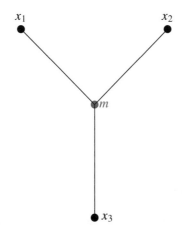

Most metric spaces are *not* tripod spaces. For instance, Riemannian manifolds of dimension > 1 do *not* satisfy the tripod property. Nevertheless, there are examples that will be important for us:

- Metric trees;
- L^∞-spaces; and more generally,
- *hyperconvex spaces* (to be defined shortly).

If such a median exists it will be a minimizer for the sum of the distances to the corresponding triple x_1, x_2, x_3. Such a point is called a Fermat point.

Our strategy will then be to quantify the deviation from the tripod property.

We get the existence of tripods if the following more general condition is satisfied. For any $x_1, x_2, x_3 \in X$ which do not lie on a geodesic, and $r_i + r_j \geq d(x_i, x_j)$, $1 \leq i < j \leq 3$,

$$\bigcap_{i=1}^{3} B(x_i, r_i) \neq \emptyset.$$

This leads to

Principle 3.1 Three balls that can intersect do intersect.

To explore this principle, and the deviation from it, we shall now introduce a 3-point analogue of (1), (2). For $x_1, x_2, x_3 \in X$ and $r_i + r_j \geq d(x_i, x_j)$,

$$\rho((x_1, x_2, x_3), (r_1, r_2, r_3)) := \inf_{x \in X} \max_{i=1,2,3} \frac{d(x_i, x)}{r_i}, \tag{4}$$

$$\rho(x_1, x_2, x_3) := \sup_{r_i + r_j \geq d(x_i, x_j), i \neq j} \rho((x_1, x_2, x_3), (r_1, r_2, r_3)). \tag{5}$$

This is uniquely solved by the *Gromov products*

$$r_1 = \frac{1}{2}(d(x_1, x_2) + d(x_1, x_3) - d(x_2, x_3)),$$

$$r_2 = \frac{1}{2}(d(x_1, x_2) + d(x_2, x_3) - d(x_1, x_3)),$$

$$r_3 = \frac{1}{2}(d(x_1, x_3) + d(x_2, x_3) - d(x_1, x_2)). \tag{6}$$

Remark It is obvious that $\rho((x_1, x_2, x_3), (r_1, r_2, r_3)) \geq 1$. Moreover, this quantity is bounded from above by 2 if X is complete.

If (with r_1, r_2, r_3 defined by (6)) $\rho(x_1, x_2, x_3) = 1$ and the infimum is attained by some m, then we have a tripod construction or equivalently a Fermat point. This implies that there exists an intermediate point through which each pair x_i, x_j can be connected.

Definition 3.2 The point m attaining the infimum in (4) is called a *weighted circumcenter*.

A weighted circumcenter solves an optimization problem in $\mathbb{R}^3$ with respect to the l_∞ norm. The larger the value of $\rho(x_1, x_2, x_3)$ is, the less optimal the weighted circumcenter as the interconnecting point will be.

We observe here

Lemma 3.1 *Weighted circumcenters exist and are unique for triangles in $CAT(0)$ spaces (Alexandrov's generalization of Riemannian manifolds of sectional curvature ≤ 0).*

4 Hyperconvexity

We shall now extend the above principle to arbitrary numbers of points.

Definition 4.1 (X, d) is *hyperconvex* if for any family $\{x_i\}_{i \in I} \subset X$ and $r_i + r_j \geq d(x_i, x_j)$ for $i, j \in I$,

$$\bigcap_{i \in I} B(x_i, r_i) \neq \emptyset.$$

In a totally convex metric space, $r_i + r_j \geq d(x_i, x_j)$ can be replaced by $B(x_i, r_i) \cap B(x_j, r_j) \neq \emptyset$ for all $i, j \in I$. Thus, when balls intersect pairwise, they also have a common intersection.

This leads to our final

Principle 4.1 Balls that can intersect do intersect.

We observe

Lemma 4.1 *Hyperconvex spaces are tripod spaces.* $\qquad\qquad\qquad\qquad\qquad$ □

We list some important properties of hyperconvex spaces.

Theorem 4.1

(a) *Hyperconvex spaces are complete and contractible to each of their points [5].*
(b) *X is hyperconvex iff every 1-Lipschitz map from a subspace of any metric space Y to X can be extended to a 1-Lipschitz map over Y [5].*
(c) *Every metric space is isometrically embedded in a hyperconvex space, called its hyperconvex hull. The hyperconvex hull of a compact space is compact and that of a finite space is a simplicial complex. [16, 29]*

We now describe the isometric embedding in part (c) and the construction of the hyperconvex hull, in order to understand the specific choice of radii in (1) and (4). By Kuratowski embedding, every bounded metric space (X, d) is isometrically embedded in the space of bounded functions on X equipped with the supremum norm, i.e. $l_\infty(X)$, via the map $x \mapsto d(x, \cdot)$, which we denote by $x \mapsto d_x$ for simplicity.

$l_\infty(X)$ contains the subspace $E(X)$ consisting of all functions f that are minimal subject to the relation

$$f(x) + f(y) \geq d(x, y), \ \forall x, y \in X. \tag{7}$$

It has been shown in [16, 29, 35] that $E(X)$ is a hyperconvex space containing the image of X under the Kuratowski embedding isometrically, and $E(X)$ is minimal in the sense that it is isometrically embeddable in any other such hyperconvex space.

The radii in (1) and (4) are functions on a 2-point space and a 3-point spaces respectively, satisfying (7).

If X is a finite metric space with $|X| = n$, the space of all functions satisfying (7) is a polyhedron in the finite vector space $\mathbb{R}^n$ obtained by the intersection of the closed half spaces $f_i + f_j \geq d(x_i, x_j)$ for $1 \leq i < j \leq n$. Therefore, the interior of every face S of this polyhedron is the intersection of some hyperplanes

$f_i + f_j = d(x_i, x_j)$. We can then define a graph $G(S)$ with vertex set X, corresponding to the symmetric relation defined by that face. More precisely, x_i is connected to x_j with an edge in $G(S)$ if for $f \in S^\circ$ we have $f_i + f_j = d(x_i, x_j)$. Now, $E(X)$ is the union of compact faces of this polyhedron and moreover the graph corresponding to each such face is a spanning graph, that is, every vertex is connected to at least one other vertex in this graph. This construction was first introduced in [16], where a combinatorial dimension for finite metric spaces was defined as the maximal dimension of a face in its hyperconvex hull. The hyperconvex hull of finite metric spaces was studied further in [6, 17] and from a different perspective in [14, 40] to obtain the *metric fan* of a finite set. In [40] a software tool was presented to visualize these hyperconvex hulls. The problem of finding faces of $E(X)$, when X is finite, as a linear programming problem was also studied in [15, 28].

In the special case $X = \{x_1, x_2\}$ with distance $d_{12} = d(x_1, x_2)$, the corresponding polyhedron is the half plane $f_1 + f_2 \geq d_{12}$ cut by the coordinate planes $f_i = 0$, $i = 1, 2$, which has only one compact face, the line segment $f_1 + f_2 = d_{12}$ connecting $(d_{12}, 0)$ to $(0, d_{12})$, i.e., $[d_{x_2}, d_{x_1}]$. Every point in this polyhedron can be reached through a ray passing this line segment and the midpoint of this segment, that is $\frac{1}{2}(d_{12}, d_{12})$ is the corresponding radius function in (3). The space of all such radius functions is illustrated in Fig. 1(a).

Similarly, one can see that for $X = \{x_1, x_2, x_3\}$, using the same notation d_{ij}, $1 \leq i, j \leq 3$ for pairwise distances, the corresponding polyhedron is the intersection of the half-spaces

$$f_i + f_j \geq d_{ij}, \ 1 \leq i < j \leq 3 \tag{8}$$

and the coordinate half spaces $f_i \geq 0$ for $i = 1, 2, 3$. Moreover, the hyperconvex hull, colored in blue in Fig. 1(b), is the union of three segments each of which connect a distance function d_{x_i} to the function $r = (r_1, r_2, r_3)$ defined in (6). For the analysis of discrete metric spaces, some variants of the notion of hyperconvexity are well suited, c.f [20, 23, 24, 33].

Definition 4.2 (X, d) is δ-hyperbolic ($\delta \geq 0$) if for any family $\{B(x_i, r_i)\}_{i \in I}$ with $r_i + r_j \geq d(x_i, x_j)$,

$$\bigcap_{i \in I} B(x_i, \delta + r_i) \neq \emptyset. \tag{9}$$

Definition 4.3 (X, d) is λ-hyperconvex ($\lambda \geq 1$) if for every family $\{B(x_i, r_i)\}_{i \in I}$ with $r_i + r_j \geq d(x_i, x_j)$,

$$\bigcap_{i \in I} B(x_i, \lambda r_i) \neq \emptyset. \tag{10}$$

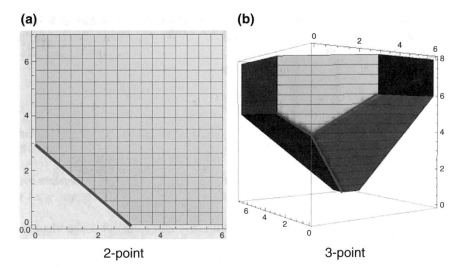

(a)   **(b)**

2-point   3-point

Fig. 1 (**a**) The yellow area is the set of all possible radius functions on 2 points and the line segment colored in blue refers to the minimal ones. (**b**) The three dimensional polyhedron is the set of all possible radius functions on 3 points and the tripod consisting of three line segments colored in blue refers to the minimal ones

Of course, 0-hyperbolicity and 1-hyperconvexity are simply hyperconvexity. For large radii, δ becomes insignificant, and the concept of δ-hyperbolicity is therefore good for asymptotic considerations. In contrast, λ-hyperconvexity is invariant under scaling the metric d, and it can therefore capture scaling invariant properties of a metric space.

The preceding concepts allow for a quantification of the deviation from hyperconvexity. The following results are known.

Theorem 4.2 *Hilbert spaces are* $\sqrt{2}$-*hyperconvex. Reflexive and dual Banach spaces are 2-hyperconvex. Therefore, for a measure space* (X, μ), $L^p(X, \mu)$, $1 < p < \infty$, *are 2-hyperconvex, and if* X *is finite,* $L^1(X, \mu)$ *is also 2-hyperconvex.* $L^\infty(X, \mu)$ *is hyperconvex [20, 33].*

5   Relation with Topological Data Analysis (TDA)

Definition 5.1 For a family $(x_i)_{i \in I}$ in a metric space (X, d) and $r > 0$, we define the *Čech complex* $\check{C}_r((x_i), X)$ containing a q-simplex whenever

$$\bigcap_{i=1,\ldots,q+1} B(x_i, r) \neq \emptyset.$$

Here $(x_i)_{i \in I}$ is called the landmark set and X is the witness set. When the witness set coincides with the landmarks, we thus define a non-empty intersection inside the sample set $(x_i)_{i \in I}$ as the criterion for a simplex. We also define the *Vietoris–Rips complex* $V R_r((x_i), X)$ containing a q-simplex whenever

$$B(x_i, r) \cap B(x_j, r) \neq \emptyset \text{ for all } i, j \in I.$$

The two structures are not as different as they might appear, as the difference between the criteria for spanning a simplex is whether the vertex set is contained in a ball of radius or of diameter r.

The principle of the important topological data analysis scheme of *persistent homology* then is to record how the homology of these complexes varies as a function of r. [12, 18, 19, 42].

Of course, every simplex of the Čech complex is also a simplex of the Vietoris–Rips complex, but not necessarily conversely unless for each simplex at least one of the balls of diameter r containing the vertex set of that simplex has a center in the witness set.

Deviation from hyperconvexity lets the Vietoris–Rips complex contain more simplices than the Čech complex, or conversely

Lemma 5.1 *In a hyperconvex space, all simplices that are filled in the Vietoris–Rips complex are also filled in the Čech complex. In particular, there is no contribution to local homology from unfilled simplices.* □

For instance, we can take a sample $(x_i)_{i \in I}$ from a geodesic metric space (X, d) and compare $V R_r((x_i), X)$ with $\check{C}_r((x_i), E(X))$. For the latter complex, we take the hyperconvex hull of X, i.e. $E(X)$, as the witness set. It is clear that $\check{C}_r((x_i), E(X)) \subset V R_r((x_i), X)$, as X is a geodesic space and hence totally convex. Conversely, every simplex in $V R_r((x_i), X)$ is defined according to the criterion that balls of radius r around its vertices intersect pairwise, which by hyperconvexity of $E(X)$ implies the existence of a common point between them in $E(X)$. In other words, the Vietoris–Rips complex of a metric family $((x_i)_{i \in I}, d)$ coincides with its Čech complex but with different witness sets. This natural principle has been used in [36] to study the metric thickening of S^1 in its hyperconvex hull. A thorough study of the Čech and the Vietoris–Rips filtration of S^1 can be found in [1, 2].

If X is a closed Riemannian manifold, for small-enough radius r depending on the injectivity radius and a curvature bound, $V R_r(X)$ is homotopy equivalent to X by a well known theorem of Hausmann [27]. On the other hand according to Nerve lemma, whenever X is a paracompact space and the family of open balls around sample points $(x_i)_{i \in I}$ with radius $r > 0$ define a cover such that the non-empty intersections of any finite number of them is contractible, the Čech complex $\check{C}_{<r}((x_i), X)$ is homotopy equivalent to the original space X, cf. [26]. Although Hausmann's theorem is restricted to the case where the original space, from which the sample is taken, is a Riemannian manifold, both constructions at some point reveal the topology of the space. However, the Vietoris–Rips filtration ignores the geometry of the space beyond the pairwise relations. The extent to which

higher order relations are overlooked by considering Vietoris–Rips complexes can be quantified by computing the deviation from hyperconvexity of different orders. This measures how much one must expand balls to obtain a simplex in the Čech complex of (x_i) with witness set X after that simplex is observed in the Čech complex of (x_i) with witness set $E(X)$. The upper bound 2 for this scale is usually stated in the TDA literature, but this bound is not sharp.

For instance, let us consider equilateral triangles of perimeter $3a$ in the Euclidean plane, in a circle and in a metric tree. That is, (x_1, x_2, x_3), (x'_1, x'_2, x'_3) and $(\bar{x}_1, \bar{x}_2, \bar{x}_3)$ are comparison triangles in the Euclidean plane, a circle and a hyperconvex space, respectively. As noted in (6), $r = \frac{a}{2}$ is the radius at which each of these triples forms a simplex in the corresponding Vietoris–Rips complex. However, we only need the upper bound of 2 in the case of (x'_1, x'_2, x'_3), where the point are sampled from a circle which has the highest deviation from hyperconvexity, for expanding the balls to obtain the simplex in the Čech complex, cf. [31].

One can also more generally let the radii of the balls be different. That is, for a vertex set $(x_i)_{i \in I}$ and a corresponding non-negative radius function r, we define the Čech complex containing a q-simplex $x_1, \ldots, x_{q+1}$ whenever

$$\bigcap_{i=1,\ldots,q+1} B(x_i, r(x_i)) \neq \emptyset.$$

The Vietoris–Rips complex is defined in a similar way. And one can then look at the resulting constructions for all such radius functions simultaneously [31].

6 Curvature

We can use the preceding concepts to compare spaces with each other, or with reference spaces, like Euclidean space. In geometry, such a comparison is quantified by the concept of *curvature*. From our abstract perspective, curvature relates intersection patterns of balls to convexity properties of distance function.

As pointed out by Klingenberg [34], the beginning of the theory of spaces of negative curvature can be dated to the work of von Mangoldt [41] in 1881 who showed that on a complete simply connected surface of negative curvature, geodesics starting at the same point diverge and can never meet again. This implies that the exponential map is a diffeomorphism. Apparently unaware of von Mangoldt's work, Hadamard [25] in 1898 proved further results about geodesics on surfaces of negative curvature. E. Cartan [13] later considered negatively curved Riemannian manifolds of any dimension. For our purposes, non-positive, as opposed to negative, curvature is the appropriate concept, as we are interested in comparison theorems.

Let us first recall a by now classical concept of non-positive curvature, introduced by Alexandrov [4].

Definition 6.1 The geodesic space (X, d) is a $CAT(0)$-*space* if for all geodesics $c_1, c_2 : [0, 1] \longrightarrow X$ with $c_1(0) = c_2(0)$

$$d(c_1(t), c_2(s)) \leq \|\bar{c}_1(t) - \bar{c}_2(s)\|, \; \forall \, t, s \in [0, 1] \tag{11}$$

where $\bar{c}_1, \bar{c}_2 : [0, 1] \longrightarrow \mathbb{R}^2$ are the sides of the Euclidean comparison triangle in $\mathbb{R}^2$ with the same side lengths as the triangle $\bigtriangleup(c_1(0), c_1(1), c_2(1))$.

According to this definition, triangles in $CAT(0)$-spaces are not thicker than Euclidean triangles with the same side lengths, cf. [3, 9, 10, 32].

There is another important concept of non-positive curvature, introduced by Busemann [11].

Definition 6.2 A geodesic space (X, d) is a *Busemann convex space* if for every two geodesics $c_1, c_2 : [0, 1] \longrightarrow X$ with $c_1(0) = c_2(0)$, the distance function $t \mapsto d(c_1(t), c_2(t))$ is convex.

Geodesics in a Busemann space diverge at least as fast as in Euclidean space.

Every $CAT(0)$ space is Busemann convex but not conversely. For complete simply connected Riemannian manifolds, however, the two definitions agree and are equivalent to non-positive sectional curvature in the sense of Riemann. In the non-simply connected case, one needs appropriate local formulations, but we do not go into that aspect here.

Several generalizations of these definitions to metric spaces that are not necessarily geodesic have been proposed, for instance [3, 7, 8]. We now present our definition from [30].

Definition 6.3 The metric space (X, d) has *non-positive curvature* if for each triple (x_1, x_2, x_3) in X with the comparison triangle $\bigtriangleup(\bar{x}_1, \bar{x}_2, \bar{x}_3)$ in $\mathbb{R}^2$, one has

$$\rho(x_1, x_2, x_3) \leq \rho(\bar{x}_1, \bar{x}_2, \bar{x}_3),$$

where $\rho(\bar{x}_1, \bar{x}_2, \bar{x}_3)$ is analogously defined by

$$\rho(\bar{x}_1, \bar{x}_2, \bar{x}_3) := \min_{x \in \mathbb{R}^2} \max_{i=1,2,3} \frac{\|x - \bar{x}_i\|}{r_i}.$$

Again, for simplicity, we present here only the global aspect which in the case of Riemannian manifolds, for instance, applies when the manifold is simply connected. Here, we do not explore local versions of our curvature condition.

According to this definition, the circumcenter of a triangle in a non-positively curved space is at least as close to the vertices as in the Euclidean case. In other words, there is a chance of finding a better intermediate point for each triple of points in such a space than in Euclidean plane.

For any triple of closed balls $\{B(x_i, r_i); \; i = 1, 2, 3\}$ with pairwise intersections, $\bigcap_{i=1,2,3} B(x_i, \rho r_i)$ is non-empty whenever $B(\bar{x}_i, \rho r_i), i = 1, 2, 3$, have a common point. Thus, balls do not need to be enlarged more than in Euclidean case to get triple intersection. Thus, we can again formulate a

Principle 6.1 *Balls intersect at least as easily as in Euclidean space.*

Examples:

- Tripod spaces have non-positive curvature in the sense of Definition 6.3, because there, $\rho = 1$, which is the smallest possible value.
- Complete $CAT(0)$ spaces have non-positive curvature in the sense of Definition 6.3. The converse is not true; in fact, our spaces need not be geodesic, nor have unique geodesics.
- An approximate version applies to discrete spaces. This is obviously important for questions in data analysis, and this in fact constitutes one of the motivations for Definition 6.3.

We also have

Theorem 6.1 *A complete, simply connected Riemannian manifold (N, g) has non-positive curvature iff it has non-positive sectional curvature, cf. [30].*

Obviously, with the same concepts and constructions, one can also define other curvature bounds than 0, by comparison with suitably scaled 2-spheres or hyperbolic planes.

7 Conclusions

The Čech construction assigns to a cover $\mathcal{U} = (U_i)_{i \in I}$ of X a simplicial complex $\Sigma(\mathcal{U})$ with vertex set I and a simplex σ_J whenever $\bigcap_{j \in J} U_j \neq \emptyset$ for $J \subset I$. When all intersections are contractible, the homology of $\Sigma(\mathcal{U})$ equals that of X (under some rather general topological conditions on X). When (X, d) is metric space, we can use covers by (open or closed) distance balls. Now, when (X, d) is a *hyperconvex* metric space, and if we use a cover $\mathcal{U}$ by distance balls, then whenever

$$\bigcap_{j \in J \setminus \{j_0\}} U_j \neq \emptyset \text{ for every } j_0 \in J, \tag{12}$$

then also

$$\bigcap_{j \in J} U_j \neq \emptyset. \tag{13}$$

In another words, whenever $\Sigma(\mathcal{U})$ contains all the boundary facets of some simplex, it also contains that simplex itself. It even satisfies the stronger condition that whenever $\Sigma(\mathcal{U})$ contains all the boundary faces of dimension 1 of some simplex, it also contains that simplex itself. This means that $\Sigma(\mathcal{U})$ is a flag complex.

Thus, there are no holes of the type of unfilled simplices, and no corresponding contributions to homology groups.

As hyperconvex spaces are contractible, then whenever non-trivial homology groups arise in Čech filtrations, the space cannot be hyperconvex, but only λ-hyperconvex for some $\lambda > 1$, since every complete metric space is λ-hyperconvex for some $1 \leq \lambda \leq 2$, cf. [24]. In the discrete case, one might work also with δ-hyperbolicity for $\delta > 0$.

From that perspective, hyperconvex spaces are the simplest model spaces, and homology can be seen as a topological measure for the deviation from such a model. However, this geometric interpretation has been dismissed in topological data analysis, by considering the Vietoris–Rips filtration instead of Čech, for the benefit of reducing computational complexity. Still, it is possible to infer topological information about a space from the Vietoris–Rips filtration, based on Hausmann's theorem. However, when one samples a metric space, this depends on how dense the sample is and the results are accurate only for small radii. For instance, the Vietoris–Rips complexes of S^1 admit holes of dimension larger than 1 as the radius increases, cf. [1].

Homology groups and Betti numbers as integer invariants are fundamental topological invariants. Geometry can provide more refined real valued invariants. And after Riemann [38, 39], the fundamental geometric invariants are curvatures. In our framework, the essential geometric content of curvature can be extracted for general metric spaces. The basic class of model spaces for curvature is given by the tripod spaces, a special class containing hyperconvex spaces. From that perspective, the geometric content of curvature in the abstract setting considered here is the deviation from the tripod condition. Euclidean spaces only have a subsidiary role, based on a normalization of curvature that assigns the value 0 to them.

Considering Euclidean spaces as model spaces is traditionally justified by the fact that spaces whose universal cover has synthetic curvature ≤ 0 in the sense of Alexandrov are homotopically trivial in the sense that their higher homotopy groups vanish. In technical terms, they are $K(\pi, 1)$ spaces, with π standing for the first homotopy group. The perspective developed here, however, is a homological and not a homotopical one, and therefore, our natural comparison spaces are tripods. We have started the investigation of these spaces in [30, 31]. A more systematic investigation of their properties should be of interest.

In order to get stronger topological properties, like those of hyperconvex spaces, which are homologically trivial, we might need conditions involving collections of more than three points.

In fact, according to [37, Theorem 4.2], if X is a tripod Banach space on which every collection of four closed balls $\{B(x_i, r_i)\}_{i=1}^4$ with non-empty pairwise intersection has a non-void intersection, then every finite family of closed balls with non-empty pairwise intersection has also a non-trivial intersection. In this case, the Vietoris–Rips and Čech complexes coincide.

One can also think about higher order relations and how they can be obtained from sub-relations (that is from the relations existing in all subsets of some smaller size). For instance, in some metric spaces, a family of n balls has a common point if every subfamily of size k in it has a non-empty intersection. [37] calls this property the (n, k)-intersection property. For instance, Helly's theorem says that Euclidean space $\mathbb{R}^d$ has the $(n, d + 1)$-intersection property for $n \geq d + 2$. For a given metric space, one can compute the deviation from such a property.

From the perspective of Čech complexes, this deviation could be quantified by the scaling parameter needed to fill an $(n - 1)$-simplex after all the faces of dimension $k - 1$ are filled. The quantitative measure we introduced provides us with the scaling function to fill a 2-simplex after its 1-dimensional boundary faces are filled.

References

1. M. Adamaszek and H. Adams. The Vietoris–Rips complexes of a circle. *Pacific Journal of Mathematics* **290**(1), 1–40 (2017).
2. M. Adamaszek, H. Adams, F. Frick, C. Peterson and C. Previte-Johnson. Nerve complexes of circular arcs. *Discrete & Computational Geometry* **56**(2), 251–273 (2016).
3. V. Alexander, V. Kapovitch and A. Petrunin. Alexandrov geometry. *arXiv:1903.08539v1* (2019).
4. A.D. Alexandrov. Über eine Verallgemeinerung der Riemannschen Geometrie. *Schr. Forschungsinst. Math. Berlin* 1, 33–84 (1957).
5. N. Aronszajn and P. Panitchpakdii. Extension of uniformly continuous transformations and hyperconvex metric spaces. *Pacific J. Math.* 6, 405–439 (1956).
6. H.-J. Bandelt and A.W.M. Dress. A canonical decomposition theory for metrics on a finite set. *Advances in mathematics* **92**(1), 47–105 (1992).
7. M. Bačak, B.B. Hua, J. Jost, M. Kell and A. Schikorra. A notion of nonpositive curvature for general metric spaces. *Diff. Geom. Appl.* 38, 22–32 (2015).
8. I.D. Berg and I.G. Nikolaev. Characterization of Aleksandrov spaces of curvature bounded above by means of the metric Cauchy–Schwarz inequality. *Michigan Math. J.* **67**, 289–332 (1993).
9. M.R. Bridson and A. Haefliger. *Metric spaces of non-positive curvature.* Grundlehren der mathematischen Wissenschaften **319**, Springer Science & Business Media (2013).
10. D. Burago, Yu. Burago and S. Ivanov. *A course in metric geometry.* AMS (2001).
11. H. Busemann. Spaces with non-positive curvature. *Acta Mathematica* **80**(1), 259–310 (1948).
12. G. Carlsson. Topology and data. *Bull. AMS* **46**, 255–308 (2009).
13. É. Cartan. *La géométrie des espaces de Riemann.* Gauthier-Villars (1925).
14. J.A. De Loera, B. Sturmfels and R. Thomas. Gröbner bases and triangulations of the second hypersimplex. *Combinatorica* **15**(3), 409–424 (1995).
15. M. Develin. Dimensions of tight spans. *Annals of Combinatorics* **10**(11), 53–61 (2006).
16. A. Dress. Trees, tight extensions of metric spaces, and the cohomological dimension of certain groups: A note on combinatorial properties of metric spaces. *Adv. Math.* 53, 321–402 (1984).
17. A. Dress, K.T. Huber and V. Moulton. An explicit computation of the injective hull of certain finite metric spaces in terms of their associated Buneman complex. *Adv. Math.* **168**, 1–28 (2002).
18. H. Edelsbrunner and J. Harer. Persistent homology – a survey. *Contemporary mathematics* **453**, 257–282 (2008).
19. H. Edelsbrunner, D. Letscher and A. Zomorodian. Topological persistence and simplification. *Foundations of Computer Science*, Proceedings, 41st Annual Symposium on IEEE (2000).

20. R. Espínola and M.A. Khamsi. Introduction to Hyperconvex Spaces. In: *Handbook of Metric Fixed Point Theory*, W.A. Kirk and B. Sims (eds.), pp. 391–435, Springer Netherlands, Dordrecht (2001). https://doi.org/10.1007/978-94-017-1748-9_13

21. M. Gromov. *Structures métriques pour les variétés riemanniennes*. Rédigé par J. Lafontaine and P. Pansu. Cedic-Nathan, Paris (1980).

22. M. Gromov. *Metric structures for Riemannian and non-Riemannian spaces*. Birkhäuser (1999).

23. B. Grünbaum. On some covering and intersection properties in Minkowski spaces. *Pacific J. Math.* **27**, 487–494 (1959).

24. B. Grünbaum. Some applications of expansion constants. *Pacific J. Math.* 10(1), 193–201 (1960).

25. J. Hadamard. Sur la forme des lignes géodésiques à l'infini et sur les géodésiques des surfaces réglées du second ordre. *Bulletin de la Société Mathématique de France* 26, 195–216 (1898).

26. A. Hatcher. *Algebraic topology*. Cambridge Univ. Press (2001).

27. J.C. Hausmann et al. On the Vietoris–Rips complexes and a cohomology theory for metric spaces. *Annals of Mathematics Studies* 138, 175–188 (1995).

28. H. Hirai. Characterization of the distance between subtrees of a tree by the associated tight span. *Annals of Combinatorics* **10**(1), 111–128 (2006).

29. J.R. Isbell. Six theorems about injective metric spaces. *Commentarii mathematici Helvetici* 39, 65–76 (1964). http://eudml.org/doc/139281

30. P. Joharinad and J. Jost. Topology and curvature of metric spaces. *Adv. Math.* **356**, 106813 (2019).

31. P. Joharinad and J. Jost. Topological representation of the geometry of metric spaces. *arXiv preprint arXiv:2001.10262* (2020).

32. J. Jost. Nonpositive curvature: Geometric and analytic aspects. Birkhäuser (1997).

33. M.A. Khamsi, H. Knaust, N.T. Nguyen and M.D. O'Neill. Λ-hyperconvexity in metric spaces. Nonlinear Anal. **43**, 21–31 (2000).

34. W. Klingenberg. *Riemannian geometry*. Walter de Gruyter (1982).

35. U. Lang. Injective hulls of certain discrete metric spaces and groups. *Journal of Topology and Analysis* **5**(3), 297–331 (2013).

36. S. Lim, F. Mémoli and O.B. Okutan. Vietoris–Rips persistent homology, injective metric spaces, and the filling radius. https://arxiv.org/abs/2001.07588v3 (2021).

37. J. Lindenstrauss. On the extension property for compact operators. *Bulletin of AMS* **68**, 484–487 (1962).

38. Ueber die Hypothesen, welche der Geometrie zu Grunde liegen. Edited with a commentary by J. Jost, Klassische Texte der Wissenschaft, Springer, Berlin (2013).

39. B. Riemann. On the hypotheses which lie at the bases of geometry Translated by W.K. Clifford, edited with a commentary by J. Jost, Classic Texts in the Sciences, Birkhäuser (2016).

40. B. Sturmfels and J. Yu. Classification of six-point metrics. The electronic Journal of Combinatorics 11(1), (2004).

41. H. von Mangoldt. Ueber diejenigen Punkte auf positiv gekrümmten Flächen, welche die Eigenschaft haben, dass die von ihnen ausgehenden geodätischen Linien nie aufhören, kürzeste Linien zu sein. *Crelle's Journal (J. Reine Angew. Math.)* **91**, 23–52 (1881).

42. A. Zomorodian and G. Carlsson. Computing persistent homology. *Discrete & Computational Geometry* 33(2), 249–274 (2005).

A Chapter About Asymptotic Geometric Analysis: Isomorphic Position of Centrally Symmetric Convex Bodies

Vitali Milman

Personal Note

In 1985 Joram Lindenstrauss and I started a new project. We began to collect and publish notes from our Israeli seminar on Geometric Aspect of Functional Analysis, GAFA in short. This was the reason why I first met a young, clever, energetic woman, Catriona Byrne. I think she had just started working for Springer. It would have been very difficult to organize and start publishing these notes without someone's help, and cooperation with a young energetic person ready to help turned out to be decisive for the success of the project.

By the way, the nickname GAFA was suggested by my then PhD student Haim Wolfson, a recently retired very famous professor in Computer Science. The actual name of the seminar (Geometric Aspect of Functional Analysis) was created to fit this nickname GAFA. Subsequently, the same nickname moved to GAFA journal, but with a different interpretation of this abbreviation (Geometric And Functional Analysis). The impact of this series turned out to be huge. To date, 13 volumes of this series have been released under the auspices of Springer, and three more under some other publishers. Under the influence and tutelage of GAFA Seminar Notes, a new direction within Functional Analysis arose, which turned into a new direction in mathematics, Asymptotic Geometric Analysis, abbreviated to AGA, but the publication of articles on this topic continued under the same traditional name GAFA Seminar Notes. For example, 50 works coauthored by Jean Bourgain were published in this series.

I thank Catriona Byrne most sincerely for helping to create and support this very important forum.

V. Milman (✉)
The School of Mathematical Sciences, Tel Aviv University, Tel Aviv, Israel
e-mail: milman@post.tau.ac.il

J.-M. Morel, B. Teissier (eds.), *Mathematics Going Forward*, Lecture Notes
in Mathematics 2313, https://doi.org/10.1007/978-3-031-12244-6_15

A Few Words on Asymptotic Geometric Analysis

Asymptotic Geometric Analysis (AGA) studies properties of geometric objects, such as normed spaces, convex bodies, or convex functions on finite-dimensional domains, when the dimensions of these objects increase to infinity. The asymptotic approach reveals many very novel phenomena which influence other fields in mathematics, especially where a large data set is of main concern, or the number of parameters becomes uncontrollably large. One of the important features of this relatively new theory is in developing tools which allow one to study high-parametric families. Among the tools developed in this theory are measure concentration, thin-shell estimates, stochastic localization, the geometry of Gaussian measures, volume inequalities for convex bodies, symmetrizations, and functional versions of geometric notions and inequalities (see [1] and [2]).

This field started on the border between geometry and functional analysis in the 80s and 90s. In this field, isometric problems that are typical for geometry in low dimensions are substituted by an "isomorphic" point of view, and an asymptotic approach (as dimension tends to infinity) is introduced. Geometry and analysis meet here in a non-trivial way. One central theme of this subject is the interaction of randomness and pattern. At first glance, life in a high dimension seems to mean the existence of multiple "possibilities", so one may expect an increase in the diversity and complexity as dimension increases. However, the concentration of measure and effects caused by convexity show that this diversity is compensated, and order and patterns are created for arbitrary convex bodies in the mixture caused by high dimensionality.

I will present now one recent development from AGA.

Isomorphic Position of a Convex Body

For simplicity of the exposition I will consider only centrally symmetric convex bodies in n-dimensional real space, i.e. convex compact K such that $K = -K$ and with non-empty interior. Of course, we may think of such K as the unit ball of some normed space X and, to emphasize this, we will write $K := K(X)$. For any non-degenerate linear map u in $\mathbb{R}^n$, of course, uK is the unit ball of isometrically the same normed space. However, geometrically it is a different body in $\mathbb{R}^n$. We call such a body a position of K. To every K there is associated with it a family of very interesting ellipsoids, which reflect, actually, different hidden symmetries in K. For example, there is a maximum volume ellipsoid inscribed in K (we call it John's ellipsoid), or a minimal volume ellipsoid containing K (the so-called Löwner's ellipsoid). There are also two ellipsoids of inertia, Legendre and Binet ellipsoids, and many others. See, e.g. [1]. We say that K is considered in John position if the John's ellipsoid is the standard Euclidean ball of the space. Similarly, we use the terminology of Löwner position of K, and if the Legendre ellipsoid of

K is the Euclidean ball we call such a position the isotropic position of K. In the asymptotic study of normed spaces and convex bodies, by increasing dimension to infinity the role of selected positions is crucial. Different remarkable properties of convex bodies (one may call them hidden symmetries) are recovered by considering them in different positions specially selected for different goals. We understand this part of the theory very well now.

However, many very central problems of the Asymptotic Geometry of high-dimensional convex bodies are still open and I would suggest here an additional "step of freedom" in attacking them. These reflections are inspired by two results. One of them has been known for a relatively long time. This is the result of Klartag from 2006. Another result is by Emanuel Milman and is very recent.

To solve some specific problem (let us call it Problem X) of the Asymptotic Theory, we will ask if there is a universal constant C such that for every dimension n and every convex body K (from our family) in $\mathbb{R}^n$ one may find another body T (from the family, i.e. centrally symmetric convex body) and the Banach–Mazur distance at most C from K and such that the Problem X would have a solution for T. Such a T we will call now the isomorphic position of K.

Isomorphic Version of Bourgain's Slicing Problem

Now we will discuss the remarkable result of Klartag from 2006. We use the survey [15], which we suggest the reader should consult for more details. The problem under discussion is the following

Problem 1 Let $K, T \subseteq \mathbb{R}^n$ be centrally-symmetric convex bodies such that $\mathrm{Vol}_{n-1}(K \cap \theta^{\perp}) \leq \mathrm{Vol}_{n-1}(T \cap \theta^{\perp})$ for all $\theta \in S^{n-1}$. Does it follow that $\mathrm{Vol}_n K \leq C \cdot \mathrm{Vol}_n T$ for some universal constant C?

The assumption of central-symmetry is not very essential, however, for simplicity of presentation we will assume it in all results. Problem 1 is in fact equivalent to the following:

Problem 2 Let $K \subset \mathbb{R}^n$ be a convex set of volume one. Does there exist a hyperplane $H \subset \mathbb{R}^n$ such that

$$\mathrm{Vol}_{n-1}(K \cap H) > 1/C$$

for some universal constant $C > 0$, independent of the dimension n?

This is known as Bourgain's slicing problem (from 1985, see [3]). A positive answer would have important consequences in convex geometry. In some sense the slicing problem, also called the hyperplane conjecture, is the "opening gate" to a better understanding of uniform measures in high dimensions. The problem is still open. Bourgain [4] provided the estimate $Cn^{(1/4)} \log n$ (instead of just C), then Klartag [12] improved it to $Cn^{(1/4)}$ and only recently Chen [8] improved it to $C(\epsilon)n^\epsilon$ for any $\epsilon > 0$.

However, Klartag found another approach to the problem, slightly modifying the question. To formulate Klartag's result we introduce the isotropic constant of the body K. Let $K \subset \mathbb{R}^n$ be a convex body. Denote by $\mathrm{Cov}(K)$ the covariance matrix of a random vector that is distributed uniformly in K. The *isotropic constant* of K is defined as

$$L_K = \frac{\det(\mathrm{Cov}(K))^{\frac{1}{2n}}}{\mathrm{Vol}_n(K)^{\frac{1}{n}}}.$$

Is it true that $L_K < C$, for some universal constant $C > 0$, independent of the dimension? This question is equivalent to Problem 2. Klartag proved the following theorem, the so-called isomorphic version of the slicing problem:

Theorem 1 (B. Klartag (2006)) *Let $K \subset \mathbb{R}^n$ be a convex body and $0 < \epsilon < 1$. Then there exists a convex body $T \subset \mathbb{R}^n$ such that*

(i) $(1 - \epsilon)T \subseteq K \subseteq (1 + \epsilon)T$.
(ii) $L_T < C/\sqrt{\epsilon}$, where $C > 0$ is a universal constant.

Later, in 2018, Klartag additionally proved that the body T from the theorem can be assumed to be a projective image of K [13].

So, the problem has a positive solution but in an isomorphic sense: there is an isomorphic position for which Problem 1 is solved.

Isomorphic Version of Log-Brunn–Minkowski Inequality

We will move now to another recent result about the Problem due to K.J. Böröczky, E. Lutwak, D. Yang, and G. Zhang (BLYZ for short). In this section we heavily use the article by Emanuel Milman [20], and also his help in composing the following text.

One important question in contemporary Brunn–Minkowski theory is that of existence and uniqueness in the L^p-Minkowski problem for $p \in (-\infty, 1)$: given a finite non-negative Borel measure μ on the Euclidean unit-sphere $\mathbb{S} = S^{n-1}$, determine conditions on μ which ensure the existence and/or uniqueness of a convex body K in $\mathbb{R}^n$ so that:

$$S_p K := h_K^{1-p} S_K = \mu. \tag{1}$$

Here h_K and S_K denote the support function and surface-area measure of K, respectively. When $h_K \in C^2(\mathbb{S})$,

$$S_K = \det(D^2 h_K)\mathrm{m},$$

where $\mathfrak{m}$ is the induced Lebesgue measure on $\mathbb{S}$, $D^2 h_K = \nabla^2_{\mathbb{S}} h_K + h_K \delta_{\mathbb{S}}$ and $\nabla_{\mathbb{S}}$ is the Levi-Civita connection on $\mathbb{S}$ with its standard Riemannian metric $\delta_{\mathbb{S}}$. Consequently, (1) is a Monge–Ampère-type equation.

The case $p = 1$ above corresponds to the classical Minkowski problem of finding a convex body with prescribed surface-area measure; when μ is not concentrated on any hemisphere and its barycenter is at the origin, existence and uniqueness (up to translation) of K were established by Minkowski, Alexandrov and Fenchel–Jessen. The extension to general p was put forth and publicized by E. Lutwak [17] as an L^p-analog of the Minkowski problem for the L^p surface-area measure $S_p K = h_K^{1-p} S_K$ which he introduced. Existence and uniqueness in the class of origin-symmetric convex bodies, when the measure μ is even and not concentrated in a hemisphere, was established for $n \neq p > 1$ by Lutwak [17] and for $p = n$ by Lutwak–Yang–Zhang [19]. A key tool in the range $p \geq 1$ is the prolific L^p-Brunn–Minkowski theory, initiated by Lutwak [17],[18] following Firey [10], and developed by Lutwak–Yang–Zhang and others, which extends the classical $p = 1$ case. Recall that the L^p-Minkowski sum $a \cdot K_0 +_p b \cdot K_1$ of $K_0, K_1 \in \mathcal{K}$ $(a, b \geq 0)$ was defined by Firey for $p \geq 1$ [10], and extended by Böröczky–Lutwak–Yang–Zhang [5, 6] to all $p \in \mathbb{R}$, as the largest convex body (with respect to inclusion) L so that:

$$h_L \leq \left(a h_{K_0}^p + b h_{K_1}^p \right)^{1/p}$$

(with the case $p = 0$ interpreted as $h_{K_0}^a h_{K_1}^b$ when $a + b = 1$). Note that for $p \geq 1$ one has equality above, that the case $p = 1$ coincides with the usual Minkowski sum, and that for $p < 1$ the resulting convex body $a \cdot K_0 +_p b \cdot K_1$ is the Alexandrov body associated to the continuous function on the right-hand-side.

The case $p < 1$ turns out to be more challenging because of the lack of an appropriate L^p-Brunn–Minkowski theory. Existence, (non-)uniqueness and regularity under various conditions on μ were studied by numerous authors when $p < 1$ (from either side of the critical exponent $p = -n$). The case $p = 0$ is of particular importance as it corresponds to the *log-Minkowski problem* for the cone-volume measure

$$V_K := \frac{1}{n} h_K S_K = \frac{1}{n} S_0 K,$$

obtained as the push-forward of the cone-measure on ∂K onto $\mathbb{S}$ via the Gauss map, and having total mass $V(K)$, the volume of K. Being a self-similar solution to the isotropic Gauss curvature flow, the case $p = 0$ and $\mu = \mathfrak{m}$ of (1) describes the ultimate fate of a worn stone in a model proposed by Firey.

Let $\mathcal{K}$ denote the collection of convex bodies in $\mathbb{R}^n$ containing the origin in their interior, and let $\mathcal{K}_e$ denote the subset of origin-symmetric elements. In [6], Böröczky–Lutwak–Yang–Zhang showed that an *even* measure μ is the cone-volume measure V_K of an *origin-symmetric* convex body $K \in \mathcal{K}_e$ if and only if it satisfies a certain subspace concentration condition, thereby completely resolving the existence part of the *even* log-Minkowski problem. As put forth by Böröczky–Lutwak–Yang–Zhang in their influential work [5, 6] and further developed in [16], the uniqueness question is intimately related to the validity of a conjectured L^0- (or log-)Brunn–Minkowski inequality for origin-symmetric convex bodies $K, L \in \mathcal{K}_e$, which would constitute a remarkable strengthening of the classical $p = 1$ case.

Specifically, the following equivalence may be shown by combining the results of [5, 7, 16]. We denote by $\mathcal{K}_{+,e}^{2,\alpha}$ the subset of $\mathcal{K}_e$ having $C^{2,\alpha}$-smooth boundary and strictly positive curvature.

Theorem 2 (After Böröczky–Lutwak–Yang–Zhang, Kolesnikov–Milman and Chen–Huang–Li–Liu) *The following statements are equivalent for any fixed $p \in (-n, 1)$:*

(1) For any $q \in (p, 1)$, uniqueness holds in the even L^q-Minkowski problem for any $K \in \mathcal{K}_{+,e}^{2,\alpha}$:

$$\forall L \in \mathcal{K}_e \ , \quad S_q L = S_q K \ \Rightarrow \ L = K. \tag{2}$$

(2) The even L^p-Brunn–Minkowski inequality holds:

$$\forall \lambda \in [0, 1] \quad \forall K, L \in \mathcal{K}_e \quad V((1 - \lambda) \cdot K +_p \lambda \cdot L)$$

$$\geq \left((1 - \lambda) V(K)^{\frac{p}{n}} + \lambda V(L)^{\frac{p}{n}} \right)^{\frac{n}{p}}. \tag{3}$$

The case $p = 0$, called the even log-Brunn–Minkowski inequality, is interpreted in the limiting sense as:

$$V((1 - \lambda) \cdot K +_0 \lambda \cdot L) \geq V(K)^{1-\lambda} V(L)^{\lambda}.$$

(3) The even L^p-Minkowski inequality holds:

$$\forall K, L \in \mathcal{K}_e \quad \frac{1}{p} \int_{\mathbb{S}} h_L^p \mathrm{d}S_p K \geq \frac{n}{p} V(K)^{1-\frac{p}{n}} V(L)^{\frac{p}{n}}. \tag{4}$$

The case $p = 0$, called the even log-Minkowski inequality, is interpreted in the limiting sense as:

$$\frac{1}{V(K)} \int_{\mathbb{S}} \log \frac{h_L}{h_K} \mathrm{d}V_K \geq \frac{1}{n} \log \frac{V(L)}{V(K)}.$$

Using Jensen's inequality in formulation (4) (or (3)), it is immediate to check that the above (equivalent) statements become stronger as p decreases. The restriction to origin-symmetric bodies is natural, and necessitated by the fact that no L^p-Brunn–Minkowski inequality nor uniqueness in the L^p-Minkowski problem can hold for general convex bodies when $p < 1$. Even when restricting to origin-symmetric bodies, it is easy to show that (3) or (4) are false for any $p < 0$, and that uniqueness in (2) does not hold for general $K, L \in \mathcal{K}_e$ and $q = 0$, as may be verified by testing two different centered parallelepipeds with appropriately chosen parallel facets.

Conjecture 1 (Böröczky–Lutwak–Yang–Zhang, "Even log-Brunn–Minkowski Conjecture") *Any (and hence all) of the above statements hold for origin-symmetric convex bodies in the "logarithmic case"* $p = 0$ *(and hence for all* $p \in [0, 1)$ *as well).*

A confirmation of this conjecture would constitute a dramatic improvement over the classical Brunn–Minkowski theory for the subfamily of origin-symmetric convex bodies, which had gone unnoticed for over a century. The conjecture is known to hold in the plane [5], but remains open in general for $n \geq 3$.

Various partial results are known regarding the BLYZ conjecture (see e.g. [16, 20]). The main result in [16] confirmed the *local* uniqueness in the even L^p-Minkowski problem (2) for all $K \in \mathcal{K}_{+,e}^{2,\alpha}$ and $p \in (p_0, 1)$ for $p_0 := 1 - \frac{c}{n^{3/2}}$. In [7], Chen–Huang–Li–Liu established a local-to-global principle for the uniqueness question, and deduced (2) and (4) for all $K \in \mathcal{K}_{+,e}^{2,\alpha}$ and $p \in (p_0, 1)$. In fact, thanks to recent progress on the KLS conjecture due to Y. Chen [8], the estimate from [16] immediately improves to $p_0 = 1 - \frac{c_\epsilon}{n^{1+\epsilon}}$ for any $\epsilon > 0$.

In [20], the following isomorphic version of the conjecture regarding uniqueness in the even log-Minkowski problem was resolved by E. Milman. We denote by $d_G(K, L)$ the geometric distance between two origin-symmetric bodies K, L, namely $d_G(K, L) := \inf\{ab > 0 \; ; \; \frac{1}{b}K \subset L \subset aK\}$.

Theorem 3 (E. Milman, Isomorphic Log-Minkowski) *For any* $\bar{K} \in \mathcal{K}_e$*, there exists a* $\tilde{K} \in \mathcal{K}_{+,e}^\infty$ *with:*

$$d_G(\bar{K}, \tilde{K}) \leq 8,$$

so that for any $T \in GL_n$*, the even log-Minkowski problem for* $K = T(\tilde{K})$ *has a unique solution:*

$$\forall L \in \mathcal{K}_e \; , \quad V_L = V_K \Rightarrow L = K,$$

and the even log-Minkowski inequality holds for K*:*

$$\forall L \in \mathcal{K}_e \quad \frac{1}{V(K)} \int_{\mathbb{S}} \log \frac{h_L}{h_K} dV_K \geq \frac{1}{n} \log \frac{V(L)}{V(K)},$$

with equality if and only if $L = cK$ *for some* $c > 0$*.*

The constant 8 obtained in the isomorphic version above is the worst case behavior for a general $\bar{K} \in \mathcal{K}_e$, when $D = d_{BM}(\bar{K}, B_2^n)$ may be as large as John's upper bound $\sqrt{n}$. However, whenever $D \ll \sqrt{n}$, a slightly finer analysis yields an *isometric* version of the above results, where one only perturbs $\bar{K}$ by at most $\gamma = 1 + \epsilon$, with $\epsilon = C\frac{\sqrt{D}}{\sqrt[4]{n}}$.

Theorem 3 is a remarkable result about the existence of an isomorphic position in a localized version of the log-Brunn-Minkowski inequality problem of BLYZ (Conjecture 1 above). At the same time Emanuel Milman does not know if the following problem has a positive solution:

Problem 3 (E. Milman) There is a universal constant C such that for every convex compact body K there is an isomorphic version K' such that $d_G(K, K') < C$ and such that for any two bodies K and L their isomorphic version K' and L' satisfy inequalities (3) for $p = 0$, i.e. the log-Brunn–Minkowski inequality.

More Isomorphic Versions of Well-Known Problems of AGA

In the article by B. Klartag and V. Milman [15] we listed a number of problems which are connected with the slicing problem of Bourgain and either follow from it, in the case of a positive solution, or imply it if they could be positively solved. In each of these problems one may ask if their isomorphic versions are true. Let me list some of them:

(i) the thin shell conjecture would imply the hyperplane conjecture (see [9]);
(ii) the Kannan, Lovasz, Simonovich (KLS) isoperimetric conjecture [11];
(iii) Mahler's conjecture on the low bound for the product of the volumes of the convex body and its polar (see [13] for connections with the slicing problem);
(iv) problems on "quick Steiner symmetrizations" (see [14])

I will not introduce these problems here. I refer the reader to [15] for their exact formulations and to the books [1] and [2] for detailed discussions of these major problems of the theory. As an example, let me just formulate some of these problems in their isomorphic form.

(ii) Isomorphic KLS Problem: Do universal constants C and C' exist such that for every centrally-symmetric convex body K there exist another centrally-symmetric body K' such that Banach–Mazur distance $d(K, K') < C$ and KLS conjecture is true for K' with a constant C' ?

(iii) Isomorphic Mahler problem: Does a universal constant C exist such that for any centrally-symmetric convex body K there is a body K' such that $d(K, K') < C$ and the Mahler volume of K' is greater than (or equal) to the Mahler volume of the cube (of the same dimension).

Many other problems of AGA may be reformulated the same way, and all of them, besides what is written above, are open.

My deep thanks to Boaz Klartag and Emanuel Milman for providing me their advice and written TeX texts on their results and to Miriam Hercberg for her help in editing the English of this article and her TeX typing.

References

1. S. Artstein-Avidan, A. Apostolos and V. Milman. *Asymptotic geometric analysis, Part I.* Mathematical Surveys and Monographs **202**, American Mathematical Society, Providence, RI (2015).
2. S. Artstein-Avidan, A. Apostolos and V. Milman. *Asymptotic geometric analysis, Part II.* Mathematical Surveys and Monographs **202**, American Mathematical Society, Providence, RI (2022).
3. J. Bourgain. On high-dimensional maximal functions associated to convex bodies. *Amer. J. Math.* **108**(6), 1467–1476 (1986).
4. J. Bourgain. On the distributions of polynomials on high-dimensional functions convex sets. In: *Geometric Aspects of Functional Analysis, Israel Seminar (1988–90)*, pp. 127–137, Springer Lecture Notes in Math. **1469** (1991).
5. K.J. Böröczky, E. Lutwak, D. Yang and G. Zhang. The log-Brunn-Minkowski inequality. *Adv. Math.* **231**(3-4), 1974–1997 (2012).
6. K.J. Böröczky, E. Lutwak, D. Yang and G. Zhang. The logarithmic Minkowski problem. *J. Amer. Math. Soc.* **26**(3), 831–852 (2013).
7. S. Chen, Y. Huang, Q.-r. Li and J. Liu. The L_p-Brunn–Minkowski inequality for $p < 1$. *Adv. Math.* **368**, 107–166 (2020).
8. Y. Chen. An almost constant lower bound of the isoperimetric coefficient in the KLS conjecture. *Geom. Func. Anal.* **31**(1), 34–61 (2021).
9. R. Eldan and B Klartag. Approximately Gaussian marginals and the hyperplane conjection. In: *Concentration, functional inequalities and isoperimetry*, pp. 55–68, Contemp. Math., Amer. Math. Soc. **545**, Providence RI (2011).
10. W.J. Firey. p-means of convex bodies. *Math. Scand.* **10**, 17–24 (1962).
11. R. Kannan, L. Lovász and M. Simonovitz. Isoperimetric problems for convex bodies and a localization lemma. *Discrete Comput. Geom.* **13**(3–4), 541–559 (1995).
12. B. Klartag. On convex perturbations with a bounded isotropic constant. *Geom. Func. Anal.* **16**(6), 1274–1290 (2006).
13. B. Klartag. Isotropic constants and the Mahler volumes. *Adv. Math.* **330**, 74–180 (2018).
14. B. Klartag and V. Milman. Rapid Steiner symmetrization of most of the convex and the slicing problem. *Combin. Probab. Comput.* **14**(5-6), 829–843 (2005).
15. B. Klartag and V. Milman. The slicing problem by Bourgain. To appear in *Analysis at Large*, a collection of articles in memory of Jean Bourgain, edited by A. Avila, M. Rassias, Y. Sinai, Springer (2022).
16. A.V. Kolesnikov and E. Milman. Local L^p-Brunn–Minkowski inequalities for $p < 1$. arxiv.org/abs/1711.01089, to appear in *Mem. Amer. Math. Soc.* (2017).
17. E. Lutwak. The Brunn–Minkowski–Firey theory. I. Mixed volumes and the Minkowski problem. *J. Differential Geom.* **38**(1), 131–150 (1993).
18. E. Lutwak. The Brunn–Minkowski–Firey theory. II. Affine and geominimal surface areas. *Adv. Math.* **118**(2), 244–294 (1996).
19. E. Lutwak, D. Yang and G. Zhang. On the L_p-Minkowski problem. *Trans. Amer. Math. Soc.* **356**(11), 4359–4370 (2004).
20. E. Milman. Centro-affine differential geometry and the log-Minkowski problem. arxiv.org/abs/2104.12408 (2021).

A Mysterious Tensor Product in Topology

Ieke Moerdijk

The students of my generation had to survive without the internet and mobile phones, and depended on books and real paper to write on. As undergraduates in mathematics, we were always carrying yellow books around, and Springer-Verlag had a big part in our mathematical development. A little later, when I was a PhD student, two Springer Lecture Notes had a lasting influence on my own mathematical work: *Homotopy Invariant Algebraic Structures on Topological Spaces* by Boardman and Vogt [1] and *The Geometry of Iterated Loop Spaces* by Peter May [6]. These two books together shaped the foundation of the theory of operads about which I will write below.

My contacts with Catriona go back to the preparation and publishing of "Sheaves in Geometry and Logic" with Saunders MacLane in the early 1990s. This period was also the beginning of the use of e-mail, and it is interesting and entertaining to see how e-mail customs and etiquette have changed over the years. I had my first e-mails to Catriona typed by a secretary, and Catriona had several people working for her who wrote on her behalf, all using an e-mail account and address under the name "Byrne". A few years after that, together with Albrecht Dold, Catriona helped me to get SLN 1616 into publishable shape. In more recent years, she remained most helpful in several matters, and I wish to thank her for that.

Now I would like to come back to Boardman and Vogt, and May, and talk about some mathematics. To begin with, let me remind you that a (coloured) *operad* P consists of a set $C = \mathrm{colours}(P)$ of *colours*, and for each sequence $(c_1, \ldots, c_n; c)$ of elements of C (where $n \geq 0$) a set of *operations* $P(c_1, \ldots, c_n; c)$, to be thought of as taking inputs of "types" $c_1, \ldots, c_n$, respectively, to an output of type c. Moreover,

I. Moerdijk (✉)
Universiteit Utrecht, Utrecht, The Netherlands
e-mail: i.moerdijk@uu.nl

© The Author(s), under exclusive license to Springer Nature Switzerland AG 2023
J.-M. Morel, B. Teissier (eds.), *Mathematics Going Forward*, Lecture Notes in Mathematics 2313, https://doi.org/10.1007/978-3-031-12244-6_16

209

P is equipped with several structure maps for symmetry and composition, such as

$$P(c_1, c_2; c) \xrightarrow{\sim} P(c_2, c_1; c)$$

$$P(c_1, c_2; c) \times P(d_1, d_2, d_3; c_1) \times P(e_1, e_2; c_2) \to P(d_1, d_2, d_3, e_1, e_2; c)$$

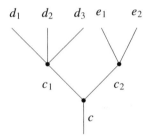

and a unit element $1_c \in P(c; c)$ for each colour c. These are to satisfy several natural conditions, such as an associativity law for composition. If $P(c_1, \ldots, c_n; c)$ is empty unless $n = 1$, this simply defines the notion of a (small) category. And if C consists of just one element $*$, one calls P uncoloured and writes $P(n)$ for $P(c_1, \ldots, c_n; c)$ where each c_i and c are necessarily $*$.

For a coloured operad P, a P-algebra A is a family of sets $\{A_c : c \in C\}$ equipped with maps

$$P(c_1, \ldots, c_n; c) \times A_{c_1} \times \cdots \times A_{c_n} \to A_c,$$

(for all sequences of colours $c_1, \ldots, c_n, c$), compatible with the structure of P mentioned above (symmetry, associativity, units). For example, if P is the uncoloured operad for which each $P(n)$ consists of a single point, a P-algebra is simply a commutative monoid, and one usually writes **Comm** for this operad. The collection Σ_n, $n \geqslant 0$ of symmetric groups also has the structure of an operad; the composition $\Sigma_n \times \Sigma_{k_1} \times \cdots \times \Sigma_{k_n} \to \Sigma_k$ for $k = k_1 + \ldots + k_n$ is defined by replacing the non-zero entries in an $n \times n$ permutation matrix representing an element of Σ_n by the permutation matrices representing given elements of $\Sigma_{k_1}, \ldots, \Sigma_{k_n}$ respectively, thus yielding a $k \times k$-permutation matrix. An algebra for this operad is an associative monoid, and one usually writes **Ass** for this operad (although Σ would obviously have been a good name as well).

These P-algebras form a category $\mathrm{Alg}(P)$, or $\mathrm{Alg}(P, \mathbf{Sets})$ to emphasize that we consider algebras in the category of sets. One can similarly define a category $\mathrm{Alg}(P, \mathcal{E})$ of algebras in any category with products, as long as expressions like "$P(c_1, \ldots, c_n; c) \times A_{c_1} \times \cdots \times A_{c_n}$" occurring in the definition of P-algebra make sense in $\mathcal{E}$. (This is the case, for example, when it is possible to view the set $P(c_1, \ldots, c_n; c)$ as an object of $\mathcal{E}$ through a suitable embedding of sets as "discrete objects" in $\mathcal{E}$.) So, if Q is another operad with set of colours $D = \text{colours}(Q)$, one can construct a category $\text{Alg}(P, \text{Alg}(Q, \textbf{Sets}))$. This category is itself a category of algebras over a new operad $P \otimes Q$ with set of colours $C \times D$; in other words,

$$\text{Alg}(P \otimes Q, \textbf{Sets}) = \text{Alg}(P, \text{Alg}(Q, \textbf{Sets})).$$

This operad $P \otimes Q$ is known as the "Boardman–Vogt tensor product" of P and Q and was first introduced in SLN [1]. It is possible to describe $P \otimes Q$ explicitly by generators and relations. In particular, if P and Q are *free* operads defined by trees S and T, i.e. $P = \text{Free}(S)$ and $Q = \text{Free}(T)$, then $P \otimes Q$ is defined by glueing free operads $\text{Free}(R)$ together, where R ranges over all the *shuffles* of the two trees S and T. A minimal example can be pictured as follows:

$$P \otimes Q = \text{Free}(R_1) \cup \text{Free}(R_2) \cup \text{Free}(R_3)$$

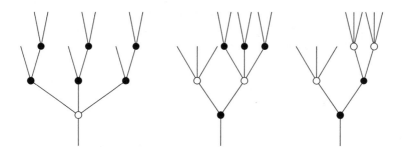

See [5] for details. The case of *categories* (viewed as operads with unary operations only, as above) corresponds to shuffling linear trees. For two trees with n and m vertices, respectively, there are $\binom{n+m}{n}$ such shuffles, as everybody who encountered products of simplicial complexes will be aware of. However, it seems impossible to find a nice closed formula for the number of shuffles of trees that aren't linear (see *loc. cit.*)

Turning to operads in topological spaces, the most famous ones are probably the (uncoloured) operads $\mathcal{C}_d$ of "little d-cubes" (for $d \geqslant 1$ the dimension of the cubes involved). Specifically, $\mathcal{C}_d(n)$ is the space of sequences of n rectilinear embeddings of a d-dimensional cube into the unit cube $[0, 1]^d$, with disjoint interiors. Here is a picture of a point in the space $C_2(3)$:

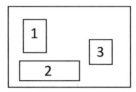

By composing such embeddings, one obtains maps

$$C_d(n) \times C_d(k_1) \times \cdots \times C_d(k_n) \to C_d(k)$$

for $k = k_1 + \ldots + k_n$, representing the composition operation of the operad C_d. These operads derive their fame from the fact that C_d-algebras (in spaces) describe d-fold loop spaces, as discussed in detail in [6]. Since the d-fold loop space of a space which is itself an e-fold loop space is evidently a $(d + e)$-fold loop space, it is natural to expect that $C_d \otimes C_e$ is closely related to C_{d+e}. This is indeed the case, as these two operads have been proved to be equivalent up to homotopy, a result known as Dunn's additivity theorem [3]. Although a positive result, Dunn's additivity theorem show at the same time that the tensor product of topological spaces behaves rather badly under weak homotopy equivalence. For example, there is a map $C_1 \to$ **Ass** or operads, assigning to a point in $C_1(n)$, i.e. a sequence of n numbered disjoint intervals in the unit interval, the permutation representing the order in which there intervals occur:

This map is weak homotopy equivalence of operads, i.e. each $C_1(n) \to$ **Ass**$(n) = \Sigma_n$ is one of spaces. On the other hand, **Ass** $\otimes$ **Ass** $=$ **Comm** by the Eckmann–Hilton trick, while $C_1 \otimes C_1 \simeq C_2$ describes double loop spaces and is very different from **Comm**.

There are variations of the operad C_d which also describe d-fold loop spaces up to homotopy, leading to a notion of "E_d-operad": An operad P is said to be an E_d-operad if it can be related to C_d by a zigzag of weak homotopy equivalences between operads,

$$C_d \leftarrow \cdot \to \cdot \leftarrow \cdot \to \ldots \leftarrow P.$$

These E_d-operads often arise as combinatorial versions of C_d, for example the one used by McClure and Smith in their proof of the Deligne conjecture [7, 8]. From a mathematical point of view, however, the notion of an E_d-operad is a rather unusual one, as it does not give any structural properties for an operad to be an E_d-operad. This becomes particularly awkward when one considers the tensor product of E_d-operads, since the tensor product is not invariant under weak equivalence, as we have just seen. So one may ask for which particular "models" of E_d-operads the additivity of Dunn holds. This is problem to which Rainer Vogt devoted much of his work (see e.g. [4]), but which remains largely unsolved.

An alternative approach is to replace the Boardman–Vogt tensor product of topological operads by a "derived" one which is invariant under weak equivalence. Denoting such a derived tensor product by $\widehat{\otimes}$, one way to construct it explicitly is as the pushout

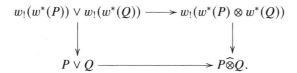

Here P and Q are topological operad with sets of colours C and D, say, and $P \vee Q$ denotes the coproduct in the category of operads with $C \times D$ as set of colours (where we first pull P and Q back along the two projections). Furthermore, the adjoint functors $w_!$ and w^* are the ones establishing a Quillen equivalence between topological operads and dendroidal sets [2]. The symbol $\otimes$ in the diagram refers to the tensor product of dendroidal sets. Since this tensor product is much better behaved, especially for "closed" operads like C_d where there is a unique nullary operation (of each colour), one can prove that $P\widehat{\otimes}Q$ is invariant under weak equivalence in each variable separately, at least for operads with free Σ_n-action on $P(n)$ (respectively $Q(n)$), for each n; see [2]. So $P\widehat{\otimes}Q$ describes quite a good tensor product, from the point of view of topology. It comes equipped with a map $P\widehat{\otimes}Q \to P \otimes Q$ expressing that $P\widehat{\otimes}Q$ is a "thick" version of the original Boardman–Vogt tensor product. It would be interesting to know for which operads this derived tensor product is equivalent to the original one of Boardman and Vogt. (This is the case for "cofibrant" operads P and Q, but cofibrant operads are hard to come by and rarely occur naturally.) It would also be interesting to know whether this derived tensor product $\widehat{\otimes}$ satisfies Dunn's additivity property for certain types of models of E_d-operads.

Thus, natural as the tensor product may seem from the point of view of algebra (remember the equation $\mathrm{Alg}(P, \mathrm{Alg}(Q, \mathbf{Sets})) = \mathrm{Alg}(P \otimes Q, \mathbf{Sets})$), it is surrounded by many unanswered questions: combinatorial ones about the number of shuffles, questions about invariance under weak equivalence, and questions about additivity.

References

1. J.M. Boardman and R.M. Vogt. *Homotopy Invariant Algebraic Structures on Topological Spaces*. Lecture Notes in Mathematics, Springer-Verlag, Berlin-Heidelberg (1973).
2. D.C. Cisinski and I. Moerdijk. Dendroidal sets and simplicial operads. *J. Topol.* **6**(3), 705–756 (2013).
3. G. Dunn. Tensor product of operads and iterated loop spaces. *J. Pure Appl. Algebra* **50**(3), 237–258 (1988). https://doi.org/10.1016/0022-4049(88)90103-X
4. Z. Fiedorowicz and R. Vogt. An Additivity Theorem for the Interchange of E_n Structures. *Adv. Math.* **273**, 421–484 (2011). https://doi.org/10.1016/j.aim.2014.10.020
5. E. Hoffbeck and I. Moerdijk. Shuffles of trees. *European J. Combin.* **71**, 55–72 (2018).
6. J.P. May. *The Geometry of Iterated Loop Spaces*. Lecture Notes in Mathematics, Springer-Verlag, Berlin-Heidelberg (1972).
7. J. McClure and J. Smith. Multivariable cochain operations and little n-cubes. *J. Amer. Math. Soc.* **16**, 681–704 (2001). https://doi.org/10.1090/S0894-0347-03-00419-3
8. J. McClure and J. Smith. Cosimplical objects and little n-cubes, I. *Amer. J. Math.* **126**, 1109–1153 (2002). https://doi.org/10.1353/ajm.2004.0038

Part V
Groups

Martin R. Bridson's article *Profinite rigidity and free groups* offers an insightful and reader-friendly presentation of results and problems related to profinitely rigid groups, those which are determined up to isomorphism by their profinite completion.

Toshiyuki Kobayashi's article *Conjectures on reductive homogeneous spaces* addresses some conjectures and problems to which the author has no solution in the areas of: Discrete series for non-symmetric homogeneous spaces G/H; discontinuous groups for G/H beyond the Riemannian setting; and analysis on non-Riemannian locally homogeneous spaces.

The article *On the algebraic K-theory of Hecke algebras* of Arthur Bartels and Wolfgang Lück concerns totally disconnected groups G which contain a normal compact open subgroup L such that G/L is cyclic. It explains how to compute the algebraic K-groups of the Hecke algebra of G from those of L, thus confirming in that case a conjecture of Farrel–Jones.

The article of Mark Pollicott and Polina Vytnova *Groups, drift, and harmonic measure* gently but firmly takes the reader from the definition of Fuchsian groups to random walks on their Cayley graphs and the associated harmonic measure on the unit circle. The authors' aim is to lead the reader to the question of the absolute continuity of such harmonic measures with respect to Lebesgue measure. This question is connected to the Hausdorff dimension of these measures, which in turn is related to their Avez random walk entropy.

The article *Groupes de Coxeter finis: centralisateurs d'involutions* of Jean-Pierre Serre is an exemplary study in the theory of finite groups. It is a description of the centralizers of involutions of finite Coxeter groups.

Conjectures on Reductive Homogeneous Spaces

Toshiyuki Kobayashi

1 Introduction

I have worked on various parts of mathematics, and "Symmetry" is a key word to create new interactions among these different disciplines.

It was when I was an undergraduate student that Springer opened an office in Tokyo within walking distance of our mathematics building. In 1989 Catriona visited Japan for seven weeks to promote editorial activities in Japan, including the forthcoming publication of the Proceedings of ICM 90 Kyoto, though I was not aware of it at that time. In 2006, the Mathematical Society of Japan relaunched the 3rd series of *Japanese Journal of Mathematics (JJM)* and started to collaborate with Springer as a global partner. Since then, as the founding Editor, I have been indebted to Springer's staff, whom Catriona mentored. She frequently visited the Takagi Lectures, and we talked about various topics over coffee. Each time, I was impressed by how much she cared about what mathematicians think really important. Her work created new interactions among mathematicians from different disciplines. I admire her for all her invaluable contributions to the mathematics community over the years.

In this paper, I would like to address some conjectures and problems in the area of "analysis of symmetries" in which I have been deeply involved, but to which I do not have a complete answer.

(1) Discrete series for non-symmetric homogeneous spaces G/H;
(2) Discontinuous groups Γ for G/H beyond the Riemannian setting;
(3) Analysis on non-Riemannian locally homogeneous spaces $\Gamma \backslash G/H$.

T. Kobayashi (✉)
Graduate School of Mathematical Sciences, The University of Tokyo, Tokyo, Japan
e-mail: toshi@ms.u-tokyo.ac.jp

© The Author(s), under exclusive license to Springer Nature Switzerland AG 2023   217
J.-M. Morel, B. Teissier (eds.), *Mathematics Going Forward*, Lecture Notes
in Mathematics 2313, https://doi.org/10.1007/978-3-031-12244-6_17

These three topics are discussed in Sects. 3–5, respectively, using a common setting which is explained in Sect. 2 with simple examples.

2 Basic Setting

Throughout this paper, our basic geometric setting will be as follows.

Setting 2.1 G is a real reductive linear Lie group, H is a closed proper subgroup which is reductive in G, and $X := G/H$.

A distinguished feature of this setting is that the manifold X carries a *pseudo-Riemannian structure* with a 'large' isometry group, namely, the reductive group G acts transitively and isometrically on X. Such a pseudo-Riemannian structure is induced from the Killing form B if G is semisimple. For reductive G, one can take a maximal compact subgroup K of G such that $H \cap K$ is a maximal compact subgroup of H, and a G-invariant symmetric bilinear form B on the Lie algebra $\mathfrak{g}$ of G such that the Cartan decomposition $\mathfrak{g} = \mathfrak{k} + \mathfrak{p}$ is an orthogonal decomposition with respect to B and that B is negative definite on $\mathfrak{k}$ and is positive definite on $\mathfrak{p}$. Then B induces a G-invariant pseudo-Riemannian structure of signature (p, q) on X, where $p + q = \dim X$ and $q = \dim K/H \cap K$.

Very special cases of homogeneous spaces in Setting 2.1 include:

Example 2.1 (Semisimple Coadjoint Orbits) For a reductive Lie group G, one can identify the Lie algebra $\mathfrak{g}$ with its dual $\mathfrak{g}^*$ via B. The coadjoint orbit $\mathcal{O}_\lambda := \mathrm{Ad}^*(G)\lambda$ is called *semisimple, elliptic, hyperbolic,* or *regular* if the element in $\mathfrak{g}$ corresponding to λ has that property. We write $\mathfrak{g}_{ss}^*$, $\mathfrak{g}_{ell}^*$, $\mathfrak{g}_{hyp}^*$, or $\mathfrak{g}_{reg}^*$ for the collection of such elements λ, respectively. By definition, $\mathfrak{g}_{ell}^*$, $\mathfrak{g}_{hyp}^* \subset \mathfrak{g}_{ss}^*$. The isotropy subgroup of λ is reductive if $\lambda \in \mathfrak{g}_{ss}^*$, hence any semisimple coadjoint orbit $\mathcal{O}_\lambda$ gives an example of Setting 2.1.

For compact G, one has $\mathfrak{g}_{ell}^* = \mathfrak{g}_{ss}^* = \mathfrak{g}^*$ and $\mathcal{O}_\lambda$ is a generalized flag variety for any $\lambda \in \mathfrak{g}^*$; $\mathcal{O}_\lambda$ is a full flag variety iff $\lambda \in \mathfrak{g}_{reg}^*$.

The subclass $\{\mathcal{O}_\lambda : \lambda \in \mathfrak{g}_{ss}^*\}$ in Setting 2.1 plays a particular role in the unitary representation theory. The orbit philosophy due to Kirillov–Kostant–Duflo–Vogan suggests an intimate relationship between the set $\mathfrak{g}^*/\mathrm{Ad}^*(G)$ of coadjoint orbits and the set of equivalence classes of irreducible unitary representations of G (the *unitary dual* $\widehat{G}$):

$$\mathfrak{g}^*/\mathrm{Ad}^*(G) \risingdotseq \widehat{G}, \qquad \mathcal{O}_\lambda \leftrightarrow \pi_\lambda. \tag{1}$$

We recall that any coadjoint orbit carries a canonical symplectic form called the *Kirillov–Kostant–Souriau form*. Then the correspondence $\mathcal{O}_\lambda \mapsto \pi_\lambda$ is supposed to be a "geometric quantization" of the Hamiltonian G-manifold $\mathcal{O}_\lambda$ if such π_λ exists. This philosophy works fairly well for $\lambda \in \mathfrak{g}_{ss}^*$ satisfying an appropriate integral condition: loosely speaking, π_λ is obtained by a unitary induction from a parabolic subgroup (real polarization of the para-Hermitian manifold $\mathcal{O}_\lambda$) for $\lambda \in \mathfrak{g}^*_{\text{hyp}}$, in a Dolbeault cohomology space on the pseudo-Kähler manifold $\mathcal{O}_\lambda$ as a generalization of the Borel–Weil–Bott theorem (complex polarization of $\mathcal{O}_\lambda$) or alternatively by a cohomological parabolic induction (e.g., Zuckerman's derived functor module $A_\mathfrak{q}(\lambda)$) for $\lambda \in \mathfrak{g}^*_{\text{ell}}$, and by the combination of these two procedures for general $\lambda \in \mathfrak{g}^*_{\text{ss}}$, although there are some delicate issues about singular λ and also about "ρ-shift" of the parameter, see [14, Chap. 2] for instance. The resulting "quantizations" π_λ of semisimple coadjoint orbits $\mathcal{O}_\lambda$ give a "large part" of the unitary dual $\widehat{G}$.

Example 2.2 (Symmetric Spaces, Real Spherical Spaces) Let σ be an automorphism of a reductive Lie group G of finite order, G^σ the fixed point subgroup of σ, and H an open subgroup of G^σ. Then H is reductive and the homogeneous space G/H provides another example of Setting 2.1. In particular, if the order of σ is two, G/H is called a (reductive) *symmetric space*. Geometrically, it is a symmetric space with respect to the Levi-Civita connection of the pseudo-Riemannian structure in the sense that all geodesic symmetries are globally defined isometries. This is a subclass of Setting 2.1 for which the L^2-analysis has been extensively studied over 60 years. Group manifolds $(G \times G)/\text{diag}(G)$, Riemannian symmetric spaces G/K and irreducible affine symmetric spaces such as $SL(p+q, \mathbb{R})/SO(p, q)$ are examples of reductive symmetric spaces. More generally, in Setting 2.1, one has

$$\{\text{symmetric spaces}\} \subset \{\text{spherical spaces}\} \subset \{\text{real spherical spaces}\},$$

where we say G/H is *spherical* if a Borel subgroup of the complexification $G_\mathbb{C}$ has an open orbit in $G_\mathbb{C}/H_\mathbb{C}$, and G/H is *real spherical* if a minimal parabolic subgroup of G has an open orbit in G/H. See Kobayashi–T. Oshima [20] for the roles that these geometric properties play in the global analysis on G/H. When H is compact, G/H is spherical if and only if it is a weakly symmetric space in the sense of Selberg.

The model space of non-zero constant sectional curvatures in pseudo-Riemannian geometry is a special case of reductive symmetric spaces:

Example 2.3 (Pseudo-Riemannian Space Form, See [32]) The hypersurface

$$X(p, q) := \{x \in \mathbb{R}^{p+q+1} : x_1^2 + \cdots + x_{p+1}^2 - x_{p+2}^2 - \cdots - x_{p+q+1}^2 = 1\}$$

in $\mathbb{R}^{p+1,q} := (\mathbb{R}^{p+q+1}, ds^2 = dx_1^2 + \cdots + dx_{p+1}^2 - dx_{p+2}^2 - \cdots - dx_{p+q}^2)$ carries a pseudo-Riemannian structure of signature (p, q) with constant sectional curvature 1. Equivalently, we may regard $X(p, q)$ as a space of constant sectional curvature -1 with respect to the pseudo-Riemannian metric of signature (q, p). If $q = 0$, $p = 0, q = 1$, or $p = 1$, then $X(p, q)$ is the sphere S^p, the hyperbolic space H^q, the de Sitter space dS^{p+1}, or the anti-de Sitter space AdS^{q+1}, respectively. For general (p, q), the generalized Lorentz group $O(p+1, q)$ acts transitively and isometrically on $X(p, q)$, and one has a diffeomorphism $X(p, q) \simeq O(p+1, q)/O(p, q)$, giving an expression of $X(p, q)$ as a reductive symmetric space of rank one.

3 Problems on Discrete Series for G/H

The 'smallest units of symmetries' defined by group actions may be

irreducible representations if the action is linear, and
homogeneous spaces if the action is smooth on a manifold.

The objects of this section are irreducible subrepresentations in $L^2(X)$ for homogeneous spaces X, that is, *discrete series representations* for X (Definition 3.2 below), which are the building blocks in global analysis on X. For instance, when X is a reductive symmetric space, parabolic inductions of discrete series representations for subsymmetric spaces yield the full spectrum in the Plancherel formula of $L^2(X)$, see [3] for instance.

This section elucidates the following problem in the generality of Setting 2.1 by using simple examples, and addresses some related conjectures.

Problem 3.1 Find all discrete series representations for G/H.

Let us fix some notation. Suppose a Lie group G acts continuously on a manifold X. Then one has a natural unitary representation (*regular representation*) of G on the Hilbert space $L^2(X)$ of L^2-sections for the half-density bundle $\mathcal{L} :=$ $(\wedge^{\dim X} T^*X \otimes \mathrm{or}_X)^{\frac{1}{2}}$ of X where or_X stands for the orientation bundle.

Definition 3.2 An irreducible unitary representation π of G is said to be a *discrete series representation* for X if there exists a non-zero continuous G-homomorphism from π to the regular representation on $L^2(X)$. In other words, discrete series representations for X are irreducible subrepresentations realized in closed subspaces of the Hilbert space $L^2(X)$.

We denote by $\mathrm{Disc}(X)$ the set of discrete series representations for X. It is a (possibly, empty) subset of the unitary dual $\widehat{G}$ of the group G.

Hereafter, suppose we are in Setting 2.1. Then there is a G-invariant Radon measure μ on the homogeneous space $X = G/H$, hence $\mathcal{L}$ is trivial as a G-equivariant bundle and $L^2(X)$ may be identified with $L^2(X, d\mu)$.

If G/H is spherical (see Example 2.2), then the ring

$$\mathbb{D}(G/H) := \{G\text{-invariant differential operators on } G/H\}$$

is commutative and the multiplicity of irreducible representations π of G in the regular representation on $C^\infty(G/H)$ is uniformly bounded, and vice versa [20]. In this case, the disintegration of the regular representation $L^2(X)$ into irreducibles (the Plancherel-type theorem) is essentially equivalent to the joint spectral decomposition for the commutative ring $\mathbb{D}(G/H)$, and Problem 3.1 highlights point spectra in $L^2(G/H)$.

Classical examples trace back to Gelfand–Graev (1962), Shintani (1967), Molchanov (1968), J. Faraut (1979), R.S. Strichartz (1983) and some others on the analysis of the space form $X(p, q)$, which we review from some modern viewpoint, see [19, Thm. 2.1] and references therein.

Example 3.1 Let $(G, H) = (O(p + 1, q), O(p, q))$ and $X = G/H$ as in Example 2.3. Then the ring $\mathbb{D}(G/H)$ is generated by the Laplacian Δ_X, which is not an elliptic differential operator if $p, q > 0$. We set

$$L^2(X)_\lambda := \{f \in L^2(X) : \Delta_X f = \lambda f \text{ in the weak sense}\}.$$

Then $L^2(X)_\lambda$ is a closed subspace in $L^2(X)$, and the resulting unitary representation of G on $L^2(X)_\lambda$ is irreducible whenever it is non-zero. Conversely, any discrete series representation for X is realized on an L^2-eigenspace $L^2(X)_\lambda$ for some eigenvalue λ. In particular, one has the equivalence:

$$\text{Disc}(G/H) = \emptyset \iff \text{there is no point spectrum of } \Delta_X \text{ in } L^2(X)$$

$$\iff p = 0 \text{ and } q \geq 1.$$

Thus there exists a point spectrum of the Laplacian Δ_X in $L^2(X)$ unless $X = X(p, q)$ is a hyperbolic space $H^q \equiv X(0, q)$. The description of the eigenspace $L^2(X)_\lambda$ for $q = 0$ is the classical theory of spherical harmonics on the sphere $S^p \equiv X(p, 0)$. For $p, q \geq 1$, one has

$$L^2(X)_\lambda \neq 0 \text{ iff } \lambda = \lambda_k \text{ for some } k \in \mathbb{Z} \text{ with } -\frac{1}{2}(p + q - 1) < k,$$

where $\lambda_k := -k(k + p + q - 1)$. The resulting irreducible unitary representation on $L^2(X)_{\lambda_k}$ is isomorphic to a 'geometric quantization' of an elliptic coadjoint orbit of minimal dimension, or alternatively in an algebraic language, it is the unitarization of Zuckerman's derived functor module $A_\mathfrak{q}(k)$ with the normalization as in [30]. This algebraic description involves delicate questions for finitely many exceptional parameters, i.e., those for $k < 0$, see Problem 3.4 below.

In the generality of Setting 2.1, we may divide Problem 3.1 into two subproblems:

(A) a characterization of the pairs (G, H) for which G/H admits at least one discrete series representation (Problem 3.3);
(B) a description of all discrete series representations for X.

We address Conjectures 3.1 and 3.2 as subproblems for (A), and formulate Conjecture 3.3 and Problem 3.4 for (B).

Problem 3.3 Find a characterization of the pairs (G, H) such that G/H admits a discrete series representation.

Similarly to the classical fact that there is no discrete spectrum of the Laplacian $\Delta_{\mathbb{R}^n}$ on $\mathbb{R}^n$ and that there is no continuous spectrum of the Laplacian $\Delta_{\mathbb{T}^n}$ on the

n-torus $\mathbb{T}^n$, the Riemannian symmetric space G/K does not admit any discrete series representation if it is of non-compact type and does not admit any continuous spectrum in the Plancherel formula if it is of compact type. The answer to Problem 3.3 is known for reductive symmetric spaces by the rank condition:

$$\text{Disc}(G/H) \neq \emptyset \iff \text{rank}\, G/H = \text{rank}\, K/H \cap K. \qquad (2)$$

The equivalence (2) was proved by Flensted–Jensen for $\Leftarrow$ and Matsuki–Oshima for $\Rightarrow$. It generalizes the Riemannian case G/K as well as Harish-Chandra's rank condition for a group manifold, see [24] and references therein.

Beyond symmetric spaces, several approaches (e.g., branching laws [12, 16], the wave front set [5], etc.) have been applied to find new families of (not necessarily, real spherical) homogeneous spaces G/H that admit discrete series representations. It is more involved to prove the converse, i.e., to prove $\text{Disc}(G/H) = \emptyset$ for non-symmetric spaces, and very little is known so far, except for certain families of spherical homogeneous spaces. For instance, one has:

Example 3.2 (Real Forms of $SL(2n + 1, \mathbb{C})/Sp(n, \mathbb{C})$, [12]) $\text{Disc}(G/H) = \emptyset$ if $G/H = SL(2n + 1, \mathbb{R})/Sp(n, \mathbb{R})$, whereas $\#\text{Disc}(G/H) = \infty$ for other real forms of $G_{\mathbb{C}}/H_{\mathbb{C}}$, i.e., $SU(2p, 2q + 1)/Sp(p, q)$ or $SU(n, n + 1)/Sp(n, \mathbb{R})$.

An optimistic solution to Problem 3.3 may be a combination of the following two conjectures:

Conjecture 3.1 ([14, Conj. 6.9]) *In Setting 2.1, one has the equivalence:*

$$\text{Disc}(G/H) \neq \emptyset \iff \#\text{Disc}(G/H) = \infty.$$

Conjecture 3.2 *In Setting 2.1, one has the following equivalence:*

$$\#\text{Disc}(G/H) = \infty \iff \mathfrak{h}^{\perp} \cap \mathfrak{g}_{\text{ell}}^* \text{ contains a non-empty open set of } \mathfrak{h}^{\perp}.$$

Both of the conjectures are true for reductive symmetric spaces G/H. In fact, Conjecture 3.2 is a reformulation of the rank condition (2) in the spirit of the orbit philosophy.

There are counterexamples for the implication $\Rightarrow$ of an analogous statement to Conjectures 3.1 and 3.2 if we drop the assumption that H is reductive, for instance, they fail when H is a parabolic subgroup and a cocompact discrete subgroup of G with rank $G >$ rank K, respectively. The implication $\Leftarrow$ in Conjecture 3.2 has recently been proved in Harris–Y. Oshima [5] without the reductivity assumption on H.

Remark 3.1 Similarly to Conjecture 3.2, one might expect the equivalence:

$$L^2(G/H) \text{ is tempered} \iff \mathfrak{h}^{\perp} \cap \mathfrak{g}_{\text{reg}}^* \text{ is dense in } \mathfrak{h}^{\perp}.$$

This is proved in Benoist–Kobayashi [1] for complex homogeneous spaces for any algebraic subgroup H without the reductivity assumption.

Once we know $\mathrm{Disc}(G/H) \neq \emptyset$, we may wish to capture *all* elements of $\mathrm{Disc}(G/H)$. We divide this exhaustion problem into two questions: one is geometric (Conjecture 3.3), and the other is algebraic (Problem 3.4).

Conjecture 3.3 *Any* $\pi \in \mathrm{Disc}(G/H)$ *is obtained as a geometric quantization of some elliptic coadjoint orbit that meets* $\mathfrak{h}^{\perp}$.

Problem 3.4 Find a necessary and sufficient condition for cohomologically parabolic induced modules $A_{\mathfrak{q}}(\lambda)$ not to vanish outside the good range of parameter λ.

Conjecture 3.3 strengthens Conjecture 3.2, and one can verify it for reductive symmetric spaces X, see [14, Ex. 2.9]. To be more precise, by using Matsuki–Oshima's theorem [24] and by using an algebraic characterization of Zuckerman's derived functor modules, one can identify any discrete series representation for G/H as a "geometric quantization" π_λ of an elliptic coadjoint orbit $\mathcal{O}_\lambda$ that meets $\mathfrak{h}^{\perp}$, with the normalization of "quantization" as in [14]. For "singular" λ, the above π_λ may or may not vanish. A missing part of Problem 3.1 in the literature for symmetric spaces is the complete proof of the precise condition on λ such that $\pi_\lambda \neq 0$, which is reduced to an algebraic question, that is, Problem 3.4. The algebraic results in [11, Chaps. 4,5] and [29] give an answer to Problem 3.4 for some classical symmetric spaces.

We examine Problem 3.4 by $X = X(p,q)$ with $p, q \geq 1$. As we saw in Example 3.1, the underlying $(\mathfrak{g}, K)$-modules (see [31, Chap. 3] for instance) of the L^2-eigenspace $L^2(X)_{\lambda_k}$ are expressed by $A_{\mathfrak{q}}(k)$. Then there are finitely many exceptional parameters $k \in \mathbb{Z}$ satisfying $-\frac{1}{2}(p + q - 1) < k < 0$, i.e., lying "outside the good range" for which the general algebraic representation theory does not guarantee the irreducibility/non-vanishing for the cohomological parabolic induction. This is the point that Problem 3.4 is concerned with.

4   Problems on Discontinuous Groups for G/H

The local to global study of geometries was a major trend of twentieth century geometry, with remarkable developments achieved particularly in Riemannian geometry. In contrast, in areas such as pseudo-Riemannian geometry, familiar to us as the space-time of relativity theory, and more generally in manifolds with indefinite metric tensor of arbitrary signature, surprisingly little is known about global properties of the geometry. For instance, the pseudo-Riemannian space form problem is unsolved, which asks the existence of a compact pseudo-Riemannian manifold M with constant sectional curvature for a given signature (p, q), see Conjecture 4.3 below.

When we highlight "homogeneous structure" as a local property, "discontinuous groups" are responsible for the global geometry. The theory of discontinuous groups beyond the Riemannian setting is a relatively "young area" in group theory that interacts with topology, differential geometry, representation theory, and number theory, among other subjects. See [13] for some background on this topic at an early stage of its development. This theme was also discussed as a new topic of future research by Kobayashi [17] and Margulis [23] on the occasion of the "World Mathematical Year 2000". For over 30 years, there have been remarkable developments which have made use of various methods ranging from topology and differential geometry to representation theory and ergodic theory, however, some important problems are still unsolved, which we illustrate in this section by using simple examples.

Beyond the Riemannian setting, we highlight a substantial difference between "discrete subgroups" and "discontinuous groups", e.g., [10].

Definition 4.1 Let G be a Lie group acting on a manifold X. A discrete subgroup Γ of G is said to be a *discontinuous group* for X if Γ acts properly discontinuously and freely on X.

The quotient space $X_\Gamma := \Gamma \backslash X$ by a discontinuous group Γ is a (Hausdorff) C^∞-manifold, and any G-invariant local geometric structure on X can be pushed forward to X_Γ via the covering map $X \to X_\Gamma$. Such quotients X_Γ are complete (G, X)-manifolds in the sense of Ehresmann and Thurston.

A classical example is a compact Riemann surface Σ_g with genus $g \geq 2$, which can be expressed as X_Γ where $\Gamma \simeq \pi_1(\Sigma_g)$ (surface group) and $X \simeq PSL(2, \mathbb{R})/PSO(2)$ by the uniformization theory. More generally, any complete affine locally symmetric space is of the form $\Gamma \backslash G/H$ where Γ is a discontinuous group for a symmetric space G/H.

Remark 4.1 The crucial assumption in Definition 4.1 is proper discontinuity of the action, and freeness is less important. In [13, Def. 2.5], we did not include the freeness assumption in the definition of discontinuous groups, allowing $X_\Gamma = \Gamma \backslash X$ to be an orbifold.

We discuss the following problems in the generality of Setting 2.1, cf. [17, Problems B and C].

Problem 4.1 Determine all pairs (G, H) such that G/H admits cocompact discontinuous groups.

Problem 4.2 (Higher Teichmüller Theory for G/H) Describe the moduli of all deformations of a discontinuous group Γ for G/H.

In the classical case where H is compact, a theorem of Borel answers Problem 4.1 in the affirmative by the existence of cocompact arithmetic discrete subgroups in G, whereas the Selberg–Weil local rigidity theorem tells us that Problem 4.2 makes sense for a cocompact Γ in a simple Lie group G only if

$\mathfrak{g} \simeq \mathfrak{sl}(2, \mathbb{R})$, and in this case the deformation of discontinuous groups gives rise to that of complex structures on the Riemann surface.

Such features change dramatically when H is non-compact: some homogeneous spaces may not admit any discontinuous group of infinite order (the Calabi–Markus phenomenon [2]), showing an obstruction to the existence of cocompact discontinuous groups for G/H. On the other hand, discontinuous groups for pseudo-Riemannian manifolds G/H tend to be "more flexible" in contrast to the classical rigidity theorems in the Riemannian case. For instance, some irreducible symmetric spaces of arbitrarily higher dimension admit cocompact discontinuous groups which are not locally rigid [7, 15], providing wide open settings for Problem 4.2.

As we mentioned, the notion "discontinuous group for G/H" is much stronger than "discreteness in G" when H is non-compact. For instance, a cocompact discrete subgroup Γ of G never acts properly discontinuously on G/H unless H is compact. Thus the existence of a lattice in G does not imply that G/H admits a cocompact discontinuous group.

We examine some related questions and conjectures to Problem 4.1. First, by relaxing the "cocompactness" assumption of Γ in Problem 4.1, one may ask the following:

Problem 4.3 Find a necessary and sufficient condition for G/H in Setting 2.1 to admit a discontinuous group Γ for G/H such that

(1) $\Gamma \simeq \mathbb{Z}$;
(2) $\Gamma \simeq$ a surface group $\pi_1(\Sigma_g)$ with $g \geq 2$.

Problem 4.3 (1) was solved in [10] in terms of the real rank condition $\text{rank}_{\mathbb{R}} G > \text{rank}_{\mathbb{R}} H$, which revealed the Calabi–Markus phenomenon [2] in the generality of Setting 2.1. Problem 4.3 (2) was solved by Okuda [26] for irreducible symmetric spaces, but is unsolved in the generality of Setting 2.1.

Cocompact discontinuous groups for G/H are much smaller than cocompact lattices of G, for instance, their cohomological dimensions are strictly smaller [10]. A simple approach to Problem 4.1 is to utilize a 'continuous analog' of discontinuous groups Γ:

Definition 4.2 (Standard Quotients $\Gamma \backslash G/H$ [8, Def. 1.4]) Suppose L is a reductive subgroup of G such that L acts properly on G/H. Then any torsion-free discrete subgroup Γ of L is a discontinuous group for G/H. The quotient space $\Gamma \backslash G/H$ is called a *standard quotient* of G/H.

If such an L acts cocompactly on G/H, then G/H admits a cocompact discontinuous group Γ by taking Γ to be a torsion-free cocompact discrete subgroup in L, which always exists by Borel's theorem. We address the following conjecture and a subproblem to Problem 4.1.

Conjecture 4.1 ([17, Conj. 4.3]) *In Setting 2.1, G/H admits a cocompact discontinuous group only if G/H admits a compact standard quotient.*

If Conjecture 4.1 were proved to be true, then Problem 4.1 would be reduced to the following one:

Problem 4.4 Classify the pairs (G, H) such that G/H admits a compact standard quotient.

This problem should be manageable because one could use the general theory of real finite-dimensional representations of semisimple Lie algebras and apply the properness criterion and the cocompactness criterion in [10, Thms 4.1 and 4.7]. See [28] for some developments.

Remark 4.2

(1) Conjecture 4.1 does not assert that any cocompact discontinuous group is a standard one. In fact, there exist triples (G, H, Γ) such that Γ is a cocompact discontinuous group for G/H and that the Zariski closure of Γ does not act properly on G/H, see [4, 7, 15].
(2) An analogous statement to Conjecture 4.1 fails if we drop the reductivity assumption on the groups G, H and L.
(3) An analogous statement to Conjecture 4.1 is proved in Okuda [26] for semisimple symmetric spaces G/H if we replace the "cocompactness" assumption with the condition that Γ is a surface group $\pi_1(\Sigma_g)$.

Special cases of Conjecture 4.1 include:

Conjecture 4.2 $SL(n, \mathbb{R})/SL(m, \mathbb{R})$ *does not admit a cocompact discontinuous group for any $n > m$.*

Conjecture 4.3 (Space Form Conjecture [17, Conj. 2.6], [21, Conj. 2.5.1]) *There exists a compact, complete, pseudo-Riemannian manifold of signature (p, q) with constant sectional curvature 1 if and only if (p, q) is in the list of Example 4.1 (4) below.*

A criterion on triples (G, H, L) of reductive Lie groups for L to act properly on $X = G/H$ was established in [10], and a list of irreducible symmetric spaces G/H admitting proper and cocompact actions of reductive subgroups L was given in [21]. Tojo [28] worked with simple Lie groups G and announced that the list in [21] exhausts all such triples (G, H, L) with L maximal, giving a solution to Problem 4.4 for symmetric spaces G/H with G simple.

A number of obstructions to the existence of cocompact discontinuous groups for G/H with H non-compact have been found over the last 30 years. One of the recent developments includes the affirmative solution to the "rank conjecture" raised by the author in 1989: it was proved in the case rank $G = $ rank H by Kobayashi–Ono (1990), and has been proved recently in the general case by Morita [25] and Tholozan [27], independently.

Conjecture 4.4 ([13, Conj. 4.15]) *If G/H admits a cocompact discontinuous group, then* rank $G + $ rank$(H \cap K) \geq $ rank $H + $ rank K.

Whereas the idea of standard quotients $\Gamma \backslash G/H$ is to replace a discrete subgroup Γ with a connected subgroup L (Definition 4.2), one may consider an "approximation" of Problem 4.1, by taking the *tangential homogeneous space* $X_\theta = G_\theta/H_\theta$ in replacement of $X = G/H$, where $G_\theta := K \ltimes \mathfrak{p}$ is the Cartan motion group of the real reductive group $G = K \exp \mathfrak{p}$ and similarly for H_θ. If G/H admits a compact standard quotient, then the tangential homogeneous space G_θ/H_θ admits a cocompact discontinuous group. The group G_θ is a compact extension of the abelian group $\mathfrak{p}$, and has a much simpler structure. We ask the following digression of Problem 4.1:

Problem 4.5 ([21]) For which pairs (G, H) in Setting 2.1 does G_θ/H_θ admit a cocompact discontinuous group?

This problem is unsolved even for symmetric spaces in general, but has a complete answer in some special settings, e.g., Example 4.1 (6) below.

We end this section with a brief summary of the state-of-art for these problems and conjectures by taking the space form $X(p, q)$ as an example.

Example 4.1 (See [2, 6, 7, 13, 15, 21, 22, 25–27]) Let $(G, H) = (O(p + 1, q), O(p, q))$, and $X = X(p, q) = G/H$ the pseudo-Riemannian space form of signature (p, q) as in Example 2.3.

(1) $X(p, q)$ admits a discontinuous group of infinite order iff $p < q$.
(2) $X(p, q)$ admits a discontinuous group which is isomorphic to a surface group iff $p + 1 < q$ or $p + 1 = q \in 2\mathbb{N}$.
(3) If $X(p, q)$ admits a cocompact discontinuous group, then $p = 0$ or "$p < q$ and $q \in 2\mathbb{N}$".
(4) $X(p, q)$ admits a cocompact discontinuous group if (p, q) is in the list below. (The converse assertion was stated as Conjecture 4.3.)

p	$\mathbb{N}$	0	1	3	7
q	0	$\mathbb{N}$	$2\mathbb{N}$	$4\mathbb{N}$	8

(5) If $(p, q) = (0, 2)$, $(1, 2)$, or $(3, 4)$, then $X(p, q)$ admits a cocompact discontinuous group which can be deformed continuously into a Zariski dense subgroup of G by keeping proper discontinuity of the action. For $(p, q) = (1, 2n)$ $(n \geq 2)$, the anti-de Sitter space $X(1, 2n)$ admits a compact quotient which has a non-trivial continuous deformation as standard quotients.
(6) The tangential homogeneous space G_θ/H_θ admits a cocompact discontinuous group if and only if $p < \rho(q)$ where $\rho(q)$ is the Radon–Hurwitz number, or equivalently, if and only if (p, q) is in the following list:

p	$\mathbb{N}$	0	1	2	3	4	5	6	7	8	9	10	11	$\cdots$
q	0	$\mathbb{N}$	$2\mathbb{N}$	$2\mathbb{N}$	$4\mathbb{N}$	$8\mathbb{N}$	$8\mathbb{N}$	$8\mathbb{N}$	$8\mathbb{N}$	$16\mathbb{N}$	$32\mathbb{N}$	$64\mathbb{N}$	$64\mathbb{N}$	$\cdots$

5 Spectral Analysis for Pseudo-Riemannian Locally Homogeneous Spaces $\Gamma \backslash G/H$

This section briefly discusses a new direction of analysis on pseudo-Riemannian locally homogeneous spaces $\Gamma \backslash G/H$.

Suppose we are in Setting 2.1. Let Γ be a discontinuous group for $X = G/H$. Then any G-invariant differential operator $D \in \mathbb{D}(G/H)$ induces a differential operator D_Γ on the quotient $X_\Gamma := \Gamma \backslash G/H$ via the covering $X \to X_\Gamma$. We think of the set $\mathbb{D}(X_\Gamma) := \{D_\Gamma : D \in \mathbb{D}(G/H)\}$ as the algebra of *intrinsic differential operators* on the locally homogeneous space X_Γ.

Example 5.1

(1) In Setting 2.1, X_Γ inherits a pseudo-Riemannian structure from X, and the Laplacian Δ_{X_Γ} belongs to $\mathbb{D}(X_\Gamma)$.
(2) For $X = X(p, q)$, $\mathbb{D}(X_\Gamma)$ is a polynomial ring in the Laplacian Δ_{X_Γ} for any discontinuous group Γ.

We address the following problem:

Problem 5.1 (See [8, 9]) For intrinsic differential operators on $X_\Gamma = \Gamma \backslash G/H$,

(1) construct joint eigenfunctions on X_Γ;
(2) find a spectral theory on $L^2(X_\Gamma)$.

In the same spirit as in Sect. 3, we highlight the "discrete spectrum".

Definition 5.1 We say $\lambda \in \mathrm{Hom}_{\mathbb{C}\text{-alg}}(\mathbb{D}(X_\Gamma), \mathbb{C})$ is a *discrete spectrum* for intrinsic differential operators on X_Γ if $L^2(X_\Gamma)_\lambda \neq \{0\}$, where we set

$$L^2(X_\Gamma)_\lambda := \{f \in L^2(X_\Gamma) : Df = \lambda(D)f \quad {}^\forall D \in \mathbb{D}(X_\Gamma)\}.$$

We write $\mathrm{Spec}_d(X_\Gamma)$ for the set of discrete spectra.

A subproblem to Problem 5.1 (1) includes:

Problem 5.2 Construct joint L^2-eigenfunctions on X_Γ.

In relation to Problem 4.2 about the deformations of a discontinuous group Γ for G/H, one may also ask the following:

Problem 5.3 Understand the behavior of $\mathrm{Spec}_d(X_\Gamma)$ under small deformations of Γ inside G.

These problems have been studied extensively in the following special settings for $X_\Gamma = \Gamma \backslash G/H$:

(1) ($H = K$). When H is a maximal compact subgroup K of G, i.e., X_Γ is a *Riemannian* locally symmetric space, a vast theory has been developed over several decades, in particular, in connection with the theory of automorphic forms when Γ is arithmetic.

(2) ($\Gamma = \{e\}$). This case is related to the topic in Section 3. In particular, Problem 5.1 has been extensively studied in the case where G/H is a reductive symmetric space and $\Gamma = \{e\}$.

(3) $G = \mathbb{R}^{p,q}$, $\Gamma = \mathbb{Z}^{p+q}$, and $H = \{0\}$. In this case, $\mathrm{Spec}_d(X_\Gamma)$ is the set of values of indefinite quadratic forms at integral points, see [18] for a discussion on Problem 5.3 in relation to the Oppenheim conjecture (proved by Margulis) in Diophantine approximation.

The situation changes drastically beyond the classical setting, namely, when H is no longer compact and $\Gamma \neq \{e\}$. New difficulties include:

(1) (Representation theory) Even when $\Gamma\backslash G/H$ is compact, the regular representation of G on $L^2(\Gamma\backslash G)$ has infinite multiplicities, as opposed to a classical theorem of Gelfand–Piatetski–Shapiro.

(2) (Analysis) In contrast to the Riemannian case where $H = K$, the Laplacian Δ_{X_Γ} is not an elliptic differential operator anymore.

As we saw in Sect. 4, if H is not compact, then not all homogeneous spaces G/H admit discontinuous groups of infinite order, but fortunately, there exist a family of reductive symmetric spaces G/H that admit "large" discontinuous groups Γ, e.g., such that $X_\Gamma = \Gamma\backslash G/H$ is compact or of finite volume. Moreover, there also exist triples (G, H, Γ) such that discontinuous groups Γ for G/H can be deformed continuously. These examples offer broad settings for Problems 5.1 and their subproblems.

For Problem 5.3, we consider two notions for stability:

Definition 5.2

(1) (stability for proper discontinuity) A discontinuous group Γ is *stable under small deformations* if the group $\varphi(\Gamma)$ acts properly discontinuously and freely on X for all $\varphi \in \mathrm{Hom}(\Gamma, G)$ in some neighbourhood $\mathcal{U}$ of the natural inclusion Γ in G.

(2) (stability for L^2-spectrum) We say $\lambda \in \mathrm{Hom}_{\mathbb{C}\text{-alg}}(\mathbb{D}(X_\Gamma), \mathbb{C})$ is a *stable spectrum* if $L^2(X_{\varphi(\Gamma)})_\lambda \neq \{0\}$ for any $\varphi \in \mathrm{Hom}(\Gamma, G)$ in some neighbourhood $\mathcal{U}$ of the natural inclusion Γ in G.

Conjecture 5.1 *Suppose that Γ is a finitely generated discontinuous group for G/H having non-trivial continuous deformations (up to inner automorphisms) with stability of proper discontinuity. Then the following conditions on the pair (G, H) are equivalent.*

(i) *There exist infinitely many stable spectra on $L^2(\Gamma\backslash G/H)$.*

(ii) *$\mathrm{Disc}(G/H) \neq \emptyset$.*

See [8] for some results in the direction (ii) $\Rightarrow$ (i), which also treat the case $\mathrm{vol}(\Gamma\backslash G/H) = \infty$.

The last section has been devoted to a "very young" topic, though special cases trace back to rich and deep classical theories. I expect this topic will create new interactions with different subjects of mathematics, and this is why I have included it as part of my article for *Mathematics Going Forward*.

Acknowledgements The author would like to express his sincere gratitude to his collaborators for the various projects mentioned in this article. This work was partially supported by Grant-in-Aid for Scientific Research (A) (18H03669), JSPS.

References

1. Y. Benoist and T. Kobayashi. Tempered homogeneous spaces IV. To appear in *J. Inst. Math. Jussieu*. Available also at *arXiv: 2009.10391*.
2. E. Calabi and L. Markus. Relativistic space forms. *Ann. of Math.* **75**, 63–76 (1962).
3. P. Delorme. Formule de Plancherel pour les espaces symétriques réductifs. *Ann. of Math. (2)* **147**, 417–452 (1998).
4. W.M. Goldman. Nonstandard Lorentz space forms. *J. Differential Geom.* **21**, 301–308 (1985).
5. B. Harris, Y. Oshima. On the asymptotic support of Plancherel measures for homogeneous spaces. *arXiv:2201.11293*.
6. K. Kannaka. Deformation of standard pseudo-Riemannian locally symmetric spaces. *In preparation*.
7. F. Kassel. Deformation of proper actions on reductive homogeneous spaces. *Math. Ann.* **353**, 599–632 (2012).
8. F. Kassel and T. Kobayashi. Poincaré series for non-Riemannian locally symmetric spaces. *Adv. Math.* **287**, 123–236 (2016).
9. F. Kassel and T. Kobayashi. Spectral analysis on standard locally homogeneous spaces. *arXiv: 1912.12601*.
10. T. Kobayashi. Proper action on a homogeneous space of reductive type. *Math. Ann.* **285**, 249–263 (1989).
11. T. Kobayashi. *Singular unitary representations and discrete series for indefinite Stiefel manifolds* $U(p, q; F)/U(p - m, q; F)$. Mem. Amer. Math. Soc. **95**, no. 462, vi+106 pp. (1992).
12. T. Kobayashi. Discrete decomposability of the restriction of $A_q(\lambda)$ with respect to reductive subgroups and its applications. *Invent. Math.* **117**, 181–205 (1994).
13. T. Kobayashi. Discontinuous groups and Clifford–Klein forms of pseudo- Riemannian homogeneous manifolds. In: *Algebraic and analytic methods in representation theory (Sønderborg, 1994)*, pp. 99–165, Perspect. Math., **17**. Academic Press, San Diego, CA (1997).
14. T. Kobayashi. Harmonic analysis on homogeneous manifolds of reductive type and unitary representation theory. In: *Translations, Series II, Selected Papers on Harmonic Analysis, Groups, and Invariants*, **183**, pp. 1–31, Amer. Math. Soc. (1998).
15. T. Kobayashi. Deformation of compact Clifford–Klein forms of indefinite-Riemannian homogeneous manifolds. *Math. Ann.* **310**, 395–409 (1998).
16. T. Kobayashi. Discrete series representations for the orbit spaces arising from two involutions of real reductive Lie groups. *J. Funct. Anal.* **152**, 100–135 (1998).
17. T. Kobayashi. Discontinuous groups for non-Riemannian homogeneous spaces. In: *Mathematics Unlimited–2001 and Beyond*, pp. 723–747, Springer, Berlin (2001).
18. T. Kobayashi. *Intrinsic sound of anti-de Sitter manifolds*. Springer Proc. Math. Stat. **191**, 83–99 (2016).
19. T. Kobayashi. Branching laws of unitary representations associated to minimal elliptic orbits for indefinite orthogonal group $O(p, q)$. *Adv. Math.* **388**, Paper No. 107862, 38pp. (2021).
20. T. Kobayashi and T. Oshima. Finite multiplicity theorems for induction and restriction. *Adv. Math.* **248**, 921–944 (2013).

21. T. Kobayashi and T. Yoshino. Compact Clifford–Klein forms of symmetric spaces—revisited. *Pure and Appl. Math. Quarterly* **1**, 603–684, Special Issue: In Memory of Armand Borel (2005).

22. R.S. Kulkarni. Proper actions and pseudo-Riemannian space forms. *Adv. Math.* **40**, 10–51 (1981).

23. G. Margulis. Problems and conjectures in rigidity theory. In: *Mathematics: frontiers and perspectives*, pp. 161–174, Amer. Math. Soc. (2000).

24. T. Matsuki and T. Oshima. A description of discrete series for semisimple symmetric spaces. *Adv. Stud. Pure Math.* **4**, 331–390 (1984).

25. Y. Morita. Proof of Kobayashi's rank conjecture on Clifford–Klein forms. *J. Math. Soc. Japan* **71**, no. 4, 1153–1171 (2019).

26. T. Okuda. Classification of semisimple symmetric spaces with proper $SL(2, \mathbb{R})$-actions. *J. Differential Geom.* **94**, 301–342 (2013).

27. N. Tholozan. Volume and non-existence of compact Clifford–Klein forms. *arXiv:1511.09448v2, preprint.*

28. K. Tojo. Classification of irreducible symmetric spaces which admit standard compact Clifford–Klein forms. *Proc. Japan Acad. Ser. A Math. Sci.* **95**, 11–15 (2019).

29. P.E. Trapa. Annihilators and associated varieties of $A_q(\lambda)$ modules for $U(p, q)$. *Compositio Math.* **129**, 1–45 (2001).

30. D.A. Vogan and G.J. Zuckerman. Unitary representations with nonzero cohomology. *Compositio Math.* **53**, 51–90 (1984).

31. N.R. Wallach. *Real Reductive Groups. I.* Pure Appl. Math. **132** Academic Press, Inc., Boston, MA (1988).

32. J.A. Wolf. *Spaces of Constant Curvature.* Sixth edition, xviii+424 pages, AMS Chelsea Publishing, Providence, RI (2011).

Profinite Rigidity and Free Groups

Martin R. Bridson

Personal Note

Many of the results that I am about to describe rely on the modern understanding of groups that act by isometries on spaces of negative and non-positive curvature, a central theme of my book with André Haefliger [10]. The success of that book project owed a great deal to the patient and thoughtful stewardship of Dr Catriona Byrne. It is with great pleasure and gratitude, therefore, that I dedicate this essay to her on the occasion of her retirement.

1 Introduction

Groups are the mathematical objects that encode symmetry in all contexts: no matter what category of objects X one may be studying, and no matter what sort of maps one may be allowing, the invertible maps from X to itself (i.e. the automorphisms of X) form a group. Thus, in all manner of contexts, one finds reasons to study groups of automorphisms $\mathrm{Aut}(X)$ in order to elucidate the nature of the underlying object X. According to one's nature, one might also be drawn to the study of groups themselves. When this is case, it is natural to reverse the passage from X to $\mathrm{Aut}(X)$: given a group Γ, one seeks objects X such that Γ acts as a group of automorphisms of X; one hopes to illuminate the nature of Γ by observing it in action. Actions on different kinds of objects provide different insights into the nature of Γ, and one

M. R. Bridson (✉)
Mathematical Institute, University of Oxford, Oxford, UK
e-mail: bridson@maths.ox.ac.uk

© The Author(s), under exclusive license to Springer Nature Switzerland AG 2023
J.-M. Morel, B. Teissier (eds.), *Mathematics Going Forward*, Lecture Notes
in Mathematics 2313, https://doi.org/10.1007/978-3-031-12244-6_18

quickly learns that the quality of the insights that one gains depends heavily on the nature of both the group and the object on which it is acting.

In all contexts, the groups that have the most unconstrained range of actions are, as the name suggests, *free groups*: associated to any set S one has the free group $F(S)$ whose elements are finite products of the elements of S (and formal inverses) subject to no constraints other than the axioms of a group. Free groups will play a central role in our discussion.

When exploring the symmetries of an object X that interests you, it is natural to grasp at a description of $\text{Aut}(X)$ by seeking (i) a set S of elementary operations that, when performed in suitable combinations, account for all of the symmetries of X, and (ii) a set of rules R describing how different combinations of these elementary operations are related: this leads to the notion of a presentation of a group $\Gamma = \langle S \mid R \rangle$. One can regard a presentation as a concise description of how Γ can be realised as a quotient of the free group $F(S)$. At the beginning of the twentieth century, mathematicians, foremost among them Max Dehn, realised that it is extremely hard to unravel the nature of a group by examining a presentation of it in isolation, even if both of the sets S and R are finite. This insight brings us back to the search for actions: rather than struggling to understand a group Γ as a quotient of a free group described by a finite presentation, one should try to unravel the nature of Γ by exploring how it can act on different kinds of objects.

The most primitive objects to consider are finite sets. Actions on finite sets capture only the finite images of groups, so the power of such actions to explain the nature of Γ is limited by the answer to the fundamental question: to what extent is Γ determined by its set of finite quotients? This compelling question has re-emerged with different emphases throughout the history of group theory, and in recent years it has been animated by a rich interplay with geometry and low-dimensional topology.

The finite images of Γ are encoded in its profinite completion $\widehat{\Gamma}$, a compact topological group that is the inverse limit of the directed system of finite quotients of Γ: if $N < M$ then $\Gamma/N \to \Gamma/M$. For finitely generated groups Γ and Λ, the set of finite images of Γ will be the same as the set of finite images of Λ if and only if $\widehat{\Lambda}$ and $\widehat{\Gamma}$ are isomorphic as topological groups; the reader unfamiliar with profinite groups will therefore lose little by reading the statement $\widehat{\Lambda} \cong \widehat{\Gamma}$ as an equality of sets of finite quotients.

If Γ has elements that do not survive in any finite quotient (see Sect. 4 for examples), then one cannot hope to recover Γ by studying $\widehat{\Gamma}$. Thus it is natural to restrict attention to *residually finite groups*, i.e. groups where every finite subset injects into some finite quotient. The most basic recognition question then becomes: which finitely generated, residually finite groups Γ are *profinitely rigid* in the sense that if Λ is finitely generated and residually finite, then $\widehat{\Lambda} \cong \widehat{\Gamma}$ implies $\Lambda \cong \Gamma$.

It is obvious that finite groups are profinitely rigid and is easy to see that finitely generated abelian groups are as well, but one quickly struggles to find further examples. The study of groups that are not profinitely rigid owes much to a paper of Serre [25] from 1964. He constructed pairs of smooth complex projective varieties that are Galois conjugate but are not homeomorphic. The fundamental groups in each pair are not isomorphic, but the profinite completions of these groups (being the étale fundamental group of their common scheme) are the same. Other illuminating examples come from the work of Stebe [26]: he described pairs of integer matrices $\varphi_1, \varphi_2 \in SL(2, \mathbb{Z})$ that are not conjugate but do have conjugate images in $SL(2, \mathbb{Z}/m)$ for every positive integer m; from this it follows easily that the mapping tori $\mathbb{Z}^2 \rtimes_{\varphi_1} \mathbb{Z}$ and $\mathbb{Z}^2 \rtimes_{\varphi_2} \mathbb{Z}$ are not isomorphic but their profinite completions are. The essence of these examples was stripped down to its bare essentials by Baumslag [5] who showed that profinite rigidity can fail even for finite extensions of $\mathbb{Z}$ (see Sect. 4).

These examples are sobering and cause one to reflect on the proof that $\mathbb{Z}^r$ is profinitely rigid. The key to this argument is the observation that if Γ satisfies a group law—in this case the law $\forall x, y \, (xy = yx)$— then $\widehat{\Lambda} \cong \widehat{\Gamma}$ will imply that Λ satisfies the same law, provided it is residually finite. In such cases, the question of absolute profinite rigidity reduces to a question of *relative* profinite rigidity, where one asks if $\widehat{\Gamma}$ distinguishes Γ from all other groups in a restricted class, for example the class of groups that satisfy a given law or a more geometric condition. Once one has manouevred into such a relative context, one might use a classification of groups in that class to identify examples of profinitely rigid groups: for example, the free nilpotent group of fixed class on a fixed number of generators is profinitely rigid (although many other nilpotent groups are not).

The pursuit of relative profinite rigidity theorems has provided a focal point for a rich body of research in recent years, particularly in geometric contexts [24]. This includes many settings in which the groups are *full-sized* in the sense that they contain non-abelian free subgroups and hence do not satisfy a law. Thus, for example, a finitely generated free group can be distinguished from any other lattice in a connected Lie group [8], or from any other residually-free group [8, 29], by its finite quotients. But such relative theorems do not lead to absolute profinite rigidity, because in the absence of a group law it is extremely difficult to rule out the possible existence of an utterly exotic Λ, finitely generated and residually finite, with $\widehat{\Lambda} \cong \widehat{\Gamma}$. One has to contend with the possibility that Λ shares few of the familiar characteristics of Γ: even if Γ is familiar to you as a group of matrices, say, why should Λ have such a representation? This is a much wilder context than that considered by Grothendieck [16], who considered pairs of residually finite groups $\iota : H \hookrightarrow G$ such that H is not isomorphic to G but ι nevertheless induces an isomorphism $\widehat{H} \to \widehat{G}$—see Sect. 4.

The paucity of our knowledge about (absolute) profinite rigidity is illustrated most starkly by the fact that the following fundamental challenge remains open.

Conjecture 1.1 *If a finitely generated, residually finite group Γ has the same finite quotients as a free group of rank r, then Γ is a free group of rank r.*

As far as I am aware, the first person to ask explicitly whether free groups are profinitely rigid was Remeslennikov [20, Question 5.48]. This remains the central challenge in the field, but in recent years there has been significant progress on related matters. In the following sections I shall describe a sample of this progress, staying close to Conjecture 1.1 and highlighting some related open problems.

2 Full-Sized Groups that Are Profinitely Rigid

The group of orientation-preserving isometries of real hyperbolic space is iso-morphic to $PSL(2, \mathbb{R})$, in dimension 2, and $PSL(2, \mathbb{C})$ in dimension 3. Thus the lattices in these Lie groups are the fundamental groups of finite-volume, orientable hyperbolic orbifolds in these dimensions; the orbifold is compact if the lattice is cocompact, and the orbifold is a manifold if the lattice has no non-trivial elements of finite order. The proof of the following theorem, which I proved with McReynolds, Reid and Spitler, relies on many aspects of the modern understanding of such orbifolds, including the arithmetic naturally associated to them and various consequences of the work of Agol and Wise showing that lattices in $PSL(2, \mathbb{C})$ act nicely on CAT(0) cube complexes (see [1]): more precisely, these groups are virtually special in the sense of [18]. Among other things, this last theorem implies that finitely generated, discrete subgroups of $PSL(2, \mathbb{C})$ are good in the sense of Serre, an important property in many results concerning profinite rigidity: Γ is *good* if for any finite $\mathbb{Z}\Gamma$-module M, the map $H^*(\widehat{\Gamma}, M) \to H^*(\Gamma, M)$ induced by $\Gamma \hookrightarrow \widehat{\Gamma}$ is an isomorphism.

Theorem 2.1 ([11, 12]) *There exist arithmetic lattices in* $PSL(2, \mathbb{C})$ *and* $PSL(2, \mathbb{R})$ *that are profinitely rigid in the absolute sense.*

For the moment, only finitely many lattices in $PSL(2, \mathbb{C})$ and $PSL(2, \mathbb{R})$ are known to be profinitely rigid. Each of the examples in $PSL(2, \mathbb{R})$ is a triangle group, i.e. the group of symmetries $\Delta(p, q, r)$ of a tiling of the hyperbolic plane by geodesic triangles with vertex angles π/p, π/q, π/r. The least-area example to which the current techniques apply is $\Delta(2, 3, 8)$. The examples in $PSL(2, \mathbb{C})$ include both cocompact and non-cocompact lattices; some of the cocompact examples have torsion and some do not. The cocompact examples include the fundamental group of the Weeks manifold, the unique compact hyperbolic 3-manifold of smallest volume. The non-cocompact examples include the Bianchi group $PSL(2, \mathbb{Z}[\omega])$, where ω is a primitive cube root of unity, and the non-cocompact lattice of minimal covolume.

Conjecture 2.1 *All lattices in* $PSL(2, \mathbb{C})$ *and* $PSL(2, \mathbb{R})$ *are profinitely rigid in the absolute sense.*

For lattices Γ_1, $\Gamma_2 < PSL(2, \mathbb{R})$ we know that $\widehat{\Gamma}_1 \cong \widehat{\Gamma}_2$ implies $\Gamma_1 \cong \Gamma_2$, since relative profinite rigidity has been established in this context [8]. For lattices in $PSL(2, \mathbb{C})$ this is unknown but Liu [21] proved that for each lattice Γ, only finitely many other lattices can have the same profinite completion as Γ, up to isomorphism: in the terminology of [17], the *profinite genus* of Γ among lattices in $PSL(2, \mathbb{C})$ is finite. It is unknown if the genus of Γ among all finitely generated (or finitely presented) residually finite groups is finite (a weaker form of Conjecture 2.1). The profinite completion distinguishes the fundamental groups of 3-manifolds that are hyperbolic from those which are not [30]. Many non-hyperbolic 3-manifolds can be distinguished from all others by the profinite completions of their fundamental groups (e.g. [28]), but the examples of Stebe described above $\mathbb{Z}^2 \rtimes_\varphi \mathbb{Z}$ show that not all 3-manifold groups enjoy this property.

Important elements of the proof of Theorem 2.1 extend to lattices in other Lie groups (cf. [22]), but other aspects, particularly the control on finitely generated subgroups, do not. Correspondingly, it is unknown whether groups such as $SL(3, \mathbb{Z})$ are profinitely rigid, and profinite rigidity is known to fail for certain lattices in other semisimple Lie groups [2], even in rank one [27].

3   Restrictions on the Nature of Profinitely-Free Groups

As we discussed earlier, a basic challenge that one faces when trying to settle a challenge such as Conjecture 1.1 is that one starts out knowing essentially nothing about the nature of Γ if $\widehat{\Gamma} \cong \widehat{F}_r$. This challenge is fertile because it forces one to find ways of extracting information about free groups from their finite quotients alone. For example, one might ask if Γ must be finitely presented, or hyperbolic (in the sense of Gromov), or linear, or residually-free? None of these properties is known. On the other hand, one can prove that Γ must have the same nilpotent quotients as F_r, that it must satisfy a version of the Freiheitsatz, and that it cannot have a finitely generated normal subgroup of infinite index other than $\{1\}$; see [13]. A theorem proved recently by Jaikin-Zapirain [19] is particularly intriguing because it is the first to place Γ in a class of groups where there is a reasonable hope of establishing a classification theorem that might allow one to resolve Conjecture 1.1: he proves that a finitely generated, residually finite group Γ with $\widehat{\Gamma} \cong \widehat{F}_r$ must be residually nilpotent, hence *parafree* in the sense of Baumslag [4]. He also proves that any finitely generated, residually finite group with the same profinite completion as a surface group must be residually nilpotent.

4   Failure of Profinite Rigidity Close to Free Groups

The relatives of free groups that we consider here are virtually free groups, hyperbolic groups, direct products of free groups, and 3-manifold groups.

It is easy to see that if N is finite then $N \times \mathbb{Z}$ is profinitely rigid. Examples of Baumslag [5] show that this rigidity fails if one replaces $N \times \mathbb{Z}$ with a semidirect product. Moreover, as explained in [14], one can modify Baumslag's construction to exhibit pairs of non-isomorphic groups H_1 and H_2 that have the same finite quotients and have the same finite index in a group $N \times \mathbb{Z}$ with N finite. To see this, we consider $G_1 = (\mathbb{Z}/25) \rtimes_\alpha \mathbb{Z}$ and $G_2 = (\mathbb{Z}/25) \rtimes_\beta \mathbb{Z}$, where, in multiplicative notation, $\alpha \in \text{Aut}(\mathbb{Z}/25)$ is $\alpha(x) = x^6$ and $\beta(x) = x^{11}$. Noting that α and β generate the same cyclic subgroup of order 5 in $\text{Aut}(\mathbb{Z}/25)$, one can prove by direct argument that $G_1 \not\cong G_2$ but $\widehat{G}_1 \cong \widehat{G}_2$.

Let $N = (\mathbb{Z}/25) \rtimes_\alpha (\mathbb{Z}/5) = \langle x, y \rangle$, where x generates the first factor and the generator y in the second factor acts as α. Let t be a generator for the second factor of $N \times \mathbb{Z}$. Then $H_1 = \langle x, yt \rangle$ and $H_2 = \langle x, y^2 t \rangle$ both have index 5 in $N \times \mathbb{Z}$, and $H_1 \cong G_1$ while $H_2 \cong G_2$.

By taking free products of copies of these groups, we see that there are virtually free groups of every finite rank that are not profinitely rigid (cf. [17]). In contrast, Conjecture 2.1 posits that free products of finite cyclic groups are profinitely rigid. Also, if Conjecture 1.1 is true, every finitely generated, residually finite group with the same finite images as a virtually free group must itself be virtually free.

Free groups are hyperbolic and 1-dimensional. Thus, when exploring the limits of Conjecture 1.1, one might wonder about hyperbolic groups of dimension 2. To explain why Conjecture 1.1 fails in this setting, we need a supply of finitely presented groups that are infinite but do not map onto any non-trivial finite group. There are various ways to construct such groups. To be explicit, we consider the following family from [9]; one knows that these groups are infinite because they are amalgamated free products of groups that have infinite abelianisation.

$$B_p = \langle a, b, \alpha, \beta \mid ba^{-p}b^{-1}a^{p+1}, \ \beta\alpha^{-p}\beta^{-1}\alpha^{p+1}, \ [bab^{-1}, a]\beta^{-1}, \ [\beta\alpha\beta^{-1}, \alpha]b^{-1} \rangle.$$

By applying the Rips construction to this presentation ([10, p. 224]), we obtain a short exact sequence $1 \to N \to \Gamma \to B_p \to 1$ with N finitely generated and Γ a residually finite, hyperbolic group with a 2-dimensional classifying space. It is easy to imagine that $\widehat{B_p} = 1$ might imply that $\widehat{N} \cong \widehat{\Gamma}$, and using the fact that $H_2(B_p, \mathbb{Z}) = 0$ one can prove that this is indeed the case (see [9]). With [7, Theorem A] in hand, one can modify this argument to prove the following result, in which I emphasise that hyperbolicity is in the sense of Gromov, in contrast to Conjecture 2.1.

Theorem 4.1 *There exist residually finite, (Gromov) hyperbolic groups Γ of dimension 2 with uncountably many non-isomorphic subgroups $\iota_H : H \hookrightarrow \Gamma$ such that $\widehat{\iota}_H : \widehat{H} \to \widehat{\Gamma}$ is an isomorphism. Moreover, one can arrange for infinitely many of these subgroups to be finitely generated.*

Because the second homology group $H_2(B_p, \mathbb{Z})$ is trivial, B_p can also serve as Q in the following criterion, which originates in the work of Platonov and Tavgen [23] and was adapted in [3] and [9].

Proposition 4.2 ([23]) *Let $f : G \to Q$ be an epimorphism of groups, with G finitely generated and Q finitely presented. Suppose that $\widehat{Q} = 1$ and $H_2(Q, \mathbb{Z}) = 0$. Then, the fibre product $P = \{(g, h) \mid f(g) = f(h)\} < G \times G$ is finitely generated and $P \hookrightarrow G \times G$ induces an isomorphism $\widehat{P} \xrightarrow{\cong} \widehat{G \times G}$.*

By applying the above criterion to epimorphisms $F \to Q$ from a free group, Platonov and Tavgen showed that *the direct product of two non-abelian free groups is not profinitely rigid*, the second part of the following theorem. The first part can be proved by applying a similar template of proof to suitable sequences of quotients $F \rightarrow Q_1 \rightarrow Q_2 \rightarrow \ldots$. The third part follows from the fact that a finitely presented subgroup of $F \times F$ must be of finite index if it maps onto both factors and intersects each non-trivially [6].

Theorem 4.3 *Let F be a finitely generated, non-abelian free group.*

(1) *There exist uncountably many non-isomorphic groups H such that $H \hookrightarrow F \times F$ induces an isomorphism $\widehat{H} \cong \widehat{F \times F}$;*
(2) *infinitely many of these groups H are finitely generated.*
(3) *There does not exist a finitely presented subgroup $H \neq F \times F$ such that $H \hookrightarrow F \times F$ induces an isomorphism $\widehat{H} \cong \widehat{F \times F}$.*

I deliberately phrased this result in a way that emphasizes the importance of *finiteness properties* in the context of profinite rigidity. For the moment, it is unclear what role finiteness properties might play in Conjecture 1.1. In particular, it is possible that the conjecture is false for finitely generated groups but true if one assumes that Γ is finitely presented. In this vein, Alan Reid, Ryan Spitler and I recently proved [15] that there exist finitely presented, residually finite groups that are profinitely rigid amongst all *finitely presented*, residually finite groups, but have infinite genus among *finitely generated*, residually finite groups. Our examples are direct products $G \times G$ where G is the fundamental group of a certain type of Seifert fibre space (a 3-manifold foliated by circles); the centre of G is infinite cyclic and $G/Z(G)$ is isomorphic to one of the triangle groups $\Delta < \mathrm{PSL}(2, \mathbb{R})$ covered by Theorem 2.1. With this recent result in mind, the reader should compare Theorem 4.3 with:

Conjecture 4.1 *Let F and F' be finitely generated free groups. If a finitely presented, residually finite group Γ has the same finite quotients as $F \times F'$, then Γ is isomorphic to $F \times F'$.*

Acknowledgements I thank my longstanding collaborator Alan Reid for the many insights that he has shared with me during our exploration of the topics discussed here, and I acknowledge with gratitude my debt to the late Fritz Grunewald, who enticed me into this field.

References

1. I. Agol. Virtual properties of 3-manifolds. In: *Proceedings of the International Congress of Mathematicians, Seoul 2014*, **1**, pp. 141–170, Kyung Moon Sa, Seoul (2014).
2. M. Aka. Arithmetic groups with isomorphic finite quotients. *J. Algebra* **352**, 322–340 (2012)
3. H. Bass and A. Lubotzky. Nonarithmetic superrigid groups: counterexamples to Platonov's conjecture. *Annals of Math.* **151**, 1151–1173 (2000).
4. G. Baumslag. Groups with the same lower central sequence as a relatively free group. I. *Trans. Amer. Math. Soc.* **129**, 308–321 (1967).
5. G. Baumslag. Residually finite groups with the same finite images. *Compositio Math.* **29**, 249–252 (1974).

6. G. Baumslag and J. Roseblade. Subgroups of direct products of free groups. *J. London Math. Soc.* **30**, 44–52 (1984).

7. M.R. Bridson. The homology of groups, profinite completions, and echoes of Gilbert Baumslag. In: *Elementary Theory of Groups and Group Rings, and Related Topics* (P. Baginski, B. Fine, A. Moldenhauer, G. Rosenberger and V. Shpilrain, eds.), pp. 11–28, De Gruyter, Berlin, Boston (2020).

8. M.R. Bridson, M. Conder and A.W. Reid. Determining Fuchsian groups by their finite quotients. *Israel J. Math.* **214**, 1–41 (2016).

9. M.R. Bridson and F. Grunewald. Grothendieck's problems concerning profinite completions and representations of groups. *Annals of Math.* **160**, 359–373 (2004).

10. M.R. Bridson and A. Haefliger. *Metric Spaces of Non-Positive Curvature.* Grund. Math. Wiss. **319**, Springer-Verlag, Heidelberg-Berlin (1999).

11. M.R. Bridson, D.B. McReynolds, A.W. Reid and R. Spitler. Absolute profinite rigidity and hyperbolic geometry. *Annals of Math.* **192**, 679–719 (2020).

12. M.R. Bridson, D.B. McReynolds, A.W. Reid and R. Spitler. On the profinite rigidity of triangle groups. *Bull. London Math. Soc.* **53**, 1849–1862 (2021).

13. M.R. Bridson and A.W. Reid. Nilpotent completions of groups, Grothendieck pairs, and four problems of Baumslag. *Int. Math. Res. Not.(IMRN)* **2015**, no. 8, 2111–2140 (2015).

14. M.R. Bridson and A.W. Reid. Profinite rigidity, Kleinian groups, and the cofinite Hopf property. *Michigan Math. J.* **72**, 25–49 (2022).

15. M.R. Bridson, A.W. Reid and R. Spitler. Absolute profinite rigidity, direct products, and finite presentability. In preparation.

16. A. Grothendieck. Representations linéaires et compatifications profinie des groupes discrets. *Manuscripta Math.* **2**, 375–396 (1970).

17. F. Grunewald and P. Zalesskii. Genus for groups. *J. Algebra* **326**, 130–168 (2011).

18. F. Haglund and D. Wise. Special cube complexes. *Geom. Funct. Anal.* **17**, 1551–1620 (2008).

19. A. Jaikin-Zapirain. The finite and soluble genus of finitely generated free and surface groups. Preprint available at https://matematicas.uam.es/~andrei.jaikin/preprints/profinitefree.pdf

20. E.I. Khukhro and V.D. Mazurov (eds.) *The Kourovka notebook. Unsolved problems in group theory.* 18th ed., Ross. Akad. Sci. Sib. Div., Inst. Math., Novosibirsk (2014).

21. Y. Liu. Finite volume hyperbolic 3-manifolds are almost determined by their finite quotients. arXiv:2011.09412.

22. D.B. McReynolds and R. Spitler. Profinite completions of linear groups and rigid representation. In preparation.

23. V.P. Platonov and O.I. Tavgen. Grothendieck's problem on profinite completions and representations of groups. *K-Theory* **4**, 89–101 (1990).

24. A.W. Reid. Profinite rigidity. In: *Proceedings of the International Congress of Mathematicians (Rio de Janeiro)*, vol. 2, pp. 1211–1234 (2018).

25. J-P. Serre. Exemples de variétés projectives conjuguées non homéomorphes. *C. R. Acad. Sci. Paris* **258**, 4194–4196 (1964).

26. P.F. Stebe. A residual property of certain groups. *Proc. Amer. Math. Soc.* **26**, 37–42 (1970).

27. M. Stover. Lattices in PU(n, 1) that are not profinitely rigid. *Proc. Amer. Math. Soc.* **147**, 5055–5062 (2019).

28. G. Wilkes. Profinite rigidity for Seifert fibre spaces. *Geom. Dedicata* **188**, 141–163 (2017).

29. H. Wilton. Essential surfaces in graph pairs. *J. Amer. Math. Soc.* **31**, 893–919 (2018).

30. H. Wilton and P.A. Zalesskii. Distinguishing geometries using finite quotients. *Geom. Topol.* **21**, 345–384 (2017).

On the Algebraic K-Theory of Hecke Algebras

Arthur Bartels and Wolfgang Lück

Personal Note

When I was invited to contribute to the Festschrift in honor of Springer's Editorial Director Dr. Catriona Byrne, I was flattered and did not hesitate to accept the invitation. I have known her since I submitted my first book with the title "Transformation groups and algebraic K-theory" to the Lecture Notes in Mathematics *in 1989. She has always been a very reliable and competent partner for all of my book projects. She is one of the few people working for a publishing house who manages to keep the right balance between the mercantile aspects and the interest of the mathematical community. One always has the impression that she cares about the contents and the authors of any submission. In particular, we became close when Andrew Ranicki and I, and Springer, had a very hard time dealing with the problems caused by the managing editor of the journal K-theory in 2007, see "Persönliches Protokoll zur Zeitschrift K-Theory" in Mitteilungen der Deutschen Mathematiker-Vereinigung, Band 15 Heft 3, 2007.*

I wish Catriona all the best for the many years to come.

A. Bartels
Westfälische Wilhelms-Universität Münster, Mathematicians Institut, Münster, Germany
e-mail: bartelsa@math.uni-muenster.de

W. Lück (✉)
Mathematicians Institut of the Rheinische Friedrich-Wilhelms-University Bonn, Bonn, Germany
e-mail: wolfgang.lueck@him.uni-bonn.de

© The Author(s), under exclusive license to Springer Nature Switzerland AG 2023
J.-M. Morel, B. Teissier (eds.), *Mathematics Going Forward*, Lecture Notes in Mathematics 2313, https://doi.org/10.1007/978-3-031-12244-6_19

1 Introduction

Let G be a td-group, i.e., a locally compact second countable totally disconnected topological Hausdorff group. Our ultimate goal is to compute the algebraic K-groups and in particular the projective class group of the (standard) Hecke algebra $\mathcal{H}(G)$ of G, which is defined in terms of locally constant functions with compact support from G to the real or complex numbers and the convolution product. We want to show that the canonical map

$$\operatorname{colim}_{K \in \operatorname{Sub}_{Cop}(G)} K_0(\mathcal{H}(K)) \overset{\cong}{\to} K_0(\mathcal{H}(G)) \tag{1}$$

is bijective. Here $\operatorname{Sub}_{Cop}(G)$ is the following category. Objects are compact open subgroups K of G, a morphism $f \colon K \to K'$ is a group homomorphism, for which there exists $g \in G$ satisfying $f(k) = gkg^{-1}$ for all $k \in K$, and we identify two such group homomorphisms $f \colon K \to K'$ and $f' \colon K \to K'$ if they differ by an inner automorphism of K'. In particular, the obvious map

$$\bigoplus_K K_0(\mathcal{H}(K)) \to K_0(\mathcal{H}(G)) \tag{2}$$

is surjective, where K runs through the compact open subgroups of G.

Dat [7, Theorem 1.6 and Corollary 4.22] showed, following ideas of Bernstein, that the map (2) is rationally surjective for a reductive p-adic group G. He used for the proof the Hattori–Stallings rank and input from the representation theory of reductive p-adic groups. Dat also asked the question whether the map (2) is surjective without rationalizing, see the sentence after [8, Proposition 1.10] and the formulation of the weaker conjecture [8, Conjecture 1.11].

The projective class group $K_0(\mathcal{H}(G))$ is interesting for the study of smooth G-representations, since every finitely generated smooth G-representation has a finite projective resolution and hence defines elements in it, see for instance [5, Theorem 29 on page 97 and Proposition 32 on page 60], [17], [18], [19], [20].

If G is discrete, the family Cop of compact open subgroups reduces to the family $\mathcal{F}in$ of finite subgroups of G and the bijectivity of the map (1) reduces to the bijectivity of the canonical map

$$\operatorname{colim}_{F \in \operatorname{Sub}_{\mathcal{F}in}(G)} K_0(\mathbb{C}F) \overset{\cong}{\to} K_0(\mathbb{C}G), \tag{3}$$

which follows from the K-theoretic Farrell–Jones Conjecture for $\mathbb{C}G$.

Our ultimate and long term goal is to the prove the version of the *K-theoretic Farrell–Jones Conjecture for the Hecke algebra of td-groups* for any closed subgroup G of any reductive p-adic group. It predicts the bijectivity of the assembly map

$$H_n^G(E_{Cop}(G); \mathbf{K}_{\mathcal{H}}^\infty) \overset{\cong}{\to} H_n^G(G/G; \mathbf{K}_{\mathcal{H}}^\infty) = K_n(\mathcal{H}(G)) \tag{4}$$

for every $n \in \mathbb{Z}$. Here the source is a smooth G-homology theory, which digests smooth G-CW-complexes and satisfies $H_n^G(G/H; \mathbf{K}_{\mathcal{H}}^\infty) = K_n(\mathcal{H}(H))$ for open subgroups $H \subseteq G$, and the smooth G-CW-complex $E_{\mathcal{C}op}(G)$ is a model for the classifying space of the family of compact open subgroups, or, equivalently the classifying space for smooth proper G-actions in the realm of G-CW-complexes. This map will be constructed in [3], where a formulation of the K-theoretic Farrell–Jones Conjecture is given for Hecke categories, which generalize the notion of a Hecke algebra.

We will not prove the K-theoretic Farrell–Jones Conjecture for Hecke categories in this paper. At least we present a direct proof of it in the special case that G is covirtually infinite cyclic, i.e., G contains a normal compact open subgroup L such that the quotient G/L is the discrete group $\mathbb{Z}$. Then the conjecture boils down to Theorem 9.1, which says that there is a Wang sequence, infinite to the left,

$$\cdots \xrightarrow{K_2(i)} K_2(\mathcal{H}(G)) \xrightarrow{\partial_2} K_1(\mathcal{H}(L)) \xrightarrow{\mathrm{id}-K_1(\varphi)} K_1(\mathcal{H}(L))$$

$$\xrightarrow{K_1(i)} K_1(\mathcal{H}(G)) \xrightarrow{\partial_1} K_0(\mathcal{H}(L)) \xrightarrow{\mathrm{id}-K_0(\varphi)} K_0(\mathcal{H}(L))$$

$$\xrightarrow{K_0(i)} K_0(\mathcal{H}(G)) \to 0,$$

where $\varphi \colon L \to L$ is the automorphism given by conjugation with some preimage of the generator of the infinite cyclic group G/L under the projection $G \to G/L$ and $i \colon L \to G$ is the inclusion, and that we have

$$K_n(\mathcal{H}(G)) = 0 \quad \text{for } n \leq -1.$$

So in this paper we can confirm the Farrell–Jones Conjecture for covirtually infinite cyclic td-groups. One may say that this paper plays the same role for the Farrell–Jones Conjecture for Hecke algebras as the papers by Farrell–Hsiang [9] and Pimsner–Voiculescu [15] did for the Farrell–Jones Conjecture for discrete groups and the Baum–Connes Conjecture. To our knowledge this paper presents the first instance of a version of the Farrell–Jones Conjecture for non-discrete groups.

One application of this paper will be that the bijectivity of (4) implies the bijectivity of (1). Moreover, Theorems 7.2 and 10.1 will be key ingredients in the part of the forthcoming proof of the Farrell–Jones Conjecture, where we will reduce the family $\mathcal{C}$vcy of (not necessarily open) covirtually cyclic subgroups to the family $\mathcal{C}$op.

We mention that we will look at more complicated Hecke algebras than the standard ones. We will allow other rings than $\mathbb{R}$ or $\mathbb{C}$. Moreover, we take a G-action on R by ring automorphisms and a normal character, which is an obvious generalization of a central character, into account. In the sequel papers we will replace the Hecke algebras by the more general notion of a Hecke category, since allowing more general coefficients will ensure the desirable inheritance to closed subgroups of the Farrell–Jones Conjecture. This is interesting in the case of reductive p-adic groups, since important subgroups such as the Borel subgroup are in general not open.

One ingredient for the main results of this paper is the Bass–Heller–Swan decompositions for additive categories and the presentation of criteria for the vanishing of the Nil-term, see Sect. 6, and [2, 13]. The second is the analysis of the filtration of the Hecke algebra of a compact td-groups in terms of approximate units, see Sect. 7.

1.1 Conventions and Notations

- A td-group is a locally compact second countable totally disconnected topological Hausdorff group. Note that for any td-group its unit has a neighborhood basis consisting of compact open subgroups.
- A subgroup is always assumed to be closed.
- A group homomorphism has closed image and is an identification onto it.
- We denote by R an associative ring, which is not necessarily commutative and does not necessarily have a unit. If a ring has a unit, it is called a *unital ring*. In almost all cases we will require for a unital ring R that $\mathbb{Q} \subseteq R$ holds, i.e., for every integer $n \geq 1$ the element $n \cdot 1 = 1 + 1 + \cdots + 1$ has a multiplicative inverse in R.
- In a ring the unit is denoted by 1. In a group the unit is denoted by e.
- For an epimorphism $p \colon S \to S'$ of sets, a *transversal* T is a subset $T \subseteq S$ such that the restriction of p to T yields a bijection $p|_T \colon T \xrightarrow{\cong} S'$. If S is a group, we always assume that the unit is in T.

2 Hecke Algebras

In this section we slightly generalize the notions of a Hecke algebra by implementing a normal character. An introduction to Hecke algebras can be found for instance in [5, 6, 10].

2.1 Basic Setup

Let R be a (not necessarily commutative) associative unital ring with $\mathbb{Q} \subseteq R$. Let G be a td-group with a normal (not necessarily open or central) subgroup $N \subseteq G$. Put $Q = G/N$. Then we obtain an extension of td-groups $1 \to N \to G \xrightarrow{\mathrm{pr}} Q \to 1$.

Consider a group homomorphism $\rho \colon G \to \mathrm{aut}(R)$, where $\mathrm{aut}(R)$ is the group of automorphism of the unital ring R. We will assume throughout the paper that the kernel of ρ is open, in other words, G acts smoothly on R.

We write $gr = \rho(g)(r)$ for $g \in G$ and $r \in R$. With this notation we get $er = r$, $g1 = 1$, $(g_1 g_2)r = g_1(g_2 r)$, $g(r_1 r_2) = (gr_1)(gr_2)$ and $g(r_1 + r_2) = gr_1 + gr_2$ for $g, g_1, g_2 \in G, r, r_1, r_2 \in R$, and the units $e \in G$ and $1 \in R$.

A *normal character* is a locally constant group homomorphism

$$\omega \colon N \to \mathrm{cent}(R)^{\times}$$

to the multiplicative group of central units of R satisfying

$$\omega(gng^{-1}) = \omega(n) \tag{5}$$

for all $n \in N$ and $g \in G$. Note that $\ker(\omega)$ is an open subgroup of N and a normal subgroup of G. We will need the following compatibility condition between the normal character and the G-action ρ on R, namely for $n \in N$, $g \in G$, and $r \in R$

$$g\omega(n) = \omega(n); \tag{6}$$

$$nr = r. \tag{7}$$

Let μ be a $\mathbb{Q}$-*valued Haar measure on Q, i.e., a Haar measure μ on Q such that for any compact open subgroup $K \subseteq Q$ we have $\mu(K) \in \mathbb{Q}^{>0}$. Given any Haar measure μ on Q, we can normalize it to a $\mathbb{Q}$-valued Haar measure by choosing a compact open subgroup $L_0 \subseteq Q$ and defining $\mu' = \frac{1}{\mu(L_0)} \cdot \mu$.

2.2 The Construction of the Hecke Algebra

An element s in the *Hecke algebra* $\mathcal{H}(G; R, \rho, \omega)_\mu$ is given by a map $s \colon G \to R$ with the following properties

- The map $s \colon G \to R$ is locally constant.
- The image of its support $\mathrm{supp}(s) := \{g \in G \mid s(g) \neq 0\} \subseteq G$ under $\mathrm{pr} \colon G \to Q$ is a compact subset of Q.
- For $n \in N$ and $g \in G$ we have

$$s(ng) = \omega(n) \cdot s(g); \tag{8}$$

$$s(gn) = s(g) \cdot \omega(n). \tag{9}$$

Definition 2.1 Let $P_{\rho,\omega}$ the set of compact open subgroups $K \subseteq G$ satisfying

$$kr = r \quad \text{for } k \in K, r \in R; \tag{10}$$

$$\omega(n) = 1 \quad \text{for } n \in N \cap K. \tag{11}$$

We abbreviate $P = P_{\rho,\omega}$ if ρ and ω are clear from the context.

We call an element $K \in P$ *admissible* for $s: G \to R$ if for all $g \in G$ and $k \in K$ we have

$$s(kg) = s(g); \tag{12}$$

$$s(gk) = s(g). \tag{13}$$

Note that the existence of an admissible element $K \in P$ is equivalent to the condition that s is locally constant, since we assume that the image of the support s under the projection $G \to Q$ is compact and the kernel of the normal character is open in N, and there exists a neighborhood basis of the unit of G consisting of compact open subgroups. Moreover, for $K \in P$, which is admissible for s, every open subgroup $K' \subseteq K$ is also admissible.

Remark 2.1 (Redundancy) Note that condition (9) follows from conditions (5) and (8) by the following calculation

$$s(gn) = s(gng^{-1}g) \overset{(8)}{=} \omega(gng^{-1}) \cdot s(g) \overset{(5)}{=} \omega(n) \cdot s(g) \overset{\omega(n) \in \mathrm{cent}(R)}{=} s(g) \cdot \omega(n).$$

Analogously condition (8) follows from conditions (5) and (9).

The sum of two elements s, s' in $\mathcal{H}(G; R, \rho, \omega)_\mu$ is defined by

$$(s + s')(g) := s(g) + s'(g) \text{ for } g \in G. \tag{14}$$

In order to define the product, choose $K \in P$ which is admissible for s and admissible for s', and a transversal T for the projection $p: G \to G/NK$, where NK is the subgroup of G given by $\{nk \mid n \in N, k \in K\}$. Define the product $s \cdot s'$ by

$$(s \cdot s')(g) := \mu(\mathrm{pr}(K)) \cdot \sum_{g' \in T} s(gg') \cdot gg's'(g'^{-1}). \tag{15}$$

Note that K may depend on s, but not on g, whereas T can depend on both s and g. The independence of the transversal follows from the following computation for $g, g' \in G, n \in N$ and $k \in K$

$$s(g(g'nk)) \cdot g(g'nk)s'((g'nk)^{-1})$$

$$= s((gg'n)k) \cdot (gg'n)ks'(k^{-1}n^{-1}g'^{-1})$$

$$\overset{(12),\,(13)}{=} s(gg'n) \cdot (gg'n)ks'(n^{-1}g'^{-1})$$

$$\overset{(10)}{=} s(gg'n) \cdot gg'ns'(n^{-1}g'^{-1})$$

$$\overset{(8),\,(9)}{=} s(gg') \cdot \omega(n) \cdot gg'n(\omega(n^{-1}) \cdot s'(g'^{-1}))$$

$$= s(gg') \cdot \omega(n) \cdot gg'n\omega(n^{-1}) \cdot gg'ns'(g'^{-1})$$

$$\overset{(6),\,(7)}{=} s(gg') \cdot \omega(n) \cdot \omega(n^{-1}) \cdot gg's'(g'^{-1})$$

$$= s(gg') \cdot \omega(n \cdot n^{-1}) \cdot gg's'(g'^{-1})$$

$$= s(gg') \cdot \omega(e) \cdot gg's'(g'^{-1}) = s(gg') \cdot gg's'(g'^{-1}).$$

We leave the elementary proof to the reader that the definition of the product (15) is independent of the choice of K and that we do get the structure of a (non-unital) ring on $\mathcal{H}(G; R, \rho, \omega)_\mu$. A more general setting including all proofs will be presented in detail in [3]. Moreover, one easily checks.

Lemma 2.2 *Consider two elements $s, s' \in \mathcal{H}(G; R, \rho, \omega)_\mu$ and compact open subgroups K, K' of G. Suppose that K is admissible for s and K' is admissible for s'.*

Then $K \cap K'$ is admissible for the product $s' \cdot s$.

2.3 Functoriality in Q

Let G, N, Q, R, ρ, ω, and μ be as in Sect. 2.1. In particular we can consider the Hecke algebra $\mathcal{H}(G; R, \rho, \omega)_\mu$ see Sect. 2.2.

Consider a (not necessarily injective or surjective) open group homomorphism $\varphi \colon G' \to G$ of td-groups. Let $N' \subseteq G'$ be a normal subgroup satisfying

$$\varphi(N') = N. \tag{16}$$

Denote by $\mathrm{pr}' \colon G' \to Q' := G'/N'$ the projection. Let $\overline{\varphi} \colon Q' \to Q$ be the open group homomorphism induced by φ. Define a group homomorphism $\rho' \colon G' \to \mathrm{aut}(R)$ and a normal character $\omega' \colon N' \to \mathrm{cent}(R)^\times$ by

$$\rho' = \rho \circ \varphi; \tag{17}$$

$$\omega'(n') = \omega(\varphi(n')) \quad \text{for } n' \in N'. \tag{18}$$

Choose a $\mathbb{Q}$-valued Haar measure on μ' on Q'. Then we can consider the Hecke algebra $\mathcal{H}(G'; R, \rho', \omega')_{\mu'}$. Next we want to construct a homomorphism of rings

$$\varphi_* \colon \mathcal{H}(G'; R, \rho', \omega')_{\mu'} \to \mathcal{H}(G; R, \rho, \omega)_\mu. \tag{19}$$

Consider an element $s' \colon G' \to R$ in $\mathcal{H}(G'; R, \rho', \omega')_{\mu'}$. Choose $K' \in P_{\rho', \omega'}$, which is admissible for s'. Then $\varphi(K') \in P_{\rho, \omega}$. Fix $g \in G$. Consider $g' \in \varphi^{-1}(g\varphi(N'K'))$. Then $\varphi(g')^{-1}g$ belongs to $\varphi(N'K')$. Choose $n' \in N'$ and $k' \in K'$ with $\varphi(g'n'k') = g$. Put

$$\widetilde{s}'(g', g) := s'(g') \cdot \omega(\varphi(n')) \in R. \tag{20}$$

One easily checks that this definition is independent of the choice of $n' \in N'$ and $k' \in K'$. Obviously we have $\widetilde{s}'(g', \varphi(g')) = s'(g')$ for $g' \in G'$. Choose a transversal T' of the projection $G' \to G'/N'K'$, which is allowed to depend on s'. Put $T'(g) = T' \cap \varphi^{-1}(g\varphi(N'K'))$. Then we define

$$\varphi_*(s')(g) = \frac{\mu'(\mathrm{pr}'(K'))}{\mu(\mathrm{pr}(\varphi(K')))} \cdot \sum_{g' \in T'(g)} \widetilde{s}'(g', g). \tag{21}$$

This is a well-defined element in $\mathcal{H}(G; R, \rho, \omega)_\mu$, which is independent of the choice of T and K'. One easily checks

Lemma 2.3

(1) *We have* $\mathrm{supp}(\varphi_*(s')) \subseteq \varphi(\mathrm{supp}(s'))$.
(2) *If* $K' \in P'_{\rho', \omega'}$ *is admissible for* s', *then* $\varphi(K')$ *is admissible for* $\varphi_*(s')$.
(3) *Suppose that* φ *is injective. Then we get*

$$\varphi_*(s')(g) = \begin{cases} \frac{\mu'(\mathrm{pr}'(K'))}{\mu(\mathrm{pr}(\varphi(K')))} \cdot s(g') & \text{if } \varphi(g') = g \text{ for some } g' \in G' \\ 0 & g' \notin \mathrm{im}(\varphi) \end{cases}$$

and

$$\mathrm{supp}_G(\varphi_*(s')) = \varphi(\mathrm{supp}_{G'}(s')).$$

(4) *The map* $\varphi_* \colon \mathcal{H}(G'; R, \rho', \omega')_{\mu'} \to \mathcal{H}(G; R, \rho, \omega)_\mu$ *is a homomorphism of (non-unital) rings.*

2.4 Approximate Units

Definition 2.4 (Rings with Approximate Units) An *approximate unit* for a ring R is a subset $\{e_i \mid i \in I\}$ of elements $e_i \in R$ indexed by some directed set I such that $e_i \cdot e_j = e_i = e_j \cdot e_i$ holds for $i \leq j$ and for every element $r \in R$ there exists an index $i \in I$ with $e_i \cdot r = r = r \cdot e_i$.

The ring R has an approximate unit if and only if there is a directed system of subrings $\{R_i \mid i \in I\}$ indexed by inclusion such that each R_i is unital and $R = \bigcup_{i \in I} R_i$. Obviously a unital ring has an approximate unit.

Note that the ring $\mathcal{H}(G; R, \rho, \omega)_\mu$ has a unit if and only if G is discrete. If G is not discrete, $\mathcal{H}(G; R, \rho, \omega)_\mu$ has at least an approximate unit by the following construction.

Lemma 2.2 implies for $K \in P$ that the subset

$$\mathcal{H}(G//K; R, \rho, \omega)_\mu \subseteq \mathcal{H}(G; R, \rho, \omega)_\mu \tag{22}$$

consisting of those elements, for which K is admissible, is closed under addition and multiplication and hence is a subring. Define an element 1_K in $\mathcal{H}(G//K; R, \rho, \omega)_\mu$ by

$$1_K(g) = \begin{cases} \frac{1}{\mu(\mathrm{pr}(K))} \cdot \omega(n) & \text{if } g \in NK, g = nk \text{ for } n \in N, k \in K; \\ 0 & \text{otherwise.} \end{cases} \tag{23}$$

Lemma 2.5 *The element 1_K is a unit in $\mathcal{H}(G//K; R, \rho, \omega)_\mu$. Moreover*

$$\mathcal{H}(G//K; R, \rho, \omega)_\mu \subseteq \mathcal{H}(G//K'; R, \rho, \omega)_\mu \quad \text{if } K' \subseteq K;$$

$$\mathcal{H}(G; R, \rho, \omega)_\mu = \bigcup_K \mathcal{H}(G//K; R, \rho, \omega)_\mu,$$

where K runs through the elements of P.

2.5 Discarding μ

In the sequel we omit the subscript μ in the notation of the Hecke algebra, since for two $\mathbb{Q}$-valued Haar measures μ and μ' on G/N there is precisely one rational number r satisfying $r > 0$ and $\mu' = r \cdot \mu$, and the map

$$\mathcal{H}(G; R, \rho, \omega)_{\mu'} \xrightarrow{\cong} \mathcal{H}(G; R, \rho, \omega)_\mu, \quad s \mapsto r \cdot s$$

is an isomorphism of rings.

3   ℤ-Categories, Additive Categories and Idempotent Completions

A ℤ-*category* is a not necessarily unital category $\mathcal{A}$ such that for every two objects A and A' in $\mathcal{A}$ the set of morphisms $\mathrm{mor}_{\mathcal{A}}(A, A')$ has the structure of a ℤ-module and composition is ℤ-bilinear. If G is a group, a *G*-ℤ-*category* $\mathcal{A}$ is a ℤ-category with a left G-action by automorphisms of ℤ-categories. Note that we do not require that $\mathcal{A}$ has identity morphisms. However, if $\mathcal{A}$ has identity morphisms, the G-action is required to respect them. Given a ring R, we denote by $\underline{R}$ the ℤ-category with precisely one object, whose ℤ-module of endomorphisms is given by R with its additive structure and composition is given by the multiplication in R. Obviously $\underline{R}$ is unital if and only if R is unital.

An *additive category* $\mathcal{A}$ is a ℤ-category with finite direct sums. Given a ring R, the category $R\text{-MOD}_{\mathrm{fgf}}$ of finitely generated free R-modules carries an obvious structure of an additive category. Note that we do not require that $\mathcal{A}$ has identity morphisms. If it does, we call it unital.

Given a ℤ-category $\mathcal{A}$, let $\mathcal{A}_{\oplus}$ be the associated additive category whose objects are finite tuples of objects in $\mathcal{A}$ and whose morphisms are given by matrices of morphisms in $\mathcal{A}$ (of the right size) and the direct sum is given by concatenation of tuples and the block sum of matrices, see for instance [13, Section 1.3]. If $\mathcal{A}$ is unital, $\mathcal{A}_{\oplus}$ is unital.

Let R be a unital ring. Then the obvious inclusion of unital additive categories

$$\underline{R}_{\oplus} \xrightarrow{\cong} R\text{-MOD}_{\mathrm{fgf}} \tag{24}$$

is an equivalence of unital additive categories.

Given an additive category $\mathcal{A}$, its *idempotent completion* $\mathrm{Idem}(\mathcal{A})$ is defined to be the following additive category. Objects are morphisms $p \colon A \to A$ in $\mathcal{A}$ satisfying $p \circ p = p$. A morphism f from $p_1 \colon A_1 \to A_1$ to $p_2 \colon A_2 \to A_2$ is a morphism $f \colon A_1 \to A_2$ in $\mathcal{A}$ satisfying $p_2 \circ f \circ p_1 = f$. Note that $\mathrm{Idem}(\mathcal{A})$ is always unital, regardless of whether $\mathcal{A}$ is unital or not. The identity of an object (A, p) is given by the morphism $p \colon (A, p) \to (A, p)$.

If $\mathcal{A}$ is unital, then there is an obvious embedding

$$\eta(\mathcal{A}) \colon \mathcal{A} \to \mathrm{Idem}(\mathcal{A})$$

sending an object A to $\mathrm{id}_A \colon A \to A$ and a morphism $f \colon A \to B$ to the morphism given by f again. A unital additive category $\mathcal{A}$ is called *idempotent complete* if $\eta(\mathcal{A}) \colon \mathcal{A} \to \mathrm{Idem}(\mathcal{A})$ is an equivalence of unital additive categories, or, equivalently, if for every idempotent $p \colon A \to A$ in $\mathcal{A}$ there are objects B and C and an isomorphism $f \colon A \xrightarrow{\cong} B \oplus C$ in $\mathcal{A}$ such that $f \circ p \circ f^{-1} \colon B \oplus C \to B \oplus C$

is given by $\begin{pmatrix} \mathrm{id}_B & 0 \\ 0 & 0 \end{pmatrix}$. The idempotent completion $\mathrm{Idem}(\mathcal{A})$ of a unital additive category $\mathcal{A}$ is idempotent complete.

Let R be unital ring. Let $R\text{-MOD}_{\mathrm{fgp}}$ be the unital additive category of finitely generated projective R-modules. We obtain an equivalence of unital additive categories $\mathrm{Idem}(R\text{-MOD}_{\mathrm{fgf}}) \xrightarrow{\simeq} R\text{-MOD}_{\mathrm{fgp}}$ by sending an object (F, p) to $\mathrm{im}(p)$. It and the functor of (24) induce an equivalence of unital additive categories

$$\theta_R \colon \mathrm{Idem}\left(\underline{R}_\oplus\right) \xrightarrow{\simeq} R\text{-MOD}_{\mathrm{fgp}}. \tag{25}$$

Let $\mathcal{A}$ be an additive category. Let $\Phi \colon \mathcal{A} \to \mathcal{A}$ be an automorphism of additive categories. Define the additive category $\mathcal{A}_\Phi[t, t^{-1}]$, called the Φ-*twisted finite Laurent category*, as follows. It has the same objects as $\mathcal{A}$. Given two objects A and B, a morphism $f \colon A \to B$ in $\mathcal{A}_\Phi[t, t^{-1}]$ is a formal sum $f = \sum_{i \in \mathbb{Z}} f_i \cdot t^i$, where $f_i \colon \Phi^i(A) \to B$ is a morphism in $\mathcal{A}$ from $\Phi^i(A)$ to B and only finitely many of the morphisms f_i are non-trivial. If $g = \sum_{j \in \mathbb{Z}} g_j \cdot t^j$ is a morphism in $\mathcal{A}_\Phi[t, t^{-1}]$ from B to C, we define the composite $g \circ f \colon A \to C$ by

$$g \circ f := \sum_{k \in \mathbb{Z}} \left(\sum_{\substack{i, j \in \mathbb{Z}, \\ i+j=k}} g_j \circ \Phi^j(f_i) \right) \cdot t^k.$$

If $\mathcal{A}$ is unital, then $\mathcal{A}_\Phi[t, t^{-1}]$ is unital again.

Let R be a (not necessarily unital) ring with an automorphism $\varphi \colon R \xrightarrow{\cong} R$ of rings. Let $R_\varphi[t, t^{-1}]$ be the ring of φ-twisted finite Laurent series with coefficients in R. We obtain from φ an automorphism $\Phi \colon \underline{R} \xrightarrow{\cong} \underline{R}$ of $\mathbb{Z}$-categories. There is an obvious isomorphism of $\mathbb{Z}$-categories

$$\underline{R}_\Phi[t, t^{-1}] \xrightarrow{\cong} \underline{R_\varphi[t, t^{-1}]}. \tag{26}$$

If R is unital, then we obtain equivalences of unital additive categories

$$(\underline{R}_\oplus)_\Phi[t, t^{-1}] \xrightarrow{\simeq} R_\varphi[t, t^{-1}]\text{-MOD}_{\mathrm{fgf}};$$

$$\mathrm{Idem}\left((\underline{R}_\oplus)_\Phi[t, t^{-1}]\right) \xrightarrow{\simeq} R_\varphi[t, t^{-1}]\text{-MOD}_{\mathrm{fgp}}. \tag{27}$$

4 The Algebraic K-Theory of $\mathbb{Z}$-Categories

Let $\mathcal{A}$ be a unital additive category. A construction of the *non-connective K-theory spectrum* $\mathbf{K}^\infty(\mathcal{A})$ of a unital additive category can be found for instance in [11] or [14]. We get from the canonical embedding $\eta(\mathcal{A}) \colon \mathcal{A} \to \mathrm{Idem}(\mathcal{A})$ a weak homotopy equivalence $\mathbf{K}^\infty(\eta(\mathcal{A})) \colon \mathbf{K}^\infty(\mathcal{A}) \to \mathbf{K}^\infty(\text{Idem}(\mathcal{A}))$ on the non-connective K-theory, see for instance [2, Lemma 3.3 (ii)].

Definition 4.1 (Algebraic K-Theory of (Not Necessarily Unital) $\mathbb{Z}$-Categories)
We will define the *algebraic K-theory spectrum* $\mathbf{K}^\infty(\mathcal{A})$ of the (not necessarily unital) $\mathbb{Z}$-category $\mathcal{A}$ to be the non-connective algebraic K-theory spectrum of the unital additive category $\text{Idem}(\mathcal{A}_\oplus)$. Define for $n \in \mathbb{Z}$

$$K_n(\mathcal{A}) := \pi_n(\mathbf{K}^\infty(\mathcal{A})).$$

Note that Definition 4.1 extends the definition of the non-connective K-theory spectrum of unital additive categories to not necessarily unital $\mathbb{Z}$-categories.

A functor $F \colon \mathcal{A} \to \mathcal{A}'$ of (not necessarily unital) $\mathbb{Z}$-categories induces a map of spectra

$$\mathbf{K}^\infty(F) \colon \mathbf{K}^\infty(\mathcal{A}) \to \mathbf{K}^\infty(\mathcal{A}'). \tag{28}$$

If the (not necessarily unital) $\mathbb{Z}$-category $\mathcal{A}$ is the directed union of (not necessarily unital) $\mathbb{Z}$-subcategories $\mathcal{A}_i$, then the canonical map

$$\text{hocolim}_{i \in I} \, \mathbf{K}^\infty(\mathcal{A}_i) \xrightarrow{\simeq} \mathbf{K}^\infty(\mathcal{A}) \tag{29}$$

is a weak homotopy equivalence and for every $n \in \mathbb{Z}$ the canonical map

$$\text{colim}_{i \in I} \, K_n(\mathcal{A}_i) \xrightarrow{\cong} K_n(\mathcal{A}) \tag{30}$$

is a bijection. We conclude (29) and (30) for instance from [11, Corollary 7.2].

If R is an associative ring (not necessarily with a unit), we define the non-connective K-theory spectrum $\mathbf{K}^\infty(R)$ to be $\mathbf{K}^\infty(\underline{R})$ and $K_n(R) := \pi_n(\mathbf{K}^\infty(R))$ for $n \in \mathbb{Z}$. If R has an approximate unit, then our definition of $K_n(R)$ agrees with the usual definition of $K_n(R)$ for a ring without unit by the kernel of the map $K_n(R_+) \to K_n(\mathbb{Z})$, where R_+ is the ring with unit associated to R. Because of Lemma 2.5 this applies to the Hecke algebra $\mathcal{H}(G; R, \rho, \omega)$.

5   Covirtually $\mathbb{Z}$ Groups

Let G, N, Q, R, ρ, P, ω, and μ be as in Sect. 2.1. In particular, we can consider the Hecke algebra $\mathcal{H}(G; R, \rho, \omega)$, see Sect. 2.2. Assume, furthermore, that we have a normal open subgroup $L \subseteq G$ satisfying:

- G/L is isomorphic to $\mathbb{Z}$;
- $N \subseteq L$;
- $M := L/N$ is compact.

Note that we get exact sequences of td-groups $1 \to L \to G \to \mathbb{Z} \to 1$ and $1 \to M \to Q \to \mathbb{Z} \to 1$, where $\mathbb{Z}$ is considered as discrete group and M is compact.

Let $g_0 \in G$ be any element which represents in G/L a generator. Let $\varphi : L \to L$ be the automorphism of L given by conjugation with g_0. Denote by $L \rtimes_{c_{g_0}} \mathbb{Z}$ the td-group given by the semi-direct product of L with the discrete group $\mathbb{Z}$ with respect to c_{g_0}. Then we get an isomorphism of td-groups

$$\alpha : L \rtimes_{c_{g_0}} \mathbb{Z} \xrightarrow{\cong} G; \quad lt^n \mapsto lg_0^n,$$

if $t \in \mathbb{Z}$ is a fixed generator. It also induces an isomorphism $\beta : M \rtimes_{c_{q_0}} \mathbb{Z} \xrightarrow{\cong} Q$, if we put $q_0 = \mathrm{pr}(g_0)$. In the sequel we identify $G = L \rtimes_{c_{g_0}} \mathbb{Z}$ and g_0 with $e_L t$ for $e_L \in L$ the unit and $Q = M \rtimes_{c_{q_0}} \mathbb{Z}$ and $g_0 N$ with $e_Q t$ for $e_Q \in Q$ the unit.

Since $L \subseteq G$ is open, the $\mathbb{Q}$-valued measure μ on G defines a $\mathbb{Q}$-valued measure on L by restriction, which we will denote by μ again. Note that we can consider the Hecke algebra $\mathcal{H}(L; R, \rho|_L, \omega)$.

Next we check that the automorphism $c_{g_0} : L \to L$ induces an automorphism of rings

$$\varphi : \mathcal{H}(L; R, \rho|_L, \omega) \xrightarrow{\cong} \mathcal{H}(L; R, \rho|_L, \omega) \tag{31}$$

by sending $s \in \mathcal{H}(L; R, \rho|_L, \omega)$ given by a function $s : L \to R$ to the element given by the function

$$\varphi(s) : L \to R, \ l \mapsto ts(t^{-1}lt). \tag{32}$$

Note that this is not just (19) applied to c_{g_0}, condition (17) is not satisfied for c_{g_0}. So we have to check that φ is well-defined and a homomorphism.

First we check that $\varphi(s)$ defines an element in $\mathcal{H}(L; R, \rho|_L, \omega)$. Obviously the image of the support of $\varphi(s)$ under $L \to L/N$ is compact, since this is true for $\mathrm{supp}(s)$ and $\mathrm{supp}(\varphi(s)) = t\,\mathrm{supp}(s)t^{-1}$.

Suppose that $K \in P$ is admissible for s. Then tKt^{-1} is admissible for $\varphi(s)$ by the following calculation for $l \in L$ and $k' \in tKt^{-1}$, if we write $k' = tkt^{-1}$ for $k \in K$

$$\varphi(s)(k'l) = ts(t^{-1}k'lt) = ts(t^{-1}tkt^{-1}lt) = ts(kt^{-1}lt) \overset{(12)}{=} ts(t^{-1}lt) = \varphi(s)(l),$$

and

$$\varphi(s)(lk') = ts(t^{-1}lk't) = ts(t^{-1}ltkt^{-1}t) = ts(t^{-1}ltk) \overset{(13)}{=} ts(t^{-1}lt) = \varphi(s)(l).$$

The following calculation shows that condition (8) is satisfied.

$$\varphi(s)(nl) = ts(t^{-1}nlt) = ts(t^{-1}ntt^{-1}lt) \overset{(8)}{=} t\big(\omega(t^{-1}nt) \cdot s(t^{-1}lt)\big)$$

$$= t\omega(t^{-1}nt) \cdot ts(t^{-1}lt) \overset{(5),\,(6)}{=} \omega(n) \cdot ts(t^{-1}lt) = \omega(n) \cdot \varphi(s)(l).$$

Recall that the condition (9) holds automatically, see Remark 2.1. Hence φ is well-defined.

It is obviously compatible with the addition. It is compatible with the multiplication by the following calculation for two elements $s, s' \in \mathcal{H}(L; R, P|_L, \omega)$ and $l \in L$, where $K \in P$ is admissible for both s and s', and T is a transversal for the projection $L \to L/NK$, and pr: $L \to M = L/N$ is the projection. We will use the fact that tTt^{-1} is a transversal for the projection $L \to L/NtKt^{-1}$ and tKt^{-1} is admissible for $\varphi(s)$ and $\varphi(s')$. Moreover, we have

$$[M : \mathrm{pr}(K)] = [tMt^{-1} : t\,\mathrm{pr}(K)t^{-1}] = [M : \mathrm{pr}(tKt^{-1})]. \tag{33}$$

We compute

$$\varphi(s \cdot s')(l) = t(s \cdot s')(t^{-1}lt)$$

$$\overset{(15)}{=} t\left(\mu(\mathrm{pr}(K)) \cdot \sum_{g' \in T} s(t^{-1}ltg') \cdot t^{-1}ltg's'(g'^{-1})\right)$$

$$= \mu(\mathrm{pr}(K)) \cdot \sum_{g' \in T} ts(t^{-1}ltg') \cdot ltg's'(g'^{-1})$$

$$= \frac{\mu(M)}{[M : \mathrm{pr}(K)]} \cdot \sum_{g' \in T} ts(t^{-1}ltg't^{-1}t) \cdot ltg't^{-1}ts'(t^{-1}tg'^{-1}t^{-1}t)$$

$$\overset{(33)}{=} \frac{\mu(M)}{[M : \mathrm{pr}(tKt^{-1})]} \cdot \sum_{g'' \in tTt^{-1}} ts(t^{-1}lg''t) \cdot lg''ts'(t^{-1}g''^{-1}t)$$

$$= \mu(\mathrm{pr}(tKt^{-1})) \cdot \sum_{g'' \in tTt^{-1}} \varphi(s)(lg'') \cdot lg''\varphi(s')(g''^{-1})$$

$$\overset{(15)}{=} (\varphi(s) \cdot \varphi(s'))(l).$$

Lemma 5.1 *There is a natural isomorphism of (non-unital) rings*

$$\Xi \colon \mathcal{H}(L; R, \rho|_L, \omega)_\varphi[t, t^{-1}] \overset{\cong}{\to} \mathcal{H}(G; R, \rho, \omega).$$

Proof Consider an element $s \in \mathcal{H}(L; R, \rho|_L, \omega)$ and an element $n \in \mathbb{Z}$. Then $\Xi(st^n)$ is defined to be the element in $\mathcal{H}(G; R, \rho, \omega)$ given by

$$G \to R, \quad (lt^m) \mapsto \begin{cases} s(l) & \text{if } m = n; \\ 0 & \text{otherwise.} \end{cases} \tag{34}$$

Obviously the image of the support of $\Xi(st^n)$ under $\mathrm{pr}\colon G \to Q$ is compact, as it is a closed subset of $t^n M$ and $M \subseteq Q$ is compact. Suppose that the compact open subgroup $K \subseteq L$ is admissible for s. Then $K \cap t^{-n} K t^n \subseteq L \subseteq G$ is admissible for $\Xi(st^n)$ by the following calculation for $l \in L$ and $k \in K \cap t^{-n} K t^n$

$$\Xi(st^n)(klt^n) \overset{(34)}{=} s(kl) \overset{(12)}{=} s(l) \overset{(34)}{=} \Xi(st^n)(lt^n),$$

and

$$\Xi(st^n)(lt^n k) = \Xi(st^n)(lt^n k t^{-n} t^n) \overset{(34)}{=} s(lt^n k t^{-n}) \overset{(13)}{=} s(l) \overset{(34)}{=} \Xi(st^n)(lt^n)$$

and the observation that we have $\Xi(st^n)(lt^m k) = \Xi(st^n)(klt^m) = \Xi(st^n)(lt^m) = 0$ for $m \in \mathbb{Z}$ with $m \neq n$. Next we verify condition (8). We get for $z \in N$ and $m \in \mathbb{Z}$ with $m \neq n$ that $\Xi(st^n)(zlt^m) = 0 = \Xi(st^n)(nlt^m)$ and

$$\Xi(st^n)(zlt^n) \overset{(34)}{=} s(zl) \overset{(8)}{=} \omega(z) \cdot s(l) \overset{(34)}{=} \omega(z) \cdot \Xi(st^n)(lt^n)$$

hold. Recall that the condition (9) holds automatically, see Remark 2.1. Thus we have shown that $\Xi(st^n)$ is a well-defined element in $\mathcal{H}(G; R, \rho, \omega)$.

Define the image under Ξ of an arbitrary element in $\mathcal{H}(L; R, \rho|_L, \omega)_\varphi[t, t^{-1}]$ given by a finite sum $\sum_{n \in \mathbb{Z}} s_n t^n$ to be the element $\sum_{n \in \mathbb{Z}} \Xi(s_n t^n)$ in $\mathcal{H}(G; R, \rho, \omega)$. Obviously Ξ is compatible with the addition. In order to show that Ξ is compatible with the multiplication, it suffices to show for $s, s' \in \mathcal{H}(L; R, \rho|_L, \omega)$, $l \in L$, and $m', n, n' \in \mathbb{Z}$

$$\left(\Xi(st^n) \cdot \Xi(s't^{n'})\right)(lt^m) = \Xi(st^n \cdot s't^{n'})(lt^m).$$

Fix a compact open subgroup $K \subseteq G$ such that K is admissible for both $\Xi(st^n)$ and $\Xi(s't^{n'})$ and $t^n K t^{-n}$ is admissible for both s and $\varphi^n(s)$. Consider a transversal T' for the projection $L \to L/NK$. Then $T = \{t^{m'} l' \mid m' \in \mathbb{Z}, l' \in T'\}$ is a transversal for the projection $G \to G/NK$ and the map $\mathbb{Z} \times T' \overset{\cong}{\to} T$ sending

(m', l') to $t^{m'}l'$ is a bijection. Moreover, $t^n T' t^{-n}$ is a transversal for the projection $L \to L/Nt^n K t^{-n}$. We have

$$\mu(\mathrm{pr}(t^n K t^{-n})) = \mu(t^n \, \mathrm{pr}(K) t^{-n})$$

$$= \frac{\mu(M)}{[M : t^n \, \mathrm{pr}(K) t^{-n}]} \overset{(33)}{=} \frac{\mu(M)}{[M : \mathrm{pr}(K)]} = \mu(\mathrm{pr}(K)). \qquad (35)$$

We compute

$$\bigl(\Xi(st^n) \cdot \Xi(s't^{n'})\bigr)(lt^m)$$

$$\overset{(15)}{=} \mu(\mathrm{pr}(K)) \cdot \sum_{g' \in T} \Xi(st^n)(lt^m g') \cdot lt^m g' \Xi(s't^{n'})(g'^{-1})$$

$$= \mu(\mathrm{pr}(K)) \cdot \sum_{l' \in T'} \sum_{m' \in \mathbb{Z}} \Xi(st^n)(lt^m t^{m'} l') \cdot lt^m t^{m'} l' \Xi(s't^{n'})((t^{m'} l')^{-1})$$

$$= \mu(\mathrm{pr}(K)) \cdot \sum_{l' \in T'} \sum_{m' \in \mathbb{Z}} \Xi(st^n)(lt^{m+m'} l' t^{-m-m'} t^{m+m'}) \cdot lt^{m+m'} l' \Xi(s't^{n'})(l'^{-1} t^{-m'})$$

$$\overset{(34)}{=} \mu(\mathrm{pr}(K)) \cdot \sum_{l' \in T'} \sum_{\substack{m' \in \mathbb{Z} \\ m+m'=n, -m'=n'}} s(lt^{m+m'} l' t^{-m-m'}) \cdot lt^{m+m'} l' s'(l'^{-1})$$

$$= \begin{cases} \mu(\mathrm{pr}(K)) \cdot \sum_{l' \in T'} s(lt^n l' t^{-n}) \cdot lt^n l' s'(l'^{-1}) & m = n + n' \\ 0 & m \neq n + n' \end{cases}$$

$$= \begin{cases} \mu(\mathrm{pr}(K)) \cdot \sum_{l' \in T'} s(lt^n l' t^{-n}) \cdot lt^n l' t^{-n} t^n s'(t^{-n} t^n l'^{-1} t^{-n} t^n) & m = n + n' \\ 0 & m \neq n + n' \end{cases}$$

$$\overset{(32)}{=} \begin{cases} \mu(\mathrm{pr}(K)) \cdot \sum_{l' \in T'} s(lt^n l' t^{-n}) \cdot lt^n l' t^{-n} \varphi^n(s')(t^n l'^{-1} t^{-n}) & m = n + n' \\ 0 & m \neq n + n' \end{cases}$$

$$\overset{(35)}{=} \begin{cases} \mu(t^n K t^{-n}) \cdot \sum_{l'' \in t^n T' t^{-n}} s(ll'') \cdot ll'' \varphi^n(s')(l''^{-1}) & m = n + n' \\ 0 & m \neq n + n' \end{cases}$$

$$\overset{(15)}{=} \begin{cases} (s \cdot \varphi^n(s'))(l) & m = n + n' \\ 0 & m \neq n + n' \end{cases}$$

$$\overset{(34)}{=} \Xi(s \cdot \varphi^n(s') \cdot t^{n+n'})(lt^m)$$

$$= \Xi(st^n \cdot s't^{n'})(lt^m).$$

Obviously Ξ is injective. It remains to show that Ξ is surjective. Any element in $\mathcal{H}(G; R, \rho, \omega)$ can be written as a sum of elements s' for which the support is contained in Lt^n for some $n \in \mathbb{Z}$. Hence it suffices to show that such s' is in the image. Define $s: L \to R$ by $s(l) = s'(lt^n)$. Choose $K \in P$ such that both K and $t^{-n} K t^n$ are admissible for s'. Obviously $K \subseteq L$ and $t^{-n} K t^n \subseteq L$. We have for $l \in L$ and $k \in K$ the equality $s'(klt^n) \overset{(12)}{=} s'(lt^n)$, which implies $s(kl) = s(l)$. We also have $s'(lkt^n) = s'(lt^n t^{-n} kt^n) \overset{(13)}{=} s'(lt^n)$ which implies $s(lk) = s(l)$. Hence K is admissible for s. Condition (8) follows from the calculation for $z \in N$.

$$s(zl) = s'(zlt^n) \overset{(8)}{=} \omega(z) \cdot s'(lt^n) = \omega(z) \cdot s(l).$$

Recall that the condition (9) holds automatically, see Remark 2.1. We conclude that s defines an element in $\mathcal{H}(L; R, \rho|_L, \omega)$ with $\Xi(st^n) = s'$. This finishes the proof of Lemma 5.1. $\square$

Lemma 5.2 *Let $\mathcal{A}$ be a (not necessarily unital) additive category which is the directed union $\mathcal{A} = \bigcup_{i \in I} \mathcal{A}_i$ of unital additive categories. Let $\Phi: \mathcal{A} \overset{\cong}{\to} \mathcal{A}$ be an automorphism of (non-unital) additive categories.*

There is an equivalence of unital additive categories

$$F: \operatorname{Idem}\left(\operatorname{Idem}(\mathcal{A})_{\operatorname{Idem}(\Phi)}[t, t^{-1}]\right) \overset{\cong}{\to} \operatorname{Idem}\left(\mathcal{A}_\Phi[t, t^{-1}]\right).$$

Proof Recall that an object in $\operatorname{Idem}(\mathcal{A})$ is given by a pair (A, p), where A is an object in $\mathcal{A}$ and $p: A \to A$ is a morphism in $\mathcal{A}$ with $p \circ p = p$. Moreover, a morphism $f: (A, p) \to (A', p')$ in $\operatorname{Idem}(\mathcal{A})_{\operatorname{Idem}(\Phi)}[t, t^{-1}]$ is given by a finite sum $f = \sum_{j \in \mathbb{Z}} f_j \cdot t^j$, where $f_j: \operatorname{Idem}(\Phi)^j(A, p) := (\Phi^j(A), \Phi^j(p)) \to (A', p')$ is a morphism in $\operatorname{Idem}(\mathcal{A})$. Hence each f_j is given by a morphism $f_j: \Phi^j(A) \to A'$ satisfying $f_j = p' \circ f_j \circ \Phi^j(p)$. We conclude that a morphism $f: (A, p) \to (A', p')$ in $\operatorname{Idem}(\mathcal{A})_{\operatorname{Idem}(\Phi)}[t, t^{-1}]$ is the same as a morphism $f: A \to A'$ in $\mathcal{A}_\Phi[t, t^{-1}]$ satisfying $(p' \cdot t^0) \circ f \circ (p \cdot t^0) = f$, since we get in $\mathcal{A}_\Phi[t, t^{-1}]$

$$(p' \cdot t^0) \circ f \circ (p \cdot t^0) = \sum_{j \in \mathbb{Z}} (p' \cdot t^0) \circ f_j \cdot t^j \circ (p \cdot t^0) = \sum_{j \in \mathbb{Z}} \left(p' \circ f_j \cdot \Phi^j(p)\right) \cdot t^j.$$

Now an object in $\operatorname{Idem}\left(\operatorname{Idem}(\mathcal{A})_{\operatorname{Idem}(\Phi)}[t, t^{-1}]\right)$ is given by $((A, p), q)$, where A is an object in $\mathcal{A}$, $p: A \to A$ is a morphism in $\mathcal{A}$ with $p \circ p = p$, and $q: (A, p) \to (A, p)$ is a morphism in $\operatorname{Idem}(\mathcal{A})_{\operatorname{Idem}(\Phi)}[t, t^{-1}]$ satisfying $q \circ q = q$. The morphism q is the same as a morphism $q: A \to A$ in $\mathcal{A}_\Phi[t, t^{-1}]$ satisfying $(p \cdot t^0) \circ q \circ (p \cdot t^0) = q$ and $q \circ q = q$. Hence we can define F on objects by

$$F((A, p), q) = (A, q).$$

Consider two objects $((A, p), q)$ and $((A', p'), q')$. A morphism $f\colon ((A, p), q) \to$ $((A', p'), q')$ in $\mathrm{Idem}\left(\mathrm{Idem}(\mathcal{A})_{\mathrm{Idem}(\Phi)}[t, t^{-1}]\right)$ is the same as a morphism $f\colon$ (A, p) $\to (A', p')$ in $\mathrm{Idem}(\mathcal{A})_{\mathrm{Idem}(\Phi)}[t, t^{-1}]$ satisfying $q' \circ f \circ q = f$ and therefore the same as a morphism $f\colon A \to A'$ in $\mathcal{A}_\Phi[t, t^{-1}]$ satisfying $(p' \cdot t^0) \circ f \circ (p \cdot t^0) = f$ and $q' \circ f \circ q = f$.

Hence we can define F on morphisms by sending the morphism $f\colon \big((A, p), q\big) \to$ $\big((A', p'), q'\big)$ in $\mathrm{Idem}\left(\mathrm{Idem}(\mathcal{A})_{\mathrm{Idem}(\Phi)}[t, t^{-1}]\right)$ to the morphism $(A, q) \to (A', q')$ in $\mathrm{Idem}\left(\mathcal{A}_\Phi[t, t^{-1}]\right)$ given by the morphism $f\colon A \to A'$ in $\mathcal{A}_\Phi[t, t^{-1}]$. One easily checks that F is compatible with composition and sends identity morphisms to identity morphisms.

Next we show that the map induced by F

$$\mathrm{mor}_{\mathrm{Idem}(\mathrm{Idem}(\mathcal{A})_{\mathrm{Idem}(\Phi)}[t,t^{-1}])}\big(((A, p), q), ((A', p'), q')\big)$$

$$\to \mathrm{mor}_{\mathrm{Idem}(\mathcal{A}_\Phi[t,t^{-1}])}\big((A, q), (A', q')\big)$$

is bijective. Obviously it is injective. In order to show surjectivity, we have to show for a morphism $f\colon (A, q) \to (A', q')$ in $\mathrm{Idem}(\mathcal{A})_{\mathrm{Idem}(\Phi)}[t, t^{-1}]$ satisfying $q' \circ f \circ q = f$ that $(p' \cdot t^0) \circ f \circ (p \cdot t^0) = f$ holds. This follows from the following computation using $(p \cdot t^0) \circ q \circ (p \cdot t^0) = q$, $q \circ q = q$, $p \circ p = p$, $(p' \cdot t^0) \circ q' \circ (p' \cdot t^0) = q'$, $q' \circ q' = q'$, and $p' \circ p' = p'$,

$$(p' \cdot t^0) \circ f \circ (p \cdot t^0) = (p' \cdot t^0) \circ q' \circ f \circ q \circ (p \cdot t^0)$$

$$= (p' \cdot t^0) \circ (p' \cdot t^0) \circ q' \circ (p' \cdot t^0) \circ f \circ (p \cdot t^0) \circ q \circ (p \cdot t^0) \circ (p \cdot t^0)$$

$$= (p' \cdot t^0) \circ q' \circ (p' \cdot t^0) \circ f \circ (p \cdot t^0) \circ q \circ (p \cdot t^0) = q' \circ f \circ q = f.$$

Finally we show that F is surjective on objects. Consider any object (A, q) in $\mathrm{Idem}\left(\mathcal{A}_\Phi[t, t^{-1}]\right)$. In order to show that (A, q) is in the image of F, we have to construct a morphism $p\colon A \to A$ in $\mathcal{A}$ such that $p \circ p = p$ holds in $\mathcal{A}$ and $(p \cdot t^0) \circ q \circ (p \cdot t^0) = q$ holds in $\mathcal{A}_\Phi[t, t^{-1}]$.

We can write q as a finite sum $q = \sum_{j \in \mathbb{Z}} q_j \cdot t^j$ for morphisms $q_j\colon \Phi^j(A) \to A$ in $\mathcal{A}$. Since $\mathcal{A}$ is the directed union $\bigcup_{i \in I} \mathcal{A}_i$ of the unital subcategories $\mathcal{A}_i$, we can find an index $i_0 \in I$ such that for each $j \in \mathbb{Z}$ with $q_j \neq 0$ and hence for all $j \in J$ the morphisms q_j and $\Phi^{-j}(q_j)$ belong to $\mathcal{A}_{i_0}$. Let $p \in \mathcal{A}_{i_0}$ be the identity morphism of the object A in $\mathcal{A}_{i_0}$. Then we get $p \circ p = p$, $p \circ q_j = q_j$, and $\Phi^{-j}(q_j) \circ p = \Phi^{-j}(q_j)$ in $\mathcal{A}$ for all $j \in \mathbb{Z}$. Now we compute

$$(p \cdot t^0) \circ q \circ (p \cdot t^0) = (p \cdot t^0) \circ \left(\sum_{j \in \mathbb{Z}} q_j \cdot t^j \right) \circ (p \cdot t^0)$$

$$= \sum_{j \in \mathbb{Z}} (p \cdot t^0) \circ (q_j \cdot t^j) \circ (pt^0) = \sum_{j \in \mathbb{Z}} (p \circ q_j \cdot \Phi^j(p)) \cdot t^j = \sum_{j \in \mathbb{Z}} (q_j \cdot \Phi^j(p)) \cdot t^j$$

$$= \sum_{j \in \mathbb{Z}} \Phi^j(\Phi^{-j}(q_j) \cdot p) \cdot t^j = \sum_{j \in \mathbb{Z}} \Phi^j(\Phi^{-j}(q_j)) \cdot t^j = \sum_{j \in \mathbb{Z}} q_j \cdot t^j = q.$$

This finishes the proof of Lemma 5.2. $\qquad\qquad\qquad\qquad\qquad\qquad\qquad$ □

The next lemma allows us to reduce the computation of the algebraic K-theory of the non-unital ring $\mathcal{H}(G; R, \rho, \omega)$ to the calculation of the algebraic K-theory of a unital additive category given by the twisted finite Laurent category of an automorphism of a unital additive category. The main advantage will be that for such a category Bass–Heller–Swan decompositions will be available.

Lemma 5.3 *There is a weak equivalence*

$$\mathbf{K}^\infty \big(\mathrm{Idem}(\underline{\mathcal{H}(L; R, \rho|_L, \omega)}_\oplus)_{\mathrm{Idem}(\varphi_\oplus)}[t, t^{-1}] \big) \xrightarrow{\simeq} \mathbf{K}^\infty \big(\mathcal{H}(G; R, \rho, \omega) \big).$$

Proof Recall that for a unital additive category $\mathcal{B}$ the obvious map $\mathbf{K}^\infty(\mathcal{B}) \to \mathbf{K}^\infty(\mathrm{Idem}(\mathcal{B}))$ is a weak homotopy equivalence. We can apply Lemma 5.2 to $\mathcal{A} = \underline{\mathcal{H}(L; R, \rho|_L, \omega)}_\oplus$ and the automorphism $\varphi_\oplus$ because of Lemma 2.5. Hence we obtain a weak equivalence

$$\mathbf{K}^\infty \big(\mathrm{Idem}(\underline{\mathcal{H}(L; R, \rho|_L, \omega)}_\oplus)_{\mathrm{Idem}(\varphi_\oplus)}[t, t^{-1}] \big)$$

$$\xrightarrow{\simeq} \mathbf{K}^\infty \big(\mathrm{Idem} \big((\underline{\mathcal{H}(L; R, \rho|_L, \omega)}_\oplus)_{\varphi_\oplus}[t, t^{-1}] \big) \big).$$

The (non-unital) additive category $(\underline{\mathcal{H}(L; R, \rho|_L, \omega)}_\oplus)_{\varphi_\oplus}[t, t^{-1}]$ is isomorphic to the (non-unital) additive category $(\underline{\mathcal{H}(L; R, \rho|_L, \omega)_\varphi[t, t^{-1}]})_\oplus$ by (26), and hence by Lemma 5.1 to the (non-unital) additive category $\underline{\mathcal{H}(G; R, \rho, \omega)}_\oplus$. Hence we obtain a weak homotopy equivalence

$$\mathbf{K}^\infty \big(\mathrm{Idem} \big((\underline{\mathcal{H}(L; R, \rho|_L, \omega)}_\oplus)_{\varphi_\oplus}[t, t^{-1}] \big) \big) \xrightarrow{\simeq} \mathbf{K}^\infty \big(\mathrm{Idem} \big(\underline{\mathcal{H}(G; R, \rho, \omega)}_\oplus \big) \big).$$

$\qquad\qquad\qquad\qquad\qquad\qquad\qquad\qquad\qquad\qquad\qquad\qquad\qquad\qquad\qquad\qquad$ □

6   A Review of the Twisted Bass–Heller–Swan Decomposition for Unital Additive Categories

In this section additive category always means a small unital additive category and functors are assumed to respect identity morphisms. The same is true for rings.

The following definitions are taken from [2, Definition 6.1].

Definition 6.1 (Regularity Properties of Rings) Let l be a natural number.

(1) We call R *Noetherian* if any R-submodule of a finitely generated R-module is again finitely generated.

(2) We call R *regular coherent* if every finitely presented R-module M is of type FP.

(3) We call R *l-uniformly regular coherent* if every finitely presented R-module M admits an l-dimensional finite projective resolution, i.e., there exists an exact sequence $0 \to P_l \to P_{l-1} \to \cdots \to P_0 \to M \to 0$ such that each P_i is finitely generated projective.

(4) We call R *regular* if it is Noetherian and regular coherent.

(5) We call R *l-uniformly regular* if it is Noetherian and l-uniformly regular coherent.

These notions are generalized to additive categories in [2, Section 6] in such a way that they reduce in the special case $\mathcal{A} = \underline{R}$ to the ones appearing in Definition 6.1. Therefore the precise definitions for additive categories are not needed to comprehend the material of this paper.

The following result follows from [2, Theorem 6.8 and Theorem 9.1].

Theorem 6.2 (The Non-connective K-Theory of Additive Categories) *Let $\mathcal{A}$ be an additive category. Suppose that $\mathcal{A}$ is regular. Consider any automorphism $\Phi\colon \mathcal{A} \xrightarrow{\cong} \mathcal{A}$ of additive categories.*

Then we get a weak homotopy equivalence of non-connective spectra

$$\mathbf{a}^{\infty}\colon \mathbf{T}_{\mathbf{K}^{\infty}(\Phi^{-1})} \xrightarrow{\simeq} \mathbf{K}^{\infty}(\mathcal{A}_{\Phi}[t, t^{-1}]),$$

where $\mathbf{T}_{\mathbf{K}^{\infty}(\Phi^{-1})}$ is the mapping torus of the map of spectra $\mathbf{K}^{\infty}(\Phi)\colon \mathbf{K}^{\infty}(\mathcal{A}) \to \mathbf{K}^{\infty}(\mathcal{A})$ induced by Φ.

7   Hecke Algebras Over Compact td-Groups and Crossed Product Rings

Let G, N, $Q := G/N$, $\mathrm{pr}\colon G \to Q$, R, P, ρ, ω, and μ be as in Sect. 2.1 and denote by $\mathcal{H}(G; R, \rho, \omega)$ the Hecke algebra which we have introduced in Sect. 2.2. Our main assumption in this section will be that Q is compact.

Definition 7.1 We call a subgroup $N \subseteq G$ *locally central* if the centralizer $C_G N$ of N in G is an open subgroup.

The main result of this section is

Theorem 7.2 *Suppose that Q is compact and N is locally central. Let l be a natural number. Let R be a unital ring with $\mathbb{Q} \subseteq R$ such that R is l-uniformly regular or regular respectively.*

Then the additive category $\mathrm{Idem}\left(\overline{\mathcal{H}(G; R, \rho, \omega)[\mathbb{Z}^m]}_{\oplus}\right)$ *is $(l + 2m)$-uniformly regular or regular respectively for all $m \geq 0$.*

For the purpose of this paper we need the conclusion of Theorem 7.2 only for the property regular, but for later applications it will be crucial to consider the property l-uniformly regular as well. The point will be that the property l-uniformly regular is compatible with infinite products of additive categories, in contrast to the property regular.

7.1   Existence of Normal $K \in P$

Lemma 7.3 *Suppose that Q is compact and N is locally central.*
Then for every compact open subgroup $K \subseteq G$ there exists a compact open subgroup $K' \subseteq G$ such that $K' \subseteq K$, $K' \subseteq C_G N$, and K' is normal in G.

Proof Put $L = K \cap C_G N$. Then L is a compact open subgroup of G satisfying $L \subseteq K$ and $L \subseteq C_G N$. Choose a transversal T of the projection $G \to G/NL = Q/\mathrm{pr}(L)$. Define $K' = \bigcap_{t \in t} t L t^{-1}$. Since pr is open, $Q/\mathrm{pr}(L)$ is compact and discrete and hence finite. This implies that the set T is finite. Hence $K' \subseteq G$ is again compact open. We get for $n \in N$ and $l \in L$

$$(tnl)L(tnl)^{-1} = tnlLl^{-1}n^{-1}t^{-1} = tnLn^{-1}t^{-1} \overset{L \subseteq C_G N}{=} tLt^{-1}.$$

This implies $K' = \bigcap_{g \in G} g L g^{-1}$. Hence $K' \subseteq G$ is a compact open normal subgroup and obviously satisfies $K' \subseteq K$ and $K' \subseteq C_G N$. $\square$

7.2   Crossed Products Rings of Finite Groups and Regularity

Let R be a unital ring and D be a (discrete) group. Recall that a *crossed product ring* $R * D$ is a unital ring which is a free left R-module with an R-basis $\{b_d \mid d \in D\}$ indexed by the elements in D such that b_e is the unit in $R * D$, for $d_1, d_2 \in D$ there is a unit $w(d_1, d_2) \in R^\times$ satisfying $b_{d_1 d_2} = w(d_1, d_2) \cdot b_{d_1} \cdot b_{d_2}$, and for $r \in R$ and $d \in D$ there exists $c_d(r) \in R$ with $c_d(r) \cdot b_d = b_d \cdot (r \cdot b_e)$, where $c_e(r) = r$ is required for $r \in R$. In particular each element b_d has an inverse b_d^{-1} in $R * D$

(which is *not* given by $b_{d^{-1}}$ in general) and there is an inclusion of rings $R \to R * D$ sending r to $r \cdot b_e$.

The notion of crossed product ring is a generalization of the notion of a twisted group ring, which is the special case where w is trivial. For more details we refer for instance to [1, Section 4] or [4, Section 6].

Lemma 7.4 *Let R be a ring with $\mathbb{Q} \subseteq R$ and D be a finite group. Let $R * D$ be a crossed product ring.*

(1) *Let M be any $R * D$-module. Let $j \colon R \to R * D$ be the canonical inclusion of rings. Then we obtain $R * D$-homomorphisms*

$$i \colon M \to R * D \otimes_R j^* M, \quad x \mapsto \sum_{d \in D} \frac{1}{|D|} \cdot b_d \otimes b_d^{-1} \cdot x;$$

$$p \colon R * D \otimes_R j^* M \to M, \quad u \otimes y \mapsto u \cdot y,$$

satisfying $p \circ i = \mathrm{id}_M$, where b_d^{-1} denotes the inverse of b_d in $R * D$.
(2) *If R is regular, then $R * D$ is regular.*
(3) *If R is l-uniformly regular, then $R * D$ is l-uniformly regular.*
(4) *If R is semi-simple, then $R * D$ is semi-simple.*

Proof

(1) We check that i is $R * D$-linear. Obviously i is compatible with addition, it remains to treat multiplication. Consider $r \in R$ and $d_0 \in D$. Note for the sequel that the element $b_d^{-1} \cdot r \cdot b_{d_0} \cdot b_{d_0^{-1} d}$ in $R * D$ belongs to R. Hence we get for $x \in M$, $r \in R$ and $d_0 \in D$

$$i(r \cdot b_{d_0} \cdot x)$$

$$= \sum_{d \in D} \frac{1}{|D|} \cdot b_d \otimes b_d^{-1} \cdot (r \cdot b_{d_0} \cdot x)$$

$$= \sum_{d \in D} \frac{1}{|D|} \cdot b_d \otimes (b_d^{-1} \cdot r \cdot b_{d_0} \cdot b_{d_0^{-1} d}) \cdot (b_{d_0^{-1} d})^{-1} \cdot x$$

$$= \sum_{d \in D} \frac{1}{|D|} \cdot b_d \cdot (b_d^{-1} \cdot r \cdot b_{d_0} \cdot b_{d_0^{-1} d}) \otimes (b_{d_0^{-1} d})^{-1} \cdot x$$

$$= \sum_{d \in D} \frac{1}{|D|} \cdot r \cdot b_{d_0} \cdot b_{d_0^{-1} d} \otimes (b_{d_0^{-1} d})^{-1} \cdot x$$

$$= r \cdot b_{d_0} \cdot \frac{1}{|D|} \cdot \sum_{d \in D} b_{d_0^{-1} d} \otimes (b_{d_0^{-1} d})^{-1} \cdot x$$

$$= r \cdot b_{d_0} \cdot \frac{1}{|D|} \cdot \sum_{d' \in D} b_{d'} \otimes (b_{d'})^{-1} \cdot x$$

$$= r \cdot b_{d_0} \cdot i(x).$$

Obviously p is a well-defined $R * D$-homomorphism satisfying $p \circ i = \mathrm{id}_M$.

(2) Since R is regular, R is in particular Noetherian. Since $R * D$ is a finitely generated R-module, $R * D$ is Noetherian as well.

It remains to show that a finitely presented $R * D$-module M is of type FP. Since R is regular and the R-module i^*M is finitely presented, i^*M is of type FP. Since $R*D$ is free as R-module and hence the functor sending an R-module N to the $R*D$-module $R*D \otimes_R N$ is flat and sends finitely generated projective R-modules to finitely generated projective $R * D$-modules, the $R * D$-module $R * D \otimes_R i^*M$ is of type FP. Since a direct summand in a module of type FP is of type FP again, the $R * D$-module M is of type FP.

(3) The proof is analogous to assertion (2), since all the statements about finite-dimension remain true if one inserts l-dimensional everywhere.

(4) A ring R is semisimple if all its modules are projective. Hence assertion (4) follows from assertion (1). This finishes the proof of Lemma 7.4.

□

7.3 The Hecke Algebra and Crossed Products

In this subsection we will assume that Q is compact.

Consider a compact open normal subgroup K of G satisfying $K \in P$. Since both K and N are normal in G, the subgroup NK of G is also normal. Put

$$D := G/NK = Q/\mathrm{pr}(K). \tag{36}$$

Note that D is a finite discrete group. We want to show

Lemma 7.5 *Suppose that Q is compact. Consider a compact open normal subgroup K of G satisfying $K \in P$.*

Then the unital ring $\mathcal{H}(G//K; R, \rho, \omega)$ is a crossed product ring $R * D$.

Proof We begin by showing that $\mathcal{H}(G; R, \rho, \omega)$ is a left R-module. Namely, define for $s \in \mathcal{H}(G; R, \rho, \omega)$ the new element $r \cdot s$ by $(r \cdot s)(g) := r \cdot s(g)$. One easily checks that $r \cdot s$ is well-defined.

Fix a set-theoretic section $\sigma: D \to G$ of the projection $p: G \to D = G/NK$ satisfying $\sigma(e_D) = e_G$. In the sequel we denote by T the transversal of p given by $T := \{\sigma(d)^{-1} \mid d \in D\}$. For $d \in D$ define $b_d \in \mathcal{H}(G//K; R, \rho, \omega)$ by the function

$$b_d: G \to R,$$

$$g \mapsto \begin{cases} \frac{1}{\mu(\mathrm{pr}(K))} \cdot \omega(n) & \text{if } p(g) = d \text{ and } g = nk\sigma(d) \text{ for } n \in N, k \in K; \\ 0 & p(g) \neq d. \end{cases} \tag{37}$$

This is independent of the choice of $n \in N$ and $k \in K$, since for $n_0, n_1 \in N$ and $k_0, k_1 \in K$ with $n_0 k_0 = n_1 k_1$ we have $n_1^{-1} n_0 = k_1 k_0^{-1} \in N \cap K$ and we compute

$$\omega(n_1) = \omega(n_1) \cdot \omega(n_1^{-1} n_0) \overset{(11)}{=} \omega(n_1) \cdot 1 = \omega(n_0). \tag{38}$$

We have to check that the required transformation formulas (12) and (13) for $g \in G$ and $k \in K$ are satisfied. If $p(g) \neq d$, then $b_d(kg) = b_d(g) = b_d(gk) = 0$ and the formulas hold. It remains to treat the case $p(g) = d$. This follows from the calculations for $g = nk\sigma(d)$ for $n \in N, k \in K$ and $k' \in K$ using $\sigma(d)k'\sigma(d)^{-1} \in K$

$$b_d(k'g) = b_d(k'nk\sigma(d)) = b_d((k'nk'^{-1})(k'k)\sigma(d))$$

$$\overset{(37)}{=} \frac{1}{\mu(\mathrm{pr}(K))} \cdot \omega(k'nk'^{-1}) \overset{(5)}{=} \frac{1}{\mu(\mathrm{pr}(K))} \cdot \omega(n) \overset{(37)}{=} b_d(g),$$

and

$$b_d(gk') = b_d(nk\sigma(d)k') = b_d(n(k\sigma(d)k'\sigma(d)^{-1})\sigma(d))$$

$$\overset{(37)}{=} \frac{1}{\mu(\mathrm{pr}(K))} \cdot \omega(n) \overset{(37)}{=} b_d(g).$$

The verification of (8) and (9) is left to the reader. This finishes the proof that b_d is a well-defined element in $\mathcal{H}(G//K; R, \rho, \omega)$.

Consider any element $s \in \mathcal{H}(G//K; R, \rho, \omega)$. Then we get

$$s = \sum_{d \in D} \mu(\mathrm{pr}(K)) \cdot s(\sigma(d)) \cdot b_d \tag{39}$$

by the following calculation for $g \in G$ with $g = nk\sigma(d)$ for $n \in N$ and $k \in K$

$$s(g) = s(nk\sigma(d)) \overset{(8),(12)}{=} \omega(n) \cdot s(\sigma(d))$$

$$\overset{\omega(n) \in \text{cent}(R)}{=} \mu(\text{pr}(K)) \cdot s(\sigma(d)) \cdot \left(\frac{1}{\mu(\text{pr}(K))} \cdot \omega(n) \right) \overset{(37)}{=} \mu(\text{pr}(K)) \cdot s(\sigma(d)) \cdot b_d(g).$$

We conclude from (39) that $\{b_d \mid d \in D\}$ is an R-basis for the left R-module $\mathcal{H}(G//K; R, \rho, \omega)$.

For d_1, d_2 in D, define an element

$$w(d_1, d_2) := \omega(n) \in R^\times \tag{40}$$

if $\sigma(d_1 d_2)\sigma(d_2)^{-1}\sigma(d_1)^{-1} = nk$ for $n \in N$ and $k \in K$. This is independent of the choice of $n \in N$ and $k \in K$ by (38). Next we want to show

$$b_{d_1} \cdot b_{d_2} = w(d_1, d_2) \cdot b_{d_1 d_2}. \tag{41}$$

Consider $d_1, d_2 \in D$ and $g \in G$. Choose elements $n \in N$ and $k \in K$ satisfying $\sigma(d_1 d_2)\sigma(d_2)^{-1}\sigma(d_1)^{-1} = nk$. If $p(g) = d_1 d_2$, we fix $n_0 \in N$ and $k_0 \in K$ satisfying $g = n_0 k_0 \sigma(d_1)\sigma(d_2)$. We compute

$$(b_{d_1} \cdot b_{d_2})(g)$$

$$\overset{(15)}{=} \mu(\text{pr}(K)) \cdot \sum_{d \in D} b_{d_1}(g\sigma(d)^{-1}) \cdot g\sigma(d)^{-1} b_{d_2}(\sigma(d))$$

$$\overset{(37)}{=} \mu(\text{pr}(K)) \cdot \sum_{d \in D, p(\sigma(d))=d_2} b_{d_1}(g\sigma(d)^{-1}) \cdot g\sigma(d)^{-1} b_{d_2}(\sigma(d))$$

$$= \mu(\text{pr}(K)) \cdot b_{d_1}(g\sigma(d_2)^{-1}) \cdot g\sigma(d_2)^{-1} b_{d_2}(\sigma(d_2))$$

$$\overset{(37)}{=} \mu(\text{pr}(K)) \cdot b_{d_1}(g\sigma(d_2)^{-1}) \cdot g\sigma(d_2)^{-1} \left(\frac{1}{\mu(\text{pr}(K))} \cdot \omega(e) \right)$$

$$= b_{d_1}(g\sigma(d_2)^{-1}) \cdot \left(g\sigma(d_2)^{-1} \cdot 1 \right)$$

$$= b_{d_1}(g\sigma(d_2)^{-1})$$

$$\overset{(37)}{=} \begin{cases} \frac{1}{\mu(\text{pr}(K))} \cdot \omega(n_0) & \text{if } p(g) = d_1 d_2; \\ = 0 & \text{if } p(g) \neq d_1 d_2. \end{cases} \tag{42}$$

Suppose for $g \in G$ that $p(g) = d_1 d_2$. We can write

$$g = n_0 k_0 \sigma(d_1)\sigma(d_2) = \left(n_0 k_0 k^{-1} n^{-1} (k_0 k^{-1})^{-1} \right)\left(k_0 k^{-1} \right)\sigma(d_1 d_2)$$

and have $n_0 k_0 k^{-1} n^{-1} (k_0 k^{-1})^{-1} \in N$ and $k_0 k^{-1} \in K$. We compute

$$
\begin{aligned}
& w(d_1, d_2) \cdot b_{d_1 d_2}(g) \\
& \overset{(40)}{=} \omega(n) \cdot b_{d_1 d_2}(g) \\
& \overset{(37)}{=} \omega(n) \cdot \frac{1}{\mu(\mathrm{pr}(K))} \cdot \omega(n_0 k_0 k^{-1} n^{-1} (k_0 k^{-1})^{-1}) \\
& = \omega(n) \cdot \frac{1}{\mu(\mathrm{pr}(K))} \cdot \omega(n_0) \cdot \omega(k_0 k^{-1} n^{-1} (k_0 k^{-1})^{-1}) \\
& \overset{(5)}{=} \omega(n) \cdot \frac{1}{\mu(\mathrm{pr}(K))} \cdot \omega(n_0) \cdot \omega(n^{-1}) \\
& \overset{\omega(n_0) \in \mathrm{cent}(R)}{=} \frac{1}{\mu(\mathrm{pr}(K))} \cdot \omega(n) \cdot \omega(n^{-1}) \cdot \omega(n_0) \\
& = \frac{1}{\mu(\mathrm{pr}(K))} \cdot \omega(n_0).
\end{aligned}
\tag{43}
$$

Since $w(d_1, d_2) \cdot b_{d_1 d_2}(g) = 0$ if $p(g) \neq d_1 d_2$, we conclude (41) from (42) and (43).

We compute for $d \in D$, $r \in R$ and $d' \in D$ using the fact that $\{\sigma(d'')^{-1} \mid d'' \in D\}$ is a transversal for $G \to G/NK = D$ and $\sigma(e_D) = e_G$

$$
\begin{aligned}
& \left(b_d \cdot (r \cdot b_{e_D}) \right)(\sigma(d')) \\
& \overset{(15)}{=} \mu(\mathrm{pr}(K)) \cdot \sum_{d'' \in D} b_d(\sigma(d')\sigma(d'')^{-1}) \cdot \sigma(d')\sigma(d'')^{-1}\left((r \cdot b_{e_D})(\sigma(d'')) \right) \\
& \overset{(37)}{=} \mu(\mathrm{pr}(K)) \cdot \sum_{d'' \in \{e_Q\}} b_d(\sigma(d')\sigma(d'')^{-1}) \cdot \sigma(d')\sigma(d'')^{-1}\left((r \cdot b_{e_D})(\sigma(d'')) \right) \\
& = \mu(\mathrm{pr}(K)) \cdot b_d(\sigma(d')e_G^{-1}) \cdot \sigma(d')e_G^{-1}\left((r \cdot b_{e_D})(e_G) \right) \\
& \overset{(37)}{=} \mu(\mathrm{pr}(K)) \cdot b_d(\sigma(d')) \cdot \sigma(d')\left(\frac{1}{\mu(\mathrm{pr}(K))} \cdot r \cdot 1 \right) \\
& = b_d(\sigma(d')) \cdot \sigma(d')r \\
& \overset{(37)}{=} \begin{cases} \frac{1}{\mu(\mathrm{pr}(K))} \cdot \sigma(d)r & \text{if } d' = d; \\ 0 & \text{otherwise.} \end{cases}
\end{aligned}
$$

This implies for $d \in D, r \in R$, and $g \in G$

$$\left(b_d \cdot (r \cdot b_{e_D})\right) \overset{(39)}{=} \sum_{d' \in D} \mu(\mathrm{pr}(K)) \cdot b_d \cdot (r \cdot b_{e_D})(\sigma(d'))b_{d'}$$

$$= \mu(\mathrm{pr}(K)) \cdot \left(\frac{1}{\mu(\mathrm{pr}(K))} \cdot \sigma(d)r\right) \cdot b_d = \sigma(d)r \cdot b_d. \qquad (44)$$

Recall from Lemma 23 that $\mathcal{H}(G//K; R, \rho, \omega)$ has a unit, namely b_{e_D}.

We conclude from (41) and (44) that the unital ring $\mathcal{H}(G//K; R, \rho, \omega)$ is the crossed product ring $R * D$ associated to (w, c) for w defined in (40) and $c_d(r) := (\rho \circ \sigma(d))(r)$. $\qquad \square$

7.4 Filtering the Hecke Algebra of a Compact Group by Normal Compact Open Subgroups

Consider a sequence $G = K_0 \supseteq K_1 \supseteq K_1 \supseteq K_2 \supseteq \cdots$ of normal compact open subgroups of G with $\bigcap_{n \geq 0} K_n = \{1\}$ such that $K_n \in P$ holds for $n \in \mathbb{N}$. It exists by Lemma 7.3 as we assume throughout this section that Q is compact and N is locally central. Let 1_{K_n} be the element in $\mathcal{H}(G; R, \rho, \omega)$ defined in (23). Then 1_{K_n} is central in $\mathcal{H}(G; R, \rho, \omega)$, since K_n is normal in G. We have $1_{K_n} \cdot 1_{K_m} = 1_{K_n} = 1_{K_m} \cdot 1_{K_n}$ for $m \leq n$. For every $s \in \mathcal{H}(G)$ there exists a natural number $n \in \mathbb{N}$ satisfying $1_{K_n} \cdot s = s = s \cdot 1_{K_n}$. In the sequel we sometimes abbreviate $1_n = 1_{K_n}$, $\mathcal{H}(G) = \mathcal{H}(G; R, \rho, \omega)$ and $\mathcal{H}(G//K_n) = \mathcal{H}(G//K_n; R, \rho, \omega)$ and put $1_{-1} = 0$. The elementary proof of the next lemma is left to the reader.

Lemma 7.6 *We have the subrings*

$$1_n \mathcal{H}(G) 1_n = \mathcal{H}(G//K_n) \text{ and } (1_n - 1_{n-1})\mathcal{H}(G)(1_n - 1_{n-1})$$

of $\mathcal{H}(G)$, which have 1_n and $(1_n - 1_{n-1})$ as unit. We get an obvious identification of rings (without unit)

$$\bigoplus_{m \geq 0} (1_m - 1_{m-1})\mathcal{H}(G)(1_m - 1_{m-1}) = \mathcal{H}(G),$$

and for $n \geq 0$ of rings with unit

$$\bigoplus_{m=0}^{n} (1_m - 1_{m-1})\mathcal{H}(G)(1_m - 1_{m-1}) = 1_n \mathcal{H}(G) 1_n.$$

Recall that a sequence $A_0 \xrightarrow{f_0} A_1 \xrightarrow{f_1} A_2$ in an additive category $\mathcal{A}$ is called *exact at* A_1 if $f_1 \circ f_0 = 0$ and for every object A and morphism $g \colon A \to A_1$ with $f_1 \circ g = 0$ there exists a morphism $\overline{g} \colon A \to A_0$ with $f_0 \circ \overline{g} = g$. For information on how this notion is related by the Yoneda embedding to the usual notion of exactness for modules we refer to [2, Lemma 5.10 and Lemma 6.3]. A functor $F \colon \mathcal{A} \to \mathcal{A}'$ of additive categories is called *faithfully flat*, provided that a sequence $A_0 \xrightarrow{f_0} A_1 \xrightarrow{f_1} A_2$ in $\mathcal{A}$ is exact, if and only if the sequence $F(A_0) \xrightarrow{F(f_0)} F(A_1) \xrightarrow{F(f_1)} F(A_2)$ in $\mathcal{A}'$ is exact.

Lemma 7.7 *Let S and T be unital rings. Let* $\mathrm{pr} \colon S \times T \to S$ *be the projection, which is a homomorphism of unital rings. Let* $i \colon S \to S \times T$ *be the inclusion sending s to $(s, 0)$, which is a homomorphism of rings (without units). Then*

(1) *There exists a diagram of unital additive categories commuting up to natural equivalence of unital additive categories*

$$
\begin{array}{ccc}
\mathrm{Idem}(\underline{S}_\oplus) & \xrightarrow{\mathrm{Idem}(i_\oplus)} & \mathrm{Idem}(\underline{S \times T}_\oplus) \\
\Theta_S \downarrow \simeq & & \simeq \downarrow \Theta_{S \times T} \\
S\text{-MOD}_{\mathrm{fgp}} & \xrightarrow[\mathrm{pr}^*]{} & S \times T\text{-MOD}_{\mathrm{fgp}}
\end{array}
$$

where the vertical arrows are the equivalences of unital additive categories of (25) and pr^* *is restriction with* pr.

(2) *The functor* $\mathrm{Idem}(i_\oplus) \colon \mathrm{Idem}(\underline{S}_\oplus) \to \mathrm{Idem}(\underline{S \times T}_\oplus)$ *has a retraction, namely* $\mathrm{Idem}(\mathrm{pr}_\oplus) \colon \mathrm{Idem}(\underline{S \times T}_\oplus) \to \mathrm{Idem}(\underline{S}_\oplus)$.

(3) *The functor* $\mathrm{Idem}(i_\oplus) \colon \mathrm{Idem}(\underline{S}_\oplus) \to \mathrm{Idem}(\underline{S \times T}_\oplus)$ *is faithfully flat.*

Proof

(1) Next we construct for every object $([l], p)$ in $\mathrm{Idem}(S_\oplus)$ an isomorphism in $S \times T\text{-MOD}_{\mathrm{fgp}}$

$$
T([l], p) \colon \mathrm{pr}^* \circ \Theta_S([l], p) \xrightarrow{\cong} \Theta_{S \times T} \circ \mathrm{Idem}(i_\oplus)([l], p).
$$

Let A be the (l, l)-matrix over S, for which $p \colon [l] \to [l]$ is given by A. If $i(A)$ is the (l, l)-matrix over $S \times T$ given by applying i to each element in A, then $\theta_{S \times T} \circ i_\oplus(p)$ is the $S \times T$-homomorphism $r_{i(A)} \colon (S \times T)^l \to (S \times T)^l$ given by right multiplication with $i(A)$. Let $i^l \colon S^l \to (S \times T)^l$ be the map sending $(x_1, x_2, \ldots, x_l)$ to $(i(x_1), i(x_2), \ldots, i(x_l))$. We obtain a commutative diagram of abelian groups

$$
\begin{array}{ccc}
S^l & \xrightarrow{\ i^l\ } & (S \times T)^l \\
\ {\scriptstyle r_A}\downarrow & & \downarrow{\scriptstyle r_{i(A)}} \\
S^l & \xrightarrow[\ i^{l'}\ ]{} & (S \times T)^l.
\end{array}
$$

Now $i^{l'}$ induces a homomorphism of abelian groups.

$$T([l], p)\colon \operatorname{im}(r_A) \to \operatorname{im}(r_{i(A)}).$$

It is injective, since i and hence i^l is injective. Next we show that $T([l], p)$ is bijective. Let y be an element of the image of $r_{i(A)}$. Choose $x = \big((s_1, t_1), \ldots, (s_l, t_l)\big)$ in $(S \times T)^l$ with $r_{i(A)}(x) = y$. Define $x' \in S$ by $x' = (s_1, \ldots, s_l)$. Then $r_{i(A)} \circ i^l(x') = r_{i(A)}(x) = y$. Hence i^l sends $r_A(x)$ to y. This finishes the proof that $T([l], p)$ is an isomorphisms of abelian groups. One easily checks that it is an isomorphism of $S \times T$-modules.

We leave it to the reader to check that the collection of the isomorphisms $T([l], p)$ defines a natural equivalence of functors $\operatorname{Idem}(\underline{S}_\oplus) \to S \times T\text{-MOD}_{\text{fgp}}$ from $\operatorname{pr}^* \circ \theta_S$ to $\theta_{S \times T} \circ \operatorname{Idem}(i_\oplus)$.

(2) This follows from $\operatorname{pr} \circ i = \operatorname{id}_S$.

(3) Since restriction is faithfully flat, the claim follows from assertion (1).

$\qquad\qquad\qquad\qquad\qquad\qquad\qquad\qquad\qquad\qquad\qquad\qquad\qquad\qquad\square$

We record for later purposes

Lemma 7.8 *Suppose that Q is compact. Consider normal compact open subgroups K and K' of G satisfying $K' \subseteq K$ and $K, K' \in P$. Let*

$$i\colon \mathcal{H}(G//K; R, \rho, \omega) \to \mathcal{H}(G//K'; R, \rho, \omega)$$

be the inclusion of rings. Let $m \geq 0$ be an integer. Denote by

$$i[\mathbb{Z}^m]\colon \mathcal{H}(G//K; R, \rho, \omega)[\mathbb{Z}^m] \to \mathcal{H}(G//K'; R, \rho, \omega)[\mathbb{Z}^m]$$

the inclusion of the (untwisted) group rings induced by i.

Then the functor

$$\operatorname{Idem}(\underline{i[\mathbb{Z}^m]}_\oplus)\colon \operatorname{Idem}\big(\underline{\mathcal{H}(G//K; R, \rho, \omega)[\mathbb{Z}^m]}_\oplus\big)$$

$$\to \operatorname{Idem}\big(\underline{\mathcal{H}(G//K'; R, \rho, \omega)[\mathbb{Z}^m]}_\oplus\big)$$

has a retraction and is faithfully flat.

Proof This follows from Lemma 7.7 and the the decomposition of unital rings

$$\mathcal{H}(G//K'; R, \rho, \omega) = \mathcal{H}(G//K; R, \rho, \omega) \oplus (1_{K'} - 1_K)\mathcal{H}(G//K'; R, \rho, \omega)(1_{K'} - 1_K),$$

cf. Lemma 7.6.   □

7.5   Proof of Theorem 7.2

Lemma 7.9 *Let* $\mathcal{A}_i$ *be a collection of additive categories. Then* $\bigoplus_{i \in I} \mathcal{A}_i$ *is* *l-uniformly regular or regular respectively if and only if each* $\mathcal{A}_i$ *is l-uniformly regular or regular respectively.*

Proof This is a consequence of the observations following from [2, Lemma 5.3], that for an object $A \in \bigoplus_{i \in I} \mathcal{A}_i$ there exists a finite subset $J \subseteq I$ with $A \in \bigoplus_{i \in J} \mathcal{A}_i$ and we have the identifications

$$\mathrm{mor}_{\bigoplus_{i \in I} \mathcal{A}_i}(?, A) = \mathrm{mor}_{\bigoplus_{i \in J} \mathcal{A}_i}(?, A);$$

$$\mathbb{Z}(\bigoplus_{i \in J} \mathcal{A}_i)\text{-MOD} = \prod_{i \in J} \mathbb{Z}\mathcal{A}_i\text{-MOD}.$$

More details of the proof can be found in [2, Section 11].   □

Consider a sequence $G = K_0 \supseteq K_1 \supseteq K_1 \supseteq K_2 \supseteq \cdots$ of normal compact open subgroups of G with $\bigcap_{n \geq 0} K_n = \{1\}$ such that $K_n \in P$ holds for $n \in \mathbb{N}$. We get from Lemma 7.6 for every natural number d identifications of additive categories

$$\bigoplus_{m \geq 0} \mathrm{Idem}\left((1_m - 1_{m-1})\mathcal{H}(G)(1_m - 1_{m-1})_\oplus[\mathbb{Z}^d]\right) = \mathrm{Idem}\left(\mathcal{H}(G)_\oplus[\mathbb{Z}^d]\right);$$

$$\bigoplus_{m=0}^{n} \mathrm{Idem}\left((1_m - 1_{m-1})\mathcal{H}(G)(1_m - 1_{m-1})_\oplus[\mathbb{Z}^d]\right) = \mathrm{Idem}\left(\mathcal{H}(G//K_n)_\oplus[\mathbb{Z}^d]\right).$$

Hence by Lemma 7.9 it suffices to show that $\mathrm{Idem}\left(\mathcal{H}(G//K_n)_\oplus[\mathbb{Z}^d]\right)$ is $(l + 2m)$-uniformly regular or regular respectively for every $n \in \mathbb{N}$.

The unital ring $\mathcal{H}(G//K_n)$ is l-uniformly regular or regular respectively, since R is l-uniformly regular or regular respectively by assumption and we have Lemmas 7.4 and 7.5. Hence $\mathrm{Idem}\left(\mathcal{H}(G//K_n)_\oplus[\mathbb{Z}^d]\right)$ is $(l + 2m)$-uniformly regular or regular respectively by [2, Corollary 6.5 and Theorem 10.1]. This finishes the proof of Theorem 7.2.

8   Negative K-Groups and the Projective Class Group of Hecke Algebras Over Compact td-Groups

Let G, N, $Q := G/N$, pr: $G \to Q$, R, $\mathcal{P}$, ρ, ω, and μ be as in Sect. 2.1 and denote by $\mathcal{H}(G; R, \rho, \omega)$ the Hecke algebra which we have introduced in Sect. 2.2. Our main assumption in this section will be that Q is compact.

Lemma 8.1 *Suppose that Q is compact and N is locally central. Suppose that the unital ring R is regular and satisfies $\mathbb{Q} \subseteq R$. Then:*

(1) *Let $\mathcal{K}$ be the set of compact open normal subgroups $K \subseteq G$ with $K \in \mathcal{P}$ directed by $K \leq K' \iff K' \subseteq K$.*
   Then we get for $n \in \mathbb{Z}$

$$K_n\big(\mathcal{H}(G; R, \rho, \omega)\big) = \mathrm{colim}_{K \in \mathcal{K}} \, K_n\big(\mathcal{H}(G//K; R, \rho, \omega)\big).$$

(2) *We get*

$$K_n\big(\mathcal{H}(G; R, \rho, \omega)\big) = 0 \quad for \; n \leq -1.$$

Proof

(1) We conclude from Lemmas 2.5 and 7.3

$$\mathcal{H}(G; R, \rho, \omega) = \bigcup_{K \in \mathcal{K}} \mathcal{H}(G//K; R, \rho, \omega).$$

Now apply (30).

(2) For $K \in \mathcal{K}$ the unital ring $\mathcal{H}(G//K; R, \rho, \omega)$ is regular by Lemma 7.4 (2) and Lemma 7.5. Hence $K_n\big(\mathcal{H}(G//K; R, \rho, \omega)\big) = \{0\}$ for $n \leq -1$, see [16, page 154]. Now apply assertion (1).

$\square$

Remark 8.1 Suppose that Q is compact and N is locally central. Because of Lemma 7.3 we can choose a nested sequence of elements in $\mathcal{K}$

$$K_0 \supseteq K_1 \supseteq K_2 \supseteq K_3 \supseteq \cdots$$

satisfying $\bigcap_{i=0}^{\infty} K_n = \{1\}$. Then for every $K \in \mathcal{K}$ there is a natural number i with $K_i \subseteq K$. Abbreviate $\mathcal{H}(G//K_i) = \mathcal{H}(G//K_i; R, \rho, \omega)$. Then the inclusion $\mathcal{H}(G//K_i) \to \mathcal{H}(G//K_{i+1})$ induces a split injection $K_n(\mathcal{H}(G//K_i)) \to K_n(\mathcal{H}(G//K_{i+1}))$ for $i \in \mathbb{N}$ and $n \in \mathbb{Z}$ by Lemma 7.8. Lemma 8.1 (1) implies that there is an isomorphism

$$K_n(\mathcal{H}(G; R, \rho, \omega))$$

$$\cong K_n(\mathcal{H}(G//K_0)) \oplus \bigoplus_{i \geq 0} \mathrm{cok}\left(K_n(\mathcal{H}(G//K_i)) \to K_n(\mathcal{H}(G//K_{i+1}))\right)$$

and $\mathrm{cok}\left(K_n(\mathcal{H}(G//K_i)) \to K_n(\mathcal{H}(G//K_{i+1}))\right)$ is isomorphic to a direct summand of $K_n(\mathcal{H}(G//K_{i+1}))$.

Now suppose additionally that R is semisimple. Then $\mathcal{H}(G//K_i)$ is semisimple and hence the abelian group $K_0(\mathcal{H}(G//K_i))$ is finitely generated free for $i \in \mathbb{N}$ by Lemma 7.4 (4) and Lemma 7.5, Hence the abelian group $K_0(\mathcal{H}(G; R, \rho, \omega))$ is free and in particular torsionfree.

9 On the Algebraic K-Theory of the Hecke Algebra of a Covirtually $\mathbb{Z}$ Totally Disconnected Group

Consider the setup of Sect. 5. In particular Q is covirtually cyclic. Denote by $\mathbf{T}_{\mathbf{K}^\infty(\varphi^{-1})}$ the mapping torus of the map

$$\mathbf{K}^\infty(\varphi^{-1}): \mathbf{K}^\infty(\mathcal{H}(L; R, \rho|_L, \omega)) \to \mathbf{K}^\infty(\mathcal{H}(L; R, \rho|_L, \omega))$$

of non-connective K-theory spectra.

Theorem 9.1 (Wang Sequence) *Suppose that the unital ring R is regular and satisfies $\mathbb{Q} \subseteq R$. Assume that N is locally central. Then:*

(1) *There is a weak homotopy equivalence of non-connective spectra*

$$\mathbf{a}^\infty: \mathbf{T}_{\mathbf{K}^\infty(\varphi^{-1})} \xrightarrow{\simeq} \mathbf{K}^\infty(\mathcal{H}(G; R, \rho, \omega)).$$

(2) *We get a long exact sequence, infinite to the left*

$$\cdots \xrightarrow{K_2(i)} K_2(\mathcal{H}(G; R, \rho, \omega)) \xrightarrow{\partial_2} K_1(\mathcal{H}(L; R, \rho|_L, \omega))$$

$$\xrightarrow{\mathrm{id}-K_1(\varphi^{-1})} K_1(\mathcal{H}(L; R, \rho|_L, \omega)) \xrightarrow{K_1(i)} K_1(\mathcal{H}(G; R, \rho, \omega))$$

$$\xrightarrow{\partial_1} K_0(\mathcal{H}(L; R, \rho|_L, \omega)) \xrightarrow{\mathrm{id}-K_0(\varphi^{-1})} K_0(\mathcal{H}(L; R, \rho|_L, \omega))$$

$$\xrightarrow{K_0(i)} K_0(\mathcal{H}(G; R, \rho, \omega)) \to 0.$$

(3) *We get for $n \leq -1$*

$$K_n(\mathcal{H}(G; R, \rho, \omega)) = 0.$$

Proof

(1) This follows from Lemma 5.3 and Theorem 6.2 applied to the additive category $\mathcal{A} = \mathrm{Idem}\left(\mathcal{H}(L; R, \rho|_L, \omega)_{\oplus}\right)$ after we have shown that the additive category $\mathrm{Idem}\left(\mathcal{H}(L; R, \rho|_L, \omega)_{\oplus}\right)$ is regular. This has already been done in Theorem 7.2.

(2) and (3) These follow from the Wang sequence associated to the left-hand side of the weak homotopy equivalence appearing in assertion (1) and Lemma 8.1 (2).

$\square$

10 Some Input for the Farrell–Jones Conjecture

Our ultimate goal is the proof and the application of the K-theoretic Farrell–Jones Conjecture for the Hecke algebra of a closed subgroup of a reductive p-adic group. For this purpose we will need Theorem 7.2 and the following Theorem 10.1 in forthcoming papers.

Consider the setup of Sect. 2.1. For the remainder of this subsection we will assume that the td-group Q is compact and N is locally central. Let $\bar{i} : Q' \to Q$ be the inclusion of a compact open subgroup of Q. Put $G' = \mathrm{pr}^{-1}(Q')$. Let $i : G' \to G$ be the inclusion. The construction in Sect. 2.3 yields a ring homomorphism

$$\mathcal{H}(i) : \mathcal{H}(G'; R, \rho', \omega) \to \mathcal{H}(G; R, \rho, \omega)$$

where $\rho' = \rho \circ i$, μ' is obtained from μ by restriction with i, and we take $N' = N$ and $\omega' = \omega$. The image $\mathcal{H}(i)(s)$ of an element $s \in \mathcal{H}(G'; R, \rho', \omega)$, which is given by an appropriate function $s : G' \to R$, is specified by the function $\mathcal{H}(i)(s) : G \to R$ sending g to $s(g)$, if $g \in G'$, and to 0, if $g \notin G'$, see Lemma 2.3 (3).

Theorem 10.1 *Suppose that Q is compact and N is locally central. Then the functor of unital additive categories*

$$\mathrm{Idem}\left(\mathcal{H}(i)_{\oplus}[\mathbb{Z}^m]\right) : \mathrm{Idem}\left(\mathcal{H}(G'; R, \rho', \omega)_{\oplus}[\mathbb{Z}^m]\right) \to \mathrm{Idem}\left(\mathcal{H}(G; R, \rho, \omega)_{\oplus}[\mathbb{Z}^m]\right)$$

is faithfully flat.

Proof Let $\mathcal{K}'$ be the directed set of normal compact open subgroups of Q which satisfy $K \subseteq Q'$, and $K \in P$, where we put $K \leq K' \Longleftrightarrow K' \subseteq K$. Note that for any compact open subgroup L of Q there exists $K \in \mathcal{K}'$ with $K \subseteq L$ by Lemma 7.3.

In the sequel we abbreviate

$$\mathcal{H}(G) := \mathcal{H}(G; R, \rho, \omega);$$

$$\mathcal{H}(G//K) := \mathcal{H}(G//K; R, \rho, \omega),$$

and analogously for G'. Next we want to show that the functor

$$\text{Idem}\left(j_{K_\oplus}[\mathbb{Z}^m]\right): \text{Idem}\left(\underline{\mathcal{H}(G//K)}_\oplus[\mathbb{Z}^m]\right) \to \text{Idem}\left(\underline{\mathcal{H}(G)}_\oplus[\mathbb{Z}^m]\right)$$

is faithfully flat for $K \in \mathcal{K}'$, where $i_K : \mathcal{H}(G//K) \to \mathcal{H}(G)$ is the inclusion. Consider morphisms $f_0 : A_0 \to A_1$ and $f_1 : A_1 \to A_2$ in $\text{Idem}\left(\underline{\mathcal{H}(G//K)}_\oplus[\mathbb{Z}^m]\right)$ with $f_1 \circ f_0 = 0$. Note that we can consider them also as morphisms in $\text{Idem}\left(\underline{\mathcal{H}(G)}_\oplus[\mathbb{Z}^m]\right)$. We have to show that it is exact in $\text{Idem}\left(\underline{\mathcal{H}(G//K)}_\oplus[\mathbb{Z}^m]\right)$ if and only if it is exact in $\text{Idem}\left(\underline{\mathcal{H}(G)}_\oplus[\mathbb{Z}^m]\right)$.

Suppose that $A_0 \xrightarrow{f_0} A_1 \xrightarrow{f_2} A_2$ is exact in $\text{Idem}\left(\underline{\mathcal{H}(G//K)}_\oplus[\mathbb{Z}^m]\right)$. In order to show that it is exact in $\text{Idem}\left(\underline{\mathcal{H}(G)}_\oplus[\mathbb{Z}^m]\right)$, we have to find for any object A and any morphism $g : A \to A_1$ in $\text{Idem}\left(\underline{\mathcal{H}(G)}_\oplus[\mathbb{Z}^m]\right)$ with $f_1 \circ g = 0$ a morphism $\overline{g} : A \to A_0$ in $\text{Idem}\left(\underline{\mathcal{H}(G)}_\oplus[\mathbb{Z}^m]\right)$ with $f_0 \circ \overline{g} = g$. We can choose an element $K' \in \mathcal{K}'$ with $K \leq K'$ such that A and g live already in $\text{Idem}\left(\underline{\mathcal{H}(G//K')}_\oplus[\mathbb{Z}^m]\right)$ by Lemma 2.5 and the first paragraph of this proof. Since the inclusion

$$\text{Idem}\left(\underline{\mathcal{H}(G//K)}_\oplus[\mathbb{Z}^m]\right) \to \text{Idem}\left(\underline{\mathcal{H}(G//K')}_\oplus[\mathbb{Z}^m]\right)$$

is faithfully flat by Lemma 7.8, we can find $\overline{g} : A \to P_0$ with $f_0 \circ \overline{g} = g$ in $\text{Idem}\left(\underline{\mathcal{H}(G//K')}_\oplus[\mathbb{Z}^m]\right)$ and hence also in $\text{Idem}\left(\underline{\mathcal{H}(G)}_\oplus[\mathbb{Z}^m]\right)$.

Suppose that $A_0 \xrightarrow{f_0} A_1 \xrightarrow{f_2} A_2$ is exact in $\text{Idem}\left(\underline{\mathcal{H}(G)}_\oplus[\mathbb{Z}^m]\right)$. In order to show that it is exact in $\text{Idem}\left(\underline{\mathcal{H}(G//K)}_\oplus[\mathbb{Z}^m]\right)$ we have to find for any object A and any morphism $g : A \to A_1$ in $\text{Idem}\left(\underline{\mathcal{H}(G//K)}_\oplus[\mathbb{Z}^m]\right)$ with $f_1 \circ g = 0$ a morphism $\overline{g} : A \to A_0$ in $\text{Idem}\left(\underline{\mathcal{H}(G//K)}_\oplus[\mathbb{Z}^m]\right)$ with $f_0 \circ \overline{g} = g$. At any rate we can find such $\overline{g} : A \to A_1$ in $\text{Idem}\left(\underline{\mathcal{H}(G)}_\oplus[\mathbb{Z}^m]\right)$. We conclude from Lemma 2.5 that there exists $K' \in \mathcal{K}'$ with $K \leq K'$ such that $\overline{g} : A \to A_1$ lies already in $\text{Idem}\left(\underline{\mathcal{H}(G//K')}_\oplus[\mathbb{Z}^m]\right)$. Recall from Lemma 7.8 that there is a retraction of the inclusion

$$\text{Idem}\left(\underline{\mathcal{H}(G//K)}_\oplus[\mathbb{Z}^m]\right) \to \text{Idem}\left(\underline{\mathcal{H}(G//K')}_\oplus[\mathbb{Z}^m]\right).$$

If we apply it to $\overline{g}$, we get a morphism $\overline{g'}\colon A \to A_1$ in $\mathrm{Idem}\left(\underline{\mathcal{H}(G//K)}_\oplus[\mathbb{Z}^m]\right)$ satisfying $f_1 \circ \overline{g'} = g$ in $\mathrm{Idem}\left(\underline{\mathcal{H}(G//K)}_\oplus[\mathbb{Z}^m]\right)$. This finishes the proof that functor $\mathrm{Idem}\left(\underline{j_K}_\oplus[\mathbb{Z}^m]\right)$ is faithfully flat. Analogously one shows that the functor $\mathrm{Idem}\left(\underline{j'_K}_\oplus[\mathbb{Z}^m]\right)\colon \mathrm{Idem}\left(\underline{\mathcal{H}(G'//K)}_\oplus[\mathbb{Z}^m]\right) \to \mathrm{Idem}\left(\underline{\mathcal{H}(G')}_\oplus[\mathbb{Z}^m]\right)$ is faithfully flat for the inclusion $j'_K\colon \mathcal{H}(G'//K) \to \mathcal{H}(G')$.

We have the following commutative diagram of functors of additive categories

$$
\begin{array}{ccc}
\mathrm{Idem}\left(\underline{\mathcal{H}(G')}_\oplus[\mathbb{Z}^m]\right) & \xrightarrow{\;\mathrm{Idem}(\underline{\mathcal{H}(i)}_\oplus[\mathbb{Z}^m])\;} & \mathrm{Idem}\left(\underline{\mathcal{H}(G)}_\oplus[\mathbb{Z}^m]\right) \\[2mm]
{\scriptstyle\mathrm{Idem}(\underline{j'_K}_\oplus[\mathbb{Z}^m])}\Big\uparrow & & \Big\uparrow{\scriptstyle\mathrm{Idem}(\underline{j'_K}_\oplus[\mathbb{Z}^m])} \\[2mm]
\mathrm{Idem}\left(\underline{\mathcal{H}(G'//K)}_\oplus[\mathbb{Z}^m]\right) & \xrightarrow{\;\mathrm{Idem}(\underline{\mathcal{H}(i//K)}_\oplus[\mathbb{Z}^m])\;} & \mathrm{Idem}\left(\underline{\mathcal{H}(G//K)}_\oplus[\mathbb{Z}^m]\right)
\end{array}
$$

whose two left vertical arrows are faithfully flat. We conclude from Lemma 2.5 that it suffices to show that the lower vertical arrow in the diagram above is faithfully flat.

We have identified $\mathcal{H}(G//K)$ and $\mathcal{H}(G'//K)$ respectively as a crossed product ring $R * F$ and $R * F'$ respectively for the finite group $F = G/K$ and $F' = G'/K$ respectively in Lemma 7.5. Moreover the inclusion $\mathcal{H}(G//K)[\mathbb{Z}^m] \to \mathcal{H}(G'//K)[\mathbb{Z}^m]$ corresponds under these identifications to the inclusions $R * F[\mathbb{Z}^m] \to R * F'[\mathbb{Z}^m]$ coming from the inclusion of finite groups $F' \to F$. The lower horizontal arrow $\mathrm{Idem}(\underline{\mathcal{H}(i//K)}_\oplus[\mathbb{Z}^m])$ becomes under the equivalences of categories of (25) and (27) the functor

$$
F\colon R*F'[\mathbb{Z}^m]\text{-}\mathrm{MOD}_{\mathrm{fgp}} \to R*F[\mathbb{Z}^m]\text{-}\mathrm{MOD}_{\mathrm{fgp}}, \qquad P \mapsto R*F[\mathbb{Z}^m]\otimes_{R*F'[\mathbb{Z}^m]} P.
$$

There is a commutative diagram

$$
\begin{array}{ccc}
R*F'[\mathbb{Z}^m]\text{-}\mathrm{MOD}_{\mathrm{fgp}} & \xrightarrow{\;F\;} & R*F[\mathbb{Z}^m]\text{-}\mathrm{MOD}_{\mathrm{fgp}} \\[2mm]
\Big\downarrow & & \Big\downarrow \\[2mm]
R[\mathbb{Z}^m]\text{-}\mathrm{MOD}_{\mathrm{fgp}} & \longrightarrow & R[\mathbb{Z}^m]\text{-}\mathrm{MOD}_{\mathrm{fgp}}
\end{array}
$$

whose vertical arrows are given by restriction from $R * F[\mathbb{Z}^m]$ or $R * F'[\mathbb{Z}^m]$ to $R[\mathbb{Z}^m]$ and whose lower vertical arrow is given by $P \mapsto \bigoplus_{i=1}^{[F:F']} P$. Since the vertical arrows and the lower horizontal arrow are obviously faithfully flat, the upper vertical arrow is faithfully flat. This finishes the proof of Lemma 10.1. $\qquad\square$

11 Characteristic p

We have assumed $\mathbb{Q} \subseteq R$, or in other words that any natural number $n \geq 1$ is invertible in R. One may wonder what happens if one drops this condition, for instance, if R is a field of prime characteristic. The following condition appearing in [6, page 9] suffices to make sense of the Hecke algebra.

Condition 11.1 *There exists a compact open subgroup K in Q such that the index* $[K : K_0]$ *of any open subgroup K_0 of K is invertible in R.*

Let Q be a reductive p-adic group. Then Condition 11.1 is satisfied if p is invertible in R, see [6, page 9].

However, this does not mean that the assertion of the Farell–Jones Conjecture or Theorem 9.1 remains true integrally. Our arguments would go through if for every compact open subgroup K in Q the index $[K : K_0]$ of any open subgroup K_0 of K is invertible in R, which is stronger than Condition 11.1.

One may hope that under Condition 11.1 the Farrell–Jones Conjecture or Theorem 9.1 remain true rationally. Let us confine ourselves to the setup of Sect. 5 and Theorem 9.1. Then we get from [13, Theorem 0.1] a weak homotopy equivalence, where we abbreviate $\mathcal{H}(G) := \mathcal{H}(G; R, \rho, \omega)$ and analogously for L

$$\mathbf{T}_{\mathbf{K}^\infty(\mathrm{Idem}(\varphi)^{-1})} \vee \mathbf{NK}^\infty(\mathrm{Idem}(\underline{\mathcal{H}(L)}_\oplus)_{\mathrm{Idem}(\underline{\varphi}_\oplus)}[t])$$

$$\vee \mathbf{NK}^\infty(\mathrm{Idem}(\underline{\mathcal{H}(L)}_\oplus)_{\mathrm{Idem}(\underline{\varphi}_\oplus)}[t^{-1}])$$

$$\xrightarrow{\simeq} \mathbf{K}^\infty(\mathcal{H}(G; R, \rho, \omega)).$$

So we need to show that the homotopy groups of the Nil-terms all vanish rationally. If L is finite, this is known to be true, see [12, Theorem 0.3 and Theorem 9.4]. Under the strong condition that there is a sequence $L \supseteq L_1 \supseteq L_2 \supseteq L_2 \cdots$ of in L normal compact open subgroups such that $\bigcap_{i \geq 0} L_i = \{1\}$ and $\varphi(L_i) = L_i$ holds for $i \geq 0$, this implies that the homotopy groups of the Nil-terms all vanish rationally. Without this strong condition we do not have a proof.

Acknowledgements The paper is funded by the ERC Advanced Grant KL2MG-interactions (no. 662400) of the second author granted by the European Research Council, by the Deutsche Forschungsgemeinschaft (DFG, German Research Foundation) under Germany's Excellence Strategy GZ 2047/1, Projekt-ID 390685813, Hausdorff Center for Mathematics at Bonn, and by the Deutsche Forschungsgemeinschaft (DFG, German Research Foundation) Project-ID 427320536 SFB 1442, as well as under Germany's Excellence Strategy EXC 2044 390685587, Mathematics Münster: DynamicsGeometryStructure.

The authors heartily thank the referee for her/his very valuable and careful report.

References

1. A. Bartels and W. Lück. On crossed product rings with twisted involutions, their module categories and L-theory. In: *Cohomology of groups and algebraic K-theory*, volume 12 of *Adv. Lect. Math. (ALM)*, pp. 1–54. Int. Press, Somerville, MA (2010).
2. A. Bartels and W. Lück. Vanishing of Nil-terms and negative K-theory for additive categories. Preprint, arXiv:2002.03412 [math.KT] (2020).
3. A. Bartels and W. Lück. Foundations for the K-theoretic Farrell-Jones Conjecture for Hecke algebras of totally disconnected groups. In preparation (2022).
4. A. Bartels and H. Reich. Coefficients for the Farrell–Jones Conjecture. *Adv. Math.* **209**(1), 337–362 (2007).
5. J. Bernstein. Draft of: Representations of p-adic groups. http//www.math.tau.ac.il/~bernstei/ Unpublished_texts/Unpublished_list.html (1992).
6. C. Blondel. Basic representation theory of reductive p-adic groups. Unpublished notes, https:// webusers.imj-prg.fr/~corinne.blondel/Blondel_Beijin.pdf (2011).
7. J.-F. Dat. On the K_0 of a p-adic group. *Invent. Math.* **140**(1), 171–226 (2000).
8. J.-F. Dat. Quelques propriétés des idempotents centraux des groupes p-adiques. *J. Reine Angew. Math.* **554**, 69–103 (2003).
9. F.T. Farrell and W.-C. Hsiang. A formula for $K_1 R_\alpha [T]$. In: *Applications of Categorical Algebra (Proc. Sympos. Pure Math., Vol. XVII, New York, 1968)*, pp. 192–218. Amer. Math. Soc., Providence, RI (1970).
10. P. Garrett. Smooth representations of totally disconnected groups. Unpublished notes (2012).
11. W. Lück and W. Steimle. Non-connective K- and Nil-spectra of additive categories. In: *An alpine expedition through algebraic topology*, volume 617 of *Contemp. Math.*, pp. 205–236, Amer. Math. Soc., Providence, RI (2014).
12. W. Lück and W. Steimle. Splitting the relative assembly map, Nil-terms and involutions. *Ann. K-Theory* **1**(4), 339–377 (2016).
13. W. Lück and W. Steimle. A twisted Bass–Heller–Swan decomposition for the algebraic K-theory of additive categories. *Forum Math.* **28**(1), 129–174 (2016).
14. E.K. Pedersen and C.A. Weibel. A non-connective delooping of algebraic K-theory. In: *Algebraic and Geometric Topology; proc. conf. Rutgers Uni., New Brunswick 1983*, volume 1126 of *Lecture Notes in Mathematics*, pp. 166–181, Springer (1985).
15. M. Pimsner and D. Voiculescu. K-groups of reduced crossed products by free groups. *J. Operator Theory* **8**(1), 131–156 (1982).
16. J. Rosenberg. *Algebraic K-theory and its applications.* Springer-Verlag, New York (1994).
17. P. Schneider and U. Stuhler. The cohomology of p-adic symmetric spaces. *Invent. Math.* **105**(1), 47–122 (1991).
18. P. Schneider and U. Stuhler. Resolutions for smooth representations of the general linear group over a local field. *J. Reine Angew. Math.* **436**, 19–32 (1993).
19. P. Schneider and U. Stuhler. Representation theory and sheaves on the Bruhat–Tits building. *Inst. Hautes Études Sci. Publ. Math.* **85**, 97–191 (1997).
20. M.-F. Vignéras. On formal dimensions for reductive p-adic groups. In: *Festschrift in honor of I.I. Piatetski-Shapiro on the occasion of his sixtieth birthday, Part I (Ramat Aviv, 1989)*, pp. 225–266, Weizmann, Jerusalem (1990).

Groupes de Coxeter finis : centralisateurs d'involutions

Jean-Pierre Serre

À Catriona Byrne, en souvenir d'une vieille amitié

Introduction.

Soit G un groupe de Coxeter fini, soit u une involution de G et soit G_u le centralisateur de u dans G. Dans certains cas, par exemple quand u est une réflexion, le groupe G_u est engendré par des réflexions de G ; en particulier, c'est un groupe de Coxeter. Il n'en est pas de même en général, mais c'est "presque" le cas. Notre but est de préciser cet énoncé (cf. th.1.1 ci-dessous), et de décrire explicitement certains invariants de G_u, pour chaque type $A_n, B_n, ..., I_2(m)$.

1. Premiers énoncés.

Rappelons d'abord quelques notations et quelques définitions.

Soit V un $\mathbf{R}$-espace vectoriel de dimension finie, et soit G un sous-groupe fini de $\mathrm{GL}(V)$ engendré par des réflexions, i.e. par des éléments d'ordre 2 fixant un hyperplan. Nous dirons, comme dans [Se 22], 1.1, que le couple (V, G) est un *couple de Coxeter*. Sauf mention expresse du contraire (cf. §4 ou 5), on supposera que V est *réduit*, c'est-à-dire ne contient aucun élément $\neq 0$ fixé par G ; cela revient à demander que $\dim V$ est égal au rang $\mathrm{rg}(G)$ de G.

Soit H un sous-groupe de G. On dit que H est un *$\mathscr{C}$-sous-groupe* de G s'il est engendré par des réflexions, i.e. si (V, H) est un couple de Coxeter. On dit que H est *parabolique* s'il existe une partie X de V telle que H soit l'ensemble des éléments de G qui fixent X ; on sait que cela entraîne que H est un $\mathscr{C}$-sous-groupe, cf. [Se 22], 1.5.

J.-P. Serre
Collège de France, Paris
email : jpserre691@gmail.com

© The Author(s), under exclusive license to Springer Nature Switzerland AG 2023
J.-M. Morel, B. Teissier (eds.), *Mathematics Going Forward*, Lecture Notes in Mathematics 2313, https://doi.org/10.1007/978-3-031-12244-6_20

Soit u une involution de G, autrement dit un élément de G de carré 1. On note V_u^+ le sous-espace vectoriel de V fixé par u, et V_u^- celui fixé par $-u$. On a $V = V_u^+ \oplus V_u^-$. Le *degré* de u est défini par $\deg(u) = \dim V_u^-$; c'est la multiplicité de -1 comme valeur propre de u ; on le note souvent d.

Théorème 1.1. *Le centralisateur G_u de u dans G est engendré par des involutions de degré 1 et 2.*

Dans le cas particulier où u est une involution de degré maximal, c'est le cor.3.18 de [Se 22].

Soit G_u^1 le sous-groupe de G_u engendré par les éléments de G_u qui sont des réflexions de G. C'est le plus grand $\mathscr{C}$-sous-groupe de G_u. Il est normal dans G_u. Notons Γ_u le quotient G_u/G_u^1 ; ce groupe précise dans quelle mesure G_u n'est pas engendré par des réflexions. Le th.1.1 équivaut à dire que Γ_u est engendré par les images des involutions de degré 2 de G_u.

Réduction au cas irréductible.

Supposons que V soit réduit. On sait (cf. [Bo 68], V3.7) que V se décompose de façon unique en une somme directe $V = \oplus V_i$ de représentations irréductibles non triviales de G, et que $G = \prod G_i$, avec $G_i \subset \mathrm{GL}(V_i)$. Les G_i sont les *composantes irréductibles* de G. On a $G_u = \prod G_{u_i}$, où les u_i sont les composantes de u. Il y a des décompositions analogues pour G_u^1, Γ_u, etc. En particulier, il suffit de démontrer le th.1.1 lorsque G est irréductible, donc de l'un des types $A, B, ..., I$; c'est ce que nous ferons dans la suite.

Théorème 1.2. (a) *Si G est irréductible non de type D_n ($n \geqslant 5$), le groupe Γ_u est isomorphe à un groupe symétrique.*

(b) *Si G est irréductible de type D_n, Γ_u est isomorphe, soit à un groupe symétrique, soit au produit d'un groupe symétrique par un groupe d'ordre 2.* [Par exemple, quand G est de type D_5, il existe une involution u de G telle que Γ_u soit abélien élémentaire de type $(2,2)$.]

Notation. Dans le cas (a), si $\Gamma_u \neq 1$, nous noterons γ_u l'unique entier $r > 1$ tel que $\Gamma_u \simeq \mathrm{Sym}_r$. Lorsque $\Gamma_u = 1$, nous écrirons tantôt $\gamma_u = 1$ et tantôt $\gamma_u = 0$, suivant le contexte.

Les théorèmes 1.1 et 1.2 seront démontrés dans les §§3-12 par une analyse cas par cas, qui donnera la structure de Γ_u, ainsi que celle des groupes de Coxeter $G_u^+, \widetilde{G}_u^+, G_u^-, \widetilde{G}_u^-$ définis ci-dessous. Nous verrons également que l'on peut choisir l'isomorphisme $\Gamma_u \to \mathrm{Sym}_{\gamma_u}$ du théorème 1.2 (a) de telle sorte que toute transposition de Sym_{γ_u} soit l'image d'une involution de degré 2 de G_u ; il y a un énoncé analogue dans le cas du type D_n, cf. §6.

2. Les groupes $G_u^+, G_u^-, \widetilde{G}_u^+, \widetilde{G}_u^-$.

L'action de G_u sur V respecte la décomposition $V = V_u^+ \oplus V_u^-$. On a donc $G_u \subset \mathrm{GL}(V_u^+) \times \mathrm{GL}(V_u^-)$, ce qui permet de définir les quatre groupes suivants :

$$G_u^+ = G_u \cap \mathrm{GL}(V_u^+),$$
$$G_u^- = G_u \cap \mathrm{GL}(V_u^-),$$
$$\widetilde{G}_u^+ = \text{image de } G_u \text{ dans } \mathrm{GL}(V_u^+) \text{ par la première projection,}$$
$$\widetilde{G}_u^- = \text{image de } G_u \text{ dans } \mathrm{GL}(V_u^-) \text{ par la seconde projection.}$$

Noter que u appartient à G_u^- ; il s'identifie à l'élément -1 de $\mathrm{GL}(V_u^-)$.

On a les inclusions :
$$G_u^+ \subset \widetilde{G}_u^+, \quad G_u^- \subset \widetilde{G}_u^- \quad \text{et} \quad G_u^+ \times G_u^- \subset G_u \subset \widetilde{G}_u^+ \times \widetilde{G}_u^-.$$

Proposition 2.1 (a) G_u^+ *et* G_u^- *sont des sous-groupes paraboliques de* G.

(b) $G_u^1 = G_u^+ \times G_u^-$.

(c) G_u^- *est engendré par les cubes de* G *d'extrémité* u.

[Rappelons que G_u^1 est le sous-groupe de G_u engendré par les réflexions de G qui commutent à u, cf. §1. Un *cube* C de G est un sous-groupe abélien engendré par des réflexions ; on appelle *extrémité* de C l'unique élément de C de degré maximum, cf. [Se 22], 4.1.]

Démonstration.

Les groupes G_u^+ et G_u^- sont des fixateurs de parties de V ; cela entraîne que ce sont des *sous-groupes paraboliques* ; d'où (a). En particulier, ils sont engendrés par des réflexions. D'après la définition de G_u^1, on a donc $G_u^1 \supset G_u^+ \times G_u^-$.

D'autre part, si $s \in G_u$ est une réflexion de G, on a $s \in G_u^+$ si $\deg(us) = \deg(u)+1$ et $s \in G_u^-$ si $\deg(us) = \deg(u)-1$; le groupe G_u^+ est donc engendré par les réflexions du premier type, et G_u^- par celles du second type. Toute réflexion de G_u est donc contenue dans $G_u^+ \times G_u^-$; comme G_u^1 est engendré par de telles réflexions, cela démontre l'inclusion $G_u^1 \subset G_u^+ \times G_u^-$. D'où (b).

Si C est un cube de G d'extrémité u, les réflexions appartenant à C sont du second type, donc appartiennent à G_u^-, d'où $C \subset G_u^-$. Inversement, toute réflexion de G_u^- appartient à un cube maximal de G_u^- ; un tel cube a pour extrémité u, puisque u est l'élément "-1" de G_u^-. D'où (c).

Corollaire 2.1. (V_u^+, G_u^+) *et* (V_u^-, G_u^-) *sont des couples de Coxeter.*

Démonstration. D'après (a), (V, G_u^+) est un couple de Coxeter. Comme $V = V_u^+ \oplus V_u^-$ et que G_u^+ opère trivialement sur V_u^-, il en est de même du couple (V_u^+, G_u^+). Le cas du couple (V_u^-, G_u^-) se traite de manière analogue.

Proposition 2.2. ([FV 05], prop.7 et [DPR 13], prop.2.2) *Le normalisateur de* G_u^- *dans* G *est égal à* G_u.

Démonstration. Il est clair que G_u^- est normal dans G_u. Inversement, soit g un élément de G normalisant G_u^-. Comme u est l'unique involution de G_u^- de degré $\deg(u)$, elle est fixée par l'automorphisme intérieur défini par g, d'où $g \in G_u$.

Remarque. L'énoncé analogue avec G_u^- remplacé par G_u^+ n'est pas toujours vrai; il se peut même que $G_u^+ = 1$ et $G_u \neq G$; c'est le cas si G est de type A_2 et u est une réflexion.

Proposition 2.3. *La suite exacte* $1 \to G_u^1 \to G_u \to \Gamma_u \to 1$ *est scindée.*

(Autrement dit, il existe un sous-groupe X_u de G_u tel que $G_u = G_u^1 \cdot X_u$ et $G_u^1 \cap X_u = 1$.)

Démonstration. Cela résulte du lemme 2 de [Ho 80], appliqué au groupe de Coxeter G_u^1. De plus, la démonstration de [Ho 80] donne une méthode pour construire un groupe X_u : on choisit une chambre $\mathscr{C}$ de G_u^1 dans V, et on lui associe le sous-groupe $H_{\mathscr{C}}$ de $\mathrm{GL}(V)$ formé des éléments qui normalisent G^1 et qui stabilisent $\mathscr{C}$. Le normalisateur de G_u^1 dans $\mathrm{GL}(V)$ est le produit semi-direct $G_u^1 . H_{\mathscr{C}}$. On prend alors $X_u = G_u^1 \cap H_{\mathscr{C}}$.

Passons maintenant aux groupes $\widetilde{G}_u^+$ et $\widetilde{G}_u^-$ définis plus haut :

Proposition 2.4. *On a des suites exactes* :

(a) $1 \to G_u^- \to G_u \to \widetilde{G}_u^+ \to 1$ *et* $1 \to G_u^+ \to G_u \to \widetilde{G}_u^- \to 1$.

(b) $1 \to G_u^- \to \widetilde{G}_u^- \to \Gamma_u \to 1$ *et* $1 \to G_u^+ \to \widetilde{G}_u^+ \to \Gamma_u \to 1$.

Les suites (b) *sont scindées.*

Démonstration. Le noyau de $G_u \to \mathrm{GL}(V_u^+)$ est G_u^-; cela entraîne la première suite exacte de (a); la seconde se prouve de la même manière.

D'après (a) on peut identifier $\widetilde{G}_u^-$ à G_u/G_u^+. L'homomorphisme $G_u \to \Gamma_u$ est trivial sur G_u^+. Il définit donc un homomorphisme $\widetilde{G}_u^- \to \Gamma_u$ qui est surjectif, et dont le noyau est $G_u^1/G_u^+ = G_u^-$, cf. prop. 2.1 (b). Cela donne la première des suites exactes (b); la seconde se prouve de manière analogue.

Le fait que ces suites soient scindées résulte du fait analogue pour la suite exacte $1 \to G_u^+ \times G_u^- \to G_u \to \Gamma_u \to 1$, cf. prop.2.4.

Remarque. Les constructions ci-dessus sont des cas particuliers de celles du *lemme de Goursat* ([Se 16], 1.4) qui décrit la structure d'un sous-groupe d'un produit de deux groupes dont les projections sur les deux facteurs sont surjectives. Ici les deux groupes sont $\widetilde{G}_u^+$ et $\widetilde{G}_u^-$; le sous-groupe est G_u.

Corollaire 2.2. *Dans la suite d'inclusions* $G_u^+ \times G_u^- \subset G_u \subset \widetilde{G}_u^+ \times \widetilde{G}_u^-$, *chaque groupe est d'indice* $|\Gamma_u|$ *dans le suivant.*

Cela résulte des suites exactes (b).

Remarque. Le plus petit des trois groupes du cor.2.6 est normal dans les deux autres. Par contre le groupe du milieu G_u est normal dans le grand seulement si Γ_u est abélien. En effet, après passage au quotient par le petit groupe, on obtient l'inclusion diagonale $1 \subset \Gamma_u \subset \Gamma_u \times \Gamma_u$. Or la diagonale n'est un sous-groupe normal que si le groupe est abélien.

Revenons à la première suite exacte (b) : $1 \to G_u^- \to \widetilde{G}_u^- \to \Gamma_u \to 1$. L'action de $\widetilde{G}_u^-$ sur G_u^- par conjugaison donne un homomorphisme $\widetilde{G}_u^- \to \mathrm{Aut}(G_u^-)$. L'image de cet homomorphisme est contenue dans le sous-groupe

$\text{Aut}_c(G_u^-)$ de $\text{Aut}(G_u^-)$ formé des automorphismes qui transforment réflexions en réflexions. Notons $\text{Out}_c(G_u^-)$ le quotient de $\text{Aut}_c(G_u^-)$ par le sous-groupe des automorphismes intérieurs. Par passage au quotient, on obtient un homomorphisme de $\Gamma_u = \widetilde{G}_u^-/G_u^-$ dans $\text{Out}_c(G_u^-)$.

Proposition 2.5. *L'homomorphisme* $\Gamma_u \to \text{Out}_c(G_u^-)$ *est injectif.*

Démonstration. On utilisera le lemme suivant :

Lemme 2.1. *Soit* (E, W) *un couple de Coxeter tel que* $-1 \in W$. *Tout élément d'ordre fini de* $\text{GL}(E)$ *qui centralise* W *appartient à* W.

Démonstration du lemme. Soit $-1 = \prod_i s_i$ une décomposition de -1 en produit de réflexions de W deux à deux distinctes et commutant entre elles. Soit D_i la droite de E sur laquelle s_i opère par -1. On a $E = \oplus_i D_i$. Soit $g \in \text{GL}(E)$ d'ordre fini et centralisant W. Puisque g commute aux s_i, il stabilise les D_i. Sa restriction à chaque D_i est une homothétie $x \mapsto \varepsilon_i x$, avec $\varepsilon_i \in \mathbf{R}^\times$ d'ordre fini, donc égal à 1 ou -1. Il en résulte que g est égal au produit des s_i tels que $\varepsilon_i = -1$. En particulier, on a $g \in W$.

Fin de la démonstration de la prop.2.7. Soit γ un élément de $\Gamma_u \to \text{Out}_c(G_u^-)$, et soit g un représentant de γ dans $\widetilde{G}_u^-$. Supposons que l'image de γ dans $\text{Out}_c(G_u^-)$ soit 1. Cela signifie qu'il existe $z \in G_u^-$ tel que $gxg^{-1} = zxz^{-1}$ pour tout $x \in G_u^-$. L'élément $z^{-1}g$ centralise G_u^-. D'après le lemme 2.8, appliqué au couple (V_u^-, G_u^-), on a $z^{-1}g \in G_u^-$, d'où $g \in G_u^-$, i.e. $\gamma = 1$.

Remarque. On définit de façon analogue un homomorphisme $\Gamma_u \to \text{Out}_c(G_u^+)$. Cet homomorphisme est injectif si $-1 \in G$: cela résulte de la prop.2.7, appliquée à $-u$; si $-1 \notin G$, il peut ne pas être injectif.

Théorème 2.1. *Si le théorème 1.1 est vrai pour* (V, G), *les couples* $(V_u^+, \widetilde{G}_u^+)$ *et* $(V_u^-, \widetilde{G}_u^-)$ *sont des couples de Coxeter.*

En particulier, $\widetilde{G}_u^+$ *et* $\widetilde{G}_u^-$ *sont des groupes de Coxeter.*
(Cela répond positivement à une question posée dans [Se 22], 3.15.)

Démonstration.

Faisons la démonstration pour $\widetilde{G}_u^+$; le cas de $\widetilde{G}_u^-$ est analogue. Soit H le sous-groupe de $\widetilde{G}_u^+$ engendré par les V_u^+-réflexions. Nous devons montrer que $H = \widetilde{G}_u^+$. Comme ce dernier groupe est un quotient de G_u, le théorème 1.1 dit qu'il est engendré par les images dans $\text{GL}(V_u^+)$ des involutions de degré 1 ou 2 dans G. Si g est une involution de G_u, notons g^+ son image dans $\text{GL}(V_u^+)$, et g^- son image dans $\text{GL}(V_u^-)$. On a $\deg(g) = \deg(g^+) + \deg(g^-)$. Si $\deg(g) \leqslant 2$, on a, soit $\deg(g^+) \leqslant 1$, soit $\deg(g^-) = 0$. Dans le premier cas, g^+ est, soit 1, soit une réflexion dans $\text{GL}(V_u^+)$, donc appartient à H. Dans le second cas, on a $g^- = 1$, i.e. g fixe V_u^-, donc g^+ appartient à G_u^+, qui est contenu dans H, on l'a vu. Cela prouve que $H = \widetilde{G}_u^+$.

---

1. Autre interprétation de $\text{Out}_c(G_u^-)$: c'est le groupe des automorphismes du graphe de Coxeter de G_u^-.

Remarque. Le théorème 2.9 était essentiellement connu, mais dans une formulation différente. On peut le déduire d'un théorème de R.B. Howlett ([Ho 80]) sur les normalisateurs de sous-groupes paraboliques, théorème qui est applicable à G_u^1 d'après la prop.2.3. Je dois cette remarque à G. Röhrle ; c'est également lui qui m'a indiqué la prop.2.3.

La démonstration de [Ho 80], comme celle donnée ici, est une vérification cas par cas. Il serait intéressant d'avoir une démonstration directe.

§3. Détermination des groupes G_u^+.

Les groupes G_u^+ s'obtiennent par une récurrence sur $\deg(u)$ qui permet de passer d'un groupe de Coxeter à un autre de rang inférieur. On se ramène ainsi au cas où u est une réflexion.

De façon plus précise :

Réduction au cas où u est une réflexion.

Proposition 3.1. *Soient v, w deux involutions de G, commutant entre elles et telles que $\deg(vw) = \deg(v) + \deg(w)$. Alors $G_{vw}^+ = (G_v^+)_w^+ = (G_w^+)_v^+$.*

Démonstration. Les hypothèses faites sur v, w équivalent à $V_{vw}^- = V_v^- \oplus V_w^-$. Un élément de G appartient à G_{vw}^+ si et seulement si il fixe V_v^- et V_w^-. D'où la proposition.

Corollaire 3.1. *Soit u une involution de degré d, et soient $s_1, ..., s_d$ des réflexions, commutant deux à deux, telles que $u = s_1 \cdots s_d$. Soit $G(i)$ $(i = 0, ..., d)$ la suite de sous-groupes de G définie par $G(0) = G$ et $G(i) = G(i - 1)_{s_i}^+$. On a $G_u^+ = G(d)$.*

Démonstration. Cela résulte de la prop. 3.1 en raisonnant par récurrence sur d.

Le cas où u est une réflexion et où G est cristallographique.

[Rappelons (cf.[Bo 68], VI.2.5) que G est dit *cristallographique* s'il stabilise un réseau de V ; cela équivaut à dire que G est le groupe de Weyl d'un système de racines de V.]

Supposons que G .soit cristallographique et irréductible. Soit R un système de racines de V dont G est le groupe de Weyl, soit $S = \{\alpha_1, ..., \alpha_n\}$ une base de R et soit X le graphe de Dynkin correspondant (celui dont l'ensemble des sommets est S).

Soit $\alpha_0 = -\tilde{\alpha}$ l'opposée de la plus grande racine de R et soit $X_0 = X \cup \{\alpha_0\}$ le graphe de Dynkin complété (cf. [Bo 68], VI.4.3). Soit Y le sous-graphe de X obtenu en supprimant les sommets $\{\alpha_i\}$ de X liés à α_0 dans X_0. Alors :

Proposition 3.2. *Soit s_0 la réflexion associée à α_0. Le groupe $G_{s_0}^+$ est égal au sous-groupe parabolique G_Y de G de base Y.*

Démonstration. Le groupe $G_{s_0}^+$ est engendré par les réflexions s_α correspondant aux racines positives orthogonales à α_0 (pour un produit scalaire défini positif et G-invariant, noté $x \cdot y$). Si l'on écrit α comme $\sum m_i \alpha_i$, on a

$\alpha \cdot \alpha_0 = \sum m_i \alpha_i \cdot \alpha_0$. Les m_i sont $\geqslant 0$ et les $\alpha_i \cdot \alpha_0$ sont $\leqslant 0$ ([Bo 68], VI.1.8, prop.8). On a donc $\alpha \cdot \alpha_0 = 0$ si et seulement si $m_i = 0$ pour tout i tel que $\alpha_i \cdot \alpha_0 \neq 0$, autrement dit pour tout i tel que $\alpha_i \notin Y$; cela revient à dire que α est une réflexion de G_Y, cf. [Bo 68], VI.1.7, cor.4 à la prop.7. D'où la proposition.

Corollaire 3.2. *Supposons que G soit de type impair. Soit s une réflexion de G. Le groupe G_s^+ est un conjugué du groupe G_Y de la prop.3.3.*

[Rappelons, cf. [Se 22] 1.13, que G est dit de type impair si tous les produits de deux réflexions sont, soit d'ordre 2, soit d'ordre impair. C'est le cas si G est de l'un des types A, D, E.]

Démonstration. Les réflexions d'un groupe de type impair sont conjuguées entre elles. Donc s est conjuguée de la réflexion s_0 de la prop.3. D'où le corollaire.

Exemples.

Voici trois exemples, qui seront utilisés dans les §§10, 11, 12 ; les notations sont celles des Tables de [Bo 68], VI.

(a) *Type E_6.*

(a_1) Le cas $\deg(u) = 1$. Dans le graphe de Dynkin étendu, le sous-diagramme Y de la prop.3.3 a pour sommets $\alpha_1, \alpha_3, \alpha_4, \alpha_5, \alpha_6$. Il est de type A_5. On a donc $G_u^+ \simeq A_5$.

(a_2) Le cas $\deg(u) = 2$. Ecrivons u comme produit de deux réflexions s_1 et s_2, commutant entre elles. D'après (a_1), le groupe $H = G_{s_1}^+$ est de type A_5 ; il contient s_2. D'après le cor.3.2, on a $G_u^+ = H_{s_2}^+$. Comme toutes les réflexions de H sont conjuguées, on en déduit que toutes les involutions de degré 2 de G sont conjuguées. En appliquant à H le cor.3.4 (ou en raisonnant directement), on voit que $H_{s_2}^+$ est de type A_3, et il en est donc de même de G_u^+.

(a_3) Le cas $\deg(u) = 3$. Un argument analogue donne à la fois le fait que toutes les involutions de degré 3 sont conjuguées et que le groupe G_u^+ est de type A_1.

(a_4) Le cas $\deg(u) = 4$. Même argument : les involutions de degré 4 sont conjuguées, et le groupe G_u^+ est 1.

On peut résumer ce qui précède par une chaîne : $E_6 \longrightarrow A_5 \longrightarrow A_3 \longrightarrow A_1 \longrightarrow 1$.

(b) *Type E_7.*

La même méthode donne la chaîne $E_7 \longrightarrow D_6 \longrightarrow A_1 \times D_4$, et montre qu'il y a une seule classe de conjugaison d'involutions de degré 2. Comme un groupe de type $A_1 \times D_4$ a deux types de réflexions, cette chaîne a deux prolongements possibles, l'un par D_4, l'autre par $(A_1)^4$; ils correspondent aux deux classes d'involutions de G de degré 3.

(c) *Type E_8.*

Le début de la chaîne est $E_8 \longrightarrow E_7$; d'après (b), elle se prolonge par D_6, puis par $A_1 \times D_4$, et puis, soit par D_4, soit par $(A_1)^4$.

4. Les types $A_1, I_2(m), H_3$ et H_4.

On suppose que le type de G est $A_1, I_2(m), H_3$ ou H_4. Soit u une involution de G. On se propose de démontrer les théorèmes 1.1 et 1.2 pour le couple (G, u), et de déterminer les groupes $G_u, \Gamma_u, G_u^+, ..., \widetilde{G}_u^-$ correspondants.

Type A_1.

Ici, G est d'ordre 2 ; l'involution u est, soit 1, soit -1. On a $G_u = G$; si $u = 1$, on a $G_u^+ = \widetilde{G}_u^+ = G$ et $G_u^- = \widetilde{G}_u^- = 1$; si $u = -1$, on a $G_u^+ = \widetilde{G}_u^+ = 1$ et $G_u^- = \widetilde{G}_u^- = G$. Dans les deux cas $\Gamma_u = 1$. Nous résumons ceci dans le tableau ci-dessous :

$\deg(u)$	$|G_u|$	G_u^-	$\widetilde{G}_u^-$	G_u^+	$\widetilde{G}_u^+$	γ_u
0	2	1	1	A_1	A_1	1
1	2	A_1	A_1	1	1	1

Type $I_2(m), m$ impair.

Le groupe G est diédral d'ordre $2m, m$ impair. Toute involution $u \neq 1$ est une réflexion et son centralisateur est $\{1, u\}$. D'où le tableau :

$\deg(u)$	$|G_u|$	G_u^-	$\widetilde{G}_u^-$	G_u^+	$\widetilde{G}_u^+$	γ_u
0	$2m$	1	1	$I_2(m)$	$I_2(m)$	1
1	2	A_1	A_1	1	1	1

Type $I_2(m), m$ pair.

Le groupe G est diédral d'ordre divisible par 4. Il contient -1. Ses réflexions forment deux classes de conjugaison, permutées par un automorphisme extérieur le centralisateur d'une réflexion est le groupe de type $(2, 2)$ engendré par cette involution et l'élément -1. On en déduit le cas $\deg(u) = 1$ du tableau ci-dessous. Le cas où $\deg(u) = 0$ (resp. 2) est immédiat, puisqu'alors $u = 1$ (resp. -1).

$\deg(u)$	$|G_u|$	G_u^-	$\widetilde{G}_u^-$	G_u^+	$\widetilde{G}_u^+$	γ_u
0	$2m$	1	1	$I_2(m)$	$I_2(m)$	1
1	4	A_1	A_1	A_1	A_1	1
2	$2m$	$I_2(m)$	$I_2(m)$	1	1	1

Type H_3.

Ici, $G = \mathrm{Alt}_5 \times \{1, -1\}$. C'est un groupe de rang 3, contenant -1; il y a une seule classe d'involutions pour chaque degré $\leqslant 3$. On a le tableau suivant :

$\deg(u)$	$|G_u|$	G_u^-	$\widetilde{G}_u^-$	G_u^+	$\widetilde{G}_u^+$	γ_u
0	$2^3 3.5$	1	1	H_3	H_3	1
1	2^3	A_1	A_1	$(A_1)^2$	$(A_1)^2$	1
2	2^3	$(A_1)^2$	$(A_1)^2$	A_1	A_1	1
3	$2^3 3.5$	H_3	H_3	1	1	1

Les lignes correspondant à $\deg(u) = 0$ ou 3 sont évidentes. Lorsque $\deg(u) = 1$, le groupe G_u est d'ordre 2^3. On a $G_u^- = A_1$ et $G_u^+ = A_1 \times A_1$; comme le produit de leurs ordres est égal à celui de G_u, cela montre que $G_u^1 = G_u$ d'où $\Gamma_u = 1$. Le cas $\deg(u) = 2$ se ramène au précédent en remplaçant u par $-u$, ce qui permute les signes "+ " et "−".

Type H_4.

C'est un groupe de rang 4, contenant -1, d'ordre $2^6 3^2 5^2$. On l'obtient par "dédoublement" à partir de Alt_5, cf. [Se 22], 5.10 et 6.12. Cette construction montre que, pour $d = 0, 1, 2, 3, 4$, le nombre des involutions de degré d est respectivement $1, 60, 450, 60, 1$, et ces involutions forment une seule classe de conjugaison. On a le tableau suivant :

$\deg(u)$	$|G_u|$	G_u^-	$\widetilde{G}_u^-$	G_u^+	$\widetilde{G}_u^+$	γ_u
0	$2^6 3^2 5^2$	1	1	H_4	H_4	1
1	$2^4 3.5$	A_1	A_1	H_3	H_3	1
2	2^5	$(A_1)^2$	B_2	$(A_1)^2$	B_2	2
3	$2^4 3.5$	H_3	H_3	A_1	A_1	1
4	$2^6 3^2 5^2$	H_4	H_4	1	1	1

Les cas $\deg(u) = 0$ et $\deg(u) = 4$ sont évidents. Le cas $\deg(u) = 1$ résulte de ce que le centralisateur d'une réflexion est de type H_3; en remplaçant u par $-u$, cela donne le cas $\deg(u) = 3$.

Lorsque $\deg(u) = 2$, les groupes G_u^+ et G_u^- sont de type $A_1 \times A_1$, car sinon ce seraient des groupes diédraux d'ordre $2m$, avec m pair $\geqslant 4$, contrairement au fait que H_4 est un groupe de type impair, au sens de [Se 22], 1.13 (variante : utiliser le cor.3.4 pour se ramener au type H_3).

Comme $|G_u| = 2^6 3^2 5^2/60 = 2^5$, et que $G_u/G_u^- \simeq G_u^+$, on a $|G_u^-| = 8$, d'où $|\Gamma_u| = 2$. Cela justifie la ligne $\deg(u) = 2$ du tableau ci-dessus. L'homomorphisme $G_u \to \Gamma_u \simeq \mathrm{Sym}_2$ est donné par l'action de Γ_u sur les deux réflexions de produit u. Il reste à montrer qu'il existe une involution g de G_u, de degré 2, dont l'image dans Γ_u est non triviale. Cela résulte d'un énoncé plus général, démontré au §8. On peut aussi faire un calcul explicite :

Notons a, x, y, z des réflexions de G réalisant le diagramme de Coxeter

$$H_4 : \quad \underset{a}{\circ}\!\!-\!\!\!-\!\!\!-\!\!\!\overset{5}{\underset{x}{\circ}}\!\!-\!\!\!-\!\!\!-\!\!\!\underset{y}{\circ}\!\!-\!\!\!-\!\!\!-\!\!\!\underset{z}{\circ} \ .$$

On a $xz = zx, xyx = yxy, yzy = zyz$. Soient $u = xz$ et $g = yuy$. Ce sont des involutions de degré 2. On a $gxg = z$; en effet, $gxg = yu.yxy.uy = yu.xyx.uy = yzyzy = y.yzy.y = z$. Ainsi, la conjugaison par g échange x et z, donc fixe u. On a $g \in G_u$, et l'image de g dans Γ_u est non triviale.

5. Type A_{n-1}.

Dans le cas du type A, il est plus commode de décrire A_{n-1} que A_n. Soit X un ensemble fini à n éléments et soit V_X un $\mathbf{R}$-espace vectoriel de base X. Le groupe $G = \mathrm{Sym}_X$ des permutations de X opère de façon fidèle sur V_X; on obtient ainsi un couple de Coxeter (V_X, G) de type A_{n-1}; les réflexions sont les transpositions de X.

L'espace "V" standard associé à G est l'hyperplan de V_X engendré par les $x - x'$ avec $x, x' \in X$.

Soit $u \in G$ une involution, autrement dit une permutation de X de carré 1. Soit d son degré. Soit $Z = X^u$ l'ensemble des points fixes de u; le groupe $\{1, u\}$ opère librement sur $X - Z$. Décomposons $X - Z$ en deux parties disjointes Y, Y' telles que $Y' = uY$. On a $d = |Y| = |Y'|$ et $n = a + 2d$, où $a = |Z|$.

Tout élément g de G_u respecte la décomposition de X en deux parties : Z et $Y \cup Y'$, donc définit une permutation g_1 de Z et une permutation g_2 de $Y \cup Y'$ commutant à u. Inversement, si l'on se donne g_1, g_2 vérifiant ces conditions, il lui correspond un élément de G_u. Le groupe formé par les g_1 est Sym_Z; il est de type A_{a-1}. Celui formé par les g_2 est de type B_d, cf. §6. On a donc :

Proposition 5.1. G_u *est isomorphe à un groupe de Coxeter de type* $A_{a-1} \times B_d$.

Soient x, x' deux éléments distincts de X. La transposition $\mathrm{tr}_{x,x'}$ appartient à G_u si et seulement si l'on a, soit $x, x' \in Z$, soit $x, x' \in Y$ et $x' = ux$. Le groupe engendré par les x, x' du premier type est Sym_Z; celui engendré par les x, x' du second type est le produit de d groupes à 2 éléments. Comme les transpositions en question engendrent G_u^1, on en déduit :

Proposition 5.2. *Le groupe* G_u^1 *est de type* $A_{a-1} \times (A_1)^d$.

Comme $B_d/(A_1)^d \simeq \mathrm{Sym}_d$, cela entraîne :

Corollaire 5.1. *On a* $\Gamma_u \simeq \mathrm{Sym}_d$.

Il reste à expliciter les groupes $G_u^-, ..., \widetilde{G}_u^+$. Si $y \in Y$, posons $y^+ = y + uy$ et $y^- = y - uy$; soit Y^+ (resp. Y^-) l'ensemble des y^+ (resp. des y^-). Alors V_u^+ a pour base $Z \cup Y^+$ et V_u^- a pour base Y^-. De plus :

(i) Les réflexions de G_u qui fixent V_u^+ sont les transpositions du type $\mathrm{tr}_{y,uy}$, avec $y \in Y$; le groupe G_u^- qu'elles engendrent est de type $(A_1)^Y \simeq (A_1)^d$.

(ii) Les réflexions de G_u qui fixent V_u^- sont les transpositions de Z ; le groupe G_u^+ qu'elles engendrent est $\mathrm{Sym}_Z \simeq \mathrm{Sym}_a$, qui est de type A_{a-1}.

(i') Les réflexions de V_u^- qui sont les restrictions d'un élément de G_u sont de deux types :

celles de (i), qui changent de signe les y^- ;

celles de la forme $s_{y_1,y_2} = \mathrm{tr}_{y_1,y_2} \mathrm{tr}_{uy_1,uy_2}$, avec $y_1, y_2 \in Y$, qui échangent y_1^- et y_2^-.

Ces réflexions engendrent un sous-groupe H de $\widetilde{G}_u^-$ qui est de type B_d, donc d'ordre $2^d d!$, cf. §6. Comme $|\widetilde{G}_u^-| = |G_u^-| \cdot |\Gamma_u| = 2^d d!$, on a $H = \widetilde{G}_u^-$. Cela montre que $(V_u^-, \widetilde{G}_u^-)$ est un couple de Coxeter de type B_d, et cela montre aussi que Γ_u est engendré par les images des s_{y_1,y_2}, donc par des images d'involutions de degré 2 de G_u. Cela achève la démonstration des th. 1.1 et 1.2 pour G.

(ii') Les réflexions de V_u^+ qui sont les restrictions d'un élément de G_u sont celles de (ii), et aussi celles de la forme s_{y_1,y_2}, cf. (i'), qui échangent y_1^+ et y_2^+. On en déduit que $\widetilde{G}_u^+$ est isomorphe à $\mathrm{Sym}_Z \times \mathrm{Sym}_{Y+}$, donc de type $A_{a-1} \times A_{d-1}$.

On obtient ainsi le tableau :

$\deg(u)$	$|G_u|$	G_u^-	$\widetilde{G}_u^-$	G_u^+	$\widetilde{G}_u^+$	γ_u
d	$2^d d! a!$	$(A_1)^d$	B_d	A_{a-1}	$A_{a-1} \times A_{d-1}$	d

[Rappelons que G est de type A_{n-1} et que a est le nombre de points fixes de u.]

Remarque. Pour certaines valeurs de d, on peut avoir $d - 1 = -1$ ou $n - 2d = -1$, ce qui introduit des facteurs A_{-1} dans G_u^+ et $\widetilde{G}_u^+$; on les interprète en convenant que $A_{-1} = A_0 = 1$, ce qui est naturel puisque le groupe des permutations d'un ensemble à 0 ou 1 élément est égal à 1. Dans les tableaux relatifs aux types B_n et D_n, on rencontre aussi B_0, B_1, D_0, D_1 ; on convient que $B_0 = 1, B_1 = A_1, D_0 = D_1 = 1$.

6. Type B_n.

Soit n un entier > 0. (On pourrait même supposer $n > 2$, car $B_1 = A_1$ et $B_2 = I_2(4)$, et ces cas ont été traités au §4.)

Rappels.

Soit Z est un ensemble fini à $2n$ éléments, et soit ε une permutation de Z de carré 1 sans point fixe. Soit Y le quotient de Z par l'action du groupe $C = \{1, \varepsilon\}$. On a $|Y| = n$. Soit $G = \mathrm{Sym}_{Z,Y}$ le groupe des permutations de Z commutant à ε, autrement dit le groupe d'automorphismes du diagramme $Z \to Y$. On a une suite exacte

$$1 \to C^Y \to G \to \mathrm{Sym}_Y \to 1,$$

où C^Y est le groupe des applications de Y dans C (i.e. un produit de n copies de C indexées par Y). Cette suite est scindée. On a $|G| = 2^n n!$.

Soit V_Z un **R**-espace vectoriel de base Z. Le groupe G opère sur V_Z, et stabilise le sous-espace formé des éléments invariants par ε, espace qui s'identifie à V_Y ; soit $V = (1 - \epsilon)V_Z$ l'espace formé par les anti-invariants de ε ; on a $V_Z = V_Y \oplus V$. L'action de G sur V est fidèle. *Le couple (V, G) est un couple de Coxeter de type B_n.*

Il y a deux classes de réflexions : les *courtes* qui sont des transpositions de la forme $\mathrm{tr}_{z,\varepsilon z}$, avec $z \in Z$, et les *longues* qui sont de la forme $\mathrm{tr}_{z,z'}\, \mathrm{tr}_{\varepsilon z, \varepsilon z'}$, avec $z, z' \in Z$ et $z' \neq z, \varepsilon z$. Les premières sont des permutations impaires de Z, et les secondes sont des permutations paires. Avec les notations de [Bo 68], VI.4.5, ces réflexions correspondent aux racines $\pm\epsilon_i$ et $\pm\epsilon_i \pm \epsilon_j$ $(i \neq j)$.

Remarque. Le groupe G, vu comme sous-groupe de Sym_Z n'est pas engendré par des transpositions si $n > 1$; mais il est engendré par des transpositions et des produits de deux transpositions, autrement dit par des involutions de $\mathrm{GL}(V_L)$ de degré 1 ou 2.

Les groupes G_u, G_u^1 et Γ_u associés à une involution u.

Soit u une involution de G, autrement dit une permutation de Z de carré 1 qui commute à ε. Lorsque $u = 1$ ou $u = \varepsilon$, on a $G_u = G$. Supposons que $u \neq 1, \varepsilon$. Soit $\Delta = \langle u, \varepsilon \rangle$ le groupe d'ordre 4 engendré par u et ε. L'action de Δ sur Z donne une partition de Z en trois sous-ensembles :

$Z_u\ \ =$ ensemble des $z \in Z$ tels que $uz = \varepsilon z$;
$Z'_u\ =$ ensemble des $z \in Z$ tels que $uz = z$;
$Z''_u =$ ensemble des $z \in Z$ tels que $uz \neq z, \varepsilon z$.

Ces ensembles sont stables par Δ, donc par ε ; soient Y_u, Y'_u, Y''_u leurs images dans Y : on obtient ainsi une partition $Y = Y_u \cup Y'_u \cup Y''_u$. L'ensemble des points de Y fixés par u est $Y_u \cup Y'_u$. Le groupe $\{1, u\}$ opère librement sur Y''_u ; soit T_u le quotient de Y''_u par cette action. Cela donne le diagramme :

$$
\begin{array}{ccccccc}
Z & = & Z_u & \cup & Z'_u & \cup & Z''_u \\
  &   & \downarrow & & \downarrow & & \downarrow \\
Y & = & Y_u & \cup & Y'_u & \cup & Y''_u \\
  &   &   &   &   &   & \downarrow \\
  &   &   &   &   &   & T_u.
\end{array}
$$

Posons $a = |Y_u|$, $a' = |Y'_u|$, $b = |T_u| = \frac{1}{2}|Y''_u|$. On a $n = a + a' + 2b$ et $\deg(u) = a + b$. Les entiers a et b caractérisent la classe de conjugaison de l'involution u, et peuvent être donnés arbitrairement pourvu que $a + 2b \leqslant n$.

Le quadruplet (Z, ε, Y, u) est réunion disjointe de trois quadruplets correspondant aux trois composantes de Y que l'on vient de définir. Cette décomposition est stable par le groupe G_u. Plus précisément, G_u est produit direct de trois facteurs :

(i) Le premier facteur est Sym_{Z_u,Y_u} est d'ordre $2^a a!$; sa contribution à G_u^- et à $\widetilde{G}_u^-$ est Sym_{Z_u,Y_u}; celle à G_u^+ et à $\widetilde{G}_u^+$ est 1; celle à Γ_u est 1.

(ii) Le second facteur est $\mathrm{Sym}_{Z'_u,Y'_u}$ est d'ordre $2^{a'} a'!$; sa contribution à G_u^- et à $\widetilde{G}_u^-$ est 1; celle à G_u^+ et à $\widetilde{G}_u^+$ est $\mathrm{Sym}_{Z'_u,Y'_u}$; celle à Γ_u est 1.

(iii) Le troisième facteur est le groupe $\mathrm{Sym}_\Delta(Z''_u)$ des permutations de Z''_u qui commutent à l'action de Δ. Il est produit semi-direct des deux groupes suivants :

• le groupe Δ^{T_u} des applications de T_u dans Δ; c'est un sous-groupe normal d'ordre 4^b;

• le groupe des Δ-automorphismes de Z''_u qui stabilisent une partie S rencontrant chaque fibre de $Z''_u \to T_u$ en un point et un seul; il est isomorphe à Sym_{T_u}.

On peut donc écrire le troisième facteur sous la forme $\Delta^{T_u}.\mathrm{Sym}_{T_u}$. Son ordre est $4^b b!$. Si $b > 1$, ce n'est pas un groupe de Coxeter.

En résumé :

Proposition 6.1. *On a* $G_u = \mathrm{Sym}_{Z_u,Y_u} \times \mathrm{Sym}_{Z'_u,Y'_u} \times \Delta^{T_u}.\mathrm{Sym}_{T_u}$.

Les deux premiers facteurs de G_u sont engendrés par des réflexions; ils sont donc contenus dans G_u^1. Il n'en est pas de même du troisième facteur si $b > 1$:

Proposition 6.2. *On a* $G_u^1 \cap \Delta^{T_u}.\mathrm{Sym}_{T_u} = \Delta^{T_u}$.

Démonstration. Il suffit de montrer que toute réflexion s de G_u qui fixe Z et Z' appartient au groupe Δ^{T_u}. Soit I_s l'ensemble des points de Z'' qui ne sont pas fixés par s; c'est un ensemble à 2 ou à 4 éléments, on l'a vu. Or I_s est stable par Δ, et les orbites de Δ dans Z'' sont d'ordre 4. Donc I_s est une orbite de Δ, ce qui entraîne que s appartient à Δ^{T_u}.

Proposition 6.3. (a) *L'action de G_u sur T_u définit par passage au quotient un isomorphisme de Γ_u sur Sym_{T_u}.*

(b) *Toute transposition de Sym_{T_u} est image d'une involution de G_u de degré 2.*

Démonstration de (a). Cela résulte des prop. 6.1 et 6.2 puisque $\Gamma_u = G_u/G_u^1$.

Démonstration de (b). Soient $t, t' \in T_u$, avec $t \neq t'$, et soient z, z' des représentants de t, t' dans Z''_u. Soient g, h les réflexions de G données par $g = \mathrm{tr}_{z,z'}\, \mathrm{tr}_{\varepsilon z, \varepsilon z'}$ et $h = \mathrm{tr}_{uz,uz'}\, \mathrm{tr}_{\varepsilon uz, \varepsilon uz'}$. Ces réflexions commutent, et l'on a $ugu = h$. On a $gh \in G_u$ et l'image de gh dans Sym_{T_u} est la transposition $\mathrm{tr}_{t,t'}$. D'où (b).

Corollaire 6.1. *Les théorèmes 1.1 et 1.2 sont vrais pour G.*

C'est clair.

On peut résumer les résultats obtenus de la façon suivante :

Proposition 6.4. *On a :*

$$G_u = \mathrm{Sym}_{Z_u,Y_u} \times \mathrm{Sym}_{Z'_u,Y'_u} \times \Delta^{T_u}.\mathrm{Sym}_{T_u} \simeq B_a \times B_{a'} \times \Delta^b.\mathrm{Sym}_b;$$

$$G_u^1 = \mathrm{Sym}_{Z_u,Y_u} \times \mathrm{Sym}_{Z'_u,Y'_u} \times \Delta^{T_u} \simeq B_a \times B_{a'} \times \Delta^b;$$

$$\Gamma_u = \mathrm{Sym}_{T_u} \simeq \mathrm{Sym}_b \,;\, \gamma_u = |T_u| = b.$$

Les groupes $G_u^+, \widetilde{G}_u^+, G_u^-, \widetilde{G}_u^-$ associés à u.

La décomposition de G en produit de trois facteurs entraîne une décomposition du même type pour les groupes $G_u^+, ..., \widetilde{G}_u^-$. Nous avons donné plus haut le cas des deux premiers facteurs. Pour le troisième facteur, on a :

Lemma 6.1. *Les troisièmes facteurs des groupes G_u^+ et G_u^- sont de type $(A_1)^b$. Ceux des groupes $\widetilde{G}_u^+$ et $\widetilde{G}_u^-$ sont de type B_b.*

Démonstration. Puisque cet énoncé ne concerne que le troisième facteur, on peut supposer que les deux premiers sont triviaux, i.e. que $a = a' = 0$ et $Z = Z''_u$. Dans ce cas, les involutions u et εu sont conjuguées, ce qui entraîne que $G_u^+ \simeq G_u^-$ et $\widetilde{G}_u^+ \simeq \widetilde{G}_u^-$. Notons ces groupes H et $\widetilde{H}$. D'après la prop.6.2, on a $H \times H \simeq \Delta^b$, ce qui entraîne que H est un groupe abélien élémentaire d'ordre 2^b ; comme c'est un groupe de Coxeter, il est isomorphe à $(A_1)^b$. Un argument analogue montre que $\widetilde{H} \times \widetilde{H} \simeq \Delta^b . \mathrm{Sym}_b$. En particulier $\widetilde{H}$ est d'ordre $2^d d!$. Or, il contient H comme sous-groupe normal. Cela entraîne que c'est un groupe de Coxeter de type B_b, en vertu du lemme suivant :

Lemme 6.2. *Soit (E, H) un couple de Coxeter. Soit $e = \dim E$. Supposons que H soit de type $(A_1)^e$. Soit H' un sous-groupe fini de $\mathrm{GL}(E)$ qui normalise H et qui est d'ordre $2^e e!$. Alors (E, H') est un couple de Coxeter de type B_e.*

Démonstration du Lemme 6.8. L'action de H décompose E en somme directe de droites $D_1, ..., D_e$. Comme H' normalise H, il permute les D_i. Soit $\langle x \cdot y \rangle$ un produit scalaire défini positif sur E invariant par H', et soit Z l'ensemble des $z \in D_1 \cup ... \cup D_e$ tels que $\langle z \cdot z \rangle = 1$. On a $|Z| = 2e$, et l'application $z \mapsto -z$ est une permutation ε d'ordre 2 de Z sans point fixe. Le groupe H' stabilise Z et commute à ε. On obtient ainsi un homomorphisme injectif de H' dans le groupe de Coxeter de type B_e défini par (Z, ε) ; comme les deux groupes ont le même ordre, cet homomorphisme est un isomorphisme.

On obtient finalement le tableau :

invariants	$|G_u|$	G_u^-	$\widetilde{G}_u^-$	G_u^+	$\widetilde{G}_u^+$	γ_u
a, a', b	$2^n a! a'! b!$	$B_a \times (A_1)^b$	$B_a \times B_b$	$B_{a'} \times (A_1)^b$	$B_{a'} \times B_b$	b

7. Type D_n.

Conservons les notations (Z, ε, Y) du §6. Soit $G' = B_{Z,Y}$ et soit $G = D_{Z,Y}$ le sous-groupe d'indice 2 de G' formé des éléments g qui sont des permutations *paires* de Z, i.e. $\mathrm{sgn}_Z(g) = 1$. Le couple $(V_{Z,Y}, G)$ est un couple de Coxeter de type D_n. Les réflexions de G sont les réflexions longues de G'.

Soit u une involution de G, et soient a, a', b ses invariants au sens du §6. Le fait que u appartienne à G équivaut à $a \equiv 0 \pmod 2$. Deux involutions de mêmes invariants sont conjuguées, sauf dans le cas $a = a' = 0$ où il y a deux classes de conjugaison.

Le groupe G_u est le sous-groupe d'indice $\leqslant 2$ de G'_u formé des éléments x tels que $\operatorname{sgn}_Z(x) = 1$. La décomposition de G'_u donnée dans la prop. 6.1 est :

$$G'_u = \operatorname{Sym}_{Z_u, Y_u} \times \operatorname{Sym}_{Z'_u, Y'_u} \times \operatorname{Aut}_\Delta(Z''_u).$$

Les éléments de $\operatorname{Aut}_\Delta(Z''_u)$ sont de signature 1. La condition $\operatorname{sgn}_Z(x) = 1$ ne porte donc que sur les deux premières composantes de x. D'où :

Proposition 7.1. *Soit H_u le sous-groupe de $\operatorname{Sym}_{Z_u, Y_u} \times \operatorname{Sym}_{Z'_u, Y'_u}$ formé des couples (g, g') tels que $\operatorname{sgn}_{Z_u}(g) = \operatorname{sgn}_{Z'_u}(g')$. On a $G_u = H_u \times \operatorname{Aut}_\Delta(Z''_u)$.* [Rappelons que $\Delta = \langle u, \varepsilon \rangle$.]

Il y a quatre possibilités pour (a, a') :

(i) $a = a' = 0$, i.e. $2b = n$. On a alors $H_u = 1$ et $G_u = \operatorname{Aut}_\Delta(Z''_u) = G'_u$. L'ordre de G_u est $2^n b!$, on a $\Gamma_u = \operatorname{Sym}_{T_u}$ et $\gamma_u = b$. Les groupes G_u^- et G_u^+ sont isomorphes à $A_1^{T_u}$; les groupes $\widetilde{G}_u^-$ et $\widetilde{G}_u^+$ sont de type B_b.

(ii) $a = 0, a' > 0$. Le premier facteur de H_u est 1 ; le second est $D(Z'_u, Y'_u)$, qui est de type $D_{a'}$. On a $G_u = D(Z'_u, Y'_u) \times \operatorname{Aut}_D(Z''_u)$. La situation est la même que pour G', avec $B_{a'}$ remplacé par $D_{a'}$. On a $\Gamma_u = \operatorname{Sym}_{T_u}$ et $\gamma_u = b$.

(iii) $a > 0$ et $a' = 0$: comme dans le type (ii), avec a et a' permutés, ainsi que (Z, Z') et (Y, Y'). Ici encore $\Gamma_u = \operatorname{Sym}_{T_u}$ et $\gamma_u = b$.

(iv) $a > 0$ et $a' > 0$. Soit $H_u^1 = D_{Z_u, Y_u} \times D_{Z'_u, Y'_u}$. C'est un sous-groupe d'indice 2 de H_u qui est engendré par des réflexions. Inversement, toute réflexion de H_u appartient à H_u^1 car c'est transposition de $Y_u \cup Y'_u$ qui stabilise à la fois Y_u et Y'_u, donc qui est une transposition, soit de Y_u, soit de Y'_u. D'autre part, le groupe engendré par les réflexions de $\operatorname{Aut}_\Delta(Z''_u)$ est le groupe Δ^{T_u}. On en conclut que $G_u^1 = H_u^1 \times \Delta^{T_u}$ et que le groupe $\Gamma_u = G_u/G_u^1$) est égal au produit de Sym_{T_u} par H_u/H_u^1 qui est d'ordre 2. C'est le cas, mentionné dans le th. 2.2, où Γ_u *n'est pas un groupe symétrique* (sauf si $b = 0$ ou 1).

La détermination des groupes $G_u^-, G_u^+, \ldots$ résulte de celle des groupes correspondants pour le type B_n. Plus précisément :

Les composantes dépendant de l'invariant " b " sont les mêmes que pour le type B_n ; dans les autres, certains groupes B_a ou $B_{a'}$ sont remplacés par D_a ou $D_{a'}$ respectivement.

On obtient ainsi le tableau :

invariants	$\lvert G_u \rvert$	G_u^-	$\widetilde{G}_u^-$	G_u^+	$\widetilde{G}_u^+$	γ_u
$0, 0, b$	$2^n b!$	$(A_1)^b$	B_b	$(A_1)^b$	B_b	b
$0, a', b \ \ a' > 0$	$2^{n-1} a'! b!$	$(A_1)^b$	B_b	$D_{a'} \times (A_1)^b$	$D_{a'} \times B_b$	b
$a, 0, b \ \ a > 0$	$2^{n-1} a! b!$	$D_a \times (A_1)^b$	$D_a \times B_b$	$(A_1)^b$	B_b	b
$a, a', b \ \ aa' > 0$	$2^{n-1} a! a'! b!$	$D_a \times (A_1)^b$	$B_a \times B_b$	$D_{a'} \times (A_1)^b$	$B_{a'} \times B_b$	$b, 2$

8. Résultats auxiliaires sur les groupes $\bar{G}_u^-$ et $\widetilde{G}_u^-$.

Ces résultats seront utilisés dans les trois sections suivantes. On note d le degré de l'involution u. On suppose que le type de G est H_4, E_6, E_7 ou E_8. On s'intéresse aux deux propriétés suivantes :

(i) u *est l'extrémité d'un seul cube*, i.e. sa décomposition en produit de d réflexions est unique, à permutation près. Cela équivaut à $\bar{G}_u^- \simeq (A_1)^d$.

(ii) *Il existe un $\mathscr{C}$-sous-groupe H de G de type A tel que $u \in H$.*

Proposition 8.1. *Si les propriétés* (i) *et* (ii) *sont satisfaites, les théorèmes 1.1 et 1.2 sont vrais pour le couple* (G,u)*, alors le groupe* $\widetilde{G}_u^-$ *est de type* B_d*, le groupe* Γ_u *est isomorphe à* Sym_d *et il est engendré par les images des involutions de* G_u *de degré* 2.

Démonstration. Soit H un sous-groupe de G satisfaisant à (ii). D'après le §5, la prop.8.5 est vraie si $G = H$. On va se ramener à ce cas. D'après (i), on a $\bar{G}_u^- = \bar{H}_u^-$. Le groupe $\widetilde{G}_u^-$ contient $\widetilde{H}_u^-$, qui est de type B_d. Cela montre que Γ_u contient un sous-groupe isomorphe à Sym_d. D'autre part Γ_u est isomorphe à un sous-groupe de $\mathrm{Out}_c(\bar{G}_u^-) \simeq \mathrm{Out}_c((A_1)^d) \simeq \mathrm{Sym}_d$, cf. prop.2.7. On a donc $\widetilde{G}_u^- = \widetilde{H}_u^-$, ce qui démontre la proposition.

Proposition 8.2. *Les propriétés* (i) *et* (ii) *sont satisfaites pour* $d \leqslant 2$ *lorsque G est de type H_4 et pour $d \leqslant 3$ lorsque G est de type E_6, E_7 ou E_8. La propriété* (ii) *est satisfaite pour $d \leqslant 4$ lorsque G est de type E_8.*

Démonstration. L'hypothèse (i) est satisfaite puisque les seuls groupes de Coxeter de rang $\leqslant 3$, de type impair, et contenant -1, sont des puissances de A_1.

Pour (ii), et G de type H_4, on remarque que le diagramme de G contient un sous-diagramme de type A_3. Or un groupe de type A_3 contient des involutions de tout degré $\leqslant 2$. Comme les involutions de G de même degré sont conjuguées entre elles, cela entraîne (ii).

Le même argument s'applique à G de type E_6, car son diagramme contient un sous-diagramme de type A_5 ; il s'applique aussi au type E_8, ainsi qu'à E_7 si $d \leqslant 2$.

Dans le cas de E_7, pour $d = 3$, il y a deux classes d'involutions, cf. [Se 22], 7.5 : celles de type *triangle* et celles de type *droite*. Pour les traiter, choisissons des réflexions $s_1, ..., s_7$ correspondant au diagramme de Coxeter de E_7 :

Posons $u = s_3 s_5 s_7$ et $u' = s_2 s_5 s_7$; ce sont des involutions de degré 3. D'après [Se 22], *loc.cit.*, un produit $s_a s_b s_c$ est du type triangle si et seulement si il existe $m \neq a, b, c$ tel que s_m soit adjacent à un et un seul des s_a, s_b, s_c. Dans le cas

---

2. Dans [Se 22], cette condition est exprimée en termes des racines $\alpha_a, \alpha_b, \alpha_c$ associées à s_a, s_b, s_c : le produit $s_a s_b s_c$ est du type triangle si et seulement si $\frac{1}{2}(\alpha_a + \alpha_b + \alpha_c)$ n'appartient pas au réseau des poids.

de u, l'entier $m = 1$ répond à cette condition ; dans le cas de u', aucun m n'est possible ; ainsi, u est du type triangle et u' du type droite. Il suffit donc de vérifier la condition (ii) pour u et pour u'. Pour u, on prend le sous-groupe de type A_5 engendré par s_3, s_4, s_5, s_6, s_7 ; pour u', on prend celui engendré par s_2, s_4, s_5, s_6, s_7.

Le cas de E_8 est analogue au précédent. Il y a deux classes d'involutions de degré 4, celles de type *rectangle* et celles de type *tétraèdre*. On les distingue de la manière suivante : on écrit l'involution u comme produit de quatre réflexions commutant deux à deux, et correspondant à des racines x, y, z, t. Alors u est de type rectangle si $x + y + z + t \in 2R$, où R désigne le réseau des racines ; sinon, u est de type tétraèdre. Soient $s_1, ..., s_8, s_0$ des réflexions correspondant au diagramme étendu de E_8 (s_0 correspondant à la plus grande racine) :

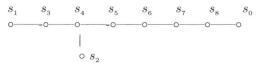

Prenons $u = s_2 s_5 s_7 s_0$ et $u' = s_1 s_3 s_5 s_7$. On vérifie par la même méthode que pour E_7 que u est du type rectangle et u' du type tétraèdre. Il suffit donc de vérifier (i) pour u et pour u' : pour u (resp. pour u') on prend le sous-groupe de type A_7 engendré par les $s_i, i \neq 1, 3$ (resp. $i \neq 0, 2$).

9. Type E_6.

C'est un groupe de rang 6 qui ne contient pas -1. Il y a une seule classe de conjugaison d'involutions pour chaque degré $d \leqslant 4$. Le tableau correspondant est :

$\deg(u)$	$|G_u|$	G_u^-	$\widetilde{G}_u^-$	G_u^+	$\widetilde{G}_u^+$	γ_u
0	$2^7 3^4 5$	1	1	E_6	E_6	1
1	$2^5 3^2 5$	A_1	A_1	A_5	A_5	1
2	$2^6 3$	$(A_1)^2$	B_2	A_3	$A_1 \times A_3$	2
3	$2^5 3$	$(A_1)^3$	B_3	A_1	$A_1 \times A_2$	3
4	$2^7 3^2$	D_4	F_4	1	A_2	3

Vérification du tableau.

Les cas $d = 0$ et $d = 1$ sont immédiats.

Pour $d = 2$ ou 3, les prop.8.1 et 8.2 montrent que le th.1.1. et le th.1.2 sont vrais pour (G, u), que $G_u^- \simeq (A_1)^d$, $\widetilde{G}_u^- \simeq B_d$ et que $\Gamma_u \simeq \mathrm{Sym}_d$.

Dans le cas $d = 2$, l'ordre de G_u est $2^6 3$ et l'on a vu au §3 que G_u^+ est de type A_3. Le groupe $\widetilde{G}_u^+$ contient G_u^+ comme sous-groupe d'indice 2 ; de plus c'est un groupe de type cristallographique ; la seule possibilité est qu'il soit de type $A_1 \times A_3$.

Dans le cas $d = 3$, un raisonnement analogue montre que l'ordre de $\widehat{G}_u^+$ est 2, donc que ce groupe est de type A_1 ; quant à $\widetilde{G}_u^+$, c'est un groupe de rang $\leqslant 3$ et d'ordre $2^2 3$, qui normalise un groupe de type A_1 ; son type est donc $A_1 \times A_2$.

Le cas $d = 4$ est traité dans [Se 22], 3.19 (et c'est lui qui est à l'origine du présent travail).

10. Type E_7.

Il y a une seule classe d'involutions de degré $0, 1, 2, 5, 6, 7$, deux classes de degré 3 et deux classes de degré 4 ; ces dernières sont notées $3, 3', 4, 4'$ dans le tableau ci-dessous. Rappelons comment on caractérise les deux classes de degré 3, cf. [Se 22], 7.5. Notons R le réseau des racines de $E7$, P le réseau des poids, et V_6 le $\mathbf{F}_2$-espace vectoriel $R/2P$; cet espace est muni d'une forme bilinéaire alternée non dégénérée. Les réflexions de E_7 correspondent bijectivement aux éléments non nuls de V_6. Soit une involution de degré 3 ; décomposons u en produit de trois réflexions s, s', s'' commutant entre elles (ce qui est unique, à permutation près) ; ces réflexions donnent trois éléments non nuls x, x', x'' de $R/2P$, deux à deux orthogonaux. Si la somme $x + x' + x''$ est 0, u est du type droite ; si elle ne l'est pas, u est du type *triangle*. Il n'est pas difficile de compter combien il y a d'involutions de chaque type : pour le cas des droites, il y a $2^6 - 1$ possibilités pour x, $2^5 - 2$ possibilités pour x' et une seule possibilité pour x''. Comme chaque u est obtenu 6 fois, le nombre des involutions du type droite est $(2^6 - 1)(2^5 - 2)/6 = 3^2 5.7$ et G_u est d'ordre $2^{10} 3^4 5.7/3^2 5.7 = 2^{10} 3^2$: c'est le type 3 du tableau ci-dessous. Un calcul analogue montre que le nombre des involutions de type triangle est $(2^6 - 1)(2^5 - 2)(2^4 - 2^2)/6 = 2^2 3^3 5.7$ et que G_u est d'ordre $2^8 3$; c'est le type $3'$ du tableau. Si u est de degré 4 (resp. $4'$), on dit que u est de type 4 si $-u$ est de type 3 (resp. $3'$).

$\deg(u)$	$\lvert G_u \rvert$	G_u^-	$\widetilde{G}_u^-$	G_u^+	$\widetilde{G}_u^+$	γ_u
0	$2^{10} 3^4 5.7$	1	1	E_7	E_7	1
1	$2^{10} 3^2 5$	A_1	A_1	D_6	D_6	1
2	$2^{10} 3$	$(A_1)^2$	B_2	$A_1 \times D_4$	$A_1 \times B_4$	2
3	$2^{10} 3^2$	$(A_1)^3$	B_3	D_4	F_4	3
$3'$	$2^8 3$	$(A_1)^3$	B_3	$(A_1)^4$	$A_1 \times B_3$	3
4	$2^{10} 3^2$	D_4	F_4	$(A_1)^3$	B_3	3
$4'$	$2^8 3$	$(A_1)^4$	$A_1 \times B_3$	$(A_1)^3$	B_3	3
5	$2^{10} 3$	$A_1 \times D_4$	$A_1 \times B_4$	$(A_1)^2$	B_2	2
6	$2^{10} 3^2 5$	D_6	D_6	A_1	A_1	1
7	$2^{10} 3^4 5.7$	E_7	E_7	1	1	1

Vérification du tableau.

Soit $d = \deg(u)$. Comme G contient -1, il nous suffit de traiter les cas où $d \leqslant 3$; les autres s'en déduisent en remplaçant u par $-u$; noter que, si u est de type 3 (resp. $3'$), $-u$ est de type 4 (resp. $4'$).

Les cas $d = 0$ et $d = 1$ sont immédiats.

Lorsque $d = 2$, les prop.8.1 et 8.2 entraînent que G_u^- est de type $(A_1)^2$, $\widetilde{G}_u^-$ est de type B_2 et $\Gamma_u \simeq \mathrm{Sym}_2$. Le groupe G_u^+ est de type $A_1 \times D_4$ d'après le §3 ; comme $\widetilde{G}_u^-$ contient G_u^+ avec indice 2, il est de type $A_1 \times B_4$.

Supposons que $d = 3$. Comme ci-dessus, les prop.8.1 et 8.2 entraînent que G_u^- est de type $(A_1)^3$, $\widetilde{G}_u^-$ est de type B_3 et $\Gamma_u \simeq \mathrm{Sym}_2$.

Quant u est de type 3, le même argument que pour $d = 2$ montre que G_u^+ est d'ordre $2^6 3$, donc de type D_4. Comme c'est un sous-groupe normal de $\widetilde{G}_u^+$, et que $\widetilde{G}_u^+/G_u^+$ est isomorphe à Sym_3, on en déduit que $\widetilde{G}_u^+$ est de type F_4.

Quand u est de type $3'$, G_u^+ est d'ordre 2^4, donc de type $(A_1)^4$, alors que $\widetilde{G}_u^+$ est d'ordre $2^5 3$, et le contient comme sous-groupe normal d'indice 6 ; cela entraîne que $\widetilde{G}_u^+$ est de type $A_1 \times B_3$. Noter que le facteur A_1 est contenu dans le noyau de la surjection $\widetilde{G}_u^+ \to \mathrm{Sym}_3$, donc est engendré par une G-réflexion ; c'est l'un des facteurs de G_u^+.

11. Type E_8.

Pour $d = 0, 1, 2, 3$ et $d = 5, 6, 7, 8$ il y a une seule classe d'involutions de degré d. Pour $d = 4$, il y en a deux : celle appelée *du type rectangle* dans [Se 22], 5.8, et celle appelée *du type tétraèdre* ; dans le tableau ci-dessous, elles correspondent aux lignes 4 et $4'$.

$\deg(u)$	$|G_u|$	G_u^-	$\widetilde{G}_u^-$	G_u^+	$\widetilde{G}_u^+$	γ_u
0	$2^{14}3^5 5^2 7$	1	1	E_8	E_8	1
1	$2^{11}3^4 5.7$	A_1	A_1	E_7	E_7	1
2	$2^{12}3^2 5$	$(A_1)^2$	B_2	D_6	B_6	2
3	$2^{11}3^2$	$(A_1)^3$	B_3	$A_1 \times D_4$	$A_1 \times F_4$	3
4	$2^{13}3^3$	D_4	F_4	D_4	F_4	3
$4'$	$2^{11}3$	$(A_1)^4$	B_4	$(A_1)^4$	B_4	4
5	$2^{11}3^2$	$A_1 \times D_4$	$A_1 \times F_4$	$(A_1)^3$	B_3	3
6	$2^{12}3^2 5$	D_6	B_6	$(A_1)^2$	B_2	2
7	$2^{11}3^4 5.7$	E_7	E_7	A_1	A_1	1
8	$2^{14}3^5 5^2 7$	E_8	E_8	1	1	1

Vérification du tableau.

On peut se borner au cas $d \leqslant 4$ puisque G contient -1. La méthode est la même que pour le type E_7. Les cas $d = 0$ et $d = 1$ sont immédiats. Lorsque $d = 2$ ou 3, les prop.8.1 et 8.2 entraînent que G_u^- est de type $(A_1)^d$, que $\Gamma_u \simeq \mathrm{Sym}_d$, que $\widetilde{G}_u^-$ est de type B_d, et que Γ_u est engendré par les images des involutions de G_u de degré 2.

Pour $d = 2$, on a vu au §3 que G_u^+ est de type D_6 ; comme $\widetilde{G}_u^+$ le contient avec indice 2, il est de type B_6. On a $|G_u| = |G_u^-| \cdot |\widetilde{G}_u^+| = 2^2 \cdot 2^6 6! = 2^{12}3^2 5$.

Pour $d = 3$, un argument analogue montre que G_u^+ est de type $A_1 \times D_4$, que $\widetilde{G}_u^+$ est de type $A_1 \times F_4$, et que $|G_u| = 2^1 13^2$.

[Noter que l'inclusion de E_7 dans E_8 transforme une involution v de type 3 de $H = E_7$ en une involution u de G de degré 3 ; le groupe $\widetilde{H}_v^+$ est donc contenu dans $\widetilde{G}_u^+$, ce qui entraîne que $\widetilde{G}_u^+$ contient F_4.]

Pour $d = 4$, on a vu au §3 que G_u^+ est, soit de type D_4, soit de type $(A_1)^4$, et que les deux cas sont possibles. C'est le premier cas que nous avons choisi de noter 4 dans le tableau ci-dessus, le second étant noté $4'$. Dans les deux cas, u et $-u$ sont conjuguées (cela résulte de la description de ces classes donnée dans [Se 22], 5.6) ; d'où $G_u^- \simeq G_u^+$.

Si u est de type $4'$, la prop.8.1 montre que $\widetilde{G}_u^-$ est de type B_4. On a $|G_u| = |G_u^-| \cdot |\widetilde{G}_u^-| = 2^4 \cdot |B_4| = 2^{11}3$, et $\Gamma_u = \mathrm{Sym}_4$, ce qui justifie la ligne ($4'$) du tableau. De plus, une comparaison avec le tableau de E_7 montre que l'involution u provient, par l'injection $E_7 \to E_8$, d'une involution de E_7 de type $4'$.

Supposons que u est de type 4. Cette involution provient d'une involution de type 4 de E_7. On en déduit que $\widetilde{G}_u^-$ contient un sous-groupe de type F_4, donc que Γ_u contient un sous-groupe isomorphe à Sym_3. Or la prop.2.7 montre que Γ_u est isomorphe à un sous-groupe de $\mathrm{Out}_c(G_u^-) \simeq \mathrm{Out}_c(D_4) \simeq \mathrm{Sym}_3$. On en conclut que $\widetilde{G}_u^- \simeq F_4$. [Autre démonstration : utiliser le fait que F_4 est un sous-groupe fini maximal de $\mathrm{GL}_4(\mathbf{Q})$, cf. [Da 65], (4.3).] On déduit de là que $|G_u| = 2^{13}3^3$ et que $\Gamma_u \simeq \mathrm{Sym}_3$, ce qui justifie la ligne 4 du tableau. De plus, cet argument montre que les éléments d'ordre 2 de Γ_u sont des images d'involutions de degré 2 de G_u.

Il reste à prouver que *les involutions de type* 4 (resp. $4'$) *sont des rectangles* (resp. *des tétraèdres*) au sens de [Se 22], 5.6. Il suffit de le faire pour le type 4, et sur un exemple explicite. Avec les notations de la fin du §8, choisissons $u = s_2 s_5 s_7 s_0$, qui est de type rectangle, on l'a vu. Posons $v = s_2 s_5 s_7$; c'est une involution de E_7 de type droite, cf. §8. Soit e l'élément -1 de E_7 ; le produit ev est une involution de E_7 de type 4. Son image dans E_8 est aussi de type 4. Mais l'image de e dans E_8 est $-s_0$; celle de ev est donc égale à $-u$. On en conclut que $-u$ est de type 4, et la même chose est vraie pour u, puisque u et $-u$ sont conjuguées.

12. Type F_4.

Ce groupe contient -1. Pour $d = 1, 2, 3$, il y a deux classes de conjugaison d'involutions de degré d.

Les réflexions de chacune des deux classes (notées L et C : "longues" et "courtes") engendrent un sous-groupe normal de G, qui est de type D_4 ; le quotient de G par ce sous-groupe est isomorphe à Sym_3. Les trois sous-groupes d'ordre 2 de Sym_3 correspondent à trois sous-groupes d'indice 3 de G, qui sont des $\mathscr{C}$-sous-groupes de type B_4. Les classes C et L sont permutées par un automorphisme d'ordre 2 de G (par exemple celui qui est évident sur le diagramme de Coxeter). Chaque classe a 12 éléments.

Le groupe G_u^+ associé à une réflexion u est de type B_3 ; lorsque u est de type L, cela résulte de la prop.3.3, car u est conjuguée de la réflexion notée s_0 dans cette proposition ; le cas où u est de type C résulte du précédent en appliquant un automorphisme de G qui permute C et L.

Les deux classes d'involutions de degré 2, notées 2 et $2'$, ont les propriétés suivantes :

Une involution u de type 2 se décompose de deux façons différentes en produit de deux réflexions s et s' ; dans l'une de ces décompositions, s et s' appartiennent à la classe C ; dans l'autre décomposition, s et s' appartiennent à la classe L. Le groupe G_u^- est de type B_2. Le nombre de telles involutions est 2.3^2.

Une involution u de type $2'$ s se décompose de façon unique en $u = ss'$, où s est une réflexion de type C et s' est une réflexion de type L. Le groupe G_u^- est de type $A_1 \times A_1$. Le nombre de ces involutions est $2^2 3^2$.

[Ces propriétés se démontrent, soit en utilisant les plongements $D_4 \subset B_4 \subset F_4$, soit en utilisant la construction de G par dédoublement à partir de Sym_4, cf. [Se 22], 6.12.]

On a le tableau suivant :

$\deg(u)$	$|G_u|$	G_u^-	$\widetilde{G_u^-}$	G_u^+	$\widetilde{G_u^+}$	γ_u
0	$2^7 3^2$	1	1	F_4	F_4	1
1 et $1'$	$2^5 3$	A_1	A_1	B_3	B_3	1
2	2^6	B_2	B_2	B_2	B_2	1
$2'$	2^4	$A_1 \times A_1$	$A_1 \times A_1$	$A_1 \times A_1$	$A_1 \times A_1$	1
3 et $3'$	$2^5 3$	B_3	B_3	A_1	A_1	1
4	$2^7 3^2$	F_4	F_4	1	1	1

Vérification du tableau.

On peut se borner au cas $d \leqslant 2$ puisque G contient -1. Les deux cas $d = 0, 1$ sont immédiats.

Lorsque u est de type 2, on a vu que G_u^- est de type B_2. Or u et $-u$ sont conjugués (puisque leurs centralisateurs ont le même ordre). Le groupe G_u^+ est donc aussi de type B_2. Comme $\widetilde{G_u^-} = G_u/G_u^+$, l'ordre de ce groupe est $2^6/2^3 = 2^3$; puisqu' il contient G_u^-, qui est d'ordre 2^3, il lui est égal. On a donc $\Gamma_u = 1$.

Le cas du type $2'$ est analogue : les groupes G_u^- et G_u^+ sont de type $A_1 \times A_1$; l'ordre de $\widetilde{G_u^-} = G_u/G_u^+$ est $2^4/2^2 = 2^2$, d'où $\widetilde{G_u^-} = G_u^-$ et $\Gamma_u = 1$.

Cette vérification termine la démonstration des énoncés du §1.

Références

[Bo 68] N. Bourbaki, *Groupes et algèbres de Lie, Chap. IV-VI*, Hermann, Paris, 1968 ; English translation, Springer-Verlag, 2002.

[Da 65] E.C. Dade, *The maximal finite groups of* 4×4 *integral matrices*, Ill. J. Math. 9 (1965), 99-122.

[DPR 13] J.M. Douglass, G. Pfeiffer & G. Röhrle, *On reflection subgroups of finite Coxeter groups*, Comm. Algebra 41 (2013), 2574–2592.

[FV 05] G. Felder & P. Veselov, *Coxeter group actions on the complement of hyperplanes and special involutions*, J. Eur. Math. Soc. 7 (2005), 101-116.

[Ho 80] R.B. Howlett, *Normalizers of parabolic subgroups of reflection groups*, J. London Math. Soc. 21 (1980), 62-80.

[Se 16] J-P. Serre, *Finite Groups : an Introduction*, International Press, Somerville, 2016 ; seconde édition corrigée, 2022.

[Se 22] J-P. Serre, *Groupes de Coxeter finis : involutions et cubes*, Ens. Math. 68 (2022), 99-133 ; arXiv :2012.03689.

Groups, Drift and Harmonic Measures

Mark Pollicott and Polina Vytnova

1 Introduction

An intriguing problem in modern geometric measure theory is the study of the harmonic measure on the unit circle which arises from a random walk on a Fuchsian group. The existing approach combines several areas of pure mathematics, such as Ergodic Theory, Probability Theory, Hyperbolic Geometry, and rigorous Numerical Analysis.

We will first recall some background. Afterwards we introduce one of the quantifiable characteristics of random walks called the *drift* and explain how it is related to properties of the harmonic measure, in particular, its Hausdorff dimension. Finally, we will draw a connection to a popular conjecture of Kaimanovich and Le Prince on the nature of the harmonic measure associated to a random walk on a Fuchsian group.

Although there are a number of partial results in special cases the general conjecture still remains open. We will offer a new perspective which covers both some known examples and some new cases. Finally, we will illustrate the question using the example of a (4, 4, 4)-triangle group which can be traced back to the works of Gauss from 1805.

M. Pollicott (✉)
Department of Mathematics, Warwick University, Coventry, UK
e-mail: masdbl@warwick.ac.uk

P. Vytnova
Department of Mathematics, University of Surrey Guildford, Surrey, UK
e-mail: P.Vytnova@surrey.ac.uk

J.-M. Morel, B. Teissier (eds.), *Mathematics Going Forward*, Lecture Notes
in Mathematics 2313, https://doi.org/10.1007/978-3-031-12244-6_21

Wahrlich[1] es ist nicht das Wissen, sondern das Lernen, nicht das Besitzen sondern das Erwerben, nicht das Da-Seyn, sondern das Hinkommen, was den grössten Genuss gewährt.

Carl Friedrich Gauß to Wolfgang Bolyai [8].

2 Preliminaries

In this section we collect together some background knowledge we need to properly formulate the problem.

2.1 Hyperbolic Geometry

We will treat Fuchsian groups as groups of isometries acting on the hyperbolic plane $\mathbb{H}$. For our considerations it will be convenient to consider the so-called Poincaré disk model of $\mathbb{H}$.

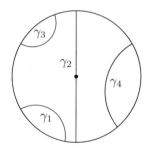

Fig. 1 The Poincaré disk where geodesics are either circular arcs which meet the boundary circle orthogonally or diameters

This is a representation of the hyperbolic plane as an open unit disk $\mathbb{D} = \{z = x + iy \colon |z| < 1\}$ equipped with the Poincaré metric

$$ds^2 = 4 \frac{dx^2 + dy^2}{(1 - (x^2 + y^2))^2}.$$

The geodesics are the extrema of the distance functional with respect to this metric. They are precisely the Euclidean diameters and circular arcs which are orthogonal to the boundary circle $\partial \mathbb{D}$, as shown in Fig. 1.

---

[1] "It is not knowledge, but the act of learning, not possession but the act of getting there, which grants the greatest enjoyment."

The orientation preserving isometries of $\mathbb{H}$ in this model are linear fractional transformations of the form

$$g(z) = \frac{az + b}{\bar{b}z + \bar{a}}, \quad \text{where } a, b \in \mathbb{C} \text{ and } |a|^2 - |b|^2 = 1.$$

As a group the orientation preserving isometries are isomorphic to the group consisting of 2×2 real matrices with determinant 1 up to multiplication by $\pm \left(\begin{smallmatrix} 1 & 0 \\ 0 & 1 \end{smallmatrix} \right)$. Namely this is the group $\mathrm{PSL}(2, \mathbb{R}) = \mathrm{SL}(2, \mathbb{R})/\{\pm I\}$.

2.2 Geometric Group Theory

In the present paper we want to consider finitely generated groups of isometries of $\mathbb{H}$. We call the group Γ *non-elementary* if it is not isomorphic to $\mathbb{Z}$. If Γ is finitely generated, then the orbit $\Gamma 0 = \{g0 : g \in \Gamma\}$ of $0 \in \mathbb{D}$ is a countable set of points in the unit disk.

Definition 2.1 We call a non-elementary group Γ a *Fuchsian group* if $\Gamma 0$ is a discrete set with respect to the Poincaré metric ds^2 introduced above.

In particular, if Γ is Fuchsian then all accumulation points with respect to the Poincaré metric must lie on the unit circle $\partial \mathbb{D} = \{z \in \mathbb{C} : |z| = 1\}$.

Let us denote the set consisting of generators and their inverses by $\Gamma_\bullet = \{g_1^{\pm 1}, \cdots, g_d^{\pm 1}\}$. We can associate to $\Gamma_\bullet$ the Cayley graph of Γ. This is an infinite graph in which the vertices can be realised as the points of $\Gamma 0 = \{g0 : g \in \Gamma\}$ and two vertices $g0$ and $h0$ are connected by an edge if and only if $gh^{-1} \in \Gamma_\bullet$.

2.3 Random Walks and the Drift

We can now introduce our main tool. Given a set of generators and their inverses $\Gamma_\bullet$ we can consider a random walk on the Cayley graph where we allow a transition from a vertex $g0$ to a neighbouring vertex $h0$ with probability $\frac{1}{2d}$. (Here we assume that $\#\Gamma_\bullet = 2d$, as above.)

Definition 2.2 Given a specific set of generators $\Gamma_\bullet$ we can associate the drift (or the rate of escape) defined by

$$\ell = \ell(\Gamma_\bullet) = \lim_{n \to +\infty} \frac{1}{(2d)^n} \sum_{g_{j_1}, \ldots, g_{j_n} \in \Gamma_\bullet} \frac{d(g_{j_1} \cdots g_{j_n} 0, 0)}{n}.$$

The limit always exists by a standard subadditivity argument and quantifies the rate at which typical points $g_{j_1} \cdots g_{j_n} 0$ escape towards the boundary circle $\partial \mathbb{D}$.

3 Some Examples

Let us now turn to specific examples of Fuchsian groups. For basic results on hyperbolic polygons we refer the reader to an excellent book by Beardon [3].

3.1 Regular Octagon Tilings

Let $P \subset \mathbb{H}$ be a regular octagon with angles $\frac{\pi}{4}$ and sides of equal length which are geodesics as shown in Fig. 2. We can consider groups of isometries generated by four transformations which identify the sides of the regular octagon. In this case the images of the octagon under the group action tile the hyperbolic plane, so that $\Gamma P = \mathbb{H}$. There are four different identifications which yield a surface of genus 2 as factor space $\mathbb{H}/\Gamma$ [15]. This property implies, in particular, that the group Γ generated by these identifications is discrete. We will consider only two identifications which lead to well-known surfaces: the Bolza surface and the Gutzwiller surface.

The Gutzwiller group [11] $\Gamma_G = \langle g_1, g_2, g_3, g_4 \rangle$ is generated by four isometries which identify the alternating sides of the regular pentagon P. They satisfy the identity $g_1 g_2^{-1} g_1 g_2^{-1} g_3 g_4^{-1} g_4 g_3^{-1} = I$ (see Fig. 2, Left).

The Bolza group [5] $\Gamma_B = \langle g_1, g_2, g_3, g_4 \rangle$ is generated by four isometries which identify the opposite sides of the regular pentagon P. They satisfy the identity $g_1 g_2^{-1} g_3 g_4^{-1} g_1^{-1} g_2 g_3^{-1} g_4 = I$ (see Fig. 2, Right).

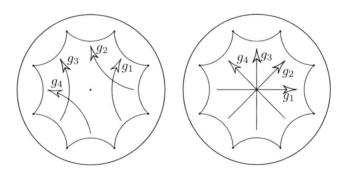

Fig. 2 Left: generators of the Gutzwiller group identify alternating sides of the regular octagon; Right: Bolza group generated by isometries identifying opposite sides of the regular octagon

It turns out the drift doesn't depend on the identification chosen, in particular, these two examples share the same value for the drift ℓ.

Theorem 3.1 *With the choice of generators specified above, the drift ℓ for the Bolza group Γ_B and for the Gutzwiller group Γ_G is the same and satisfies*

$$1.690771 < \ell < 1.691313.$$

The method we use for estimating ℓ involves looking at the action of Γ on $\partial \mathbb{D}$ and computing the maximal Lyapunov exponent. This is achieved by obtaining estimates on the spectral radius of transfer operators acting on the space of α-Hölder continuous functions $\mathcal{L}_t : C^\alpha(\partial \mathbb{D}) \to C^\alpha(\partial \mathbb{D})$ defined by $[\mathcal{L}_t f](z) = \frac{1}{2n} \sum_{g \in \Gamma_*} |g'(z)|^t f(gz)$ for t close to 1 and suitably small $\alpha > 0$. A more detailed exposition of the technical computer-assisted argument will appear elsewhere.

3.2   Hyperbolic Triangle Groups

Another class of interesting examples is perhaps the class of Coxeter groups generated by reflections in the sides of a hyperbolic triangle, the so-called triangle groups. It is easy to see that the group is discrete if and only if all angles of the triangle are rational multipliers of π. We will restrict our considerations to the case when triangle has angles $\frac{\pi}{k}$, $\frac{\pi}{l}$ and $\frac{\pi}{m}$ where k, l and m are integers. Recall that sum of the angles of the hyperbolic triangle is strictly less than π, and therefore k, l, m should satisfy the inequality $\frac{1}{k} + \frac{1}{l} + \frac{1}{m} < 1$.

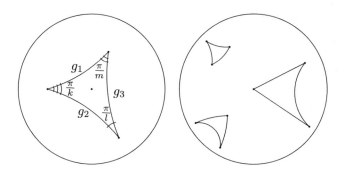

Fig. 3 Left: a triangle in $\mathbb{D}$ with boundaries which are geodesics with respect to the Poincaré metric and internal angles $\frac{\pi}{k}$, $\frac{\pi}{l}$ and $\frac{\pi}{m}$, containing the centre of the disk in its interior. Right: hyperbolic triangles with different shape

Definition 3.2 The (k, l, m)-triangle group is a group generated by reflections in the three sides of a hyperbolic triangle with angles $\frac{\pi}{k}$, $\frac{\pi}{l}$ and $\frac{\pi}{m}$.

It follows from the properties of reflections, that the generators g_1, g_2, g_3 of the (k, l, m)-triangle group, shown in Fig. 3 satisfy the following relations

$$g_1^2 = g_2^2 = g_3^2 = I \quad \text{and} \quad (g_1 g_2)^k = (g_2 g_3)^l = (g_3 g_1)^m = I.$$

These are the defining relations in the sense that any group with three generators which satisfy these condition is the (k, l, m)-triangle group. Evidently, this group is also cocompact. Furthermore, similarly to the case of the regular octagon, the images of the original triangle with respect to the group form a tessellation of the hyperbolic plane.

The study of the groups generated by reflections with respect to the sides of curvilinear triangles can be traced back to the works of Gauss. Bolyai, commenting on Gauss' work, suggests that in a drawing from "Cereri Palladi Junoni Sacrum" dated February 1805 Gauss introduced the idea of reflection with respect to the circle. A copy of the drawing, taken from [9, p. 104], is shown in Fig. 4 on the left. On the right we see a tessellation of the hyperbolic plane generated by the $(4, 4, 4)$-triangle group. The difference between the two drawings is due to the choice of the location of the original triangle. In Gauss' drawing the centre of the disk is one of the vertices. In our drawing, the centre of the disk is the barycentre of the triangle. Despite the appearance of a tessellation of the Poincaré disk in Gauss' drawing it was written 49 years before the birth of Poincaré! Lobachevsky laid the foundations of the hyperbolic geometry in 1823 (see, e.g. [16], chapter 18).

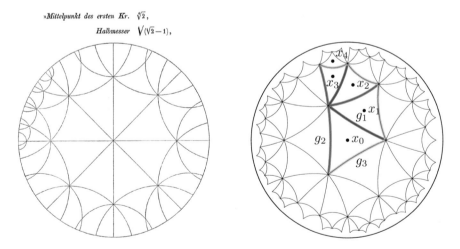

Fig. 4 Left: an original drawing by Gauss of a part of the $(4, 4, 4)$-triangle group tessellation; Right: an orbit of a random walk corresponding to the $(4, 4, 4)$-triangle group: $x_1 = g_1 x_0$, $x_2 = g_1 g_2 x_0$, $x_3 = g_1 g_2 g_1 x_0$, and finally $x_4 = g_1 g_2 g_1 g_3 x_0$

Using the same machinery as in the case of the regular octagon we can estimate the drift. In Table 1 we list different examples of triangle groups and give upper and lower bounds on the associated drift. In a special case of the (4, 4, 4)-triangle group, we have the following result.

Theorem 3.3 *The drift of the random walk on the* (4, 4, 4)-*triangle group with the choice of generators specified above satisfies*

$$0.1282273 < \ell < 0.1282264.$$

Table 1 Upper and lower bounds on the drift for some (k, l, m)-hyperbolic triangle groups

k	l	m	Lower bound on ℓ	Upper bound on ℓ
3	7	2	0.009936413804542	0.009974294432083
3	8	2	0.016242376981342	0.016295700460901
3	9	2	0.020422904820936	0.020508218335138
4	5	2	0.024263195172778	0.024341830945392
4	6	2	0.037765501277040	0.037870175386186
4	8	2	0.050724918174930	0.050934249274956
5	5	2	0.046019792084900	0.046155635941842
5	6	2	0.058159239428682	0.058334985605960
5	7	2	0.065329026703739	0.065563197936118
6	6	2	0.069559814745121	0.069846131636394
4	3	3	0.046694831446660	0.046816105401585
5	3	3	0.069435926662536	0.069689191304812
6	3	3	0.081515978567027	0.081925767935374
7	3	3	0.088431558608918	0.089059709051931
3	4	4	0.088752444507380	0.088919437571219
3	6	6	0.148515148139248	0.149179933451390
4	4	4	0.128086862380309	0.128344145942091
5	5	5	0.182618423778876	0.183286144055414
6	6	6	0.209779208475952	0.211031605163552
7	7	7	0.224864828238411	0.228908301867331
8	8	8	0.232248419011566	0.238574707256068
9	9	9	0.236782098913020	0.247054233672500
10	10	10	0.240409132283172	0.252180931190328

4 Two Problems

After a brief discussion of the groups we are concerned with we continue by introducing one of the central objects of the theory—*the harmonic measure*.

4.1 The Harmonic Measure on the Unit Circle

As we have seen already, a typical orbit of the random walk associated to a Fuchsian group converges to a point on the boundary circle with respect to the Euclidean and the hyperbolic metric. The distribution ν of the limit points on the boundary $\partial\mathbb{D}$ defines a probability measure. More precisely, given a generating set $\Gamma_\bullet$ of a Fuchsian group we can define a family of probability measures on $\mathbb{D}$ by

$$\nu_n = \left(\frac{1}{2d}\right)^n \sum_{g_{j_1},\cdots,g_{j_n} \in \Gamma_\bullet} \delta_{g_{j_1}\cdots g_{j_n}0},$$

where $\delta_{g_{j_1}\cdots g_{j_n}0}$ is the Dirac measure supported at $g_{j_1}\cdots g_{j_n}0$. The measures ν_n converge in the weak star topology (on the closed unit disk) to a probability measure ν on $\partial\mathbb{D}$.

Definition 4.1 The measure ν is called the *harmonic measure* or the *hitting measure*.

We can denote by $\Lambda \subset \partial\mathbb{D}$ the support of this measure (i.e., the smallest closed set of the full measure). It is known that either $\Lambda = \partial\mathbb{D}$ or $\Lambda \subsetneq \partial\mathbb{D}$ is a Cantor set.

4.2 Singularity of the Harmonic Measure

The following natural question was posed by Kaimanovich and Le Prince [12]:

Question 4.1 Can we characterise Fuchsian groups for which the associate harmonic measure is absolutely continuous with respect to the Lebesgue measure?

Of course, if the support of the harmonic measure is a Cantor set then the measure is singular with respect to Lebesgue measure. Therefore, we will only consider the case that $\Lambda = \partial\mathbb{D}$. In the special case when one of the generators in $\Gamma_\bullet$ is parabolic it was shown by Gadre, Maher, and Tiozzo that the harmonic measure is always singular [7].

Furthermore, there are examples of non-discrete groups due to Bourgain for which ν is absolutely continuous [6] (see also [2]). In the more general setting when the weights in the random walk differ the measure ν may be singular [12].

In the setting of the surface groups, this question has been intensively studied [13, 14]. One set of examples is Fuchsian groups generated by isometries identifying the sides of hyperbolic polygons. However, many of the examples with more than four sides have harmonic measures that are singular (see [13, Theorem 1]).

4.3 Dimension of the Harmonic Measure

An important quantitative characteristic of the measure is its Hausdorff dimension.

Definition 4.2 The Hausdorff dimension of the measure is the infimum of Hausdorff dimensions of sets of the full measure:

$$\dim_H(v) = \inf\{\dim_H(X) \mid X \subset \partial \mathbb{D} \text{ Borel and } v(X) = 1\}.$$

There is a useful result due to Tanaka which relates the question of absolute continuity of the harmonic measure to the numerical value of its Hausdorff dimension [17].

Proposition 4.3 (Tanaka) *The harmonic measure v is absolutely continuous if and only if $\dim_H(v) = 1$.*

This leads to the following stronger version of Question 21.1.

Question 4.2 Assuming the harmonic measure is not absolutely continuous with respect to Lebesgue measure, can we estimate its Hausdorff dimension?

We now return to our examples.

It follows from the result of Kosenko [13, Theorem 1.2] (see also [14]) that the harmonic measure v_B associated to *the Bolza group* Γ_B is singular. We can improve on this result.

Theorem 4.4 *The dimension of the harmonic measure v_B for the Bolza group satisfies*

$$\dim_H(v_B) \leq 0.86116.$$

The dimension of the harmonic measure v_G for the Gutzwiller group Γ_G satisfies

$$\dim_H(v_G) \leq 0.86317.$$

In particular, the harmonic measure v_G is also singular.

In order to explain the proof, we need one extra ingredient.

4.4 Relation to the Avez Entropy

We introduce another numerical characteristic of the random walk which is commonly used to estimate the dimension of a measure.

Definition 4.5 We can associate to a harmonic measure ν the *Avez random walk entropy* defined by

$$h_A(\nu) = \lim_{n \to +\infty} \frac{1}{n} H(\nu^{*n})$$

where $H(\cdot)$ is the usual Shannon entropy function and ν^{*n} denotes the n-fold convolution [1].

The limit always exists by subadditivity. The dimension, entropy and drift are related using the following identity [4, 10, 17].

Proposition 4.6 *For the harmonic measure ν we have that* $\dim_H(\nu) = h_A(\nu)/\ell(\nu)$.

Now we are ready to prove Theorem 4.4.

Proof of Theorem 4.4 Combining Propositions 4.3 and 4.6 we see that in order to establish that the harmonic measure is singular it is sufficient to show that $h(\nu) \geq \ell(\nu)$. In particular, we need to establish an upper bound on the entropy and a lower bound on the drift. For the Gutzwiller group an estimate on the entropy is given in a beautiful paper [10]. Even the most basic bound they give in Example 2.3 $h(\nu_B) \leq 1.46$ in combination with the estimate on the drift $\ell(\nu_B)$ from Theorem 3.1 allows us to deduce that the measure in singular. In the case of the Bolza group, we can use the estimate on the drift $\ell(\nu_G)$ from Theorem 3.1 and the upper bound on the entropy of $h(\nu_G) \leq \frac{3}{4}\log 7$ coming from the free group on four generators. $\square$

4.5 Final Remarks

Should one wish to apply the same approach to show that the harmonic measures for the triangle groups are singular it will be necessary to obtain an effective upper bound on the Avez entropy. Unfortunately, the naive bound of $\frac{1}{3}\log 2 \approx 0.231049\ldots$ corresponding to the Avez entropy of the random walk on the free product $\mathbb{Z}_2 \times \mathbb{Z}_2 \times \mathbb{Z}_2$ isn't quite low enough to show that the measure is singular for most of triangle groups listed in Table 1. Nevertheless in the case of $(8, 8, 8)$, $(9, 9, 9)$, and $(10, 10, 10)$-triangle groups we may conclude that the measure is singular. It is reasonable to suggest that the drift for (k, k, k)-triangle group is monotone increasing as $k \to \infty$. This would imply that the harmonic measure is singular for $k \geq 8$.

Acknowledgements The first author is partly supported by ERC-Advanced Grant 833802-Resonances and EPSRC grant EP/T001674/1 the second author is partly supported by EPSRC grant EP/T001674/1.

References

1. A. Avez. Théoréme de Choquet–Deny pour les groupes à croissance non exponentielle. *C. R. Acad. Sci. Paris Sér. A* **279:2528** (1974).
2. B. Barany, M. Pollicott and K. Simon. Stationary measures for projective transformations. *Journal of Statistical Physics* **148**(3), 393–421 (2012).
3. A.F. Beardon. *The geometry of discrete groups.* Corrected reprint of the 1983 original. Graduate Texts in Mathematics **91**. Springer-Verlag, New York (1995).
4. S. Blachére, P. Haissinsky and P. Mathieu. Harmonic measures versus quasiconformal measures for hyperbolic groups. *Ann. Sci. Ec. Norm. Sup.* **44**(4), 683–721 (2011).
5. O. Bolza. On binary sextics with linear transformations into themselves. *Amer. J. Math.* **10** no. 1, 47–70 (1887).
6. J. Bourgain. On the Furstenberg measure and density of states for the Anderson–Bernoulli model at small disorder. *Journal d'Analyse Mathématique* **117**, 273–295 (2012).
7. V. Gadre, J. Maher and G. Tiozzo. Word length statistics for Teichmüler geodesics and singularity of harmonic measure. *Comment. Math. Helv.* **92**, no. 1, 1–36 (2017).
8. Schreiben Gauss an Wolfgang Bolyai, Göttingen, 2. 9. 1808. In: Franz Schmidt, Paul Stäckel (Hrsg.): *Briefwechsel zwischen Carl Friedrich Gauss und Wolfgang Bolyai*, B.G. Teubner, Leipzig, S. 94 (1899).
9. C.F. Gauß. *Werke.* Band VIII. (German) [Collected works. Vol. VIII] Reprint of the 1900 original. Georg Olms Verlag, Hildesheim (1973).
10. S. Gouëzel, F. Mathéus and F. Maucourant. Sharp lower bounds for the asymptotic entropy of symmetric random walks. *Groups, Geometry, and Dynamics* **9**, 711–735 (2015).
11. H. Ninnemann. Gutzwiller's octagon and the triangular billiard $T^*(2, 3, 8)$ as models for the quantization of chaotic systems by Selberg's trace formula. *Internat. J. Modern Phys. B* **9**, no. 13–14, 1647–1753 (1995).
12. V. Kaimanovich and V. Le Prince. Matrix random products with singular harmonic measure. *Geometriae Dedicata* **150**, 257–279 (2011).
13. P. Kosenko. Fundamental inequality for hyperbolic Coxeter and Fuchsian groups equipped with geometric distances. *arXiv preprint*: arXiv:1911:00801 (2019).
14. P. Kosenko and G. Tiozzo. The fundamental inequality for cocompact fuchsian groups. *arXiv preprint*: arXiv:2012:07417 (2020).
15. T. Kuusalo and M. Näätänen. On arithmetic genus 2 subgroups of triangle groups. In: *Extremal Riemann surfaces (San Francisco, CA, 1995)*, 21–28, Contemp. Math. **201**, Amer. Math. Soc., Providence, RI (1997).
16. J. Stillwell. Mathematics and Its History. Springer, Berlin (2010).
17. R. Tanaka. Dimension of harmonic measures in hyperbolic spaces. *Ergod. Th. and Dynam. Sys.* **39**, 474–499 (2019).

Part VI
History of Mathematics

This part presents three very different aspects of the historical viewpoint on mathematics.

The article *Some Problems in the modern history of mathematics* by Jeremy Gray begins with a concise, critical, and insightful presentation of the historiography of modern mathematics in the West. The author then proposes five fundamental projects for going forward: a socio-historical book on the history of mathematics, a book on the history of applied mathematics, a history of differential equations, and a book on mathematics of the eighteenth century. The motivations for each are of course of great interest even for non-historians. The same is true of the paragraph making the case for a new philosophical perspective in the history of mathematics.

Norbert Schappacher's article *Mathematics going backwards? A logological encounter between mathematics and archeology* begins with a description of Borromean links appearing in mosaics on the floor of the Villa del Casale in Sicily. It continues with a vivid description of the relationship between mathematicians and archeologists centered around knots and links and ends with a meditation on the possible use of links as symbols of cohesion in mathematics.

David Rowe's text *Max Dehn as a historian of mathematics* presents a fascinating example of a mathematician deeply versed in the history of his discipline. It explains how this fits in with Dehn's view of mathematics as an important part of human culture, and with the activities of his very influential seminar. The article also points to the differences between Dehn's approach to mathematics and that of the Göttingen school and more generally details Dehn's interactions with his mathematical environment. Finally there are insightful presentations of Dehn's work on the history of geometry and on the Newton–Leibniz controversy.

Some Problems in the History of Modern Mathematics

Jeremy Gray

Personal Note

It is a great pleasure to take this opportunity to thank Catriona Byrne for the many ways she has helped the history of mathematics community, and for the numerous encouraging conversations I have had with her over the years, which have helped shape how I have thought about mathematics and its history. Of particular note is the work she did to bring about Springer's contribution to the Prize awarded by the European Mathematical Society. More personally, I thank her for helping to open Springer's doors to what became my four volumes on the history of mathematics in the SUMS series, which are now ably looked after by Remi Lodh.

1   A Brief Historiography

I shall restrict my attention to mathematics in the West in the period from 1600 to 2000 (the modern period). The word 'mathematics' in this essay will always mean the mathematics of those four centuries. Before addressing my main theme of problems for future historians of mathematics, I would like briefly to set the context with a few historiographical remarks.

Science and technology had played a hugely significant role on both the Allied and Axis sides in the second World War and then throughout the Cold War, with implications and opportunities that needed to be understood. In the English-speaking world there was a boom in history and philosophy of science, and in logic,

J. Gray (✉)
The Open University, Milton Keynes, UK
e-mail: jeremy.gray@open.ac.uk

© The Author(s), under exclusive license to Springer Nature Switzerland AG 2023
J.-M. Morel, B. Teissier (eds.), *Mathematics Going Forward*, Lecture Notes
in Mathematics 2313, https://doi.org/10.1007/978-3-031-12244-6_22

often focussed on what the key ingredients of progress might be: what was special about science, what was science anyway? Enthusiasm and funding eventually waned, but two books stand out as survivors: Tom Kuhn's [38] *The Structure of Scientific Revolutions* and Imré Lakatos's [39] *Proofs and Refutations*. Whatever their strengths or weaknesses, they both offered a point of view that many outside the field could relate to; Kuhn's paradigm shifts and Lakatos's monster barring and other ideas seemed much fresher than a trudge through the names and dates of the great scientists and mathematicians, or raking over the coals of formalism, intuitionism, and logicism.

The return to original research in the history of mathematics based on archival documents was quieter. In the U.K. there were the examples of Ivor Grattan-Guinness [23] and Tom Whiteside [57]; in the USA Carl Boyer and Morris Kline [36]; in France René Taton [55]. In East and West Germany there were a number of long-standing Chairs in the subject, and in the Soviet Union there were some good links between mathematicians and historians of mathematics (for example, Kolmogorov and Youschkevitch [37]), doubtless complicated by a desire of the communist system to produce Marxist 'history'. However, the English-speaking world also saw a growing separation between history of mathematics and history of science, as historians of science sought to orient their work more closely with historians of other topics, in whose department they worked. Scientific rationality and the experimental method became only a part of arguments about status, social acceptability, funding, and national priorities. This was not a fertile ground for historians of mathematics of any kind, for whom there are aspects of mathematics already pointed out by philosophers of the 1950s and by Leo Corry more recently in lectures, which make the writing of interesting history of mathematics difficult; chiefly, what can be said in mathematics is very tightly constrained by the standards of rigour of its time. But nor, with a few exceptions, did philosophers of mathematics sustain an interest in history, preferring set-theoretic reductionism.

Where is the history of modern mathematics today? There is no consensus about research in the field, nor should there be. Among the substantial editorial achievements that have altered our picture of modern mathematics, the most significant for the early twentieth century is the ten volumes of Hausdorff's *Werke* [31]. For the nineteenth century there have been a number of well-researched biographies, and Thomas Hawkins's exceptional book [32] *Emergence of the Theory of Lie Groups*. For the eighteenth century there is the creation of the online Euler Archive, and for the seventeenth century I would single out Henk Bos's [3] *Redefining Geometrical Exactness: Descartes' Transformation of the Early Modern Concept of Construction*, which taught us how to read Descartes with fresh eyes, and Niccolò Guicciardini's *Isaac Newton on Mathematical Certainty and Method*, which does the same for Newton. A recent, two-volume, source-based account for students by June Barrow-Green et al. [2] may serve as an introduction to much of this material and more.

Although there has been a welcome turn towards historiographical studies (Guicciardini [28], Remmert, Schneider, and Sørensen [46]) and a methodological stiffening, most openly in the groups around Catherine Goldstein and Karine

Chemla in Paris, the history of modern mathematics can too easily project a sense of worthiness that shades into seeming insignificance. That should be set against the fact that science, especially physics, celestial mechanics, and mathematics are inextricable; that on many occasions philosophy has turned to mathematics and mathematical physics to renew its pursuit of its own questions; that mathematics is a key element in the educational systems of the modern world, and that modern mathematics can be seen as a species of modernism (Gray [24]). Without detracting at all from anything that has been done, historians of mathematics might profitably reassert where their work stands at the nexus of several vital concerns in the shaping of the modern world in the last four or more centuries. Perhaps surprisingly, a way forward might begin by considering what would be involved in a social-historical approach (so often a domain bereft of serious mathematics).

2 The Problems

2.1 Write a Social-Historical Book in the History of Mathematics

We currently lack a book in the history of mathematics rooted in its contemporary social developments, although the forthcoming six-volume *Bloomsbury Cultural History of Mathematics* edited by J.W. Dauben and D.E. Rowe should be an excellent resource for many audiences (full disclosure, I have an essay in volume 5). What does it mean that mathematics was the province of a few gifted individuals in the seventeenth century, often outside the small university world? Or that these small numbers persisted into the eighteenth century, the age of the Academies and the Republic of Letters? Or that matters then passed to the ever-expanding universities of the nineteenth and twentieth centuries, concurrent with the rise of modern capitalism, and the number of mathematicians grew by perhaps a factor of 50? Soon, reform of the school educational system became a feature of every country that aspired to advanced mathematics, but there is no recent study of the connection between the rise of the universities after 1800 and any truly interesting account of the effect this had on mathematics, although there was a flurry of interest in neohumanism in Germany (see e.g. Pyenson [45]), there have been close studies of mathematics in nineteenth century Cambridge (Craik [9] and (Warwick [56]), and more recently there have been arguments about the first century of the École Polytechnique. Rowe's [48], a look at the Göttingen tradition, grounded in archival sources, offers a fresh indication of how things could be done, as do the dense and valuable books by Reinhard Siegmund-Schultze (for example, his [51] *Mathematicians Fleeing from Nazi Germany: Individual Fates and Global Impact*). Even if there are no significant connections between advanced pure mathematics and any social need, that would be a valuable part of the story (see Harris [30]).

It is time for the history of statistics to be integrated with the rest of the history of mathematics, despite the fact that in many universities mathematics and statistics remain in different departments. There are two good books on the history of statistics (Porter [44], Stigler [52]), but right now the subject seems to be lying fallow. There are two books on measure theory and the axiomatic approach to probability of Kolmogorov and others (Hochkirchen [33], von Plato [43]). Measure theory went in two directions, one towards probability (in the work of Doob) and the other towards functional analysis and ultimately also quantum mechanics. We now live at a time when probabilistic thinking is enriching several domains of analysis (notably number theory and partial differential equations) and while it would be absurd to imagine writing a history of such a fast-moving field today this might create the opportunity for a fresh look at the history of probabilistic thinking in mathematics. A start could be made with a fresh look at the history of thermodynamics.

2.2 Write a Book on the History of Applied Mathematics

Most fundamentally, we need to rethink the term 'applied mathematics', not just because it has various meanings in different countries even today, but because it emerged as a term only at the start of the nineteenth century; previously there had been a division into pure and mixed mathematics. Truesdell spoke of the rational mechanics of the eighteenth century, referring to the striking absence of experimental work in science in the period and the reliance on untested mathematics (inevitable, of course, in the dominant field of celestial mechanics). If it is true that experimental physics only took off at the start of the nineteenth century, perhaps with the study of electricity, and theoretical physics only came later (see Jungnickel and McCormmach [34]), there still needs to be a historical investigation of the associated mathematics in this context.

We lack a book on the complicated relationship between modern mathematics and modern physics, although it is widely believed that this was strangely attenuated in the early years of quantum mechanics. There is much we know about each side, see (Schneider [50]) and e.g., from an enormous literature on Einstein, (Renn [47]), but less on the interaction between the fields. There has been no shortage of solid work in the history of applied mathematics, much of it concentrated on the nineteenth century, and much of it in the form of biographies. We also have the pioneering books by Olivier Darrigol on electrodynamics [10], hydrodynamics [11], and optics [12]. More recently there has been at least two books on mathematics and the first world war (Aubin and Goldstein [1], Royle [49]), and books on mathematics and the early history of flight. But, for example, we still lack a book on the major British applied mathematicians that Klein so appreciated in his [35] *Vorlesungen über die Entwicklung der Mathematik im 19. Jahrhundert (1928)*. From the physics side, we have Buchwald's [6] and [7] concerning responses to Maxwellian physics and the rise of the wave theory of light.

Any such book would have to consider such topics as Maxwell's equations, the Einstein field equations, Schrödinger's equation and other partial differential equations, which underlines the importance of the next problem.

2.3 A History of Differential Equations

There is a number of accounts of one or another equation, and an extensive mathematical folklore, but my attempt to say something in a book (Gray [25]) aimed at final-year students convinced me that there is a need for a more thorough history with a more sophisticated methodology and less constrained by mathematical difficulty. It seems very likely that a good treatment of the subject will change our ideas about the growth of mathematical analysis, and move it away from an over-emphasis on rigour towards what might be called rigour for good reasons. A history of partial differential equations would be particularly valuable, especially if it could engage with the fundamental changes introduced by 1920. After that, the field becomes highly diverse and complicated and may defy historians for some time (see, however, Lützen [40]). Two topics that I had to omit stand out: Maxwell's equations (but see Buchwald [6]) and perturbation theory, so important in celestial mechanics and other fields. The twentieth century also saw the introduction, following ideas of Poincaré, of an abstract theory of flows, probabilistic ideas, and the ergodic theorems of Birkhoff and von Neumann.

A few isolated equations aside, theories of ordinary and partial differential equations began with Euler, Laplace, Lagrange, and Monge in the eighteenth century, and that leads into the next problem.

2.4 A Book on the History of Mathematics in the Eighteenth Century

Researching the history of mathematics in the eighteenth century sits uneasily between the preferences of mathematics and history of science departments. Two intertwined themes are the advances in celestial mechanics, and the reformulations of the calculus. Laplace's intimidating *Mécanique Céleste* was widely taken to have removed all doubts about the workings of the solar system, but the history of planetary astronomy still needs to be properly included in the history of mathematics, despite Gillispie's [22] *Pierre-Simon Laplace 1749–1827* and the work of Curtis Wilson; see, e.g. his [58] and [59]. The Euler Archive is a valuable initiative in this direction.

Scholarship on calculus in the century is almost bracketed by Guicciardini's two books [26, 27] on Newton and various studies of aspects of Cauchy's rigorization of analysis. In between, Euler brought about a shift in the foundations of the subject by successfully introducing the concept of a function, in whatever limited a form (see Ferraro [16]), but attempts by Lagrange to provide rigorous foundations of the calculus failed, as Ferraro and Panza [17] have shown. Another figure on whom scholarship has just begun, is Johann Heinrich Lambert, a member of every section of the Académie Royale des Sciences in Berlin, who wrote on many subjects and about whom we have several fragmented accounts.

Recently, Andrea del Centina (see e.g. his paper of 2020 [8]), and Jean-Yves Briend and Marie Anglade (see a series of papers starting with their paper [4] in 2017), have been revising our understanding of what falls under the heading of projective geometry in the seventeenth century. But we lack recent accounts of geometry in the eighteenth century, although there is (Bruneau [5]) on MacLaurin, and De Risi's innovative work on the foundations of geometry and attitudes to Euclid's *Elements*, much of which is still to appear (but for an early yet valuable work see his [14]). Presently, it seems as if, books by Euler and Cramer notwithstanding, geometry went into something of a decline. There was, for example, surprisingly little differential geometry of surfaces in the eighteenth century.

Historians and other intellectuals have given the idea of a progressive Enlightenment a rough time in the last 20 years, but apart from Hankins' book [29] there has been very little written on the involvement of mathematicians in the Enlightenment project and, for example, the production of the great *Encyclopédie*. As one example to be integrated into the history of mathematics, there is the controversy between d'Alembert, Rameau, and Rousseau about the new theory of music. This is only one reason for a fresh examination of mathematics of the age of the Academies.

2.5 The History of Mathematics from a New Philosophical Perspective

Such were the crises of mathematics around 1900 that Hilbert was driven to say that to solve their problems mathematicians had to become philosophers. The highest standards were required of the rigour, reliability, and perhaps meaning of mathematics, and three families of ideas emerged: intuitionism, formalism, and logicism. More than anyone else, Gödel and Tarski answered many of the questions raised at the start of the twentieth century. More recently, the idea that mathematics is about axiomatically defined structures in various inter-relations (structuralism) has been a rival to a feeling that, at base, mathematics is an outgrowth of set theory and logic, a view that leaves most mathematicians cold. In the last 10 years, however, in the work of Mancosu [41, 42], Tappenden, in his [53] and [54] (to appear), and others, attempts have been made to engage philosophically with questions that mathematicians do ask themselves: What makes an idea fruitful? What is the right definition of a new concept? What is meant by purity of method and why is it valuable? What characterises a 'right' proof? How does advanced mathematics emerge from the seemingly incontrovertible elementary arithmetic, and with what consequences (here, see Ferreirós [19])? We now have a variety of sophisticated tools, such as Epple's epistemic objects (Epple [15]).

All this refocuses history of mathematics on how are discoveries made, which has a weakness of assuming too easily that some discoveries just get made, as if they were ordained in advance. There is a need for histories of mathematics that deliberately play down the eventual successes in favour of the seeking after new results, and recent developments in the philosophy of mathematical practice may provide the tools for such a thing.

It is time to take seriously the questions of why mathematics matters and why it is convincing, as opposed to taking these questions for granted. From a social perspective, and for the nineteenth century, a lot might hang on the opinions of astronomers and certain kinds of physicists, but the question is worth asking philosophically. To what extent is rigour a spur to mathematical discovery? This raises the question of what various mathematicians have been trying to do, why, under what constraints, and with what success. Such a book would be more Lakatosian than Kuhnian. Ultimately, we need a methodological approach to the mathematics of the previous centuries that is not a survey of results that reads like *Mathematics Reviews* for the past, and a philosophy of mathematical practice may well offer such a thing. Aside from the books mentioned above, one could also draw inspiration from Ferreirós's [18], and two books [20, 21] by Marcus Giaquinto.

3 Concluding Remarks

I heard recently of a 600-page history of modern Germany that mentioned Bach precisely once. This was not to find fault, but merely to indicate how much necessarily gets left out of such a book, never mind how much Bach's legacy remains more alive than a lot of what was included. I haven't been able to check, but I doubt if Euler's name was mentioned at all in the book, and that is the problem of the history of mathematics, as it is of music and art. The challenge is to find places to stand, and organising principles, that make the history of mathematics not only accessible but vital.[1]

References

1. D. Aubin and C. Goldstein (eds.) *The War of Guns and Mathematics*. AMS HMath **42** (2014).
2. J.E. Barrow-Green, J.J. Gray and R. Wilson. *The History of Mathematics – a Source-Based Approach*. American Mathematical Society, Vol. 1 (2019), Vol. 2 (2022).

---

[1] I would like to thank Tom Archibald, David Rowe, and José Ferreirós for helpful comments on earlier drafts of this essay.

3. H.J.M. Bos. *Redefining Geometrical Exactness: Descartes' Transformation of the Early Modern Concept of Construction.* Springer (2001).

4. J.-Y. Briend and M. Anglade. La notion d'involution dans le Brouillon Project de Girard Desargues. *Archive for History of Exact Sciences* **71**, 543–588 (2017).

5. O. Bruneau. *Colin Maclaurin, ou l'Obstination mathématicienne d'un Newtonien.* Presses Universitaires de Nancy (2011).

6. J.Z. Buchwald. *From Maxwell to Microphysics.* Chicago U.P. (1985).

7. J.Z. Buchwald. *The Rise of the Wave Theory of Light.* Chicago U.P. (1989).

8. A. del Centina. Pascal's mystic hexagram, and a conjectural restoration of his lost treatise on conic sections. *Archive for History of Exact Sciences* **74**, 469–521 (2020).

9. A.D.D. Craik. *Mr Hopkins' Men: Cambridge Reform and British Mathematics in the 19th Century.* Springer (2007).

10. O. Darrigol. *Electrodynamics from Ampère to Einstein.* Oxford U.P. (2000).

11. O. Darrigol. *Worlds of Flow.* Oxford U.P. (2005).

12. O. Darrigol. *A History of Optics from Greek Antiquity to the Nineteenth Century.* Oxford U.P. (2012).

13. J.W. Dauben and D.E. Rowe (eds.) *The Bloomsbury Cultural History of Mathematics.* 6 vols (forthcoming).

14. V. De Risi. *Geometry and Monadology: Leibniz's Analysis Situs and Philosophy of Space.* Birkhäuser (2007).

15. M. Epple. Between Timelessness and Historiality: On the Dynamics of the Epistemic Objects of Mathematics. *Isis* **102**, 481–493 (2011).

16. G. Ferraro. *The Rise and Development of the Theory of Series up to the Early 1820s.* Springer (2008).

17. G. Ferraro and M. Panza. Lagrange's Theory of Analytical Functions and his Ideal of Purity of Method. *Archive for History of Exact Sciences* **66**, 95–197 (2012).

18. J. Ferreirós. *Labyrinth of Thought: A History of Set Theory and its Role in Modern Mathematics.* Birkhäuser (2007).

19. J. Ferreirós. *Mathematical Knowledge and the Interplay of Practices.* Princeton U.P. (2015).

20. M. Giaquinto. *The Search for Certainty: A Philosophical Account of Foundations of Mathematics.* Clarendon Press, Oxford (2002).

21. M. Giaquinto. *Visual Thinking in Mathematics: An Epistemological Study.* Oxford U.P. (2007).

22. C.G. Gillispie. *Pierre-Simon Laplace 1749–1827: A Life in Exact Science.* Princeton University Press (1997).

23. I. Grattan-Guinness. *Convolutions in French Mathematics, 1800–1840.* 3 vols., Birkhäuser (1990).

24. J.J. Gray. *Plato's Ghost: The Modernist Transformation of Mathematics.* Princeton U.P. (2008).

25. J.J. Gray. *Change and Variations: A History of Differential Equations to 1900.* Springer (2021).

26. N. Guicciardini. *Reading the* Principia. *The Debate on Newton's Mathematical Methods for Natural Philosophy from 1687 to 1736.* Cambridge U.P. (1999).

27. N. Guicciardini. *Isaac Newton on Mathematical Certainty and Method.* MIT Press (2009).

28. N. Guicciardini (ed.) *Anachronisms in the History of Mathematics.* Cambridge U.P. (2021).

29. T.L. Hankins. *Science and the Enlightenment.* Cambridge U.P. (1970).

30. M. Harris. *Mathematics without apologies: portrait of a problematic vocation.* Princeton U.P. (2015).

31. F. Hausdorff. *Gesammelte Werke.* E. Brieskorn, M. Epple, W. Purkert, E. Scholz, et al. (eds.) 10 vols. Springer (2013–2021).

32. T. Hawkins. *Emergence of the Theory of Lie Groups.* Springer (2000).

33. T. Hochkirchen. *Die Axiomatisierung der Wahrscheinlichkeitsrechnung und ihre Kontexte: Von Hilberts sechstem Problem zu Kolmogoroffs Grundbegriffen.* Vandenhoeck & Ruprecht (1999)..

34. C. Jungnickel and R. McCormmach. *The Intellectual Mastery of Nature.* 2 vols., Chicago U.P. (1986).

35. C.F. Klein. *Vorlesungen über die Entwicklung der Mathematik im 19. Jahrhundert.* Original 1928. Chelsea repr. 2 vols. in 1 (1967).
36. M. Kline. *Mathematical Thought from Ancient to Modern Times.* Oxford U.P. (1972).
37. A.N. Kolmogorov and A.P. Youschkevitch (eds.) *Mathematics of the 19th Century,* Birkhäuser (1992).
38. T.S. Kuhn. *The Structure of Scientific Revolutions.* Chicago U.P. (1962).
39. I. Lakatos. *Proofs and Refutations.* Cambridge U.P. (1976).
40. J. Lützen. *The Prehistory of the Theory of Distributions.* Springer (1982).
41. P. Mancosu (ed.) *The Philosophy of Mathematical Practice.* Oxford U.P. (2008).
42. P. Mancosu. *The Adventure of Reason. Interplay between philosophy of mathematics and mathematical logic: 1900–1940.* Oxford U.P. (2010).
43. J. von Plato. *Creating Modern Probability: Its Mathematics, Physics, and Philosophy in Historical Perspective.* Cambridge U.P. (1994).
44. T.M. Porter. *The Rise of Statistical Thinking, 1820–1920.* Princeton U.P. (1986).
45. L. Pyenson. *Neo-humanism and the Persistence of Pure Mathematics in Wilhelmian Germany.* American Philosophical Society (1983).
46. V. Remmert, M.R. Schneider and H.K. Sørensen (eds.) *Historiography of Mathematics in the 19th and 20th Centuries.* Springer (2016).
47. J. Renn (with H. Gutfreund). *The Road to Relativity.* Princeton U.P. (2015).
48. D.E. Rowe. *A Richer Picture of Mathematics: The Göttingen Tradition and Beyond.* Springer (2018).
49. T. Royle. *The Flying Mathematicians of World War I.* McGill-Queen's U.P. (2021).
50. M. Schneider. *Zwischen zwei Disziplinen: B.L. van der Waerden und die Entwicklung der Quantenmechanik.* Springer (2011).
51. R. Siegmund-Schultze. *Mathematicians Fleeing from Nazi Germany: Individual Fates and Global Impact.* Princeton U.P. (2009).
52. S.M. Stigler. *The History of Statistics: The Measurement of Uncertainty before 1900.* Harvard U.P. (1986).
53. J. Tappenden. The Riemannian Background to Frege's Philosophy. In: *The Architecture of Modern Mathematics: Essays in History and Philosophy,* J. Ferreirós and J.J. Gray (eds.), Oxford U.P. (2006).
54. J. Tappenden. *Philosophy and the Origins of Contemporary Mathematics: Frege and his Mathematical Context.* Oxford U.P. (to appear).
55. R. Taton. *L'Oeuvre Scientifique de Gaspard Monge.* Presses Universitaires de France (1951).
56. A. Warwick. *Masters of Theory: Cambridge and the Rise of Mathematical Physics.* Chicago U.P. (2003).
57. D.T. Whiteside (ed.) *The Mathematical Papers of Isaac Newton.* 8 vols., Cambridge U.P. (1967–1981).
58. C. Wilson. The work of Lagrange in celestial mechanics. In: *Planetary Astronomy from the Renaissance to the Rise of Astrophysics, Part B: The Eighteenth and Nineteenth Centuries,* 108–130, R. Taton and C. Wilson (eds.), Cambridge U.P. (1995).
59. C. Wilson. Newton and Celestial Mechanics. In: *The Cambridge Companion to Newton,* I.B. Cohen and G.E. Smith (eds.), 203–226, Cambridge U.P. (2002).

Mathematics Going Backward?
A Logological Encounter Between
Mathematics and Archaeology

Norbert Schappacher

I dedicate these freewheeling reflections to Catriona Byrne. We have met time and again over the years, usually at mathematical conferences. There were always exciting book projects to discuss. Catriona also participated at the Strasbourg IMU meeting that triggered my logological thoughts.

1 The Archaeologist and the Mathematician

The story goes back to May/June 2019. My partner and I toured the center and the Western part of Sicily, starting with the famous Roman villa in the Casale district of Piazza Armerina. As good, grateful tourists, we were duly overwhelmed by the opulent mosaic floors, especially the lavish hunts and the cute scenes with children.

On the next day we met the Canadian archaeologist Roger J. Wilson from Vancouver and his wife at an active excavation site near Gerace, not far from Piazza Armerina. Wilson is the author of a book about the Villa del Casale [9], which appeared some time ago. However, when we met in 2019, we only talked about some of his more recent findings and publications.

Little did I know at the time that my Strasbourg colleague Thomas Delzant and his wife were also touring Sicily at about the same time, naturally visiting the Roman Villa del Casale on their trip as well. Being a staunch mathematician, Thomas Delzant stopped in the small Room 18, in the North wing of the ancient

N. Schappacher (✉)
IRMA, Université de Strasbourg, Strasbourg, France
e-mail: n.schappacher@unistra.fr

Roman villa,[1] before moving on to its well-known mosaic highlights. Only part of the mosaic floor in Room 18 has survived. It is purely decorative, without any representative element. What made him pause here—in spite of all the lush scenes that he knew were waiting for him in the other parts of the complex—was a very neat representation of the so-called Borromean rings—see Fig. 1. This brings us back to the IMU logo (Fig. 2).

Fig. 1 The Borromean link in the mosaic floor of Room 18 of the Roman Villa del Casale near Piazza Armerina, Sicily. On the left, overview of Room 18; photographer Tyler Bell. On the right, detail; photographer Thomas Delzant

Thomas told me about his discovery during the IMU event staged in Strasbourg in September 2021.[2] He also asked me whether this mosaic might be the earliest occurrence of the Borromean rings in history. I replied with the usual caution about "earliest occurrences" that every historian of mathematics—and indeed every historian—is familiar with: not only is priority often hard to prove, but the fuss about priority tends to obscure historically meaningful nuances and contexts that distinguish recurring occurrences of "the same" phenomenon. Nevertheless, Thomas's discovery prompted me to start an email exchange with Roger Wilson. This is how I learnt about the archaeological sources presented below.

---

[1] We stick to the numbering of the rooms originally proposed by Gino Vinicio Gentili and followed in [9]. Brigitte Steger uses a different numbering in [8], for reasons probably best known to her.

[2] See https://imucentennial.math.unistra.fr (consulted 22 March 2022).

Fig. 2 The logo of the IMU (Courtesy IMU)

2 The Mathematician and the Archaeologist

Mathematicians like to know precisely what they are talking about. To achieve this they are even prepared, at least since the late nineteenth century, to replace concrete objects by ways of talking about them. Thus they may pass, for instance, from permutations or symmetries to group theory.

Thomas Delzant stopped in Room 18 of the Villa del Casale because he encountered the shape of a *Brunnian link*: the three closed curves, or loops, shown in the design are inseparably linked, but if any one of them is cut open and removed, the remaining two curves are no longer entangled; they can be separated from each other without cutting. Our "Borromean rings"—so called because such a configuration, wrongly suggesting it could be made of three flat rings, occurs more than a thousand years later in the code of arms of the House of Borromeo from Milan—are the simplest non-trivial example of a Brunnian link.

The name alludes to Hermann Brunn (1862–1939), a geometer and Arabist from Munich, who thought about knots and links between 1887 and 1897 and published four texts, the last one being the talk he gave at the very first International Congress of Mathematicians (ICM) held in Zürich in 1897. These papers contain little more than various suggestions to measure the complexity of a given knot.[3] Brunn's enthusiasm for knots and links may have been fanned by memories of his father, who was an archaeologist—a fact that incidentally explains why Brunn was born in Rome.

Independently of the concept of Brunnian link that orients the mathematician's appraisal of the pattern and sees it as a variant of the IMU logo, the mosaic in Room 18 is in the first place a concrete piece of craftsmanship, and a fine one indeed. As such, every detail of its execution matters in principle: the precise contours, the structure of each loop, the colors, the octagonal frame, the uniform background, brightly colored, and so forth. The mosaics at the Villa del Casale were in all likelihood composed in the second quarter of the fourth century CE by North African specialists from the Carthage region.[4] The particular motif that

---

[3] See the discussion of Brunn's work in [2], pp. 180–182.

drew Thomas Delzant's attention was evidently part of their repertoire. We know nothing about the significance it may have had for them, or for those who may have specifically ordered it.

This being said, there are (at least) two respects in which the craftsmanship of decorative mosaic floors differs from other pictorial art forms—think of Rembrandt's paintings, for example—where each individual tableau stands alone as a unique, valuable original work of its master, which is irreplaceable.

The first difference is particularly easy to explain to mathematicians: Contrary to a painting, any pixelization of which can be reasonably claimed to be inadequate, a mosaic pattern is basically a discrete object—even if the precise shape of the individual stones, and certain nuances of their colors, may be said to require very fine scales. In other words, any mosaic can in principle be coded as a finite sequence of data listing the pieces (stones) it consists of, their properties and their positions with respect to a given geometrical framework. In this sense, mosaics could provide an intriguing example of a hybrid art form situated somewhere between notational and non-notational arts in the sense of Nelson Goodman [3]. The catch with this observation, which renders it rather philosophical, is that nothing much seems to be known about the usage of such coding practices for mosaics, especially in antiquity.

The second difference—potentially more relevant for our purpose—is that one may try and catalogue standard patterns occurring repeatedly, say, in non-representational, decorative mosaic pavements from various parts of the Roman Empire throughout its long history. Such a project grew out of French initiatives proposed by archaeologists since the 1960s, and was finally published in two impressive volumes [1]. The thousands of black-and-white drawings catalogued there were painstakingly executed by hand in a coherent style, by Richard Prudhomme for the first volume, and after his death by Marie-Pat Raynaud for the second. Most of them are based on specific archaeological sites, which are duly indicated in footnotes. Yet, the drawings go a long way towards reducing the information to a basic discrete code underlying the concrete pavement in question, as envisaged above.

For example, the frame pattern in which our Brunnian link appears in Room 18 of the Villa del Casale is listed in Plate 177, motif **e** of [1], vol. I—see Fig. 3 below. The source given for this framework pattern is a pavement from Djemilla, Algeria.

However, the choice of the editors of this momentous work [1] is to order all these patterns, and to classify them in terms of labels that are formulated in a semi-terminological language. These descriptions are given first in French, and then translated into English, German, Italian, and Spanish. This produces rather elaborate descriptions. The floor patterns **d** and **e** shown in Fig. 3, for instance, are described in English like this:

> **d.** Polychrome orthogonal pattern of eight-pointed stars of two interlaced squares of simple guilloche, tangent at one angle, the cross-shaped interspaces bearing an inscribed eight-lozenge star (forming squares).

---

[4] This excludes the so-called bikini girls mosaic, which was installed later.

Fig. 3 R. Prudhomme's drawing of framepatterns, Plate 177, patterns **d** through **f**. Sample **e** is the type encountered in Fig. 1. Source: [1], vol. I, p. 275

e. Polychrome variety of **d**, the cross-shaped interspaces consisting of a central poised square ensconced in the angles, between four pairs of adjacent lozenges (creating the effect of an orthogonal pattern of adjacent octagons).

Such elaborate descriptions of mosaic patterns raise high expectations for the treatment that the authors reserve for our favorite Brunnian link in the second volume of [1], which is dedicated to "centered ornaments" (*décors centrés*). Alas, these expectations are frustrated because the early predecessor of the IMU's logo is listed on p. 43 of vol. II of [1] as one of the very last "elementary motifs." This whole list of *elementary motifs* is made up of 89 "simple" [1, pp. 34–38] and 132 "complex motifs" [1, pp. 38–43]. Setting such allegedly *elementary motifs* apart at the very beginning of the volume, the authors take the freedom of depicting them more like plain geometric patterns, rather than insisting on the mosaic texture. Accordingly, no indication of existing pavement sites is given where these *elementary motifs* can be found. Also the frequency of their respective occurrences is not discussed—see Fig. 4.

Furthermore, the Brunnian link is addressed there as a "knot of three figures of eight." This is devastating for the mathematician, whose whole appreciation of the motif is naturally based on three simple loops entangled in three-space in a particular way, whereas a single figure 8 is already the union of two loops. Testing a few friends (non-mathematicians), I did encounter the same confusion between a simple loop and a figure 8 when I showed them Fig. 1. Still, the least one can say is that introducing our way of writing the number eight introduces an anachronistic element into the description of this ancient motif from a Roman mosaic pavement. On the positive side, the description proposed in [1], vol. II, does recognize our link as a kind of knot.

Various other archaeologists have tried to improve on the formal classification of all those motifs. One of the co-authors of [1], Anne-Marie Gumier-Sorbets, independently proposed a computer-based classification of mosaic ornaments in

Nœud en lyre (nœud de S
affrontés)
  Knoten in Form einer Lyra
  (Knoten aus 2 gegenstän-
  digen S)
  Lyre-shaped knot (knot of
  facing S's)
  Nudo en lira (nudo de S
  afrontadas)
  Nodo a lira

Nœud de deux triangles
curvilignes
  Knoten aus 2 gekurvten
  Dreiecken
  Knot of two curvilinear
  triangles
  Nudo de dos triángulos
  curvilíneos
  Nodo di due triangoli curvi-
  linei

Nœud de trois huit
  Knoten aus 3 achtförmigen
  Gebilden
  Knot of three figures of eight
  Nudo de tres ochos
  Nodo di tre otto

Nœud de quatre huit
  Knoten aus 4 achtförmigen
  Gebilden
  Knot of four figures of eight
  Nudo de cuatro ochos
  Nodo di quattro otto

Nœud de deux trilobes
  Knoten aus 2 Dreipässen
  Knot of two trilobes
  Nudo de dos trilóbulos
  Nodo di due trilobi

Coquille
  Muschel
  Scallop
  Concha
  Conchiglia

« Plat d'argent »
  « Silberplatte »
  « Silver-plate motif »
  «Plato de plata»
  Motivo gemmato

Fig. 4 The last seven from the list of *elementary motifs*: [1], vol. II, p. 43

her thesis. Klaus Schmelzeisen in his thesis [7] tried to be more formal than his various predecessors. It was written with a view to accounting for a fixed corpus of Roman mosaics from North Africa. Schmelzeisen invented a new system, in which our Brunnian logo received the label 14L.03. We refer the interested reader to this thesis and to the literature cited there.

3   A Link That Everyone Recognizes as a Knot

A peculiarity of the Brunnian link that I noticed when I showed Fig. 1 to people, lies in the fact that it is not immediately seen by everyone as a knot.[5] As far as links represented in ancient mosaics are concerned, a rewarding comparison is provided by what has been known for centuries as *Salomon's Knot*, its variants and some of its generalizations. Salomon's Knot has just two loops—they may be drawn as circles or slightly deformed, sometimes into shapes with vertices—that are doubly interlocked, so that it is virtually impossible to form it with a finger and the thumb of each of one's hands. See Fig. 5 for several versions of this motif. It is present over the centuries and in many civilizations, possibly starting with a Mesopotamian clay tablet from Shurukapp dated to about 2600 BCE, where the knotted strings are represented as snakes, and continuing well beyond the Roman Empire.[6]

Eventually, not only the symbol but also its name must have been so well-known that it entered world literature as the emblematic representative of whatever is tightly tangled. Thus it is referred to in Dante Alighieri's *Tenzone* exchange of voluntarily insulting sonnets with his childhood friend Forese Donati. The first text, proposed by Forese, describes how its author had seen, the night before, Dante's (already dead) father tied up in a "knot of which I don't know the name; was it Salomon's or that of another sage. . . " In his riposte, Dante picks up this image: *Ben ti faranno il nodo Salamone . . .* , to explain that what was going to tie up and ruin Forese were in fact the partridge breasts and other yummy things he was devouring. Indeed, Forese will reappear in Dante's *Divine Comedy* on the sixth terrace of the Purgatory, where the emaciated gluttonous are punished with the repeated craving for delicacies that is stirred by the trickling of water on a tree and the odour of a blossom.[7]

The twofold link in Salomon's Knot is the fundamental difference between this motif and our logo. This difference is basic, both from the mathematical point of view and when it comes to describing things plainly. Given any two of the three loops in Fig. 1, they cross over each other four times, and each time it is the same loop which passes above, and the same which slips underneath the other. In Salomon's Knot, however, the relative position of the two components alternates from one crossing to the next.

---

[5] Just as the archaeological books quoted above, I use here the colloquial meaning of the word 'knot', not the mathematical term, which would signify the continuous image in space of a single line. The colloquial usage does not differentiate between knots and links.

[6] Cf. the introductory chapters of [5], especially pp. 23–31. Trying to deal with the ubiquitous presence of the motif, this book attempts to interpret it as a deeply rooted "symbol and archetype of the idea of union", by appealing for instance to Jungian psychoanalysis as well as to many other ideas, from ethnology to art history.

[7] See Dante's *Divine Comedy*, Purgatory, Canto 23; see also Canto 24.

Nœud de trois boucles
Knoten aus 3 Schlaufen
Knot of three loops
Nudo de tres bucles
Nodo triplice

Nœud de trois boucles à œillet
Knoten aus 3 ösenförmigen
Schlaufen
Knot of three loops with
eyelets
Nudo de tres bucles con
ojete
Nodo triplice a occhielli

Nœud en trèfle
Knoten in Kleeblattform
Trefoil knot
Nudo en trébol
Nodo trilobo

Nœud en *hedera*
Knoten in Herzblattform
Hedera knot
Nudo en *hedera*
Nodo a *hedera*

Nœud de Salomon
Flechtbandknoten (Salo-
monsknoten)
Solomon knot
Nudo de Salomón
Nodo di Salomone

Var., en boucles lâches
Variante, lockere Ausführung
Var., loosely tied
Var., de bucles sueltos
Nodo di Salomone allentato

Var., en boucles carrées
Variante, eckig gebildete
Schlaufen
Var., with squared loops
Var., de bucles cuadrados
Nodo di Salomone
quadrato

Double nœud de Salomon
Doppelter Salomonsknoten
Double Solomon knot
Doble nudo de Salomón
Doppio nodo di Salomone

Carré de sparterie (nœud de Salomon et carré à sommets arrondis, entrelacés)
Geflochtenes Quadrat
(Salomonsknoten und
Quadrat mit abgerundeten
Ecken, ineinandergeflochten)
Square of interlaced bands
(Solomon knot and square
with rounded points, inter-
looped)
Cuadrado entretejido (nudo
de Salomón y cuadrado con
vértices redondeados,
entrelazados)
Quadrato a stuoia (Nodo di
Salomone intrecciato a un
quadrato con spigoli arro-
tondati)

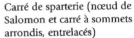

Carré de sparterie en boucles rondes et carrées (nœud de Salomon en boucles carrées et carré à sommets arrondis, entrelacés).
Geflochtenes Quadrat mit
gerundeten und quadrati-
schen Schlaufen
(Salomonsknoten in quadra-
tischer Schlaufenform und
Quadrat mit abgerundeten
Ecken, ineinandergeflochten)
Square of interlaced bands
with rounded and squared
loops (Solomon knot of
squared loops and square
with rounded points, inter-
looped)
Cuadrado entretejido de
bucles redondos y cuadra-
dos (nudo de Salomón de
bucles cuadrados y
cuadrado con vértices
redondeados, entrelazados)
Nodo di Salomone quadrato
intrecciato ad un quadrato
con spigoli arrotondati

Fig. 5 Variations of Salomon's Knot, from the list of *elementary motifs*: [1], vol. II, p. 42

Given the wide circulation of Salomon's Knot, especially in Roman mosaics from various periods and parts of the Empire, could this concrete mathematical difference between the two designs explain the relative rarity of our Brunnian link and logo? In our email exchange, Roger Wilson pointed to three occurrences in mosaics from Jordan, which date from the sixth, resp. the fourth century CE.[8] Each of these altogether four mosaic representations of our pattern, from Sicily and Jordan, predate the earliest occurrences that have made it into Wikipedia so far: the triangular Viking version of the link from Lärbro, Sweden, and the Christian usage of a ring arrangement to represent the Holy Trinity.[9]

4   The Brunnian Link and the Pomegranate

The most difficult question regarding our pattern is asked last: What does our motif signify, for those who created and those who commissioned these mosaics at the time, and what does it mean for us today? On the IMU website quoted above, we read that the new IMU logo "represents the interconnectedness not only of the various fields of mathematics, but also of the mathematical community around the world." This is certainly well intentioned. It also remains rather vague. And one immediately wonders how to read the salient feature of the Brunnian link, which is the reason why mathematicians are fascinated by it: that it suffices to cut just one loop in order to make it fall apart.

Maybe one should interpret each loop as a separate path of promoting international cohesion in mathematics. Then one loop could stand for the history of the International Congresses of Mathematicians (ICMs), which started as early as 1897 in Zürich. Another loop would represent the IMU, which was first founded in 1920 and fell apart shortly afterwards, as a consequence of the Bologna ICM in 1928.[10] The new IMU founded at the beginning of the 1950s still exists today and has successfully established itself as the dominating actor of the global mathematical community, controlling the ICMs as well as the procedures to choose the winners of the Fields Medals and several other prizes.

---

[8] See [4], No. 136 (p. 125) from the crypt of Saint Elianus in Madaba, Jordan (see Fig. 6 below); No. 566 (p. 295) from the Church of Bishop Isaiah at Jerash, Jordan; and No. 684 (p. 328) from the Byzantine baths at Gadara, Jordan. The last one is the one among these three mosaics that dates from the fourth century.

[9] See https://en.wikipedia.org/wiki/Borromean_rings (consulted 25 April 2022); cf. [5], Fig. 143 (p. 119) and Fig. 196 (p. 164).

[10] See [6], Sec. 4.4.3.

And the remaining third loop would reflect the structural evolution of the sciences at large, and mathematics in particular, which has made the functioning of the international community possible. This includes the professionalization of mathematics in the universities during the nineteenth century, the growth and structural evolution of the mathematical publications, but also, for instance, the influence of American philanthropy, especially through the Rockefeller Foundation, for international connections in the field of mathematics in the 1920s and 1930s. Only when all three axes are active is the link intact.

But what can we learn from looking back? The Brunnian link in Room 18 of the Villa del Casale—at least if we just look at this motif, and not at the ambient pavement—seems as pure and crystalline as possible, not suggesting any significance beyond the intertwining of the three differently colored loops. Compared to this, the version from the crypt of Saint Elianus in Madaba, Jordan, shown in Fig. 6 is much more generous: it places various fruits in the open spaces left by the design, with a pomegranate proudly placed in the center. The whole motif now looks rather like a basket full of fruit, braided with the Brunnian link as its basic element.

The pomegranate has a long history as a symbol with many meanings. Thus it happened, for instance, at the end of the last century that the "pomegranate was chosen as the logo for the Millennium Festival of Medicine from a shortlist that included DNA, the human body, and a heart beat. Not only has the pomegranate been revered through the ages for its medicinal properties but it also features in the heraldic crests of several medical institutions involved in the organisation of the festival."[11]

Avoiding such vindications that refer to other logos, and sticking to the pictorial evidence of old mosaics, I would like to suggest that the Brunnian link in its version from Madaba should be understood as a delicate structure that holds together valuable fruits, i.e., results obtained, as well as ever more seeds for further growth.

---

[11] See https://www.ncbi.nlm.nih.gov/pmc/articles/PMC1118911/ (consulted 27 April 2022).

Fig. 6 Pavement from the crypt of Saint Elianus in the Church of the Prophet Elias at Madaba, Jordan. Source: [4, p. 125]

References

1. C. Balmelle, M. Blanchard-Lemée, J. Christophe, J.-P. Darmon, A.-M. Gumier-Sorbets, H. Lavagne, R. Prudhomme and H. Stern. *Le décor géométrique de la mosaique romaine.* Vol. I: *Répertoire graphique et descriptif des compositions linéaires et isotropes.* Vol. II *Répertoire graphique et descriptif des décors centrés.* Picard Paris (1985) (vol. I), (2002) (vol. II).
2. M. Epple. *Die Entstehung der Knotentheorie. Kontexte und Konstruktionen einer modernen mathematischen Theorie.* Vieweg Braunschweig (1999).
3. N. Goodman. *Languages of Art. An Approach to a Theory of Symbols.* The Bobbs-Merrill Co. Indianapolis & New York (1968).
4. M. Piccirillo, P.M. Bikai and T.A. Dailey. *The Mosaics of Jordan.* American Center of Oriental Research, Amman (1993).
5. U. Sansoni. *Il nodo di Salomone. Simbolo e archetipo d'alleanza.* Electa Milan (1998).
6. N. Schappacher. *Framing Global Mathematics. The International Mathematical Union Between Theorems and Politics.* Springer Cham (2022). https://doi.org/10.1007/978-3-030-95683-7
7. K. Schmelzeisen. *Römische Mosaiken der Africa Proconsularis. Studien zu Ornamenten, Datierungen und Werkstätten.* Peter Lang Frankfurt (1992).
8. B. Steger. *Piazza Armerina : la villa romaine du Casale en Sicile.* Antiqua 17. Picard Paris (2017).
9. R.J.A. Wilson. *Piazza Armerina.* Grafton London (1983).

Max Dehn as Historian of Mathematics

David E. Rowe

Personal Note

It gives me pleasure to dedicate this paper to Catriona Byrne for her many years of engagement on behalf of mathematics and its history.

1 Introduction: Biography and History

Compared with nearly any other field of knowledge, mathematics has an extraordinarily long and rich history. From time to time scholars have also avidly studied the mathematics of the past, and in some cases they took inspiration from it to invent something novel. A particularly striking example came about after 1588 when the eight books of the *Collection* of Pappus of Alexandria were published in Venice in the Latin edition prepared by Federigo Commandino. Pappus lived around 300 A.D., so some 500 years after the high water mark of ancient Greek mathematics and, as Thomas Little Heath remarked, his compendium was "obviously written with the object of reviving classical Greek geometry" [17, 2: 357]. Aside from the major works of Euclid, Archimedes, and Apollonius, the *Collection* is the most important mathematical text we possess from the ancient classical world. Indeed, without this text and its commentaries historians would never have been able to imagine the scope of the Greek tradition of geometrical problem solving. Pappus, to be sure, was not an inventive mathematician on the level of his predecessors; in fact, it would be more apt to think of him as an early historian of mathematics. Several of

D. E. Rowe (✉)
Mathematics Institute, Mainz University, Mainz, Germany
e-mail: rowe@mathematik.uni-mainz.de

© The Author(s), under exclusive license to Springer Nature Switzerland AG 2023   337
J.-M. Morel, B. Teissier (eds.), *Mathematics Going Forward*, Lecture Notes in Mathematics 2313, https://doi.org/10.1007/978-3-031-12244-6_24

those who read him, on the other hand, were very formidable mathematicians, two of them being Descartes and Newton. To immediately appreciate the significance of Pappus's *Collection* for the flourishing of European mathematics in the seventeenth century, one needs only to read Henk Bos's insightful study of Descartes's *La Géométrie* [1].

During the early twentieth century, a resurgence of interest in history of mathematics came to fruition within the German mathematical community. Otto Neugebauer, who largely managed Richard Courant's Mathematics Institute in Göttingen, undertook pioneering research on Babylonian mathematics and astronomy. He and Otto Toeplitz, who taught in Kiel before moving to Bonn, founded the Springer journal *Quellen und Studien zur Geschichte der Mathematik Astronomie und Physik* in 1929. Meanwhile, in Frankfurt, Max Dehn was running a weekly seminar that studied ancient and early modern mathematical texts in their original languages. These sorts of studies took place during the short-lived era that saw the flowering of Weimar culture, which ended all too abruptly in 1933 when Hitler came to power. Neugebauer, Toeplitz, and Dehn all fled Nazi Germany, the latter two under threat to their lives. Historical research in mathematics continued in Germany, but much of it was thereafter colored by a nationalist or even explicitly racist agenda, led by the efforts of the Berlin mathematician Ludwig Bieberbach (Segal [39, pp. 334–417]). Although he lost all his positions after the fall of the NS-regime, Bieberbach maintained certain connections with influential figures who shared his interest in promoting historical studies of mathematics in Germany.

During the postwar era, the heyday of Bourbaki, a new wave of historical interest arose in the West. Axiomatization, rigor, purism, structuralist concepts—these watchwords of modern mathematics deeply affected the way mathematicians came to see but also to judge the mathematics of the past. The Bourbaki project itself had modest beginnings, but with time its goal was to canonize the fundamental structures in those theories which the group considered the established core theories of modern mathematics. In this sense, Bourbaki was only incidentally interested to look backward and identify when these ideas first arose. It would appear doubtful that the historical notes, which Bourbaki included in the volumes of *Éléments de mathématique* [2] and which were later gathered together in [3], generated great interest among mathematicians or historians of mathematics. Their intent, after all, was essentially just to provide a larger account of the intellectual context connected with the topics covered in the *Éléments*. For students of mathematics with a modicum of interest in the subject as intellectual history, many of these notes are still well worth reading today. That goes without saying, of course, for André Weil's history of number theory [47].

By the 1970s, though, a handful of scholars who pursued history of mathematics from other perspectives began to publish work that Weil, in particular, found distasteful or worse. Those partisan battles from long ago need not concern us here, but one aspect has real significance for the theme of this essay. Bourbaki represented a purist movement that hoped to canonize a certain body of mathematics, which contemporary mathematicians—or those who considered themselves to be well-rounded—would acknowledge as core knowledge. Jean Dieudonné described this objective very clearly in a lecture he delivered in 1968, later translated in [11]. This canonization naturally lent itself to a highly selective view of the past, a style of historiography that Ivor Grattan-Guinness called "the royal road to me." In that respect, it is worth noting that Dieudonné thought quite highly of Klein's lectures on nineteenth-century mathematics [24], as did Neugebauer and others (Rowe [34, pp. 32–33]). They generally lauded his account, in particular Klein's strikingly subjective remarks laced with autobiographical anecdotes. The contrast with the dry factual information in Bourbaki [3] is striking, but of course the name index in the back of that volume reveals a very clear Bourbakian image of the history of mathematics. To exaggerate only a little, this style of historiography judged the past almost exclusively from the standpoint of the present. Moreover, the names that appeared throughout were those credited with an important new idea or result. History of mathematics was thus reduced to a certain impressive chain of disembodied ideas. Who produced those ideas and why they put them into circulation were questions that went largely unasked, and these "mathematical people" never emerged from the shadows. Otto Neugebauer, I'm quite sure, thought that serious history of mathematics had nothing to do with the personal lives of mathematicians, but all that began to change in the 1970s.

It was also during that decade that three editors at Springer—Alice and Klaus Peters and Walter Kaufmann-Bühler—launched a newsletter they called "The Mathematical Intelligencer" (on its history, see Senechal [40]). The world of mathematics was still rather small in those days, but large enough that Springer's mailing list put "The Old Intelligencer" (as it was later called) into the hands of a few thousand mathematicians. The rest, as they say, is history, and today's glossy magazine bears practically no resemblance to those early issues. Throughout its nearly 50-years, *MI* has kept pace with new trends, emerging and older communities, subcultures, crossovers with the arts and sciences, etc., etc. History and biography played a large part as well, all part of a complex unfolding of varied interests in the realm of mathematical culture.

Here I'd like to offer some brief reflections on the life and work of Max Dehn (1878–1952), stressing, in particular, his interests in the history of mathematics. Dehn was remembered often in the pages of *The Mathematical Intelligencer*, beginning with an essay about the man and his work written by his former student, Wilhelm Magnus [27]. In the 1980s, John Stillwell translated Dehn's most important papers and published these with commentary in [10]. Since that time, Dehn's name and fame have only grown. This essay is adapted from parts of a forthcoming book, *Max Dehn: A Polyphonic Portrait* (Lorenat et al. [26]).

2 Dehn in Frankfurt

In 1921, following complex negotiations, Max Dehn assumed the professorship formerly held by Ludwig Bieberbach in Frankfurt. Founded in 1914 as a privately endowed institution (*Stiftungsuniversität*), Frankfurt University hired many more scholars of Jewish background than nearly all the older German universities. In mathematics, Frankfurt's senior mathematician, Arthur Schoenflies, took full advantage of this situation during the first years of the Weimar Republic. When Dehn joined the faculty as its second full professor in mathematics, three others held positions as associate professors: Ernst Hellinger (appointed in 1914), Otto Szász, and Paul Epstein. These five Frankfurt mathematicians were all ethnic Jews, though after Schoenflies's retirement in 1922 his chair went to Carl Ludwig Siegel, the only non-Jew in this tightly-knit group.

In 1924, Dehn became head of the Frankfurt Mathematics Seminar, a post he held up until 1935 when his position was terminated, forcing him into early retirement. What unfolded under his leadership was a community of scholars who worked together in an atmosphere largely free from the competition and rivalries typical at other leading universities. In Frankfurt, the watchwords were cooperation and harmony. Those idyllic years were later memorialized by Carl Siegel, the community's last surviving member in [42], a lecture he delivered on 13 June 1964 in the Frankfurt Mathematics Seminar.

Soon after his arrival, Dehn decided to launch a private reading circle, a *Lesekränzchen*, that would long be remembered by all who attended. This group devoted its attention to the study of classical mathematical texts in their original languages, in particular Greek and Latin works written by, among others, Euclid, Bombelli, Cavalieri, Kepler, Roberval, Wallis, Huygens, Barrow, Newton, Leibniz, and Euler. The entire mathematics faculty took part in these gatherings as a truly communal undertaking, even though Dehn was its acknowledged *spiritus rector*.

Shortly before the Nazi Party came to power, Dehn gave a lecture to the German Mathematical Society, "Problems in Post-Secondary Teaching of Mathematics" [8]. This gave him the opportunity to speak about some of the unique features of the Frankfurt program, in particular its history of mathematics seminar. Among its several qualities, he laid stress on a humanistic virtue, namely, the sense of humility one gains through a deeper appreciation of the intellectual achievements of one's forebears. "Studying the development of mathematics, steadily, deeply, and without haste together with close colleagues," he wrote, "makes every mathematician more mature and fills him with a more human love of his science." At the same time, Dehn had no illusions about the effectiveness of this special seminar as a teaching tool. Most students lacked the necessary linguistic skills, but even more, the intellectual patience required to delve into difficult texts. He also noted very aptly that mathematical and historical thinking tend to run in opposite directions. Over the course of 10 years, he doubted whether more than a half-dozen students had gained anything of lasting value from the seminar. This telling remark clearly suggests that its true purpose was *Fortbildung*, i.e., cultural enrichment for the faculty and a few older teachers from the surrounding community.

Students were naturally encouraged to participate in the Frankfurt historical seminars as well, though not many possessed the requisite language skills to do so. Dehn thought only a rare few truly profited from the experience. A young astronomy student named Willy Hartner, who later founded Frankfurt's Institute for History of Science, recalled in 1981 how much he regretted never having participated regularly in Dehn's seminar. Hartner possessed the necessary prerequisites—that unusual mixture of philological and mathematical talents—but he admitted that in 1922, the year he first met Dehn, he had not yet discovered his interest in history (he was only 17 at the time). Nevertheless, he shared some vivid memories of the contrasting styles of Dehn and Hellinger as teachers:

> Anyone who, like me, ever heard Ernst Hellinger's differential and integral calculus and other lectures will have remembered well into old age his almost unequaled mastery. Today educational methods are very much in fashion, but I am sure Hellinger never bothered with such theories; with him it was as if a friendly fairy had put that in his cradle.
>
> Max Dehn embodied a completely different type of brilliance. In contrast to Hellinger, he loved to improvise and abandon himself to the overflow of thoughts storming through him. With all due acknowledgment of his mastery, this proved a bit difficult for us, his inexperienced listeners. Feeling very despondent, I asked him for a brief interview. It lasted a good two hours spent in the professors' cafeteria, where one drank miserable inflationary coffee at a price of about a billion marks a cup. I was pleasantly surprised that Dehn responded to my request without any sign of annoyance. The rest of the conversation was about very different things—art, music, languages, classical and modern, about history, and finally also about the political situation. It was the beginning of a lifelong friendship that we preserved in even more difficult times. (Burde, Schwarz and Wolfart [4, pp. 23–24])

Among the students who regularly attended this mathematics history seminar, one in particular stood out from all the rest—Adolf Prag, whose later career in some ways mirrored Dehn's. Not only did Prag's life crisscross with those of Max Dehn and his two daughters, Eva and Maria, but he also went on to play a singular role in historical studies devoted to the mathematics of the seventeenth century.

Prag was born in 1906 in a small village on the edge of the Black Forest, but soon thereafter his family moved to Frankfurt, where he attended the humanistic Goethe Gymnasium. There he acquired a solid grounding in classical languages that he would cultivate throughout his life. From 1925 to 1929 he studied mathematics at Frankfurt University, where he became a mainstay in Dehn's history of mathematics seminar. As Christoph Scriba later imagined the situation:

> Dehn, with his wide historical and philosophical interests, must have sparked a congenial vein in Prag. In addition, the outstanding linguistic abilities of this student, who was able to translate Latin and even Greek texts fluently into the German language, were a welcome asset for the discussions of this circle. [38, p. 410]

During this time, a lifelong friendship developed between Prag and two of Dehn's best students, Ruth Moufang and Wilhelm Magnus.

After completing his studies, Prag still needed to pass the state examination for teaching candidates and submit a thesis (*Staatsexamensarbeit*). For a topic he went to Dehn, who suggested that he write about the Oxford mathematician John Wallis, whose work Prag had studied in the seminar. The resulting thesis was so impressive that Dehn sent it to Otto Neugebauer, who published it in his new series *Quellen und Studien zur Geschichte der Mathematik* [31].[1] Years later, Christoph Scriba took up

---

[1] The published version, however, omitted a chapter on the Pell equation.

research on Wallis, a project that led to a close personal connection with Prag (see below).

As a Jew, Prag had no chance of gaining a position at a state-run school, so in 1931 he accepted a post at a private Jewish school in Herrlingen (Württemberg) run by Anna Essinger, a remarkable educator. Sensing early on that her undertaking had no future in Germany, she obtained support from Quaker organizations in 1933 to move the school to Kent, England. There at Bunce Court, a large house near Faversham, Prag continued teaching, and he later became deputy head of the school. The two daughters of Max and Toni Dehn attended this same school, where their father also taught from January to April 1938. In the spring of 1937, Frede Warburg, daughter of the well-known art historian Aby Warburg, joined the school staff, and the following year she and Adolf Prag wed. They survived the difficult times that lay ahead and died only months apart 65 years later in 2004. In the final section, I will briefly discuss Prag's singular role in the historiography of early modern mathematics.

Occasionally, visitors attended the Frankfurt historical seminar, one being André Weil, who vividly recalled the impression Dehn left on him:

> A humanistic mathematician who saw mathematics as one chapter—certainly not the least important—in the history of human thought, Dehn could not fail to make an original contribution to the historical study of mathematics, and to involve his colleagues and students in the project. This contribution, or rather this creation, was the historical seminar of the Frankfurt mathematics institute. Nothing could have seemed simpler or less pretentious. A text would be chosen and read in the original, with an effort to follow closely not only the superficial lines but also the thrust of the underlying ideas. ... It was only later that I attended it, on subsequent visits to Frankfurt, a place I made a point of visiting as often as I could. I am not sure whether it was already in the summer semester of 1926 that, during a seminar session devoted to Cavalieri, Dehn showed how this text had to be read from the viewpoint of the author, taking into account both what was commonly accepted in his lifetime and the new ideas that Cavalieri was trying to the best of his ability to implement. Everyone participated in the discussion, contributing what he could to the group effort. [46, p. 52]

Weil was also very struck by the radically different atmosphere in Richard Courant's Göttingen [46, pp. 52–53]. He recalled, in particular, how he learned very little in conversations with those in Courant's own group. Nearly every time he got talking with one of them, the exchange would end rather abruptly with a remark like, "sorry, I have to go write a chapter for Courant's book" [46, p. 51]. There was a distinct awareness in Göttingen that Max Dehn and Carl Ludwig Siegel, both of whom thought of mathematics as an art form, were cultivating an approach to research in Frankfurt that stood in conscious opposition to the Göttingen model. Siegel's main hobby during these years was painting, especially impressionistic landscapes. After coming to Frankfurt, he lived at first with the painter Fritz Wucherer and his family in Kronberg, a wealthy town in the idyllic Taunus region northwest of Frankfurt. Wucherer owned an impressive villa and belonged to an artists' colony in Kronberg. He was well known for his landscape paintings, and for some time Siegel took lessons from him.

Otto Neugebauer, who served as Courant's "floor manager" at the Mathematics Institute in Göttingen, was certainly sensitive to the implicit criticism coming from

Frankfurt.[2] Neugebauer played a central role in designing the institute's new quarters, built with funding from the Rockefeller Foundation. When it opened in December 1929, Hermann Weyl delivered a lecture honoring Felix Klein, who had long dreamed of housing mathematics in such a building. Neugebauer, on the other hand, was eager to describe the physical arrangements as an inviting place for teaching staff and students to gather and meet. "We hope and believe," he wrote, "that the new mathematics institute will *not* provide new impetus for the "mechanization" of science, as so often prophesied, ... but rather will offer a workplace, where one can *enjoy* teaching and learning and, above all, the pursuit of pure science" [29, p. 4].

Courant's Göttingen was a multi-faceted enterprise, but at its heart flourished a "publish or perish" culture that stood as the antithesis of the one cultivated in Frankfurt. Indeed, one of the striking features of the latter was how little Dehn and Siegel chose to publish once they began working together. This hardly meant that they were unproductive, however; nor did they lack ambition. In fact, their decision to withdraw from this arena stemmed from a shared understanding that "more was not better"—real progress would take place outside the "mathematical factories," which were for producing and disseminating such an abundance of new results that contemporary mathematicians found themselves drowning in their own literature.

André Weil remembered Dehn invoking just this image when he visited Frankfurt around Christmas of 1926. Mathematics, Dehn told him,

> was in danger of drowning in the endless streams of publications; but this flood had its source in a small number of ideas, each of which could be exploited only up to a certain point. If the originators of such ideas stopped publishing them, the streams would run dry; then a fresh start could be made. To this purpose, Dehn and his colleagues refrained from publishing. (Weil [46, p. 53])

This view probably comes closer to Siegel's attitude than to Dehn's, if only because the latter was a born teacher and collaborator, famous for his generosity in sharing fresh ideas to help others.

Dehn's seminar proved to be deeply inspirational for Siegel, whose singular ability to attack truly formidable problems in number theory was becoming legendary (Yandell [51, p. 208]). He was surely long intrigued with the mysterious results Riemann had communicated in his 8-page paper on the zeta-function, which no one had been able to unravel. With the assistance of his friend, Erich Bessel-Hagen, he set to work studying Riemann's unpublished notes related to the distribution of primes, a question that Riemann's teachers, Gauss and Dirichlet, had studied before him. Siegel worked on this topic, off and on, for several years.

On 6 November 1927, he composed a 10-page manuscript that dealt with Riemann's ideas, though he clearly never intended this text for publication. Instead, he gave it to Max Dehn, no doubt as a birthday present, as Dehn turned 49 on 13 November of that year. At the end of the manuscript, he even added a humanistic touch to fit the occasion. Figure 1 shows Siegel's portrait of Riemann along with

---

[2] On Neugebauer's early career, see Rowe [34].

some lines from a famous poem, "Friede mit der Welt" (Peace with the World) by
Friedrich Rückert,[3] which he found among Riemann's notes:

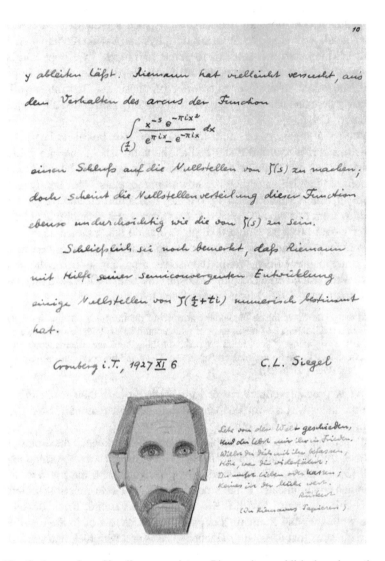

Fig. 1 The final page from Siegel's manuscript on Riemann's unpublished work on the zeta-
function. Dehn Papers, Dolph Briscoe Center for American History, University of Texas at Austin

---

[3] Rückert's poetry was set to music by numerous famous composers; best known among these
works are the "Kindertotenlieder" in the composition by Gustav Mahler.

Lebe von der Welt geschieden,
Und du lebst mit ihr in Frieden.
Willst du dich mit ihr befassen,
Höre, was dir widerfährt!
Du musst lieben oder hassen;
Keines ist der Mühe wert.

(Live apart from the world,
And you live with her in peace.
Should you want to engage with her,
Hear, what shall befall you!
You must love or hate;
Neither is worth the effort.)

Siegel's research project eventually led to his reconstruction of the Riemann-Siegel Formula, published in *Quellen und Studien* [41]. H.M. Edwards summed up his accomplishment with these words:

> The difficulty of Siegel's undertaking could scarcely be exaggerated. Several first-rate mathematicians before him had tried to decipher Riemann's disconnected jottings, but all had been discouraged either by the complete lack of explanation for any of the formulas, or by the apparent chaos in their arrangement, or by the analytical skill needed to understand them. One wonders whether anyone else would ever have unearthed this treasure if Siegel had not. [12, p. 136]

In January 1928, Max Dehn addressed a large audience at Frankfurt University when he spoke about "The Mentality of the Mathematician" [6], a speech Abe Shenitzer later translated for readers of *Mathematical Intelligencer* [9]. Dehn spoke on a ceremonial occasion, namely the annual celebration of the founding of the modern German nation in January 18, 1871. Since he had to approach this topic from some higher plane, though, he chose to illustrate what he hoped to convey by appealing to history, even going back to ancient times.

Certainly the views Dehn expressed in "The Mentality of the Mathematician" cast considerable light on the speaker's own quite unique way of thinking. His first and most immediate task was to assure his listeners that mathematicians were engaged in a creative activity. For "the layman often thinks that mathematics is by now a closed science, and gives little thought to the origin of the discipline he is familiar with from school." Dehn spoke of the sense of divine inspiration that ancient Greek mathematicians felt after making a profound discovery, and how "Eratosthenes and Perseus, in the manner of winners in an Olympic competition, made votive offerings out of joy at attaining their goals." Turning to early modern times, he talked about Cardano's wild urge to work out all the various types of solutions of cubic equations in his *Ars Magna*, but he also made clear that mathematical knowledge had to be clarified and communicated to have a decisive impact. This was particularly evident in the case of Descartes, who fashioned himself as having made a great new discovery—a method for systematically solving geometrical problems by reducing them to algebraic equations—when, in fact, he had mainly brought forth a known method with exceptional clarity.

Dehn's admiration for Descartes' accomplishments did not extend to his person, however, as this great French thinker was extremely impressed by his own sense of superiority and gloated over what he had accomplished. For Max Dehn, Gerolamo Cardano was a far more sympathetic figure, as can be seen from this passage:

> Cardano, who died in 1576 at the age of 75, was a typical man of the Renaissance. In view of our present topic—the creative power of the mathematician—Cardano is of special interest to us. His productivity was unbelievably extensive. Ninety years after his death, ten large folios of his work appeared, and the publisher assured readers that this was only half of what Cardano had written. There is no area between heaven and earth that he left untreated. He wrote about all the natural sciences, medicine, astrology, theology, philosophy and history. His autobiography—which Goethe compared to Benvenuto Cellini's—has great charm. In it he describes with touching ingenuousness a life afflicted with manifold misfortunes. At times we are strongly reminded of Rousseau's *Confessions*. Goethe writes at length about Cardano in his history of the science of color—about his talent, his passion, his wild and confused state that always comes to the fore .... [9, p. 20]

Turning to Dehn's seminar, one can easily see that the choice of texts was largely confined to classical antiquity and the period in early modern Europe leading up to the emergence of the calculus in the works of Newton and Leibniz.[4] Siegel thus recalled spending a number of semesters studying works by Euclid and Archimedes. Another block of texts dealt with developments in algebra and geometry from Leonardo of Pisa and Cardano to Viète, Descartes, and Desargues. Finally, the seminar looked carefully at texts documenting the emergence of infinitesimal calculus over the course of the seventeenth century, especially key authors associated with the British tradition: Wallis, Gregory, Barrow, and Newton. This overall plan was thus entirely conventional; yet even so, knowing in advance what one expected to find in an older mathematical text was usually of little help when it came to reading and *actually understanding* such works *in detail*.

Some three decades later, when Carl Siegel returned to Frankfurt to speak about the times he shared with his former colleagues there, he had this to say about their history of mathematics seminar:

> As I look back now, those communal hours in the seminar are some of the happiest memories of my life. Even then I enjoyed the activity which brought us together each Thursday afternoon from four to six. And later, when we had been scattered over the globe, I learned through disillusioning experiences elsewhere what rare good fortune it is to have academic colleagues working unselfishly together without thought to personal ambition, instead of just issuing directives from their lofty positions. [42, p. 226]

---

[4] Protocol books from Dehn's seminar are in the possession of the Frankfurt University Archives.

3 Dehn on the History of Geometry

Max Dehn's contributions to the literature on history of mathematics came mainly in the form of essays and occasional articles. His single most impressive piece was a six-part appendix to the third edition of Moritz Pasch's classic monograph *Vorlesungen über die neuere Geometrie* [30]. The second edition of Pasch's book, published by Teubner in 1912, had long been out of print. During the postwar era, after Springer assumed Teubner's former role as the leading German publisher of mathematical texts, Courant's "yellow series" often published older standard works in an updated form. Pasch was already approaching 80, so he was in no position to produce a substantially new edition, but Courant was surely more than pleased when Dehn agreed to write an appropriate supplementary appendix.

Some 5 years later, Courant turned once again to Dehn to request a supplement for a new edition of Arthur Schoenflies' textbook on analytic geometry [35]. Dehn's six appendices to [7, pp. 298–411] not only offered an overview of foundations and a modern treatment of linear algebra, it also contained a brief historical overview as well as a section on still unsolved problems in analytic geometry. In short, this material made the book far more than simply an elementary textbook. Here, as well as in the case of Pasch's book, Dehn drew on material he had developed for his courses in Frankfurt. This circumstance is reflected in his preface to Pasch [30], where he wrote: "The appendix corresponds approximately to a two-hour, one-semester lecture course, in which the instructor reports on what he considers to be all the more important questions, discussing the most important problems in detail, and above all seeking to stimulate independent study and the reading of classical works" [5, p. viii].

Among the classics in the history of geometry that Dehn had in mind, two were preeminent: Euclid's *Elements* and Hilbert's *Grundlagen der Geometrie* [18], which in 1922 was published in a 5th edition containing several new supplements. As for the significance of Pasch's original text from 1882, Dehn described this as marking the end of a quest to derive projective geometry from purely elementary principles, formulated in a complete system of axioms that avoids appealing to congruence properties or notions of continuity, such as the Axiom of Archimedes [5, p. 188]. Hilbert's axiom system, on the other hand, stood closer to the original system of Euclid, which made it possible to analyze which parts of geometry were susceptible to an elementary treatment and which were not.

Dehn's approach in this survey was largely systematic, though he added footnotes containing brief historical remarks coupled with references. The first question he raises is the role of the parallel postulate in ancient Greek geometry, a problem compounded by philological difficulties. In most of the extant manuscripts this postulate appears under the "common notions," which textbook authors usually referred to as axioms to distinguish these from the strictly geometrical postulates. The "parallel postulate" was then given as Axiom 11 in these texts (in the English tradition, following Robert Simson, it was Axiom 12). The Danish historian of mathematics Johann Heiberg argued that this was due to the editorial intervention of Theon of Alexandria who, according to Heiberg, had removed the fifth postulate from its original position and placed it under the "common notions." In preparing the modern Greek/Latin edition, Heiberg restored the postulate to what he believed was its original place. He found it listed as the fifth postulate in an older non-Theonine manuscript housed in the Vatican Library, which he took as his principal *Urtext* in preparing the modern edition. The English translation published afterward by T.L. Heath [17] follows Heiberg's edition almost without exception. In Dehn's day, these were very recent events, though today few realize that the parallel postulate has only been called Euclid's fifth for little more than a century. In several places, when discussing Greek mathematics, Dehn made similar comments about difficulties arising from a dearth of historical source material.

The logical or mathematical status of the parallel postulate long remained one of the most famous of all geometrical mysteries. Pasch's work put the last touches on projective geometry, a theory in which the properties of parallel lines play no role. Alongside those developments, however, more subversive thinkers—Lobachevsky and Bolyai—staked out arguments for a new theory of geometry in which parallel lines no longer satisfy Euclid's fifth postulate. Although it took several decades for mathematicians to embrace non-Euclidean geometry, once they did so, the contingent status of the parallel postulate became clear: Euclidean geometry was only a special case. Indeed, among the infinitely many possible spaces of constant curvature, Euclidean geometry was the one in which that constant was zero. Dehn's discussion took up the connection between non-Euclidean geometries and projective geometry, an insight Felix Klein recognized once he learned about the possibility of obtaining a general projective metric, a technique Arthur Cayley used to derive Euclidean geometry. Dehn also briefly noted how Riemann's notion of a manifold with local curvature properties led to the natural question of the various possible global extensions, a problem that led to Clifford–Klein space forms.

Dehn sketched these various topics quite rapidly before turning to problems underlying the foundations of projective geometry. Here he focused on the difficulty of providing a logically sound and complete construction of coordinate systems in projective geometry. Dehn distinguished between an older, more intuitive approach that depended on the Archimedean axiom and the purely projective methods developed by Pasch. From Desargues theorem—which follows immediately from the incidence axioms for points, lines, and planes in space—one can easily generate a network of rational points in the plane by iterating the construction of a fourth harmonic point for every triple. Pasch then found a way to extend this construction to irrational points by invoking a projective substitute for the Archimedean axiom.

These brief remarks then led over to Dehn's main topic, which begins with a modernized account of Hilbert's approach to segment arithmetic based on the two lines theorem of Pascal (Pappus's theorem) and the theorem of Desargues. His treatment of these, however, draws on elementary group theory for geometric transformations, leading to a proof that the fundamental theorem of projective geometry entails both theorems, Pascal as well as Desargues. Dehn also gave a proof of Hessenberg's theorem, namely that the planar theorem of Desargues follows from Pascal's. The individual achievements of others (Staudt, Wiener, Hilbert) are only mentioned in a footnote, and Dehn caps off this section with a schematic chart providing an overview of the relative dependence of the various axioms and fundamental theorems. All of this reflects Hilbertian interests, except for the appeal to group theory, where for details he points to Schwan [37]. The author of this study was a Gymnasium teacher in Düsseldorf, who went on to write his dissertation under Max Dehn.[5]

Following this overview, Dehn presents a section containing proofs of the key theorems. He emphasizes that one must first prove Pascal's theorem without recourse to continuity, and he begins with a synopsis of the original proof given by Friedrich Schur in [36]. This proof made essential use of a beautiful idea first discovered by Germinal Pierre Dandelin in connection with conics that lie on a hyperboloid of one sheet, thus a quadric surface generated by two systems of lines. Dandelin showed that a spatial hexagon obtained by connecting 6 points along corresponding generators, as these alternate between the two families, leads to a so-called Brianchon point, the common intersection point of the 3 diagonals.[6] The dual incidence relation follows as well, and taking a plane section of the quadric then leads to a conic with an inscribed hexagon that satisfies Pascal's theorem. Dehn not only credited Hermann Wiener with having brought out the significance of the theorems of Desargues and Pascal for foundations of geometry, he also emphasized how Schur's proof of Pascal's theorem was inspired by Dandelin's older ideas. These enabled Schur to prove the two-line version of Pascal's theorem, the case required for a commutative segment arithmetic (Dehn [5, pp. 228–232]).

Turning back to his earlier discussion of Euclid's *Elements*, Dehn underscored what Schur had achieved, namely the very first purely synthetic introduction of a segment arithmetic without any appeal to continuity or the parallel postulate. He thought this work, and not Saccheri's, could more fittingly have borne the title "Euclidis ab omni naevo vindicatus" (Euclid freed of every flaw). A century earlier, the English mathematician Henry Saville had pointed out two major flaws in the classical presentation: the opaque use of the parallel postulate in Book I and the glaring break in Book V, where Euclid inserted a general theory of ratio and proportion before applying it to develop the theory of similar rectilinear figures in Book VI. The cornerstone concept in Book V was the famous Definition V.5 that provides a theoretical criterion for determining when two ratios will be equal. Euclid merely needed to invoke that definition once, in the first proposition of Book VI, after which everything fell easily into place.

Dehn seemed to be saying that this historical development—from Saville to Saccheri and Lambert, passing through the discovery of non-Euclidean geometry and Pasch's grounding of projective geometry, and then the rigorous coordinatization of elementary synthetic geometry with Schur's work—represents a story that was already essentially closed when Hilbert stepped onto the scene. What

---

[5] Wilhelm Schwan, "Extensive Größe, Raum und Zahl," Diss. Frankfurt University, 1923.

[6] Pictures illustrating this argument for this case of Brianchon's theorem can be found in Hilbert and Cohn-Vossen [19, pp. 92–93].

he wrote immediately afterward, though, fully clarifies why Hilbert's *Grundlagen der Geometrie* occupies such a significant place in this chain of developments. Indeed, in surveying what had transpired up until 1899, Dehn described the series of highways and byways that led to important stations, but in such a complicated fashion that one could hardly view these as more than a collection of significant results that fell well short of constituting a unified theory. Hilbert, on the other hand, was the first to recognize the validity of "exotic geometries," as for example, plane geometries in which the theorem of Desargues fails to hold. This finding went hand in hand with one of his central insights: *The validity of the plane theorem of Desargues is the necessary and sufficient condition for deciding whether the plane can be embedded in space.* Schur's proof of the Pascal theorem made essential use of spatial geometry, whereas Hilbert sought to reveal the possibilities for building a theory of geometry in the plane by exploiting the power of the parallel postulate. After spelling out this motivation, Dehn proceeded to give Hilbert's planar proof of Pascal's theorem.

In the closing section on projective geometry, Dehn describes some of the simple consequences of arithmetization, illustrating the theorems of Desargues and Pascal by means of incidence configurations for points and lines in the plane. Hilbert sometimes called these closure theorems, since they lead to closed figures that lie in special position in the plane. The Desargues theorem leads to a $(10, 3)$ configuration, whereas Pascal is a $(9, 3)$ (thus 9 points and 9 lines that are incident in triples). In the first case, one has 30 linear equations, three for each of the 10 lines whose equations are satisfied by substituting the coordinates of the 3 points that lie on them. But since these linear relations are not independent, translating the theorem into algebra leads to the result that one can deduce the final relation from 29 of them. Similarly for the Pascal theorem, as both are examples of *Schnittpunktsätze*, as Hilbert described in *Grundlagen der Geometrie*. In this setting, duality follows immediately from the fact that point and line coordinates enter symmetrically in systems of linear equations.

In the remaining parts of his survey, Dehn took up several topics closely related to Hilbert's researches as well as his own. He addressed here the problem of proving the absolute consistency of an axiomatic system, as Hilbert long claimed must be possible. Like Henri Poincaré before him, Dehn doubted that the principle of complete induction could ever be reduced to a consistency argument (Dehn [5, pp. 260–262]). The focal point of the Hilbert–Bernays program to formalize mathematics was their effort to prove the consistency of the axioms for arithmetic. Hilbert regarded this as the first step toward solving his second Paris problem, which required doing the same for the real numbers. Thus, by the mid-1920s, Dehn publicly doubted the feasibility of Hilbert's formalist program. At the time he spoke in Frankfurt, L.E.J. Brouwer was trying to topple formalism, while pressing for a new approach to foundations based on his philosophy of intuitionism. Only 2 years later, Kurt Gödel would demonstrate the power of Hilbert's proof theory by using it to demonstrate that the consistency of arithmetic was a formally *unprovable* proposition.

Pasch had no time to study Dehn's essay in any detail when he received the page proofs; his failing eyesight likely hindered him from doing more that glancing through the text. Still, he sent his congratulations to Dehn, while expressing his delight over the sheer volume of material his survey contained as well as the careful handling of it.[7] He only added his wish that Dehn somewhere mention the term "Pasch's Axiom" in his text, since several writers had used this terminology. Hilbert himself had acknowledged in a footnote that Axiom II.4, the last among his axioms of order, was first introduced by Pasch in 1882.

Dehn's survey was intended as an overview of historical developments from antiquity to modern times, not, of course, as a detailed historical study. In all likelihood, he wrote this text without having to undertake any substantial amount of research. After all, this topic had long been an integral part of his own teaching and research. As noted earlier, he viewed the Frankfurt reading circle as a vehicle for intellectual enrichment, not as a training ground for future historians of mathematics. One of his star students, though, continued in that direction, a largely unknown story with some surprising wrinkles.

4   Historical Studies on Leibniz and Newton

Adolf Prag never lost his passion for history of mathematics, and in some symbolic sense one could say that Prag played an important role in resolving one of the most contentious issues in the history of mathematics. This concerned the famous priority dispute over the invention of the calculus between the followers of Newton and Leibniz.[8] In fact, this was only one of an entire series of conflicting issues that divided Newton and Leibniz, who held starkly opposing views regarding God's place in the world He created. Leaving all else aside, it remains difficult to say whether Prag took great interest in the calculus controversy, which was both prolonged and vicious. Like Max Dehn, he took a deep interest in the British tradition, but there seems to be no evidence that the Frankfurt seminar paid close attention to the latest iteration of the Newton vs. Leibniz squabble in contemporary revisionist literature. Prag only entered this story through a back door decades after World War II. To appreciate the context, though, requires glancing further backward to the years after the First World War.

First, though, a few words about the circumstances that led to this controversy. Newton was a secretive and mistrustful personality, so very few knew anything about his early mathematical work from the mid-1660s, including those parts related to the calculus. This was still the case when Leibniz visited London in

---

[7] Moritz Pasch to Max Dehn, 7 July 1926, Dehn Papers, Dolph Briscoe Center for American History, University of Texas at Austin.

[8] To be sure, historians have continued to grapple with the issues at stake in this conflict up to the present day. A particularly thoughtful analysis can be found in Guicciardini [15, pp. 329–384]. See also Westfall's account in [48].

the mid-1670s. After he heard something about this work, Leibniz made inquires, which Newton answered in two letters. These passed through the hands of Henry Oldenburg, secretary of the Royal Society, and they would later be used as evidence against Leibniz. He and Newton never personally met, so this incidental exchange evidently went unnoticed at the time.

A decade later in 1684, when Leibniz published the basic rules for the differential calculus, he made no mention of Newton's mathematical work, which still remained unpublished. Three years later, Newton published his *Principia*, which required mathematical methods rooted in calculus. Newton could have derived some of the main results using his theory of fluxions, but if he did so, he left no trace of this in the text. Instead, he dressed up all his results in a geometrical garb, which avoided infinitesimals by invoking the method of first and last ratios. However, hints that Newton might have anticipated Leibniz's invention began to surface in the late 1690s. Around that time, insiders gradually learned that Sir Isaac claimed ownership of the essential methods Leibniz had put into print. These rumors eventually turned into charges of foul play, and in 1712 a committee of the Royal Society, under Newton's presidency, undertook an investigation of the matter. Its report (prepared essentially by Newton himself) concluded, not surprisingly, that Leibniz stole the calculus from Newton and later pretended that he alone had invented it. Since the events surrounding this whole story have been dealt with many times and are far too complicated to describe here, let me skip over them entirely in order to describe a later chapter in this controversy, one which has only rarely been discussed in the historical literature.[9]

Interest in the conflict between Newton and Leibniz had much to do with the fact that partisans for the two sides drenched the matter in blatantly nationalistic rhetoric. This aspect was hardly forgotten when the old debate broke forth in a new form soon after the Great War ended. In 1920, the English mathematician J.M. Child leaped into the fray after undertaking a careful study of the work of Isaac Barrow, who preceded Newton as the first Lucasian Professor of Mathematics at Trinity College, Cambridge. Since most of the relevant texts from the period were in Latin, Child published English translations in [25], together with critical remarks directed at the commentaries written by Carl Immanuel Gerhardt, the nineteenth-century editor of Leibniz's mathematical writings. Gerhardt discovered Leibniz's own account of his path to the calculus, "Historia et origo calculi differentialis," which he published already back in 1846. This marks the beginning of Gerhardt's efforts to reclaim Leibniz's place in the history of mathematics, work that Child sharply criticized. Whereas Newton and his acolytes had charged Leibniz with appropriating Newtonian methods, Child disagreed with this claim, arguing that the brilliant German had instead obtained his main ideas from Isaac Barrow.[10] Child's

---

[9] For a detailed account of the original controversy, see Hall [16].

[10] An attempt to support the case of Barrow was made in Feingold [13]. For an analysis of more recent scholarship relating to the possibility that Leibniz was influenced by Barrow's work, see Probst [32].

translations of the "Historia et origo" and other texts relating to Leibniz's early works received praise in the English-speaking world, although David Eugene Smith made a point of condemning the polemical manner in which Child defended his claims in support of Barrow [43].

In Germany, on the other hand, leading scholars associated with the ongoing Leibniz Edition in Hannover sought to refute Child's claims, some of which were based on speculation. To a large extent, Child was forced to argue in the dark, owing to the fact that he had no knowledge of Leibniz's writings beyond those contained in Gerhardt's publications. The latter had alluded to notes Leibniz had made in his copy of Barrow's *Lectiones Geometricae*, but Child had no access to this book when he published [25]. Heinrich Wieleitner recovered Leibniz's copy of Barrow's *Lectiones* along with a number of other manuscripts that Gerhardt had overlooked. Wieleiter urged Dietrich Mahnke, then a young philosopher who had studied under Husserl and Hilbert in Göttingen, to investigate these sources. This led to Mahnke's *Habilitationsschrift*, published as [28], which contained a lengthy rebuttal of Child's claims. One year later, Wieleitner habilitated in history of mathematics at Munich University. One of the theses he proposed to defend at that time read: "It is totally unjustified to accuse our Leibniz of untruthfulness (or even only forgetfulness) in regard to the reporting about the course of his invention of the differential and integral calculus" (Hofmann [20, p. 211]).

At his death in 1716, Leibniz left behind a vast collection of writings and correspondence. As part of an effort to organize these documents for publication, the Berlin Academy established a Leibniz Commission with leading figures from a variety of fields, including the physicist Max Planck, the mathematician Ludwig Bieberbach, and the philosopher Nicolai Hartmann. Mahnke and Conrad Müller were appointed as mathematical editors. Shortly before the outbreak of the war, however, the Academy aligned itself with the government by appointing Theodor Vahlen, a high-level Nazi mathematician, as its president (Thiel [44]). Vahlen was a natural choice, owing to his close alliance with Ludwig Bieberbach, secretary of the Academy's Mathematics and Natural Sciences Division.

Three years earlier, Bieberbach had founded the journal *Deutsche Mathematik* with the support of government funds. Both he and Vahlen took considerable interest in promoting Leibniz's mathematical reputation. After Mahnke was killed in a car accident in 1939, Bieberbach invited Joseph Ehrenfried Hofmann to take Mahnke's place in the project. Vahlen even appointed him head of the entire work group in Berlin, thereby signaling that publication of Leibniz's mathematical work and correspondence held highest priority. Soon thereafter, Hofmann published an essay in *Deutsche Mathematik* setting forth his approach to history of mathematics [21].

Together with his wife Josepha, Hofmann worked on the Leibniz papers up until 1943, when their house in Berlin was destroyed in a bombing raid. Thereafter, they moved to the small town of Ichenhausen in the Swabian region of Bavaria. Hofmann lost most of his private library, but not the manuscript material in his possession. After the war, the Berlin Academy terminated Hofmann's position with the Leibniz Edition, though he refused to accept this decision or to cooperate with the new staff charged with editing Leibniz's mathematical papers and correspondence. With exclusive access to this material, Hofmann began writing a monograph that gave the first detailed account of Leibniz's early mathematical journey during the years 1672–1676. Although completed in 1946, this text was first published 3 years later in [22]. Seen against the backdrop of earlier events, this book represents the (almost) final response of German historians of mathematics to charges that Leibniz owed a major intellectual debt either to Newton or to Barrow. It takes little imagination to realize, however, that practically no one living in Germany in the year 1949 would have been interested to read about such arcane matters. Later, though, following the resurgence of interest in history of mathematics in the 1960s and 70s, Hofmann's work found many readers, thanks in large part to the efforts of Adolf Prag. Through his translation of [22] into the updated English edition [23], Prag played a major role in making this story available to a wider audience.

After the war ended, Hofmann maintained his ties with Bieberbach, who lost his professorship in Berlin. As a notorious spokesman for Nazi principles, Bieberbach's fate was sealed the moment his case came up during the denazification procedures. Others, on the other hand, came away unscathed. Freiburg's Wilhelm Süss, who had assumed a leadership role in the German mathematical community during the Nazi era, was quickly reinstalled in his former professorship. During the period under French occupation, he converted his former center for war-related research in the Black Forest into a conference center, today the internationally renowned Mathematics Research Institute in Oberwolfach. Hofmann enjoyed good relations with Süss, who in the wake of the war invited him to spend several months in Oberwolfach writing his book on Leibniz's mathematical development [22]. He also supported Süss's main project at the time, namely preparation of a volume on pure mathematics for the series FIAT Reviews of German Science (Remmert [33, pp. 142–145]).

Beginning in 1954, Hofmann organized yearly workshops on the history of mathematics in Oberwolfach. Many who attended these were senior mathematicians or teachers at secondary schools, but they also attracted young historians. One of these was Christoph J. Scriba, who eventually became Hofmann's collaborator and later his successor as a workshop organizer. Scriba also served as a key figure for building bridges to England, where he spent 2 years as a post-doc in Oxford during the early 1960s. During his stay, he struck up a warm friendship with Adolf Prag. Since Scriba's research project was devoted to studying the papers and correspondence of John Wallis, he clearly had good reason to make this personal connection, having read Prag's paper on Wallis [31]. In 1965, he invited Prag to attend a workshop on history of mathematics in Oberwolfach, and in subsequent years Adolf Prag was often among those who participated at these events. During one of these meetings in the 1960s, he and Hofmann discussed a plan to bring out an English translation of [22].

After his retirement from teaching in 1966, Prag lived in Oxford, which gave him the chance to work in the Bodleian Library. He had already struck up a friendly cooperation with Tom Whiteside, who was beginning work on his voluminous edition of Newton's mathematical papers. Prag had served as an external examiner for Whiteside's Cambridge doctoral exam, which took place in 1959. The latter's thesis on "Patterns of Mathematical Thought in the Later Seventeenth Century" [49] fell directly into the field of studies that Prag first took up more than three decades before in Frankfurt under Max Dehn. He now began to play an important supporting role in Whiteside's work, which led to the publication of *The Mathematical Papers of Isaac Newton* in eight volumes [50].[11]

When the first volume appeared in 1967, Prag brought a copy with him to Oberwolfach, just as he did in 1981 when he spoke there about the eighth. In the preface to that final volume, Whiteside wrote:

> It is wholly just that my old friend and colleague Adolf Prag endures to share the title-page of this final volume of Newton's mathematical papers with me. In his seventies he remains the ever-willing, near omniscient helper that he has always been, and without his furnishing and correction of a wide spectrum of matters literary, technical and historical this edition would have been much the poorer in its detail. (Scriba [38, pp. 409–410])

Whiteside's former mentor, Michael Hoskin, played a major part in lining up funding for this Newton project, but also in persuading Cambridge University Press to publish it. Once it was underway, he perhaps also had a hand in the delicate negotiations with CUP over the translation of Hofmann's biography of Leibniz. This was only completed the year after Hofmann died in 1973. Thus, Adolf Prag not only served as a kind of ambassador for Whiteside's Newton during his trips to Germany, his translation [23] gave the English-speaking world a full account of what German scholarship had to say about Leibniz's early mathematical career.

André Weil, one of those who read it very carefully, had also attended Max Dehn's Frankfurt seminar in the 1920s. When he reviewed the book for the American Mathematical Society, however, his words of praise were mixed with a general sense of disappointment. For Weil, the charges leveled against Leibniz had been refuted long ago, which left him puzzled why Hofmann wrote at such length about the priority debate:

> Perhaps the reader of this volume would have been spared a great deal of dull material if the author, at the outset, had made up his mind whether to write the "grand synthesis" he seemed to promise us or to appear as the lawyer for the defense in the absurd prosecution for plagiarism launched against Leibniz in the early years of the eighteenth century by Sir Isaac's sycophants and eventually by Sir Isaac himself. Even if there could ever have been a case against Leibniz, C.I. Gerhardt's excellent publications seemed to have closed it long ago. But we find Hofmann constantly on the defensive ... [45, p. 680]

Naturally, Weil took no interest in the nationalistic motives on both sides of this controversy, but how else can one explain all the ink various writers have spilled over a peculiar priority dispute? In the book's preface, Hofmann indicated how he realized, after Prag and Whiteside had approached him with idea of preparing a translation, "that the original text would require thorough revision" [23, p. ix]. Yet, as Weil rightly pointed out in his review, the text itself is virtually identical to [22],

---

[11] In his obituary for D.T. Whiteside, Niccolò Guicciardini duly noted Prag's importance for the success of this momentous undertaking [14, p. 5].

the original German version. André Weil clearly preferred Max Dehn's approach to history, which took a loftier view of earlier mathematical accomplishments rather than dwelling on petty squabbles over intellectual property rights.

References

1. H. Bos. *Redefining Geometrical Exactness: Descartes' Transformation of the Early Modern Concept of Construction.* Heidelberg: Springer (2001).
2. N. Bourbaki. *Éléments de mathématique.* 10 vols., Paris: Hermann (1939–1998).
3. N. Bourbaki. *Éléments d'histoire des mathématiques.* 2nd ed., Paris: Hermann (1960/1984).
4. G. Burde, W. Schwarz and J. Wolfart. Max Dehn und das mathematische Seminar. *Preprint* (2002). https://www.yumpu.com/de/document/view/6453522/max-dehn-und-das-mathematische-seminar-institut-fur-mathematik
5. M. Dehn. Die Grundlegung der Geometrie in historischer Entwicklung. In: [30, pp. 185–271] (1926).
6. M. Dehn. Über die geistige Eigenart des Mathematikers. Rede anlässlich der Gründungsfeier des Deutschen Reiches am 18. Januar 1928. *Frankfurter Universitätsreden* Nr. 28 (1928).
7. M. Dehn. Sechs Anhänge. In: [35, pp. 208–411] (1930).
8. M. Dehn. Probleme des Hochschulunterrichts in der Mathematik. *Jahresbericht der Deutschen Mathematiker-Vereinigung* **43**, 71–79 (1932).
9. M. Dehn. The Mentality of the Mathematician: a Characterization. Trans. of [6] by A. Shenitzer, *Mathematical Intelligencer* **5**(2), 18–26 (1983).
10. M. Dehn. *Papers on Group Theory and Topology.* Translated and introduced by J. Stillwell, New York: Springer (1987).
11. J. Dieudonné. The work of Nicolas Bourbaki. *American Mathematical Monthly* **77**, 134–145 (1970).
12. H.M. Edwards. *Riemann's Zeta Function.* New York: Academic Press (1974).
13. M. Feingold. Newton, Leibniz, and Barrow, Too. An Attempt at a Reinterpretation. *Isis* **84**, 310–338 (1993).
14. N. Guicciardini. In Memorium. Derek Thomas Whiteside (1932–2008). *Historia Mathematica* **36**, 4–9 (2009).
15. N. Guicciardini. *Isaac Newton on Mathematical Certainty and Method.* Cambridge MA: MIT Press (2009).
16. A.R. Hall. *Philosophers at war. The quarrel between Newton and Leibniz.* Cambridge: Cambridge University Press (1980).
17. T.L. Heath. *A History of Greek Mathematics.* 2 vols., Oxford: Clarendon Press; reprinted by Dover, New York (1921/1981).
18. D. Hilbert. Grundlagen der Geometrie. In: *Festschrift zur Einweihung des Göttinger Gauss-Weber Denkmals,* Leipzig: Teubner (2. Aufl., 1903); *Grundlagen der Geometrie (Festschrift 1899),* Klaus Volkert, ed., Heidelberg: Springer (2015).
19. D. Hilbert and S. Cohn-Vossen. *Anschauliche Geometrie.* Berlin: Springer (1932).
20. J.E. Hofmann. Heinrich Wieleitner. *Jahresbericht der Deutschen Mathematiker- Vereinigung* **42**, 199–223 (1933).
21. J.E. Hofmann. Über Ziele und Wege mathematikgeschichtlicher Forschung. *Deutsche Mathematik* **5**, 150–157 (1940).
22. J.E. Hofmann. *Die Entwicklungsgeschichte der Leibnizschen Mathematik während des Aufenthaltes in Paris (1672– 1676).* München: Leibniz Verlag (1949).
23. J.E. Hofmann. *Leibniz in Paris, 1672–1676: His Growth to Mathematical Maturity.* Trans. Adolf Prag and D.T. Whiteside, Cambridge: Cambridge University Press (1974).
24. F. Klein. *Die Entwicklung der Mathematik im 19. Jahrhundert.* Bd. 1. Berlin: Julius Springer (1926).

25. G.W. Leibniz. *The Early Mathematical Manuscripts of Leibniz.* Translated from the Latin Texts Published by Carl Immanuel Gerhardt With Critical and Historical Notes by J.M. Child. Chicago: Open Court (1920).
26. J. Lorenat, J. McCleary, V.R. Remmert, D.E. Rowe, M. Senechal, eds. *Max Dehn: A Polyphonic Portrait.* Providence RI: American Mathematical Society (2023).
27. W. Magnus. Max Dehn. *Mathematical Intelligencer* **1**(3), 132–143 (1978/79).
28. D. Mahnke. *Neue Einblicke in die Entdeckungsgeschichte der höheren Analysis.* Abhandlungen der preußischen Akademie der Wissenschaften, Physikalisch-mathematische Klasse, Nr. 1, Jahrgang 1925, Berlin: Akademie-Verlag (1926).
29. O. Neugebauer. Das mathematische Institut der Universität Göttingen. *Die Naturwissenschaften* **18**(1), 1–4 (1930).
30. M. Pasch. *Vorlesungen über die neuere Geometrie. Mit einem Anhang von Max Dehn: Die Grundlegung der Geometrie in historischer Entwicklung.* Berlin: Springer (1882/1926).
31. A. Prag. John Wallis (1616–1703). Zur Ideengeschichte der Mathematik im 17. Jahrhundert, *Quellen und Studien zur Geschichte der Mathematik,* Abt. B, **1**, 381–412 (1929).
32. S. Probst. Leibniz as Reader and Second Inventor: The Cases of Barrow and Mengoli. In: *G.W. Leibniz, Interrelations between Mathematics and Philosophy,* N.B. Goethe, P. Beeley, D. Rabouin, eds., Heidelberg: Springer (2015).
33. V.R. Remmert. Oberwolfach in the French occupation zone: 1945 to early 1950s. *Revue d'Histoire des Mathématiques* **26**(2), 121–172 (2020).
34. D.E. Rowe. From Graz to Göttingen: Neugebauer's Early Intellectual Journey. In: *A Mathematician's Journeys: Otto Neugebauer and Modern Transformations of Ancient Science,* A. Jones, C. Proust and J. Steele, eds., pp. 1–59, Archimedes, New York: Springer (2016).
35. A. Schoenflies and M. Dehn. *Einführung in die analytische Geometie,* 2te Auflage, Berlin: Springer (1930).
36. F. Schur. Über den Fundamentalsatz der projektiven Geometrie. *Mathematische Annalen* **51**, 401–409 (1898).
37. W. Schwan. Streckenrechnung und Gruppentheorie. *Mathematische Zeitschrift* **3**, 11–28 (1919).
38. C.J. Scriba. In Memoriam Adolf Prag (1906–2004). *Historia Mathematica* **31**, 409–413 (2004).
39. S.L. Segal. *Mathematicians under the Nazis.* Princeton: Princeton University Press (2003).
40. M. Senechal. Happy Birthday! *Mathematical Intelligencer* **30**(1), 6–19 (2008).
41. C.L. Siegel. Über Riemanns Nachlass zur analytischen Zahlentheorie. *Quellen und Studien zur Geschichte der Mathematik, Astronomie und Physik,* Abt. B: Studien 2, 45–80 (1932).
42. C.L. Siegel. On the History of the Frankfurt Mathematics Seminar. *Mathematical Intelligencer* **1**(4), 223–230 (1979).
43. D.E. Smith. The Early Mathematical Manuscripts of Leibniz. *Bulletin of the American Mathematical Society* **27**, 31–35 (1920).
44. J. Thiel. Leibniz-Tag, Leibniz-Medaillen, Leibniz-Kommission, Leibniz-Ausgabe – Die Preußische Akademie der Wissenschaften und ihr Ahnherr im 'Dritten Reich'. In: Wenchao Li and Hartmut Rudolf, eds., *'Leibniz' in der Zeit des Nationalsozialismus,* Stuttgart: Franz Steiner (2013).
45. A. Weil. Review of [23]. *Bulletin of the American Mathematical Society* **27**, 676–688 (1975).
46. A. Weil. *The Apprenticeship of a Mathematician.* Basel: Birkhäuser (1992).
47. A. Weil. *Number theory, an Approach through History from Hammurapi to Legendre.* Basel: Birkhäuser (2001).
48. R.S. Westfall. *Never at Rest. A Biography of Isaac Newton.* Cambridge: Cambridge University Press (1980).
49. D.T. Whiteside. Patterns of Mathematical Thought in the Later Seventeenth Century. *Archive for History of Exact Sciences* **1**, 179–388 (1961).
50. D.T. Whiteside. *The Mathematical Papers of Isaac Newton.* 8 vols., Cambridge: Cambridge University Press (1967–1981).
51. B.H. Yandell. *The Honors Class. Hilbert's Problems and their Solvers.* Natick, Mass.: AK Peters (2002).

Part VII
Information Theory

In *Multiterminal statistical inference: An unsolved problem*, Shun-ichi Amari describes a remarkably simple statistical inference problem involving two correlated signals and integrating information theory and statistics, on which more than 40 years have passed without significant progress. The problem is elucidated with simple examples.

Lampros Gavalakis and Ioannis Kontoyiannis' article *Information in probability: Another information-theoretic proof of a finite de Finetti theorem* is a really original contribution matching both requirements that were made for this volume. First, there is a neat and very complete review of the rich use of information-theoretic arguments in other fields, particularly dynamical systems and probability theory. The second section provides an illustration with an original proof of De Finetti's theorem using information theoretical tools.

Multiterminal Statistical Inference: An Unsolved Problem

Shun-ichi Amari

1 Introduction

I have worked on various subjects in the mathematical sciences, including information geometry and mathematical neuroscience (the theory of artificial neural networks). When working on these subjects it is important to be acquainted with a wide range of methods and topics.

It was in the middle of the 1980s when Catriona Byrne visited me at the University of Tokyo. We had lots of pleasant discussions concerning current trends and methods. From her, for example, I studied wavelet analysis, which was not yet popular in Japan at that time. She helped to open my eyes to the world. She is a guardian, not only encouraging but also fostering mathematical scientists. Due to her long years of encouragement and support, I authored a monograph, *Information Geometry and Its Applications* [5], and took the initiative to inaugurate a new journal, *Information Geometry*. I have no sufficient words to thank her.

In this short paper, I would like to present a problem which I have been deeply involved in, but cannot solve. It looks like a simple problem concerning the estimation of correlations between two random variables x and y (for simplicity's sake, let them be binary). There are n repeated iid observations $x = (x_1, \ldots, x_n)$ and $y = (y_1, \ldots, y_n)$. However, x and y are observed at different terminals, so they should be sent to a common terminal, where statistical inference takes place. When there are rate restrictions on information transmission, say only rn bits of information are sent, $0 < r < 1$, among n bits each of x and y, we need to encode x and y into nr-bit codewords $c_X(x)$ and $c_Y(y)$, respectively. What is the optimal

S. Amari (✉)
Teikyo University, Advanced Comprehensive Research Organization, Tokyo, Japan

RIKEN Center for Brain Science, Wako, Japan
e-mail: shun-ichi.amari@riken.jp

encoding scheme that maximizes the Fisher information included in $c_X(x)$ and $c_Y(y)$, which characterizes the quality of statistical inference? This problem, which connects multiterminal information theory and statistical inference, was proposed by T. Berger [7] more than 40 years ago. Since then many first-class information scientists have attempted to solve the problem [1–4, 6, 8–14], but without significant success.

We analyze the problem by using a simple but typical model and present interesting intuitive ideas for its solution.

2 Formulation of the Problem

We illustrate the problem with a simple example.

Let (x, y) be a pair of random variables. We assume x and y are binary variables, taking values 0 and 1. A sequence of iid pairs (x_i, y_i), $i = 1, \ldots, n$, are observed. However, the x's and y's are observed at different terminals X and Y, respectively, which might be separated by a long distance. Let

$$M = \{p(x, y; \theta)\}$$

be a family of probability distributions specified by the parameter θ. We consider the situation when the marginal distributions $p(x, \theta)$ and $p(y, \theta)$ do not depend on θ, so we cannot estimate θ from $x = (x_1, \ldots, x_n)$ only or $y = (y_1, \ldots, y_n)$ only, without knowing the other. The parameter θ represents the intensity of correlation between x and y. We consider the case when the probability distributions are given by

$$p(x, y; \theta) = \frac{1}{2} \left\{ \theta \delta_{xy} + (1 - \theta) \left(1 - \delta_{xy} \right) \right\},$$

where δ_{xy} is the Kronecker delta. The expectations of x and y are both $1/2$. The Fisher information given by

$$G(\theta) = \mathrm{E} \left[\left(\frac{\partial \log p(x, y, \theta)}{\partial \theta} \right)^2 \right],$$

where E is expectation, represents how well we can perform statistical inference concerning θ, for example, its estimation or statistical hypothesis testing: $\theta = \theta_0$.

When we observe all data x and y, the problem reduces to ordinary statistical inference. The multiterminal situation assumes that X and Y are separated and we cannot combine x and y. Instead, we send the observed data to a common receiving terminal R, where statistical inference takes place. However, there are restrictions on the transmission channels, and only rn bits of information can be sent from X to R and Y to R. See Fig. 1.

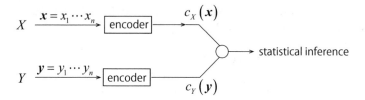

Fig. 1 Scheme of multiterminal statistical inference

Let $c_X(x)$ and $c_Y(x)$ be codewords into which x and y are encoded. The transmitted signals are $c_X(x)$ and $c_Y(y)$. Because of the rate restrictions, the cardinalities of the sets of codewords $C_X = \{c_X(x)\}$ and $C_Y = \{c_Y(y)\}$ are limited to

$$|C_X| = |C_Y| = 2^{rn}.$$

The Fisher information included in the random variables $c_X(x)$ and $c_Y(y)$ is calculated from the probability of messages $c_X(x)$ and $c_Y(y)$,

$$p\{c_X, c_Y, \theta\} = \sum_{x \in C_X, y \in C_Y} p(x, y, \theta),$$

where

$$p(x, y, \theta) = \prod_{i=1}^{n} p(x_i, y_i, \theta).$$

The problem is to find the optimal encoding scheme that maximizes the Fisher information.

A trivial encoding scheme is to pick out first rn bits of x and y, respectively. Then, the codewords are simply

$$c_X = x_1 \cdots x_{rn},$$

$$c_Y = y_1 \cdots y_{rn}.$$

The Fisher information is

$$G_C(\theta) = \frac{r}{n} G(\theta),$$

where $G(\theta)$ is the Fisher information included in a pair (x, y). An interesting problem is if there exist encoding schemes better than the trivial one.

To focus on the problem, we consider a very special case where only one bit of information is allowed to be sent, that is, both $c_X(x)$ and $c_Y(y)$ are binary. This encoding scheme divides the sets $\{x\}$ and $\{y\}$ into two parts C_X and its complement $\bar{C}_X$ (C_Y and $\bar{C}_Y$), respectively, such that

$$C_X = \{x \,|\, c_X(x) = 1\}, \quad C_Y = \{y \,|\, c_Y(y) = 1\}.$$

That is, $c_X(x) = 1$, when $x \in C_X$ and 0, otherwise. Similarly for $c_Y(y)$.

The trivial encoding scheme is just to choose any i, $i = 1, \ldots, n$, and put

$$c_X(x) = x_i,$$
$$c_Y(y) = y_i.$$

Since x_i is correlated to y_i and independent of other y_j's ($j \neq i$), our intuition suggests that there are no encoding schemes better than the trivial one.

Obviously, it is difficult to estimate θ from only one bit of each of $c_X(x)$ and $c_Y(y)$. However, we use a local encoding scheme such that $c_X(x)$ depends only on the first s bits $x_1 \cdots x_s$ of x. (Similarly for $c_Y(y)$.) In such a case, we divide the entire n series x into n/s blocks, each block including s letters. We encode each block in this manner. Then, we have n/s independent observations of $c_X(x)$'s. Similarly for y. The entire Fisher information is n/s times the Fisher information G per bit of the original problem. However, note that this strategy is possible only for $r < 1/2$. We search for the optimal s for this encoding scheme, which may depend on θ. When $s = 1$, this reduces to the trivial encoding scheme. Interestingly, there exist encoding schemes which are better than the trivial one. Amari [4] proved that, when θ is in a small neighborhood of $\theta = 1/2$, that is, x and y are nearly independent, the trivial scheme is the best. But when θ takes much larger or smaller values than $1/2$, there exist better schemes. We describe two types of promising encoding schemes.

The first is the s-parity encoding scheme given by

$$c_X(x) = x_1 \oplus \cdots \oplus x_s,$$
$$c_Y(y) = y_1 \oplus \cdots \oplus y_s.$$

We can calculate the Fisher information $G_s(\theta)$ for this scheme. Let s^* be

$$s^*(\theta) = \arg\max_s G_s(\theta).$$

Then we have the local s^*-parity encoding, which depends on θ. Even though this depends on θ, it is useful for hypothesis testing: $\theta = \theta_0$ or not. For estimation, if we know an approximate $\hat{p}$, $s^*(\hat{p})$ encoding is useful. K. Kobayashi (private communication) showed that this scheme is optimal for $s = 1, \ldots, 5$ by an exhaustive search.

A second encoding scheme is s-majority encoding. Let

$$a(x) = \sum_{i=1}^{s} x_i.$$

We define

$$c_X(x) = \begin{cases} 1, & \text{when } a(x) > \frac{s}{2}, \\ 0, & \text{when } a(x) < \frac{s}{2}, \end{cases}$$

where we assume that s is odd for simplicity. We use the same s-majority encoding for y. We can analyze this scheme, giving a promising Fisher information of $G_s^{\text{maj}}(\theta)$, where s is chosen such that it maximizes $G_s^{\text{maj}}(\theta)$, given θ.

The parity encoding gives a good achievable lower bound of the optimal Fisher information, although we cannot yet prove that it is the best among those that summarize s bits into a 1 bit encoding scheme. However, it cannot be used for $r > 1/2$.

The problem is seemingly simple, but still unsolved.

Conclusions

We have described a problem connecting multiterminal information theory and statistical inference. Elucidating the problem by using a simple model, we presented a number of local encoding schemes, showing that there exist non-trivial encoding schemes better than the trivial one which give achievable lower bounds. However, the general problem remains unsolved.

References

1. R. Ahlswede and M. Burnashev. On minimax estimation in the presence of side information about remote data. *Ann. Statist.*, vol. 18, no. 1, 141–171 (1990).
2. R. Ahlswede and I. Csiszár. Hypothesis testing with communication constraints. *IEEE Trans. Inform. Theory*, vol. IT-32, 533–542, July (1986).
3. S. Amari. Fisher information under restriction of Shannon information in multi-terminal situations. *Ann. Inst. Statist. Math.*, vol. 41, no. 4, 623–648 (1989).
4. S. Amari. On optimal data compression in multiterminal statistical inference. *IEEE Trans. Inform. Theory*, vol. 57, 5577–5587 (2011).
5. S. Amari. *Information Geometry and Its Applications.* Springer (2016).
6. S. Amari and T.S. Han. Statistical inference under multiterminal rate restrictions: A differential geometric approach. *IEEE Trans. Inform. Theory*, vol. 35, 217–227, Mar. (1989).
7. T. Berger. Decentralized estimation and decision theory. Presented at the IEEE 7th Spring Workshop on Information Theory, Mt. Kisco, NY, Sept. (1979).

8. T.S. Han. Hypothesis testing with multiterminal data compression. *IEEE Trans. Inform. Theory*, vol. IT-33, 759–772, Nov. (1987).

9. T.S. Han and S. Amari. Parameter estimation with multiterminal data compression. *IEEE Trans. Inform. Theory*, vol. 41, 1802–1833, Nov. (1995).

10. T.S. Han and S. Amari. Statistical Inference Under Multiterminal Data Compression. *IEEE Trans. Inform. Theory*, vol. 44, no. 6, 2300–2324 (1998).

11. T.S. Han and K. Kobayashi. Exponential-type error probabilities for multiterminal hypothesis testing. *IEEE Trans. Inform. Theory*, vol. 35, 2–14, Jan. (1989).

12. H.M.H. Shalaby and A. Papamarcou. Multiterminal detection with zero-rate data compression. *IEEE Trans. Inform. Theory*, vol. 38, 254–267, Mar. (1992).

13. H.M.H. Shalaby and A. Papamarcou. Error exponent for distributed detection of Markov sources. *IEEE Trans. Inform. Theory*, vol. 40, 397–408, Mar. (1994).

14. Z. Zhang and T. Berger. Estimation via compressed information, *IEEE Trans. Inform. Theory*, vol. 34, 198–211, Mar. (1988).

Information in Probability: Another Information-Theoretic Proof of a Finite de Finetti Theorem

Lampros Gavalakis and Ioannis Kontoyiannis

1 Entropy and Information in Probability

Shannon's landmark 1948 paper [76] founded the field of information theory and ignited the fuse that led to much of the subsequent explosive development of communications theory and engineering in the twentieth century. At the same time, it also led to a wave of applications of information theory to numerous other branches of science. Of those, some, e.g. those in bioinformatics and neuroscience, were successful, while some others, despite Shannon's "bandwagon" warning [77], much less so.

Within mathematics, information-theoretic ideas have had a major impact along several directions, perhaps most notably (although certainly not exclusively) in connection with probability theory. For our present purposes, the most relevant line of work is based on the idea of utilising information-theoretic tools and ideas in order to prove core probabilistic results. Over the past 55 years, a great number of such proofs have appeared. These are often accompanied by new interpretations and rich intuition, thus providing new ways of understanding why fundamental probabilistic theorems are true, and sometimes also giving stronger versions of the original results.

In the rest of this introduction we describe some of the main landmarks along this path, and we indicate directions of current and likely near-future activity. This brief survey is necessarily incomplete and biased, due to our own subjective taste

L. Gavalakis
Department of Engineering, University of Cambridge, Cambridge, UK
e-mail: lg560@cam.ac.uk

I. Kontoyiannis (✉)
Statistical Laboratory, DPMMS, University of Cambridge, Centre for Mathematical Sciences, Cambridge, UK
e-mail: ik355@cam.ac.uk

© The Author(s), under exclusive license to Springer Nature Switzerland AG 2023
J.-M. Morel, B. Teissier (eds.), *Mathematics Going Forward*, Lecture Notes in Mathematics 2313, https://doi.org/10.1007/978-3-031-12244-6_26

and bounded knowledge. Then in Sect. 2 we state and prove a new finite version of de Finetti's classical representation theorem for finite-valued exchangeable random variables.

The first appearance of information-theoretic ideas in the proof of a genuinely probabilistic result was in 1958, when Hájek [41, 42] proved that the laws μ and ν of any two Gaussian processes are either absolutely continuous with respect to each other, or singular. Hájek exploited the implications of $D(\mu\|\nu) + D(\nu\|\mu)$ being finite or infinite, where $D(\mu\|\nu)$ denotes the *relative entropy* or *Kullback–Leibler divergence* between μ and ν,

$$D(\mu\|\nu) := \begin{cases} \int \log \frac{d\mu}{d\nu}\, d\mu, & \text{if } \frac{d\mu}{d\nu} \text{ exists} \\ +\infty, & \text{otherwise.} \end{cases} \tag{1}$$

[Throughout, log denotes the natural logarithm.] In the same year, Kolmogorov [56] introduced *entropy* in ergodic theory. He provided a way to calculate the entropy of a transformation to conclude that Bernoulli shifts of different entropies are not metrically isomorphic. The importance of entropy in ergodic theory was also highlighted more than a decade later, when Ornstein [70–72] proved that Bernoulli shifts with the same entropy are necessarily isomorphic.

The following year, 1959, Linnik [63] gave an information-theoretic proof of the central limit theorem (CLT), showing that the law of the standardised sum $S_n = (1/\sigma\sqrt{n})\sum_{i=1}^{n} X_i$ of n independent and identically distributed (i.i.d.) random variables $X_1, \ldots, X_n$ with variance σ^2 (or, more generally, of independent random variables satisfying the Lindeberg condition) converges in distribution to a Gaussian. Linnik's connection between the CLT and information-theoretic ideas was the first in a long series of works, along a path that remains active until today. Indeed, in a sequence of papers, including works by Shimizu [78], Brown [13], Barron [5], Johnson [48], Artstein et al. [2], Tulino and Verdú [82], and Madiman and Barron [65], it was shown that the *differential entropy* $h(S_n)$ of the standardised sums in fact *increases* with n, and its limiting value is the entropy $h(Z)$ of the standard Gaussian $Z \sim N(0, 1)$. This monotonic convergence in combination with the fact that the Gaussian has maximum entropy among all random variables with variance σ^2, presents an appealing analogy between the CLT and the second law of thermodynamics.

In the above discussion, the differential entropy of a continuous random variable X with density f is given by $h(X) = h(f) = -\int f \log f$, and the relative entropy between two probability measures μ, ν on $\mathbb{R}$ with densities f, g is $D(\mu\|\nu) = D(f\|g) = \int f \log(f/g)$. A simple computation shows that the "entropic CLT" just described can equivalently be stated as, $D(f_n\|\varphi) \downarrow 0$ as $n \to \infty$, where f_n is the density of S_n and φ the standard normal density. This convergence in the sense of relative entropy implies convergence in L^1: *Pinsker's inequality*, established by Csiszár [21], Kullback [59] and Kemperman [54], states that:

$$D(\mu\|\nu) \geq \frac{1}{2\log 2}\|\mu - \nu\|_1^2. \tag{2}$$

Subsequent work along these lines includes Carlen and Soffer's dynamical systems approach [17], Johnson's convergence to Haar measure on compact groups [53], Johnson and Barron's rates of convergence in the entropic CLT [51], Bubeck and Ganguly's entropic CLT for Wishart random matrices [15], and, most recently, an information-theoretic CLT for discrete random variables [36].

A year after Linnik's paper made the first entropy-CLT connection, in 1960, Rényi [74] examined the convergence of Markov chains to equilibrium from an information-theoretic point of view, thus initiating another path of information-theoretic investigation in probability. Rényi showed that the relative entropy $D(P_n \| \pi)$ between the time-n distribution P_n of a finite-state chain with an all-positive transition matrix and its unique invariant distribution π, decreases to zero as $n \to \infty$. Similar and slightly more general results were established independently by Csiszár [20] in 1963, who also employed Rényi's notion of f-divergence, an important generalisation of relative entropy. In the same year, Kendall [55] extended Rényi's techniques and results, to include certain countable state space chains. A significant advance came with Fritz's 1973 work [35], where he studied the asymptotic behavior of reversible Markov kernels and established their weak convergence to equilibrium. Barron in 2000 [7] extended Fritz's result to convergence in relative entropy, and in 2009 Harremoës and Holst [44] used ideas related to information projections to further extend and generalise those earlier results.

The problem of Poisson approximation and convergence was first examined through the lens of information theory around 20 years ago, leading to a development analogous to that of the entropic CLT. Harremoës in 2001 identified the Poisson as the maximum entropy distribution among all laws that arise from sums of independent Bernoulli random variables with a fixed mean [43]. This characterisation was extended in 2007 by Johnson [50] to the class of ultra log-concave laws on the nonnegative integers. Meanwhile, in 2005 Kontoyiannis et al. [57] derived convergence results and nonasymptotic Poisson approximation bounds using entropy-theoretic methods. Interestingly, some of those results were based, in part, on a discrete modified logarithmic Sobolev inequality for the entropy established by Bobkov and Ledoux [11]. In a related direction, Harremoës et al. [45, 46] obtained Poisson approximation results under the thinning operation.

A similar program was carried out in the case of compound Poisson approximation. Compound Poisson laws on the integers were again given a natural maximum-entropy interpretation by Johnson et al. [52] and Yu [89], and compound Poisson approximation bounds and convergence results were established via information-theoretic techniques by Madiman et al. [66] and Barbour et al. [4]. Interestingly, in some cases the resulting nonasymptotic bounds give the best results to date.

The Method of Types and Large Deviations Suppose $\{X_n\}$ are i.i.d. random variables with common probability mass function (PMF) Q on a finite alphabet A of size $m = |A|$. The *type* $\hat{P}_n$ of a string $x_1^n = (x_1, \ldots, x_n) \in A^n$ is simply the empirical PMF induced by x_1^n on A. Let $\mathcal{P}_n$ denote the collection of all n-types on A, namely, all PMFs that arise as types of strings of length n. Then, e.g., we have the obvious bound,

$$|\mathcal{P}_n| \leq (n+1)^m, \tag{3}$$

and direct computation also shows that, for any $x_1^n \in A^n$,

$$Q^n(x_1^n) = e^{-n[H(\hat{P}_n)+D(\hat{P}_n\|Q)]}. \tag{4}$$

Here, $H(P) := -\sum_{x \in A} P(x) \log P(x)$ is the (discrete Shannon) *entropy* of a PMF P on A, and the definition (1) of the relative entropy $D(P\|Q)$ between two PMFs P, Q on the same discrete alphabet becomes $D(P\|Q) = \sum_{x \in A} P(x) \log[P(x)/Q(x)]$. Slightly more involved calculations lead to interesting and useful bounds. For example, for an n-type P, let $T(P)$ denote the *type class* of P, consisting of all $x_1^n \in A^n$ with type P. Then the cardinality and probability of $T(P)$ satisfy,

$$(n+1)^{-m} e^{nH(P)} \leq |T(P)| \leq e^{nH(P)} \tag{5}$$

$$(n+1)^{-m} e^{-nD(P\|Q)} \leq Q^n(T(P)) \leq e^{-nD(P\|Q)}. \tag{6}$$

The method of types is a collection of combinatorial estimates for probabilities associated with discrete i.i.d. random variables and memoryless channels, of which the examples in (3)–(6) above are the starting point. Based in part on preliminary ideas of Wolfowitz [88], the method of types was fully developed in 1981 by Csiszár and Körner [28]. As described in Csiszár's review [25], the method of types has been employed very widely and with great success in numerous information-theoretic problems arising from different communication-theoretic scenarios.

Based in part on the method of types, and also building on ideas from related work by Groenebook et at. [38], Csiszár was able to establish a series of important results in large deviations. In 1975 [22] he identified the exponent in Sanov's theorem [75] as an extremum of relative entropies, and in 1984 [23] he proved a general, strong version of Sanov's theorem, by a combination of the method of types, discretisation arguments, and a general Pythagorean inequality for the relative entropy established by Topsøe [81]. He also gave a simpler proof along the same lines in his 2006 paper [26].

Moreover, in the same paper [23] Csiszár established a version of the *Gibbs conditioning principle* (also know as the *conditional limit theorem*) using the same tools. This was further extended by Csiszár et al. in 1987 [27] to the case of Markov conditioning, and by Algoet et al. in 1992 [1] to Markov types.

The method of types and the Gibbs conditioning principle will both play an important role in our proof of the finite de Finetti theorem in Sect. 2.

Exchangeability Suppose $\{X_n\}$ are i.i.d. random variables, and let $\mathcal{E}$ denote the *exchangeable* σ-algebra, that is, the sub-σ-algebra of $\sigma(\{X_n\})$ that consisting of those events that are invariant under finite permutations of the indices in the sequence $\{X_n\}$. In 2000, O'Connell [69] gave a beautiful, elementary information-theoretic proof the Hewitt-Savage 0-1 law [47]: $\mathcal{E}$ is trivial, in that all events in $\mathcal{E}$ have probability either zero ore one.

Another aspect of exchangeability comes up in connection with de Finetti's theorem. Let $\{X_n\}$ be an exchangeable sequence of random variables with values in the same finite alphabet A. Here, exchangeability means that, for any n and any permutation π on $\{1, 2, \ldots, n\}$, the distribution of the random variables $(X_{\pi(1)}, X_{\pi(2)}, \ldots, X_{\pi(n)})$ is the same as that of $(X_1, X_2, \ldots, X_n)$. De Finetti's theorem [30, 31] states that $\{X_n\}$ is exchangeable if and only if it is a mixture of i.i.d. sequences, that is, if and only if there is a measure $\bar{\mu}$ on the simplex $\mathcal{P}$ of probability distributions on A, such that, for any $k \geq 1$ and any $x_1^k = (x_1, \ldots, x_k) \in A^k$,

$$\mathbb{P}(X_1^k = x_1^k) = M_{\bar{\mu},k}(x_1^k) := \int_{\mathcal{P}} Q^k(x_1^k)\mathrm{d}\bar{\mu}(Q). \tag{7}$$

De Finetti's theorem plays an important role in the foundations of subjective probability and Bayesian statistics, see, e.g., the discussions in [8, 33]. But arguments about its practical relevance are limited by the fact that, as is well known [34], the representation (7) fails in general if it is only assumed that a finite collection of random variables $(X_1, \ldots, X_n)$ is exchangeable for some fixed n. Nevertheless, approximate versions of (7) remain valid in this case [33, 34]. Such a 'finite' version of de Finetti's theorem for binary random variables was recently established in [37], using information-theoretic ideas: It was shown that there is a mixing measure μ on $\mathcal{P}$ such that, for any $1 \leq k \leq n$, the distribution P_k of $(X_1, \ldots, X_k)$ is close to $M_{\mu,k}$ in the precise sense that:

$$D(P_k \| M_{\mu,k}) \leq \frac{5k^2 \log n}{n - k}. \tag{8}$$

A different information-theoretic proof of a different finite version of de Finetti's theorem is given in Sect. 2.

Further Connections There are numerous other directions along which information-theoretic methods have been employed to establish either known or new probabilistic results. We briefly mention only a few more from the long list of relevant works, some of which go beyond probability theory. The interested reader may also consult Barron's reviews of information-theoretic proofs and connections with statistics and learning [6, 7], Csiszár's review of information-theoretic methods in probability [24], and Johnson's text [49].

A natural and powerful connection has been drawn between information theory and concentration of measure inequalities, through what has come to be known as the *entropy method*. Often attributed Herbst [29], the entropy method was primarily developed by Ledoux [60–62]. Marton's 1996 work [68] had an early and significant influence in this direction as well. Entropy also appears naturally in connection with related work on transportation theory [10, 83]. Book-length accounts of measure concentration and related inequalities, including the entropy method, are given in [12] and [73].

A fascinating and multifaceted series of connections between information-theoretic ideas and functional inequalities started with Shannon's entropy power inequality (EPI), stated in his original 1948 paper [76] and later proved by Stam [79] and Blachman [9]. Much of the relevant literature up to 1991 is summarised in Dembo et al.'s review [32], including the connection with Gross' celebrated Gaussian logarithmic Sobolev inequality [39]. This paper also contains an early discussion of the strong ties between entropy inequalities and high-dimensional convex geometry, starting with Costa and Cover's 1984 observation [18] that the Brunn–Minkowski inequality can be viewed as a special case of a generalised EPI.

Building on the technical ideas of Stam and Blachman, Bakry and Émery in a very influential 1985 paper [3] derived an important representation of the derivative of the relative entropy $D(P_t \| Q_t)$ of the time-t distributions of a diffusion with different initial conditions. Under appropriate assumptions, strong connections were established with logarithmic Sobolev inequalities, generalising the earlier connection between the EPI and Gross' Gaussian logarithmic Sobolev inequality, and facilitating the study of the long-term behaviour of the underlying diffusion. An important observation, independently re-discovered by Barron [6], is that this derivative can be expressed as a "relative" Fisher information, which also admits an interpretation as a minimum mean squared error. This interpretation had been promoted earlier in work by Brown, see e.g. [14], and it was re-framed in more information theoretic terms by Guo et al. in 2005 [40], leading to a variety of subsequent developments.

More recently, a remarkable equivalence between the subadditivity property of entropy and the classical Brascamp–Lieb inequality was pointed out by Carlen and Cordero-Erausquin [16], and a unified information-theoretic treatment was given by Liu et al. [64]. In yet another direction, Tao in 2010 [80] developed a series of discrete entropy inequalities motivated by sumset and inverse sumset bounds in additive combinatorics, also leading to a discrete version of the EPI. More recent work in this direction includes [58, 67].

Finally, we mention that a natural analog of the entropy in free probability was introduced in a series of papers by Voiculescu [84–87], where several properties of the free entropy are established, including a free version of the EPI. In related work, convergence results to maximum free entropy distributions is considered by Johnson in [49, Chapter 8].

2 Information-Theoretic Proof of a Finite de Finetti Theorem

Suppose $X_1^n = (X_1, \ldots, X_n)$, for some fixed n, are exchangeable, discrete random variables, with values in a finite alphabet A of $m = |A|$ elements. Let $\hat{P}_{X_1^n}$ denote the (random) type of X_1^n, and let the measure $\mu = \mu_n$ denote the law of $\hat{P}_{X_1^n}$ on the probability simplex $\mathcal{P}$. In this section we provide an information-theoretic proof of the following:

Theorem 2.1 (Finite de Finetti Theorem) *For any* $1 \leq k \leq n$, *let* P_k *denote the distribution of* $X_1^k = (X_1, \ldots, X_k)$ *and* $M_{\mu,k}$ *denote the mixture-of-i.i.d.s:*

$$M_{\mu,k}(x_1^k) = \int_{\mathcal{P}} Q^k(x_1^k) d\mu(Q), \qquad x_1^k \in A^k.$$

For any $1 \leq k \leq (n/100)^{1/3}$, *we have,*

$$D(P_k \| M_{\mu,k}) \leq \epsilon(n, k) := 2\delta + ke^{-\frac{n}{k}\delta}\left(\frac{n}{k} + 1\right)^{2m^k} \log n, \qquad (9)$$

with $\alpha = \alpha_{n,k} = \left[\frac{2k}{\sqrt{n}}\left(\frac{1+2k}{\sqrt{n}} + 1\right)\right]^{1/2}$ *and* $\delta = \delta_{n,k} = \alpha \log(m^k/\alpha)$.

Before giving the proof of the theorem, some remarks are in order:

(1) It can be seen from (9) that, if k stays bounded as $n \to \infty$, then:

$$\epsilon(n, k) = O(\delta_{n,k}) = O\left(\left(\frac{k}{\sqrt{n}}\right)^{1/2} \log \frac{n}{k}\right) \to 0.$$

Moreover, in order for $\epsilon(n, k)$ to vanish, k can grow at most logarithmically with n. This is, at least asymptotically, weaker than the bound (8) given in [37] for the binary case $m = 2$. What's more, the proof of (9) given below is longer and more involved that the corresponding proof of (8) in [37]. So why bother? The reason is that the proof given here follows a completely different information-theoretic path than that in [37], and that path consists of an appealing sequence of steps making interesting connections. So we first present a heuristic outline, and then give the actual proof. In fact, as will be seen from the proof (especially Lemma 2.2), it is easy to improve the bound $\epsilon(n, k)$, but our purpose here is to illustrate the ideas rather than to obtain optimal results.

(2) We have cheated slightly in the statement of the theorem, in that the proof below is only given for the case when n is a multiple of k. However, this is only a minor technical inconvenience; for example, we can replace n with an integer multiple of k which is no less than $n - k$, leading to the same bound with $\epsilon(n - k, k)$ in place of $\epsilon(n, k)$.

(3) De Finetti's original theorem (7) easily follows from (9) by an application of Pinsker's inequality (2) and a standard weak convergence argument.

Heuristic Proof of de Finetti's Theorem (7)

Step 1: Since the sequence $\{X_n\}$ is exchangeable it is also stationary, therefore, by the ergodic theorem $\hat{P}_{X_1^n}$ converges as $n \to \infty$ a.s. to a (random) P on A. Let $\bar{\mu}$ denote the law of P, and let $\{Y_n\}$ be i.i.d. random variables uniformly distributed on A. Then, by exchangeability, we clearly have for any n, any $k \le n$, any n-type Q_n, and any $a_1^k \in A^k$,

$$\mathbb{P}(X_1^k = a_1^k | \hat{P}_{X_1^n} = Q_n) = \mathbb{P}(Y_1^k = a_1^k | \hat{P}_{Y_1^n} = Q_n). \tag{10}$$

Step 2: Choose and fix any one of the almost all realisations $\{Q_n\}$ along which $\hat{P}_{X_1^n}$ converges to some Q as $n \to \infty$. By (10) and symmetry we have,

$$\mathbb{P}(X_1 = a | \hat{P}_{X_1^n} = Q_n) = \mathbb{E}\Big(\mathbb{I}_{\{Y_1=a\}} \Big| \hat{P}_{Y_1^n} = Q_n\Big) = \mathbb{E}\Big(\frac{1}{n}\sum_{i=1}^n \mathbb{I}_{\{Y_i=a\}} \Big| \hat{P}_{Y_1^n} = Q_n\Big),$$

so that,

$$\mathbb{P}(X_1 = a | \hat{P}_{X_1^n} = Q_n) = \mathbb{E}\Big(\hat{P}_{Y_1^n}(a) \Big| \hat{P}_{Y_1^n} = Q_n\Big) = Q_n(a),$$

for any $a \in A$, and letting $n \to \infty$ yields,

$$\lim_{n\to\infty} \mathbb{P}(X_1 = a | \hat{P}_{Y_1^n} = Q_n) = Q(a). \tag{11}$$

Step 3: Next we generalise (11) to blocks of random variables. As before, choose and fix any one of the almost all realisations $\{Q_n\}$ of the random $\hat{P}_{X_1^n}$ such that $Q_n \to$ some Q as $n \to \infty$. Define a new sequence of i.i.d. random variables $Z_n = (Y_{2n-1}, Y_{2n})$, $n \ge 1$, so that each Z_n is uniformly distributed on $A \times A$. From (10), taking $k = 2$ and an arbitrary even $n = 2\ell$,

$$\mathbb{P}\big((X_1, X_2) = (a_1, a_2) | \hat{P}_{X_1^{2\ell}} = Q_{2\ell}\big) = \mathbb{P}\big(Z_1 = (a_1, a_2) | \hat{P}_{Z_1^\ell} \in E(Q_{2\ell})\big), \tag{12}$$

where $E(Q)$ denotes the set of probability distributions W on $A \times A$ with the property that the average of the two marginals W_1 and W_2 of W equals Q,

$$E(Q) = \Big\{W \text{ on } A \times A : \frac{W_1 + W_2}{2} = Q\Big\}.$$

If we write U for the uniform distribution on $A \times A$, it is easy to check that the distribution W_ℓ^* that uniquely achieves the $\min_{W \in E(Q)} D(W \| U)$ is simply $Q \times Q$.

At this point, we would wish to apply the conditional limit theorem [19] to the i.i.d. process $\{Z_n\}$, to obtain that,

$$\lim_{\ell \to \infty} \mathbb{P}\big(Z_1 = (a_1, a_2)\big| \hat{P}_{Z_1^\ell} \in E(Q_{2\ell})\big) = \lim_{\ell \to \infty} W_\ell^*(a_1, a_2)$$

$$= \lim_{\ell \to \infty} Q_{2\ell}(a_1) Q_{2\ell}(a_2)$$

$$= Q(a_1) Q(a_2),$$

and combining this with (12) would yield:

$$\mathbb{P}\big((X_1, X_2) = (a_1, a_2)\big| \hat{P}_{X_1^{2\ell}} = Q_{2\ell}\big) \to Q(a_1) Q(a_2), \quad \ell \to \infty.$$

The same argument can be used without difficulty to show that for any $k \geq 1$ and any $a_1^k \in A^k$,

$$\lim_{\ell \to \infty} \mathbb{P}(X_1^k = a_1^k | \hat{P}_{X_1^{k\ell}} = Q_{k\ell}) = Q^k(a_1^k). \tag{13}$$

Step 4: Since (13) holds for almost every sequence $\{Q_n\}$, letting $\ell \to \infty$, by the bounded converge theorem we have,

$$\mathbb{P}(X_1^k = a_1^k) = \mathbb{E}\Big(\mathbb{P}(X_1^k = a_1^k | \hat{P}_{X_1^{k\ell}})\Big) \to \int_{\mathcal{P}} Q^k(a_1^k) \mathrm{d}\bar{\mu}(Q),$$

as required.   □

The only problem with the above argument is that the set $E(Q)$ has an empty interior so that the conditional limit theorem is not directly applicable. Nevertheless, in the next section where we take a finite-n approach, we are able to 'imitate' the proof of the conditional limit theorem and replace the step where the non-empty interior assumption is used with a different argument.

2.1 Proof

Recall the notation and terminology for types described in the Introduction. Let $\mu = \mu_n$ denote the law of $\hat{P}_{X_1^n}$ on $\mathcal{P}$, and let $\{Y_n\}$ be i.i.d. random variables uniformly distributed on A. For any $k \leq n$, any n-type Q_n, and any $a_1^k \in A^k$,

$$\mathbb{P}(X_1^k = a_1^k | \hat{P}_{X_1^n} = Q_n) = \mathbb{P}(Y_1^k = a_1^k | \hat{P}_{Y_1^n} = Q_n).$$

Foe $k = 1$ and any $a \in A$, by symmetry we have,

$$\mathbb{P}(X_1 = a | \hat{P}_{X_1^n}) = \hat{P}_{X_1^n}(a),$$

and taking the expectation of both sides with respect to μ shows that in fact $P_1 = M_{\mu,1}$.

For general $1 \leq k \leq n$ with $n = k\ell$, for any n-type Q we have,

$$\mathbb{P}(X_1^k = a_1^k | \hat{P}_{X_1^{k\ell}} = Q) = \mathbb{P}(Z_1 = a_1^k | \hat{P}_{Z_1^\ell} \in E_k(Q))$$

$$= \mathbb{E}(\hat{P}_{Z_1^\ell}(a_1^k) | \hat{P}_{Z_1^\ell} \in E_k(Q)),$$

where now $\{Z_n\}$ is a sequence of i.i.d. random variables uniformly distributed on A^k, and $E_k(Q)$ consists of all probability distributions W on A^k with the property that the average of the k one-dimensional marginals of W equals Q. Taking expectations with respect to $\hat{P}_{X_1^{k\ell}} = \hat{P}_{X_1^n} \sim \mu = \mu_n$,

$$\mathbb{P}(X_1^k = a_1^k) = \int \mathbb{E}(\hat{P}_{Z_1^\ell}(a_1^k) | \hat{P}_{Z_1^\ell} \in E_k(Q)) \mathrm{d}\mu(Q),$$

and by the joint convexity of relative entropy,

$$D(P_k \| M_{\mu,k}) = D\left(\int \mathbb{E}(\hat{P}_{Z_1^\ell} | \hat{P}_{Z_1^\ell} \in E_k(Q)) \mathrm{d}\mu(Q) \middle\| \int Q^k \, \mathrm{d}\mu(Q) \right)$$

$$\leq \int D\left(\mathbb{E}(\hat{P}_{Z_1^\ell} | \hat{P}_{Z_1^\ell} \in E_k(Q)) \middle\| Q^k \right) \mathrm{d}\mu(Q)$$

$$\leq \int \mathbb{E}\left(D(\hat{P}_{Z_1^\ell} \| Q^k) \middle| \hat{P}_{Z_1^\ell} \in E_k(Q) \right) \mathrm{d}\mu(Q). \tag{14}$$

We will obtain an explicit bound for the relative entropy in (14). First, we construct a joint ℓ-type W with desirable properties. Let $\mathcal{P}_\ell$ denote the set of ℓ-types on A^k.

Lemma 2.2 *For any $\ell > k \geq 1$ and any n-type Q, there is a $W \in E_k(Q) \cap \mathcal{P}_\ell$ with:*

$$\max_{a_1^k} |W(a_1^k) - Q^k(a_1^k)| \leq M := \left[\frac{2}{\ell} + \frac{4k}{\ell} + 2\sqrt{\frac{k}{\ell}} \right]^{1/2}.$$

Moreover, for $2 \leq k \leq \sqrt{\ell}/10$,

$$|H(W) - H(Q^k)| \leq -M \log \frac{M}{m^k}.$$

Proof Let $x_1^{k\ell} \in A^{k\ell}$ have type Q, let $V_1^{k\ell}$ be a random permutation of $x_1^{k\ell}$, and let $\widehat{W}$ denote its (random) ℓ-type. Obviously we have that $\widehat{W} \in E_k(Q)$ by construction, and we will also show that $\widehat{W}$ satisfies the statement of the lemma with positive probability. Taking any $k \leq \ell$ and $\gamma > 0$ arbitrary,

$$
\mathbb{P}\left(\max_{a_1^k} |\widehat{W}(a_1^k) - Q^k(a_1^k)| > \gamma \right)
$$

$$
\leq \sum_{a_1^k} \mathbb{P}\left(|\widehat{W}(a_1^k) - Q^k(a_1^k)| > \gamma \right)
$$

$$
\leq \sum_{a_1^k} \gamma^{-2} \mathbb{E}\left[\left(\widehat{W}(a_1^k) - Q^k(a_1^k) \right)^2 \right]
$$

$$
= \gamma^{-2} \sum_{a_1^k} \left[\rho_2(a_1^k) - 2Q^k(a_1^k)\rho_1(a_1^k) + Q^k(a_1^k)^2 \right], \tag{15}
$$

where $\rho_2(a_1^k) = \mathbb{E}\left[\widehat{W}(a_1^k)^2 \right]$ and $\rho_1(a_1^k) = \mathbb{E}\left[\widehat{W}(a_1^k) \right]$.

Now we find appropriate bounds so that the above probability is < 1. To get an upper bound on $\rho_1(a_1^k)$ for some fixed a_1^k note that,

$$
\rho_1(a_1^k) = \mathbb{P}(V_1^k = a_1^k) \leq \prod_{i=1}^{k} \frac{n(a_i)}{\ell k - i + 1},
$$

where $n(a_i)$ is the number of appearances of a_i in $x_1^{k\ell}$, and hence,

$$
\rho_1(a_1^k) \leq \prod_{i=1}^{k} \frac{n(a_i)}{\ell k} \frac{\ell k}{\ell k - i + 1}
$$

$$
\leq Q^k(a_1^k) \left(\frac{\ell k}{\ell k - k} \right)^k
$$

$$
\leq Q^k(a_1^k) \left(1 - \frac{1}{\ell} \right)^{-k}
$$

$$
\leq Q^k(a_1^k)(1 + k/\ell),
$$

since $(1 - x)^{-k} \leq 1 + kx$ for $x \in [0, 1)$. Similarly, writing $a_1^k * a_1^k$ for the concatenation of a_1^k with itself, we can estimate,

$$
\mathbb{P}(V_1^{2k} = a_1^k * a_1^k) \leq Q^k(a_1^k)^2(1 + 4k/\ell),
$$

so that,

$$\rho_2(a_1^k) = \ell \frac{1}{\ell^2} \rho_1(a_1^k) + \ell(\ell-1)\frac{1}{\ell^2} \mathbb{P}(V_1^{2k} = a_1^k * a_1^k)$$

$$\leq (1/\ell)(1 + k/\ell)Q^k(a_1^k) + (1 + 4k/\ell)Q^k(a_1^k)^2$$

$$\leq (2/\ell)Q^k(a_1^k) + (1 + 4k/\ell)Q^k(a_1^k)^2. \tag{16}$$

Substituting the bound (16) in (15) we have,

$$\mathbb{P}\left(\max_{a_1^k} |W(a_1^k) - Q^k(a_1^k)| > \gamma\right)$$

$$\leq \gamma^{-2} \sum_{a_1^k} Q^k(a_1^k)\left[\frac{2}{\ell} + \left(2 + \frac{4k}{\ell}\right)Q^k(a_1^k) - 2\rho_1(a_1^k)\right]$$

$$\leq \gamma^{-2} \max_{a_1^k}\left[\frac{2}{\ell} + \left(2 + \frac{4k}{\ell}\right)Q^k(a_1^k) - 2\rho_1(a_1^k)\right]. \tag{17}$$

Finally, we get a lower bound on $\rho_1(a_1^k)$. In the case where for some $\beta > 0$ (to be chosen later), $Q(a_i) > \beta$ for all a_i, we have,

$$\rho_1(a_1^k) \geq \prod_{i=1}^{k} \frac{n(a_i) - i + 1}{\ell k - i + 1} \geq \prod_{i=1}^{k} \frac{n(a_i) - k + 1}{\ell k} \geq Q^k(a_1^k) \prod_{i=1}^{k}\left(1 - \frac{1}{\ell Q(a_i)}\right),$$

so that,

$$\rho_1(a_1^k) > Q^k(a_1^k)\left(1 - \frac{1}{\ell\beta}\right)^k \geq Q^k(a_1^k)\left(1 - \frac{k}{\ell\beta}\right), \tag{18}$$

assuming $\beta \geq 1/2\ell$, since $(1-x)^k \geq 1 - kx$ for all $k \geq 1$ and all $x \in [0, 2]$. For all such a_1^k, using (18) we can bound the expression in the maximum in (17) by

$$\left[\frac{2}{\ell} + \left(2 + \frac{4k}{\ell}\right)Q^k(a_1^k) - 2\rho_1(a_1^k)\right] < \frac{2}{\ell} + \frac{4k}{\ell} + \frac{2k}{\ell\beta}.$$

And in the case when at least one a_i has $Q(a_i) \leq \beta$, simply omitting the negative term and noting that $Q^k(a_1^k) \leq \beta$ we bound the same term above by,

$$\left[\frac{2}{\ell} + \left(2 + \frac{4k}{\ell}\right)Q^k(a_1^k)\right] \leq \frac{2}{\ell} + \left(2 + \frac{4k}{\ell}\right)\beta.$$

Combining the last three bounds,

$$\mathbb{P}\Big(\max_{a_1^k}|W(a_1^k) - Q^k(a_1^k)| > \gamma\Big) \leq \gamma^{-2}\Big[\frac{2}{\ell} + \max\Big\{\frac{2k}{\ell}\Big(2 + \frac{1}{\beta}\Big),\ \Big(2 + \frac{4k}{\ell}\Big)\beta\Big\}\Big],$$

where the inequality is strict when the first term dominates the maximum. To obtain a good bound we take for β a value approximately equal to the minimiser of the above expression: We set $\beta^* = \sqrt{k/\ell}$. Note that for this β^* it can be easily verified that the first term strictly dominates the maximum, giving,

$$\mathbb{P}\Big(\max_{a_1^k}|W(a_1^k) - Q^k(a_1^k)| > \gamma\Big) < \gamma^{-2}\Big[\frac{2}{\ell} + \frac{4k}{\ell} + 2\sqrt{\frac{k}{\ell}}\Big],$$

and taking $\gamma = M$ as in the lemma, completes the proof of the first statement.

For the second part, noting that for $2 \leq k \leq \sqrt{\ell}/10$ we have $M < 1/2$, the result follows from [28, Lemma 2.7]. $\qquad\square$

Next we obtain an upper bound on the conditional expectation in (14).

Lemma 2.3 *Suppose* $n = \ell k$, *with* $2 \leq k \leq \sqrt{\ell}/10$. *For any n-type Q we have:*

$$\mathbb{E}\Big(D(\hat{P}_{Z_1^\ell}\|Q^k)\Big|\hat{P}_{Z_1^\ell} \in E_k(Q)\Big) \leq \epsilon(n, k).$$

Proof We follow the same steps as in the proof of the conditional limit theorem in [19]. Recall that if we write U_k for the uniform distribution on A^k, then the W_k^* that uniquely achieves $D^* = \min_{W \in E_k(Q)} D(W\|U_k)$ is $W_k^* = Q^k$. We partition $E_k(Q)$ into $B_{2\delta}$ and $C = E_k(Q) - B_{2\delta}$, where $B_{2\delta} = \{W \in E_k(Q) : D(W\|U_k) \leq D^* + 2\delta\}$, with $\delta = \delta_{n,k}$. Then, writing ν_ℓ for the distribution of $\hat{P}_{Z_1^\ell}$,

$$\nu_\ell(C|E_k(Q)) = \frac{\nu_\ell(C \cap E_k(Q))}{\nu_\ell(E_k(Q))} \leq \frac{\nu_\ell(C)}{\nu_\ell(B_{2\delta})}.$$

Next we bound the above numerator and denominator. For the numerator, writing again $\mathcal{P}_\ell$ for the set of ℓ-types on A^k,

$$\nu_\ell(C) \overset{(a)}{=} \sum_{W \in C \cap \mathcal{P}_\ell} U_k^\ell(T(W))$$

$$\overset{(b)}{\leq} \sum_{W \in C \cap \mathcal{P}_\ell} e^{-\ell D(W\|U_k)}$$

$$\overset{(c)}{\leq} |E_k(Q) \cap \mathcal{P}_\ell| e^{-\ell(D^* + 2\delta)}$$

$$\overset{(d)}{\leq} (\ell + 1)^{m^k} e^{-\ell(D^* + 2\delta)},$$

where $T(W)$ in (a) denotes the type class of all strings of length ℓ in A^k with type W, (b) is a standard property [19], (c) follows from the definition of C and the fact that $E_k(Q) \cap \mathcal{P}_\ell \subset E_k(Q)$, and (d) follows from the standard observation that $|E_k(Q) \cap \mathcal{P}_\ell| \leq |\mathcal{P}_\ell| \leq (\ell + 1)^{m^k}$. Similarly, letting W_0 denote the type from Lemma 2.2,

$$
\begin{aligned}
\nu_\ell(B_{2\delta}) &\geq \nu_\ell(B_\delta) \\
&= \sum_{W \in B_\delta \cap \mathcal{P}_\ell} U_k^\ell(T(W)) \\
&\geq U_k^\ell(T(W_0)) \\
&\geq (\ell + 1)^{-m^k} e^{-\ell D(W_0 \| U)} \\
&\geq (\ell + 1)^{-m^k} e^{-\ell(D^* + \delta)}.
\end{aligned}
$$

Combining these bounds, we obtain,

$$
\mathbb{P}\big(\hat{P}_{Z_1^\ell} \in C \big| \hat{P}_{Z_1^\ell} \in E_k(Q)\big) \leq (\ell + 1)^{2m^k} e^{-\ell \delta},
$$

or,

$$
\mathbb{P}\Big(D(\hat{P}_{Z_1^\ell} \| U_k) > D^* + 2\delta \Big| \hat{P}_{Z_1^\ell} \in E_k(Q)\Big) \leq (\ell + 1)^{2m^k} e^{-\ell \delta}.
$$

Since the set $E_k(Q)$ is closed and convex, we may apply the Pythagorean identity for relative entropy [19] to conclude that:

$$
\mathbb{P}\Big(D(\hat{P}_{Z_1^\ell} \| Q^k) > 2\delta | \hat{P}_{Z_1^\ell} \in E_k(Q)\Big) \leq (\ell + 1)^{2m^k} e^{-\ell \delta}.
$$

Thus,

$$
\mathbb{E}D(\hat{P}_{Z_1^\ell} \| Q^k) \big| \hat{P}_{Z_1^\ell} \in E_\ell(Q) \leq (\ell + 1)^{2m^k} e^{-\ell \delta} \max_{P \in E_k(Q)} D(P \| Q^k) + 2\delta.
$$

The claimed bound now follows by Lemma 2.4 on taking $\ell = k/n$. $\qquad\square$

Lemma 2.4 *For any n-type Q, $\max_{W \in E_k(Q)} D(W \| Q^k) \leq k \log n$.*

Proof If $a_1^k \in A^k$ is such that $Q^k(a_1^k) = \prod_{i=1}^k Q(a_i) = 0$, then $Q(a_{i_0}) = 0$ for some i_0. Since $Q(a_{i_0}) = \frac{1}{k} \sum_{j=1}^k W_j(a_{i_0})$, we must have $W_1(a_{i_0}) = \cdots = W_k(a_{i_0}) = 0$, which implies that $W(a_1^k) = 0$.

On the other hand, if $Q^k(a_1^k) > 0$ then $Q^k(a_1^k) \geq \frac{1}{n^k}$. Thus, for any $W \in E_k(Q)$,

$$
\begin{aligned}
D(W \| Q^k) &= \sum_{a_1^k \in A^k} W(a_1^k) \log \frac{W(a_1^k)}{Q^k(a_1^k)} \\
&\leq \sum_{a_1^k \in A^k} W(a_1^k) \log \frac{W(a_1^k)}{(\frac{1}{n})^k} \\
&= k \log n - H(W) \\
&\leq k \log n,
\end{aligned}
$$

as required. □

Theorem 2.1 follows from (14) combined with Lemma 2.3.

Acknowledgements We wish to thank Andrew Barron for pointing out several useful references for the historical development outlined in our Introduction.

References

1. P.H. Algoet and B.H. Marcus. Large deviation theorems for empirical types of Markov chains constrained to thin sets. *IEEE Trans. Inform. Theory* **38**(4), 1276–1291, July (1992).
2. S. Artstein, K. Ball, F. Barthe and A. Naor. Solution of Shannon's problem on the monotonicity of entropy. *J. Amer. Math. Soc.* **17**(4), 975–982 (2004).
3. D. Bakry and M. Émery. Diffusions hypercontractives. In: *Seminaire de probabilités XIX 1983/84*, pp. 177–206, Springer (1985).
4. A.D. Barbour, O. Johnson, I. Kontoyiannis and M. Madiman. Compound Poisson approximation via information functionals. *Electron. J. Probab.* **15**, 1344–1369 (2010).
5. A.R. Barron. Entropy and the central limit theorem. *Ann. Probab.* **14**(1), 336–342 January (1986).
6. A.R. Barron. Information theory in probability, statistics, learning, and neural nets. In: Y. Freund and R.E. Schapire, editors, *Proceedings of the Tenth Annual Conference on Computational Learning Theory (COLT)*, Nashville, Tennessee, July 1997. Available at: http://www.stat.yale.edu/~arb4/publications_files/COLT97.pdf
7. A.R. Barron. Limits of information, Markov chains, and projection. In: *2000 IEEE International Symposium on Information Theory (ISIT)*, Sorrento, Italy, June (2000).
8. M. Bayarri and J.O. Berger. The interplay of Bayesian and frequentist analysis. *Statistical Science* **19**(1), 58–80 (2004).
9. N.M. Blachman. The convolution inequality for entropy powers. *IEEE Trans. Inform. Theory* **11**(2), 267–271 April (1965).
10. S.G. Bobkov and F. Götze. Exponential integrability and transportation cost related to logarithmic Sobolev inequalities. *J. Funct. Anal.* **163**(1), 1–28 (1999).
11. S.G. Bobkov and M. Ledoux. On modified logarithmic Sobolev inequalities for Bernoulli and Poisson measures. *J. Funct. Anal.* **156**(2), 347–365 (1998).

12. S. Boucheron, G. Lugosi and P. Massart. *Concentration inequalities: A nonasymptotic theory of independence.* Oxford University Press, Oxford, U.K. (2013).
13. L.D. Brown. A proof of the central limit theorem motivated by the Cramér–Rao inequality. In: *Statistics and Probability: Essays in Honor of C.R. Rao*, pp. 141–148, North-Holland, Amsterdam (1982).
14. L.D. Brown. *Fundamentals of statistical exponential families: With applications in statistical decision theory.* Volume 9 of *IMS Lecture Notes Monograph Series*, Institute of Mathematical Statistics, Hayward, CA (1986).
15. S. Bubeck and S. Ganguly. Entropic CLT and phase transition in high-dimensional Wishart matrices. *International Mathematics Research Notices* **2018**(2), 588–606 (2018).
16. E.A Carlen and D. Cordero-Erausquin. Subadditivity of the entropy and its relation to Brascamp–Lieb type inequalities. *Geometric & Functional Analysis* **19**(2), 373–405 (2009).
17. E.A. Carlen and A. Soffer. Entropy production by block variable summation and central limit theorems. *Comm. Math. Phys.* **140**(2), 339–371 (1991).
18. M.H.M. Costa and T.M. Cover. On the similarity of the entropy power inequality and the Brunn–Minkowski inequality. *IEEE Trans. Inform. Theory* **30**(6), 837–839 November (1984).
19. T.M. Cover and J.A. Thomas. *Elements of information theory.* J. Wiley & Sons, New York, second edition (2012).
20. I. Csiszár. Eine informationstheoretische ungleichung und ihre anwendung auf den beweis der ergodizität von Markoffschen ketten. *Publ. Math. Inst. Hungar. Acad. Sci.* **8**, 85–108 (1963).
21. I. Csiszár. Information-type measures of difference of probability distributions and indirect observations. *Studia Sci. Math. Hungar.* **2**, 299–318 (1967).
22. I. Csiszár. *I*-divergence geometry of probability distributions and minimization problems. *Ann. Probab.* **3**(1), 146–158 February (1975).
23. I. Csiszár. Sanov property, generalized *I*-projection and a conditional limit theorem. *Ann. Probab.* **12**(3), 768–793 August (1984).
24. I Csiszár. Information theoretic methods in probability and statistics. In: *1997 IEEE International Symposium on Information Theory (ISIT)*, Ulm, Germany, June (1997).
25. I. Csiszár. The method of types. *IEEE Trans. Inform. Theory* **44**(6), 2505–2523, October (1998).
26. I. Csiszár. A simple proof of Sanov's theorem. *Bulletin of the Brazilian Mathematical Society* **37**(4) (2006).
27. I. Csiszár, T.M. Cover and B.S. Choi. Conditional limit theorems under Markov conditioning. *IEEE Trans. Inform. Theory* **33**(6), 788–801 November (1987).
28. I. Csiszár and J. Körner. *Information theory: Coding theorems for discrete memoryless systems.* Academic Press, New York (1981).
29. E.B. Davies and B. Simon. Ultracontractivity and the heat kernel for Schrödinger operators and Dirichlet Laplacians. *J. Funct. Anal.* **59**(2), 335–395 (1984).
30. B. De Finetti. Sul significato soggettivo della probabilita. *Fundamenta Mathematicae* **17**(1), 298–329 (1931).
31. B. De Finetti. La prévision: ses lois logiques, ses sources subjectives. *Ann. Inst. Henri Poincaré* **7**(1), 1–68 (1937).
32. A. Dembo, T.M. Cover and J.A. Thomas. Information-theoretic inequalities. *IEEE Trans. Inform. Theory* **37**(6), 1501–1518 November (1991).
33. P. Diaconis. Finite forms of de Finetti's theorem on exchangeability. *Synthese* **36**(2), 271–281 (1977).
34. P. Diaconis and D.A. Freedman. Finite exchangeable sequences. *Ann. Probab.* **8**(4), 745–764 (1980).
35. J. Fritz. An information-theoretical proof of limit theorems for reversible Markov processes. In: *Transactions of the Sixth Prague Conference on Information Theory, Statistical Decision Functions, Random Processes (Tech. Univ., Prague, 1971; dedicated to the memory of Antonín Špaček)*, pp. 183–197, Academia, Prague (1973).
36. L. Gavalakis and I. Kontoyiannis. Entropy and the discrete central limit theorem. *arXiv e-prints*, 2106.00514 [math.PR], June (2021).

37. L. Gavalakis and I. Kontoyiannis. An information-theoretic proof of a finite de Finetti theorem. *Electron. Comm. Probab.* **26**, 1–5 (2021).
38. P. Groeneboom, J. Oosterhoff and F.H. Ruymgaart. Large deviation theorems for empirical probability measures. *Ann. Probab.* **7**(4), 553–586 August (1979).
39. L. Gross. Logarithmic Sobolev inequalities. *Amer. J. Math.* **97**(4), 1061–1083 (1975).
40. D. Guo, S. Shamai and S. Verdú. Mutual information and minimum mean-square error in Gaussian channels. *IEEE Trans. Inform. Theory* **51**(4), 1261–1282 April (2005).
41. J. Hájek. On a property of normal distributions of any stochastic process. *Czechoslovak Math. J.* **8**(83), 610–618 (1958).
42. J. Hájek. A property of J-divergences of marginal probability distributions. *Czechoslovak Math. J.* **8**(3), 460–463 (1958).
43. P. Harremoës. Binomial and Poisson distributions as maximum entropy distributions. *IEEE Trans. Inform. Theory* **47**(5), 2039–2041 July (2001).
44. P. Harremoës and K.K. Holst. Convergence of Markov chains in information divergence. *J. Theoret. Probab.* **22**(1), 186–202 March (2009).
45. P. Harremoës, O. Johnson and I. Kontoyiannis. Thinning and the law of small numbers. In: *2007 IEEE International Symposium on Information Theory (ISIT)*, pp. 1491–1495, Nice, France, June (2007).
46. P. Harremoës, O. Johnson and I. Kontoyiannis. Thinning, entropy, and the law of thin numbers. *IEEE Trans. Inform. Theory* **56**(9), 4228–4244 (2010).
47. E. Hewitt and L.J. Savage. Symmetric measures on Cartesian products. *Trans. Amer. Math. Soc.* **80**(2), 470–501 (1955).
48. O. Johnson. Entropy inequalities and the central limit theorem. *Stoch. Proc. Appl.* **88**(2), 291–304 (2000).
49. O. Johnson. *Information theory and the central limit theorem.* Imperial College Press, London, U.K. (2004).
50. O. Johnson. Log-concavity and the maximum entropy property of the Poisson distribution. *Stoch. Proc. Appl.* **117**(6), 791–802 June (2007).
51. O. Johnson and A.R. Barron. Fisher information inequalities and the central limit theorem. *Probab. Theory Related Fields* **129**(3), 391–409 (2004).
52. O. Johnson, I. Kontoyiannis and M. Madiman. Log-concavity, ultra-log-concavity, and a maximum entropy property of discrete compound Poisson measures. *Discrete Applied Mathematics* **161**(9), 1232–1250 (2013).
53. O. Johnson and Yu.M. Suhov. Entropy and convergence on compact groups. *J. Theoret. Probab.* **13**(3), 843–857 (2000).
54. J.H.B. Kemperman. On the optimum rate of transmitting information. In: *Probability and information theory*, pp. 126–169. Springer-Verlag (1969).
55. D.G. Kendall. Information theory and the limit-theorem for Markov chains and processes with a countable infinity of states. *Ann. Inst. Statist. Math.* **15**(1), 137–143 May (1963).
56. A.N. Kolmogorov. A new metric invariant of transitive dynamical systems and Lebesgue space automorphisms. *Dokl. Acad. Sci. USSR*, volume 119, 861–864 (1958).
57. I. Kontoyiannis, P. Harremoës and O. Johnson. Entropy and the law of small numbers. *IEEE Trans. Inform. Theory* **51**(2), 466–472 February (2005).
58. I. Kontoyiannis and M. Madiman. Sumset and inverse sumset inequalities for differential entropy and mutual information. *IEEE Trans. Inform. Theory* **60**(8), 4503–4514 August (2014).
59. S. Kullback. A lower bound for discrimination information in terms of variation. *IEEE Trans. Inform. Theory* **13**(1), 126–127 January (1967).
60. M. Ledoux. Isoperimetry and Gaussian analysis. In: *Lectures on probability theory and statistics*, volume 1648 of *Lecture Notes in Mathematics*, pp. 165–294, Springer, Berlin (1996).
61. M. Ledoux. On Talagrand's deviation inequalities for product measures. *ESAIM Probab. Statist.* **1**, 63–87 (electronic) (1997).
62. M. Ledoux. *The concentration of measure phenomenon.* American Mathematical Society, Providence, RI (2001).

63. Ju.V. Linnik. An information-theoretic proof of the central limit theorem with Lindeberg conditions. *Theory Probab. Appl.* **4**, 288–299 (1959).
64. J. Liu, T.A. Courtade, P. Cuff and S. Verdú. Brascamp–Lieb inequality and its reverse: An information theoretic view. In: *2016 IEEE International Symposium on Information Theory (ISIT)*, pp. 1048–1052, Barcelona, Spain, July (2016).
65. M. Madiman and A.R. Barron. Generalized entropy power inequalities and monotonicity properties of information. *IEEE Trans. Inform. Theory* **53**(7), 2317–2329 July (2007).
66. M. Madiman, O. Johnson and I. Kontoyiannis. Fisher information, compound Poisson approximation, and the Poisson channel. In: *2007 IEEE International Symposium on Information Theory (ISIT)*, pp. 976–980, Nice, France, June (2007).
67. M. Madiman and I. Kontoyiannis. Entropy bounds on abelian groups and the Ruzsa divergence. *IEEE Trans. Inform. Theory* **64**(1), 77–92, January (2016).
68. K. Marton. Bounding $\bar{d}$-distance by informational divergence: A method to prove measure concentration. *Ann. Probab.* **24**(2), 857–866 April (1996).
69. N. O'Connell. Information-theoretic proof of the Hewitt–Savage zero-one law. Technical report, Hewlett-Packard Laboratories, Bristol, U.K., June (2000).
70. D.S. Ornstein. Bernoulli shifts with the same entropy are isomorphic. *Advances in Mathematics* **4**(3), 337–352 (1970).
71. D.S. Ornstein. Imbedding Bernoulli shifts in flows. In: *Contributions to ergodic theory and probability*, pp. 178–218, Springer (1970).
72. D.S. Ornstein. Two Bernoulli shifts with infinite entropy are isomorphic. *Advances in Mathematics* **5**(3), 339–348 (1970).
73. M. Raginsky and I. Sason. *Concentration of Measure Inequalities in Information Theory, Communications and Coding.* Foundations and Trends in Communications and Information Theory. NOW Publishers, Boston, MA (2018).
74. A. Rényi. On measures of entropy and information. In: *Proc. 4th Berkeley Sympos. Math. Statist. and Prob., Vol. I*, pp. 547–561, Univ. California Press, Berkeley, Calif. (1961).
75. I.N. Sanov. On the probability of large deviations of random variables. *Mat. Sb.* **42**, 11–44 (1957). English translation in *Sel. Transl. Math. Statist. Probab.* **1**, 213–244 (1961).
76. C.E. Shannon. A mathematical theory of communication. *Bell System Tech. J.* **27**(3), 379–423, 623–656 (1948).
77. C.E. Shannon. The bandwagon. *IRE Trans. Inform. Theory* **2**(1), 3, March (1956).
78. R. Shimizu. On Fisher's amount of information for location family. In: G.P. Patil, S. Kotz and J.K. Ord, editors, *A Modern Course on Statistical Distributions in Scientific Work*, pp. 305–312, Springer, Dordrecth, Netherlands (1975).
79. A.J. Stam. Some inequalities satisfied by the quantities of information of Fisher and Shannon. *Information and Control* **2**(2), 101–112 (1959).
80. T. Tao. Sumset and inverse sumset theory for Shannon entropy. *Combinatorics, Probability and Computing* **19**(4), 603–639 July (2010).
81. F. Topsøe. Information-theoretical optimization techniques. *Kybernetika* **15**(1), 8–27 (1979).
82. A.M. Tulino and S. Verdú. Monotonic decrease of the non-Gaussianness of the sum of independent random variables: A simple proof. *IEEE Trans. Inform. Theory* **52**(9), 4295–4297 September (2006).
83. C. Villani. *Optimal transport: Old and new.* Springer, Berlin (2009).
84. D. Voiculescu. The analogues of entropy and of Fisher's information measure in free probability theory, I. *Comm. Math. Phys.* **155**(1), 71–92 (1993).
85. D. Voiculescu. The analogues of entropy and of Fisher's information measure in free probability theory, II. *Inventiones Mathematicae* **118**(1), 411–440 (1994).
86. D. Voiculescu. The analogues of entropy and of Fisher's information measure in free probability theory III: The absence of Cartan subalgebras. *Geometric & Functional Analysis* **6**(1), 172–199 (1996).
87. D. Voiculescu. The analogues of entropy and of Fisher's information measure in free probability theory, IV: Maximum entropy and freeness, in free probability theory. *Fields Inst. Commun.* **12**, 293–302 (1997).

88. J. Wolfowitz. *Coding theorems of information theory*. Springer, Berlin (1961).
89. Y. Yu. On the entropy of compound distributions on nonnegative integers. *IEEE Trans. Inform. Theory* **55**(8), 3645–3650 August (2009).

Part VIII
Logic

It would seem that the title of Angus Macintyre's article *Between the rings $\mathbb{Z}/p^n\mathbb{Z}$ and the ring $\mathbb{Z}_p$: Issues of axiomatizability, definability and decidability* says it all. However, one can add that it is also a historical and insightful introduction to a fundamental area of the intersection of logic and arithmetic.

Between the Rings $\mathbb{Z}/p^n\mathbb{Z}$ and the Ring $\mathbb{Z}_p$: Issues of Axiomatizability, Definability and Decidability

Angus J. Macintyre

1 Introduction

This paper is about the metamathematics of the rings $\mathbb{Z}$, $\mathbb{Z}/p^n\mathbb{Z}$, and $\mathbb{Z}_p$, the p-adic completion of $\mathbb{Z}$. The mathematical study of the first two rings goes back to the early days of our subject, and flourishes still. The definition of $\mathbb{Z}_p$, by Hensel, at the very end of the nineteenth century, has enriched more and more of mathematics in the intervening 120 years (John Tate's thesis being an example of great beauty).

Metamathematics did not exist when $\mathbb{Z}_p$ was introduced, and it took a long time for its central ideas, namely axiomatisation, definability and decidability, to become attractive to a variety of mathematicians. However, from the beginning exceptional mathematicians began to contribute (for example Hilbert, Skolem, von Neumann, Gödel, Tarski), and by the 1930s Tarski had established marvellous results on the metamathematics of the real field $\mathbb{R}$, which would prove in the ensuing century to generalise to methods extremely powerful in diophantine geometry. However, for $\mathbb{Z}$ the analysis initiated by Gödel in the 1930s revealed barriers to our understanding of the metamathematics of $\mathbb{Z}$, and the culminating result in 1970 (when Matjasevic completed a daring programme of Davis, Putnam and Robinson) is beyond doubt one of the supreme achievements of mathematical logic, showing the unsolvability of Hilbert's 10th problem.

The present paper concerns the metamathematics of the $\mathbb{Z}_p$, uniformly in p as far as possible. Some 6 years before Matjasevic, independent work by Ax and Kochen, and Ersov, solved positively the axiomatisation and decidability problems. The definability problems were solved by combination of ideas of Paul Cohen, Denef

A. J. Macintyre (✉)
University of Edinburgh, Edinburgh, UK

Queen Mary University of London, London, UK
e-mail: a.macintyre@qmul.ac.uk

© The Author(s), under exclusive license to Springer Nature Switzerland AG 2023
J.-M. Morel, B. Teissier (eds.), *Mathematics Going Forward*, Lecture Notes in Mathematics 2313, https://doi.org/10.1007/978-3-031-12244-6_27

and Macintyre. Uniformity problems, including decidability, were solved by Ax. But many problems remain concerning the computational complexity of $\mathbb{Z}_p$. These matters are all described in the paper.

Finally, the metamathematical issues for the class of all $\mathbb{Z}/p^n\mathbb{Z}$ are discussed. For an *individual* $\mathbb{Z}/p^n\mathbb{Z}$ the issues are trivial, as these are explicitly given finite rings. But to understand the situation as either the n or the p, or both, vary, is a very substantial matter. If we fix $n = 1$ and let p vary, Ax solved all the problems, using major results from number theory. But if p is *fixed* and n varies, the analysis is hard, and depends on refined analysis of the $\mathbb{Z}_p$ case. This is an example of the always fascinating situations where uniformities in finite settings get understood by passing to some kind of completion.

1.1 The Diagrams and Some Axioms Connected to Them

The title refers to the diagrams of commutative rings with 1:

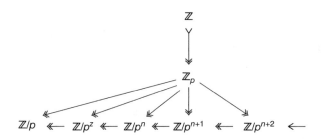

L_{rings} is the standard first-order language for ring theory, with primitives $+, \cdot, -, 0, 1$ [34]. The structures in the diagram above are all models of the axioms (in L_{rings}) for commutative rings with 1. All but $\mathbb{Z}$ are local rings. The class of local rings is axiomatizable by a single sentence of L_{rings} (see the very lucid [40]). The residue field is uniformly interpretable [12]. In the diagram(s) above it is $\mathbb{F}_p$. This property is also given by an axiom, but not uniformly in p. In the above diagram all the maximal ideals are principal, and this can be said by a single sentence uniformly for local rings. Finally, a first-order property common to all the local rings above is that the set of principal ideals is linearly ordered by reverse inclusion (i.e. the rings are so-called chain rings), with order $<$ given by

$$(1) \subset (p) \subset (p^2) \cdots \subset (0)$$

This chain will be crucial below, in particular in showing that we have no need to appeal to any language for valued rings. Note that in the diagram $\mathbb{Z}_p$ and $\mathbb{Z}/p\mathbb{Z}$ are the only local domains. Some of the most outstanding work in so-called "applied model theory" has been concerned been concerned with uniformities in p for these structures [1–4, 22]. The present paper is concerned with neglected issues, and I have studied those issues in much joint work both with Paola D'Aquino and with Jamshid Derakhshan [16, 21].

1.2   Tarski's Legacy

One may fairly credit Tarski [38] with founding a deep mathematical theory of truth, satisfaction, definability and (with the arrival of computability theory) decidability. His work on the field $\mathbb{R}$ ([37], but with publication long delayed by war) is a masterpiece with decisive and inspirational results on axiomatization, structure of definable sets, and decidability. Combined with later, quite different, work by Abraham Robinson and Anatoli Malcev on similar issues for a variety of classical (and also novel) structures, this provided the basic repertoire for people entering "applied model theory" from the late 1950s onward. This paper is concerned with issues arising in the 1960s work of Ax–Kochen and Ersov on Henselian fields. In real settings, a shift of emphasis on the structure of definable sets in higher dimension by van den Dries, in the late 1970s, led to the beautiful and amazingly fertile notion of o-minimality [39], and Wilkie's proof of o-minimality of the real exponential field initiated an astonishing ongoing interaction between logic, real analytic geometry and diophantine geometry [42]. As regards decidability, Wilkie and I in [33] proved the decidability of the real exponential field, under the assumption that Schanuel's Conjecture is merely true (some assumption of this kind seems to me unavoidable). I want to point out that our "algorithm" is not very informative, involving a standard "wait and see" procedure in which two recursive enumerations are at work. Alas, often this is all we get when we look for decidability. The "wait and see" method allows us to show that the structure under analysis is not Gödelian, but cannot be expected to yield much detailed algorithmic knowledge. I return to this later in Sect. 3.3.

1.3   Metamathematics of $\mathbb{Z}$

The theory $Th(M)$ of an L-structure M is defined as the set of L-sentences true in M. $\equiv$, elementary equivalence, is the relation (between L-structures) of having the same theory.

$Th(\mathbb{Z})$ is only dimly understood. $\mathbb{Z}$ is the fundamental Gödelian ring, and has an undecidable theory. In particular, $\mathbb{Z}$ has no computable complete set of axioms, and the set of polynomials over $\mathbb{Z}$ solvable in $\mathbb{Z}$ is not computable [17].

The definable relations over $\mathbb{Z}$, the so-called arithmetic sets, have a beautiful hierarchical structure, in which more and more quantifier alternations are needed to exhaust all definitions, but most of the individual arithmetic sets have no intelligible structure. This is in total contrast to what Tarski [37] showed for $\mathbb{R}$, where the definable sets need no quantifiers if $<$ is taken as a primitive. Another contrasting result is that of Wilkie [42] for the real exponential, where only existential formulas are needed.

1.4   Metamathematics of PA

PA is a natural recursive set of axioms [28] true in $\mathbb{Z}$, basically saying that any nonempty definable set bounded below has a least element (there are some isolated, simple axioms needed too). PA is incomplete by Gödel (and even for sentences expressing unsolvability of diophantine equations, but it is nevertheless very powerful, and it is notoriously hard to find mathematically interesting sentences true in $\mathbb{Z}$ but not provable in PA. I will look later at the strength of PA in connection with the structures in the main diagram. For example, I will connect the theory of class of quotient rings M/q, where M is a nonstandard model of PA and q is a (possibly) nonstandard prime power in M, and the theory of $\{\mathbb{Z}/p^n\mathbb{Z} : p, n \text{ varying}\}$. The moral will be that Gödel Incompleteness fades away in residue rings. This will be discussed in Sect. 3.

2   Classes $\mathbb{C}$ of L-Structures

Although one is often dealing with an individual structure, like $\mathbb{R}$, $\mathbb{Z}$ or $\mathbb{Z}_p$, there are times when one has to deal with classes of structures, with a view to uniformity results. Of special importance for the present paper are the classes

  (i) $\{\mathbb{Z}_p : p \text{ prime}\}$
  (ii) $\{\mathbb{F}_p : p \text{ prime}\}$
  (iii) For fixed p, $\{\mathbb{Z}/p^n : n = 0, 1, 2...\}$
  (iv) $\{\mathbb{Z}/p^n\mathbb{Z} : p, n \text{ varying}\}$

Thus one is led to define $Th(\mathbb{C})$ as $\bigcap_{\mathcal{M} \in \mathbb{C}} Th(\mathcal{M})$.

For fixed p, the individual $\mathbb{Z}/p^n\mathbb{Z}$, for $n \geq 0$ are not interesting since they are finite, finitely axiomatizable and decidable. The structure of definable relations is more subtle, but then we look for uniformity in n. Thus we look at $\mathbb{C}_p$ defined as the class of all $\mathbb{Z}/p^n\mathbb{Z}$ for $n \geq 0$. Now serious mathematics comes in when we ask about axioms for $Th(\mathbb{C}_p)$, decidability, and the fine structure of definability.

2.1   Let p Vary

Now let p, n vary, and let $\mathbb{C}$ be the union of all the $\mathbb{C}_p$. The problems set in the preceding subsection become much more demanding.

2.2 Ax

Most important of all the classes here are two classes of fields:

(1) $\mathbb{C}_{prime}$, the class of all $\mathbb{Z}/p$
(2) $\mathbb{C}_{Fin}$, the class of all finite fields.

Note that fields which are not prime do not occur in the diagram.

Ax's work [1] gave us our fundamental understanding of axioms, and decidability in models of the theories of the two classes just defined. Ax used ideas around two great theorems of number theory, Weil's Riemann Hypothesis for curves over finite fields, and Cebotarev's Theorem on decomposition groups of primes in number fields [1]. They fit marvellously with the metamathematical task Ax set himself.

3 The Metamathematical Analysis of $\mathbb{Z}_p$

In the mid 1960s Ax and Kochen [2–4] and Ersov [22] established a new research area, the model theory of Henselian fields, which has expanded greatly in the subsequent 60 years, and has led to significant interaction between model theory and number theory. There are currently some hopes that this work can link to recent profound work in p-adic Hodge Theory, due to Fargues and Fontaine [23].

The logic component of the research is very lucidly explained in van den Dries' Cetraro Notes [40]. These notes are mainly concerned with Henselian valued fields with residue field of characteristic 0, and with finitely ramified mixed characteristic Henselian fields.

In this paper I work with the valuation rings $\mathcal{O}_K$ of such fields K. Thus I work with $\mathbb{Z}_p$ rather than $\mathbb{Q}_p$. Since $\mathbb{Q}_p$ is interpretable in $\mathbb{Z}_p$ uniformly in p, and $\mathbb{Z}_p$ is definable in $\mathbb{Q}_p$ uniformly in p (even by a low complexity formula) [12], the two formulations are equivalent.

$\mathbb{Z}_p$ is natural in this paper, since we want to connect the metamathematics of the finite $\mathbb{Z}/p^n\mathbb{Z}$ to that of the projective limit of the $\mathbb{Z}/p^n\mathbb{Z}$, i.e. $\mathbb{Z}_p$. Note that $\mathbb{Z}/p\mathbb{Z}$ and $\mathbb{Z}_p$ are domains, but the remaining $\mathbb{Z}/\{\mathbb{Z}/p^n\}$ are not.

3.1 The Axioms for $\mathbb{Z}_p$

The following is an adaptation of what was given in [3], but here formulated in L_{rings}. To avoid taking a primitive for valuation, we work with the chain Γ of nonzero principal ideals $\langle x \rangle$ (cf. Sect. 1). This carries the structure of an ordered semigroup via

$$\langle x \rangle \oplus \langle y \rangle = \langle xy \rangle.$$

Define $v(x)$ as $\langle x \rangle$, even when $x = 0$, and treat $\langle 0 \rangle$ as ∞, greater than any element of Γ.

Now we formulate the Axioms for $\mathbb{Z}_p$, en bloc, assuming the preceding notation.

AXIOMS R *is a local domain, of characteristic 0, with maximal ideal generated by p, and residue field of cardinality p. The set of nonzero principal ideals forms a chain Γ under reverse inclusion. With respect to the $\oplus$ introduced above, Γ becomes an ordered semigroup which is naturally isomorphic to the nonnegative part of a model of Presburger arithmetic. If we adjoin $\langle 0 \rangle$ as ∞, and define v as above, then v is a valuation on Γ. $\langle p \rangle$ is the least positive element of Γ. R is henselian with respect to v.*

Theorem 3.1 AXIOMS *is a complete set of axioms for $\mathbb{Z}_p$. $\mathbb{Z}_p$ is decidable.*

The reader may well prefer the more standard axiomatizations in the literature. The point of the version above is that it adapts well to axiomatizations for the $\mathbb{Z}/p^n\mathbb{Z}$.

Note The axiomatization is entirely explicit, analogous to Tarski's. But the decidability proof is lacking in any explicit information, and merely shows that $\mathbb{Z}_p$ is not Gödelian. I turn now to the issue of definable sets, where deep results have been obtained.

3.2 Definable Sets

Though [2–4] and [22] obtained useful information about definable sets in both $\mathbb{Q}_p$ and $\mathbb{Z}_p$, they did not get far enough to obtain a clear picture of the landscape of definable sets. In 1974 [31], by reflecting on Tarski's Elimination for $\mathbb{R}$, and on the solvability of the absolute Galois group of $\mathbb{Q}_p$, I found an Elimination Theorem which nearly a decade later combined beautifully with ideas of Paul Cohen [13] and Jan Denef [18] allowing Denef to make dramatic advances in our understanding of p-adic Poincaré series.

Tarski had used, and needed, a primitive for $>$ to get his quantifier-elimination. $>$ is of course definable in $\mathbb{R}$ in L_{rings}, via

$$x > 0 \iff x \neq 0 \wedge (\exists y)(y^2 = x).$$

In any ring R we can define $P_n(x)$ as

$$x \neq 0 \wedge (\exists y)(y^n = x).$$

Tarski used P_2.

I proved that for either $\mathbb{Q}_p$ or $\mathbb{Z}_p$ one has quantifier-elimination in the definitional extension got by enriching L_{rings} by new predicate symbols $P_2, P_3, \ldots, P_k \ldots$, for $k > 1$.

Note that some writers use P_n simply for the nth powers, rather than for the nonzero n^{th} powers. We stick with the convention given above. Of course the two formalisms are equivalent for definability results.

It is to be noted that the elimination is not uniform in p. In addition, even for a single p, it is known to be computable (since one already had decidability for $\mathbb{Q}_p$ and one could appeal to a "wait and see" argument) but no one has convincingly pushed the algorithm down to elementary recursive. However, Weisspfenning has given a primitive recursive elimination [41]. But in general primitive recursive procedures do not give useful bounds.

Note that one can deduce that in either the compact ring $\mathbb{Z}_p$, or in the locally compact $\mathbb{Q}_p$, definable sets are Haar measurable. Denef used his method to prove rationality results (in p^{-s}) for integrals of the form

$$\int_A |f(\bar{x})|^s \, d\bar{x}$$

where A is a definable subset of $\mathbb{Q}^n$ or $\mathbb{Z}^n$, f is a definable function on A, $|\cdot|$ is the standard p-adic absolute value, and s is a real number chosen in a suitable range [18].

A standard procedure allows these rationality results to be used to prove rationality in T of a huge variety of formal power series

$$\sum_{n=0}^{\infty} c_n T^n$$

arising in number theory, publicized by Borevich and Shafarevich [8].

Denef's original work did not yield any kind of uniformity in p, in situations where one might hope for this, e.g. when A and f are defined over $\mathbb{Q}$ or $\mathbb{Z}$. But later Denef, Pas and I proved substantial uniformity results [19, 32, 36], giving complicated proofs. Some aspects of this work were later given a much deeper analysis, in the superb work of Denef and Loeser [20] on motivic integration. I use the uniformities of Denef, Pas and myself to get explicit axioms for $Th(\mathbb{C}_p)$, and I know no other way to do this. It remains mysterious to me that one may need the much more sophisticated uniform p-adic analysis to get axioms for the $\mathbb{C}_p$, whose members are very accessible and congenial to beginners in number theory. Our analysis also gives axioms for $Th(\mathbb{C})$ where both p and n vary. It also leads to axioms for $Th(\mathbb{Z}/m\mathbb{Z})$ via [15].

3.3   Decidability

Both [2–4] and [22] proved the decidability of $\mathbb{Z}_p$ and $\mathbb{Q}_p$, but by different methods. Ax and Kochen [2–4] went via a complete, computable axiomatization, proved using infinitistic methods involving ultraproducts, and Ersov [22] used methods of Abraham Robinson. Neither method is constructive.

Paul Cohen, in 1969 [13], gave a primitive recursive decision problem , using a cell-decomposition method proved by a brilliant and ingenious induction, using Hensel's Lemma only (and no fancy model theory). His bare-hands method was preceded by an analogous proof for the real case (also published in [13]). Cell-decomposition in o-minimality had not yet emerged, though real geometers were already involved with related special cases. In the p-adic case, Cohen's cell-decomposition was transformed by Denef [20] into a central technique of p-adic definability.

It remains a very challenging problem to get better bounds for the complexity of $\mathbb{Z}_p$ and $\mathbb{Q}_p$. I believe that there should be elementary recursive bounds (towers of exponentials) for the central constructions such as cell-decompositions, quantifier-eliminations and the like. Such bounds have been found in the real case, where there is an extensive literature and deep results [5]. For the real field there are double-exponential upper bounds (in the appropriate parameters), for which I advise the reader to consult the literature. For Presburger arithmetic there are lower bounds which are greater than the upper bounds for the ordered field $\mathbb{R}$ [24]. From this it follows that the basic decision procedures for the p-adic case are strictly harder than for the real case.

I have made two serious attempts (once with Lavinia Egidi, and once with Raf Cluckers) to get elementary recursive bounds, but neither succeeded. I believe that I have now been able to reach the final goal, but this is not yet published. The reader may wish to consult Scott Brown's [10] for a striking iterated exponential bound arising from the original Ax–Kochen transfer principle.

One should establish bounds for Denef's version of Cell-Decomposition. My 1989 paper on uniform bounds [32] does quite a bit in this direction, and Denef and Pas have excellent uniformities without explicit bounds.

4 Ax on Finite Fields and p-Adic Uniformities

4.1 $\mathbb{C}_{Fin}$

The problem of decidability of the class $\mathbb{C}_{Fin}$ had been around for a long time, until in 1968 Ax in [1] solved it positively in a tour de force. He used, as well as the kind of model theory used in [2] (ultraproducts, for example), major twentieth century results from number theory, around Weil's Riemann Hypothesis for curves over finite fields [35] and Cebotarev's Theorem on the distribution of primes [11]. Weil was relevant mainly for axiomatization, and Cebotarev for definability and uniformities in p. The method which puts these together was devised a bit later by Fried and Sacerdote [25], and called "Galois Stratification". It gives an Elimination Method much better than general recursive.

A general strategy of Abraham Robinson is to begin any analysis of any $Th(\mathbb{C})$ by looking at systems of equations and inequations

$$\sum(\bar{x}, \tilde{\alpha}) \quad (\tilde{\alpha} \text{ parameters})$$

and finding tractable conditions $\Theta(\tilde{\alpha})$ on the $\tilde{\alpha}$ such that the $\tilde{\alpha}$ satisfying Θ in some $R \in \mathbb{C}$ are the $\tilde{\alpha}$ for which the system is solvable in R.

This admits numerous variations, e.g. involving definitional expansions of L_{rings}, as in the quantifier-eliminations of Tarski and me.

What to do when $\mathbb{C}$ is the class of finite fields? Ax uses the Lang–Weil estimates [35] for the number of $\mathbb{F}_q$-points on an absolutely irreducible variety V over $\mathbb{F}_q$. (There may be none, but there are notions of the complexity of V such that if $q >$ complexity of V then there are points). It then follows that any infinite ultraproduct K of finite fields satisfies:

(PAC) *If V is an absolutely irreducible variety over K then V has a K-valued point.*

In fact, any PAC field (pseudo-algebraically closed field) has to be infinite, but need not be algebraically closed. An infinite ultraproduct of finite fields is never algebraically closed, since every finite field satisfies the set of elementary conditions saying that it has a unique extension of dimension n for all n. Moreover, these ultraproducts are perfect, since all finite fields are, and this can be expressed by an infinite set of axioms, one for each prime p.

4.2 Pseudofinite Fields

Definition 4.1 K is pseudofinite if and only if K is PAC, perfect, and has a unique extension of each degree.

This is a set of first-order conditions, consistent by the preceding ultraproduct arguments. None of the K constructed in this way is algebraic, but Jarden [27] showed that:

$$\{\sigma \in \text{Gal}(\mathbb{Q}) : \text{Fix}(\sigma) \text{ is pseudofinite}\} \text{ has measure 1 in the usual}$$

Haar measure on $\text{Gal}(\mathbb{Q})$.

Here $\text{Gal}(\mathbb{Q})$ is $\text{Aut}(\mathbb{Q}^{alg}|\mathbb{Q}$, where $\mathbb{Q}^{alg}$ is the algebraic closure of $\mathbb{Q}$.

For much nice detail on PAC fields one should consult [26].

4.3 How Does Pseudofinite Relate to Finite?

Here are some of Ax's main theorems from the first part of [1]. The proofs of all these results, except Theorem 4.4, are in [1].

Theorem 4.2 *K is pseudofinite iff $K \equiv L$, where L is an infinite ultraproduct of finite fields.*

Note that K can be of characteristic 0.

Theorem 4.3 *A pseudofinite field K is characterized up to elementary equivalence by*

$$\{f \in Z[x] : f \text{ has a root in } K\}.$$

This is a special case of an Elimination Theorem, involving "solvability formulas" long familiar in the model theory of fields. These are the formulas $\mathrm{Sol}_n(x_1, ...x_n))$, defined as

$$(\exists y) \, (x^n + c_1 x^{n-1} + \cdots + c_n = 0).$$

Note that these include the members of the set $\{f \in Z[x] : f \text{ has a root in } K\}$.

Theorem 4.4 ([29]) *Uniformly for all pseudofinite fields any formula is equivalent to a Boolean combination of Sol_n formulas.*

Theorem 4.5 *A pseudofinite field K is determined up to elementary equivalence by a conjugacy class of procyclic subgroups of the absolute Galois group of the prime subfield of K.*

Theorem 4.6 *A sentence θ holds in all pseudofinite fields iff θ holds in all but finitely many finite fields.*

And, finally, a startling one, deeply related to Cebotarev's Theorem.

Theorem 4.7 *A pseudofinite field of characteristic 0 is elementarily equivalent to an (infinite) ultraproduct of finite prime fields.*

All these theorems are relevant for the analysis of the class of all $Z/p^n Z$ as p, n vary.

5 Towards Decidability Uniform in p

The main ideas here relate to Cebotarev's Theorem. A central notion, with no reference to infinite fields, is the Ax Boolean algebra. To any sentence θ of L_{rings} one associate the set

$$\Delta(\theta)$$

defined as the set of prime powers p^n such that $\theta \in Th(\mathbb{F}_{p^n})$.

In Section 9 of [1] a conventional argument from algebraic number theory, which is not merely computable but even elementary recursive, allows one to decide, except for finitely many primes p, the structure of $\Delta(\theta)$ for θ a solvability sentence. In addition it computes, in an elementary recursive way, the finite set of exceptional primes, and the contribution their powers make to $\Delta(\theta)$. The structure is given very explicitly in terms of arithmetic progressions, and hints at periodicities such as those occurring in the theory of recurrence relations [6, 9]. The argument extends to Boolean combinations of solvability sentences explicitly given, and still in an elementary recursive way. BUT it is not an elementary recursive algorithm if our data is given as an arbitrary sentence. The reason is that the elimination argument that shows every sentence is equivalent to a Boolean combination of solvability sentences is merely computable, and not obviously elementary recursive. The Turing computability comes from a standard "wait and see" argument, which typically yields no specific algorithmic information.

The preceding use of Cebotarev leads, in a fairly straightforward way, to several major decidability results, namely:

Theorem 5.1

(1) *The theory of pseudofinite fields is decidable,*
(2) *The theory of pseudofinite fields of specified characteristic is decidable.*
(3) *The theory of the class of all fields $\mathbb{F}_p$ is decidable.*
(4) *The theory of all finite fields is decidable.*

Clearly the proof of (4) comes from that of (1) by taking account computably of finite exceptional sets of primes as in Ax's use of Cebotarev.

Note In none of the four cases does Ax give an elementary recursive decision procedure. To my knowledge, the only improvement, say for (4), obtained in the last 50 years is that of Fried–Sacerdote [25], where primitive recursive bounds are sketched via Galois stratification.

However, the situation is much better for Axiomatization. In all cases there exists an elementary recursive axiomatization, using bounds in the Riemann Hypothesis for curves [16, 31].

6   Uniformity for the $\mathbb{Z}_p$

Ax readily combined his work on ultraproducts of finite fields with the earlier work on Henselian fields ([2–4] and [22]) to get both axiomatizability and decidability results uniformly for the fields $\mathbb{Q}_p$ and the rings $\mathbb{Z}_p$. We will use the latter case. For definability one has to work harder, putting Kiefe's work into the mix, and we do not pursue this here.

The main result, for our purposes is:

Theorem 6.1 *The class of all $\mathbb{Z}_p$ is decidable, and has an elementary recursive axiomatization.*

This result is in [1], but in reference to elementary recursive axiomatisation.

Notes

1. The results of the preceding theorem do not modify routinely, because of problems with characteristic of the residue field.

      In Theorem 5.1 one has not yet done better than a computable axiomatization.
2. In Theorem 5.1 one does have an elementary recursive axiomatization in characteristic 0.
3. It is not immediately obvious that one can replace "decidable" in Theorem 5.1 by "primitive recursive".

We are finally ready to look at the issues for the class of all $\mathbb{Z}/p^n\mathbb{Z}$, both for fixed p and varying n, and for p and n both varying.

7   What Do We Know About the $\mathbb{Z}/p^n\mathbb{Z}$?

7.1   Some Uninformative Decidability Results

For now we fix p. The $\mathbb{Z}/p^n\mathbb{Z}$ are the $\mathbb{Z}_p/\langle x \rangle$ where x is a nonzero element of $\mathbb{Z}_p$. The case $x = 1$ (i.e. $n = 0$) is of course uninteresting, and we just ignore it. Now, the family of all the $\mathbb{Z}_p/\langle x \rangle$ is interpretable in $\mathbb{Z}_p$ and so we get instantly

Theorem 7.1 *The theory of the class of all $\mathbb{Z}/p^n\mathbb{Z}$ is decidable.*

Proof $\mathbb{Z}_p$ is decidable, by Ax.   □

Note This is far short of what we want to know. The proof provides no explicit algorithm. All we get is that we are not in a Gödelian situation, and given the importance of these residue rings we expect to get a lot more positive information. Also, we want an explicit axiomatization (and we do not get one simply from the explicit axiomatization of $\mathbb{Z}_p$!), and some normal form for definitions. Before turning to this, we state another easy theorem, related to the preceding.

Theorem 7.2 *The set of sentences true in all but finitely many $\mathbb{Z}/p^n\mathbb{Z}$ is decidable.*

Proof This also comes from the interpretation, since the order on the set of principal ideals in $\mathbb{Z}_p$ is interpretable via divisibility. □

Note The "algorithm" is again uninformative.

7.2 Generalizing "Pseudofinite"

One can use the theorem (or rather its proof) to develop some work on ultraproducts of the $\mathbb{Z}/p^n\mathbb{Z}$, and on generalizations of the notion of "pseudofinite" in the style of Ax, but I leave this for another occasion.

7.3 Digression on PA

It is natural to ask if there is any trace of the Gödel phenomenon in modular arithmetic. This leads one to look at modular arithmetic in models M of PA. Because of the proof of unsolvability of Hilbert's 10th Problem one knows a wide range of pathology concerning diophantine equations in such M. However, no trace of Incompleteness concerning residue rings was found.

Ax showed that the set of finite systems of polynomials Σ over $\mathbb{Z}$ solvable in all residue rings $\mathbb{Z}/n\mathbb{Z}$ is decidable (and this was also a folklore result in the number theory community). Ax's result was not trivial even modulo the big results in the main part of his paper, and he left as an open problem whether or not the theory of the class of all $\mathbb{Z}/n\mathbb{Z}$ is decidable. Many years later Jamshid Derakhshan and I gave a positive answer, using serious results on definability in the ring of adeles over $\mathbb{Q}$ [21].

All this raises the question as to how much can be done in PA. In 1978 [31] I studied the structure of residue rings of models M of PA modulo prime elements and (even nonstandard) powers of such primes. The nonstandard M are not principal ideal domains, but definable ideals are principal. The standard primes p of $\mathbb{Z}$ remain prime in M and have quotients $\mathbb{F}_p$, but if M is nonstandard there are always nonstandard p. By using Bombieri's elementary proof of Weil's Riemann Hypothesis for curves [7] and rewriting it in PA I showed

Theorem 7.3 *The following are equivalent:*

(1) K *is a pseudofinite field of characteristic 0 or a field* $\mathbb{F}_p$, p *in* $\mathbb{Z}$
(2) K *is elementarily equivalent to some* M/rM, *where* M *is a model of PA and* r *is a prime in* M.
(3) K *is elementarily equivalent to an ultraproduct of prime finite fields.*

Much more recently, Paola D'Aquino and I [16] extended the analysis to q which
are prime powers in a model M of PA, obtaining

Theorem 7.4 *The following are equivalent:*

(1) *R is elementarily equivalent to some M/qM, where M is a model of PA and q
   is a prime power in M.*
(2) *R is elementarily equivalent to an ultraproduct of rings $\mathbb{Z}/p^n$ (so p, n standard)
   where p and n vary.*
(3) *Either R is a chain ring, with residue field either a standard $\mathbb{F}_p$, and maximal
   ideal $\langle p \rangle$ (i.e. R unramified), elementarily equivalent to some $S/\langle \alpha \rangle$, where S
   is elementarily equivalent to $\mathbb{Z}_p$ and α is not 0,*

   or

   *R is elementarily equivalent to a proper quotient by a principal ideal of the
   valuation ring of some Henselian valued field with residue field pseudofinite of
   characteristic 0 and value group a Presburger group.*

One moral from these theorems is that PA is as strong as $Th(\mathbb{Z})$ as far as the
basic algebra of the residue rings is concerned, so that there is no obvious Gödelian
phenomenon in this area.

8   How to Get at Axioms for the $\mathbb{Z}/p^n\mathbb{Z}$

Fix p for now. Let R be one of the $\mathbb{Z}/p^n\mathbb{Z}$. We know that R is a local ring, a chain
ring, with residue field $\mathbb{F}_p$ and maximal ideal $\langle p \rangle$. R is Henselian. The valuation
structure of R is interpreted in ring theory via the chain of principal ideals. Any
ultraproduct J of the $\mathbb{Z}/p^n\mathbb{Z}$ has the preceding properties, and the chain of principal
ideals is what is called a Presburger TOAG in [14]. This is an initial segment (with
last element) of a model of Presburger arithmetic. From the theorems quoted in
the preceding subsection, and inspection of the [4] proof for $\mathbb{Q}_p$ one can readily
prove that the elementary theory of J is determined by the Presburger type of the
penultimate element γ of the TOAG, and by quantifier elimination for Presburger
that type is determined by the truth values of each of

$$\gamma = 0, \gamma = 1, \gamma = 2, \ldots \text{ and } \gamma \equiv k \pmod{m}, \text{ for } m = 1, 2, \ldots \text{ and } k < m.$$

Here our addition is $\oplus$.

Any Presburger type as above arises from a J as above. The infinite J get
characterized up to elementary equivalence merely by the congruence conditions
(and so the elementary types are classified by the quantifier-free Presburger types).
The finite J are each characterized by a single $\gamma = k$.

Note that the TOAG of a model R of $\mathbb{C}_p$ satisfies first-order induction for the
corresponding TOAG for any predicate definable in R. This is an axiom a bit
more explicit than one expressing mere decidability, but it is definitely not too

illuminating qua axiom. We are looking for much more explicit "algebraic" axioms, and more "algebraic" definability theory.

8.1 Definability Theory for the $\mathbb{Z}/p^n\mathbb{Z}_p$

As usual for now, p is fixed. In [16] one proves decidability for the theory of $\mathbb{C}_p$ via a uniform interpretability in $\mathbb{Z}_p$ (and the result holds even in the setting of primes in models of PA). Ax had, long before in [1], proved th decidability of $\mathbb{C}_p$. As remarked earlier, this result has little detail per se. But what about consequences of the method of [16] for definability? We will see that there is an interesting quantifier elimination here, but with the defect that we do not know if it is elementary recursive.

Let Θ be a sentence of ring theory. Consider the formula (x, Θ) in the free variable x which expresses in a model R of $Th(\mathbb{Z}_p)$ that $x \neq 0$ and Θ is true in $R/\langle x \rangle$.

Recall that $\langle x \rangle$ is the principal ideal generated by x. We construe $\langle x \rangle$ as an element of the TOAG (or value semigroup) of R.

Now, by my quantifier-elimination [30], (x, Θ) is equivalent in $\mathbb{Z}_p$ to a Boolean combination of formulas of the form

$$P_{n_j}((g_{n_j})(x))$$

for polynomials g_{n_j} in x with coefficients in $\mathbb{Z}$.

This leads us to consider some basic preservation properties for the projection

$$R \rightarrow R/\langle a \rangle,$$

where as before R is a model of $Th(\mathbb{Z}_p)$, and $a \neq 0$.

We are interested in formulas of the form $P_n(g(x))$, where $g \in \mathbb{Z}[x]$.

8.1.1 The Simplest Case

Assume g is a constant $m \in \mathbb{Z}$, and $g \neq 0$ in R.

Then $g = p^k \cdot r$ where r is prime to p. In R, g is an nth power iff n divides k and r is an nth power modulo $p^{2v(n)+1}$. The latter clause comes from Hensel–Rychlik.

Note that g will be 0 in $R/\langle a \rangle$ iff a divides g in $\mathbb{R}$ iff $v(a) \leq v(g)$ iff $v(a) \leq k$). So, our assumption that $g \neq 0$ in R is equivalent to $k > v(a)$.

Collecting the above remarks, we have:

Lemma 8.1 *Let R be a model of $Th(\mathbb{Z}_p)$, and let S be $R/\langle a \rangle$ for a nonzero element a of R. Let n, g be positive integers, with $g \neq 0$ in S. Suppose $g = p^k \cdot r$, where r*

is prime to p. Then g satisfies P_n in R iff n divides k and r is an nth power modulo $p^{2v(n)+1}$.

So we have proved that if $v(a) > 2v(n) + 1$, and $k > v(a)$, and n divides k then $P_n(g)$ holds in R iff it holds in S. The constraint that r is an nth power modulo $p^{2v(n)+1}$ is a blemish that we will now remove (or, rather, formulate internally in S).

About S we know that it is isomorphic to $\mathbb{R}/\langle a\rangle$ for some R elementarily equivalent to $\mathbb{Z}_p$. For standard n (and p fixed), $v(n)$ is independent of R and so the finite residue ring $R/p^{2v(n)+1}R$ is independent of R. Moreover, it is naturally isomorphic to the ring of remainders modulo $p^{2v(n)+1}$ in $\mathbb{Z}$. So, in addition for integers r prime to p, and arbitrary m, the condition that r be an mth power in the ring of remainders is independent of R.

We now have the prototype for a basic definability result.

Lemma 8.2 *Let R be a model of Th($\mathbb{Z}_p$), and let S be $R/\langle a\rangle$ for a nonzero and noninvertible element a of R. Let n, g be positive integers, with $g \neq 0$ in S. Suppose $g = p^k \cdot r$, where r is prime to p. Then g satisfies P_n in R iff g satisfies P_n in S.*

8.2 Quantifier Elimination

We will sketch a quantifier elimination for the class of all $\mathbb{Z}/p^k\mathbb{Z}$, or equivalently for all the S in the preceding analysis. The preceding analysis involving g was intended as a warm-up for the technical work we must now do.

We fix a formula ψ of ring theory, with free variables $y_0, ..., y_l$, and we consider the sets it defines in the various S as above. We fix S, and an R, as in the preceding discussion. As usual S is $R/\langle a\rangle$, which is neither 0 nor a unit of R. The variables y are intended to range over R, but, because of the natural definable surjection from R to S, we will also construe the y as ranging over R (in some sense replacing y by $y + \langle a\rangle$, or when a starts to vary, as x, by $y + \langle x\rangle$). The point is to transform the formula $\psi(y + \langle x\rangle)$, and what it defines in S, to an R formula ψ^*, with the y and the single x as its free variables, so that $\psi^*(y, x)$ expresses in R exactly what $\psi(y+\langle x\rangle)$ expresses in $R/\langle x\rangle$. This is completely routine, an exercise in interpretability.

Now we can apply to $\psi^*(y, x)$ my quantifier elimination for $\mathbb{Z}_p$, to get (even primitive recursively by Weisspfenning's [41]) a Boolean combination of formulas of the form $P_n(g(y, x)$, with the g polynomials over $\mathbb{Z}$, with that Boolean combination being equivalent to $\psi^*(y, x)$ over R. Now we want to replace (as we did for the illustrative special case above) $P_n(g(y, x)$ by an equivalent "Macintyre language formula" over S. As remarked earlier the y will be used interchangeably with the $y + \langle x\rangle$, and the x in the polynomial $g(y, x)$ will be replaced by 0. The first thing to sort out is when $g(y, 0)$ is 0 in S. The answer is obvious when x divides $g(y, 0)$ in R, and this is equivalent to $v(x) \leq v(g(y, 0))$. We are going to work under the assumption that $g(y, 0) \neq 0$ in R, and that is equivalent to assuming that $v(x) > v(g(y, 0))$.

Now we continue along the lines of the argument given earlier for sentences. Clearly if $v(x) > v(g(y, 0))$, then if $P_n(g(y, x))$ holds in R then $P_n((g(y, 0))$ holds in S. Conversely, still assuming $v(x) > v(g(y, 0))$ in R, let $g(y, 0) = p^{k_y} \cdot r_y$, where r_y is a unit of R. Then $P_n((g(y, 0))$ holds in § iff n divides k_y and r_y is an nth power in S. Now we reach the Hensel–Rychlik argument used in the special case done earlier. r_y will be an nth power in R iff r_y is a nth power modulo $p^{2v(n)+1}$ in R. So there is a finite set of standard integers E_n so that r_y is an nth power in R iff r_y is congruent modulo $p^{2v(n)+1}$ to an element of E_n. So we have

Lemma 8.3 *Suppose $g(y, x)$ is a polynomial over $\mathbb{Z}$, and a is a nonzero nonunit of R. Let n be a positive integer, and suppose $v(a) > v(g(b, 0)$ for some b in R. Let S be $R/\langle a \rangle$. Then $P_n(g(b, 0)$ holds in S iff $P_n(g(b, a))$ holds in R.*

If $v(g(b, 0)) \geq v(a)$ then $P_n(g(b, a))$ fails in S since $g(b, 0) = 0$. But $P_n(g(b, a))$ may still hold in R.

Note We do not pursue this argument further. It will be elaborated in forthcoming work with Paola D'Aquino.

9 Axioms for the Class of All $\mathbb{Z}/p^n\mathbb{Z}$

This is a surprisingly tricky business, and is connected to the issue of possible iterated exponential bounds in some natural algorithmic problems in the analysis. The axioms we will produce are probably not optimal, but our methods give uniformity in p and a basic periodicity result for truth of sentences.

Let α be a sentence of ring theory "defined over $\mathbb{Z}$" (so that it makes sense over any $\mathbb{Z}_p$ or $\mathbb{Z}_p/p^n\mathbb{Z}_p$. Let $c_{n,p}(\alpha) = 1$, if α holds in $\mathbb{Z}_p/p^n\mathbb{Z}_p$, and $= 0$ if α is false in $\mathbb{Z}_p/p^n\mathbb{Z}_p$. Then for fixed p the Poincaré series

$$\sum c_{n,p}(\alpha) \cdot T^n$$

is a rational function $P(p, \alpha, T)/Q(p, \alpha, T)$ over $\mathbb{Z}$ in T. The discussion in 7.6 of Denef's 1984 paper [18] allows one to extract from this a useful linear recurrence for the $c_{n,p}(\alpha)$, from which we will reach the axioms we seek.

A rather difficult result (using Denef's fundamental work [18]) shows that there are integers $B_P(\alpha)$ and $B_Q(\alpha)$, independent of p, bounding the degrees of $P(p, \alpha, T)$ and $Q(p, \alpha, T)$ respectively. The coefficients of numerator and denominator do depend on p.

We now apply some of the basic algebra of recurrence relations, and we find it suffices to refer to [6, 9]. We will give only a hint of what we do, as Paola D'Aquino and I plan to use that material in [16].

Let $k = B_P(\alpha) + 1$. Then $k > B_P(\alpha)$, and then the general theory of recurrence relations gives us

$$\text{for } n \geq k, c_{n,p}(\alpha) = a_1 \cdot c_{n-1,p}(\alpha) + \cdots + a_k C_{n-k,p}(\alpha)$$

where

$$Q(p, \alpha, T) = 1 - a_1 T - \cdots - a_k T^k.$$

We then use Bright's paper [9] on modular recurrence (modulo 2), and the fact that our Poincaré series has all coefficients either 0 or 1, to get nonnegative integers $\lambda > 0$ and $\mu \geq 0$, so that $\lambda + \mu \leq 2^k < 2^{B_P(\alpha)+1}$, and for all $j \geq \mu$.

For each $0 \leq m \leq k$, α holds in $\mathbb{Z}_p/p^{j+m}\mathbb{Z}_p$ iff α holds in $\mathbb{Z}_p/p^{j+\lambda+m}\mathbb{Z}_p$.

Choose the lexicographically minimal (μ, λ) satisfying this periodicity. If we fix p we write μ_p, λ_p, k_p for the corresponding μ, λ, k. Since we have to consider rings different from the standard finite $\mathbb{Z}_p/p^n\mathbb{Z}_p$, we define for a general R satisfying $Th(\mathbb{Z}_p/p^n\mathbb{Z}_p)$ for some (p, n), and for a nonzero x in R

$c_x(\alpha) = 1$ if α holds in $R/\langle x \rangle$, and $= 0$ if α is false in $R/\langle x \rangle$.

Now we consider the class of all ultraproducts of the $\mathbb{Z}_p/p^n\mathbb{Z}_p$. D'Aquino and Macintyre [16] listed various axioms for these, namely:

Commutative chain rings, where the chain, under reverse inclusion, is a model of the nonnegative part of a model of Presburger arithmetic, and $v(x) = \langle x \rangle$. They are local rings, with residue fields a model of FinPrim, well known axioms for the theory of finite prime fields [1, 16]. The maximal ideal is principal, and if the residue field is finite it is $\mathbb{F}_p$ for a prime p and in that case p generates the maximal ideal. The TOAG satisfies a first-order version of induction with respect to a definable subset of the ring. If R/μ is finite of characteristic p, then R is (naturally) isomorphic to a unique $\mathbb{Z}_p/p^n\mathbb{Z}_p$.

Now we come to the remaining axioms for the class of R elementarily equivalent to such ultraproducts. Mainly there will be an axiom for each α, and it will typically involve data like $B_Q(\alpha)$. We will simplify the notational demands by writing d_α for this. (Note that this is independent of p.)

Axiom 1 for α

If the TOAG of R has at least 2^{d_α} elements, then there are

$$\mu, \mu + \lambda \leq 2^{d_\alpha}$$

in the TOAG such that $\lambda > 0$ and $c_{\mu+m}(\alpha) = c_{\mu+\lambda+m}(\alpha)$, for each $0 \leq m \leq 2^{d_\alpha}$, i.e. α has the same truth value in $\mathbb{R}/\mu + m$ and $\mathbb{R}/(\mu+\lambda+m)$ for each $0 \leq m \leq 2^{d_\alpha}$

Note

- This is true by an argument in [9].
- When the TOAG is finite, and $n < 2^{d_\alpha}$ then, by the preceding remarks about axioms listed by D'Aquino and me, a finite set of "numerical" conditions enumerates what happens.

Axiom 2 for α

If the TOAG of R has at least 2^{d_α} elements, and μ is the least element for which there exists λ as in Axiom 1, then if λ_0 is the least such λ, then for all $\delta \geq \mu$, α has the same truth value in $R/\delta + m$ and $R/(\delta + \lambda_0 + m)$, for all $0 \leq m \leq 2^{d_\alpha}$

Note

- This is also true by an argument in [9].

Axiom 3 for α

If $\langle x \rangle \geq 2^{d_\alpha}$, and $D = \mathrm{lcm}(j \leq 2^{d_\alpha})$, then

$$R/\langle x \rangle \models \alpha \iff R/\langle x \cdot D \rangle \models \alpha.$$

Note

- This is the real periodicity.
- What if $\langle x \rangle < 2^{d_\alpha}$? Assume first R/μ is of characteristic p. Then it is easily proved that $\langle x \rangle = m$ for some standard m, and R/mR is naturally isomorphic to $\mathbb{Z}/p^m$ and so the truth or falsity of α is determined by m, and p

Next assume R/μ is of characteristic 0. Just replace p by any generator t of the maximal ideal. Then it is easily seen that $\langle x \rangle = m$ as in preceding paragraph, and so $\langle x \rangle = \langle t^m \rangle$. But then $R/\langle t^m \rangle$ is elementarily equivalent [1] to an infinite ultraproduct of various $\mathbb{F}_p[[t]]/p^m$. Now the truth of α is determined by the ultrafilter.

10 Conclusion

I do not give here the proof that the axioms listed above (all evidently true in all $\mathbb{Z}/p^n\mathbb{Z}$) prove all sentences true in all $\mathbb{Z}/p^n\mathbb{Z}$. The proof will go into [16]. The periodicity plays a crucial role. The axioms form a primitive recursive set, by [41]. To get an equivalent iterated exponential (Kalmar elementary) bound one has to get the uniform degree bounds elementary recursive. This is a daunting task of unwinding in Kreisel style. But I am optimistic.

What I wanted to stress here is how hard it seems to be merely to get an explicit set of axioms, and how wide-open is the problem of elementary recursive axiomatization. Very little has been achieved in this direction in the last 50 years.

References

1. J. Ax. The elementary theory of finite fields. *Annals of Mathematics (2)* **88**(2), 239–271 (1968).
2. J. Ax and S. Kochen. Diophantine problems over local fields I. *American Journal of Mathematics* **87**(3), 605–630 (1965).
3. J. Ax and S. Kochen. Diophantine problems over local fields II. A complete set of axioms for p-adic number theory. *American Journal of Mathematics* **87**(3), 631–648 (1965).
4. J. Ax and S. Kochen. Diophantine problems over local fields: III. Decidable fields. *Annals of Mathematics (2)* **83**(3), 437–456 (1966).
5. M. Ben-Or, D. Kozen and J. Reif. The complexity of elementary algebra and geometry. In: *Proceedings of the sixteenth annual ACM symposium on Theory of computing*, pp. 457–464, Association for Computing Machinery, New York (1984).
6. J. Berstel and M. Mignotte. Deux propriétés décidables des suites récurrentes linéaires. *Bulletin de la Societe mathematique de France* **104**, 175–184 (1976).
7. E. Bombieri. Hilbert's eighth problem: an analogue. In: *Proceedings of Symposia in Pure Mathematics*, AMS (1976).
8. Z.I. Borevich and I.R. Shafarevich. *Number theory*. Academic Press (1986).
9. C. Bright. Modular periodicity of linear recurrence sequences. Technical report, University of Waterloo (2008).
10. S.S. Brown. Bounds on transfer principles for algebraically closed and complete discretely valued fields. Memoirs of the American Mathematical Society, volume **204**, AMS (1978).
11. J.W.S. Cassels and A. Fröhlich. *Algebraic number theory*. London Mathematical Society, London (2010).
12. R. Cluckers, J. Derakhshan, E. Leenknegt and A. Macintyre. Uniformly defining valuation rings in Henselian valued fields with finite or pseudo-finite residue fields. *Annals of Pure and Applied Logic* **164**(12), 1236–1246 (2013).
13. P.J. Cohen. Decision procedures for real and p-adic fields. *Communications on pure and applied mathematics* **22**(2), 131–151 (1969).
14. P. D'Aquino, J. Derakhshan and A.J. Macintyre. Truncations of ordered abelian groups. *Algebra universalis* **82**(2), 1–19 (2021).
15. P. D'Aquino and A. Macintyre. Model theory of some local rings. *IfCoLog Journal of Logics and their Applications* **4**(4), 885–899 (2017).
16. P. D'Aquino and A. Macintyre. The model theory of residue rings of models of Peano arithmetic: The prime power case. *arXiv preprint arXiv:2102.00295* (2021).
17. M. Davis, Y. Matijasevich and J. Robinson. Hilbert's tenth problem: Diophantine equations, positive aspects of a negative solution. In: *Proceedings of Symposia in Pure Mathematics*, AMS (1976).
18. J. Denef. The rationality of the Poincaré series associated to the *p*-adic points on a variety. *Invent. math.* **77**(1), 1–23 (1984).
19. J. Denef. On the degree of Igusa's local zeta function. *American Journal of Mathematics* **109**(6), 991–1008 (1987).
20. J. Denef and F. Loeser. Definable sets, motives and p-adic integrals. *Journal of the American Mathematical Society* **14**(2), 429–469 (2001).
21. J. Derakhshan and A. Macintyre. Model theory of adeles I. *Annals of Pure and Applied Logic* **173**(3), 103074 (2022).
22. Y.L. Ershov. On the elementary theory of maximal normed fields. *Doklad. Ak. Nauk. USSR* **6**, 1390–1393 (1965).
23. L. Fargues and J.M. Fontaine. *Courbes et fibrés vectoriels en théorie de Hodge p-adique*. Société mathématique de France (2018).
24. M.J. Fischer and M.O. Rabin. Super-exponential complexity of presburger arithmetic. In: *Quantifier Elimination and Cylindrical Algebraic Decomposition*, pp. 122–135, Springer (1998).

25. M. Fried and G. Sacerdote. Solving diophantine problems over all residue class fields of a number field and all finite fields. *Annals of Mathematics* **104**(2), 203–233 (1976).

26. M.D. Fried and M. Jarden. *Field arithmetic*. Ergenbisse der Mathematik und ihrer Grenzgebiete, vol. **11**, Springer (2008).

27. M. Jarden. Elementary statements over large algebraic fields. *Transactions of the American Mathematical Society* **164**, 67–91 (1972).

28. R. Kaye. *Models of Peano Arithmetic*. Oxford Logic Guides **15**, Clarendon Press (1991).

29. C. Kiefe. Sets definable over finite fields: their zeta-functions. *Transactions of the American Mathematical Society* **223**, 45–59 (1976).

30. A. Macintyre. On definable subsets of p-adic fields. *The Journal of Symbolic Logic* **41**(3), 605–610 (1976).

31. A. Macintyre. Residue fields of models of P. In: *Studies in Logic and the Foundations of Mathematics*, vol. **104**, pp. 193–206, Elsevier (1982).

32. A. Macintyre. Rationality of p-adic Poincaré series: Uniformity in p. *Annals of pure and applied logic* **49**(1), 31–74 (1990).

33. A. Macintyre and A.J. Wilkie. On the decidability of the real exponential field. In: *Kreiseliana, About and around George Kreisel*, pp. 441–467, A.K. Peters (1996).

34. D. Marker. *Model theory: an introduction*. Graduate Texts in Mathematics vol. **217**, Springer Science & Business Media (2006).

35. B. Osserman. The Weil conjectures. In: *The Princeton Companion to Mathematics*, pp. 729–732, Princeton University Press (2008).

36. J. Pas. Uniform p-adic cell decomposition and local zeta functions. *J. reine angew. Math.* **399**, 137–172 (1989).

37. A. Tarski. *A decision method for elementary algebra and geometry*. Prepared for publication with the assistance of JCC McKinsey. University of California Press (1951).

38. A. Tarski. *Logic, semantics, metamathematics: papers from 1923 to 1938*. Hackett Publishing (1983).

39. L. Van den Dries. *Tame topology and o-minimal structures*. London Mathematical Society Lecture Note Series, vol. **248**, Cambridge University Press (1998).

40. L. van den Dries. Lectures on the model theory of valued fields. In: *Model Theory in Algebra, Analysis and Arithmetic – Cetraro, Italy 2012*, Chapter 5, Springer (2014).

41. V. Weispfenning. Quantifier elimination and decision procedures for valued fields. In: *Models and Sets*, pp. 419–472. Springer (1984).

42. A.J. Wilkie. Model completeness results for expansions of the ordered field of real numbers, by restricted Pfaffian functions and the exponential function. *Journal of the American Mathematical Society* **9**(4), 1051–1094 (1996).

Part IX
Mathematical Models

In his programmatic essay *Mathematical biology: Looking back and going forward*, Phillip Maini describes how "mathematical biology has grown over the past 50 years from a niche subject, pursued by a small number of visionary pioneers, to a core subdiscipline of mathematics that is becoming increasingly inter- and intra-disciplinary, and playing a key role in many areas within ecology, epidemiology and the life and medical sciences". Citing Joel E. Cohen, he concludes that "Mathematics is biology's next microscope, only better; Biology is mathematics' next physics, only better".

In *Kermack and McKendrick models on a two-scale network and connections to the Boltzmann equations* Stephan Luckhaus and Angela Stevens deliver casual notes about a famous epidemiological model, stressing the fact that it was inspired by Boltzmann, yet its study might bring back some new insight on Boltzmann's equations. They explain a paradox: lockdown measures in a pandemic can actually have the opposite effect in a first phase.

Ivar Ekeland's contribution *The Pygmalion syndrome, or how to fall in love with your model* is a philosophical dream about how mathematics could be used more critically in economic theory. It concludes with the statement of a theorem (Ekeland's variational principle), epitomizing the author's love of pure mathematics.

Morel's title *Can we teach functions to an artificial intelligence by just showing it much "ground truth"* is self-explanatory and addresses one of the major pitfalls of artificial intelligence, which is currently identified with deep neural networks.

Mathematical Biology: Looking Back and Going Forward

Philip K. Maini

Personal Note

I would like to start by thanking Catriona Byrne for all the work she has done for mathematics. In particular, I am most grateful for the help and support that she, and Ute McCrory, gave me during my time as Editor for the Springer Lecture Notes in Mathematics Subseries, Biology.

1 What Is Mathematical Biology?

Mathematical biology is the name given to the study of biological phenomena through the use of mechanistic mathematical models. Here, the term "biology" includes ecology, epidemiology and medicine, and "mechanistic" is used to distinguish this field from purely statistical and data analysis approaches, such as bioinformatics. While the latter discover correlations between phenomena, the former is the study of *why* things happen. While such a mechanistic approach yields a deeper understanding of the science and is potentially more powerful than statistical/bioinformatics approaches, the latter are more tractable at present. But that is changing.

P. K. Maini (✉)
University of Oxford, Mathematical Institute, Oxford, UK
e-mail: philip.maini@maths.ox.ac.uk

© The Author(s), under exclusive license to Springer Nature Switzerland AG 2023
J.-M. Morel, B. Teissier (eds.), *Mathematics Going Forward*, Lecture Notes
in Mathematics 2313, https://doi.org/10.1007/978-3-031-12244-6_28

2   How Has Mathematical Biology Changed?

Mathematical biology has always moved with the data. In the early days, data were largely very coarse-grained and at a macroscopic level—for example, simple visual observations such as the stripes on a zebra, or spatially-averaged temporal data, such as the number of people infected by a disease. Accordingly, models tended to be composed of coupled systems of discrete or ordinary/partial differential equations that would generate large scale dynamics, such as patterns resembling those observed on animal coats, or predictions of the temporal evolution of a disease through a population, to name but a few examples (see, for example [3, 7, 9, 12–16]). This allowed us to move to testing (and generating) hypotheses for systems which, due to their complexity, are not understandable by verbal reasoning alone. In many cases, the models were built with analytical tractability in mind, and results were generated that agreed, by visual inspection, with the data available. Of course, this is only the first step in model validation but further progress was usually hampered, partly due to limitations in biotechnology, but also to the fact that it was uncommon for theoreticians and experimentalists to work together.

The 1990s saw major advances in biotechnology, generating data at the gene level as well as across scales (cell and tissue level) and, together with increasing computational power, the sub-disciplines of bioinformatics and systems biology (including multiscale modelling) were born. While the former focussed on statistical approaches for correlation analysis, the latter continued the mechanistically-driven approach that is at the core of mathematical biology. Released from the constraints of analytical tractability, mathematical models became increasingly complex, not only in terms of the number and nonlinear complexity of equations, but also in the form models took. Thus, to bridge across scales, hybrid agent-based models were used, in which some variables are modelled discretely, others as continua. In this way, intracellular dynamics could feed into intercellular interactions, leading to macroscopic level behaviour (see, for example, [6]). While, in the case of epidemiology, multi-scale models allowed us to couple within-host infection dynamics with population level spread [11]. Other mathematical approaches also came into use, for example, Boolean algebra, topology, graph and network theory [2] to analyse large, complex interaction networks (for example, gene regulatory networks); stochastic modelling to account for the effects of noise and small numbers, when the continuum limit breaks down, came increasingly into focus [8].

These more complex models typically "outgrew" the experimental observations on which they were built in the sense that it was unusual to be able to fully, and accurately, parametrise them with data and then make testable experimental predictions.

3   Going Forward

We have now reached another critical step-change in the field of mathematical modelling, as ever-increasing amounts of dynamical data (time series, spatial dynamics) are now becoming available. For certain cases, sufficient data are available for machine learning (ML) and artificial intelligence (AI) approaches but, for other cases, data are too sparse and/or noisy. The challenges now are: can we use a mathematical biology approach to mechanistically understand the results from ML and AI [1, 4] and, in the cases where ML and AI cannot be used, how do we leverage sufficient information from the data, in terms of parameter estimation, identifiability etc? At the same time, it is being increasingly recognised by biologists that mathematics can be used to help, not only in identifying correlations, but in understanding mechanism. This is now leading to interdisciplinary research in which biological hypotheses are being translated into mathematical models, allowing us to generate predictions which are then tested experimentally, enabling us to refine our models and continue on the predict-test-refine-predict cycle. This leads to the challenge of which summary statistics are most informative for our modelling, requiring advances in statistics and in pure mathematics, such as topological data analysis.

Another major challenge in the field consists of bringing together mechanics [10] and biochemistry. These fields have, in the past, developed separately, but it is clear that in biology, these processes are intrinsically coupled through the phenomena of mechanotransduction, growth, geometry etc.

We all knew that biology was complicated but now we are on the verge of having the tools, both experimental and theoretical, necessary to dig deeper than ever before into acquiring a mechanistic understanding of this complexity. This will require bringing these tools together through research that is not only interdisciplinary, but also, intradisciplinary. For example, in mathematics, this means bringing together the sub-disciplines of mathematical modelling, applied analysis, stochastic analysis, numerical analysis and computation, network and graph theory, topology, algebra and statistics etc. This leads to a team science approach where the complexity of biology defines new problems that will require technical advances in these sub-disciplines, as well as finding creative and original ways of combining tools from across a large spectrum of mathematics to solve problems driven by the science. This, in turn, will lead to new biology. To borrow the title of Joel E. Cohen's 2004 paper [5] "Mathematics is biology's next microscope, only better; Biology is mathematics' next physics, only better".

References

1. M. Alber, A.B. Tepole, W.R. Cannon, S. De, S. Dura-Bernal, K. Garikipati, G. Karniadakis, W.W. Lytton, P. Perdikaris, L. Petzold and E. Kuhl. Integrating machine learning and multiscale modeling – perspectives, challenges, and opportunities in the biological, biomedical, and behavioral sciences. *npj Digital Medicine* **2**, 115 (2019). https://doi.org/10.1038/s41746-019-0193-y
2. R. Albert and H.G. Othmer. The topology of the regulatory interactions predicts the expression pattern of the segment polarity genes in Drosophila melanogaster. *Journal of Theoretical Biology* **223**, 1–18 (2003).
3. R.M. Anderson and R.M. May. *Infectious Diseases of Humans: Dynamics and Control.* Oxford University Press (1992).
4. J.E. Baker, J.-M. Peña, J. Jayamohan and A. Jérusalem. Mechanistic models versus machine learning, a fight worth fighting for the biological community? *Biology Letters* **14**, 20170660 (2018). https://doi.org/10.1098/rsbl.2017.0660
5. J.E. Cohen. Mathematics is biology's next microscope, only better; Biology is mathematics' next physics, only better. *PLoS Biology* **2(12)**, e439 (2004). https://doi.org/10.1371/journal.pbio.0020439
6. J.C. Dallon and H.G. Othmer. A discrete cell model with adaptive signalling for aggegration of *Dictyostelimu discoideum*. *Philosophical Transactions of the Royal Society London B* **352(1351)**, 391–417 (1997).
7. L. Edelstein-Keshet. *Mathematical Models in Biology.* Classics in Applied Mathematics, SIAM (2005) (First published by Random House, New York, NY, 1988).
8. R.E. Erban and S.J. Chapman. *Stochastic Modelling of Reaction-Diffusion Processes.* Cambridge University Press (2020).
9. A. Goldbeter. *Biochemical Oscillations and Cellular Rhythms.* Cambridge University Press (1997).
10. A. Goriely. *The Mathematics and Mechanics of Biological Growth.* Springer (2017).
11. W.S. Hart, P.K. Maini, C.A. Yates and R.N. Thompson. A theoretical framework for transitioning from patient-level to population-scale epidemiological dynamics: influenza A as a case study. *Journal of the Royal Society Interface* **17**, 20200230 (2020). https://doi.org/10.1098/rsif.2020.0230
12. J. Keener and J. Sneyd. *Mathematical Physiology I: Cellular Physiology* (2nd Edition). Springer (2009).
13. J. Keener and J. Sneyd. *Mathematical Physiology II: Systems Physiology* (2nd Edition). Springer (2009).
14. H. Meinhardt. *The Algorithmic Beauty of Sea Shells.* Springer-Verlag, Berlin-Heidelberg (1995)
15. J.D. Murray. *Mathematical Biology I: An Introduction* (3rd Edition). Springer (2002).
16. J.D. Murray. *Mathematical Biology II: Spatial Models and Biomedical Applications* (3rd Edition). Springer (2003).

Kermack and McKendrick Models on a Two-Scale Network and Connections to the Boltzmann Equations

Stephan Luckhaus and Angela Stevens

Dedicated to Catriona Byrne by two friends

On this special occasion, we consider it befitting to refer to a Scottish hero of applied mathematics. Actually, Anderson Gray McKendrick was not a mathematician at all, but a doctor in the Indian Medical Service.

1 Introduction

In 1914 McKendrick [12], inspired by L. Boltzmann, came up with a method to describe the evolution of statistical distributions in an—in principle continuous—parameter space, also for the life sciences and for the social sciences. The method was developed in the von Mises framework and it leads to partial differential equations. One has to discretize the parameters, and time too, use the law of large numbers, and when refining the discretization, one has to assume some form of uniform convergence for the transition probabilities.

The invention of measure theory by Lebesgue [8], Carathéodory [1], and Kolmogoroff [6] was a huge step forward mathematically. But in practical terms, if one wants to determine transition probabilities, one still has to discretize parameter space and count transition frequencies. Therefore, the statistical problem of balancing the coarseness of the discretization with the number of available observations is not overcome. We mention this, though we do not deal in this paper with the derivation of kinetic equations as a hydrodynamic limit of stochastic processes. This was done independently already by Feller [3], who was not aware of McKendrick's work at the time. McKendrick's ideas were not limited to applications in medicine,

S. Luckhaus
University of Leipzig, Mathematical Institute, Leipzig, Germany
e-mail: Stephan.Luckhaus@math.uni-leipzig.de

A. Stevens (✉)
University of Münster, Institute for Analysis and Numerics, Münster, Germany
e-mail: angela.stevens@wwu.de

© The Author(s), under exclusive license to Springer Nature Switzerland AG 2023
J.-M. Morel, B. Teissier (eds.), *Mathematics Going Forward*, Lecture Notes in Mathematics 2313, https://doi.org/10.1007/978-3-031-12244-6_29

but have gained interest lately, not least because of the Covid-19 pandemic. That is also true for the authors. The link to a series of lectures dealing with Covid-19 data specifically can be found among the references (Luckhaus [9]).

Often the Kermack–McKendrick models are misinterpreted solely as the well-known *SIR-ODE*-system for the dynamics of susceptibles, infectious and removed during an epidemic (Kermack and McKendrick [4, 5], Diekmann [2]). But McKendrick's equations are by far more general. Here we explain

- how his systems can be adapted to cover a small world-large world scenario, where the small world has a different infection mechanism,
- how they can be adapted to several variants of viruses competing, and
- how even the classical Boltzmann equations can be written in McKendrick form.

Mathematically, already in the *ODE*-systems one observes a difference between a system with mass action law for the infection and a system with more general types of infection mechanisms. One probably has to look at such more general infection mechanisms also for Covid, instead of laws of mass action. After all, for any aerosol infection the probability of getting infected depends in a sigmoidal fashion on the amount of virus inhaled.

The contact process model of infections leads to a mass action law, but this does not explain the threshold phenomena that were observed for Covid. For instance, in Switzerland, in April 2020, there was a sizeable epidemic wave in Geneva, whereas in Zurich there wasn't. In a mass action type law for infections both regions should have behaved identically. Either one is in a stable situation, so that the imported infections are amplified by a finite rate, or there is a (small) positive probability that one single infection will cause an epidemic. In real life it seems that in order for the Covid epidemic to really take off, a certain not so small incidence of infections is needed. Such a metastability phenomenon does not occur for a pure mass action law. To describe it one has to modify the infection rates, working for example with two rates, a lower rate below a certain threshold for the incidence of infections and a higher rate above.

Mass action laws have the crucial advantage that one can derive invariants for the respective equations. These invariants are the only chance to estimate the global behavior of an epidemic as opposed to its local behavior. In kinetic Kermack–McKendrick equations, also for structured populations, determining where heteroclinic trajectories end is purely a question of the invariants (Luckhaus [10, 11], Rass and Radcliffe [15]). The local behavior, on the other hand, is always a question for the linearized equations and not for the full system. Introducing an infection rate depending on i, the incidence of infections, in a piecewise constant way could be seen as a minimal modification, retaining invariants, but with jumping weights, to cover metastability.

The next big difference appears, when looking at an *SIR* or an *SI* model with or without recovery, i.e. when removed individuals can or cannot re-enter the class of susceptibles again. Like in the *ODE*-system, also the full McKendrick equations without recovery have a Krasnoselsky monotone (or order preserving) structure

(Krasnoselsky [7]), which is lost once you have recovery. In many cases it is possible to rewrite the system with recovery as a system without recovery, simply by introducing more variables. This way of representing the equations is illustrated in Sect. 5 in the case of the Boltzmann equations which of course have recovery but can artificially—by counting the number of collisions—be written as a system without recovery.

Whether the McKendrick representation is useful also in the Boltzmann system itself is not so clear. But it might be that it allows us to decouple the effect of the free flight and the mixing properties of the collision.

Here we will concentrate on the application of McKendrick's method to epidemiology, as it was described in his papers [13, 14].

2 The Basic Kermack–McKendrick Equation

Suppose we are given an average evolution of infectiousness $\alpha(a)$ of an infected individual. Here a denotes the time elapsed since infection, and a_0 denotes the maximal duration of infectiousness. Let $\beta(a)$ denote a removal probability by death or quarantine measures. Then the basic Kermack–McKendrick equations for susceptibles, infected, and removed read

$$\partial_t s(t) = -i(0, t) = -s(t) \int_0^{a_0} \alpha(a)i(a, t)\, da$$

$$(\partial_t + \partial_a)i(a, t) = -\beta(a)i(a, t)$$

$$\partial_t r(t) = i(a_0, t) + \int_0^{a_0} \beta(a)i(a, t)\, da.$$

This system of equations has the structure of an integral operator equation, whose linearization is a Krasnoselsky monotone operator (order preserving operator), a property widely used in population dynamics, see e.g. Thieme [16, 17]. It ensures, for example, that the leading eigenvalues of stationary points are real. There are also two invariants, namely

$$I_1 = s(t) + \int_0^{a_0} i(a, t)\, da + r(t) \quad \text{and}$$

$$I_2 = \log s(t) + \int_0^{a_0} i(a, t)\mu(a)\, da + r(t)\mu(a_0) \quad \text{where}$$

$$\mu(a) = \int_0^{a_0} \alpha(\sigma) \exp\left(-\int_0^\sigma \beta(\tau)d\tau\right) d\sigma - \int_a^{a_0} \alpha(\sigma) \exp\left(-\int_a^\sigma \beta(\tau)d\tau\right) d\sigma.$$

The evolution of $s(t)$ and the infection rate $i(0, t)$ is identical to that of the following s-$\tilde{i}$-system, with $\tilde{i}(a, t) = \exp\left(\int_0^a \beta(\tau)\, d\tau\right) i(a, t)$, namely

$$\partial_t s(t) = -i(0, t) = -\tilde{i}(0, t) = -s(t) \int_0^\infty \bar{\alpha}(a)\, \tilde{i}(a, t)\, da$$

$$(\partial_t + \partial_a)\tilde{i}(a, t) = 0 \quad \text{where}$$

$$\bar{\alpha}(a) = \exp\left(-\int_0^a \beta(\tau)\, d\tau\right) \alpha(a) \quad \text{if } a < a_0 \quad \text{and} \quad \bar{\alpha}(a) = 0 \quad \text{otherwise.}$$

Now the reason for Krasnoselsky monotonicity is that in this model there is no recovery, i.e. the individuals progress from one stage to the next, but not back. Already for the *ODE-SIR*-system with recovery, i.e. $\varepsilon > 0$,

$$\partial_t s = -\alpha i s + \varepsilon r$$
$$\partial_t i = \alpha i s - \beta i \tag{1}$$
$$\partial_t r = \beta i - \varepsilon r$$

this monotonicity property is lost. The stable stationary point has a complex eigenvalue with negative real part.

The reason for the existence of invariants, on the other hand, is the mass action law $\alpha i s$ for infection. For the *ODE*-system without recovery

$$\partial_t s = -s f(i)$$
$$\partial_t i = s f(i) - \beta i$$
$$\partial_t r = \beta i$$

one retains monotonicity but loses the invariants, if f is non-linear, e.g. a sigmoidal function.

The classical model (1) without recovery, i.e. $\varepsilon = 0$, is almost explicitly solvable. First we can rewrite the equations as

$$\partial_t(s + i + r) = 0 \quad , \quad \partial_t i = i(\alpha s - \beta) \quad , \quad \partial_t\left(\log s + \frac{\alpha}{\beta} r\right) = 0 .$$

There are two invariants: $s + i + r = 1$ and $\partial_t\left(\log s - \frac{\alpha}{\beta} s - \frac{\alpha}{\beta} i\right) = 0$.

As a consequence everything reduces to a single equation

$$-\alpha i s = \partial_t s = -\beta\, s \log s + \alpha\, s^2 + \beta\, \gamma(0)\, s ,$$

$$\text{where } \beta\, \gamma(0) = \beta \log s(0) - \alpha\, s(0) - \alpha\, i(0) .$$

If $s(0) \leq \beta/\alpha$, since s is decreasing, i is also decreasing, and no epidemic outbreak happens. If $s(0) > \beta/\alpha$, then i initially increases, until s has decreased to β/α, and an epidemic outbreak happens with maximum at $s = \beta/\alpha$. Finally $i \to 0$ for $t \to \infty$.

What is the final size of the epidemic in this setting?

The asymptotic limit for large times, $t = \infty$, is just given by the identities

$$i(\infty) = 0 \quad \text{and} \quad \beta \log s(\infty) - \alpha\, s(\infty) = \beta \, \log s(0) - \alpha\, s(0) - \alpha\, i(0) ,$$

with the additional information, that $\partial_s \left(\log s - \frac{\alpha}{\beta} s \right)(\infty) = \frac{1}{s(\infty)} - \frac{\alpha}{\beta} \geq 0$, since s and i will both eventually decrease. So all depends on the simple curve $\beta \log s - \alpha s = \phi(s)$, which is strictly concave with global maximum at $s = \beta/\alpha$.

Starting with any $s(0), i(0) > 0$ for large times, the fraction of removed will approach the unique solution of $\beta \log s - \alpha s = \beta \log s(0) - \alpha s(0) - \alpha\, i(0)$ with $s < \frac{\beta}{\alpha}$. This value $\frac{\beta}{\alpha}$ is called herd immunity. It has two interpretations:

1. The maximal possible ratio of the population which has escaped the infection during the whole course of the epidemic.
2. The value of the ratio of susceptibles in the population at which the number of infected starts to decrease.

Within the context of this simple model, the message for disease control is equally simple. Suppose the epidemic starts with small $i(0)$ and $s(0) > \frac{\beta}{\alpha}$, and suppose you are able, but only temporarily, to decrease the infection rate α. How far should you decrease α? Well, you know in any case, that $s(\infty) \leq \frac{\beta}{\alpha}$ eventually. So to get there quickly and in order not to overshoot, the optimal choice is

$$\tilde{\alpha} = \alpha\beta \, \frac{\log \beta - \log \alpha - \log s(0)}{\beta - \alpha[s(0) + i(0)]} .$$

Reducing the contact ratio more will get you only to an $s(T), i(T)$ at the time T of lifting the temporary restrictions, which is the starting point of a new epidemic. Such new epidemic waves were observed in Europe in the fall/winter season of 2020/2021.

3 The Two-Scale s-i-Model

The main idea of this paper is to explain the following phenomenon, which has been observed for example during the Covid epidemic. When lockdowns were imposed, the effect on the overall number of infections was minimal, because the infections in small closed groups increased, thus partially offsetting the reduced infection risk in public transport, pubs, concerts, etc. If one wants to describe this effect in an epidemiological model, the in-group infection rate must be modeled explicitly. Here we present models which do that. They have a two-scale structure.

In our first model we start from the Kermack–McKendrick equations in its s-i-formulation but replace individuals by finite size groups. In terms of a contact graph this corresponds to a finite size complete graph with finite contact probability attached to the vertices of a large scale complete graph with an $O\left(\frac{1}{n}\right)$ contact probability. Individuals get infected only once. But the groups have suffered k infections at different times. For an illustration see Fig. 1.

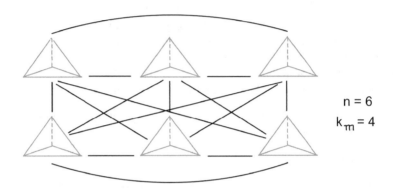

Fig. 1 Example of a two scale contact graph. Grey edges represent possible contacts within a group of size $k_m = 4$ and black edges and curves the possible contacts between the 6 different groups

What we get is
$$\partial_t s(t) = -s(t) \sum_l \int_{\Omega_l} \alpha_{0,l}(\vec{b}) i_l(\vec{b}, t) \, d\vec{b}$$

$$(\partial_t + \partial_{\vec{a}}) \, i_k(\vec{a}, t) = -i_k(\vec{a}, t) \left[\sum_l \int_{\Omega_l} \alpha_{k,l}(\vec{b}) i_l(\vec{b}, t) \, d\vec{b} + \beta_k(\vec{a}) \right]$$

$$i_k(0, a_2, \ldots, a_k, t) = -\left[(\partial_t + \partial_{\vec{a}}) i_{k-1}\right](a_2, \ldots, a_k, t).$$

Here $\vec{a} \in \Omega_k = \{(a_1, a_2, \ldots, a_k) \mid 0 \le a_1 \le a_2 \le \ldots \le a_k\}$ denotes the time elapsed since the successive k infections within the population of groups, who have suffered k infections, and $\partial_{\vec{a}} := \sum_{j=1}^{k} \partial_{a_j}$. Further, i_k denotes the percentage of these groups among all groups. Here β_k are the rates of infection within the groups having already undergone k infections. The $\alpha_{k,l}(\vec{b})$ are the rates depending on the times $b_1, \ldots, b_l$ elapsed since infection with which the l infectious in the groups having suffered l infections infect another individual in the groups having suffered k infections, and k is less than or equal to k_m, the group size. Again there are two invariants. The obvious one

$$s(t) + \sum_k \int_{\Omega_k} i_k(\vec{a}, t) \, d\vec{a} \, ,$$

and the additional logarithmic one

$$\log(s(t)) + \sum_k \int_{\Omega_k} \mu_k(\vec{a}) i_k(\vec{a}, t) \, d\vec{a} \, ,$$

where the weights μ_k can be obtained as follows.
The identity obtained by integration by parts

$$\sum_k \int_{\Omega_k} \mu_k(\vec{a}) \partial_t i_k(\vec{a}, t) \, d\vec{a} = \sum_k \int_{\Omega_k} \partial_{\vec{a}} \mu_k(\vec{a}) i_k(\vec{a}, t) \, d\vec{a}$$

$$+ \sum_k \int_{\Omega_{k-1}} \mu_k(0, \vec{b}) i_k(0, \vec{b}, t) \, d\vec{b} + \sum_k \int_{\Omega_k} \mu_k(\vec{a}) (\partial_t + \partial_{\vec{a}}) i_k(\vec{a}, t) \, d\vec{a}$$

leads to the equations

$$\partial_{\vec{a}} \mu_k(\vec{a}) = \alpha_{0,k}(\vec{a}) \quad , \quad \mu_k(0, \vec{b}) = \mu_{k-1}(\vec{b}) \quad , \quad \mu_0 = 0 \, ,$$

$$\mu_k(a_1, \dots, a_k) = \mu_{k-1}(a_2 - a_1, \dots, a_k - a_1) + \int_0^{a_1} \alpha_{0,k}(a_1 - t, \dots, a_k - t) \, dt.$$

So far we have only one additional invariant, not more, and μ_k does not depend on β_k. The influence of β_k is indirect. If β_k is large, that feeds back into $\log s(t)$ via the difference $\mu_k(a_1, \dots, a_k) - \mu_{k-1}(a_2 - a_1, \dots, a_k - a_1)$. This can be used to estimate the influence of in-group infections. The $s-i$-models formally do not have stationary points. To describe the possible endpoints of an epidemic, one has to split again

$$i_k = i_k \, \chi_{a_1 < a_0} + r_k$$

and linearize at all points

$$i_k \, \chi_{a_1 < a_0} \equiv 0 \, ,$$

where a_0 is chosen as the maximal time of infectiousness. If β_k is large, this entails an explicit estimate of $\int_{\Omega_k} r_k$ by $\int_{\Omega_k} i_k$, and so an estimate of $\int_{\Omega_k} r_k$ by $r := \int_{\Omega_{k_m}} r_{k_m}$, where k_m is the maximal k, the group size. It is no longer true that one has a simple equation determining (s_∞, r_∞) in terms of $(s(0), r(0))$, but one has estimates.

4   More General Models of McKendrick Type with Several Competing Infectious Agents

Even if individuals can be infected several times, one can artificially represent the process as a process without recovery, by keeping track of the successive infections. As in dynamic programming or in normal-form game representation, this leads to a rooted tree whose vertices are the discrete infection stages. An example is given at the end of this section.

For each l let T_l be a rooted tree, V_l be the collection of its vertices, E_l be the collection of its directed edges, and v_{0l} be its root. Then

$$T_l = (v_{0l}, V_l, E_l) \quad , \quad E_l \subset \{(v, v') \mid v, v' \in V_l\} \quad , \quad v_{0l} \in V_l.$$

The edges represent infections, the vertices the infection states. As in the previous section, to each vertex v (except the root) one has associated a number k_v and the domain $\Omega_v = \{0 \le a_1 \le \dots \le a_{k_v}\}$, with the property that $k_{v'} = k_v + 1$ for $(v, v') \in E_l$, and a differential equation

$$(\partial_t + \partial_{\vec{a}_v}) i_v(\vec{a}_v, t) = -i_v(\vec{a}_v, t) \sum_{j=1}^{m} \sum_{e=(v,v')\in E_l} f_{je}(I(t,.)).$$

Here $j = 1, \dots, m$ are the indices of m meeting places, and $f_{je}(I(t,.))$ is the probability for an individual (or group) in the lth subpopulation to progress from stage v to stage v' by an infection at the place j.

$I(t,.)$ is a map from $\Pi_{\cup(V_l \setminus v_{0l})} \, \Omega_v \to \mathbb{R}^+$ with product structure $I(t, (\vec{a}_v)_v) = \Pi_v \, i_v(\vec{a}_v, t)$ and f_{je} is an integral operator.

The initial condition for the *PDE* is again

$$i_{v'}(0, \vec{a}_v, t) = i_v(\vec{a}_v, t) \sum_{j} f_{je}(I(t,.)) , \quad \text{for } e = (v, v') \in E_l.$$

The system is still monotone in the sense of Krasnoselsky, and if f, i.e. the matrix of integral operators f_{je}, is linear, there exist again logarithmic invariants, one for each tree. But this last property is lost in the case of typical threshold dynamics for f, and it seems that we observe such a dynamic for human to human aerosol infections such as COVID-19.

To give an example of such a tree model, suppose an individual can be infected by two variants of SARS-CoV-2. That means, it can progress from the state of susceptibility first to an infection with the first variant, then to the state where it has undergone both infections, or first to an infection with variant two. The five vertices would be denoted by (0), (1), (2), $(1, 2)$, $(2, 1)$, and the four directed edges by $[(0), (1)]$, $[(1), (1, 2)]$, $[(0), (2)]$, $[(2), (2, 1)]$.

5   Boltzmann's Equation in McKendrick Form

As a curiosity we write down the Boltzmann system in the McKendrick form. One does this by introducing a memory of past collisions.

Let $f_k(t, x, v, a_1, \ldots, a_k, w_1, \ldots, w_k)$ denote the density of particles which underwent k collisions at times $t - a_1, \ldots, t - a_k$ and whose velocities were w_l in the interval of time $(t - a_{l-1}, t - a_l)$. If $K(w, w', v)$ denotes the Boltzmann kernel, giving the probability of a particle with velocity w to change its velocity to v, after colliding with another particle of velocity w', the equations read

$$(\partial_t + \partial_{\vec{a}} + v\partial_x)\, f_k(t, x, v, \vec{a}, \vec{w}) = -f_k(t, x, v, \vec{a}, \vec{w}) \int \tilde{K}(v, w') F(t, x, w')\mathrm{d}w',$$

where $\quad f_k : \mathbb{R}^+ \times \mathbb{R}^3 \times \mathbb{R}^3 \times \Omega_k \times \mathbb{R}^{3k} \to \mathbb{R}^+ \quad$ with initial conditions

$$f_k(t, x, v, 0, \vec{a}, \vec{w}, w_k) = f_{k-1}(t, x, w_k, \vec{a}, \vec{w}) \int K(w_k, w', v) F(t, x, w')\, \mathrm{d}w'$$

for $\quad \vec{a} \in \Omega_{k-1}$, and $\quad \vec{w} \in \mathbb{R}^{3(k-1)}$.

Here $\displaystyle F(t, x, w') = \sum_{k \in \mathbb{N}} \int_{\Omega_k \times \mathbb{R}^{3k}} f_k(t, x, w', \vec{a}, \vec{w})\, \mathrm{d}\vec{w}\, \mathrm{d}\vec{a}$

and $\displaystyle \tilde{K}(w, w') = \int K(w, w', v)\, \mathrm{d}v.$

We get the same type of invariants, including the logarithmic one, as we have obtained for the two-scale Kermack–McKendrick model. Let us repeat the calculation.

$$(\partial_t + v\partial_x) \int_{\mathbb{R}^3 \times \Omega_k \times \mathbb{R}^{3k}} \mu_k(v, w, \vec{a}, \vec{w}) f_k(t, x, w, \vec{a}, \vec{w})\, \mathrm{d}\vec{w}\, \mathrm{d}\vec{a}\, \mathrm{d}w$$

$$= J_{1,k} + J_{2,k} - J_{3,k}$$

$$J_{1,k} = \int_{\mathbb{R}^3 \times \Omega_k \times \mathbb{R}^{3k}} \partial_{\vec{a}} \mu_k(v, w, \vec{a}, \vec{w}) f_k(t, x, w, \vec{a}, \vec{w})\, \mathrm{d}\vec{w}\, \mathrm{d}\vec{a}\, \mathrm{d}w$$

$$J_{2,k} = \int_{\mathbb{R}^3 \times \Omega_{k-1} \times \mathbb{R}^{3k}} \mu_k(v, w, 0, \vec{b}, \vec{w}) f_k(t, x, w, 0, \vec{b}, \vec{w})\, \mathrm{d}\vec{w}\, \mathrm{d}\vec{b}\, \mathrm{d}w$$

$$J_{3,k} =$$

$$\int_{\mathbb{R}^3 \times \Omega_k \times \mathbb{R}^{3k} \times \mathbb{R}^3} \mu_k(v, w, \vec{a}, \vec{w}) f_k(t, x, w, \vec{a}, \vec{w}) \tilde{K}(w, w') F(t, x, w')\, \mathrm{d}w'\, \mathrm{d}\vec{w}\, \mathrm{d}\vec{a}\, \mathrm{d}w$$

$$\text{for } k \geq 1$$

$$(\partial_t + v\partial_x) \log f_0 = -\int_{\mathbb{R}^3} \tilde{K}(v, w') F(t, x, w')\mathrm{d}w'.$$

So again, in order to balance

$$\sum_k J_{1,k}$$

one has to set

$$\partial_{\vec{a}} \mu_k(v, w, \vec{a}, \vec{w}) = \tilde{K}(v, w).$$

And to arrive at

$$\sum_{k \geq 1} J_{2,k} = \sum_{k \geq 0} J_{3,k}$$

one has to set

$$\mu_k(v, w, 0, \vec{a}, \vec{w}, w_k) = \mu_{k-1}(v, w_k, \vec{a}, \vec{w})$$

for all $(\vec{a}, \vec{w}) \in \Omega_{k-1} \times \mathbb{R}^{3(k-1)}$, $k \geq 1$ and $\mu_0 = 0$.

It should be noted that this logarithmic invariant is not connected to the logarithmic entropy of Boltzmann, which is the logarithmic entropy w.r.t. an invariant measure. Here the evolution is $f_k \rightarrow 0$ for all k.

Whether the McKendrick representation is useful for the Boltzmann system is not so clear. But it might make it possible to decouple the effect of the free flight and the mixing properties of the collision.

References

1. C. Carathéodory. *Vorlesung über reelle Funktionen*. Teubner Verlag, Leipzig, Berlin (1918).
2. O. Diekmann. The 1927 epidemic model of Kermack and McKendrik: a success story or a tragicomedy? *Newsletter of the JSMB* **92**, 8–11 (2020).
3. W. Feller. Die Grundlagen der Volterraschen Theorie des Kampfes ums Dasein in wahrscheinlichkeitstheoretischer Behandlung. *Acta Bioth. Ser. A* **5**, 11–40 (1939).
4. W.O. Kermack and A.G. McKendrick. A Contribution to the Mathematical Theory of Epidemics. *Proc. R. Soc. Lond. A, Cont. Pap. Math. Phys. Character* **115**, no. 772, 700–721 (1927).
5. W.O. Kermack and A.G. McKendrick. The Solution of Sets of Simultaneous Integral Equations Related to the Equation of Volterra. *Proc. London Math. Soc. (2)* **41**, no. 6, 462–482 (1936).
6. A.N. Kolmogoroff. Beiträge zur Maßtheorie. *Math. Ann.* **107**, no. 1, 351–366 (1933).
7. M. Krasnoselsky. *Positive Solutions of Operator Equations*. Noordhoff, Groningen (1964).
8. H.L. Lebesgue. Intégrale, Longueur, Aire. *Annali di Matematica Pura ed Applicata, Serie III* **7**, 231–359 (1902).
9. S. Luckhaus. Corona "Herd-Immunity" and all that. Lecture Series (2020). Part 1: https://www.youtube.com/watch?v=SZ4dIEb2ttM; Part 2: https://www.youtube.com/watch?v=M5zVoa6dtzw; Part 3: https://www.youtube.com/watch?v=O79g6_e00Ro; Part 4: https://www.youtube.com/watch?v=TjCZPHjIhso; Part 5: https://www.youtube.com/watch?

v=hhKXQmfVY_Q; Part 6: https://www.youtube.com/watch?v=BV9lswFvZx0; Part 7: https://www.youtube.com/watch?v=dyLa1VgPYzY

10. S. Luckhaus. Mathematical Epidemiology, SIR Models and Covid-19. MPI for Mathematics in the Sciences, Preprint 60 (2020).
11. S. Luckhaus. Corona, Mathematical Epidemiology, Herd Immunity, and Data. MPI for Mathematics in the Sciences, Preprint 105 (2020).
12. A.G. McKendrick. Studies on the Theory of Continuous Probabilities, with Special Reference to its Bearing on Natural Phenomena of a Progressive Nature. *Proc. London Math. Soc. (2)* **13**, 401–416 (1914).
13. A.G. McKendrick. Application of Kinetic Theory of Gas to Vital Phenomena. *Indian J. of Medical Research* **3**, 667 (1916).
14. A.G. McKendrick. Application of Mathematics to Medical Problems. *Proc. R. Soc. Edinburgh* **44**, 98–130 (1926).
15. L. Rass and J. Radcliffe. *Spatial deterministic epidemics.* American Mathematical Society (2003).
16. H.R. Thieme. *Mathematics in Population Biology.* Princeton Univ. Press (2003)
17. H.R. Thieme. Eigenvectors of homogeneous order-bounded order-preserving maps. *Discrete Contin. Dyn. Syst. Ser. B* **22**, no. 3, 1073–1097 (2017).

The Pygmalion Syndrome, or How to Fall in Love with Your Model

Ivar Ekeland

1 A Missed Opportunity?

Two years ago I was approached by a colleague—let us call him Prospero—who suggested we write a book together (doubtless to be published by Springer Verlag, with some help by Catriona Byrne). The title was still to be found, but the book was to describe how, in half a century, mathematics had progressed from describing the natural world to describing the social world. Modern science had begun with Galileo, modern mathematics with Descartes, Newton and Leibniz. Prospero claimed that our generation, the one that started working after WWII, had the enormous privilege of extending the scope of mathematics to the social order and the human mind itself. Economists, using mathematical models, had turned Adam Smith's vague intuitions about an invisible hand into a global market which at the present time rules the world, and financiers had created an enormous industry from an abstruse mathematical formula related to the heat equation. Is it not an incredible achievement, which shows that the power of mathematics extends far beyond the physical world, into the inner recesses of the human mind?

Our experience as teachers and researchers supported this view. When I was recruited at the recently created Université de Paris-Dauphine in 1970, we had to set up a mathematics department in a university where there were no natural sciences, no physics, no chemistry, no biology. This was a predicament, because mathematics is a language, and language must be about something. Traditionally, it was about the natural sciences: mathematicians went to mechanics and physics to find problems, and the major outlet for mathematical education (aside from teaching) was engineering. But in 1970 the situation had changed, and mathematical models

I. Ekeland (✉)

CEREMADE, Université Paris-Dauphine, Paris, France

e-mail: ekeland@math.ubc.ca

had entered economics and management: Arrow and Debreu had solved the major problems of general equilibrium [1], and firms were starting to use optimization methods in management. This was the heyday of linear programming, and of all the optimization methods which went under the generic name of operations research. So we decided to create a mathematics curriculum where the students would learn no natural sciences, but economics, finance, statistics, and, crucially, computer science. We never looked back. Today, business, insurance and banking have replaced engineering as the major outlet for mathematical degrees.

This transformation of mathematics was made possible by the development of economic theory, and its expansion into a general theory of human behaviour. There are of course several schools of thought in economics, and Adam Smith has more than one heir. Marx and Keynes still have followers, but the second half of the twentieth century witnessed the triumph of the neoclassical school, led by Friedrich von Hayek and Milton Friedman, and epitomized in the classical textbook of Paul Samuelson, *Economics*, first published in 1948, which is now in its 19th edition and has sold four million copies worldwide. Samuelson considered mathematics to be the natural language of economics, as it is in the natural sciences, and popularized a mathematical model of human behaviour which is known as *homo oeconomicus*. In this model, men (and women, and children) are born optimizers: they classify linearly all options available to them, and they pick the best one subject to budget constraints. There is no limit to the scope of this model, which goes far beyond economics, understood as the study of the production, exchange and consumption of material goods: it is no less than a general theory of human behaviour. This program was stated explicitly by Gary Becker [2] and he carried in through in studying social institutions, such as the family [3], racial discrimination [4], crime [5], and addiction [6]. In later years, others followed, developing formal theories of just about every institution, firms, churches or states, and every social norm, from witch-burning in Africa to sumo competitions in Japan.

In 1973, Fischer Black and Myron Scholes published a paper entitled *The pricing of options and corporate liabilities*. An option on a stock is a bet on the price of the stock at some future date, and the paper did exactly what the title said, that is, Black and Scholes gave a formula for pricing the option simply by looking at the past performance of the stock. This was a revolution. One would have thought that to evaluate the odds one would have to go into the past performance and the future outlook of the issuing company, but Black and Scholes showed that one could be content with looking at the past performance of the stock. This formula, together with the arrival of computers powerful enough and fast enough to store the information and perform the calculations, was the birth of the global financial industry as we know it today. Of course, to extend the Black and Scholes idea to other situations, where more complex assets were to be priced, more mathematics was needed, and the community was happy to provide them. Stochastic analysis turned out to be the right tool, another illustration of the classical miracle, whereby some piece of mathematics developed for purely theoretical purposes turns out to be the right tool to solve some practical problem.

Where will it all stop, asked Prospero? We are on the way of building a mathematical theory of society from the mathematical theory of the individual, as our predecessors built (or are trying to build) a mathematical theory of solids from the mathematical theory of atoms and molecules. And as our predecessors have proved their worth by building machines or instruments, automobiles or lasers, contemporary mathematicians have created an enormous institution, more powerful than states, the global financial market. Of course, the complexities of building a global model are enormous, so much so in fact that certain phenomena can emerge at the macro level which have no equivalent at the micro level, but in principle it is possible, perhaps by a deeper mind than ours and more computing power than we can ever manage. All the information is there, in the basic model of the individual or of the atom, just waiting to be developed. The twentieth century has set mathematics on course to explore the human world, just as the 16th had set it on course to conquer the natural world. We should really write a book about it!

2 Testing Microeconomic Theory

I declined. Prospero builds castles in the air. Physicists and biologists try to understand a hard reality that would continue to exist if mankind disappeared from this planet. Economists and social scientists study the rules of games in which they participate. To quote the anthropologist David Graeber, *The ultimate, hidden truth of the world is that it is something that we make and could just as easily make differently.* This is certainly not true about the natural world, there is no way we could change the laws of gravitation if we tried, but it just as certainly is for the social world: in the span of a human life, laws, religions, tastes and state boundaries seem as immutable as the laws of physics, but at the scale of millennia they come and go with bewildering speed. The change can be sudden, precipitated by the action of a few activists, as was the case with the French Revolution, or the outcome of a long process of decay, with little or no human agency, as for the fall of the Roman Empire. Willingly or not, consciously or not, the rules of the games change with time, and it is extremely hard to believe that there would be a universal theory of human behaviour, independent of the particular game these humans would be engaged in. To be precise, the economic approach to human behaviour, as advocated by Gary Becker and the utilitarian school, applies at best to individuals schooled in modern capitalistic societies, where the market rules with the help of a powerful state

Let me give an example. The basic assumption of neoclassical microeconomics, is that individuals have the ability to order linearly, from better to worse, all options open to them, and that they choose the best one. There are a few others, to deal with situations when the outcomes are uncertain, or deferred, but this is really the fundamental fact about *homo oeconomicus*. But is it true?

According to Popper, a theory is scientific if and only if one can devise an experiment, the result of which is different according to whether the theory is true (that is, correctly predicts the outcome) or not. If such an experiment does not exist, the theory is outside science. If it exists, and the result is negative (that is, the outcome is not the one predicted by the theory), the theory is disproved and should be discarded. If the result is the one predicted, the theory is not disproved, which does not mean that it is true (more experiments would be needed to assert that conclusion), but that it can be relied upon until some new experiment is devised that disproves it. The typical example is the famous experiment of Michelson and Morley (1887), which disproved the aether theory of light and led eventually to Einstein's special relativity.

Does there exist such an *experimentum crucis* in economic theory? There is, and it has been known for a long time, even though it has only been understood and performed very recently. Let us first restate the basic assumption, namely the fact that every individual can order linearly all the options open to him/her and chooses the best one. We represent these options as points in $\mathbb{R}^n$. With a few minor assumptions, such as continuity, the individual preferences can be represented by a function $u : \mathbb{R}^n \to \mathbb{R}$, in the sense that the individual prefers x to y if and only if $u(x) > u(y)$. An individual going shopping with an amount w to spend, and observing the prevailing prices (good i has price p_i, $1 \le i \le n$), will choose the quantities $x^i \ge 0$ in order to maximise $u(x^1, \ldots, x^n)$ subject to the budget constraint $\sum x^i p_i \le w$

$$\max u(x) \qquad\qquad\qquad (\mathcal{P})$$

$$px \le w$$

Problem ($\mathcal{P}$) is known as the consumer's problem in microeconomics. To make sense, it is necessary that the solution is unique: indeed, if there were several solutions, the decision problem would still be open, that is, the theory would not tell us what the consumer actually chooses. For this reason, it is assumed that the consumer's utility function is strictly concave, and even that it is C^2 with $u''(x)$ positive definite. We can then restate our basic question: is it true that individuals have a strictly concave utility function, and choose their consumption by solving problem ($\mathcal{P}$)?

Note that we are not asking whether mathematicians have a concave utility functions, but whether truckers, rockers, teenagers, housewives and nurses, who probably never have heard of a function, concave or not, have a concave utility function. Very improbable, is it not? We cannot ask them: we would be laughed out of court, and anyway the utility function cannot be observed, even a mathematician cannot tell what his utility function is. The only thing which is observable is what people actually buy, that is, the solution $x(p)$ of problem ($\mathcal{P}$). Hence the idea of the *experimentum crucis*: observe $x(p)$, that is, observe how consumption changes with prices, and deduce information on u.

Note that $x\,(p)$ is a map from R^n into itself. Introduce the following matrix:

$$S_{ij} = \frac{\partial x^i}{\partial p_j} - \sum p_n \frac{\partial x^i}{\partial p_n} x^j$$

It was first established by Antonelli (1886), then forgotten, and later rediscovered by Slutsky (1915), that:

Proposition 2.1 *There is a function u (x) such that x (p) solves problem ($\mathcal{P}$) if and only if the matrix S is symmetric:*

$$S_{ij} = S_{ji} \tag{$\mathcal{E}$}$$

Some comments are in order. The result is local in x, of course, and if you want $u\,(x)$ to be concave there are some positivity conditions on S to be satisfied, which we are not going into. The function $u\,(x)$ is not unique: if u describes a certain set of preferences, so does $h \circ u$ for any increasing function $h : \mathbb{R} \to \mathbb{R}$. To get a feeling for the condition ($\mathcal{E}$), note that if the second term was not present, the condition $S_{ij} = S_{ji}$ would reduce to $\frac{\partial x^i}{\partial p_j} = \frac{\partial x^j}{\partial p_i}$, equality of cross-derivatives, the classical condition for $x\,(p)$ to be a gradient. Finally, I challenge the reader to take a piece of paper and write down three functions $x_i\,(p_1, p_2, p_3)$, $i = 1, 2, 3$ satisfying ($\mathcal{E}$), which is a system of three nonlinear PDEs of the first order.

It turns out that what mathematicians cannot do, truckers, rockers, teenagers, housewives and nurses do every day. In fact, large-scale experiments show that the consumption patterns in Canada satisfy equation ($\mathcal{E}$). More precisely, if one estimates $x\,(p)$ for various types of consumers in Canada, and substitute in the Slutsky matrix S, one finds that $S_{ij} = S_{ji}$. This was done in seminal work by Browning and Chiappori [7]. Many economists before them had tried to check the Slutsky conditions on consumption data, without succes. These consumption data concern households, and typically do not satisfy the symmetry condition. Browning and Chiappori were the first ones to separate the data concerning singles, and found that they do satisfy the symmetry condition. Two-person households do not, simply because there are two, not one, utility functions, and choices cannot be made by maximising two functions simultaneously. They satisfy another, less stringent condition, which was discovered and found to be necessary and sufficient by Chiappori and Ekeland [9]

To sketch the mathematical content, I will just say that the problem can be rephrased in terms of the exterior differential calculus of Elie Cartan. Introducing the 1-form $\omega = \sum x^i \mathrm{d} p_i$, equation ($\mathcal{E}$) can be rewritten as $\omega \wedge \mathrm{d}\omega = 0$. If there are 2 persons in the household, the consumption $x\,(p)$ has to satisfy the condition:

$$\omega \wedge \mathrm{d}\omega \wedge \mathrm{d}\omega = 0$$

and if you want the utility functions $u_1\,(x)$ and $u_2\,(x)$ of the two members of the household to be concave, some positivity conditions have to be satisfied, and everything turns out to be necessary and sufficient. It is a beautiful piece of mathematics, culminating in the Cartan–Kähler theorem, which is used to characterize, not individual demand, not household demand, but market demand.

3 The Model as an Icon

On the face of it, we have an unmitigated success: the microeconomic theory of the consumer has been put to the test and prevailed. We have shown that ordinary people unconsciously solve nonlinear systems of first-order PDEs, and from observing their consumption patterns of households, we can find out how many they are. It all uses beautiful mathematics, which has fallen into disuse nowadays, but which is superb in the way it combines analysis and algebra.

However, the experiment checks only part of the theory, the one that concerns decisions with immediate and certain outcomes. Many decisions have deferred and uncertain outcomes, such as saving for one's retirement or picking a job. Problem ($\mathcal{P}$) does not cover such cases, and the theory has to be extended. The basic utility for an uncertain flow delivering x_t at time t is then given by:

$$E\left[\int_0^\infty e^{-rt} u\left(x_t\right) dt\right]$$

Note the exponential discount rate r and the expectation E. I know of no experiment which has comforted this model the way the Browning–Chiappori experiment has comforted consumer theory. On the other hand, I know of much psychological evidence which challenges both the expected utility and exponential discount model: clearly people put more weight on losses than on gains, and on the immediate future than on the far one.

But let us forget about the inner workings of the *homo oeconomicus* model, and ask ourselves what does it represent exactly? Human beings or the social situation they are put in? The Browning–Chiappori experiment shows that households in Canada, when shopping, to a large extent behave as consumer theory predicts. But is it innate behaviour, or acquired behaviour? After all, in North America, from the early childhood, individuals are exposed to advertisement and taught to maximise their consumption within their budget constraint: the ubiquitous mantra is *Why pay more?*. For those who want higher education, colleges and universities carry courses in "home economics", and for the really gifted, there are courses in economics and finance. Experiments carried out with economic students tend to show that they behave in more selfish ways than students in other fields [8]—whether it is a result of their training or whether they self-select into that field is an open question.

In other words, the *homo oeconomicus* model is very much like an icon, an idealized description of a human being, set as an example to the multitude which should strive to conform its behaviour to the story it is told, and which is helped towards that worthy goal by a powerful institution, namely the global market, and comforted by a well-developed theory, understood only by a few *cognoscenti* who lend their moral support to the powerful.

4 The Right Use of Mathematics

What I have said up to now should be uncontroversial. It can even be found within the theory itself: the whole idea is that individuals can rank options linearly, but where does that ranking come from? To state problem ($\mathcal{P}$), the fundamental problem of consumer theory, one needs the utility function $u(x)$, but where does that function come from? It may be that *homo oeconomicus* has neither mother nor father, and is born with an innate utility function, but real people begin life as infants, and it takes about 20 years to turn them into adult members of society, responsible for their own choices. It takes enormous efforts and resources to shape their preferences. One could hardly see why society would devote considerable resources to education and why the advertising industry would exist if preferences were innate. If one is interested in building a mathematical model of human behaviour, the fundamental question should be: what are the right preferences to have?

One possible view is that anything goes: with the proper education and training, human beings can have any preferences whatsoever. There is no behaviour so divinely good or so devilishly bad that human beings cannot adopt as a way of life. Although much of human history tends to favour this view, we will discard it for it simply states that no general theory is possible.

Another view is that human behaviour should be rational. It has been upheld by Western philosophy since the Greeks, and it is often cited in support of the current economic approach to human behaviour, according to which "being rational" means "choosing the best available option among all possible ones". This is not only an extremely arrogant approach, since it means that everyone who disagrees is irrational, in other words crazy, but it is also contradicted by anthropology, which studies scores of societies where men live by different rules, by history, which unearthes in our own societies the roots of the present ideology, by sociology, which points out the institutions that enable *homo oeconomicus* to thrive, not to mention psychology and philosophy. It is a sociologist, Max Weber, who made the celebrated distinction between two kinds of rationality, rationality of means (*Zwecksrationalität*) and rationality of purpose (*Wertrationalität*). Rationality of purpose consists in choosing the right ends to pursue: what is it that makes life worth living? That is the classical philosophical question of the good life, and it is no concern of the economist. Once the goals have been decided on, there is the question of how best to achieve them: that is rationality of means, and this is where the economist comes in.

What is the role of mathematics? It has shown itself to be a worthy ally to economists in exploring rationality of means. Could it be helpful in exploring rationality of purpose? Not directly, it is all about values, which are uncommensurable and the essence of mathematics is to measure and compare: setting a price for human life, for instance, is common practice in economics, when investigating whether to improve a road or deciding on public health policy, and unacceptable in individual behaviour (except for certain kind of professionals). Some mathematicians have tried to formalize values judgment, Garrett Birkhoff for instance, who wrote a book on *Aesthetic Measure* (1933), but the result is less than convincing, to put it kindly.

However, mathematics could be very helpful in developing the consequences of certain choices: what if one instituted a carbon tax? what if one restricted high-frequency trading? what if one did away with limited liability for stockholders? what if one introduced money and banks in economic models? I am sick and tired of reading papers on executive compensation, concluding that they should be paid more and more to keep them interested in doing their job: should one not question the governance rules which lead to this kind of result, and suggest others which would lead to more palatable results? As we mentioned in the beginning, the rules by which markets and firms operate are not set in stone, and can be changed for better ones. Unfortunately, there is very little research in that direction. The bulk of research in mathematical finance, for instance, is devoted to exploring to the very end the consequences of no-arbitrage theory, and pays no attention to the way payments are made, banks operate and money is created. Similarly, the model mainstream economists use to study global warming is so simplified as to be irrelevant: there is no biosphere and the diversity of human populations is compressed into a single "representative" individual. There is ample room for making the model more realistic, and there is some activity in that direction, but it is overshadowed by the Nobel prize of William Nordhaus, who devoted his life to that model, and concludes that a warming of 3 °C is "optimal", in direct contradiction to the unanimous conclusions of biologists, physicists and other scientists.

What I suggest here, to use mathematics in a critical way, in order to question the rules of the prevailing game, is of course more difficult than to respond to the demand of the financial industry or the global markets. This is the difference between engineers and scientists: the first ones solve practical problems which are brought to them, the second ones try to understand the world they live in. There is also a third dimension available to mathematicians, namely poetry, or rather art for art's sake. The world we live in is not pretty, and the more one understands it the uglier it looks. For this reason, one needs sometime to escape in another one, like the protagonists of that wonderful and terrible movie "Brazil": this is what "pure" mathematics does for you, at least what it has done for me. I have taken incredible pleasure in the work of some masters, like John Milnor and René Thom, and the muse of mathematics has been very kind to me. Let me conclude this by quoting one of her presents, which accompanies me to this day:

Theorem 4.1 *Let (X, d) be a complete metric space and $f : X \rightarrow R \cup \{+\infty\}$ a lower semi-continuous function, bounded from below: $f \geq 0$. For every x_0 and every $\lambda > 0$ there exists some x_1 such that:*

- $f(x_1) \leq f(x_0) - \lambda d(x_0, x_1)$
- $d(x_0, x_1) \leq \frac{1}{\lambda} f(x_0)$
- $\forall x, \ f(x) \geq f(x_1) - \lambda d(x, x_1)$

References

1. K. Arrow and G. Debreu. Existence of an equilibrium in a competitive economy. *Econometrica* **22**(3), 265–290 (1954).
2. G. Becker. *The economic approach to human behaviour.* The University of Chicago Press (1976).
3. G. Becker. *A treatise on the family.* Harvard University Press (1981).
4. G. Becker. *The economics of discrimination.* The University of Chicago Press (1971).
5. G. Becker. *Essays in the economics of crime and punishment.* National Bureau of Economic Research: distributed by Columbia University Press (1974).
6. G. Becker. *A theory of rational addiction.* The University of Chicago Press (1984).
7. M. Browning and P.-A. Chiappori. Efficient intra-household allocations: a general characterization and empirical tests. *Econometrica* **66**(6), 1241–1278 (1998).
8. J. Carter and M. Irons. Are economists different, and if so, why?. *Journal of Economic Perspectives* **5**(2), 171–177 (1991).
9. P.-A. Chiappori and I. Ekeland. The economics and mathematics of aggregation: formal models of group behavior. *Foundations and Trends in Microeconomics* **5**(1–2), 1–151 (2009).

Can We Teach Functions to an Artificial Intelligence by Just Showing It Enough "Ground Truth"?

Adrien Courtois, Thibaud Ehret, Pablo Arias, and Jean-Michel Morel

1 Introduction

Artificial neural networks (NNs) are complex non-linear functions $y = \mathcal{N}(w; x)$ obtained by combining multiple simple units, in a structure that is reminiscent of how neurons are organized in the brain. NNs are parameterized by a vector w of millions, billions, or sometimes even trillions of parameters.

A neural network $\mathcal{N}$ can be described as a family of functions $(f_i)_{i \in [\![1,D]\!]}$ and a vector of weights $w \in \mathbb{R}^d$ such that the network is defined as

$$\mathcal{N}(x; w) = f_D(f_{D-1}(\ldots(f_2(f_1(x; w); w)\ldots; w); w).$$

Notably, each function f_i is differentiable almost everywhere and is parameterized by weights stored in $w \in \mathbb{R}^d$. During a supervised training, neural networks are trained using large number of examples pairs (x, y) of inputs and the desired outputs (e.g. noisy and noiseless image pairs) stored in a dataset $\mathcal{D} := (x_n, y_n)_{n \in [\![1,N]\!]} \in (\mathbb{R}^p \times \mathbb{R}^q)^N$. Supervised training is to be distinguished from self-supervised and unsupervised training, which do not involve ground truth. The objective of training is to minimize the risk, which is the expected value of the loss

$$\mathcal{L}(w) = \int L(\mathcal{N}(x; w), y) d\mathbb{P}(x, y), \tag{1}$$

A. Courtois · T. Ehret · P. Arias
ENS Paris-Saclay, Gif-sur-Yvette, France

J.-M. Morel (✉)
Ecole Normale Supérieure Paris-Saclay, Paris, France

where $\mathbb{P}$ is the probability distribution of the data, and w denotes the vector of all parameters of the network. The distribution $\mathbb{P}$ is usually unknown and the empirical distribution of a collection of training samples $\mathbb{P} = \frac{1}{N} \sum_{n=1}^{N} \delta_{(x_n, y_n)}$ is used instead, so that

$$\mathcal{L}(w) = \frac{1}{N} \sum_{n=1}^{N} L(\mathcal{N}(x_n; w), y_n). \tag{2}$$

To minimize this risk—or this loss—a gradient descent algorithm is used, which therefore reaches a local minimum, as the functional is generally not convex. The gradient $\nabla_w \mathcal{L}$ of the loss over the training data is computed with respect to the set of parameters w. In practice, due to computational limitations, the stochastic gradient descent is used and more sophisticated optimization methods are becoming common [27, 41]. The dominant learning structure is called deep learning and refers to "deep" multi-layer neural networks who are being trained on massive data to learn the "right" answer to a request. In supervised learning, the training is sometimes led by human "ground truth", obtained by manual annotation. In other examples, the training is obtained from a large set of observed data.

Most tasks proposed to deep learning are highly complex and hardly definable in mathematical terms. The hope is that by accumulating enough ground truth data and feeding it to a neural network, the right operator will be obtained. This has two implications: the operator being learned is often not mathematically or formally defined; it is defined "in extenso" by its application to many examples.

The goal of data-driven techniques is that the network trained on a dataset will "learn" to solve the task and generalize to new unseen data. To test this generalization ability, the dataset is split into training and test datasets. Then, the network is trained using data sampled from the training set, but the goal is to optimize the performance on the test set. However, while networks for a given task generalize well between splits of a same dataset, *they often fail when applied on a different dataset* [53]. This leads to a well-known failing of deep learning algorithms, called "statistical overfitting" or "dataset bias" [52]: e.g. the learning task is not necessarily extensible outside of the dataset it has learned from. The fact that neural networks struggle to generalize outside of the training domain is actually well-known [52, 53]. As pointed out in this last paper,

> Some datasets, that started out as data capture efforts aimed at representing the visual world, have become closed worlds unto themselves.

The testing set has, legitimately enough, exactly the same origin as the learning set, being generally extracted for the same original larger set by uniform random sampling. In short, the problem supposedly resolved by learning parameters on the training data set has no other definition than the dataset itself. And the verification of the network's capacity on a testing set statistically identical to the training set is a congenital defect of machine learning. But what else could be done? To solve this problem, different approaches have been studied in the literature. Among those, we can extract the four main ideas: uncertainty detection [1], domain transfer [56, 63], self-supervised training [23, 47] and unsupervised training [47].

Uncertainty detection is about making the network output its confidence score to influence its calibration [12, 18, 39, 40, 54]. This is not an easy task as neural networks tend to be overconfident and badly calibrated [19].

Domain transfer means making a network trained on a certain dataset work on another dataset. There are three main types of domain transfer algorithms. The first one consists in modifying the images of the new dataset so it matches the statistics of the images of the training dataset [22, 34, 64]. The second one consists in training the network with an additional loss term which enforces learning features that have the same statistics across different datasets [46, 48, 49]. Lastly, some approaches propose to leverage the network's features to generate pseudo-labels which will be used to train a new network [13, 26]. This can be useful when the new dataset is not fully labeled. Data augmentation can be seen as a domain transfer technique. It consists in applying random transformations to the training samples to artificially increase the diversity and size of the dataset.

Self-supervised learning has received a lot of attention in the past few years as it allows to train a network without any label. This partially cures the problem of wrong labels and alleviates the cost of labeling an entire dataset. In some of those approaches, the distance between the representation of similar data samples is reduced while it is increased for dissimilar samples [15]. For some others, no dissimilarity is used [25]. When used, the notion of similarity is to be redefined for each task [59, 62]. While it has its own caveats, this technique is considered to be very promising. However, neural networks trained in a self-supervised fashion are still trained on a dataset and they can still be subject to dataset biases, contradictions and ambiguities.

Unsupervised learning, on the other hand, aims at exhibiting the structure underlying a dataset. For instance, it can find clusters of data samples sharing the same properties [5, 6]. However, this domain is not receiving as much attention as self-supervision.

An approach that has not been pursued extensively however, is changing the properties of the training dataset itself to limit contradictions or generate a fully artificial dataset in which there is an unbiased ground truth. In this paper, we will illustrate the problems of supervised learning approaches through two examples where neural networks trained on "natural" datasets learn what is actually a geometrically definable property. Through those two examples, we will show that by enforcing these geometric properties on the training dataset, or directly on images, we can drastically change the outcomes of machine learning. The two geometric questions we shall consider are monocular depth estimation (MDE) from a single image ("monocular vision") and line segment detection (LSD) in natural images.

In Sect. 2, we contrast a natural monocular depth estimation dataset with a synthetic *Rectangle Depth Estimation Dataset (RDE)*. While synthetic, the RDE dataset retains the main difficulties of the depth estimation problem, but has an unambiguous ground truth where reaching 100% accuracy is theoretically possible.

In Sect. 3 we address the very basic line segment detection in images and oppose two groups of methods that we describe briefly: the first approach is a geometric statistical definition of line segments. It is represented by two methods. The other approach is represented by not less than five deep neural networks that were taught to recognize segments on a human annotated dataset. We make a simple experimental comparison of these methods, as their results are easily interpreted and linked to their methodology. Section 4 is a final discussion.

2 Depth Estimation and Its Impossible Ground Truth

Fig. 1 Depth estimation (right) of a neural network [45] trained on natural images for the Library image (left)

A classical example of recent success in machine learning has been to train a network to "see in depth" by giving it a large number of images coupled with their depth map. The network is then asked to deduce from a single image its depth map, an operation called monocular depth perception. Depth estimation datasets [8, 33, 57, 61] consist in the association between many images and their associated depth map measured by range laser. In other terms, each image pixel has two attributes, color and range (or depth). No faulty human intervention can be found in such real ground truth. Yet, the problem of deducing depth from color is obviously ambiguous: if a flat green image is shown, what is the distance of green? Thus, networks are at first sight requested to perform an impossible task. Yet, the hope is that the very large size of the dataset ensures that the network's decision will somehow be akin to our imprecise but useful perception of perspective. The result on images similar to the indoor scenes of the training set is impressive: See the experiment of Fig. 1, made on an "in domain" image.

Does it mean that these networks resolve perspective vision? Note that they are trained to find a normalized depth (between 0 and 1), not the actual depth measured in, say, meters. But monocular perspective perception is not to be identified with finding real distances, even if normalized.

The human monocular perception theory of perspective [36, 37] tries to explain how our perception orders objects in space. There are two classes of cues helping decide if an object is farther away than another. The first ones are of optical-geometric nature, like atmospheric perspective, organization of straight lines toward vanishing points, shape from shading, shape from texture, and occlusion, that is signaled by T-junctions. The other ones are empirical and rely on our experience of the spatial organization of recognizable objects natural scenes, where in addition we know in advance the size of these objects, humans, animals, trees, vehicles, buildings, etc. One can deduce their approximate distance from their apparent size in an image.

The artificial monocular depth estimators that we shall consider [38, 45] are not being trained with a loss that is invariant to any increasing function applied to the measured depth. They are not requested to provide an order in space, but directly a real depth.

In order to get a clearer view of at least one element of depth perception, we shall rely on Kanizsa's perception of depth. Kanizsa [24] discovered that a particular organization of visible boundaries in images in T-junctions was the predominant cue to organize objects in space from the farthest to the closest ones. (See Fig. 2 for an illustration of the role of T-junctions: they signal which rectangle is in front of another.) We defined in [9] a fully unambiguous depth dataset to evaluate if and how the relative position of objects could be evaluated from Kanizsa's theory alone. The synthetic dataset was made of images like the one in Fig. 2.

Other synthetic datasets [43, 51] have been proposed to analyze and quantify the effect of certain layers or training methods, allowing one to discover effects that would otherwise be impossible to unveil [43]. Notably, synthetic datasets are commonly used for image quality evaluation [29]. The vast majority of the literature on depth estimation uses datasets of real images labeled using lidar lasers. Part of these works aim at improving already-existing networks used for depth estimation. For instance, some focus on designing better losses [30, 31, 33, 45, 55, 58] while others [38, 44] devise post-processing strategies based on already-trained networks to make them work on new cases.

Our dataset is inspired by the dead leaves model [14] and poses a very simple depth estimation task where objects are replaced by simple rectangles. The rectangles can overlap one another, creating a spatial organization that naturally puts objects of top of others. The ground truth is the ordering of rectangles in increasing integer order from the black (0) background to the closest rectangle (up to 10). Each order is represented by a color. In other terms, the goal of the algorithm is, given the input image to output similar images where colors have been replaced by the new fixed colors in correspondence to the depth order from 0 to 10. The task given to a neural network is closely related to real-world depth estimation as it accurately reflects one of its main difficulties. When an object is partially occluded by others that divide it into several components, the network must regroup the parts that have been separated, which can only be done by recognizing the same color and/or

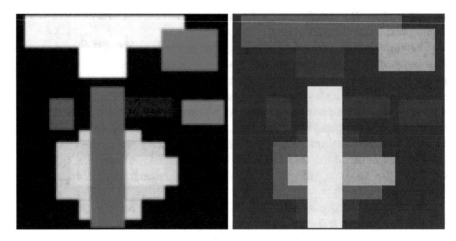

Fig. 2 An example image from the RDE dataset and the associated ground truth. The brighter the color, the higher the number of rectangles that are beneath it. The unambiguous images of the dataset are made of up to ten superposed rectangles with fixed colors. The ground truth is the ordering from 0 to 10 and it can be deduced visually and by a simple algorithm. (Each color on the right is associated with a class from 0 to 10). T-junctions, namely points where a region border stops on another border, are here the key shape ordering indicators

detecting edge alignment. The ground truth is unambiguous. Indeed, there exists a deterministic reconstruction algorithm based on three nonlocal cues: a) color similarity (all rectangles are monochromatic, thus can be recovered nonlocally); b) T-junctions, a local cue that propagates nonlocally, c) convexity, which leads to decisions about the overlap of a convex region with the one surrounding it. A full description of the algorithm recovering depth (identified with the overlap order) is described in [9]. An example can be found in Fig. 2.

We can now evaluate if, indeed, a neural network trained on a set deprived of any ambiguity surpasses in performance the same network trained on the very same dataset but were the carefully removed ambiguous cases have been added back to the dataset. The result of this experiment is given in Table 2, where different variants of the RDE dataset were investigated. We used the exact same architecture for every training; only the training and test set were modified. The specificity of each variant is described in Table 1. Various measurements of the performance of the networks are shown. In particular, when training on a dataset where all the ambiguities were retained (second line), we notice that the last three measurements are multiplied by a factor 3 to 4. When changing the blur kernel, the results are catastrophic. This shows a strong dataset bias even in this over-simplified modeling of depth estimation.

In Table 2, the test loss refers to an average error (as measured by the loss) on the test set. The root mean square error passes from 5 to 2 roughly. The three metrics we report in this table are three of the most commonly employed metrics for Monocular Depth Estimation tasks [4, 7, 38, 45]: the Root Mean Square Error

(RMSE), the $\delta_{1.25}$ and the Ord metric. The RMSE is defined as

$$\text{RMSE}(\hat{y}, y) := \sqrt{\frac{1}{HW} \sum_{i,j} (\hat{y}_{i,j} - y_{i,j})^2},$$

where $\hat{y}$ is the prediction and y the ground-truth. This measures the error between the network's prediction and the prediction it was supposed to make. The $\delta_{1.25}$ metric is the percentage of pixels such that

$$\delta_{1.25} := \frac{1}{HW} \sum_{i,j} \mathbf{1}_{\{\max\left(\frac{\hat{y}_{i,j}}{y_{i,j}}, \frac{y_{i,j}}{\hat{y}_{i,j}}\right) > 1.25\}}.$$

In other terms, it corresponds to the portion of pixel where the network's prediction erred by more than 25%. The ordinal loss consists in sampling 50,000 pairs of pixels $((i_1, j_1), (i_2, j_2))$ and for each of those pairs, compute

$$l_i = \begin{cases} +1, & \text{if } y_{i_1, j_1}/y_{i_2, j_2} \geq 1 + \tau \\ -1, & \text{if } y_{i_1, j_1}/y_{i_2, j_2} \leq \frac{1}{1+\tau} \\ 0, & \text{otherwise.} \end{cases}$$

Using the same pairs, the equivalent quantity $\hat{l}$ is computed for the prediction. The ordinal loss is given by

$$\text{Ord} := \frac{1}{|\mathcal{P}|} \sum_{i \in \mathcal{P}} \mathbf{1}_{\{l_i \neq \hat{l}_i\}}.$$

In practice, we used $\tau = 0.03$ and all the networks were evaluated using the same set of pairs of pixels when computing the ordinal loss. Said differently, this metric measures if the network predicted the correct ordering of the scene i.e. if the value of one pixel is above another one in the ground truth, this ordering must prevail in the network's prediction regardless of the exact value.

The reader should be curious at this point to see what a big and successful neural network trained on millions of natural images gives on our synthetic images. All in all, we should have expected that such a network gives a constant depth on each rectangle with constant color, and that it orders correctly each pair of rectangles that overlap. The result is shown in Fig. 3. In the second column we see the depth estimated by very powerful neural networks trained on millions of image-depth pairs. They react to this abstract scene as if it were a natural perspective scene, where generally the more distant objects are situated at the top and the closest ones at the bottom. In other terms, the networks have learned "average" perspective laws and apply them everywhere. There is no surprise in this result, it is "dataset bias"

Table 1 List of the different variants of the RDE dataset we considered. Both Gaussian and uniform blur have the same kernel size. The noise is centered white Gaussian noise of standard deviation $\sigma = \frac{2}{256}$

Dataset name	Ambiguous GT	Gaussian blur	Uniform blur	Noise
Unambiguous	✗	✓	✗	✗
Ambiguous	✓	✓	✗	✗
No blur	✗	✗	✗	✗
Noisy	✗	✓	✗	✓
Uniform blur	✗	✗	✓	✗

Table 2 Results on the RDE dataset and variants, as described in Table 1. All metrics are multiplied by 100 for readability. For these classic performance metrics, the lower the better

Train dataset	Test dataset	Test loss ↓	Ord ↓	$\delta_{1.25}$ ↓	RMSE ↓
Unambiguous	Unambiguous	0.72	1.79	1.98	1.95
Ambiguous	Unambiguous	1.78	6.03	7.36	4.91
Unambiguous	No blur	7.70	4.53	5.48	5.13
Unambiguous	Noisy	1.21	2.56	6.36	3.76
Unambiguous	Uniform blur	32.0	10.4	14.5	11.3

on a plausible, yet "out of domain" image. But, certainly, this result differs much from what humans would propose.

Some Partial Conclusions

The proposed image dataset and its ground truth were designed to follow a series of requirements:

Unambiguous ground-truth, Well-posedness: The input contains enough information to solve the task;

Focus on a specific required network property: The network must be able to deduce the exact ground truth from the input image only if it has the assessed property. In our case this property was the nonlocal propagation of the depth property to all pixels having the same color, and of the order indicated by each T-junction to the two concerned rectangles.

Perceptual and physical validity: Although the images are not natural, a correct interpretation of the image must be realizable by a quick visual inspection. In our case it could be realized with rectangular colored paper sheets disposed on a table.

These properties are not attainable with natural datasets, as they contain many statistical cues that help compensate for a structural deficiency of the network. We conjecture that using such fully controlled datasets, but very similar to a classic artificial intelligence task, might be used for neural network design, and for explaining the observed properties and flaws of neural networks trained on "natural" datasets. We have just verified in the above example that eliminating ambiguities boosts the overall network performance.

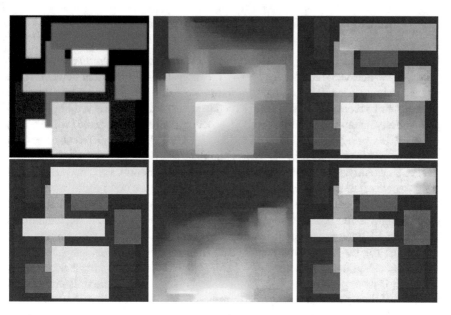

Fig. 3 Results of state-of-the-art networks when evaluated on our task without retraining. First column: input and ground truth. Second column: result of MiDaS [45], result of MergeNet [38]. Third column: result of the baseline U-Net, result of our best network, both being trained on our synthetic dataset. The disappointing results of SOTA networks on a visually interpretable image show that these networks are guided by hidden natural statistics, much more than by nonlocal geometric reasoning. By a nonlocal association failure, the baseline U-Net also errs on the large disconnected brown rectangle

3 Human Ground Truth and Line Segment Detection

Line segment detection in images has innumerable applications, as most human made objects contain straight edges that cause straight image edges, namely straight segments across which the image intensity changes drastically, due to different plane orientations toward the light [28, 35]. In this section we shall contrast two methodologies for the detection in images of an object that might have claims to be defined in purely mathematical terms: line segments. The first methodology (two algorithms will be presented) is purely geometric and statistical, and it does not require a single learning example. The second methodology builds sophisticated neural networks that learn from a human-annotated dataset associating images with their line segments.

3.1   Line Segments Detected by Statistical Testing

LSD [16, 50] and EDlines [2, 3] are based on the *non-accidentalness* statistical principle, which we shall briefly illustrate in the case of alignment detection. Its simple guiding idea is to count the number of aligned pixels on any possible segment and accept the set of pixels as a line segment if the observed alignment is perceptually meaningful, namely could not occur in a fully disordered (white noise) image. This principle originates in Gestalt theory [11]. The difference between LSD and EDlines resides in the way candidate segments are extracted from the image. (It would be inconvenient to test all possible segments). LSD finds connected components of pixels sharing the same gradient orientation up to some precision. If the shape of the component is approximately a rectangle, then the medial axis of the rectangle is the detected segment, provided it satisfies the non-accidentalness principle. EDlines proceeds differently and generates candidates by joining seed candidates for the tips of the segments. Then these candidates are selected by the non-accidentalness principle. More quantitatively, a line segment detection is validated if its expectation in white noise is low [11]. In a white noise image, all pixels (and their gradient orientations) are independent and uniformly distributed. Let us define the "Number of False Alarms (NFA)" of a line segment as follows. Let A be a discretized segment comprised of n pixels on the image domain. A valid line segment of the image should have "many" pixel gradient directions aligned to the direction normal to the segment A. Suppose A has at least k points with their directions aligned (up to an error $\pm p/2$) with the normal direction to A in an image of size $N \times N$. Define the NFA of A as

$$NFA(n, k) = N^4 \sum_{i=k}^{n} \binom{n}{i} p^i (1 - p)^{n-i}, \tag{3}$$

where N^4 is the number of potential line segments in an $N \times N$ image. The probability p used in the computation of the binomial tail is the accuracy of the line direction. An event (a line segment in this case) is called ε-meaningful if its $NFA(n, k) \leq \varepsilon$. The authors advise setting ε to 1, which corresponds to one false detection per image. Given these definitions, a line segment of length n, with k aligned pixels k is considered valid if $NFA(n, k) \leq 1$. Otherwise the line is rejected.

Fig. 4 Middle: two images from the *wireframe* dataset. Top: their line segment ground truth in the *wireframe* dataset. Bottom: their interpretation by LSD. This experiment illustrates the frailty of human annotation: many obvious line segments are missing in the ground truth; some are partially occluded

3.2 Learning by Examples What a Line Segment Is

We shall compare the above geometric-statistical methods LSD and EDlines with not less than five methods created by machine learning, SOLD2 [42], M-LSD [17], TP-LSD [21], ULSD [32], and LETR [60]. These five deep neural networks are almost contemporary and very recent. They can be compared because they have been trained (and tested) on the same dataset, *Wireframes*. In architecture design, a wireframe is often referred to as a line drawing of a building or a scene on paper. The authors of [20] have built this very large new dataset of over 5000 images with wireframes thoroughly labeled by humans. Note that a wireframe is not just a set of line segments; it is a set of line segments aimed at describing an architecture. Hence, its line segments mostly correspond to edges and corners of structures (walls, buildings, furniture, etc.), and they mostly end on T-junctions or

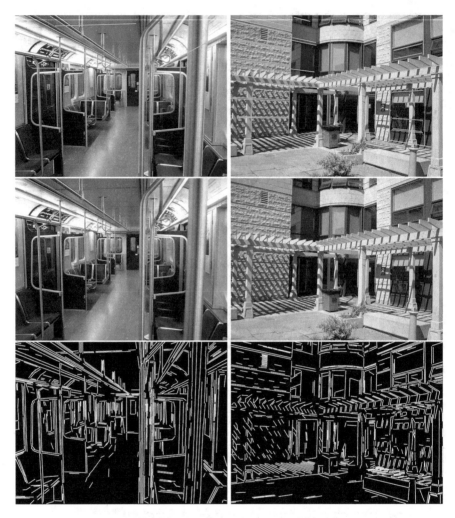

Fig. 5 Middle: two images from the *YorkUrban* dataset. Top: their line segment ground truth in the *YorkUrban* dataset. Bottom: their interpretation by LSD. This experiment illustrates the bias of human annotation in Wireframe dataset: an architect is only interested in architecturally meaningful separations. There are also lines drawn that do not correspond to an actual line on the image. Finally, many obvious line segments are missing and some are partially occluded

corners. This led annotators to neglect many segments that can be conspicuous but have a minor explanatory role to understand the architecture, for example the dark rays separating planks. For all five neural networks we shall consider, training was made on this dataset and validated on its testing part and on another wireframe test dataset, the YorkUrban dataset [10]. All of these methods focus on predicting different parts of the lines annotated in the dataset. In particular, they try to predict the junctions (i.e. the endpoints of the lines), the line itself as a heatmap, or a root point defined as the central point of the line as well as the displacements from this root point to other defining points of the line. Additionally, the methods also include a classification module that learns whether a predicted line corresponds to an actual line in the dataset.

It is to be noted that some of the methods we describe have been trained following an unsupervised or self-supervised scheme. This further points out that the downfalls of "statistical overfitting" and "ground truths" are not exclusive to supervised learning.

3.3 Comparison of All Methods

Fig. 6 Comparison of seven line segment detectors on a photograph of Le Pirée: original image "Le Pirée", results of LSD (2008), EDlines (2011), TP-LSD (2020), ULSD (2021), LETR (2021), M-LSD (2021), SOLD2 (2021). The first two detectors are handcrafted and based on edge growing followed by an a contrario detection threshold. The last five are obtained by sophisticated mainly unsupervised deep learning methods

In Fig. 6 we compare the results of the seven above presented line segment detectors on an architectural photograph. The experiment displays the original image "Le Pirée", followed by results of LSD (2008), EDlines (2011), TP-LSD (2020), ULSD (2021), LETR (2021), M-LSD (2021), SOLD2 (2021). As we commented, the first two detectors are handcrafted and based on edge growing followed by an a contrario detection threshold. A detailed comparison of both images shows that most segments present in one image are present in a similar position in the other. Sometimes a segment detected by EDlines is actually split in two smaller segments with a gap in the LSD result. This corresponds to a slightly different heuristic exploration of the image gradient field by both methods: LSD requires connectedness for a segment to be detected, while EDlines allows for some gap in a segment and therefore sometimes presents a single segment where LSD found two. All in all, the detection maps are very similar because they obey the same statistical definition of a meaningful segment. No machine learning is involved in this definition and therefore no dataset dependence: these detectors are agnostic.

Something radically different is at stake with the last five detectors obtained by sophisticated deep learning methods learning primarily from the *Wireframe* dataset. Unsurprisingly, the line segment interpretation given by machine learning taught on wireframes are widely different from those proposed by the agnostic statistical definition.

This is confirmed by taking a look at images from the *wireframe* dataset [20], see Fig. 4. The middle row shows two images from the *wireframe* dataset and their "ground truth" (first row) which is a human annotation. This annotation is actually a mental reconstruction of the architectural sketch of the scene. It presents segments that are hardly visible, neglects many that are conspicuous, and even partially occluded ones. This is in strong contrast with the result of the agnostic LSD detection in the bottom row showing hundreds of line segments bordering the ground planks on the left image. On the right-hand image, the ground truth interprets the muntins of the windows as single line segments while these are thick and bounded by two straight sides. This experiment illustrates perhaps the bias of human annotation: many obvious line segments are missing because they were not considered meaningful for interpreting the scene. The five neural networks were also evaluated (and compared to LSD and EDlines) on the YorkUrban dataset [10]. In Fig. 5 we display results in the same format as for the Wireframe images. On the middle row, two images from this dataset. On top of them their "wireframe" interpretation. On the bottom row, the line segments found by LSD.

Returning to Fig. 6, one can notice some agreement between the results of the four first mentioned networks trained on *wireframe*, TP-LSD, ULSD, LETR, and MLSD, while the result of SOLD2 is puzzling. The four first results give a sketchy but coherent view of the building visible on the foreground. The input image belongs to the "domain" on which the neural network is competent.

Yet, on a slightly "out of domain" image, the results obtained by neural networks trained on *wireframe* diverge considerably more, as evident in the results compared in Fig. 7. The scene is actually an indoor scene, but containing just and only aligned chairs in perspective. The networks perform a "wireframe" analysis. Hence, the rounded angles of the chairs are replaced by corners; more strikingly, most chair borders are visually ridges, not edges, thus delimited by two parallel straight lines. These pairs are found by LSD and EDlines, but systematically replaced by a single line in the wire frame interpretation.

In the experiments of Figs. 6 and 7 the algorithms depend on parameters. For LSD and EDlines, by a classic principle of a contrario detection methods, one false alarm (in expectation) is allowed per image. This principle makes sense in detection tasks where multiple detections are expected. Unfortunately for neural networks no such principle is available or accessible to learning.

The considered machine learning methods depend on one or two parameters. One is common to all: the score or detection threshold ranging from 0 to 1. But an additional threshold may be used to enforce accuracy like in ML-LSD endowed with a distance threshold, and SOLD2, which has an inlier threshold ranging in [0, 1]. LSD and EDlines have both a single threshold, the NFA, which is fixed to 1, thus actually not an active user parameter.

In Fig. 8 we display the results obtained when varying the parameters of each method, particularly the score threshold. Although the median score 0.5 would be expected to give the best result on the learning dataset, this threshold is clearly no longer valid on an "out of domain" image.

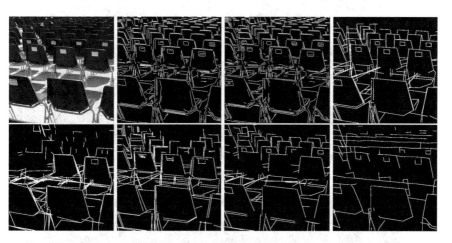

Fig. 7 Comparison of seven line segment detectors on an "out of domain" image, a photograph of chairs: original image "chairs" results of LSD (2008), EDlines (2011), TP-LSD (2020), ULSD (2021), LETR (2021), M-LSD (2021), SOLD2 (2021). The first two detectors are handcrafted and based on edge growing followed by an a contrario detection threshold. The last five are obtained by sophisticated mainly unsupervised deep learning methods

4 Conclusion

Neural networks are generally given ambiguous or contradictory training examples and given a proposed task that is not even definable. In the two use cases we examined, the problem that neural networks are asked to solve is not defined by anything but by the dataset. We picked these two examples for several reasons. First of all, the question to be resolved by these neural networks has a clear physical meaning (for the depth problem) and in the second example an intuitive geometric meaning. Indeed, line segments correspond mostly to physical edges of buildings and other man-made objects.

Fig. 8 Parameter dependence of five line segment detectors obtained by deep learning on an "out of domain" image, a photograph of chairs. First row: results of LETR with score thresholds 0.7, 0.3 and 0.0 respectively. Second row results of TP-LSD with score thresholds 0.25, 0.1, and an alternative version (TP-LSD-320) with score threshold 0.1. Third row: ULSD with score thresholds 0, 0.1 and 0.2 respectively. Fourth row M-LSD with parameter pairs (0.2, 1), (0,20) and (0.5, 10) respectively. Fifth row SOLD2 with parameter pairs (0.1, 0.5), (0.1, 0.99), (0.5, 0.99) respectively. The parameters are chosen so the number of segments decreases from over-detection to under-detection; the parameters in the middle column are the more plausible ones

We tried to argue that defining a problem by a "natural" dataset may become a dead end. First, because no external verification is possible. Or, rather, in the cases we examined, an external verification using sound but slightly "out of domain" data yields questionable or unstable results, which is alarming. The AI movement is carried by the hope that neural networks have or will have a "generalization power". The two examples we examined rather suggest that we should find workarounds to force neural networks to generalize by fully controlling the structure of well-defined synthetic datasets, rather than relying on uncontrolled "ground truth". Avoiding ambiguities and contradictions in what we teach neural networks seems to be a sound pedagogic recommendation. The notions of well-posedness, and geometric invariance for example, have been powerful requirements in physics and applied mathematics. Such axiomatic requirements should be applied to the training datasets and on the loss. For example, we saw that the loss in monocular depth perception has not the obvious geometrically required invariance. We should perhaps explore the limits of ad hoc synthetic datasets, with mathematically definable ground truth, before we start even using "ground truth in the wild". Last but not least, our examination of line segment detection suggests that the variability of networks trained on the very same dataset should be a major concern and actually perhaps an a posteriori criterion to evaluate each proposed dataset. Similarly, measuring the variance of the result for different training sessions of the same neural network trained on the same dataset should also be an obvious reliability criterion.

Neural networks are like Plato's prisoners, shackled in a cave and shown shadows of objects on a wall. Because they know of nothing else, they accept anything shown as the whole reality and cannot imagine anything beyond it. Contrary to what a mathematically naive approach would suggest, training an artificial intelligence requires a teaching strategy that avoids contradictions and ambiguities, and the task should be feasible in the end, much like when teaching humans.

References

1. M. Abdar, F. Pourpanah, S. Hussain, D. Rezazadegan, L. Liu, M. Ghavamzadeh, P. Fieguth, X. Cao, A. Khosravi,U. Rajendra Acharya, et al. A review of uncertainty quantification in deep learning: Techniques, applications and challenges. *Information Fusion* **76**, 243–297 (2021).
2. C. Akinlar and C. Topal. Edlines: Real-time line segment detection by edge drawing (ed). In: *2011 18th IEEE International Conference on Image Processing*, pp. 2837–2840, IEEE (2011).
3. C. Akinlar and C. Topal. EDLines: A real-time line segment detector with a false detection control. *Pattern Recognition Letters* **32**(13), 1633–1642 (2011).
4. I. Alhashim and P. Wonka. High quality monocular depth estimation via transfer learning. *arXiv preprint* arXiv:1812.11941 (2018).
5. M. Caron, P. Bojanowski, A. Joulin and M. Douze. Deep clustering for unsupervised learning of visual features. In: *Proceedings of the European conference on computer vision (ECCV)*, pp. 132–149, Springer (2018).

6. M. Caron, I. Misra, J. Mairal, P. Goyal, P. Bojanowski and A. Joulin. Unsupervised learning of visual features by contrasting cluster assignments. *Advances in Neural Information Processing Systems* **33**, 9912–9924 (2020).
7. W. Chen, Z. Fu, D. Yang and J. Deng. Single-image depth perception in the wild. *Advances in neural information processing systems* **29**, 730–738 (2016).
8. W. Chen, S. Qian, D. Fan, N. Kojima, M. Hamilton and J. Deng. Oasis: A large-scale dataset for single image 3d in the wild, In: *Proceedings of the IEEE/CVF Conference on Computer Vision and Pattern Recognition*, pp. 679–688, Springer (2020).
9. A. Courtois, J.-M. Morel and P. Arias. Investigating Neural Architectures by Synthetic Dataset Design. *arXiv preprint arXiv:2204.11045* (2022).
10. P. Denis, J.H. Elder and F.J. Estrada. Efficient edge-based methods for estimating Manhattan frames in urban imagery. In: *Proceedings of the European conference on computer vision (ECCV)*, pp. 197–210, Springer (2008).
11. A. Desolneux, L. Moisan and J.-M. Morel. *From Gestalt Theory to Image Analysis: A Probabilistic Approach.* Interdisciplinary Applied Mathematics **34**, Springer Science & Business Media (2007).
12. T. DeVries and G.W. Taylor. Learning confidence for out-of-distribution detection in neural networks. *arXiv preprint arXiv:1802.04865* (208).
13. H. Feng, M. Chen, J. Hu, D. Shen, H. Liu and D. Cai. Complementary pseudo labels for unsupervised domain adaptation on person re-identification. *IEEE Transactions on Image Processing* **30**, 2898–2907 (2021).
14. Y. Gousseau and F. Roueff. The dead leaves model: general results and limits at small scales. *arXiv preprint arXiv:math/0312035* (2003).
15. J.-B. Grill, F. Strub, F. Altché, C. Tallec, P. Richemond, E. Buchatskaya, C. Doersch, B. Avila Pires, Z. Guo, M. Gheshlaghi Azar et al. Bootstrap your own latent: A new approach to self-supervised learning. *Advances in Neural Information Processing Systems* **33**, 21271–21284 (2020).
16. R. Grompone von Gioi, J. Jakubowicz, J.-M. Morel and G. Randall. LSD: A fast line segment detector with a false detection control. *IEEE transactions on pattern analysis and machine intelligence* **32**(4), 722–732 (2008).
17. G. Gu, B. Ko, SH. Go, S.-H. Lee, J. Lee, M. Shin. Towards light-weight and real-time line segment detection. *arXiv preprint arXiv:2106.00186* (2022).
18. C. Guo, G. Pleiss, Y. Sun and K.Q. Weinberger. On calibration of modern neural networks. In: *International Conference on Machine Learning*, pp. 1321–1330, PRML (2017).
19. M. Hein, M. Andriushchenko and J. Bitterwolf. Why ReLU networks yield high-confidence predictions far away from the training data and how to mitigate the problem. In: *Proceedings of the IEEE/CVF Conference on Computer Vision and Pattern Recognition*, pp. 41–50, IEEE (2019).
20. K. Huang, Y, Wang, Z. Zhou, T. Ding, S. Gao and Y. Ma. Learning to parse wireframes in images of man-made environments. In: *Proceedings of the IEEE Conference on Computer Vision and Pattern Recognition*, pp. 626–635, IEEE (2018).
21. S. Huang, F. Qin, P. Xiong, N. Ding, Y. He and X. Liu. Tp-lsd: Tri-points based line segment detector. In: *European Conference on Computer Vision*, pp. 770–785, Springer (2020).
22. P. Isola, J.-Y. Zhu, T. Zhou and A.A. Efros. Image-to-image translation with conditional adversarial networks. In: *Proceedings of the IEEE conference on computer vision and pattern recognition*, pp. 1125–1134, IEEE (2017).
23. A. Jaiswal, A.R. Babu, M.Z. Zadeh, D. Banerjee and F. Makedon. A survey on contrastive self-supervised learning. *Technologies* **9**(1), 2 (2020).
24. G. Kanizsa. *Organization in vision: Essays on Gestalt perception.* Praeger Publishers (1979).
25. P. Khosla, P. Teterwak, C. Wang, A. Sarna, Y. Tian, P. Isola, A. Maschinot, C. Liu and D. Krishnan. Supervised contrastive learning. *Advances in Neural Information Processing Systems* **33**, 18661–18673 (2020).

26. Y. Kim and C. Kim. Semi-Supervised Domain Adaptation via Selective Pseudo Labeling and Progressive Self-Training. In: *2020 25th International Conference on Pattern Recognition (ICPR)*, pp. 1059–1066, IEEE (2021).

27. D.P. Kingma and J. Ba. Adam: A method for stochastic optimization. *arXiv preprint* arXiv:1412.6980 (2014).

28. Z. Kou, Z. Shi and L. Liu. Airport detection based on line segment detector. In: *2012 International Conference on Computer Vision in Remote Sensing*, pp. 72–77, IEEE (2012).

29. D. Kundu, L.K. Choi, A.C. Bovik and B.L. Evans. Perceptual quality evaluation of synthetic pictures distorted by compression and transmission. *Signal Processing: Image Communication* **61**, 54–72 (2018).

30. J.H. Lee, M.-K. Han, D.W. Ko and I.H. Suh. From big to small: Multi-scale local planar guidance for monocular depth estimation. *arXiv preprint* arXiv:1907.10326 (2019).

31. J.-H. Lee and C.-S. Kim. Multi-loss rebalancing algorithm for monocular depth estimation. In: *Computer Vision–ECCV 2020: 16th European Conference, Glasgow, UK, August 23–28, 2020, Proceedings, Part XVII 16*, pp. 785–801, Springer (2020).

32. H. Li, H. Yu, J. Wang, W. Yang, L. Yu and S. Scherer. ULSD: Unified line segment detection across pinhole, fisheye, and spherical cameras. *ISPRS Journal of Photogrammetry and Remote Sensing* **178**, 187–202 (2021).

33. Z. Li and N. Snavely. Megadepth: Learning single-view depth prediction from internet photos. In: *Proceedings of the IEEE Conference on Computer Vision and Pattern Recognition*, pp. 2041–2050, IEEE (2018).

34. M.-Y. Liu and O. Tuzel. Coupled generative adversarial networks. In: *Advances in neural information processing systems* **29**, Curran Associates, Inc. (2016).

35. X. Lu, J. Yao, K. Li and L. Li. Cannylines: A parameter-free line segment detector. In: *2015 IEEE International Conference on Image Processing (ICIP)*, pp. 507–511, IEEE (2015).

36. W. Metzger. *Gesetze des Sehens* [1975]. Verlag Waldemar Kramer (2008).

37. W. Metzger. *Laws of Seeing*. English translation of "Gesetze des Sehens" (1936, first German edition), by Lothar Spillmann, Steven Lehar, Mimsey Stromeyer and Michael Wertheimer. Cambridge: MIT Press (2006).

38. S.M. Miangoleh, S. Dille, L. Mai, S. Paris and Y. Aksoy. Boosting Monocular Depth Estimation Models to High-Resolution via Content-Adaptive Multi-Resolution Merging. In: *Proceedings of the IEEE/CVF Conference on Computer Vision and Pattern Recognition*, pp. 9685–9694, IEEE (2021).

39. M. Minderer, J. Djolonga, R. Romijnders, F. Hubis, X. Zhai, N. Houlsby, D. Tran and M. Lucic. Revisiting the calibration of modern neural networks. *Advances in Neural Information Processing Systems* **34**, pp. 15682–15694, Curran Associates, Inc. (2021).

40. J. Mukhoti, J. van Amersfoort, P.H.S. Torr and Y. Gal. Deep Deterministic Uncertainty for Semantic Segmentation. *arXiv preprint* arXiv:2111.00079 (2021).

41. Y.E. Nesterov. A method for solving the convex programming problem with convergence rate $O(1/k^2)$. *Dokl. akad. nauk SSSR* **269**, 543–547 (1983).

42. R. Pautrat, J.-T. Lin, V. Larsson, M.R. Oswald and M. Pollefeys. SOLD2: Self-supervised occlusion-aware line description and detection. In: *Proceedings of the IEEE/CVF Conference on Computer Vision and Pattern Recognition*, pp. 11368–11378, IEEE (2021).

43. A. Power, Y. Burda, H. Edwards, I. Babuschkin and V. Misra. Grokking: Generalization Beyond Overfitting on Small Algorithmic Datasets. *ICLR MATH-AI Workshop* (2021).

44. M. Ramamonjisoa, Y. Du and V. Lepetit. Predicting sharp and accurate occlusion boundaries in monocular depth estimation using displacement fields. In: *Proceedings of the IEEE/CVF Conference on Computer Vision and Pattern Recognition*, pp. 14648–14657, IEEE (2020).

45. R. Ranftl, K. Lasinger, D. Hafner, K. Schindler and V. Koltun. Towards robust monocular depth estimation: Mixing datasets for zero-shot cross-dataset transfer. *IEEE transactions on pattern analysis and machine intelligence* **44**(3), 1623–1637 (2020).

46. A. Rozantsev, M. Salzmann and P. Fua. Beyond sharing weights for deep domain adaptation. *IEEE transactions on pattern analysis and machine intelligence* **41**(4), 801–814 (2018).

47. L. Schmarje, M. Santarossa, S.-M. Schröder and R. Koch. A survey on semi-, self- and unsupervised learning for image classification. *IEEE Access* **9**, 82146–82168 (2021).
48. B. Sun, J. Feng and K. Saenko. Return of frustratingly easy domain adaptation. In: *Proceedings of the AAAI Conference on Artificial Intelligence* **30**(1) (2016).
49. B. Sun and K. Saenko. Deep coral: Correlation alignment for deep domain adaptation. In: *European conference on computer vision*, pp. 443–450, Springer (2016).
50. R. Grompone von Gioi, J. Jakubowicz, J.-M. Morel and G. Randall. On straight line segment detection. *Journal of Mathematical Imaging and Vision* **32**(3), 313–347 (2008).
51. Y. Tay, M. Dehghani, S. Abnar, Y. Shen, D. Bahri, P. Pham, J. Rao, L. Yang, S. Ruder and D. Metzler. Long Range Arena: A Benchmark for Efficient Transformers. In: *International Conference on Learning Representations* (2021). https://openreview.net/forum?id=qVyew-grC2k
52. T. Tommasi, N. Patricia, B. Caputo and T. Tuytelaars. A deeper look at dataset bias. In: *Domain adaptation in computer vision applications*, pp. 37–55, Springer (2017).
53. A. Torralba and A.A. Efros. Unbiased look at dataset bias. In: *Proceedings of the IEEE Conference on Computer Vision and Pattern Recognition 2011*, pp. 1521–1528, IEEE (2011).
54. D.-B. Wang, L. Feng and M.-L. Zhang. Rethinking Calibration of Deep Neural Networks: Do Not Be Afraid of Overconfidence. In: *Advances in Neural Information Processing Systems* **34**, pp. 11809–11820, Curran Associates, Inc. (2011).
55. L. Wang, J. Zhang, Y. Wang, H. Lu and X. Ruan. Cliffnet for monocular depth estimation with hierarchical embedding loss. In: *European Conference on Computer Vision*, pp. 316–331, Springer (2020).
56. M. Wang and W. Deng. Deep visual domain adaptation: A survey. *Neurocomputing* **312**, 135–153 (2018).
57. K. Xian, C. Shen, Z. Cao, H. Lu, Y. Xiao, R. Li and Z. Luo. Monocular relative depth perception with web stereo data supervision. In: *Proceedings of the IEEE Conference on Computer Vision and Pattern Recognition*, pp. 311–320, IEEE (2018).
58. K. Xian, J. Zhang, O. Wang, L. Mai, Z. Lin and Z. Cao. Structure-guided ranking loss for single image depth prediction. In: *Proceedings of the IEEE/CVF Conference on Computer Vision and Pattern Recognition*, pp. 611–620, IEEE (2020).
59. Z. Xie, Y. Lin, Z. Zhang, Y. Cao, S. Lin and H. Hu. Propagate yourself: Exploring pixel-level consistency for unsupervised visual representation learning. In: *Proceedings of the IEEE/CVF Conference on Computer Vision and Pattern Recognition*, pp. 16684–16693, IEEE (2021).
60. Y. Xu, W. Xu, D. Cheung and Z. Tu. Line segment detection using transformers without edges. In: *Proceedings of the IEEE/CVF Conference on Computer Vision and Pattern Recognition*, pp. 4257–4266, IEEE (2021).
61. H. Yu, N. Xu, Z. Huang, Y. Zhou and H. Shi. High-Resolution Deep Image Matting. *Proceedings of the AAAI Conference on Artificial Intelligence* **35**(4), 3217–3224 (2021).
62. F. Zhang, P. Torr, R. Ranftl and S. Richter. Looking Beyond Single Images for Contrastive Semantic Segmentation Learning. In: *Advances in Neural Information Processing Systems* **34**, pp. 3285–3297, Curran Associates, Inc. (2021).
63. L. Zhang and X. Gao. Transfer adaptation learning: A decade survey. *arXiv preprint* arXiv:1903.04687 (2019).
64. J.-Y. Zhu, T. Park, P. Isola and A.A. Efros. Unpaired image-to-image translation using cycle-consistent adversarial networks. In: *Proceedings of the IEEE international conference on computer vision*, pp. 2223–2232, IEEE (2017).

Part X
Mathematical Physics and PDEs

This part is mainly devoted to a programmatic vision and conjectures of several leading experts in analysis and partial differential equations.

In *Divergence-free tensors and cofactors in geometry and fluid dynamics*, Denis Serre, the inventor of compensated integrability, presents connections between Div-free/Div-BV symmetric tensors and geometrical topics such as convex bodies and minimal surfaces. In passing, he establishes results on the cofactor map and the geometric mean of positive definite matrices.

Kostas Dafermos' *Hyperbolic conservation laws: Past, present and future* provides a concise birds-eye view of the vast field of hyperbolic conservation laws. The author has managed to give an introduction to the field, where it came from, what were major developments, and where it could/should be heading, in a mere 10 pages, as only a true master could. The article focuses on general trends of research rather than single advances, this way one is able to get a very good impression of the subject.

In *Which nuclear shape generates the strongest attraction on a relativistic electron? An open problem in relativistic quantum mechanics*, Maria J. Esteban, Mathieu Lewin, and Eric Séré formulate several conjectures concerning the lowest eigenvalue of a Dirac operator with an external electrostatic potential. The latter describes a relativistic quantum electron moving in the field of some (pointwise or extended) nuclei. The main question is whether the eigenvalue is minimal when the nuclear charge is concentrated at a single point. This well-known property in nonrelativistic quantum mechanics has escaped all attempts of proof in the relativistic case.

In *Strong singularities of solutions to nonlinear elliptic equations*, V. Maz'ya, already the author of a paper offering *Seventy Five (Thousand) Unsolved Problems in Analysis and Partial Differential Equations*, presents a puzzling conjecture on the asymptotic behavior at infinity of the Riccati equation in a strip.

In their beautiful programmatic paper, S. Albeverio, F.C. De Vecchi and S. Ugolini describe connections between stochastic mechanics, optimal control, and nonlinear Schrödinger equations. They relate, for large N, the quantum mechanics of N particles to certain nonlinear Schrödinger equations, the latter of which are used also to describe the physical effect of Bose–Einstein condensation. A newer stochastic optimal control approach to Bose–Einstein condensation is presented with a sketch of future research lines in the different areas of mathematics involved.

Divergence-Free Tensors and Cofactors in Geometry and Fluid Dynamics

Denis Serre

Dedicated to Catriona Byrne, whose action in favour of the dissemination of mathematics was as much efficient as discreet.

Hommage

Retiring is one of the few delicate moments in a life. A lot of examples, within the mathematical community, prove that it can turn to the best. Keep your interest on what you liked, Catriona, and use your new free time to discover or to achieve something else. Enrich your mind !

Notations

The space of $n \times n$ symmetric matrices is $\mathbf{Sym}_n$. The open cone of real positive definite symmetric matrices is $\mathbf{SPD}_n$, its closure being $\mathbf{Sym}_n^+$. The latter cone defines the natural order $\prec$ in $\mathbf{Sym}_n$. Transposition of matrices and vectors is written $A \mapsto A^T$. The cofactor matrix of $M \in \mathbf{M}_n(\mathbb{R})$ is $\widehat{M}$; we recall that $\widehat{M} M^T = M^T \widehat{M} = (\det M) I_n$, and that $\det \widehat{M} = (\det M)^{n-1}$.

Partial derivatives in $\mathbb{R}^n$ are denoted $\partial_1, \ldots, \partial_n$. The inner product of $\mathbb{R}^n$ is $x \cdot y$ and the unit sphere is $\mathbb{S}^{n-1}$. If K is a convex body in $\mathbb{R}^n$, its support function, which is positively homogeneous of degree one, is

$$h_K(x) = \sup\{x \cdot \xi \mid \xi \in K\}.$$

D. Serre (✉)
UMPA, Ecole Normale Supérieure de Lyon, Lyon, France
e-mail: denis.serre@ens-lyon.fr

© The Author(s), under exclusive license to Springer Nature Switzerland AG 2023
J.-M. Morel, B. Teissier (eds.), *Mathematics Going Forward*, Lecture Notes
in Mathematics 2313, https://doi.org/10.1007/978-3-031-12244-6_32

1   Divergence-Free Symmetric Tensors

A symmetric tensor is a map $x \mapsto A(x) \in \mathbf{Sym}_n(\mathbb{R})$, where x runs over an open domain $U \subset \mathbb{R}^n$. When the entries a_{ij} are distributions, we define the row-wise Divergence operator

$$\mathrm{Div}\, A = \left(\sum_{j=1}^{n} \partial_j a_{ij} \right)_{1 \leq i \leq n}.$$

Definition 1.1 We say that A is Div-BV if each entry a_{ij} and each coordinate $(\mathrm{Div}\, A)_i$ is a finite Radon measure over U. We say that A is Div-free if actually $\mathrm{Div}\, A \equiv 0$.

The definition was motivated initially by the study of gas dynamics, whose conservation of mass and linear momentum can be written in terms of a Div-free tensor in a space-time domain; see Sect. 2.1. The BV context is justified on the one hand by the requirement of positiveness, and on the other hand by the role of the mass of Div A in the functional inequality; see Theorem 2.1.

1.1   Special Div-Free Tensors

A central result of the mathematical theory of elasticity is the Piola identity

$$\mathrm{Div}\, \widehat{\nabla f} \equiv 0$$

where f is a $W_{loc}^{1,n-1}$-vector field. Several proofs of this fact have been elaborated, some of them being especially elegant; we refer to [5] for a review.

When f is itself the gradient map of some potential $\theta \in W_{loc}^{2,n-1}(U)$, we infer that the cofactor matrix $\widehat{D^2\theta}$ of the Hessian is symmetric, locally integrable and Div-free. If moreover θ is convex, this tensor is positive semi-definite; we call it a *special* Div-free tensor. When $n = 2$, this construction exhausts the cone of positive integrable Div-free tensors, though it does not when $n \geq 3$.

1.1.1   Multi-linearization

We recall that the map $\Phi \mapsto \phi$ defined by $\phi(e) = \Phi(e, \ldots, e)$ is an isomorphism from the space of d-linear symmetric maps $\Phi : E^d \to F$ onto that of homogeneous polynomial maps $\phi : E \to F$ of degree d. Hereabove, E, F are vector spaces over a field of characteristic 0. Applying this to $E = F = \mathbf{M}_n(\mathbb{R})$ and to the cofactor map $M \mapsto \widehat{M}$, we are led to the

Definition 1.2 Cof_n is the unique $(n-1)$-linear symmetric map from $\mathbf{M}_n(\mathbb{R})$ into itself such that

$$\widehat{M} = \mathrm{Cof}_n(M, \dots, M), \qquad \forall M \in \mathbf{M}_n(\mathbb{R}).$$

We warn the reader that the index n refers to the size $n \times n$ of the matrices, but that Cof_n has only $n-1$ arguments, because $M \mapsto \widehat{M}$ is polynomial of degree $n-1$.

An alternate definition is that $\mathrm{Cof}_n(A_1, \dots, A_{n-1})$ is the coefficient of the monomial $X_1 \cdots X_{n-1}$ in the polynomial map

$$(X_1, \dots, X_{n-1}) \mapsto \frac{1}{(n-1)!} \widehat{A(X)}, \qquad A(X) = X_1 A_1 + \cdots + X_{n-1} A_{n-1}.$$

For instance, if $n = 3$, then Cof_3 is bilinear, defined by

$$\mathrm{Cof}_3(A, B) = \frac{1}{2} \left(\widehat{A+B} - \widehat{A} - \widehat{B} \right).$$

More generally, the (ℓ, m)-entry of $\mathrm{Cof}_n(A_1, \dots, A_{n-1})$ equals $(-1)^{\ell+m} \mathrm{Det}_{n-1}$ $(M_1, \dots, M_{n-1})$ where Det_{n-1} is the multi-linearization of the homogenous polynomial $\mathbf{M}_{n-1}(\mathbb{R}) \overset{\mathrm{det}}{\to} \mathbb{R}$, and M_j is the sub-matrix of A_j obtained by deleting the ℓth row and the mth column.

The multi-linearization allows us to extend the construction of special tensors:

Proposition 1.1 *Given the functions* $\theta_1, \dots, \theta_{n-1} \in W^{2,n-1}(U)$, *the tensor*

$$\mathrm{Cof}_n(\mathrm{D}^2 \theta_1, \dots, \mathrm{D}^2 \theta_{n-1}) \in L^1(U; \mathbf{Sym}_n)$$

is Div-free.

If the θ_j*'s are convex, then this tensor is positive semi-definite (see Proposition 3.1).*

Remark that even though this generalization provides much more Div-free tensors than just the special ones if $n \geq 3$, it still does not exhaust the linear cone of positive Div-free symmetric tensors. Actually, we expect that the linear space of Div-free tensors depends upon $\binom{n+1}{2} - n = \binom{n}{2}$ independent functions, counting the entries a_{ij} and the differential relations to which they obey to. This is larger than the number $n-1$ of the potentials involved in Proposition 1.1.

1.2 The Homogeneous Case: The Minkowski Problem

Let $\theta(x)$ be a positively homogeneous function of degree 1, say of class $W^{2,n-1}_{loc}$. Its Hessian is therefore homogeneous of degree -1 and the corresponding Div-free tensor is homogeneous of degree $1-n$. Because of the Euler identity $\mathrm{D}^2\theta(x)x \equiv 0$, we actually have

$$\widehat{D^2\theta} = \mu_\theta(x)\frac{xx^T}{|x|^{n+1}}$$

where μ_θ is positively homogeneous of degree 0. The identity $\mathrm{Div}\,\widehat{D^2\theta} = 0$ amounts to saying that the center of mass of μ_θ lies at the origin (see [10]):

$$\int_{\mathbb{S}^{n-1}} \mu_\theta(\xi)\xi\,ds(\xi) = 0. \tag{1}$$

Conversely, every tensor of the form

$$A = \mu(x)\frac{xx^T}{|x|^{n+1}} \tag{2}$$

for some positively homogeneous function μ of degree 0, satisfies the identity

$$\mathrm{Div}\,A = V_\mu \delta_{x=0}, \qquad V_\mu := \int_{\mathbb{S}^{n-1}} \mu(\xi)\xi\,ds(\xi),$$

hence is Div-free if and only if μ satisfies (1). If μ is positive, then A turns out to be special, according to a theorem of Pogorelov [8]. In other words, there exists a convex function θ such that $\mu = \mu_\theta$; this function is positively homogeneous of degree 1. Pogorelov's theorem actually applies when μ is a positive measure over $\mathbb{S}^{n-1}$ satisfying (1), provided that μ is not supported by a great sphere (the intersection of $\mathbb{S}^{n-1}$ with a linear hyperplane).

Since positively homogeneous functions of degree 1 are support functions of convex bodies, Pogorelov's result is usually written in geometrical terms. Given a convex body K, $\mu_{\theta_K}(x)$ is the inverse of the Gauß curvature of ∂K at the point x at which the outward normal is ξ. Thus Pogorelov's Theorem solves the famous Minkowski's problem:

> Given the Gauß curvature $\xi \mapsto \frac{1}{\mu(\xi)}$ as a function of the outward normal, satisfying (1), find the convex body.

For us, it says that, given a positive μ satisfying (1), there exists a convex body K such that $\mu = \mu_\theta$ for $\theta = \theta_K$. We refer to [10] for the details. Mind that θ is unique up to an affine additive factor ; equivalently, K is unique up to a translation.

1.3 Mixed Convex Bodies

A curious situation happens when the potentials θ_j of Proposition 1.1 are convex and positively homogeneous functions of degree 1, thus are support functions of convex bodies K_j. Then the Hessians $x \mapsto D^2\theta_j$ are homogeneous of degree -1, and the Div-free tensor $A = \mathrm{Cof}_n(D^2\theta_1, \ldots, D^2\theta_{n-1})$ is homogeneous of degree $1 - n$. Actually, A is the coefficient of $X_1 \cdots X_{n-1}$ in the polynomial $X \mapsto \widehat{D^2\Theta_X}$, where $\Theta_X = X_1\theta_1 + \cdots + X_{n-1}\theta_{n-1}$. Because of $D^2\Theta_X x \equiv 0$, one has an expression of the form

$$\widehat{D^2\Theta_X} = \mu_X(x)\frac{xx^T}{|x|^{n+1}}$$

where μ_X is homogeneous of degree 0. This implies that A has the form (2). Since A is Div-free, the function μ satisfies (1). Because it is positive, Pogorelov's Theorem tells us that it is the cofactor matrix $\widehat{D^2\theta}$ associated with the support function of some convex body K. Since positive Hessians are associated with classes of convex bodies modulo translation, we obtain that the correspondence

$$(\theta_1, \ldots, \theta_{n-1}) \longmapsto \mathrm{Cof}_n(D^2\theta_1, \ldots, D^2\theta_{n-1}),$$

restricted to convex positively homogeneous functions of degree 1, rewrites as a map

$$(K_1, \ldots, K_{n-1}) \mapsto K$$

where the convex bodies are understood up to a translation. This construction is equivalent to that of the *mixed body* $K = [K_1, \ldots, K_{n-1}]$ by Firey [2] (see also Lutwak [7]) ; in other words, the measure $\mu(x)dx$ coincides with the *mixed area* $S(K_1, \ldots, K_{n-1})$.

The properties of the mixing map include symmetry, homogeneity and consistency:

- For every $\rho \in \mathfrak{S}_{n-1}$, we have $[K_{\rho(1)}, \ldots, K_{\rho(n-1)}] = [K_1, \ldots, K_{n-1}]$.
- For every $\lambda > 0$, we have $[\lambda K_1, \ldots, K_{n-1}] = \lambda^{\frac{1}{n-1}}[K_1, \ldots, K_{n-1}]$.
- For every convex body K, we have $[K, \ldots, K] = K$.

If each K_j is balanced (that is the unit ball of some norm $\theta_j = \|\cdot\|_j$), then K is the class of a balanced body. The map above thus induces a map

$$(\|\cdot\|_1, \ldots, \|\cdot\|_{n-1}) \overset{\mathcal{N}_n}{\longmapsto} \|\cdot\|$$

over the space of norms. It is symmetric, of partial degree of homogeneity $\frac{1}{n-1}$, and satisfies

$$\mathcal{N}_n(\|\cdot\|, \ldots, \|\cdot\|) = \|\cdot\|.$$

1.4   The Euclidian Case

Suppose that each K_j is an ellipsoid, that is $\|x\|_j = \sqrt{x^T S_j x}$ for some $S_j \in \mathbf{SPD}_n$. Then the resulting norm is Euclidian too, $\|x\| = \sqrt{x^T S x}$ where $S \in \mathbf{SPD}_n$ is implicitly defined through its cofactor

$$\widehat{S} = \mathrm{Cof}_n(S_1, \ldots, S_{n-1}).$$

The resulting body $[K_1, \ldots, K_{n-1}]$ is thus an ellipsoid too.

Since $\mathrm{vol}\, K_j = \omega_n \sqrt{\det S_j}$ where ω_n is the volume of the unit ball, Corollary 3.1 and the identity $\det \widehat{S} = (\det S)^{n-1}$ tell us that

$$\mathrm{vol}\,[K_1, \ldots, K_{n-1}] \geq \left(\prod_1^{n-1} \mathrm{vol}\, K_j\right)^{\frac{1}{n-1}}. \tag{3}$$

This inequality is a special case of that for general convex bodies (Theorem 4.1 of [7]).

2   Compensated Integrability

Compensated Integrability deals with positive semi-definite Div-BV tensors. The positivity means that for every $\xi \in \mathbb{R}^n$, the distribution

$$A(\xi) := \sum_{i,j=1}^n \xi_i \xi_j a_{ij}$$

is non-negative, hence a Radon measure. See Proposition 1.1 for some examples of this situation.

When A is positive semi-definite, the expression $(\det A)^{\frac{1}{n}}$ produces a well-defined Radon measure, which is absolutely continuous with respect to the trace $\mathrm{Tr}\, A$ because of the AM-GM inequality

$$0 \leq (\det A)^{\frac{1}{n}} \leq \frac{1}{n} \mathrm{Tr}\, A.$$

Compensated Integrability tells us that positiveness, associated with the control the divergence, imply an enhanced integrability:

Theorem 2.1 (Serre [9, 10]) *Let A be a positive semi-definite Div-BV tensor over $\mathbb{R}^n$. Then the measure $(\det A)^{\frac{1}{n}}$ actually belongs to $L^{\frac{n}{n-1}}(\mathbb{R}^n)$, and we have the* *functional inequality*

$$\left\| (\det A)^{\frac{1}{n}} \right\|_{\frac{n}{n-1}} \leq c_n \left\| \mathrm{Div}\, A \right\|, \tag{4}$$

where the norm in the right-hand side is the total mass

$$\sup \left\{ \left\langle \mathrm{Div}\, A, \vec{\phi} \right\rangle \mid \vec{\phi} \in C_K(\mathbb{R}^n), \ \sup_x |\vec{\phi}(x)| \leq 1 \right\}.$$

The constant c_n is sharp: the equality in (4) is achieved with $A = \chi_B I_n$ when χ_B is the characteristic function of a ball.

The statement above generalizes several well-known inequalities from Functional Analysis, say:

- The Gagliardo inequality [4]. Take $A = \chi_K \mathrm{diag}(f_1, \ldots, f_n)$ where χ_K is the characteristic function of a hypercube, and each f_j ignores the coordinate x_j ($\partial_j f_j \equiv 0$).
- The Sobolev–Gagliardo–Nirenberg inequality—the choice $A = f(x)I_n$ gives

$$\|f\|_{\frac{n}{n-1}} \leq c \|\nabla f\|_1, \qquad \forall f \in W^{1,1}(\mathbb{R}^n).$$

- The Isoperimetric inequality—taking $A = \chi_D I_n$ where χ_D is a characteristic function yields

$$\left(\frac{\mathrm{vol}(D)}{\mathrm{vol}(B)} \right)^{\frac{1}{n}} \leq \left(\frac{\mathrm{area}(\partial D)}{\mathrm{area}(\partial B)} \right)^{\frac{1}{n-1}}$$

where B is ball in $\mathbb{R}^n$.

2.1 The Role of Div-Free Tensors in Mathematical Physics

A paradigmatic positive Div-free tensor comes from the Euler equations of gas dynamics. Here $n = 1 + d$ where d is the space dimension, and $x = (t, y)$ where t stands for the time variable. The conservation laws of mass and momentum,

$$\partial_t \rho + \mathrm{div}_y(\rho v) = 0,$$

$$\partial_t(\rho v) + \mathrm{Div}_y(\rho v \otimes v) + \nabla_y p = 0$$

can be recast as $\mathrm{Div}_{t,y} T = 0$ where

$$T = \begin{pmatrix} \rho & \rho v^T \\ \rho v & \rho v v^T + p I_d \end{pmatrix}.$$

The tensor above is obviously symmetric, positive semi-definite whenever the pressure p is non-negative, a rather natural assumption.

Symmetric Div-free tensors turn out to be ubiquitous in Mathematical Physics, as explained in [12]. They express the conservation laws associated with the invariance of some Lagrangian, under changes of coordinates, through Noether's Theorem. The group of invariance contains space and time shifts, which imply the conservation of momentum and of energy. It also contains the orthogonal group of some quadratic form (either dx^2 or $c^2 dt^2 - dy^2$), which ensures the symmetry of the tensor.

Compensated Integrability yields new estimates that depend only upon first principles. In the canonical example of gas dynamics in the full space $\mathbb{R}^d$, we thus obtain a Strichartz-like estimate

$$\int_0^{+\infty} dt \int_{\mathbb{R}^d} \rho^{\frac{1}{d}} p \, dy \leq c_d' M^{\frac{1}{d}} \left(\frac{1}{4} \int \int_{\mathbb{R}^d \times \mathbb{R}^d} \rho_0(y) \rho_0(z) |v_0(z) - v_0(y)|^2 dz \, dy \right. $$

$$\left. + M \int_{\mathbb{R}^d} e_0(y) \, dy \right)^{\frac{1}{2}},$$

where ρ_0, e_0, v_0 are the density, internal energy and velocity at initial time, and $M = \int \rho_0 dy$ is the total mass. The double integral in the right-hand side is a Galilean-invariant form of the product of the mass with the (initial) kinetic energy.

Other examples occur in various models of continuous or discrete mechanics, for instance Vlasov-type equations [10] or kinetic (Boltzman) equations [9]. The corresponding tensor is positive semi-definite whenever the interaction between particles is repulsive; this excludes for instance viscous forces, or attractive Coulomb force. The Maxwell's system, whether it be linear or nonlinear, provides also a symmetric Div-free tensor, expressing the conservation of energy and momentum, but this one is always indefinite [10].

3   Div-BV Tensors and the Curvature of Hypersurfaces

Another application, to hard-spheres dynamics, is presented in [11]. It involves a non-trivial version of Compensated Integrability for Div-BV tensors supported by a graph. More generally, there are interesting situations in Differential Geometry or in Mathematical Physics, where a Div-BV tensor is supported by a lower-dimensional subset, say a submanifold. A remarkable example is the following.

Theorem 3.1 *Let $\mathcal{M}$ be a C^2-hypersurface in $\mathbb{R}^n$, the tangent subspace at $x \in \mathcal{M}$ being $T_x \mathcal{M}$. Define the tensor $K_{\mathcal{M}}$ by*

$$K_{\mathcal{M}} = \pi_x \, \mathcal{H}_{n-1}|_{\mathcal{M}}$$

where $\mathcal{H}_{n-1}$ is the $(n-1)$-dimensional Hausdorff measure, and π_x is the orthogonal projection to $\vec{T}_x\mathcal{M}$. Then we have

$$\text{Div } K_{\mathcal{M}} = (n-1)\kappa_x\vec{v}_x \; \mathcal{H}_{n-1}|_{\mathcal{M}} \tag{5}$$

where κ_x is the mean curvature at $x \in \mathcal{M}$, and $\vec{v}_x$ is the unit normal vector. In particular $K_{\mathcal{M}}$ is Div-free if and only if $\mathcal{M}$ is a minimal surface.

Comments

1. Notice that since the entries of $K_{\mathcal{M}}$ are measures, we expected *a priori* that Div $K_{\mathcal{M}}$ be a distribution of order -1. The reason why it is actually a measure, is that $K_{\mathcal{M}}$ vanishes in the normal direction.
2. Compensated Integrability, in its classical form (that is, in the ambient space), is useless in such a context, because the rank of the tensor is bounded by the dimension of the support; in the example above, π_x has rank $n-1$. Hence the determinant vanishes identically and (4) becomes trivial. But it seems likely that a version written within the variety (mind that the dimension occurs explicitly in our functional inequality through exponents) must be valid; this is left as an open problem. Of course the functional inequality will involve the ambient dimension $n-1$ instead of n. Our expectation is supported by Serre [11], where we showed the role played by the vertices of the graph; there, the tensor is special, associated with the support function of a convex body, whose volume enters in the functional inequality. Another supporting argument is that CI in $\mathbb{R}^{n-1}$ can easily be derived from CI in $\mathbb{R}^n$.
3. An interesting consequence of Identity (5) is that, given a BV-function f over $\mathcal{M}$, we have

$$\text{Div}\,(fK_{\mathcal{M}}) = (\nabla f(x) + (n-1)\kappa_x f(x)\vec{v}_x) \; \mathcal{H}_{n-1}|_{\mathcal{M}}\,.$$

Therefore the tangential gradient $\nabla f(x)$, and the normal component $(n-1)\kappa_x f(x)$, appear as the head and tail of the same coin, that of the Divergence of an isotropic tensor. This suggests that the expected variant of Compensated Integrability would give a new proof of the Sobolev–Gagliardo–Nirenberg inequality on submanifolds, first established by Michael and Simon [6] for general manifolds under the assumption of non-negative Ricci curvature. Here, it expresses that for compactly supported functions f over $\mathcal{M}$,

$$\|f\|_{\frac{n-1}{n-2}} \leq c_{n-1} \int_{\mathcal{M}} (|\nabla f(x)| + |\kappa_x f(x)|) \; ds(x).$$

4. The connection between the mean curvature and the projection operator π_x has been observed for a long time. A formula equivalent to (6) below, though not stated in terms of distributional Divergence, was used by Allard [1] as a definition of *generalized mean curvature and inner normal* in the context of varifolds.

The proof below covers the more general case where the hypersurface $\mathcal{M}$ admits a boundary. The complete formula is

$$\text{Div } K_{\mathcal{M}} = (n-1)\kappa_x \vec{v}_x \ \mathcal{H}_{n-1}|_{\mathcal{M}} - \vec{N}_x \ \mathcal{H}_{n-2}|_{\partial \mathcal{M}}, \tag{6}$$

where $\vec{N}_x \in \vec{T}_x \mathcal{M}$ is the outer unit normal vector to $\partial \mathcal{M}$ at x.
Proof Let Ω be an open neighbourhood of $\mathcal{M}$. By definition, $K_{\mathcal{M}}$ acts against test tensors $\mathcal{S} \in \mathcal{D}(\Omega)^{n^2}$ through the formula

$$\langle K_{\mathcal{M}}, \mathcal{S} \rangle = \int_{\mathcal{M}} \text{Tr}\,(\pi_x \mathcal{S}(x)) \, \mathcal{H}_{n-1}.$$

The distributional derivative $\text{Div } K_{\mathcal{M}}$ acts over vector-valued test functions $\vec{\phi} \in \mathcal{D}(\Omega)^n$ and is defined as usual by duality

$$\langle \text{Div } K_{\mathcal{M}}, \vec{\phi} \rangle = -\langle K_{\mathcal{M}}, \nabla \vec{\phi} \rangle = -\int_{\mathcal{M}} \text{Tr}\,(\pi_x \nabla \phi(x)) \, \mathcal{H}_{n-1}.$$

Because differential operators act locally, we content ourselves to work with test fields whose support is suitably small. For a given point $m \in \mathcal{M}$, we therefore consider a ball B containing m, such that $\mathcal{M} \cap B$ is the graph of a function $y \mapsto w(y)$ in an orthonormal system of coordinates $(y_1, \ldots, y_{n-1}, z)$ of $\mathbb{R}^n$. The function is of class C^2 over an open subset $\omega \subset \mathbb{R}^{n-1}$. The boundary $B \cap \partial \mathcal{M}$ is the image $w(\gamma)$, where γ is a part of $\partial \omega$, and w is of class C^1 up to γ. All the calculations below are done in the coordinates (y, z) ; vectors are written blockwise accordingly.
If $\text{Supp}\,\vec{\phi} \subset B$, its restriction to $\mathcal{M} \cap B$ defines a field $\vec{\psi}$ by $\vec{\psi}(y) := \vec{\phi}(y, w(y))$. We have

$$\nabla \vec{\psi} = \nabla_y \vec{\phi} + \partial_z \vec{\phi} \nabla w. \tag{7}$$

In the following calculations, we use the formulæ

$$\vec{v}_x = \frac{1}{\sqrt{1 + |\nabla w|^2}} \begin{pmatrix} -\nabla w \\ 1 \end{pmatrix}, \qquad \pi_x = I_n - \vec{v}_x \otimes \vec{v}_x, \qquad \mathcal{H}_{n-1}|_{\mathcal{M}} \sim \sqrt{1 + |\nabla w|^2}\, dy.$$

From (7), we get

$$\pi_x \nabla \vec{\phi} = \begin{pmatrix} \nabla \vec{\psi} \\ 0 \end{pmatrix} - \frac{1}{1 + |\nabla w|^2} \begin{pmatrix} ((\nabla w \cdot \nabla)\vec{\psi})\nabla w \\ -(\nabla w \cdot \nabla)\vec{\psi} \end{pmatrix}.$$

Defining a vector field $\vec{q} := (\psi_1, \ldots, \psi_{n-1})$, we thus have

$$\text{Tr}\,(\pi_x \nabla \vec{\phi}) = \text{div}\,\vec{q} - \frac{1}{1 + |\nabla w|^2}\,(\nabla w^T \nabla \vec{q}\, \nabla w - \nabla w \cdot \nabla \psi_n).$$

so that

$$\langle \text{Div}\,K_{\mathcal{M}}, \vec{\phi} \rangle = -\int_\omega \left\{ \sqrt{1 + |\nabla w|^2}\, \text{div}\,\vec{q} - \frac{1}{\sqrt{1 + |\nabla w|^2}}\,(\nabla w^T \nabla \vec{q}\, \nabla w - \nabla w \cdot \nabla \psi_n) \right\} dy.$$

Integrating by parts, this yields

$$\langle \text{Div}\,K_{\mathcal{M}}, \vec{\phi} \rangle = \int_\omega \vec{q} \cdot \left(\nabla \sqrt{1 + |\nabla w|^2} - \text{Div}\,\frac{\nabla w \otimes \nabla w}{\sqrt{1 + |\nabla w|^2}} \right) dy + \int_\omega \psi_n \text{div}\,\frac{\nabla w}{\sqrt{1 + |\nabla w|^2}}\,dy$$

$$- \int_\gamma \vec{q} \cdot \left(\sqrt{1 + |\nabla w|^2}\, n_y - \frac{n_y \cdot \nabla w}{\sqrt{1 + |\nabla w|^2}}\,\nabla w \right) ds - \int_\gamma \psi_n \frac{n_y \cdot \nabla w}{\sqrt{1 + |\nabla w|^2}}\,ds,$$

$$(8)$$

where ds is the element of area over γ and n_y is the outer unit normal to γ.

We recall the well-known formula

$$\text{div}\,\frac{\nabla w}{\sqrt{1 + |\nabla w|^2}} = (n-1)\kappa_x.$$

Noticing on the one hand that

$$\nabla \sqrt{1 + |\nabla w|^2} - \text{Div}\,\frac{\nabla w \otimes \nabla w}{\sqrt{1 + |\nabla w|^2}} = -(n-1)\kappa_x \nabla w, \qquad (9)$$

the bulk integrals simplify into

$$\int_\omega (n-1)\kappa_x(\psi_n - \vec{q} \cdot \nabla w)\,dy = \int_\omega (n-1)\kappa_x \vec{\psi} \cdot \left(\sqrt{1 + |\nabla w|^2}\, \vec{\nu}_x \right) dy$$

$$= \int_{\mathcal{M} \cap B} (n-1)\kappa_x \vec{\phi} \cdot \vec{\nu}_x \mathcal{H}_{n-1}. \qquad (10)$$

On the other hand, when $x \in \partial \mathcal{M}$, the vector

$$V := \pi_x \begin{pmatrix} n_y \\ 0 \end{pmatrix}$$

obviously belongs to $\vec{T}_x\mathcal{M}$ and is normal to $\vec{T}_x\partial M$, thus is parallel to N_x. Both V and N_x point actually in the same direction, so that $V = |V|N_x$. Denoting ∇_τ the tangential (to γ) part of the gradient, we have also

$$|V|^2 = 1 - \frac{(n_y \cdot \nabla w)^2}{1 + |\nabla w|^2} = \frac{1 + |\nabla_\tau w|^2}{1 + |\nabla w|^2},$$

and therefore

$$V = \sqrt{\frac{1 + |\nabla_\tau w|^2}{1 + |\nabla w|^2}}\, N_x.$$

We deduce that the boundary integral in (8) equals

$$-\int_\gamma \vec{\psi} \cdot V \sqrt{1 + |\nabla w|^2}\, ds = -\int_\gamma \vec{\psi} \cdot N_x \sqrt{1 + |\nabla_\tau w|^2}\, ds$$

$$= -\int_{\partial\mathcal{M} \cap B} \vec{\phi} \cdot N_x \mathcal{H}_{n-2}. \tag{11}$$

Assembling (8), (10) and (11), we conclude

$$\langle \mathrm{Div}\, K_\mathcal{M}, \vec{\phi} \rangle = \int_{\mathcal{M} \cap B} (n-1) H_x \vec{\phi} \cdot \vec{v}_x \mathcal{H}_{n-1} - \int_{\partial\mathcal{M} \cap B} \vec{\phi} \cdot N_x \mathcal{H}_{n-2},$$

which gives the desired formula (6).   □

A nice consequence of Formula (6) concerns self-intersecting minimal hypersurfaces:

Theorem 3.2 *Let $\mathcal{M}_j$ ($j = 1, 2, 3$) be C^2-hypersurfaces, meeting along their common boundary Γ, a codimension-2 submanifold. Define $\mathcal{M} = \mathcal{M}_1 \cup \mathcal{M}_2 \cup \mathcal{M}_3$ and the tensor $K_\mathcal{M}$ as in Theorem 3.1.*
Then $\mathcal{M}$ is a minimal hypersurface if, and only if, $\mathrm{Div}\, K_\mathcal{M} = 0$.

Proof A self-intersecting minimal hypersurface is a configuration which satisfies Plateau's laws:

- each film $\mathcal{M}_j$ is a smooth hypersurface with vanishing mean curvature,
- along the common boundary, the films make pairwise an angle of $\frac{2\pi}{3}$.

Since

$$\mathrm{Div}\, K_\mathcal{M} = (n-1) \sum_{j=1}^{3} \kappa_x \vec{v}_x\, \mathcal{H}_{n-1}|_{\mathcal{M}_j} - \left(\sum_{j=1}^{3} \vec{N}_{jx} \right) \mathcal{H}_{n-2}|_\Gamma ,$$

we see that $K_\mathcal{M}$ is Divergence-free if, and only if, each of the $\mathcal{M}_j$'s have vanishing mean curvature (whence the first law), and the unit outer normal vectors N_{jx} sum up to zero along Γ. But three unit vectors sum up to zero if and only if they are coplanar (which they are, being normal to Γ) and make pairwise an angle of $\frac{2\pi}{3}$ (whence the second law). □

Appendix: Cofactors and Geometric Mean in SPD$_n$

We gather here some results, possibly new, concerning the $(n-1)$-linear map Cof_n. We point out that the map $S \mapsto \widehat{S}$ is $\mathbf{O}_n$-equivariant, and thus its multi-linearisation is so:

$$\forall Q \in \mathbf{O}_n, \forall A \in \mathbf{M}_n(\mathbb{R}), \qquad \widehat{Q^T A Q} = Q^T \widehat{A} Q, \tag{12}$$

$$\forall A_1, \ldots, A_{n-1} \in \mathbf{M}_n(\mathbb{R}), \qquad \mathrm{Cof}_n(Q^T A_1 Q, \ldots, Q^T A_{n-1} Q) = Q^T \mathrm{Cof}_n(A_1, \ldots, A_{n-1}) Q.$$

Recall that if A is symmetric and invertible, then $\widehat{A} = (\det A) A^{-1}$. In particular the cone $\mathbf{SPD}_n$ is left invariant under the cofactor map. Our first remark is that this fact persists after multi-linearization:

Proposition 3.1 *If $A_1, \ldots A_{n-1} \in \mathbf{SPD}_n$, then $\mathrm{Cof}_n(A_1, \ldots, A_{n-1}) \in \mathbf{SPD}_n$.*

By continuity, the same result holds true when the open cone $\mathbf{SPD}_n$ is replaced by its closure $\mathbf{Sym}_n^+$.

Proof We shall prove that for every unit vector x, we have $x^T \mathrm{Cof}_n(A_1, \ldots A_{n-1}) x > 0$. Because of (12), and since $\mathbf{O}_n$ acts transitively over the unit sphere, is suffices to consider the case of the first vector e_1 of the canonical basis. Thus we need only to consider the upper-left entry of $\mathrm{Cof}_n(A_1, \ldots A_{n-1})$ and prove that it is positive.

This entry is the multi-linearization of the map $A \mapsto (\widehat{A})_{11} = \det A'$, where A' is the principal submatrix obtained by deleting the first row and column. Thus

$$e_1^T \mathrm{Cof}_n(A_1, \ldots A_{n-1}) e_1 = \mathrm{Det}_{n-1}(A_1', \ldots, A_{n-1}'),$$

where Det_{n-1} is the multi-linearization of the determinant over $\mathbf{Sym}_{n-1}$.

Let us recall that the determinant, a homogeneous polynomial, is hyperbolic over $\mathbf{Sym}_{n-1}$ in the sense of Gårding [3]. Its forward cone is $\mathbf{SPD}_{n-1}$, to which the principal submatrices $A_1', \ldots, A_{n-1}'$ belong.

When P is a hyperbolic polynomial, homogeneous of degree d, with forward cone Γ, its multi-linearization Φ satisfies a "reverse-Hölder" inequality:

$$\Phi(z_1, \ldots, z_d) \geq \left(\prod_{j=1}^{d} P(z_j) \right)^{\frac{1}{d}}, \qquad \forall z_1, \ldots, z_d \in \Gamma. \tag{13}$$

When applied to the determinant, (13) gives

$$\text{Det}_{n-1}(A'_1, \ldots, A'_{n-1}) \geq \left(\prod_{j=1}^{n-1} \det A'_j\right)^{\frac{1}{n-1}},$$

which implies

$$e_1^T \, \text{Cof}_n(A_1, \ldots A_{n-1}) \, e_1 > 0.$$

□

Our next result involves the notion of geometrical mean of r elements $B_1, \ldots, B_r$ of $\textbf{SPD}_n$. It is defined in terms of the distance

$$d(B, C) = \| \log(B^{-1/2} C B^{-1/2}) \|_F = \left(\sum_\alpha (\log \mu_\alpha)^2\right)^{1/2},$$

where the μ_α's are the eigenvalues of $C B^{-1}$. This distance induces a Riemannian structure, which enjoys the properties of a symmetric cone:

- Congruences $B \mapsto M^T B M$ (where $M \in \textbf{GL}_n(\mathbb{R})$) are isometries.
- The inversion $B \mapsto B^{-1}$ is an isometry.

The geometric mean $\mathfrak{G}(B_1, \ldots, B_r)$ is the unique matrix $X \in \textbf{SPD}_n$ which minimizes

$$B \mapsto \sum_{i=1}^r d(B, B_i)^2.$$

It is characterized by the equation

$$\sum_{i=1}^r \log(X^{-1/2} B_i X^{-1/2}) = 0_n, \qquad X \in \textbf{SPD}_n. \tag{14}$$

When $r = 2$, it is rather denoted $B_1 \sharp B_2$, and we have an explicit formula

$$B \sharp C = B^{1/2} (B^{-1/2} C B^{-1/2})^{1/2} B^{1/2} = C^{1/2} (C^{-1/2} B C^{-1/2})^{1/2} C^{1/2}.$$

Taking the trace of (14), we obtain the elementary property

Proposition 3.2 *The determinant of the geometrical mean equals the geometrical mean of the determinants:*

$$\det \mathfrak{G}(B_1, \ldots, B_r) = \left(\prod_{i=1}^r \det B_i \right)^{1/r}.$$

Another useful fact is the

Proposition 3.3 *Given* $B_1, \ldots, B_r \in \mathbf{SPD}_n$ *and* $v \in \mathbb{R}^n$, *we have*

$$v^T \mathfrak{G}(B_1, \ldots, B_r)v \leq \left(\prod_{i=1}^r v^T B_i v \right)^{1/r}.$$

In other words, the evaluation (as a quadratic form) of the geometrical mean of positive definite matrices, is bounded above by the geometric mean of the evaluations.

Proof We first observe that if $\Lambda > 0$ is a diagonal matrix, and $w \in \mathbb{R}^n$, then by concavity of the logarithm

$$w^T (\log \Lambda) w = \sum_\alpha w_\alpha^2 \log \lambda_\alpha \leq |w|^2 \log \left(\frac{1}{|w|^2} \sum_\alpha w_\alpha^2 \lambda_\alpha \right) = |w|^2 \log \frac{w^T \Lambda w}{|w|^2}.$$

Since every $S \in \mathbf{SPD}_n$ is unitarily similar to a positive diagonal matrix, we infer

$$\forall S \in \mathbf{SPD}_n, \forall w \in \mathbb{R}^n, \qquad w^T (\log S) w \leq |w|^2 \log \frac{w^T S w}{|w|^2}.$$

Let X be the geometrical mean. Evaluating (14) at w and applying the inequality above, we have

$$0 = \sum_{i=1}^r w^T \log(X^{-1/2} B_i X^{-1/2}) w \leq |w|^2 \sum_{i=1}^r \log \frac{w^T X^{-1/2} B_i X^{-1/2} w}{|w|^2}.$$

Substituting $v = X^{-1/2} w$, this rewrites as

$$0 \leq \sum_{i=1}^r \log \frac{v^T B_i v}{v^T X v},$$

which gives the result. □

The mean inherits the symmetry properties of the distance d. For instance it is equivariant under congruences:

$$\forall M \in \mathbf{GL}_n(\mathbb{R}), \forall B_1, \ldots, B_r \in \mathbf{SPD}_n, \qquad \mathfrak{G}(M^T B_1 M, \ldots, M^T B_r M) = M^T \mathfrak{G}(B_1, \ldots, B_r) M.$$

Likewise, it is invariant under inversion:

$$\forall B_1, \ldots, B_r \in \mathbf{SPD}_n, \qquad \mathfrak{G}(B_1, \ldots, B_r)^{-1} = \mathfrak{G}(B_1^{-1}, \ldots, B_r^{-1}).$$

It is also homogeneous of degree $\frac{1}{r}$ in each of its arguments:

$$\forall a_1, \ldots, a_r \in \mathbb{R}, \forall B_1, \ldots, B_r \in \mathbf{SPD}_n, \qquad \mathfrak{G}(a_1 B_1, \ldots, a_r B_r) = (a_1 \cdots a_r)^{1/r} \mathfrak{G}(B_1, \ldots, B_r).$$

From the two last properties, plus the identity $\widehat{A} = (\det A)A^{-1}$, we deduce that it commutes with the cofactor map:

$$\widehat{\mathfrak{G}}(B_1, \ldots, B_r) = \mathfrak{G}(\widehat{B_1}, \ldots, \widehat{B_r}).$$

We now state our main result. This is a sort of generalization of (13), where the cofactor map $S \mapsto \widehat{S}$ plays the role of P and the geometric mean in the right-hand side is understood in the sense discussed above.

Theorem 3.3 *For every* $A_1, \ldots, A_{n-1} \in \mathbf{SPD}_n$, *we have*

$$\mathfrak{G}(\widehat{A_1} \ldots, \widehat{A_{n-1}}) \prec \mathrm{Cof}_n(A_1 \ldots, A_{n-1}).$$

Proof We have to compare the evaluations of both sides at an arbitrary vector $v \neq 0$. By homogeneity and $\mathbf{O}_n$-equivariance, it suffices to compare them when $v = e_1$ is the first element of the canonical basis, that is to compare the $(1, 1)$-entries of both matrices.

On the one hand, we have (Proposition 3.3)

$$e_1^T \mathfrak{G}(\widehat{A_1} \ldots, \widehat{A_{n-1}}) e_1 \leq \left(\prod_1^{n-1} e_1^T \widehat{A_i} e_1 \right)^{\frac{1}{n-1}}.$$

Writing blockwise the matrices

$$A_i = \begin{pmatrix} \cdot & \cdot \\ \cdot & A_i' \end{pmatrix}, \qquad A_i' \in \mathbf{SPD}_{n-1},$$

we have $e_1^T \widehat{A_i} e_1 = \det A_i'$, whence

$$e_1^T \mathfrak{G}(\widehat{A_1} \ldots, \widehat{A_{n-1}}) e_1 \leq \left(\prod_1^{n-1} \det A_i' \right)^{\frac{1}{n-1}}.$$

Now, because the determinant is a hyperbolic polynomial over $\mathbf{SPD}_{n-1}$, we infer

$$e_1^T \mathfrak{G}(\widehat{A_1} \ldots, \widehat{A_{n-1}}) e_1 \leq \mathrm{Det}_{n-1}(A_1', \ldots, A_{n-1}').$$

We conclude by remarking that the right-hand side above is nothing but $e_1^T \operatorname{Cof}_n(A_1, \ldots, A_{n-1}) e_1$.   □

Corollary 3.1 *For every* $A_1, \ldots, A_{n-1} \in \mathbf{Sym}_n^+$, *we have*

$$\prod_{i=1}^{n-1} \det A_i \leq \det \operatorname{Cof}_n(A_1 \ldots, A_{n-1}).$$

Proof By density and continuity, we may assume that every A_i is positive definite. From Proposition 3.2 and Theorem 3.3, we have

$$\left(\prod_1^{n-1} \det \widehat{A_i} \right)^{\frac{1}{n-1}} = \det \mathfrak{G}(\widehat{A_1} \ldots, \widehat{A_{n-1}}) \leq \det \operatorname{Cof}_n(A_1 \ldots, A_{n-1}).$$

We conclude by using $\det \widehat{A} = (\det A)^{n-1}$.   □

Acknowledgements I am indebted to the anonymous referee, who let me know a relevant piece of literature.

References

1. W. K. Allard. On the first variation of a varifold. *Annals of Math.*, **95** (1972), pp 417–491. On the first variation of a varifold: Boundary behaviour. *Annals of Math.*, **101** (1975), pp 418–446.
2. W. J. Firey. Blaschke sums of convex bodies and mixed bodies. *Proc. Colloq. Convexity* (Copenhagen 1965), Københavns Univ. Mat. Inst., 1967, pp 94–101.
3. L. Gårding. An inequality for hyperbolic polynomials. *J. Math. Mech.*, **8** (1959), pp 957–965.
4. E. Gagliardo. Proprietà di alcune classi di funzioni in più variabili. *Ricerche Mat.*, **7** (1958), pp 102–137.
5. R. Kupferman, A. Schachar. A geometric perspective on the Piola identity in Riemannian settings. *J. Geom. Mech.* **11** (2019), 59–76.
6. J. H. Michael, L. M. Simon. Sobolev and mean value inequalities on generalized submanifolds of $\mathbb{R}^n$. *Comm. Pure & Appl. Math.*, **26** (1973), pp 361–379.
7. E. Lutwak. Volume of mixed bodies. *Trans. of the Amer. Math. Soc.*, **294** (1986), pp 487–500.
8. A. V. Pogorelov. *The Minkowski multidimensional problem.* Scripta Series in Mathematics. V. H. Winston & Sons, Washington, D.C.; Halsted Press (John Wiley & Sons), New York–Toronto–London (1978).
9. D. Serre. Divergence-free positive symmetric tensors and fluid dynamics. *Annales de l'Institut Henri Poincaré (analyse non linéaire)*, **35** (2018), pp 1209–1234.
10. D. Serre. Compensated integrability. Applications to the Vlasov–Poisson equation and other models of mathematical physics. *J. Math. Pures & Appl.*, **127** (2019), pp 67–88.
11. D. Serre. Hard spheres dynamics: Weak *vs* strong collisions. *Arch. Rat. Mech. Anal.*, **240** (2021), pp 243–264.
12. D. Serre. Symmetric Divergence-free tensors in the Calculus of Variations. *Comptes Rendus, Mathématique*, **360** (2022), pp. 653–663.

Hyperbolic Conservation Laws: Past, Present, Future

Constantine M. Dafermos

1 Introduction

Hyperbolic conservation laws is the term commonly used for quasilinear first-order hyperbolic systems of partial differential equations in divergence form. In classical continuum physics, such systems express the conservation laws for continuous media in the absence of diffusion induced by viscosity, heat flux, etc.

The oldest, and still most important, example is provided by the Euler equations

$$\begin{cases} \partial_t \rho + \operatorname{div}(\rho v) = 0 \\ \\ \partial_t (\rho v) + \operatorname{div}(\rho v \otimes v) + \operatorname{grad} p(\rho) = 0, \end{cases} \tag{1}$$

which express the conservation of mass and momentum in the barotropic flow of an inviscid gas. Because of its central importance in classical physics and technology, this system has been studied intensively over the past three centuries by mathematicians, physicists and engineers. The amount and variety of the amassed information is enormous and yet, from the standpoint of mathematical analysis, the fundamental questions are still open. This equally describes the state of affairs of the entire field of hyperbolic conservation laws: a long history, a wealth of information, recent fruitful activity and major open questions.

The aim of this brief article is to present a bird's-eye view of the area, with an eye to the challenges posed by open problems that are currently under investigation. It does not purport to provide a review of the field—that would be a much more ambitious project. In particular, attributing each development to its contributor(s) would require hundreds of references. In their place, the bibliography consists of

C. M. Dafermos (✉)
Division of Applied Mathematics, Brown University, Providence, RI, USA

© The Author(s), under exclusive license to Springer Nature Switzerland AG 2023
J.-M. Morel, B. Teissier (eds.), *Mathematics Going Forward*, Lecture Notes
in Mathematics 2313, https://doi.org/10.1007/978-3-031-12244-6_33

a collection of texts and monographs, where the reader may find comprehensive expositions, together with a sample of seminal papers, of historical importance or recent vintage, which have set, or are setting, directions in the development of the field.

The author has struggled to gain an overview of the field in the course of his efforts to prepare an encyclopedic presentation of the subject. The resulting book, *Hyperbolic Conservation Laws in Continuum Physics*, was published by Springer in the year 2000, and in order to keep up with current developments has undergone four editions, with a fifth in preparation. Throughout this 25 year endeavor, the encouragement and assistance of Catriona Byrne has been invaluable and indispensable. This article is a small token of gratitude for her support.

2 The Past

This section spans two centuries, between 1750 and 1950. The following was accomplished during that period:

The origins of the subject lie in classical physics. Mathematical physicists formulated the major field theories that are governed by hyperbolic systems of conservation laws. This was achieved initially within the framework of mechanics, where mass and momentum are conserved. The development of thermodynamics in the mid-nineteenth century introduced the energy conservation law together with the notions of entropy, viscosity and heat conductivity, which play a major role in the subject. The construction of the edifice of classical physics was completed with the development of electrodynamics, which became yet another source for systems of hyperbolic conservation laws.

The earliest efforts in the analysis of hyperbolic conservation laws were focused on the Euler equations in one space dimension:

$$
\begin{cases}
\partial_t \rho + \partial_x(\rho v) = 0 \\[2mm]
\partial_t(\rho v) + \partial_x(\rho v^2) + \partial_x p(\rho) = 0,
\end{cases}
\tag{2}
$$

where ρ is the density, v is the velocity and p denotes the pressure of the gas. In the course of the study of (2) it was soon realized that the characteristic property of nonlinear hyperbolic conservation laws is that the lifespan of classical solutions terminates in finite time because of wave breaking. This prompted Stokes to introduce the notion of a shock wave, which manifests one of the earliest examples of a weak solution to a partial differential equation. In his groundbreaking study of (2), Riemann discovered the presence of what are now called *Riemann invariants* in classical solutions and constructed self-similar weak solutions, solving the *Riemann problem*.

Appending the energy conservation law to the Euler equations yields the system

$$\begin{cases} \partial_t u - \partial_x v = 0 \\ \partial_t v + \partial_x p(u, s) = 0 \\ \partial_t \left[\varepsilon(u, s) + \frac{1}{2}v^2\right] + \partial_x \left[p(u, s)v\right] = 0, \end{cases} \tag{3}$$

which is here written in Lagrangian coordinates. In comparison to (2), (3) introduces the new fields of entropy s and internal energy ε. The symbol u denotes the specific volume, the inverse of density. The structure of shock waves for this important system was first investigated by Hugoniot.

The introduction of solutions with shocks brought out a number of questions on uniqueness and stability that had to be addressed by a combination of physical and mathematical arguments.

Intensive investigation of the systems of conservation laws of gas dynamics and other specific hyperbolic conservation laws was conducted during the first half of the twentieth century so as to meet the demands for technological developments in the aerospace and chemical industries. By mid century, intimate familiarity with these equations had been gained and a great number of particular solutions had been constructed, especially for steady-state, irrotational transonic flow. The list of contributors to that research effort included applied mathematicians, aerodynamicists and other engineers. However, because of the formidable obstacles posed by the analysis, the bulk of these results had been derived by heuristic arguments or formal asymptotics. Experiments in wind tunnels also played an important role. A systematic formulation of the general theory and a rigorous mathematical treatment were still lacking.

3 The Present

A major turn in the direction of research took place in the early 1950s. Following the general trends in partial differential equations, the study of the qualitative theory of hyperbolic conservation laws was initiated. Motivated by a pioneering paper on the Burgers equation

$$\partial_t u + \partial_x \left(\frac{1}{2}u^2\right) = 0 \tag{4}$$

by Hopf, Lax coined the term "hyperbolic systems of conservation laws" and provided a codification of the subject based on a distillation of the earlier works by mathematicians, physicists and engineers. In consequence, hyperbolic conservation laws became an important part of the theory of partial differential equations. Major progress has been made in the following directions.

The scalar conservation law has been thoroughly investigated. Even the one-space dimensional case,

$$\partial_t u + \partial_x f(u) = 0, \tag{5}$$

manifests fascinating structure. In the multi-space dimensional situation,

$$\partial_t u + \sum_{i=1}^{m} \partial_{x_i} f_i(u) = 0, \tag{6}$$

a major development was the insightful definition of admissible weak solutions by Kruzhkov, inducing contraction in L^1 that renders the Cauchy problem well-posed. An alternative, equivalent formulation deemed "kinetic," introduced by Lions, Perthame and Tadmor, provides a powerful tool that leads to a precise description of the fine properties of solutions. The status of the theory of the scalar conservation law is by now definitive, but still new interesting properties continue to be discovered.

Great progress has also been achieved in the study of systems of conservation laws, but the theory is far from approaching completeness.

Beginning with the one-space dimensional case,

$$\partial_t U + \partial_x F(U) = 0, \tag{7}$$

U in $\mathbb{R}^n$, a thorough investigation of self-similar solutions, $U(x, t) = V(x/t)$, to the Riemann problem has contributed to clearing the issues of admissibility and uniqueness of solutions.

A major step was the development of the *random choice* method of Glimm, which constructs BV solutions to the Cauchy problem for (7), under initial data with small total variation, by using solutions to the Riemann problem as building blocks. This methodology enabled DiPerna, T.-P. Liu and many other authors to derive a detailed description of the structure and long time behavior of BV solutions. An alternative but parallel approach, termed *front tracking*, devised by the Italian school led by Bressan, has established both the existence and the uniqueness of solutions. Even sharper results on existence, uniqueness and stability were obtained by Bianchini and Bressan via the method of *vanishing viscosity*, which derives solutions of (7) as limits of solutions to the parabolic system

$$\partial_t U + \partial_x F(U) = \varepsilon \, \partial_x^2 U, \tag{8}$$

as the artificial viscosity ε tends to zero. Thus, the basic theory of the Cauchy problem for (7) under initial data with small total variation is essentially complete. However, the case of initial data with large variation is still wide open.

Progress has also been made in the theory of balance laws

$$\partial_t U + \partial_x F(U) = G(U), \tag{9}$$

with source terms $G(U)$ of relaxation type.

For a special, but important, class of systems, weaker solutions, in L^∞, were constructed by means of the functional analytic method of compensated compactness.

Modeling with systems of conservation laws in one-space dimension has proliferated and now extends beyond physics to traffic theory and several other areas in science and engineering.

We next turn to the case of systems

$$\partial_t U + \sum_{i=1}^{m} \partial_{x_i} F_i(U) = 0 \tag{10}$$

in several space dimensions. After a long period of stagnation, significant progress has been made in recent years.

The investigation of the Euler equations (1) has yielded notable breakthroughs over the past few years. To begin with, it has been verified that classical solutions break down in finite time with the development of shocks. Furthermore, following pioneering work by De Lellis and Szekelyhidi, the method of *convex integration* has been employed for constructing weak solutions to the Cauchy problem which will be termed *exotic*. In particular, it was shown that infinitely many exotic solutions exist for a dense class of initial data. This raises the issue of uniqueness and admissibility of weak solutions, which has not yet been addressed in a satisfactory manner.

In a different direction, a successful research program launched by Gui-Qiang Chen and coauthors has established the existence of steady solutions to the Euler equations, in two-space dimensions, in the range where the equations change type, from elliptic to hyperbolic. As noted in the previous section, such solutions, which govern transonic gas flow, had been anticipated, but not rigorously established, for the needs of the aerospace industry.

The methodology for constructing weak solutions via convex integration has also been applied to certain systems of conservation laws related to the Euler equations but not to general systems (10), for which the existence of weak solutions to the Cauchy problem is still open. As a partial remedy, research has been redirected to constructing *measure-valued solutions*. This notion of solution, introduced by DiPerna, is very weak, which raises the issues of admissibility and uniqueness. As a minimum, it has been shown that, so long as they exist, classical solutions to the Cauchy problem are unique and stable within the class of weak or measure-valued solutions that dissipate the aggregate entropy.

Progress in the qualitative theory of hyperbolic conservation laws has proceeded hand in hand with advances in the numerical analysis of solutions. Indeed, the art and science of scientific computation in the area of gas dynamics has flourished over the past 50 years, as a result of sophisticated algorithms combined with the use of fast computers and the overall gains in experience with computing.

4   The Future

Any attempt to forecast the future carries risks, but at least it looks reasonable to expect developments in the following directions, which were set in the previous section.

In regard to systems (7), in one spatial dimension, the fundamental question whether the Cauchy problem in the BV setting is well-posed for initial data with large total variation remains wide open. Systems have been designed in which the variation, and even the L^∞ norm, of solutions blow up in finite time, but it is not clear whether this pathological behavior is limited to special systems or is generic. Efforts to demonstrate either global existence or finite-time blowup of BV solutions have failed to reach a conclusion in either direction, even for the simplest systems, such as (2). The difficulty lies in that wave interactions may generate such an enormous variety of complex wave patterns that it looks impossible to classify and analyze them. Computer-assisted proofs may lend a helping hand for that purpose. It is conceivable that even for systems as simple as (2) the variation may blow up, albeit only for solutions with special, nongeneric initial data. Perhaps, it may prove easier to demonstrate breakdown of the variation by designing resonating source terms, with no regard to the initial data. In any case, the general theory will remain in limbo until this issue is settled.

Staying in the realm of one spatial dimension, questions of central importance on the Cauchy problem for the system (9) of balance laws remain unanswered, even when the initial data have small variation. Indeed, success in controlling growth of the variation in systems of conservation laws (7) hinges on exploiting the dispersion effect: after colliding, two waves of distinct characteristic families move away from each other, never to interact again. By contrast, in the case of solutions to systems of balance laws (9), the source term induces scattering that redistributes the wave strength among the various characteristic families and may thus offset the beneficial effects of dispersion. The investigation of the role of scattering by the source is a necessary prerequisite for completing the theory of systems of balance laws.

Turning to systems of conservation laws in several space dimensions, it is fair to say that the area is still terra incognita, albeit on the verge of being explored. An important component that is missing in the theory of both weak and measure-valued solutions is an admissibility condition that would render the Cauchy problem well-posed. Experimentation in that direction has begun, drawing on the experience already gained in the one-space dimensional case. Arguments based on entropy dissipation or vanishing viscosity have been tried, but the results are far from conclusive at the present time. This is a major challenge that must be met.

The recent activity on constructing exotic weak solutions to the Euler equations via convex integration, reported in the previous section, has left a number of open questions. To begin with, it will be important to see how exotic solutions look, as the only information available at the present time is that they are highly oscillatory. In particular, the conjectured relation between the existence of exotic solutions and the phenomenon of turbulence must be investigated.

It has been verified experimentally that weak solutions that are constant in all but one spatial direction, such as planar shocks or fans thereof, which are determined by solving Riemann problems, are physically relevant in gas dynamics. This must be explained in the face of the existence of a multitude of exotic solutions that start out with the same initial values but then vary in all spatial directions and dissipate entropy at a higher rate. A possible explanation is that we are missing the proper admissibility condition on weak solutions. Alternatively, it may turn out that exotic solutions are not observed in nature because they are not generic. This issue must be resolved by future research.

Still another important question is whether the existence of the rich family of exotic solutions is peculiar to the Euler equations, resulting from a particular lack of determinacy in their structure, or whether it extends to more general classes of systems of conservation laws (10). To that end, the natural candidate for serving as a test case is the system of conservation laws of elastodynamics.

The prediction is that the quest for a general theory will continue for a long time.

The investigation of the Cauchy problem for hyperbolic conservation laws under random initial data is at an early stage of development and is expected to grow in the near future. Perhaps it will shed some light on the question of uniqueness by identifying which solutions are likely to be observed in nature.

Great progress should also be expected in the art and science of scientific computation of solutions, in the following directions.

On the theoretical side, the supporting numerical analysis of the algorithms employed in the solution of systems is based to a great extent on the theory of the scalar conservation law. The expectation is that progress will be made by bringing closer the numerical and the theoretical analysis of systems.

A new challenge to scientific computing is posed by the exotic solutions of the Euler equations constructed via convex integration. Their computation will not be an easy task as they are highly oscillatory. It should be noted that the success of such a project would also provide valuable assistance to the theory by revealing the structure of these solutions, which is currently unknown. An equally challenging task is the development of methodologies for computing measure-valued solutions.

Computational projects combining traditional algorithms, based on the equations, with data science will become widespread.

The range of applications will also be expanded. Examples of topics in which there is current activity in modeling with hyperbolic conservation laws include traffic in congested transportation networks, gas flow in pipe networks and control theory. Research in these areas will undoubtedly continue and new areas will open up.

References

1. S. Benzoni-Gavage and D. Serre. *Multi-dimensional Hyperbolic Partial Differential Equations.* Oxford University Press (2007).
2. S. Bianchini and A. Bressan. Vanishing viscosity solutions of nonlinear hyperbolic systems. *Ann. of Math.* **161**, 223–342 (2005).
3. A. Bressan. *Hyperbolic Systems of Conservation Laws. The One-dimensional Cauchy Problem.* Oxford University Press (2000).
4. Gui-Qiang Chen and M. Feldman. *The Mathematics of Shock Reflection-Diffraction and von Neumann's Conjecture.* Princeton University Press (2018).
5. Robin Ming Chen, A.F. Vasseur and Yu Cheng. Global ill-posedness for a dense set of initial data to the isentropic system of gas dynamics. *Adv. Math.* **393**, 108057 (2021).
6. R. Courant and K.O. Friedrichs. *Supersonic Flow and Shock Waves.* Wiley-Interscience (1948).
7. D. Christodoulou. *The Formation of Shocks in 3-Dimensional Fluids.* European Mathematical Society (2007).
8. C.M. Dafermos. *Hyperbolic Conservation Laws in Continuum Physics,* 4th edition. Springer (2016).
9. C. De Lellis and L. Szekelyhidi. On admissibility criteria for weak solutions of the Euler equations. *Arch. Rational Mech. Anal.* **195**, 225–260 (2010).
10. R.J. DiPerna. Measure-valued solutions to conservation laws. *Arch. Rational Mech. Anal.* **88**, 223–270 (1985).
11. L. Euler. Principes géneraux du mouvement des fluides. *Mémoires de l' académie des Sciences de Berlin* **11**, 274–315 (1755).
12. J. Glimm. Solutions in the large for nonlinear hyperbolic systems of equations. *Comm. Pure Appl. Math.* **18**, 697–715 (1965).
13. H. Holden and N.H. Risebro. *Front Tracking for Hyperbolic Conservation Laws,* 2nd edition. Springer (2015).
14. E. Hopf. The partial differential equation $u_t + uu_x = \mu u_{xx}$. *Comm Pure Appl. Math.* **3**, 201–230 (1950).
15. H. Hugoniot. Sur la propagation du mouvement dans les corps et spécialement dans les gaz parfaits. *J. Ecole Polytechnique* **57**, 3–97 (1887); **58**, 1–125 (1889).
16. S.N. Kruzhkov. First order quasilinear equations with several space variables. *Mat. Sbornik* **10**, 228–255 (1970).
17. P.D. Lax. Hyperbolic systems of conservation laws. *Comm. Pure Appl. Math.* **10**, 537–566 (1957).
18. P.G. LeFloch. *Hyperbolic Systems of Conservation Laws.* Birkhäuser (2002).
19. R.J. LeVeque. *Numerical Methods for Conservation Laws,* 2nd edition. Birkhäuser (1992).
20. P.-L. Lions, B. Perthame and E. Tadmor. Kinetic formulation for the isentropic dynamics and p-systems. *Comm. Math. Phys.* **163**, 415–431 (1994).
21. Tai-Ping Liu. *Shock Waves.* American Mathematical Society (2021).
22. A. Majda. *Compressible Fluid Flow and Systems of Conservation Laws in Several Space Variables.* Springer (1984).
23. B. Riemann. Ueber die Fortpflanzung ebener Luftwellen von endlicher Scwingungsweite. *Gött. Abh. Math. Cl.* **8**, 43–65 (1860).
24. D. Serre. *Systems of Conservation Laws,* Vols. 1–2. Cambridge University Press (1999).
25. J.A. Smoller. *Shock Waves and Reaction-Diffusion Equations,* 2nd edition. Springer (1994).
26. G.G. Stokes. On a difficulty in the theory of sound. *Philos. Magazine,* Ser. 3, **33**, 349–356 (1848).
27. L.C. Tartar. Compensated compactness and applications to partial differential equations. In: *Nonlinear Analysis and Mechanics: Heriot-Watt symposium,* Vol IV, pp. 136–212, Pitman (1979).

Which Nuclear Shape Generates the Strongest Attraction on a Relativistic Electron? An Open Problem in Relativistic Quantum Mechanics

Maria J. Esteban, Mathieu Lewin, and Éric Séré

This article is dedicated to Catriona Byrne on the occasion of her retirement. Her extremely good knowledge of the mathematical community and profession and her kindness made her presence in mathematical events always enjoyable and very useful.

1 A Conjecture for Relativistic Electrons

In this note we describe some conjectures which we recently coined in [13, 14], concerning the effect of a nuclear charge on a relativistic electron. We first describe the main conjecture somewhat informally, before we discuss its proper mathematical formulation more thoroughly. Consider a non-negative finite Borel measure μ on $\mathbb{R}^3$ and the corresponding linear Schrödinger operator

$$-\frac{\Delta}{2} - \mu * \frac{1}{|x|}, \tag{1}$$

which describes a non-relativistic electron moving in the Coulomb potential generated by the positive charge distribution μ, in atomic units. The lowest (negative)

M. J. Esteban (✉) · M. Lewin
CEREMADE, CNRS, Université Paris-Dauphine, PSL Research University, Paris, France
e-mail: esteban@ceremade.dauphine.fr; mathieu.lewin@math.cnrs.fr

É. Séré
CEREMADE, Université Paris-Dauphine, PSL Research University, CNRS, Paris, France
e-mail: sere@ceremade.dauphine.fr

eigenvalue of this operator is given by the variational principle [26]

$$\lambda_1 \left(-\frac{\Delta}{2} - \mu * \frac{1}{|x|} \right)$$

$$= \inf_{\substack{\varphi \in H^1(\mathbb{R}^3) \\ \int_{\mathbb{R}^3} |\varphi|^2 = 1}} \left\{ \frac{1}{2} \int_{\mathbb{R}^3} |\nabla \varphi(x)|^2 \, dx - \int_{\mathbb{R}^3} \left(\mu * \frac{1}{|\cdot|} \right) (x) \, |\varphi(x)|^2 \, dx \right\}. \quad (2)$$

Since this is an infimum over affine functions of μ, we deduce immediately that the eigenvalue is a *concave* function of μ. Therefore, it is minimized, at fixed mass $\mu(\mathbb{R}^3)$, when μ is proportional to a delta and we have

$$\lambda_1 \left(-\frac{\Delta}{2} - \mu * \frac{1}{|x|} \right) \geq \lambda_1 \left(-\frac{\Delta}{2} - \frac{\mu(\mathbb{R}^3)}{|x|} \right) = -\frac{\mu(\mathbb{R}^3)^2}{2} \quad (3)$$

for every $\mu \geq 0$. The interpretation is that the lowest possible electronic energy is reached by taking the most concentrated charge distribution, at fixed total charge $\mu(\mathbb{R}^3)$. In fact, in [25, 27] it is proved that the eigenvalue decreases when μ is deformed using an arbitrary contraction, for instance a dilation $\alpha^3 \mu(\alpha \cdot)$ with $\alpha \geq 1$. This was generalized to molecular systems in [19, 25, 27], where it is proved that the electronic part of the ground state energy decreases when all the distances between the nuclei are decreased.

Relativistic effects play an important role in the description of quantum electrons in molecules containing heavy nuclei, even for not so large values of the nuclear charge. A proper description of such systems is based on the Dirac operator [11, 38]. This is a first-order differential operator which has very different properties compared to its non-relativistic counterpart $-\Delta/2$ in (1). For instance the spectrum of the free Dirac operator is not semi-bounded, which prevents us from giving an unambiguous definition of a "ground state" and turns out to be related to the existence of the positron [11]. In addition, because of its scaling properties, the Dirac operator has a critical behavior with respect to the Coulomb potential $1/|x|$ which gives a bound $Z \leq 137$ on the highest possible charge of atoms in the periodic table, for point nuclei.

In atomic units for which $m = c = \hbar = 1$, the free Dirac operator D_0 can be written as

$$D_0 = -i\boldsymbol{\alpha} \cdot \nabla + \beta = -i \sum_{k=1}^{3} \alpha_k \partial_{x_k} + \boldsymbol{\beta}, \quad (4)$$

where α_1, α_2, α_3 and β are 4×4 Hermitian matrices which satisfy the following anticommutation relations:

$$\begin{cases} \alpha_k \alpha_\ell + \alpha_\ell \alpha_k = 2\delta_{k\ell} \, \mathbb{1}, \\ \alpha_k \beta + \beta \alpha_k = 0, \\ \beta^2 = \mathbb{1}. \end{cases}$$

The usual representation in 2×2 blocks is given by

$$\beta = \begin{pmatrix} I_2 & 0 \\ 0 & -I_2 \end{pmatrix}, \qquad \alpha_k = \begin{pmatrix} 0 & \sigma_k \\ \sigma_k & 0 \end{pmatrix} \qquad (k = 1, 2, 3),$$

where the Pauli matrices are defined as

$$\sigma_1 = \begin{pmatrix} 0 & 1 \\ 1 & 0 \end{pmatrix}, \qquad \sigma_2 = \begin{pmatrix} 0 & -i \\ i & 0 \end{pmatrix}, \qquad \sigma_3 = \begin{pmatrix} 1 & 0 \\ 0 & -1 \end{pmatrix}. \tag{5}$$

The operator D_0 is self-adjoint on the domain $H^1(\mathbb{R}^3, \mathbb{C}^4)$ in the Hilbert space $L^2(\mathbb{R}^3, \mathbb{C}^4)$ and its spectrum is $\sigma(D_0) = (-\infty, -1] \cup [1, \infty)$ [38]. Moreover, $(D_0)^2 = -\Delta + 1$.

A relativistic electron in the presence of the nuclear charge μ is described by the Dirac–Coulomb operator

$$D_0 - \mu * \frac{1}{|x|} \tag{6}$$

in place of the non-relativistic operator (1). In our units μ represents the nuclear charge multiplied by the fine-structure constant $\alpha \simeq 1/137$. We defer the precise definition of the Dirac–Coulomb operator to the next section. Eigenvalues in the gap $(-1, 1)$ physically correspond to stationary states of the relativistic electron. Therefore it seems natural to expect that the lowest eigenvalue in $(-1, 1)$ will again be minimized for the Dirac measure $\mu(\mathbb{R}^3)\delta_0$, like in the Schrödinger case (3). This is the conjecture which we recently made in [13, 14].

Conjecture 1.1 (General Charges [13, 14]) *For any non-negative Borel measure μ such that $\mu(\mathbb{R}^3) \leqslant 1$, the lowest eigenvalue in the gap $(-1, 1)$ satisfies*

$$\lambda_1\left(D_0 - \mu * \frac{1}{|x|}\right) \geq \lambda_1\left(D_0 - \frac{\mu(\mathbb{R}^3)}{|x|}\right) = \sqrt{1 - \mu(\mathbb{R}^3)^2}. \tag{7}$$

In relativistic quantum chemistry one often relies on extended nuclear charges, hence the interest of looking at any possible μ. If we restrict our attention to pointwise nuclei, then we have $\mu = \sum_m \theta_m \delta_{R_m}$ and the conjecture becomes

Conjecture 1.2 (Multi-Center Potentials [13, 14]) *We have*

$$\lambda_1 \left(D_0 - \sum_{m=1}^{M} \frac{\theta_m}{|x - R_m|} \right) \geq \lambda_1 \left(D_0 - \frac{\sum_{m=1}^{M} \theta_m}{|x|} \right) = \sqrt{1 - \left(\sum_{m=1}^{M} \theta_m \right)^2} \quad (8)$$

for all $M \geq 2$, all $R_1, \ldots, R_M \in \mathbb{R}^3$ and all $\theta_m \geq 0$ so that $\sum_{m=1}^{M} \theta_m \leqslant 1$,

Since any μ can be approximated by a combination of Dirac deltas for the narrow topology, Conjecture 1.2 is equivalent to Conjecture 1.1. Indeed λ_1 is continuous for this topology [14, Lemma 12].

The case $M = 2$ was conjectured by Klaus in [23, p. 478] and by Briet–Hogreve in [3, Sec. 2.4]. Numerical simulations from [2, 30] seem to confirm the conjecture for $M = 2$, even for large values of the nuclear charges. In [13, 14] and here we make the stronger conjecture that the same holds for any M. Note that the numerical simulations seem to indicate that λ_1 decreases when the Euclidean distance between nuclear charges is decreased, a property proved by Lieb and Simon [25, 27] in the non-relativistic case. This leads to a third conjecture:

Conjecture 1.3 (Monotonicity) *Let μ be a non-negative Borel measure such that $\mu(\mathbb{R}^3) \leqslant 1$ and let $f : \mathbb{R}^3 \to \mathbb{R}^3$ be a contraction for the Euclidean norm of $\mathbb{R}^3$. Then, denoting by $f_*\mu$ the pushforward of μ by f, we have*

$$\lambda_1 \left(D_0 - \mu * \frac{1}{|x|} \right) \geq \lambda_1 \left(D_0 - (f_*\mu) * \frac{1}{|x|} \right). \quad (9)$$

Conjecture 1.1 is a special case of Conjecture 1.3, as can be seen by taking $f = 0$. In this note we only discuss Conjecture 1.1, which is already far from obvious. The main difficulty is that the lowest Dirac eigenvalue in the gap $(-1, 1)$ is not given by a minimum like in (2). In fact, as quickly explained below, it is given by a min-max formula [7, 13, 17, 35]. Unfortunately, it does not seem easy to derive a concavity property of $\lambda_1(D_0 - \mu * |x|^{-1})$ from this variational characterization, and this prevents us from using the same argument as in the nonrelativistic case. However the min-max formula implies that $\lambda_1(D_0 + V)$ is monotone in V, so that Conjecture 1.1 holds true if one restricts it to *radially symmetric measures* μ. Indeed, for such measures we have the *pointwise* bound

$$\left(\mu * \frac{1}{|\cdot|} \right)(x) \leqslant \frac{\mu(\mathbb{R}^3)}{|x|},$$

by Newton's theorem [26] and (7) follows. If one only considers radial contractions f, Conjecture 1.3 is also true for radially symmetric measures μ. No other case seems to have been proved in the literature.

In the next section we discuss the proper definition of the Dirac operator $D_0 - \mu * |x|^{-1}$ in (6) and the exact meaning of the "lowest eigenvalue in the gap" $\lambda_1 (D_0 - \mu * |x|^{-1})$ appearing in the conjecture.

2 Dirac Operator with External Charges

2.1 Self-adjointness

For Coulomb-like potentials V, it is not an easy task to define $D_0 + V$ as a self-adjoint operator. The reason is that $1/|x|$ has the same homogeneity as the differential part $\boldsymbol{\alpha} \cdot \boldsymbol{\nabla}$ of the free Dirac operator. In the pure Coulomb case $\mu = \nu\delta_0$, everything is explicit. The operator $D_0 - \nu|x|^{-1}$ has a unique self-adjoint realization for $\nu \leqslant \sqrt{3}/2$ and infinitely many for $\nu > \sqrt{3}/2$. For $\nu \in (\sqrt{3}/2, 1]$ one self-adjoint extension is special, with the corresponding eigenfunctions being the least singular at the origin. It is called the "distinguished" extension. For $\nu > 1$ all the self-adjoint realizations look the same, with eigenfunctions having similar oscillations near the origin [20]. For $\nu \in [0, 1]$ it is known that the lowest eigenvalue of the distinguished extension in the gap $(-1, 1)$ equals $\sqrt{1 - \nu^2}$ and therefore remains positive. The formula for this eigenvalue was already used on the right side of (7).

Many works have been devoted to the case of a general Coulomb-type potential V since the 1970s [22–24, 33, 34, 36, 40–42]. Various methods were introduced to prove that there also exists a unique "distinguished" self-adjoint extension. The results typically cover any potential V satisfying the pointwise inequality

$$0 \geq V(x) \geq -\frac{\nu}{|x|}, \qquad \nu \in (0, 1).$$

In this case, "distinguished" can have several possible meanings, which were all eventually shown to be equivalent. One requirement was that the domain of the operator be a subspace of $H^{1/2}(\mathbb{R}^3, \mathbb{C}^4)$, so that the energy is well defined. Another natural property was that the operator is the norm-resolvent limit of the Dirac operator with a regularized potential. Using a quite different approach Esteban and Loss proved more recently in [15, 16] that a distinguished self-adjoint extension could also be defined in the critical case $\nu = 1$.

For small values of ν, the domain of self-adjointness is just the Sobolev space $H^1(\mathbb{R}^3, \mathbb{C}^4)$ but for larger values of ν, the domain was not explicit in most of the above-cited works. The recent articles [12, 35] contain a more detailed analysis of the domain.

In [13] all the previous works were generalized to cover the case of potentials $V = -\mu * |x|^{-1}$. The existence of a "distinguished" extension was shown under the sole assumption that μ is a non-negative finite measure which has no atom of mass larger than or equal to 1. This gave a clear definition to the operator $D_0 - \mu * |x|^{-1}$ in (6), describing one electron in the presence of a nuclear charge μ.

2.2   Dirac Eigenvalues in the Gap

Once the operators have been well defined, the next question is how to find and characterize the stationary states, that is, the eigenvalues in the spectral gap $(-1, 1)$. This has also attracted a lot of attention in spectral theory and mathematical physics in the last two decades [7–10, 12, 17, 31, 32, 35]. We are not going to state the precise result here, but the conclusion is that one can characterize the eigenvalues in the spectral gap using non-standard min-max variational methods. Potentials of the form $V = -\mu * |x|^{-1}$ were not covered by most of the existing results but they were handled in [13], following the method in [7, 12, 35].

Let us emphasize that there is some difficulty in defining what it means to be the "lowest eigenvalue in the gap $(-1, 1)$", as in our two Conjectures 1.1–1.3. If we have a well-behaved (e.g. bounded) negative potential V, then the eigenvalues of $D_0 + tV$ will be close to 1 for small $t > 0$ and will all decrease when t is increased. The lowest eigenvalue will eventually touch the lower spectrum at -1, at a certain finite value of t, and dissolve in the continuum. Then the second eigenvalue in the gap becomes the lowest one. We do not wish to look at these pathological discontinuities and want to be sure that the lowest eigenvalue remains so for all $t \leqslant 1$.

In fact, should our Conjectures 1.1 and 1.2 hold true, they would imply that

$$\lambda_1\big(D_0 - t\mu * |x|^{-1}\big) \geq 0, \qquad \forall t \in (0, 1).$$

In particular, when we turn on the potential $V = -\mu * |x|^{-1}$ by means of the parameter t, the lowest eigenvalue will always be non-negative and there will be no spectrum in the lower half of the gap $(-1, 0)$. No eigenvalue will dive into the negative continuum, which justifies considering the lowest one.

Since we do not know how to prove the conjecture, a natural first step was to investigate which measure μ can have eigenvalues approaching the negative threshold -1. In [14], we defined a critical charge ν_1 as the largest positive number for which

$$\lambda_1(D_0 - \mu * |x|^{-1}) > -1 \text{ for all } 0 < \mu(\mathbb{R}^3) < \nu_1 .$$

For measures with $\mu(\mathbb{R}^3) < \nu_1$ there is thus no ambiguity of what it means to be the "lowest eigenvalue". Our Conjectures 1.1 and 1.2 contain the statement that $\nu_1 = 1$. The following was shown in [14].

Theorem 2.1 (The Critical Charge ν_1 [14]) *The critical number ν_1 satisfies*

$$0.9 \simeq \frac{2}{\pi/2 + 2/\pi} \leqslant \nu_1 \leqslant 1. \tag{10}$$

It is also the best constant in the Hardy-type inequality

$$\int_{\mathbb{R}^3} \frac{|\sigma \cdot \nabla \varphi|^2}{\mu * |x|^{-1}} \, dx \geq \frac{\nu_1^2}{\mu(\mathbb{R}^3)^2} \int_{\mathbb{R}^3} \left(\mu * \frac{1}{|x|} \right) |\varphi|^2 \, dx \tag{11}$$

for every $\varphi \in C_c^\infty(\mathbb{R}^3, \mathbb{C}^2)$ and every finite non-negative measure $\mu \geq 0$, where $\sigma_1, \sigma_2, \sigma_3$ are the 2×2 Pauli matrices defined above in (5).

The estimate (10) was proved using an inequality due to Tix [39], whereas the link with the Hardy inequality (11) comes from the variational characterization of the first eigenvalue. Such inequalities have played an important role in the study of Dirac operators [1, 4–7].

3 Two Results from [13, 14]

In this last section we mention two results from [13, 14] which are related to our Conjectures 1.1–1.3.

3.1 Existence of an Optimal Measure μ

Even if we do not know that concentrating all the mass at one point gives the lowest eigenvalue, we could at least prove that there exists an optimizer μ for a fixed mass $\mu(\mathbb{R}^3) = \nu < \nu_1$ and that it has a very small support.

Theorem 3.1 (Existence of an Optimal Measure [14]) *For any $\nu \in [0, \nu_1)$, there exists a positive Borel measure μ_ν with $\mu_\nu(\mathbb{R}^3) = \nu$ so that*

$$\lambda_1 \left(D_0 - \mu_\nu * \frac{1}{|x|} \right) = \min_{\substack{\mu: \\ \mu(\mathbb{R}^3) = \nu}} \lambda_1 \left(D_0 - \mu * |x|^{-1} \right).$$

The support of any such minimiser μ_ν is a compact set of zero Lebesgue measure.

The theorem is proved in [14] by a rather delicate adaptation of techniques from nonlinear analysis to the context of Dirac operators. The first eigenvalue is a highly nonlinear function of the measure μ, even if the operator only depends linearly on μ. The main "enemy" is the action of the non-compact group of space translations, which is controlled using Lions' concentration-compactness method [28, 29]. The main difficulty was to prove that the problem is locally compact under the assumption that $0 \leqslant \nu < \nu_1$ and this is another reason why the critical mass ν_1 plays a central role. In spirit, the local compactness holds true because the eigenvalue cannot dive into the lower continuous spectrum by definition of ν_1. But the actual proof is rather involved and relies on variational arguments using the min-max characterization of the first eigenvalue. That the support has zero Lebesgue measure was shown in [14] by means of a unique continuation principle for Dirac operators, which extends famous results in the Schrödinger case [21, 37].

3.2 The Potential Energy Surface

In quantum chemistry one is interested in the *potential energy surface* which, by definition, is the graph of the first eigenvalue of the multi-center Dirac–Coulomb operator, seen as a function of the locations of the nuclei, including the nuclear repulsion:

$$(R_1, \ldots, R_M) \mapsto \lambda_1 \left(D_0 - \sum_{m=1}^{M} \frac{\theta_m}{|x - R_m|} \right) + \sum_{1 \leqslant m < \ell \leqslant M} \frac{\theta_m \theta_\ell}{|R_m - R_\ell|}.$$

For the case $M = 2$ the properties of the above function were analyzed in [3, 18, 23] in the case of subcritical singularities with charge $\theta_m < 1$. In [13] we extended these results to cover the case $M > 2$ and also to include the critical case of nuclear charge equal to 1. We proved the following

Theorem 3.2 (The Potential Energy Surface [13]) *Let* $0 < \theta_1, \ldots, \theta_M \leqslant 1$.

(i) *The map* $(R_1, \ldots, R_M) \mapsto \lambda_1 \big(D_0 - \sum_{m=1}^{M} \theta_m |x - R_m|^{-1} \big)$ *is continuous on the open set*

$$\Omega = \Big\{ (R_1, \ldots, R_M) \in (\mathbb{R}^3)^M \ : \ R_m \neq R_\ell \, for \, m \neq \ell$$

$$\lambda_1 \left(D_0 - \sum_{m=1}^{M} \frac{\theta_m}{|x - R_m|} \right) > -1 \Big\}.$$

(ii) *Moreover,*

$$\lim_{\min_{k \neq \ell} |R_k - R_\ell| \to \infty} \lambda_1 \left(D_0 - \sum_{m=1}^{M} \frac{\theta_m}{|x - R_m|} \right) = \sqrt{1 - \max_{m} \theta_m^2}. \tag{12}$$

(iii) *If in addition* $\sum_{m=1}^{M} \theta_m < \nu_1$ *then*

$$\lim_{\max_{k \neq \ell} |R_k - R_\ell| \to 0} \lambda_1 \left(D_0 - \sum_{m=1}^{M} \frac{\theta_m}{|x - R_m|} \right) = \sqrt{1 - \left(\sum_{m=1}^{M} \theta_m \right)^2}. \tag{13}$$

By (ii) we see that Conjecture 1.2 is valid when the nuclei are infinitely far apart. On the other hand, (iii) says that the lowest eigenvalue is continuous when all the nuclei are merged to one point. Conjecture 1.2 says that the limit (13) should be from above and it would be interesting to try to prove the conjecture when the nuclei are very close to each other. The limit (13) was also stated for $M = 2$ and $\nu_1 = \nu_2 < 1/2$ in [3] but we could not fill all the details of the argument of the proof.

The properties of Dirac–Coulomb operators are fascinating and much more involved than the non-relativistic Schrödinger case. Many tools (such as min-max methods) have been developed to better deal with Dirac operators. Our Conjectures 1.1, 1.2 and 1.3 are strongly supported by numerical results in the physics and chemistry literature, but their proof will probably require introducing new techniques.

Acknowledgements This project has received funding from the European Research Council (ERC) under the European Union's Horizon 2020 research and innovation programme (grant agreement MDFT No 725528 of M.L.), and from the Agence Nationale de la Recherche (grant agreement molQED).

References

1. N. Arrizabalaga, J. Duoandikoetxea and L. Vega. Self-adjoint extensions of Dirac operators with Coulomb type singularity. *J. Math. Phys.* **54**, 041504 (2013).
2. A.N. Artemyev, A. Surzhykov, P. Indelicato, G. Plunien and T. Stöhlker. Finite basis set approach to the two-centre Dirac problem in Cassini coordinates. *J. Phys. B* **43**, 235207 (2010).
3. P. Briet and H. Hogreve. Two-centre Dirac–Coulomb operators: regularity and bonding properties. *Ann. Phys.* **306**, 159–192 (2003).
4. B. Cassano, F. Pizzichillo and L. Vega. A Hardy-type inequality and some spectral characterizations for the Dirac–Coulomb operator. *Rev. Mat. Complut.* **33**, 1–18 (2020).
5. J. Dolbeault, M.J. Esteban, J. Duoandikoetxea and L. Vega. Hardy-type estimates for Dirac operators. *Ann. Sci. École Norm. Sup.* **40**, 885–900 (2007).
6. J. Dolbeault, M.J. Esteban M. Loss and L. Vega. An analytical proof of Hardy-like inequalities related to the Dirac operator. *J. Funct. Anal.* **216**, 1–21 (2004).
7. J. Dolbeault, M.J. Esteban and É. Séré. On the eigenvalues of operators with gaps. Application to Dirac operators. *J. Funct. Anal.* **174**, 208–226 (2000).
8. J. Dolbeault, M.J. Esteban and É. Séré. Variational characterization for eigenvalues of Dirac operators. *Calc. Var. Partial Differ. Equ.* **10**, 321–347 (2000).
9. J. Dolbeault, M.J. Esteban and É. Séré. A variational method for relativistic computations in atomic and molecular physics. *Int. J. Quantum Chem.* **93**, 149–155 (2003).

10. J. Dolbeault, M.J. Esteban and É. Séré. General results on the eigenvalues of operators with gaps, arising from both ends of the gaps. Application to Dirac operators. *J. Eur. Math. Soc. (JEMS)* **8**, 243–251 (2006).

11. M.J. Esteban, M. Lewin and É. Séré. Variational methods in relativistic quantum mechanics. *Bull. Amer. Math. Soc. (N.S.)* **45**, 535–593 (2008).

12. M.J. Esteban, M. Lewin and É. Séré. Domains for Dirac–Coulomb min-max levels. *Rev. Mat. Iberoam.* **35**, 877–924 (2019).

13. M.J. Esteban, M. Lewin and É. Séré. Dirac–Coulomb operators with general charge distribution. I. Distinguished extension and min-max formulas. *Ann. Henri Lebesgue* **4**, 1421–1456 (2021).

14. M.J. Esteban, M. Lewin and É. Séré. Dirac–Coulomb operators with general charge distribution. II. The lowest eigenvalue. *Proc. London Math. Soc.* **123**, 345–383 (2021).

15. M.J. Esteban and M. Loss. Self-adjointness for Dirac operators via Hardy–Dirac inequalities. *J. Math. Phys.* **48**, 112107 (2007).

16. M.J. Esteban and M. Loss. Self-adjointness via partial Hardy-like inequalities. In: *Mathematical results in quantum mechanics*, pp. 41–47, World Sci. Publ., Hackensack, NJ (2008).

17. M. Griesemer and H. Siedentop. A minimax principle for the eigenvalues in spectral gaps. *J. London Math. Soc. (2)* **60**, 490–500 (1999).

18. E.M. Harrell and M. Klaus. On the double-well problem for Dirac operators. *Ann. Inst. H. Poincaré Sect. A (N.S.)* **38**, 153–166 (1983).

19. T. Hoffmann-Ostenhof. A comparison theorem for differential inequalities with applications in quantum mechanics. *J. Phys. A* **13**, 417–424 (1980).

20. H. Hogreve. The overcritical Dirac–Coulomb operator. *J. Phys. A, Math. Gen.* **46**, 025301 (2013).

21. D. Jerison and C.E. Kenig. Unique continuation and absence of positive eigenvalues for Schrödinger operators. With an appendix by E.M. Stein. *Ann. of Math. (2)* **121**, 463–494 (1985).

22. T. Kato. Holomorphic families of Dirac operators. *Math. Z.* **183**, 399–406 (1983).

23. M. Klaus. Dirac operators with several Coulomb singularities. *Helv. Phys. Acta* **53**, 463–482 (1980).

24. M. Klaus and R. Wüst. Characterization and uniqueness of distinguished selfadjoint extensions of Dirac operators. *Commun. Math. Phys.* **64**, 171–176 (1978/79).

25. E.H. Lieb. Monotonicity of the molecular electronic energy in the nuclear coordinates. *J. Phys. B* **15**, L63–L66 (1982).

26. E.H. Lieb and M. Loss. *Analysis.* Vol. 14 of Graduate Studies in Mathematics, American Mathematical Society, Providence, RI, 2nd ed. (2001).

27. E.H. Lieb and B. Simon. Monotonicity of the electronic contribution to the Born–Oppenheimer energy. *J. Phys. B* **11**, L537–L542 (1978).

28. P.-L. Lions. The concentration-compactness principle in the calculus of variations. The locally compact case, Part I. *Ann. Inst. H. Poincaré Anal. Non Linéaire* **1**, 109–149 (1984).

29. P.-L. Lions. The concentration-compactness principle in the calculus of variations. The locally compact case, Part II. *Ann. Inst. H. Poincaré Anal. Non Linéaire* **1**, 223–283 (1984).

30. S.R. McConnell. *Two centre problems in relativistic atomic physics.* PhD thesis, University of Heidelberg (2013).

31. S. Morozov and D. Müller, On the minimax principle for Coulomb–Dirac operators. *Math. Z.* **280**, 733–747 (2015).

32. D. Müller. Minimax principles, Hardy–Dirac inequalities, and operator cores for two and three dimensional Coulomb–Dirac operators. *Doc. Math.* **21**, 1151–1169 (2016).

33. G. Nenciu. Self-adjointness and invariance of the essential spectrum for Dirac operators defined as quadratic forms. *Commun. Math. Phys.* **48**, 235–247 (1976).

34. G. Nenciu. Distinguished self-adjoint extension for Dirac operator with potential dominated by multicenter Coulomb potentials. *Helv. Phys. Acta* **50**, 1–3 (1977).

35. L. Schimmer, J.P. Solovej and S. Tokus. Friedrichs Extension and Min-Max Principle for Operators with a Gap. *Ann. Henri Poincaré* **21**, 327–357 (2020).

36. U.-W. Schmincke. Distinguished selfadjoint extensions of Dirac operators. *Math. Z.* **129**, 335–349 (1972).

37. E.M. Stein. Appendix to "Unique Continuation" by Jerison and Kdolbeaultenig. *Annals of Math.* **121**, 489–494 (1985).

38. B. Thaller. *The Dirac equation.* Texts and Monographs in Physics, Springer-Verlag, Berlin (1992).

39. C. Tix. Strict positivity of a relativistic Hamiltonian due to Brown and Ravenhall. *Bull. London Math. Soc.* **30**, 283–290 (1998).

40. R. Wüst. A convergence theorem for selfadjoint operators applicable to Dirac operators with cutoff potentials. *Math. Z.* **131**, 339–349 (1973).

41. R. Wüst. Distinguished self-adjoint extensions of Dirac operators constructed by means of cut-off potentials. *Math. Z.* **141**, 93–98 (1975).

42. R. Wüst. Dirac operations with strongly singular potentials. Distinguished self-adjoint extensions constructed with a spectral gap theorem and cut-off potentials. *Math. Z.* **152**, 259–271 (1977).

Strong Singularities of Solutions to Nonlinear Elliptic Equations

Vladimir Maz'ya

Personal Note

I am happy to contribute to the Festschrift in honour of Springer's Editorial Director Dr. Catriona Byrne. My short paper is dedicated to an area in the Theory of Nonlinear Elliptic Partial Differential Equations which is not sufficiently explored at present and is promising for future interesting discoveries. Before turning to the mathematical stuff I wish to share some memories related to my collaboration with Dr. Byrne and perhaps the best way to do that is to reproduce my letter written on July 28, 2006.

Dear Catriona,

I congratulate you with your quarter of the century with Springer Verlag. During this period of time I always felt your friendly and qualified support. I remember quite well your assistance in publishing "Sobolev Spaces", 1985, my first Springer book. Another example is your editing of my English in the Lecture Notes volume of 1997 which is far beyond your formal duties. More recently you gave me an invaluable help with publication of the volume "Differential Equations with Operator Coefficients", 1999. I am fortunate that our collaboration continues even now.

I also congratulate Springer Verlag with having such an excellent member of staff, devoted, charming, highly efficient and full of energy.

The author has been supported by the RUDN University Strategic Academic Leadership Program.

V. Maz'ya (✉)
Department of Mathematics, Linköping University, Linköping, Sweden

Peoples' Friendship University of Russia (RUDN University), Moscow, Russian Federation
e-mail: vladimir.mazya@liu.se

© The Author(s), under exclusive license to Springer Nature Switzerland AG 2023
J.-M. Morel, B. Teissier (eds.), *Mathematics Going Forward*, Lecture Notes
in Mathematics 2313, https://doi.org/10.1007/978-3-031-12244-6_35

I wish you, dear Catriona, good health, happiness and many years of fruitful work for the benefit of the world mathematical community.
Yours truly,
Vladimir Maz'ya

The intensity of my collaboration with Catriona and with Springer Verlag in general can be illustrated by the following list of books.

1. V. Maz'ya, *Sobolev Spaces*, 1985.
2. V. Kozlov and V. Maz'ya, *Theory of Higher-Order Sturm–Liouville Equations*, Lecture Notes, No. 1659, 1997.
3. V. Kozlov and V. Maz'ya, *Differential Equations with Operator Coefficients*, 1999.
4. G. Kresin, V. Maz'ya, *Sharp Real-Part Theorems*, Lecture Notes, No. 1903, 2007.
5. V. Maz'ya, T.O. Shaposhnikova, *Theory of Sobolev Multipliers with Applications to Differential and Integral Operators*, 2009.
6. V. Maz'ya, *Sobolev Spaces with Applications to Elliptic Partial Differential Equations*, 2011.
7. V. Maz'ya, A. Movchan, M. Nieves, *Green's Kernels and Meso-Scale Approximations in Perforated Domains*, Lecture Notes, No. 2077, 2013.

1   An Open Problem

The construction of asymptotic formulae for solutions to linear elliptic boundary value problems in strips, cylinders or domains with angular and conic boundary points has been developed in numerous publications (see, for instance [1, 5, 7, 8, 21] and the bibliography there). Less attention has been paid to the asymptotics of solutions to nonlinear boundary value problems. In more detail, properties of solutions to the p-Laplace equation were investigated (see [2, 4, 9–12, 16, 17, 22, 23]). In the case of weak singularities, when the problem can be linearized at infinity or near a singular point, boundary value problems for semilinear and more general quasilinear equations were considered in [14, 15, 18]. As for solutions with strong singularities, the situation is quite different. Since the principal terms of the asymptotics depend on the nonlinear operator as a whole, direct linearization is impossible. This case was dealt with in [3, 19, 20]. A formal asymptotic representation of solutions to the Dirichlet problem for the Riccati equation near an angular point was given in [16]. A description of asymptotic behavior of all solutions to the Neumann problem for the two-dimensional Riccati equation near an angular point was obtained in [6].

Now I turn to an open problem concerning the Riccati equation in a strip. By u we denote an arbitrary solution of the quasi-linear equation

$$u_{xx} + u_{yy} + au_x^2 + bu_y^2 = 0, \quad a, b = const > 0 \tag{1}$$

in the half-strip $\Omega = \{(x, y) : x > 1, 0 < y < l\}$ which is continuous in $\overline{\Omega}$ and subject to the boundary conditions

$$u(x, 0) = u(x, l) = 0. \tag{2}$$

We are interested in the asymptotic behavior of u at infinity. A priori assuming that the solution is uniformly bounded, one can show by a standard linearization argument that it is asymptotically equivalent to

$$C \exp\left(-\frac{k\pi x}{l}\right) \sin \frac{k\pi y}{l}, \qquad k = 1, 2, \ldots,$$

as $x \to +\infty$, where $C = const$. One can show that if the Dirichlet data are prescribed for $x = 1$, $y \in [0, l]$, the uniformly bounded solution u is unique.

The hypothesis I propose to justify is as follows. Suppose that a function u, solving the problem (1)–(2), is not bounded at infinity. Show that the uniform asymptotic formula holds

$$u(x, y) = \frac{1}{b} \log\left[\exp\left(\sqrt{\frac{b}{a}}\frac{\pi x}{l}\right) \sin \frac{\pi y}{l} + \frac{\cos\left(\sqrt{\frac{b-a}{a}}\left(\frac{\pi y}{l} - \frac{\pi}{2}\right)\right)}{\cos\left(\sqrt{\frac{b-a}{a}}\frac{\pi}{2}\right)} \right] + O\left(\exp(-\delta x)\right), \tag{3}$$

where $\delta = const > 0$ and

$$(b - a)a^{-1} \neq m^2, \qquad m = 1, 2, \ldots. \tag{4}$$

Clearly, the logarithmic term in (3) vanishes on the horizontal parts of $\partial\Omega$. Obviously, here the main term of the asymptotics depends both on linear and nonlinear parts of the elliptic operator.

Note that (3) is global in the sense that it does not contain boundary layer terms, unlike the asymptotics obtained in [16]. (Obviously, the angle in [16] should be replaced by the strip using a conformal mapping.)

There are other interesting questions related to solutions of (1) and (2). For example, what happens with the asymptotics without the restriction (4)? Also, what will the asymptotics look like if a and b in (1) are of different signs or if they depend on the point (x, y)? The answers to these questions are far from being obvious and, as far as I know, there exist no general methods for their treatment at present.

In conclusion I would like to attract reader's attention to other unsolved problems in analysis and partial differential equations collected in [13].

References

1. S. Agmon and L. Nirenberg. Properties of solutions of ordinary differential equations in Banach space. *Comm. Pure Appl.* **16**, 121–239 (1963).
2. M. Dobrowolski. On quasilinear elliptic equations in domains with conical boundary points. *J. reine angew. Math.* **394**, 186–195 (1989).
3. J. Huentutripay, M. Jazar and L. Véron. A dynamical system approach to the construction of singular solutions of some degenerate elliptic equations. *J. of Diff. Equations* **195**(1), 175–193 (2003).
4. T. Iwaniec and J.J. Manfredi. Regularity of p-harmonic functions on the plane. *Rev. Math. Iberoamericana* **5**(1–2), 1–19 (1989).
5. V.A. Kondratiev. Boundary value problems for elliptic equations in domains with conic or angular points. *Trudy Mosk. Mat. Obshch.* **16**, 209–292 (1967), English translation: *Moscow Math. Soc.* **10**, 227–313 (1967).
6. V. Kozlov and V. Maz'ya. Angle singularities of solutions to the Neumann problem for the two-dimensional Riccati equation. *Asymp. Anal.* **19**(1), 57–79 (1999).
7. V.A. Kozlov, V.G. Maz'ya and J. Rossmann. *Elliptic Boundary Value Problems in Domains with Point Singularities.* Amer. Math. Society (1997).
8. V.A. Kozlov, V.G. Maz'ya and J. Rossmann. *Spectral Problems Associated with Corner Singularities of Solutions to Elliptic Equations.* Amer. Math. Society (2000).
9. I.N. Krol. The solutions of equation $D_{x_i}(|Du|^{p-2}D_{x_i}u) = 0$ with a singularity at a boundary point. *Trudy Mat. Inst. Steklov* **125**, 127–139 (1973).
10. I.N. Krol and V.G. Maz'ya. The absence of continuity and H?older continuity of solutions of quasilinear elliptic equations near a nonregular boundary. *Trudy Moskov. Mat. Obšč.* **26**, 75–94 (1972), English translation: *Transactions of the Moscow Math. Soc., for the year 1972*, American Math., Soc. Providence RI (1974).
11. V.G. Maz'ya. The continuity at a boundary point of the solutions of quasi-linear elliptic equations. (Russian) *Vestnik Leningr. Univ.* **25**, no. 13, 42–55 (1970); English translation in *Vestnik Leningr. Univ. Math.* **3**, 225–242 (1976).
12. V.G. Maz'ya. *Boundary Behavior of Solutions to Elliptic Equations in General Domains.* Tracts in Math. **30**, European Mathematical Society (2018).
13. V. Maz'ya. Seventy five (thousand) unsolved problems in analysis and partial differential equations. *Integr. Equ. Oper. Theory* **90**(25), 1–44 (2018).
14. V.G. Maz'ya and B.A. Plamenevskii. *The behavior of solutions to quasilinear elliptic boundary value problems in a neighbourhood of a conical point. Zapiski Nauchn. Sem. Leningrad Otdel. Mat. Inst. Steklov (LOMI)* **38**, 94–97 (1973).
15. V.G. Maz'ya, B.A. Plamenevskii and N.F. Morozov. On nonlinear bending of a plate with a crack. *Amer. Math. Soc. Transl.* **123**, 125–139 (1984).
16. V.G. Maz'ya and A.S. Slutskii. Asymptotic solution to the Dirichlet problem for a two-dimensional Ricatti's type equation near a conic point. *Asymptot. Anal.* **39**, no. 2, 169–185 (2004).
17. V.G. Maz'ya, A.S. Slutskii and V.A. Fomin. Asymptotic behavior of the stress function near the vertex of a crack in the problem of torsion. *Izv. AN SSSR, Mekhanika Tverdogo Tela* **21**(4), 170–176 (1986).
18. E. Miersemann. Asymptotic expansion of solutions of the Dirichlet problem for quasilinear elliptic equations of second order near a conical point. *Math. Nachr.* **135**, 239–274 (1988).
19. S.A. Nazarov. Asymptotic behavior near an angular point of the boundary of the solution to a nonlinear equation. *Math. Notes* **31**(3–4), 211–216 (1982).
20. S.A. Nazarov and K.I. Piletskas. Asymptotic behavior of the solution of the nonlinear Dirichlet problem with a strong singularity near a corner point. *Math. USSR, Izv.* **48**(6), 531–550 (1984).
21. A. Pazy. Asymptotic expansions of solutions of differential equations in Hilbert space. *Arch. Rat. Mech. Anal.* **24**, 193–218 (1967).

22. J.B. Serrin. Isolated singularities od solutions to quasi-linear equations. *Acta Math.* **113**, 219–246 (1965).
23. P. Tolksdorf. On the Dirichlet problem for quasilinear equations in domain with conical boundary point. *Comm. Partial Differential Equations* **8**, 773–817 (1983).

Some Connections Between Stochastic Mechanics, Optimal Control, and Nonlinear Schrödinger Equations

Sergio Albeverio, Francesco Carlo De Vecchi, and Stefania Ugolini

1 Introduction

Dedication
We are happy to have been given the opportunity to contribute a little paper to a publication in honor of Catriona Byrne. The first named author had the great luck of meeting Catriona already in the early 80s and has discussed with her many issues concerning our common passion, mathematics in all its multiform and fascinating aspects. She is a very special communicative person, full of enthusiasm. It is always a great pleasure to meet her and share with her impressions about not only mathematics but also the world of arts. She saw at an early stage how the still rather scattered attempts to create more bridges between probability theory, rather abstract aspects of the theory of stochastic processes and infinite-dimensional analysis on one hand, and apparently distant other areas of mathematics, from number theory to geometry and non-standard analysis on the other, could be enhanced, also through interactions coming from mathematical physics (especially quantum theory). Catriona joined, directly or through her coworkers, several scientific meetings, in particular those where S.A. was in some way involved (from Bielefeld, Bochum, Bonn and Oberwolfach to Levico, Verona, Warwick), and the informal discussions with her produced new interconnections between the participants. S.A. also remembers with gratitude the encouraging interest she expressed in work he was pursuing on the theory of Feynman path integrals. This

S. Albeverio (✉) · F. C. De Vecchi
Department for Applied Mathematics, Universität Bonn, Bonn, Germany
e-mail: albeverio@iam.uni-bonn.de; francesco.devecchi@uni-bonn.de

S. Ugolini
Dipartimento di Matematica, Università di Milano, Milano, Italy
e-mail: stefania.ugolini@unimi.it

led in particular to a second edition of a *Lecture Notes in Mathematics* [11], with the late Raphael Høegh-Krohn and Sonia Mazzucchi as coauthors. Also on her initiative four lectures in the series "Saint-Flour Seminars in Probability" were republished in a Springer book with the title *Mathematical physics at Saint-Flour* [8] with, besides S.A., Hans Föllmer, Leonard Gross and Ed Nelson as authors. It is not by chance that these lectures happen to have a strong component in analysis and probability theory besides one in mathematical physics. The present paper relates to a number of arguments that have their roots in the topics treated in that book initiated by Catriona. Her influence in fact is also recognizable in many books of Proceedings edited by S.A. in various collaborations, in particular those emanating from activities of the Research Center BiBos. It is a great pleasure for all authors to thank Catriona for all she has done for the mathematical community and in particular for our areas of work. We do this with our heartfelt wishes for good health, happiness, enjoyment and success in all her future undertakings (it is really difficult to imagine a future for her not full of beautiful activities!).

The Topics Discussed
The topics we shall present in the present paper are connected, in several ways, with those of concern in the above mentioned books [8, 11]. The main motivations come from questions that arose in physics, namely on how to better understand certain phenomena appearing in nature, as manifestations of an underlying "quantum world".

In the first part of this exposé (Sects. 2–6) we shall concentrate on attempts to understand in a mathematical way some aspects of the particular complex phenomenon of Bose–Einstein condensation (BEC). In the second part (Sect. 7) we shall mention and briefly discuss future possible developments in this connection, but also more general issues connected with multiform and fascinating relations between quantum evolution and probabilistic evolution that still have not been brought to light.

Let us start by briefly mentioning what the physical phenomena of BEC is. BEC might be characterized by saying that it happens when a sufficiently diluted gas of bosons (i.e. consisting of identical particles with integer spin, in the case we shall consider with spin zero, called "bosons") confined to a box, is cooled down in an appropriate way to "very low temperatures" (close to absolute zero). In this case a large fraction of the number of bosons of the gas happen to get into the same lowest energy quantum state ("ground state"), and behaves as a single quantum object. Since the cooled down gas is often macroscopic, we have then a macroscopic system exhibiting quantum behavior. The phenomenon was predicted in the sense of theoretical physics for an "ideal boson gas" (without any interaction), using quantum statistical considerations, by S.N. Bose and A. Einstein already in 1924–1925. Its experimental verification for a "real gas" had to wait until 1995 (for this experimental work E. Cornell, W. Ketteler and C. Wiemann received the Nobel prize in 2001). Present day experimental techniques have been developed very much since then, and permit us to establish many detailed properties of BEC.

Of particular interest for us is that the quantum state associated with a BE condensate can be well described by a single quantum mechanical wave function satisfying a nonlinear Schrödinger equation with a cubic nonlinearity called the Gross–Pitaevskii equation (see below and for references, e.g., [64, 65] and [80]). The nonlinear term in this equation expresses a local self-interaction of each particle of the condensate and depends on the density of the particles (it is the collective result of the presence of 2-particle interactions between the particles of the gas).

The mathematical derivation of the Gross–Pitaevskii (GP) equation and other related equations from a quantum mechanical N-particle system, described by a Hamiltonian H_N (see (1) below), usually with a confining potential V and with two particle interactions given by a potential v_N, taking the limit as N tends to infinity, has been an important issue in mathematical physics for many years and there is still much research going on, as we shall indicate.

The derivation involves the choice of particular 2-particle interactions, scaled in a certain way depending on N and the dimension n of the underlying space in which particles move (here we shall mainly consider the case $n = 3$, but other values of n have been examined by similar methods). As we shall mention in detail in Sect. 2, in the case $n = 3$ essentially three choices of scaling, characterized by a parameter $0 \leq \beta \leq 1$, have been discussed: the mean-field one for $\beta = 0$ (where the limit equation, called the mean-field or Hartree equation, contains a cubic nonlinear and non-local term with a "good kernel"); the intermediate one for $0 < \beta < 1$ (where the limit equation is a nonlinear Schrödinger equation with a cubic local nonlinearity with a constant factor in front involving the integral of the original 2-particle potential); and for the value $\beta = 1$ the GP equation (with a local cubic nonlinearity and a constant in front depending on the scattering length of the 2-particle potential).

As we mentioned above, the GP regime ($\beta = 1$) is the most used in the study of BEC, but it is also the one that is most mathematically complex. The major results were obtained in a series of papers by Lieb, Seiringer and Yngvason (see, e.g., [63, 65] and the book [64], see also [24, 43]). The choice of the 2-particle, translation invariant, potentials is a point interaction one, that heuristically permits certain explicit calculations (typical of point interactions, see, e.g., [9, 13]) leading in particular to the presence of a local nonlinearity, but also already presents for $n = 3$ intriguing mathematical problems in the choice of the starting Hamiltonian (connected with the theory of self-adjoint extensions of symmetric operators and renormalization theory; these problems also arise in physical phenomena like the Efimov and Thomas effects, and not by chance their study, both theoretical and experimental, has strong connections with the work on BEC: see, e.g., [7], [42] and also the excellent exposition in [47], we shall say a bit more on this in Sect. 7). For a detailed explanation of the mean-field scaling limit $\beta = 0$ and its applications see, e.g., [61, 62]. For the intermediate case $0 < \beta < 1$ see, e.g., [83].

In our presentation in the first part of the present paper we shall stress a new approach to this circle of problems developed in the last decade, starting from [72], based on ideas of Nelson's stochastic mechanics (see, e.g., [23, 40, 76, 78, 79]), associating to a solution of the N-particle Schrödinger equation related to the N-body Hamiltonian H_N a certain diffusion process on $\mathbb{R}^{nN}$ having invariant measure whose drift is the logarithmic derivative of the solution of the original Schrödinger equation. It was shown in [72] that in the GP-limit one gets a process with drift depending on the wave function of the BE condensate. A further discussion can be found in [16]. Progress associated with this state is discussed and a probabilistic counterpart of the asymptotic localization of the interaction energy has been shown in [73] and chaotic properties have been established in [86] for this scaling limit. Other developments in this setting are discussed in Sect. 7. We shall also present, in Sect. 4, original results on the mean field limit. For this we shall use a new variational approach that is inspired by previous work of K. Yasue [87] and Guerra–Morato [51], starting from an N-particle approximation of the relative mean-field stochastic optimal control problem introduced in [4].

In Sect. 5 we present a Markovian N-particle approximation (based on our work in [4]) to the stochastic optimal control discussed in Sect. 4. With a suitable choice of potentials we prove two convergence results: one involving the invariant measure of the optimal controlled N-particle process, the other concerning the law of the process on the whole path space $C^0([0, T], \mathbb{R}^{nN})$ (for any arbitrary $T > 0$ fixed). In Theorem 5.1 the convergence to zero of the $\frac{1}{N}$-multiple of the entropy of $\rho_{0,N}$ (the invariant measure of the optimally controlled N-particle system) relative to $\rho_0^{\otimes N}$ (the tensor product of the invariant measure of the optimally controlled mean-field system) is proven. A corresponding result, Theorem 5.2, holds for the conditional entropy (with respect to the k-partial marginals, for any $k \in \mathbb{N}$) on the path space.

In Sect. 6 the case of a variational problem with a convoluted delta potential is studied for all values of $\beta \in (0, 1]$. The optimal control is discussed in relation to the methods used in Sects. 4 and 5 for the case $\beta = 0$. In the case $0 < \beta < 1$ both the convergence of the "value function" and the probability measure on the path space, with respect to the relative entropy, are considered (see Theorems 6.1 and 6.2 respectively), using the methods of [5]. In the case $\beta = 1$, a weak convergence result of the probability law on the path space, obtained in [6], is also mentioned.

In Sect. 7 we first discuss possible extensions of the work presented in the previous sections on stochastic optimal control, especially to a time-dependent case (rather than the stationary case studied before). We also broaden the perspective to other problems where the relations between hyperbolic problems and parabolic ones play an important role, e.g., we mention the truly infinite-dimensional problems one meets when one replaces particle quantum mechanics with relativistic quantum field theory. Here new problems arise and very little is known about extending optimal stochastic control to this area. We observe that much success in the study of quantum fields has been obtained by taking a "Euclidean, Wiener-like, path integral" method instead of the "hyperbolic path integral" (Feynman path integral). The latter corresponds in a sense to taking imaginary time in the Euclidean path integral, followed by an analytic continuation procedure. More direct methods have been developed to extend the existing rigorous mathematical work of Feynman path integrals (see [10, 11, 68]) from the "finite-dimensional case" of non-relativistic quantum mechanics to the "infinite-dimensional case" of quantum field theory. Additional connections between probability, analysis, and geometry are also briefly mentioned.

2 Quantum Mechanics and Bose–Einstein Condensation

For the sake of simplicity hereafter we consider the quantum mechanical description of $N \in \mathbb{N}$ identical Bosons of mass $m > 0$. More precisely, the N-body Hamiltonian used in the description of the experiments on Bose–Einstein Condensation (BEC) [37, 56, 69] is of the type

$$H_N = \sum_{i=1}^{N} \left(-\frac{\hbar^2}{2m} \Delta_i + V(\mathbf{r_i}) \right) + \sum_{1 \le i < j \le N} v_N(\mathbf{r_i} - \mathbf{r_j}), \tag{1}$$

where $V : \mathbb{R}^3 \to \mathbb{R}$ is a confining potential, v_N a pair-wise repulsive (rotation invariant) interaction potential and $\mathbf{r}_i \in \mathbb{R}^3$, $i = 1, \ldots, N$. H_N is realized (under suitable assumptions) as a self-adjoint operator in the complex $L_s^2(\mathbb{R}^{3N})$-space of permutation symmetric square-integrable functions ("wave functions"). We denote the scalar product in this space by $(\cdot, \cdot)$ and the norm by $||\cdot||$. $\hbar$ denotes the (reduced) planck's constant.

The state of the system is described by the wave function $\Psi_{N,t}$ solving the Schrödinger equation

$$i\hbar \partial_t \Psi_{N,t} = H_N \Psi_{N,t} = \sum_{i=1}^{N} \left(-\frac{\hbar^2}{2m} \Delta_i \Psi_{N,t} + V(\mathbf{r_i}) \Psi_{N,t} \right) + \sum_{1 \le i < j \le N} v_N(\mathbf{r_i} - \mathbf{r_j}) \Psi_{N,t}.$$
$$\tag{2}$$

with the initial condition $\Psi_{N,0} \in L^2(\mathbb{R}^{3N})$, whose modulus square $\rho_t^N(\mathbf{r}) = |\Psi_{N,t}(\mathbf{r})|^2$, $\mathbf{r} \in \mathbb{R}^{3N}$ gives (by Born's interpretation) the probability density (with respect to Lebesgue measure) associated with the system of N-particles. In the following we will focus on the stationary case, more precisely the ground state, i.e. the wave function $\Psi_{N,t}$ does not depend on t, and it is the eigenfunction of the lowest eigenvalue of H_N. In the study of BEC, the ground state plays the main role (physically this is due to the fact that the BEC phenomenon happens at very low temperatures). Let us characterize the ground state denoted by Ψ_0, by a variational principle: consider the functional

$$E[\Psi] := \frac{1}{2} \int_{\mathbb{R}^{3N}} \Psi(\mathbf{r}) H_N \Psi(\mathbf{r}) d\mathbf{r} = T_\Psi + \Phi_\Psi. \tag{3}$$

$E[\Psi]$ is the mean quantum mechanical energy, where

$$T_{\Psi} := \sum_{i=1}^{N} \int_{\mathbb{R}^{3N}} |\nabla_i \Psi|^2 d\mathbf{r}_1 \cdots d\mathbf{r}_N$$

is the *"(mean) kinetic energy"* and

$$\Phi_{\Psi} = \sum_{i=1}^{N} \int_{\mathbb{R}^{3N}} V(\mathbf{r}_i)|\Psi|^2 d\mathbf{r}_1 \cdots d\mathbf{r}_N + \frac{1}{2}\sum_{i=2}^{N} \int v_N(\mathbf{r}_1 - \mathbf{r}_i)|\Psi|^2 d\mathbf{r}_1 \cdots d\mathbf{r}_N$$

the *(mean) potential energy* associated with $\Psi \in L_s^2(\mathbb{R}^{3N})$. If there exists a minimizing function Ψ_N^0 of $E[\Psi]$ with respect to the complex-valued functions Ψ in $L_s^2(\mathbb{R}^{3N})$ subject to the constraint $\|\Psi\|_2 = 1$, it is called a variational *ground state*. The corresponding energy $E[\Psi_N^0]$ given by

$$E[\Psi_N^0] := \inf\left\{ E(\Psi) : \|\Psi\|^2 = 1 \right\},$$

where Ψ in the previous set belongs to $L_s^2(\mathbb{R}^{3N})$, is called *ground state energy*. Under suitable assumptions on the potentials V and v one can prove the existence and uniqueness of the ground state Ψ_N^0 for (1). By the minimax principle (see, e.g., [81, Thm. XIII.1]) one has $H_N\Psi_0^N = E[\Psi_N^0]\Psi_0^N$, i.e. Ψ_0^N is the eigenfunction corresponding to the lowest eigenvalue $E[\Psi_N^0]$ of H_N, as a self-adjoint operator acting in $L_s^2(\mathbb{R}^{3N})$.

Remark 2.1 Uniqueness of the ground state is to be understood as uniqueness apart from an *overall phase*. Regularity conditions on V and v implying the strict positivity and the continuous differentiability of the ground state (wave function) are well known (indeed they follow by a suitable adaptation of the arguments in [81] (Thm.XIII.46 and XIII.47) and [81] (Thm.XIII.11)), respectively.

The mathematical notion of the quantum phenomenon of Bose–Einstein condensation can be introduced in quantum theory by starting from the one-particle density matrix, i.e. the operator in $L^2(\mathbb{R}^3)$ having kernel:

$$\gamma(\mathbf{r}, \mathbf{r}') = \int \Psi_N^0(\mathbf{r}, \mathbf{r}_2, \dots \mathbf{r}_N) \cdot \Psi_N^0(\mathbf{r}', \mathbf{r}_2, \dots \mathbf{r}_N) d\mathbf{r}_2 \cdots d\mathbf{r}_N,$$

where Ψ_N^0 denotes the wave function of the ground state.

Definition 2.1 Complete BEC is defined by the property that

$$\lim_{N \uparrow \infty} \gamma(\mathbf{r}, \mathbf{r}') = \varphi(\mathbf{r})\varphi(\mathbf{r}')$$

for some $\varphi \in L^2(\mathbb{R}^3)$ and in some topology for density matrices.

One of the main problems in the mathematical physics literature of the subject consists in justifying the various non-linear one-particle approximation models for describing the Bose–Einstein condensate. This goal is pursued, in the ground state framework, starting from the N-body Hamiltonian for N Bose particles (1) and by performing a suitable limit of an infinite number of particles.

Under certain assumptions on V and v_N, it has been shown that, for $N \to +\infty$, there is a limit wave function φ in $L^2(\mathbb{R}^3)$ of norm 1 solving a suitable (nonlinear) Schrödinger equation with Hamiltonian of the form

$$H_{BE}(\varphi) = -\frac{\hbar^2}{2m}\Delta + V(\mathbf{r}) + \tilde{v}(|\varphi|^2, \mathbf{r}), \quad \mathbf{r} \in \mathbb{R}^3, \tag{4}$$

where $\tilde{v}(|\varphi|^2, \mathbf{r})$ is an $L^2(\mathbb{R}^3)$-operator depending on the probability density $|\varphi|^2$ on $\mathbb{R}^3$. The related energy functional is given by the expression

$$E_{BE}(\varphi) = \int_{\mathbb{R}^3} \left(\frac{1}{2m}|\nabla\varphi(\mathbf{r})|^2 + \frac{1}{2}V(\mathbf{r})|\varphi(\mathbf{r})|^2 + \frac{1}{4}\tilde{v}(|\varphi|^2, \mathbf{r})|\varphi(\mathbf{r})|^2 \right) d\mathbf{r}. \tag{5}$$

The precise form of the operator $\tilde{v}$ is strongly dependent on the kind of scaling limit of the original interaction potential v_N. If we take v_N (in (1)) of the form

$$v_N(\mathbf{r}) = \frac{N^{3\beta}}{N-1}v_0(N^{\beta}\mathbf{r}), \quad 0 \le \beta \le 1 \tag{6}$$

we can distinguish three regimes:

1. the *mean-field regime (also called Hartree)*, that is $\beta = 0$, in which

$$\tilde{v}(|\varphi|^2, \mathbf{r}) = \int_{\mathbb{R}^3} v_0(\mathbf{r} - y)|\varphi|^2(y)dy = (v_0 * |\varphi|^2)(\mathbf{r});$$

2. the *intermediate regime* (also called nonlinear Schrödinger), i.e. $0 < \beta < 1$, in which

$$\tilde{v}(|\varphi|^2, \mathbf{r}) = \left(\int_{\mathbb{R}^3} v_0(y)dy \right) |\varphi|^2(\mathbf{r}) = \left(\int_{\mathbb{R}^3} v_0(y)dy \right) (\delta_0 * |\varphi|^2)(\mathbf{r})$$

(where δ_0 is the Dirac delta in 0);

3. the *Gross–Pitaevskii regime*, i.e. $\beta = 1$, in which $\tilde{v}(|\varphi|^2, \mathbf{r}) = \frac{4\pi\hbar^2 a}{m}(\delta_0 * |\varphi|^2)(\mathbf{r})$, where a is the scattering length of the potential v_0 (see, e.g., [64, Appendix C] for the definition of scattering length, and see also [9] for other physical contexts where it plays an important role).

We remark that when $\beta = 0$, which corresponds properly to the mean-field approximation, the potential range is fixed and the intensity of the interaction potential decreases as $1/N$ for $N \to \infty$. In the regime corresponding to $0 < \beta < 1$

the interaction potential goes to a delta function in the sense of the convergence of measures. This intermediate (or general) mean-field case is not very well studied and it is usually called the nonlinear Schrödinger limit. There are many results both for the mean-field and for the intermediate case. For the latter there are some quantitative estimates of the convergence rate for small values of β (see [62, 82] and references therein). In [4] the general mean-field convergence problem ($0 < \beta < 1$) is faced by using the hard results for the case $\beta = 1$ and the convergence of the one-particle ground-state energy to the ground-state energy of the nonlinear Schrödinger functional for the case of purely repulsive interaction potential is proved.

We finally stress that the case $\beta = 1$ cannot be considered as a mean-field regime and it involves the scattering length of the interaction potential. The convergence of the ground state energy in this setting has been provided by Lieb and Seiringer [63] and Lieb et al. [65] and, recently, in [75].

In the time-dependent framework one of the main problems is that of controlling whether the Bose–Einstein condensation is preserved by the time evolution, that is, whether at time $t > 0$ for N large enough the one-particle density $\gamma_{N,t}^1$ is, in some approximation, equal to $|\varphi_t|^2$, where φ_t is the solution of the nonlinear (time-dependent) Schrödinger or Gross–Pitaevskii equation. More precisely, starting from a factorized initial wave function for the N-body Hamiltonian (1) and introducing the time evolution $\Psi_{N,t}$ of the initial wave function, the goal is to prove that the one-particle density associated to $\Psi_{N,t}$ converges to $|\varphi_t|^2$, with φ_t playing the role of the time-dependent wave function of the Bose–Einstein condensate (see, e.g., [1, 20, 27, 43]). The techniques used in the time-dependent setting are different from those of the stationary one, in particular instead of the mean quantum energies the Schrödinger hierarchies are used. Many other problems, such as the study of the fluctuations around the limit, are actually faced in the more general time-dependent framework, see for instance [26, 28].

3 Nelson's Stochastic Mechanics

One of the main problems in giving a stochastic representation of solutions to the Schrödinger equation is the reversibility in time of the quantum evolution (which is given by a one-parameter unitary group, and not by a contraction semigroup). Indeed the time marginal probability of, for example, a diffusion Markov process is a solution to the Fokker–Planck equation, which is a parabolic (and thus non-time-reversible) equation. A possible solution to this problem, at least in the one particle case, is given by Nelson's Stochastic Mechanics, introduced by Edward Nelson in 1966. It intends to study certain quantum phenomena using a well-determined class of diffusion processes (see [23, 30, 40, 76, 78, 79]). See [29] for a relatively recent review on Nelson's Stochastic Mechanics. Here we recall only the basic elements of the theory in order to present a suggestive variational approach (due to K. Yasue, see [87], and F. Guerra and L. Morato, see [51], see below and Sect. 7 for other references) that motivated our own approach in Sect. 4.

Consider a quantum particle of mass m moving on $\mathbb{R}^n$, subject to a force of potential V (and thus having Hamiltonian $H = -\frac{\hbar^2}{2m}\Delta + V(x)$). Nelson associates to it a Markovian process which is a solution to the following SDE:

$$dX_t = b(X_t, t)dt + \nu dW_t, \tag{7}$$

where $\nu = \sqrt{\frac{\hbar}{m}}$, W is a standard Wiener process in $\mathbb{R}^n$, and $b : \mathbb{R}^n \times \mathbb{R}_+ \to \mathbb{R}^n$ is a measurable vector field whose regularity will be made more precise below. The core of the kinetic part of the theory is the fundamental pair of stochastic derivatives. The forward stochastic derivative is on smooth real functions f on $\mathbb{R}^n$ defined by:

$$Df(X_t) := \lim_{h\downarrow 0}\mathbb{E}_t\left[\frac{f(X_{t+h}) - f(X_t)}{h}\right]$$

(where $\mathbb{E}_t$ is the conditional expectation with respect to X_t) and has the property that:

$$DX_t = b(X_t, t).$$

Nelson also introduced the backward stochastic derivative:

$$D_* f(X_t) := \lim_{h\downarrow 0}\mathbb{E}_t\left[\frac{f(X_t) - f(X_{t-h})}{h}\right]$$

which gives:

$$D_* X_t = b_*(X_t, t),$$

for a certain vector field b_* on $\mathbb{R}^n \times \mathbb{R}_+$. The literature on time reversal of diffusion processes is quite large (see, e.g., [71] and references therein, see also [74]). Foellmer [46] individuated, in the context of Stochastic Mechanics presented here, a sufficient condition for the existence of the backward derivative: $\mathbb{E}[|b(X_t, t)|^2] < \infty$. If the vector field b is such that the probability density ρ_t for the solution X_t to the SDE (7) is strictly positive and differentiable, we have the relation

$$b_*(x, t) = b(x, t) - \frac{\nu^2}{2}\nabla \log \rho_t(x). \tag{8}$$

As for the dynamic, Nelson introduced the stochastic Newton equation

$$\frac{1}{2}[DD_* + D_*D]X_t = -\nabla V(X_t), \tag{9}$$

where V is the potential in which the particle of mass m is moving. Using the relation between b_* and b, writing $u(x, t) = \frac{1}{2}(b(x, t) - b_*(x, t))$ and $v(x, t) = \frac{1}{2}(b_*(x, t) + b(x, t))$ we get

$$\partial_t v = -\nabla V(x) + u \cdot \nabla u + v \cdot \nabla v + \frac{v^2}{2} \Delta u, \tag{10}$$

$$\partial_t u = -\nabla(u \cdot v) - \frac{v^2}{2} \nabla(\nabla \cdot v), \tag{11}$$

which are reversible (in time) equations. It is possible to prove, by choosing the initial conditions in a suitable way, that the previous system of PDEs (10), (11) admits solutions which, by an important result of Carlen (see [29–32], permit us to solve Eq. (7) and to associate a stochastic process to the quantum system. The solution process is then associated with the Schrödinger equation with the potential V (appearing in Eq. (9)). In the stationary case Eq. (10) reduces to an equation for u which is of the form $V_u = \frac{1}{2}(|u|^2 + v^2 \mathrm{div}(u))$.

K. Yasue initiated a heuristic variational formulation of the association of X_t to the Schrödinger equation by introducing a Lagrangian function $\mathcal{L}$ associated with the quantum Hamiltonian H

$$\mathcal{L}(\mathrm{D}X_t, \mathrm{D}^* X_t, X_t) = \frac{1}{4} \left(|\mathrm{D}^* X_t|^2 + |\mathrm{D}X_t|^2 \right) - V(X_t). \tag{12}$$

An alternative action functional, proposed by Guerra and Morato, is given by the expression

$$\tilde{\mathcal{L}}(\mathrm{D}X_t, \mathrm{D}^* X_t, X_t) = \frac{1}{2} \left(\mathrm{D}^* X_t \cdot \mathrm{D}X_t \right) - V(X_t). \tag{13}$$

By the relations (8) and (9) the Lagrangian (12) can be thought of as a function of the vector field b, the process X_t and the probability density ρ_t associated with it. Thanks to this observation we can use the Lagrangian $\mathcal{L}$ to formulate an optimal control problem for the controlled SDE (7) (where the vector field b plays the role of control parameter). We consider the finite horizon optimal control problem and the ergodic control problem associated with the Lagrangian $\mathcal{L}$, i.e. we have the respective cost functions

$$J^{\mathrm{fh}}(\rho_0, b, \rho_t) = \mathbb{E}_{X_0 \sim \rho_0} \left[\int_0^T \mathcal{L}(\mathrm{D}X_t, \mathrm{D}^* X_t, X_t) \mathrm{d}t \right], \tag{14}$$

$$J^{\mathrm{e}}(\rho_0, b) = \mathbb{E}_{X_0 \sim \rho_0} \left[\limsup_{T \to +\infty} \frac{1}{T} \int_0^T \mathcal{L}(\mathrm{D}X_t, \mathrm{D}^* X_t, X_t) \mathrm{d}t \right] \tag{15}$$

(the suffixes fh and e in J stand for "finite horizon" and "ergodic", respectively, see, e.g., [44, 45] for a reference on stochastic optimal control i the notation $X_0 \sim \rho_0$ stands for X_0 having law ρ_0). The reason for the choice of the cost functionals is that the optimal controls of the previous problems satisfy the Schrödinger equation. The same optimal control problems can be obtained replacing the Lagrangian $\mathcal{L}$ with the functional $\tilde{\mathcal{L}}$, given in Eq. (13), in the definition of the cost functionals J^{fh} and J^{e}. More precisely, if b is an optimal control to the problem (14) where the optimal solution process X_t has density ρ_t, then there is a unique (up to a complex multiplicative constant) function $\Psi_t : \mathbb{R}^n \times \mathbb{R}_+ \to \mathbb{C}$ such that

$$b(x, t) = \mathrm{Re}\left(\frac{\nabla \Psi_t(x)}{\Psi_t(x)} \right) + \mathrm{Im}\left(\frac{\nabla \Psi_t(x)}{\Psi_t(x)} \right), \quad \rho_t(x) = |\Psi_t(x)|^2,$$

and the function Ψ_t satisfies the Schrödinger associated to the Hamiltonian H. Carlen proved the existence of Nelson diffusions also in the general case in which there are nodes of the wave function [29], [30] under a finiteness condition on the Fisher information. In the ergodic case the optimal control b and the related probability density ρ *do not depend on the time* t and they are of the form

$$b(x) = \nabla \log(\Psi^0(x)), \quad \rho(x) = |\Psi^0(x)|^2,$$

where the function Ψ^0 is the (real) ground state of the Hamiltonian H. Let us mention, finally, that the variational formulations by Yasue, as well as by Guerra–Morato, have important connections with the entropic optimal transport problem (see [38] and [35] for studies on this connection in a rigorous probabilistic setting related to the heat rather than the Schrödinger equation). See also Sect. 7 for other variational approaches.

4 Non-linear Stochastic Mechanics

We want to take inspiration from the above sketched variational formulation of stochastic mechanics and the methods used in the convergence proof of the Bose–Einstein condensation to study some stochastic optimal control problems of McKean–Vlasov type (namely where the cost function depends not only on the solution process X_t to the controlled equation but also on its law).

Following [4], let us start with the autonomous stochastic differential equation (SDE)

$$dX_t = b(X_t)dt + \nu dW_t; \quad t \geq 0, \tag{16}$$

where b is a C^1 function from $\mathbb{R}^n$ to $\mathbb{R}^n$, $\nu > 0$ is a constant, and W_t, $t \geq 0$, is an n-dimensional standard Brownian motion. The starting point for X_t at $t = 0$ is $x_0 \in \mathbb{R}^n$. We look here at b as a "control vector field" and we associate to (16) the following "cost functional"

$$J(b, x_0) = \limsup_{T \to +\infty} \frac{1}{T} \left(\int_0^T \mathbb{E}_{x_0} \left[\frac{|b(X_t)|^2}{2} + \mathcal{V}(X_t, \text{Law}(X_t)) \right] dt \right), \quad x_0 \in \mathbb{R}^n.$$
(17)

Let $\mathcal{P}(\mathbb{R}^n)$ be the space of probability measures on $\mathbb{R}^n$ endowed with the topology given by weak convergence, $\mathcal{V} : \mathbb{R}^n \times \mathcal{P}(\mathbb{R}^n) \to \mathbb{R}$ where $\mathcal{P}(\mathbb{R}^n)$ is the set of probability measures on $\mathbb{R}^n$ is a regular function (hereafter called "potential") satisfying some technical hypotheses (see Hypothesis $\mathcal{V}$ below) and $\mathbb{E}_{x_0}$ denotes the expectation with respect to the solution X_t to the SDE (16) such that $X_0 = x_0 \in \mathbb{R}^n$.

In [4] we proved existence and uniqueness of the optimal control $b \in C^1(\mathbb{R}^n, \mathbb{R}^n)$ for the problem given by (16) and (17). Here we give a simplified proof of these results. We remark that the action functional explicitly depends on the law of X_t through the potential $\mathcal{V}$ but we find that the optimal control itself can be expressed in terms of the same law.

We define the following "value function":

$$\mathfrak{J} := \text{ess sup}_{x_0 \in \mathbb{R}^n} \left(\inf_{b \in C^1(\mathbb{R}^n, \mathbb{R}^n)} J(b, x_0) \right),$$
(18)

where ess sup is the essential supremum over $x_0 \in \mathbb{R}^n$ and J is the cost functional (17).

Remark 4.1 There are two important observations to make about the initial conditions chosen in the definition of the value function (18). The first one is that the function $x_0 \longmapsto \inf_{\alpha \in C^1(\mathbb{R}^n, \mathbb{R}^n)} J(\alpha, x_0)$ is almost surely constant in x_0 with respect to the Lebesgue measure (see Theorem 4.1 below). This means that the ess $\sup_{x_0 \in \mathbb{R}^n}$ is used only to exclude a set of measure zero with respect to x_0.

The second observation is that it is possible to extend our analysis by considering

$$\bar{J}(b, \rho) := \limsup_{T \to +\infty} \frac{1}{T} \left(\int_0^T \mathbb{E}_{X_0 \sim \rho(x)dx} \left[\frac{|b(X_t)|^2}{2} + \mathcal{V}(X_t, \text{Law}(X_t)) \right] dt \right),$$
(19)

where the process X_t now has an initial probability law $\text{Law}(X_0)$ which is absolutely continuous with respect to Lebesgue measure of the form $\rho(x)dx$. Indeed in both Theorem 4.1 and Lemma 4.1, below, we can replace the deterministic initial condition with a random one, of the previous type, obtaining the corresponding statement. This fact proves that

$$\mathfrak{J} = \inf_{b \in C^1(\mathbb{R}^n, \mathbb{R}^n)} \bar{J}(b, \rho),$$

for any $\rho \in L^1(\mathbb{R}^n)$. In this paper we consider deterministic initial conditions in order to simplify the treatment of the general problem.

Definition 4.1 If $\mathcal{K} : \mathcal{P}(\mathbb{R}^n) \to \mathbb{R}$ is a function we say that $\mathcal{K}$ is Gâteaux differentiable if for any $\mu, \mu' \in \mathcal{P}(\mathbb{R}^n)$ there exists a bounded continuous function $\partial_\mu \mathcal{K}(\cdot, \mu) : \mathbb{R}^n \to \mathbb{R}$ such that

$$\lim_{\epsilon \to 0^+} \frac{\mathcal{K}(\mu + \epsilon(\mu - \mu')) - \mathcal{K}(\mu)}{\epsilon} = \int_{\mathbb{R}^n} \partial_\mu \mathcal{K}(y, \mu)(\mu(dy) - \mu'(dy)), \qquad (20)$$

and we can choose the normalization condition given by

$$\int_{\mathbb{R}^n} (\partial_\mu \mathcal{K})(y, \mu)\mu(dy) = 0.$$

When a function $\bar{\mathcal{K}} : \mathbb{R}^n \times \mathcal{P}(\mathbb{R}^n) \to \mathbb{R}$ depends also on $x \in \mathbb{R}^n$ we say that $\bar{\mathcal{K}}$ is Gâteaux differentiable if $\bar{\mathcal{K}}(x, \cdot)$ is Gâteaux differentiable for any $x \in \mathbb{R}^n$. In this case we write

$$\lim_{\epsilon \to 0^+} \frac{\bar{\mathcal{K}}(x, \mu + \epsilon(\mu - \mu')) - \bar{\mathcal{K}}(x, \mu)}{\epsilon} = \int_{\mathbb{R}^n} \partial_\mu \bar{\mathcal{K}}(x, y, \mu)(\mu(dy) - \mu'(dy)).$$

After these remarks, let us make precise the hypothesis on the functional $\mathcal{V}$ entering in the cost functional (17):

- **Hypotheses $\mathcal{V}$:**

  (i) *The map $\mathcal{V}$ is continuous from $\mathbb{R}^n \times \mathcal{P}(\mathbb{R}^n)$ to $\mathbb{R}$ (where we recall that $\mathcal{P}(\mathbb{R}^n)$ is equipped with the weak topology of convergence of measures).*

  (ii) *There is a positive function V and there are three positive constants c_1, c_2, c_3, with $c_2 > 0$, such that for any $\mu \in \mathcal{P}(\mathbb{R}^n)$:*

$$V(x) - c_1 \le \mathcal{V}(x, \mu) \le c_2 V(x) + c_3, \quad x \in \mathbb{R}^n. \qquad (21)$$

  Furthermore, we assume that V is such that

$$|\partial^\alpha V(x)| \le C_\alpha V(x) \qquad V(x) \le C_1 V(y) \exp(C_2|x - y|), \quad x, y \in \mathbb{R}^n \qquad (22)$$

  where $\alpha \in \mathbb{N}^n$ is a multiindex of length at most $|\alpha| \le 2$, and C_α, C_1 and C_2 are positive constants; V is also assumed to be growing to $+\infty$ as $|x| \to +\infty$.

  (iii) *The map $\mathcal{V}$ is Gâteaux differentiable and $\partial_\mu \mathcal{V}(x, y, \mu)$ is uniformly bounded from below and we have*

$$\partial_\mu \mathcal{V}(x, y, \mu) \le D_1 + D_2 V(x)V(y), \quad x, y \in \mathbb{R}^n \qquad (23)$$

for some constants D_1, $D_2 \geq 0$. Furthermore, whenever

$$\partial_\mu \tilde{\mathcal{V}}(y, \mu) = \mathcal{V}(y, \mu) + \int_{\mathbb{R}^n} \partial_\mu \mathcal{V}(x, y, \mu) \mu(\mathrm{d}x)$$

is well defined (namely when $\int_{\mathbb{R}^n} V(x) \mu(\mathrm{d}x) < +\infty$), we require that $\partial_\mu \tilde{\mathcal{V}}(\cdot, \mu)$ is a $C^{\frac{n}{2}+\delta}(\mathbb{R}^n, \mathbb{R})$ Hölder function for some $\delta > 0$.

- **Hypothesis C$\mathcal{V}$** *the functional $\tilde{\mathcal{V}}$ is convex.*
- **Hypothesis QV**: *the function V, in Hypotheses $\mathcal{V}$, is radially symmetric $V(x) = \bar{V}(|x|)$, where $\bar{V}$ is a $C^1(\mathbb{R}_+, \mathbb{R})$ increasing function for which there are constants $e_1, \epsilon > 0, e_2, e_3 \geq 0$ such that:*

(i) $\bar{V}(r) \geq e_1 r^{2+\epsilon} - e_2$,
(ii) $\bar{V}'(r) \leq e_3 (\bar{V}(r))^{\frac{3}{2}}, r := |x|$.

Remark 4.2 The previous hypotheses on the functional $\mathcal{V}$ cover the mean-field scaling regime of the interacting potential v_0 in the Bose–Einstein Condensation (BEC) (see Eq. (6) for $\beta = 0$). Indeed in the mean-field BEC the functional $\mathcal{V}$ in Eq. (17)) has the form

$$\mathcal{V}(x, \mu) = V_0(x) + \int_{\mathbb{R}^n} v_0(x - y) \mu(\mathrm{d}y), \tag{24}$$

where $V_0, v_0 \in C^{\frac{n}{2}+\epsilon}(\mathbb{R}^n), \epsilon > 0$ and $\mu \in \mathcal{M}_c(\mathbb{R}^n)$ (where $\mathcal{M}_c(\mathbb{R}^n)$ is the space of signed measures on $\mathbb{R}^n$ having total mass less than $c \in \mathbb{R}_+$). Furthermore, we require that V_0 grows to plus infinity as $|x| \to +\infty$, and there is a function V, satisfying the relation (22) and Hypothesis QV, such that $V_0(x) \sim V(x)$ as $|x| \to +\infty$ (where $\sim$ stands for $V_0(x)$ is bounded from above and below by positive constants times $V(x)$ as $|x| \to +\infty$). We also assume that v_0 is bounded, reflection symmetric, i.e., $v_0(x) = v_0(-x)$, and that there exists a positive measure π on $\mathbb{R}^n$ such that, for any $x \in \mathbb{R}^n$, $v_1(x) = \int_{\mathbb{R}^n} e^{-ikx} \pi(\mathrm{d}k)$ (i.e. v_1 is the Fourier transform of a positive measure). The class of functionals (24) satisfy the above Hypotheses $\mathcal{V}$ and C$\mathcal{V}$.

First we have that if the vector field b in Eq. (16) is such that $J(b, x_0) < +\infty$, then there is a unique invariant measure ρ_b of Eq. (16).

Lemma 4.1 *Under hypotheses $\mathcal{V}(i)$ and $\mathcal{V}(ii)$, if $J(b, x_0)$ as given by (17) (with $b \in C^1$) is not equal to $+\infty$ there exists an unique and ergodic invariant probability density measure $\rho_b \in W^{1, \frac{n}{2}}(\mathbb{R}^n)$ for the SDE (16) so that $\mu_b(\mathrm{d}x) = \rho_b(x)\mathrm{d}x$ is the invariant ergodic probability measure for the SDE (16). Furthermore, we have*

$$\tilde{J}(b, \rho_b) \leq J(b, x_0)$$

for almost all $x_0 \in \mathbb{R}^n$ with respect to Lebesgue measure, where

$$\tilde{J}(b, \rho_b) := \int_{\mathbb{R}^n} \left(\frac{|b(x)|^2}{2} + \mathcal{V}(x, \rho_b) \right) \rho_b(x) \mathrm{d}x. \tag{25}$$

Proof The proof is given in [4].   □

In order to minimize the cost functional (25) with respect to ρ, for $\rho \in W^{1,\frac{n}{2}}(\mathbb{R}^n)$, $\rho(x) \geq 0$ and $\int_{\mathbb{R}^n} \rho(x)\mathrm{d}x = 1$, we set

$$\mathcal{C}_\rho = \{b \in C^1(\mathbb{R}^n, \mathbb{R}^n), \ L_b^*(\rho) = 0 \text{ and } |\tilde{J}(b, \rho)| < +\infty\}. \tag{26}$$

Then $\mathcal{C}_\rho$ is the subset of $C^1(\mathbb{R}^n, \mathbb{R}^n)$ vector fields $b_\rho \in \mathcal{C}_\rho$ such that $L_{b_\rho}^*(\rho) = 0$ (where $L_{b_\rho}^*$ is the adjoint of the infinitesimal generator L_{b_ρ} for the solution process X_t of Eq. (16) and the equality, in the definition of $\mathcal{C}_\rho$ in (26), is understood in a distributional sense) and $|\tilde{J}(b_\rho, \rho)| < +\infty$.

Remark 4.3 Suppose that $b \in C^1(\mathbb{R}^n, \mathbb{R}^n)$ such that $J(b, x_0) < +\infty$, then by Lemma 4.1 there is a unique positive probability density ρ_b which is invariant and thus, since $b \in C^1(\mathbb{R}^n, \mathbb{R}^n)$ by well-known results (see Proposition 3.1 in [4]), it satisfies the equation $L_b^*(\rho_b) = 0$. This implies that $b \in \mathcal{C}_{\rho_b}$, where $\mathcal{C}_{\rho_b}$ is defined by Eq. (26) with $\rho = \rho_b$.

We now introduce the following energy functional, for $\rho \in W^{1,\frac{n}{2}}(\mathbb{R}^n)$,

$$\mathcal{E}(\rho) := \mathcal{E}_K(\rho) + \mathcal{E}_P(\rho) = \int_{\mathbb{R}^n} \frac{|\nabla\rho|^2}{2\rho} \mathrm{d}x + \int_{\mathbb{R}^n} \mathcal{V}(x, \rho)\rho(x)\mathrm{d}x, \tag{27}$$

where the two terms on the right-hand side correspond by definition to the kinetic $\mathcal{E}_K(\rho)$ and potential $\mathcal{E}_P(\rho)$ energies, respectively. The kinetic term is also called the Fisher information.

The next lemma states a useful monotonicity property of the cost functional $\tilde{J}$.

Lemma 4.2 *For any given $\rho \in W^{1,\frac{n}{2}}(\mathbb{R}^n)$ we have*

$$\mathcal{E}(\rho) = \tilde{J}\left(\frac{\nabla\rho}{\rho}, \rho \right) \leq \inf_{b \in \mathcal{C}_\rho} \tilde{J}(b, \rho),$$

where $\tilde{J}(b, \rho)$ is defined in (25).

Proof By [25, Chapter 3, Theorem 3.1.2], if ρ is the density of the invariant measure of the SDE (16) we have that

$$\int_{\mathbb{R}^n} \frac{|\nabla\rho(x)|^2}{\rho^2(x)} \rho(x)\mathrm{d}x \leq \int_{\mathbb{R}^n} |b(x)|^2 \rho(x)\mathrm{d}x,$$

for any $b \in C_\rho$, with the equality holding if and only if $b = \frac{\nabla \rho}{2\rho}$. Since $\int_{\mathbb{R}^n} \mathcal{V}(x, \mu) \rho(x) \mathrm{d}x$ depends only on the invariant measure $\rho(x) \mathrm{d}x$, the lemma is proved. $\square$

Let us now minimize the function $\mathcal{E}(\rho)$ given by (27) under the condition $\int_{\mathbb{R}^n} \rho(x) \mathrm{d}x = 1$. Introducing the variable $\varphi = \sqrt{\rho}$ the energy functional (27) becomes

$$\mathcal{E}(\varphi^2) = \int_{\mathbb{R}^n} \left(\frac{|\nabla \varphi|^2}{2} + \mathcal{V}(x, \varphi^2)\varphi^2(x) \right) \mathrm{d}x, \tag{28}$$

with $\varphi \in L^2(\mathbb{R}^n)$ satisfying the condition $\int_{\mathbb{R}^n} \varphi^2(x) \mathrm{d}x = 1$.

Remark 4.4 The above well-known energy functional admits a unique minimizer which is strictly positive (see [4] and references therein). Furthermore, in the case where $\mathcal{V}$ is of the form (24), it coincides with the functional (5) of Bose–Einstein condensation in the mean field regime ($\beta = 0$ in relation (6)).

Lemma 4.3 *Under hypotheses $\mathcal{V}$ and $C\mathcal{V}$ the variational problem (27), with $\varphi \in L^2(\mathbb{R}^n)$ satisfying the condition $\int_{\mathbb{R}^n} \varphi^2(x) \mathrm{d}x = 1$, admits a unique minimizer $\rho_0 = \varphi_0^2$. Furthermore, φ_0 is $C^{2+\epsilon}(\mathbb{R}^n)$ for some $\epsilon > 0$, it is strictly positive and satisfies (weakly) the equation*

$$-\Delta \varphi_0(x) + 2\mathcal{V}(x, \varphi_0^2)\varphi_0(x) + 2 \int_{\mathbb{R}^n} \partial_\mu \mathcal{V}(y, x, \varphi_0^2)\varphi_0^2(y) \mathrm{d}y \varphi_0(x) = \mu_0 \varphi_0(x), \tag{29}$$

where the uniquely determined constant μ_0 is given by

$$\mu_0 = 2\mathcal{E}(\varphi_0^2) + \int_{\mathbb{R}^n} \partial_\mu \mathcal{V}(y, x, \varphi_0^2)\varphi_0^2(y)\varphi_0^2(x) \mathrm{d}y \mathrm{d}x. \tag{30}$$

Remark 4.5 Under Hypotheses $\mathcal{V}$ and $C\mathcal{V}$ we have that $J\left(\frac{\nabla \rho_0}{\rho_0}, x_0 \right) = \mathcal{E}(\rho_0)$, where $\rho_0 = \varphi_0^2$ is the unique minimizer of $\mathcal{E}$.

Finally we obtain our generalization, via an optimal control approach, of stochastic mechanics versus non-linear quantum models:

Theorem 4.1 *Under Hypotheses $\mathcal{V}$ and $C\mathcal{V}$, the logarithmic gradient of the unique minimizer $\rho_0 = \varphi_0^2$ of $\mathcal{E}$, that is $b = \frac{\nabla \rho_0}{2\rho_0}$, is the optimal control for the problem (17) for almost every $x_0 \in \mathbb{R}^n$ with respect to the Lebesgue measure.*

Proof (of Theorem 4.1) By Remark 4.5, and the definition of $\mathfrak{J}$ (given in Eq. (18)) we have that

$$\mathfrak{J} \leq \text{ess sup}_{x_0 \in \mathbb{R}^n} J\left(\frac{\nabla \rho_0}{\rho_0}, x_0\right) = \mathcal{E}(\rho_0). \tag{31}$$

In order to prove the statement of the theorem, it is sufficient to prove that $\mathcal{E}(\rho_0) \leq \mathfrak{J}$, indeed, by Lemma 4.5 and inequality (31), this implies that

$$\mathfrak{J} \leq \text{ess sup}_{x_0 \in \mathbb{R}^n} J\left(\frac{\nabla \rho_0}{\rho_0}, x_0\right)$$

and thus the thesis. By Lemma 4.1, we have $\tilde{J}(b, \rho_b) \leq J(b, x_0)$ and by Lemma 4.2, and since, by Remark 4.3, $b \in \mathcal{C}_{\rho_b}$, we get, for any fixed $b \in C^1(\mathbb{R}^n, \mathbb{R}^n)$ such that $J(b, x_0) < +\infty$,

$$\mathcal{E}(\rho_b) = \tilde{J}\left(\frac{\nabla \rho_b}{\rho_b}, \rho_b\right) \leq \inf_{\hat{b} \in \mathcal{C}_{\rho_b}} \tilde{J}(\hat{b}, \rho_b) \leq \tilde{J}(b, \rho_b).$$

Combining the previous two inequalities and Lemma 4.3, we obtain that, for any $b \in C^1(\mathbb{R}^n, \mathbb{R}^n)$ such that $J(b, x_0) < +\infty$,

$$\mathcal{E}(\rho_0) \leq \mathcal{E}(\rho_b) \leq \tilde{J}(b, \rho_b) \leq \text{ess sup}_{x_0 \in \mathbb{R}^n} J(b, x_0).$$

Taking the inf over $b \in C^1(\mathbb{R}^n, \mathbb{R}^n)$ from the previous inequality we get $\mathcal{E}(\rho_0) \leq \mathfrak{J}$. $\qquad\square$

Remark 4.6 An important consequence of Theorem (4.1) is that under Hypotheses $\mathcal{V}$ and $C\mathcal{V}$ we have that

$$\mathfrak{J} = \mathcal{E}(\rho_0) = \inf_{\varphi \in H^1(\mathbb{R}^n), \int \varphi^2 dx = 1} \mathcal{E}(\varphi^2),$$

where $\mathfrak{J}$ is the value function associated with the problem (16) and the cost functional (17), defined by (18).

Summarizing we proved that the (stationary) Nelson diffusion with drift of gradient type solves the ergodic optimal control problem with cost functional (17), with $\mathcal{V}$ satisfying Hypothesis $\mathcal{V}$, $C\mathcal{V}$ and $Q\mathcal{V}$. In particular our result contains the mean-field nonlinear Schrödinger model for the Bose–Einstein condensate (in this case where the potential $\mathcal{V}$ in the cost functional (17) is given by (24)).

5 Convergence of Markovian N-Particle Approximation

We are interested in studying a Markovian N-particle approximation to the stochastic optimal control problem given by (16) and (17). This approximation is inspired by the variational version of stochastic mechanics presented in Sect. 3.

We consider the process $X_{N,t} = (X_{N,t}^1, \ldots, X_{N,t}^N) \in \mathbb{R}^{nN}$ satisfying the SDE

$$dX_{N,t}^i = b_N^i(X_{N,t})dt + \nu dW_t^i, \quad i = 1, \ldots, N \tag{32}$$

where $b_N := (b_N^1, \ldots, b_N^N) : \mathbb{R}^{nN} \to \mathbb{R}^{nN}$ is a $C^{1+\epsilon}$ function, for some $\epsilon > 0$, and the $W_t^i, i = 1, \ldots, N$ are independent Brownian motions taking values in $\mathbb{R}^n$. If $\mathcal{V}$ is a functional satisfying Hypotheses $\mathcal{V}$, we introduce the sequence

$$\mathcal{V}_N(x) = \sum_{i=1}^N \mathcal{V}\left(x_i, \frac{1}{N-1}\sum_{k=1,k\neq i}^N \delta_{x^i}\right),$$

where $x = (x_1, \ldots, x_N) \in \mathbb{R}^{nN}$, $N \geq 2$, and δ_{x_i} is a Dirac delta measure in $x_i \in \mathbb{R}^n$. We consider the ergodic control problem (normalized with respect to the number N of particles)

$$J_N(b_N, x_0) = \limsup_{T \to +\infty} \frac{1}{NT} \int_0^T \mathbb{E}_{x_0}\left[\frac{|b_N(X_t)|^2}{2} + \mathcal{V}_N(X_t)\right]dt. \tag{33}$$

The corresponding (normalized) energy functional (analogous to the one defined in (27)) is

$$\mathcal{E}_N(\rho_N) = \mathcal{E}_{K,N}(\rho_N) + \mathcal{E}_{P,N}(\rho_N)$$
$$= \frac{1}{N}\left(\int_{\mathbb{R}^{nN}} \frac{|\nabla \rho_N|^2}{2\rho_N}dx + \int_{\mathbb{R}^{nN}} \mathcal{V}_N(x)\rho_N(x)dx\right), \tag{34}$$

where ρ_N is a positive Lebesgue integrable function such that $\int_{\mathbb{R}^{nN}} \rho_N(x)dx = 1$. We also consider the value function

$$\mathfrak{I}_N = \operatorname{ess\,sup}_{x_0 \in \mathbb{R}^n}\left(\inf_{b_N \in C^1(\mathbb{R}^{nN}, \mathbb{R}^{nN})} J_N(b_N, x_0)\right). \tag{35}$$

Remark 5.1 It is important to note that in the case where $\mathcal{V}$ is of the form (24) (i.e. for the mean-field BEC), from the definition of $\mathcal{V}_N$, we get:

$$\mathcal{V}_N(x_1, \ldots, x_N) = \sum_{k=1}^N V(x_k) + \frac{1}{N-1}\sum_{k,h=1}^N v_0(x_k - x_h),$$

which is exactly the total potential of the Hamiltonian H_N in (1), where v_N is the mean field scaling limit of v_0 (see Eq. (6) in the case $\beta = 0$).

In the setting described above we are able to prove two convergence results: the first one involves the invariant measure $\rho_{0,N}$ of the optimal controlled process $X_{N,t}$, the second concerns the law of the process $X_{N,t}$ on the whole path space $C^0([0, T], \mathbb{R}^{nN})$. In order to state the two convergence results we introduce the relative entropy between probability densities ρ_N, ρ_N' on $\mathbb{R}^{nN}$ resp. between probability measures $\mathbb{P}, \mathbb{Q}$ on the path space $C^0([0, T], \mathbb{R}^n)$ as

$$H_N(\rho_N|\rho_N')) := \begin{cases} \int_{\mathbb{R}^{nN}} \log\left(\frac{\rho_N(x)}{\rho_N'(x)}\right) \rho_N(x)\mathrm{d}x & \text{if supp}(\rho_N) \subset \text{supp}(\rho_N') \\ +\infty & \text{elsewhere} \end{cases},$$

$$\mathcal{H}_{C^0([0,T],\mathbb{R}^n)}(\mathbb{P}|\mathbb{Q}) := \int_\Omega \log\left(\frac{\mathrm{d}\mathbb{P}}{\mathrm{d}\mathbb{Q}}(\omega)\right) \mathbb{P}(\mathrm{d}\omega).$$

Theorem 5.1 *Suppose that $\mathcal{V}$ satisfies hypotheses $\mathcal{V}$, $C\mathcal{V}$ and $Q\mathcal{V}$ and let $\rho_{0,N}$ and ρ_0 be the unique minimizers of the energies* (34) *and* (27) *respectively, then we have as $N \to +\infty$*

$$\frac{1}{N} H_N(\rho_{0,N}|\rho_0^{\otimes N}) \to 0. \tag{36}$$

Remark 5.2 It is important to note that if $\rho_{0,N}^{(k)}(x_1, \ldots, x_k)$ denotes the marginal of the measure $\rho_{0,N}(x_1, \ldots, x_N)$ with respect to the first k variables, then Theorem 5.1 implies $H_k(\rho_{0,N}^{(k)}|\rho_0^{\otimes k}) \to 0$ (see [4] for the details). By the Csiszar–Kullback inequality [39, 57], this implies that $\rho_{0,N}^{(k)}$ converges to $\rho_0^{\otimes k}$ in $L^1(\mathbb{R}^{nk})$ and it also means that the N-particle system (32) (when evaluated at the optimal control $b_N = \frac{1}{2}\nabla \log \rho_{N,0}$) is Kac and entropy chaotic (see, e.g., [53] for the definition of these properties).

Let $\mathbb{P}_{0,N}$ be the probability law of the process $X_{N,t}$, on the path space $C^0([0, T], \mathbb{R}^{nN})$, for the case where $b_N(x) = \frac{1}{2}\nabla \log(\rho_{0,N}(x))$, $x \in \mathbb{R}^{nN}$ is the optimal control and the law of the initial condition $X_{N,0}$ is the invariant (optimal) measure $\rho_{0,N}$. We denote by $\mathbb{P}_{0,N}^{(k)}$ the marginal of $\mathbb{P}_{0,N}$ on the path space of the first k particles $C^0([0, T], \mathbb{R}^{nk})$. Finally, let $\mathbb{P}_0$ be the law of the process X_t solution to Eq. (16), with the optimal control $b(x_1) = \frac{1}{2}\nabla \log(\rho_0(x_1))$, $x_1 \in \mathbb{R}^n$, and starting at the invariant measure ρ_0.

Theorem 5.2 *Under hypotheses $\mathcal{V}$, $C\mathcal{V}$ and $Q\mathcal{V}$ we have that for any $k \in \mathbb{N}$*

$$\lim_{N\uparrow+\infty} \mathcal{H}_{C^0([0,T],\mathbb{R}^{nk})}(\mathbb{P}_{0,N}^{(k)}|\mathbb{P}_0^{\otimes k}) = 0. \tag{37}$$

Proof (Idea of the Proof of Theorem 5.2) One first proves that the "value function" $\frac{1}{N}\mathfrak{I}_N$ of the N-particle system converges, as $N \to +\infty$, to $\mathfrak{I}$, i.e. the "value function" of the limit problem given by (16) and (17). A stronger result holds: the kinetic part of the energy of the N-particle system converges to the kinetic energy of the limit problem, namely

$$\lim_{N\to+\infty} \frac{1}{N} \int_{\mathbb{R}^{nN}} \frac{|\nabla\rho_{0,N}(x_1,\ldots,x_N)|^2}{2\rho_{0,N}(x_1,\ldots,x_N)} dx_1 \ldots dx_N = \int_{\mathbb{R}^n} \frac{|\nabla\rho_0(x_1)|^2}{2\rho_0(x_1)} dx_1,$$

(38)

(see [4, Theorem 5.1]). Furthermore, if $b_N(x) = \frac{1}{2}\nabla\log(\rho_{0,N}(x))$ and $b(x_1) = \frac{1}{2}\nabla\log(\rho_0(x_1))$, by Girsanov's theorem we get that

$$\frac{1}{N}\mathcal{H}_{C^0([0,T],\mathbb{R}^{nN})}(\mathbb{P}_{0,N}|\mathbb{P}_0^{\otimes N}) = \mathbb{E}_{\mathbb{P}_{0,N}}[|b_N^1(X_{N,t}) - b(X_{N,t}^1)|^2],$$

(with the upper index 1 indicating the first $\mathbb{R}^n$ component of the corresponding vector in $\mathbb{R}^{nN}$). Using Eq. (29) we get

$$\frac{1}{N}\mathbb{E}_{\mathbb{P}_{0,N}}[|b_N^1(X_s) - b(X_s^1)|^2] = \int_{\mathbb{R}^{nN}} \frac{|\nabla_1\varphi_{0,N}(x)|^2}{2} dx - \mu_0$$

$$+ \int_{\mathbb{R}^{nN}} 2\left(\mathcal{V}(x^1,\rho_0) - \int_{\mathbb{R}^n} \partial_\mu\mathcal{V}(y,x^1,\rho_0)\rho(y)dy\right)\varphi_{0,N}^2(x)dx,$$

(39)

where $\varphi_{0,N} = \sqrt{\rho_{0,N}}$. Exploiting the explicit formula (30) for μ_0, the convergence of $\frac{1}{N}\mathfrak{I}_N$ to $\mathfrak{I}$, for $N \to +\infty$, and the limit (38) we obtain that $\lim_{N\to+\infty} \frac{1}{N}\mathcal{H}_{C^0([0,T],\mathbb{R}^{nN})}(\mathbb{P}_{0,N}|\mathbb{P}_0^{\otimes N}) = 0$. Finally, by the inequality

$$\mathcal{H}_{C^0([0,T],\mathbb{R}^{nk})}(\mathbb{P}_{0,N}^{(k)}|\mathbb{P}_0^{\otimes k}) \leq \frac{k}{N}\mathcal{H}_{C^0([0,T],\mathbb{R}^{nN})}(\mathbb{P}_{0,N}|\mathbb{P}_0^{\otimes N}), k = 1,\ldots,N,$$

see [4], we get the thesis.   □

6 The Case of the Dirac Delta Potential

In this section we propose to the reader a potential $\mathcal{V}$ of the following form

$$\mathcal{V}_\delta(x,\mu) = V_0(x) + g\delta_x * \mu,$$

(40)

where V_0 is a regular positive function growing at infinity ("trapping potential"), δ_x is the Dirac delta centered at $x \in \mathbb{R}^n$, $g \in \mathbb{R}_+$ is a strictly positive constant, $*$ stands for convolution, and μ is a probability measure. The potential $\mathcal{V}_\delta$ does not satisfy the regularity Hypotheses $\mathcal{V}(i)$ and $\mathcal{V}(iii)$. On the other hand it satisfies Hypothesis $\mathcal{V}(ii)$ and $C\mathcal{V}$, and (whenever the Gâteaux derivative is well defined) we have $\partial_\mu^2(\tilde{\mathcal{V}}_\delta) = 2\delta_{x-y}$, where $\tilde{\mathcal{V}}_\delta = \int_{\mathbb{R}^n} \mathcal{V}(x,\mu)\mu(dx)$, which is a positive definite distribution.

Here we do not consider the problem of proving that the optimal control ergodic problem has a unique optimal control (i.e. we do not prove here the equivalent of Theorem 4.1 for the potential (40)). We suppose that there exists a family $\mathcal{C}_{V_0} \subset C^1(\mathbb{R}^n, \mathbb{R}^n)$ of vector fields b on $\mathbb{R}^n$ (in general we expect that it can depend on the trapping potential V_0 in (40)) such that

$$
\inf_{b \in \mathcal{C}_{V_0}} \left(\limsup_{T \to +\infty} \frac{1}{T} \left(\int_0^T \mathbb{E}_{x_0} \left[\frac{|b(X_t)|^2}{2} + V_0(X_t) + g\rho_{x_0,\alpha,t}(X_t) \right] dt \right) \right)
$$
$$
= \mathbb{E}_{X_0 \sim \rho_0(x)dx} \left[\frac{|\nabla \rho_0(X_t)|^2}{4\rho_0^2(X_t)} + V_0(X_t) + g\rho_0(X_t) \right], \tag{41}
$$

where $\rho_{x_0,b,t}$ is the probability density of the law of the solution to the SDE (16) starting at $x_0 \in \mathbb{R}^n$ evaluated at time t, and ρ_0 is the density of the probability distribution minimizing the functional

$$
\mathcal{E}_\delta(\rho) = \mathcal{E}_K(\rho) + \mathcal{E}_{\delta,P}(\rho) = \int_{\mathbb{R}^n} \left(\frac{|\nabla \rho(x)|^2}{4\rho(x)} + V_0(x)\rho(x) + g(\rho(x))^2 \right) dx. \tag{42}
$$

In other words we suppose that in the set $\mathcal{C}_\rho$ (introduced in (26)) the optimal control for the problem (16) with cost functional (17) and potential $\mathcal{V}_\delta$ (see [16] for an alternative derivation of a stochastic process associated with the above cost functional) exists and it is given by $b = \frac{\nabla \rho_0}{2\rho_0}$. What we want to consider here is an N-particle problem converging to the solution of the optimal control ergodic problem just described (namely we are looking for an analogous of Theorem 5.2 for the case where $\mathcal{V}$ is given by $\mathcal{V}_\delta$ in (40)).

In general, since $\mathcal{V}_\delta$ is not well-defined for positive measures μ that are not absolutely continuous measures, let us then consider an approximating potential of the form

$$
\mathcal{V}_{\delta,N}(x,\mu) = V_0(x) + \int_{\mathbb{R}^n} v_N(x-y)\mu(dy),
$$

where $v_N : \mathbb{R}^n \to \mathbb{R}$ is a sequence of positive functions converging in the sense of distributions to a Dirac delta δ_0 when $N \to \infty$. Let us choose a specific sequence of the following form

$$
v_N(\mathbf{r}) = \frac{N^{3\beta}}{N-1} v_0(N^\beta \mathbf{r}), \qquad x \in \mathbb{R}^n \tag{43}
$$

for $\beta > 0$, where v_0 is a positive smooth radially symmetric function with compact support (as in formula (6)). We take the N-particle approximation having the control $b_N(x_1, \ldots, x_N)$ given by the logarithmic derivative of $\rho_{0,N}$, that is the minimal probability density of the energy functional $\mathcal{E}_\delta$ associated with $\mathcal{V}_{\delta,N}$, namely

$$\mathcal{E}_{\delta,N}(\rho) = \mathcal{E}_{K,N}(\rho) + \mathcal{E}_{\delta,P,N}(\rho)$$

$$= \frac{1}{N} \sum_{i=1}^{N} \left(\int_{\mathbb{R}^{Nn}} \left(\frac{|\nabla_i \rho|^2}{4\rho} + V_0(x_i)\rho \right) dx \right.$$

$$\left. + \frac{1}{N-1} \sum_{j=1,\ldots,N, j \neq i} \int_{\mathbb{R}^{Nn}} v_N(x_i - x_j)\rho dx \right).$$

In the rest of the paper we show how the results on Bose–Einstein condensation (mainly for $n = 3$, see, e.g., [61–65, 69, 75, 82]) can be used to study the convergence of the N-particles approximation of the control problem with potential (40). For this reason hereafter we shall limit our discussion to the case $n = 3$.

6.1   The Intermediate Scaling Limit

The case $0 < \beta < 1$, where β is the parameter used in the rescaling (43), which is known as intermediate scaling limit, is very similar to the regular case ($\beta = 0$) that we discussed in Sect. 5. Indeed, in this case we can prove the following theorem.

Theorem 6.1 *Under the previous hypotheses and notations, if $0 < \beta < 1$ we have, as $N \to +\infty$, the convergence statements $\mathcal{E}_{\delta,N}(\rho_{0,N}) \to \mathcal{E}_\delta(\rho_0)$, $\mathcal{E}_{\delta,P,N}(\rho_{0,N}) \to \mathcal{E}_{\delta,P}(\rho_0)$ and $\rho_{0,N}^{(1)} \to \rho_0$ (where the last convergence is in the weak L^1 sense) with the constant $g = \int_{\mathbb{R}^3} v_0(x) dx$ (where $g \in \mathbb{R}_+$ is the constant appearing in Eqs. (40) and (41)).*

Proof The proof of the theorem can be found in [62] for $0 \leq \beta < \frac{1}{3}$ (for any n and a more general class of potentials v_0 than the one considered here) and in [5] for $0 \leq \beta < 1$ (for $n = 3$ and a positive-definite interaction potential v_0). See also [83]. □

Theorem 6.1 is the analogue of the results mentioned in the proof of Theorem 5.2 in this context. Thanks to Theorem 6.1 we can repeat the reasoning performed in the proof of Theorem 5.2, obtaining:

Theorem 6.2 *Under the previous hypotheses and notations, if $0 < \beta < 1$ we have that the law $\mathbb{P}_{0,N}^{(k)}$ of the first k particles satisfying the system (32), with $\mathcal{V}$ replaced* *by $\mathcal{V}_\delta$, converges in total variation on the path space $C^0([0, T], \mathbb{R}^{3k})$ to $\mathbb{P}_0^{\otimes k}$ (where $\mathbb{P}_0$ is the law on $C^0([0, T], \mathbb{R}^3)$ of the system (16) associated with (40)).*

Proof The proof can be found in [5].   □

6.2 The Gross–Pitaevskii Scaling Limit

The case $\beta = 1$ is completely different from the previous ones. The main difference between the cases $0 < \beta < 1$ and $\beta = 1$ is that in this latter case the value function convergence result of Theorem 6.1 does not hold.

Theorem 6.3 *Under the previous hypotheses and notations, if $\beta = 1$ we have that, as $N \to \infty \, \mathcal{E}_{\delta,N}(\rho_{0,N}) \to \mathcal{E}_\delta(\rho_0)$ and $\rho_{0,N}^{(1)} \to \rho_0$ (where the latter convergence is in the weak sense in L^1) for $g = 4\pi a$ (where $g \in \mathbb{R}_+$ is the constant appearing in Eqs. (40) and (41), and $a > 0$ is the scattering length of the interaction potential v_0 (see [64])). Furthermore, putting $\hat{s} = \frac{1}{g} \int_{\mathbb{R}^3} \frac{|\nabla \rho_0|^2}{\rho_0} dx \in (0, 1)$ we have, as $N \to +\infty$:*

$$\mathcal{E}_{K,N}(\rho_{0,N}) \to \mathcal{E}_{\delta,K}(\rho_0) + g\hat{s} \int_{\mathbb{R}^3} \rho_0^2(x)dx.$$

Proof The proof of the first part of the theorem is a well-known result proven in [63, 65, 75]. The second part is proved in [65].   □

In this case we cannot repeat the reasoning of Theorem 5.2 since we are not able to prove that the relative entropy $\mathcal{H}(\mathbb{P}_{0,N}^{(k)} | \mathbb{P}_0^{\otimes k})$ converges to 0 (in fact we do not know whether the relative entropy converges to 0 or to another value). On the other hand it is possible to prove a weaker result for $\beta = 1$, we have namely:

Theorem 6.4 *Under the previous hypotheses and notations, if $\beta = 1$ we have that the law $\mathbb{P}_{0,N}^{(k)}$ converges weakly on the path space $C^0([0, T], \mathbb{R}^{3k})$ to $\mathbb{P}_0^{\otimes k}$.*

Proof The proof can be found in [6].   □

Remark 6.1 In [72] a different kind of convergence is proven and in [86] a transition to chaos result for the particle system related to the control problem is obtained.

7 Future Research Lines

We plan to extend our stochastic approach to BEC in three different directions. First it would be interesting to face the more difficult Gross–Pitaevskii scaling limit ($\beta = 1$) with a similar optimal control approach. Since this scaling limit gives rise to a singular action functional we could try to extend stochastic mechanics to this non-linear singular Schrödinger model on one hand and obtain the solution to a singular optimal control problem. Since our ergodic control problem can be looked upon as being of McKean–Vlasov type, both the drift of the SDE and the potential depending on the probability density of the invariant measure, that is, on the law of the stochastic process, we hope to be able to prove the complete BEC (in the sense of Definition 2.1) and its justification by taking advantage of the advanced stochastic techniques developed in connection with the well-studied McKean–Vlasov optimal control problem (see, e.g., [21, 33, 58], and also [19, 85]). Finally, a big effort would be needed to extend our stochastic approach to the general time-dependent setting. This is not a direct consequence of the ground state case, even in the mathematical physics approach. Indeed, the proof of BEC with a time-dependent wave function requires different techniques (see, e.g., [1, 20, 27, 28, 43]).

Let us close with a look back to the basic problems underlying the study in this article, in order to insert them into a "future research line" prospective. Since the Enlightenment, much of the development of mathematics has been influenced by problems that arose in connection with the investigation of nature, in particular physical phenomena. Especially in the last century and into the present one, the description and interpretations of quantum phenomena have played an important role (for instance, in relation to classical deterministic and stochastic dynamical systems, among other examples). For the description of quantum phenomena ideas and methods coming from the theory of infinite-dimensional spaces, and operators on them, play a central role on the "abstract level" (accompanied by a more "concrete level", like the Schrödinger equation of non-relativistic quantum mechanics). But this is certainly not the whole story, as is seen from the early steps of quantum mechanics itself, where other areas of mathematics entered and got enriched in one way or the other, e.g., the representation theory of Lie groups (to express transformation properties of observables and conservation laws), see, e.g., [66]. Variational principles also played a founding role, inasmuch as quantum mechanics can be seen as a deformation of classical mechanics, and reciprocally (see, e.g., [89]), and the most important variational principles have originated in classical mechanics (see, e.g., [2, 22] and references therein). These influences are certainly present in the genesis of R. Feynman's reformulation of quantum mechanics in terms of the famous heuristic "Feynman path integral", that became quite important both in physics and mathematics inspired by physics. We recall that Feynman's original approach consisted in describing the quantum mechanical evolution by an "integral kernel" of the form $e^{\frac{i}{\hbar} S(\gamma)}$ (i being the imaginary unit, $\hbar$ Plank's constant and $S(\gamma)$ the action functional, i.e., a time integral of a Lagrangian, for a path γ in a space of continuous paths). A heuristic variational principle would then permit us to get classical orbits from integrals involving $e^{\frac{i}{\hbar} S(\gamma)}$ expressing, for example, the solutions of Schrödinger equation, in the "semiclassical limit" (where $\hbar$ is considered to be very small). This general programme has found some mathematical realization in single cases, e.g. non-relativistic quantum mechanics in flat space (see, e.g., [11, 15, 68]) and also in some more geometric settings and for quantum fields (see, e.g, [10, 60]). But the mathematical and physical potentiality is much richer, even at the non-relativistic level, see the deep discussions of this issue in the work of J.-C. Zambrini (e.g. [88, 89]). Another aspect of mathematics developed in connection with quantum mechanics is stochastic analysis. It has its origins in the 1923 work of N. Wiener on Brownian motion, where the heat semigroup kernel plays a role similar to the above Feynman kernel. It yields the solutions of the heat equation, the parabolic analogue of the Schrödinger equation (but Wiener himself was also interested in quantum mechanics, see [67]). The "Wiener path integral" and its transformations play an important role in stochastic differential equations, invented by mathematicians like S. Bernstein (1932) and K. Itô (1948). It is emblematic that the same K. Itô who founded the probabilistic Itô calculus also gave a first approach to the Feynman path integral ([54, 55]; for further developments, see, e.g., the references [10, 11, 68] cited above). As we mentioned in Sect. 3, Nelson's stochastic mechanics is a probabilistic approach to quantum mechanics; Euclidean methods in quantum field theory (also strongly influenced by Nelson as a tool for the construction of relativistic, hence hyperbolic, thus Feynman's type) quantum fields are also ways to bring together Feynman's methods and probabilistic methods; there is a further approach, put forward by J.-C. Zambrini, and called by him "Euclidean stochastic mechanics", that exports to the world of probability structures that are somewhat hidden inside Feynman's hyperbolic formalism. Zambrini actually took much inspiration from Schrödinger's work, based on a time symmetric view of the heat equation (rather than Schrödinger's equation, see, e.g., [89]). Our point is to observe that there is an immense amount of work to be done in mathematics to better understand all these interwoven and fascinating structures.

The relations between Nelson's stochastic mechanics and variational principles in the study of certain quantum mechanical problems ([51, 87] mentioned in Sect. 3) have also generated a lot of interest in the study of the "Schrödinger probabilistic problem". Here a lot of activity has recently been developed, e.g., in [17, 35, 46, 59, 74, 88]. In this line there are also connections with probabilistic and analytic works on optimal transport (see, e.g., [36, 59, 70]) that deserve much further attention. The world of systems of many quantum particles, and their limits (see [19, 85]), proper of quantum statistical mechanics, is another area of application where such methods should be very useful. This also would imply applications to other areas of science like biology, mathematical finance and game theory (see for instance [33, 34, 41]).

A final comment concerns the developments of similar constructions and connections in the world of fields (and strings) instead of particles. This can be seen as an infinite-dimensional extension of the work we just discussed in relation to non-relativistic quantum theory, as the path γ at a fixed time t in the above Feynman approach, instead of taking values in a finite-dimensional space, would take values in infinite-dimensional spaces. Here the constraints of invariance with respect to transformation groups, imposed by the Poincaré invariance of relativistic quantum fields, causes in addition worse local singularities for the paths (much stronger than the non-differentiability of the Wiener paths), and forces regularizations and limits much more involved and challenging than those involved in the non-relativistic world. The Euclidean methods of constructive quantum field theory based on a construction at imaginary time ("heat equation world") followed by an analytic continuation of the relevant correlation functions to real time ("Schrödinger equation world") have been useful for the construction and the study of models in space-time dimensions up to 3 (see, e.g., [48, 77, 84], for similar methods for path integrals in quantum statistical mechanics see, e.g., [12]). Recently new constructive methods based on a singular stochastic partial differential equation (of the Parisi–Wu stochastic quantization type) initiated by Hairer and Gubinelli–Imkeller–Perkowski have been developed, see for instance [14, 49, 50, 52]. Also here there are relations to variational principles [18] and elliptic methods [3]. However a fully fledged transposition to the field case of the methods related to stochastic mechanics and the corresponding variational methods has still to be elaborated.

Another aspect that might be useful to examine more closely is that in a certain limit (like the non-relativistic one starting from relativistic models with polynomial interactions) the Hilbert space becomes a direct sum of spaces with a fixed number N of particles with point interactions, similar to the non-relativistic models used in Sects. 2–6 for deriving asymptotically, for $N \to +\infty$, the Gross–Pitaevskii equation. The N-scaling used there for $\beta = 1$ is a prototype of the quantum field renormalization procedures and has been used to give a meaning to the Hamiltonian (1) in the case where v_0 is a Dirac delta distribution, see references [7, 9, 42, 47] (where interesting connections with the Efimov and Thomas atomic physics effects are discussed). It also gives a meaning to the non-relativistic limit of the mentioned quantum theoretical models at least in space time dimension 2 (see the references in [7]). Similar methods might also be helpful in dimension 3 and 4, but more work has definitely to be done.

In conclusion, we mentioned some open problems involving processes with finite-dimensional resp. infinite-dimensional state space, and both using infinite-dimensional analysis: there is indeed plenty of room for new developments out there, and many more beautiful flowers to be found!

Acknowledgements The first named author thanks the DFG for the financial support via the grant AL 214/50-1 "Invariant measures for SPDEs and Asymptotics". The first and second named authors are also funded by the DFG under Germany's Excellence Strategy—GZ 2047/1, project-id 390685813.

References

1. R. Adami, F. Golse and A. Teta. Rigorous derivation of the cubic NLS in dimension one. *Journal of Statistical Physics* **127**, no. 6, 1193–1220 (2007).
2. S. Albeverio, A.B. Cruzeiro and D. Holm, editors. *Stochastic geometric mechanics*. Volume 202 of Springer Proceedings in Mathematics & Statistics. Papers from the Research Semester "Geometric Mechanics—Variational and Stochastic Methods" held at the Centre Interfacultaire Bernoulli (CIB), Ecole Polytechnique Fédérale de Lausanne, Lausanne, January–June, 2015. Springer, Cham (2017).

3. S. Albeverio, F. De Vecchi and M. Gubinelli. The elliptic stochastic quantization of some two dimensional Euclidean QFTs. In: *Annales de l'Institut Henri Poincaré, Probabilités et Statistiques*, volume 57, pp. 2372–2414, Institut Henri Poincaré (2021).

4. S. Albeverio, F. De Vecchi, A. Romano and S. Ugolini. Mean-field limit for a class of stochastic ergodic control problems. *SIAM J. Control Optim.* **60**(1), 479–504 (2022).

5. S. Albeverio, F.C. De Vecchi, A. Romano and S. Ugolini. Strong Kac's chaos in the mean-field Bose–Einstein condensation. *Stoch. Dyn.* **20**(5):2050031, 21 (2020).

6. S. Albeverio, F.C. De Vecchi and S. Ugolini. Entropy chaos and Bose–Einstein condensation. *J. Stat. Phys.* **168**(3), 483–507 (2017).

7. S. Albeverio and R. Figari. Quantum fields and point interactions. *Rend. Mat. Appl. (7)* **39**(2), 161–180 (2018).

8. S. Albeverio, H. Föllmer, L. Gross and E. Nelson. *Mathematical physics at Saint-Flour. Probability at Saint-Flour.* Reprints of selected lectures from the Summer Schools. Springer, Heidelberg (2012).

9. S. Albeverio, F. Gesztesy, R. Hoegh-Krohn and H. Holden. *Solvable models in quantum mechanics.* Springer Science & Business Media (2012).

10. S. Albeverio, A. Hahn and A.N. Sengupta. Rigorous Feynman path integrals, with applications to quantum theory, gauge fields, and topological invariants. In: *Stochastic analysis and mathematical physics (SAMP/ANESTOC 2002)*, pp. 1–60, World Sci. Publ., River Edge, NJ (2004).

11. S. Albeverio, R. Høegh-Krohn and S. Mazzucchi. *Mathematical theory of Feynman path integrals. An introduction.* Volume 523 of Lecture Notes in Mathematics. Springer-Verlag, Berlin, second edition (2008).

12. S. Albeverio, Y. Kondratiev, Y. Kozitsky and M. Roeckner, The statistical mechanics of quantum lattice systems. A path integral approach, *Eur. Math. Soc.* (2009)

13. S. Albeverio and P. Kurasov. *Singular perturbations of differential operators. Solvable Schrödinger type operators.* Volume 271 of London Mathematical Society Lecture Note Series. Cambridge University Press, Cambridge (2000).

14. S. Albeverio and S. Kusuoka. The invariant measure and the flow associated to the Φ_3^4-quantum field model. *Ann. Sc. Norm. Super. Pisa Cl. Sci. (5)*, **20**(4), 1359–1427 (2020).

15. S. Albeverio and S. Mazzucchi. Path integral: mathematical aspects. *Scholarpedia* **6**(1), 8832 (2011).

16. S. Albeverio and S. Ugolini. A Doob h-transform of the Gross–Pitaevskii Hamiltonian. *J. Stat. Phys.* **161**(2), 486–508 (2015).

17. J. Backhoff, G. Conforti, I. Gentil and C. Léonard. The mean field Schrödinger problem: ergodic behavior, entropy estimates and functional inequalities. *Probab. Theory Related Fields* **178**(1–2), 475–530 (2020).

18. N. Barashkov and M. Gubinelli. A variational method for Φ_3^4. *Duke Math. J.* **169**(17), 3339–3415 (2020).

19. C. Bardos, L. Erdős, F. Golse, N. Mauser and H.-T. Yau. Derivation of the Schrödinger–Poisson equation from the quantum N-body problem. *C. R. Math. Acad. Sci. Paris* **334**(6), 515–520 (2002).

20. C. Bardos, F. Golse and N. Mauser. Weak coupling limit of the N-particle Schrödinger equation. *Methods and Applications of Analysis* **7**(2), 275–294 (2000).

21. E. Bayraktar, A. Cosso, and H. Pham. Randomized dynamic programming principle and Feynman–Kac representation for optimal control of McKean–Vlasov dynamics. *Trans. Amer. Math. Soc.* **370**(3), 2115–2160 (2018).

22. P. Blanchard and E. Brüning. *Mathematical methods in physics. Distributions, Hilbert space operators, variational methods, and applications in quantum physics.* Volume 69 of Progress in Mathematical Physics. Birkhäuser/Springer, Cham, second edition (2015).

23. P. Blanchard, P. Combe and W. Zheng. *Mathematical and physical aspects of stochastic mechanics.* Lecture Notes in Physics Volume 281. Springer (1987).

24. C. Boccato, C. Brennecke, S. Cenatiempo and B. Schlein. Complete Bose–Einstein condensation in the Gross–Pitaevskii regime. *Comm. Math. Phys.* **359**(3), 975–1026 (2018).

25. V.I. Bogachev, N.V. Krylov, M. Röckner and S.V. Shaposhnikov. *Fokker–Planck–Kolmogorov equations.* Volume 207 of Mathematical Surveys and Monographs. American Mathematical Society, Providence, RI (2015).

26. C. Brennecke, P. Thành Nam, M. Napiórkowski and B. Schlein. Fluctuations of N-particle quantum dynamics around the nonlinear Schrödinger equation. In: *Annales de l'Institut Henri Poincaré (C) Analyse Non Linéaire*, volume 36, pp. 1201–1235, Elsevier (2019).

27. C. Brennecke and B. Schlein. Gross–Pitaevskii dynamics for Bose–Einstein condensates. *Anal. PDE* **12**(6), 1513–1596 (2019).

28. M. Caporaletti, A. Deuchert and B. Schlein. Dynamics of mean-field bosons at positive temperature. *arXiv preprint* arXiv:2203.17204 (2022).

29. E. Carlen. Stochastic mechanics: a look back and a look ahead. In: *Diffusion, quantum theory, and radically elementary mathematics*, volume 47 of Math. Notes, pp. 117–139, Princeton Univ. Press, Princeton, NJ (2006).

30. E.A. Carlen. Conservative diffusions. *Comm. Math. Phys.* **94**(3), 293–315 (1984).

31. E.A. Carlen. Existence and sample path properties of the diffusions in Nelson's stochastic mechanics. In: *Stochastic processes–mathematics and physics (Bielefeld, 1984)*, volume 1158 of Lecture Notes in Math., pp. 25–51, Springer, Berlin (1986).

32. E.A. Carlen. Progress and problems in stochastic mechanics. In: *Stochastic methods in mathematics and physics (Karpacz, 1988)*, pp. 3–31, World Sci. Publ., Teaneck, NJ (1989).

33. R. Carmona and F. Delarue. *Probabilistic theory of mean field games with applications. I. Mean field FBSDEs, control, and games.* Volume 83 of Probability Theory and Stochastic Modelling. Springer, Cham (2018).

34. R. Carmona and F. Delarue. *Probabilistic theory of mean field games with applications. II. Mean field games with common noise and master equations.* Volume 84 of Probability Theory and Stochastic Modelling. Springer, Cham (2018).

35. P. Cattiaux, G. Conforti, I. Gentil and C. Léonard. Time reversal of diffusion processes under a finite entropy condition. *arXiv preprint* arXiv:2104.07708 (2021).

36. Y. Chen, T. Georgiou and M. Pavon. On the relation between optimal transport and Schrödinger bridges: a stochastic control viewpoint. *J. Optim. Theory Appl.* **169**(2), 671–691 (2016).

37. E.A. Cornell and C.E. Wieman. Bose–Einstein condensation in a dilute gas: the first 70 years and some recent experiments (Nobel lecture). *Chemphyschem* **3**(6), 476–493 (2002).

38. A.B. Cruzeiro, L. Wu, and J.C. Zambrini. Bernstein processes associated with a Markov process. In: *Stochastic Analysis and Mathematical Physics*, pp. 41–72, Springer (2000).

39. I. Csiszár. Information-type measures of difference of probability distributions and indirect observations. *Studia Sci. Math. Hungar.* **2**, 299–318 (1967).

40. N. Cufaro Petroni and L. Morato. Entangled states in stochastic mechanics. *J. Phys. A* **33**(33), 5833–5848 (2000).

41. P. Dai Pra. Stochastic mean-field dynamics and applications to life sciences. In: *Stochastic dynamics out of equilibrium*, volume 282 of Springer Proc. Math. Stat., pp. 3–27, Springer, Cham (2019).

42. G.F. Dell'Antonio, R. Figari and A. Teta. Hamiltonians for systems of N particles interacting through point interactions. *Ann. Inst. H. Poincaré Phys. Théor.* **60**(3), 253–290 (1994).

43. L. Erdős, B. Schlein and H.-T. Yau. Rigorous derivation of the Gross–Pitaevskii equation. *Physical review letters* **98**(4), 040404 (2007).

44. G. Fabbri, F. Gozzi and A. Swiech. *Stochastic optimal control in infinite dimension. Dynamic programming and HJB equations.* With a contribution by Marco Fuhrman and Gianmario Tessitore. Volume 82 of Probability Theory and Stochastic Modelling, Springer, Cham (2017).

45. W.H. Fleming and H.M. Soner. Controlled Markov processes and viscosity solutions. Stochastic Modelling and Applied Probability, Volume 25. Springer Science & Business Media (2006).

46. H. Föllmer. Random fields and diffusion processes. In: *École d' Été de Probabilités de Saint-Flour XV–XVII, 1985–87*, pp. 101–203, Springer, Berlin (1988).

47. M. Gallone and A. Michelangeli. Self-adjoint extension schemes and modern applications to quantum Hamiltonians. *arXiv preprint* arXiv:2201.10205 (2022).

48. J. Glimm and A. Jaffe. *Quantum physics. A functional integral point of view.* Springer-Verlag, New York-Berlin (1981).
49. M. Gubinelli and M. Hofmanová. A PDE construction of the Euclidean φ_3^4 quantum field theory. *Comm. Math. Phys.* **384**(1), 1–75 (2021).
50. M. Gubinelli, P. Imkeller and N. Perkowski. Paracontrolled distributions and singular PDEs. *Forum Math. Pi* 3:e6, 75 (2015).
51. F. Guerra and L.M. Morato. Quantization of dynamical systems and stochastic control theory. *Phys. Rev. D (3)* **27**(8), 1774–1786 (1983).
52. M. Hairer. A theory of regularity structures. *Invent. Math.* **198**(2), 269–504 (2014).
53. M. Hauray and S. Mischler. On Kac's chaos and related problems. *J. Funct. Anal.* **266**(10), 6055–6157 (2014).
54. K. Itô. Wiener integral and Feynman integral. In: *Proceedings of the fourth Berkeley symposium on mathematical statistics and probability*, volume 4, pp. 227–238, Univ. of California Press (1961).
55. K. Itô. Generalized uniform complex measures in the Hilbertian metric space with their application to the Feynman integral. In: *Proc. Fifth Berkeley Sympos. Math. Statist. and Probability (Berkeley, Calif., 1965/66), Vol. II: Contributions to Probability Theory, Part 1*, pp. 145–161, Univ. California Press, Berkeley, Calif. (1967).
56. W. Ketterle and N.J. Van Druten. Evaporative cooling of trapped atoms. In: *Advances in atomic, molecular, and optical physics*, volume 37, pages 181–236, Elsevier (1996).
57. S. Kullback. A lower bound for discrimination information in terms of variation (Corresp.). *IEEE transactions on Information Theory* **13**(1), 126–127 (1967).
58. D. Lacker. Limit theory for controlled McKean–Vlasov dynamics. *SIAM J. Control Optim.* **55**(3), 1641–1672 (2017).
59. C. Léonard. A survey of the Schrödinger problem and some of its connections with optimal transport. *Discrete Contin. Dyn. Syst.* **34**(4), 1533–1574 (2014).
60. T. Lévy and A. Sengupta. Four chapters on low-dimensional gauge theories. In: *Stochastic geometric mechanics*, volume 202 of Springer Proc. Math. Stat., pp. 115–167, Springer, Cham (2017).
61. M. Lewin. Mean-field limit of Bose systems: rigorous results. In: Proceedings of the international Congress Assoc. Math. Phys. *arXiv preprint* arXiv:1510.04407 (2015).
62. M. Lewin, P. Nam and N. Rougerie. The mean-field approximation and the non-linear Schrödinger functional for trapped Bose gases. *Transactions of the American Mathematical Society* **368**(9), 6131–6157 (2016).
63. E.H. Lieb and R. Seiringer. Proof of Bose–Einstein condensation for dilute trapped gases. *Physical review letters* **88**(17), 170409 (2002).
64. E.H. Lieb, R. Seiringer, J.P. Solovej and J. Yngvason. *The mathematics of the Bose gas and its condensation*, volume 34 of Oberwolfach Seminars. Birkhäuser Verlag, Basel (2005).
65. E.H. Lieb, R. Seiringer and J. Yngvason. Bosons in a trap: A rigorous derivation of the Gross–Pitaevskii energy functional. In: *The Stability of Matter: From Atoms to Stars*, pp. 685–697, Springer (2001).
66. G. Mackey. *Mathematical foundations of quantum mechanics.* With a foreword by A.S. Wightman. Reprint of the 1963 original. Dover Publications, Inc., Mineola, NY (2004).
67. P.R. Masani. Norbert Wiener and the future of cybernetics. In: *Proceedings of the Norbert Wiener Centenary Congress, 1994 (East Lansing, MI, 1994)*, volume 52 of Proc. Sympos. Appl. Math., pp. 473–503, Amer. Math. Soc., Providence, RI (1997).
68. S. Mazzucchi. *Mathematical Feynman path integrals and their applications.* World Scientific Publishing Co. Pte. Ltd., Hackensack, NJ (2009).
69. A. Michelangeli. *Bose–Einstein condensation: analysis of problems and rigorous results.* PhD thesis, SISSA (2007).
70. T. Mikami. *Stochastic optimal transportation: stochastic control with fixed marginals.* Springer Nature (2021).
71. A. Millet, D. Nualart and M. Sanz. Integration by parts and time reversal for diffusion processes. *Ann. Probab.* **17**(1), 208–238 (1989).

72. L.M. Morato and S. Ugolini. Stochastic description of a Bose–Einstein condensate. *Ann. Henri Poincaré* **12**(8), 1601–1612 (2011).
73. L.M. Morato and S. Ugolini. Localization of relative entropy in Bose–Einstein condensation of trapped interacting bosons. In: *Seminar on Stochastic Analysis, Random Fields and Applications VII*, volume 67 of Progr. Probab., pp. 197–210, Birkhäuser/Springer, Basel (2013).
74. M. Nagasawa. Stochastic variational principle of Schrödinger processes. In: *Seminar on Stochastic Processes, 1989 (San Diego, CA, 1989)*, volume 18 of Progr. Probab., pp. 165–175, Birkhäuser Boston, Boston, MA (1990).
75. P.T. Nam, N. Rougerie and R. Seiringer. Ground states of large bosonic systems: the Gross–Pitaevskii limit revisited. Analysis & PDE **9**(2), 459–485 (2016).
76. E. Nelson. *Dynamical theories of Brownian motion*, volume 3. Princeton university press (1967).
77. E. Nelson. The free Markoff field. *J. Functional Analysis* **12**, 211–227 (1973).
78. E. Nelson. *Quantum fluctuations*. Princeton University Press (1985).
79. E. Nelson. Stochastic mechanics and random fields. In: *École d' Été de Probabilités de Saint-Flour XV–XVII, 1985–87*, volume 1362 of Lecture Notes in Math., pp. 427–450, Springer, Berlin (1988).
80. L. Pitaevskii and S. Stringari. *Bose–Einstein condensation and superfluidity*. International series of monographs on physics, volume 164. Oxford University Press, 2016.
81. M. Reed and B. Simon. *Methods of modern mathematical physics. IV. Analysis of operators.* Academic Press [Harcourt Brace Jovanovich, Publishers], New York-London (1978).
82. N. Rougerie. De Finetti theorems, mean-field limits and Bose–Einstein condensation. LMV Lecture Notes. *arXiv preprint* arXiv:1506.05263 (2015).
83. N. Rougerie. Scaling limits of bosonic ground states, from many-body to non-linear Schrödinger. *EMS Surveys in Mathematical Sciences* **7**(2), 253–408 (2021).
84. B. Simon. *The $P(\varphi)_2$ Euclidean (quantum) field theory*. Princeton Series in Physics. Princeton University Press, Princeton, N.J. (1974).
85. H. Spohn. Kinetic equations from Hamiltonian dynamics: Markovian limits. *Rev. Modern Phys.* **52**(3), 569–615 (1980).
86. S. Ugolini. Bose–Einstein condensation: a transition to chaos result. *Commun. Stoch. Anal.* **6**(4), 565–587 (2012).
87. K. Yasue. Stochastic calculus of variations. *J. Functional Analysis* **41**(3), 327–340 (1981).
88. J.-C. Zambrini. Variational processes and stochastic versions of mechanics. *J. Math. Phys.* **27**(9), 2307–2330 (1986).
89. J.-C. Zambrini. The research program of stochastic deformation (with a view toward geometric mechanics). In: *Stochastic analysis: a series of lectures*, volume 68 of Progr. Probab., pp. 359–393, Birkhäuser/Springer, Basel (2015).

Part XI
Number Theory

The article *Can we dream of a 1-adic Langlands correspondence?* by Xavier Caruso, Agnès David and Ariane Mézard presents arguments for the existence of a 1-adic Langlands correspondence which would, among other things, explain uniformity with respect to p of results in the p-adic Langlands correspondence. A link with the field with one element is mentioned.

True to its title, Henri Cohen's article *Computational number theory, past, present, and future* first provides a comprehensive history of the subject. This is followed by a very complete description of the state of the art and the results obtained, particularly in the topics of Algebraic Number Fields, L-Functions and Automorphic Forms, and Numerical Methods. The paper ends with problems and suggestions for the future.

Michel Waldschmidt's article *The four exponentials problem and Schanuel's conjecture* is an in-depth and unifying presentation of a set of problems and conjectures related to the algebraic independence of values of the exponential function or of values of the logarithm function, as well as problems of a similar nature.

Can We Dream of a 1-Adic Langlands Correspondence?

Xavier Caruso, Agnès David, and Ariane Mézard

To Catriona Byrne

The Langlands programme is a far-reaching and influential web of theorems and conjectures which has motivated a lot of research in Number Theory and Arithmetic Geometry for more than 50 years. Very roughly, it stipulates a profound and meaningful correspondence between representations of Galois groups on the one hand and representations of reductive groups on the other hand. Many variations on this theme are actually possible, depending on which base field (number field, function field, p-adic field, etc.) and which category of coefficients we are working with.

The pioneering works in the Langlands programme were mostly concerned with $\mathbb{C}$-valued representations. However, since the beginning of the twenty-first century, a purely p-adic version of Langlands correspondence has emerged under the impulsion of Breuil. Nowadays, this p-adic correspondence is fully established for 2-dimensional representations of $\mathrm{Gal}(\overline{\mathbb{Q}}_p/\mathbb{Q}_p)$ but, beyond this, little is known. Many examples have however been worked out and several conjectures have been proposed—and sometimes proved—throughout the years. One of them is the Breuil–Mézard conjecture, which predicts that the geometrical properties of some Galois deformation spaces are directly related to the decomposition properties of some representations of the corresponding reductive group.

X. Caruso
CNRS; IMB, Université de Bordeaux, Talence, France
e-mail: xavier.caruso@normalesup.org

A. David
LMB, Université de Franche-Comté, Besançon, France
e-mail: David.Agnes@math.cnrs.fr

A. Mézard
DMA, École Normale Supérieure PSL, Paris, France
e-mail: Ariane.Mezard@ens.fr

© The Author(s), under exclusive license to Springer Nature Switzerland AG 2023
J.-M. Morel, B. Teissier (eds.), *Mathematics Going Forward*, Lecture Notes in Mathematics 2313, https://doi.org/10.1007/978-3-031-12244-6_37

Looking more carefully into the aforementioned works, we notice that, in many cases, the underlying prime number p often plays a figurative role in the calculations. Typically, the relevant reductive groups are usually defined over $\mathbb{Z}$ and a significant part of the constructions and arguments can be carried out at this level. On the Galois side, this constancy is not so obvious but it is nevertheless visible; indeed, even though a prime number needs to be fixed from the very beginning, we often observe, at the end of the day, that the results of the computations are mostly independent from it.

We make the hypothesis that these strong properties of uniformity with respect to p could have a deep meaning and all be the consequences of a new type of Langlands correspondence, which should be considered as the common denominator of the p-adic Langlands correspondences when p varies. We call this new hypothetic correspondence the 1-*adic Langlands correspondence* because we believe that the natural language to formulate it is the mysterious theory of characteristic one whose main protagonist is the famous field with one element.

The aim of this note is to bring the reader to the agreement that our hypothesis is not crazy but has conceivable foundations and deserves consideration. We start our argumentation by reviewing in Sect. 1 some recent developments towards the Breuil–Mézard conjecture, with the objective to highlight the places where the arguments and/or the notions take a combinatorial flavour in which we have the feeling that the underlying prime number p plays a secondary role. Then, in Sect. 2, we briefly recall the philosophy of the field with one element and show that it is incredibly appropriate for interpreting many objects and carrying out many constructions encountered in Sect. 1. Finally, in order to give more substance to our dream, we conclude this article with an appendix in which we share some thoughts towards the development of a Galois theory in characteristic one, which is certainly a prerequisite for a 1-adic Langlands correspondence.

Grant–The three authors are supported by the ANR project clap-clap (ANR-18-CE40-0026-01)

1   Combinatorics Around the Breuil–Mézard Conjecture

Let $p > 2$ be a prime number. Throughout this section, we fix a finite extension K of $\mathbb{Q}_p$ and write $G_K = \mathrm{Gal}(\overline{\mathbb{Q}}_p/K)$ for its absolute Galois group. The Breuil–Mézard conjecture is a concrete statement relating deformation spaces of representations of G_K, on the one hand, and representations of p-adic reductive groups, on the other hand. The aim of this section is, firstly, to recall the formulation of this conjecture and, secondly, to emphasize that, in many cases, it can be approached using combinatorial arguments and constructions. These observations will be the key to build bridges with the field with one element in Sect. 2.

1.1   Review on the Breuil–Mézard Conjecture

We denote by $\mathcal{O}_K$ (resp. k_K) the ring of integers (resp. the residue field) of K. Let $\overline{\rho} : G_K \to \mathrm{GL}_n(\overline{\mathbb{F}}_p)$ be a continuous $\overline{\mathbb{F}}_p$-representation of G_K of dimension n and let $R_{\overline{\rho}}$ denote the $\mathbb{Z}_p$-algebra parametrizing the deformations of $\overline{\rho}$. In [31], Kisin proved that $R_{\overline{\rho}}$ admits quotients with strong arithmetical interest. More precisely, given in addition the two following data:

- a *Hodge type* λ, that is, by definition, the datum of a tuple $(\lambda_1, \ldots, \lambda_n) \in \mathbb{Z}^n$ with $\lambda_1 \geq \cdots \geq \lambda_n$ for all embeddings $\iota : k_K \hookrightarrow \overline{\mathbb{F}}_p$,
- an *inertial type* t, that is, by definition, a finite-dimensional $\overline{\mathbb{Q}}_p$-representation of the inertia subgroup $I_K \subset G_K$ having open kernel and admitting an extension to G_K,

Kisin constructed a surjective morphism of $\mathbb{Z}_p$-algebras $R_{\overline{\rho}} \to R_{\overline{\rho}}^{\lambda,\mathrm{t}}$ that parametrizes the lifts of $\overline{\rho}$ which are potentially crystalline with Hodge–Tate weights $(\lambda_i + n - i)_{i,\iota}$ and inertial type t.

The Breuil–Mézard conjecture is a numerical relation between the Hilbert–Samuel multiplicity of the special fibre of $R_{\overline{\rho}}^{\lambda,\mathrm{t}}$, denoted by $e(R_{\overline{\rho}}^{\lambda,\mathrm{t}} \otimes_{\mathbb{Z}_p} \overline{\mathbb{F}}_p)$, and invariants coming from the representation theory of GL_n. Precisely, let L_λ be the irreducible algebraic $\mathbb{Z}_p$-representation of GL_n of highest weight λ. After [14, 25, 43], we know that there is a finite-dimensional smooth irreducible $\overline{\mathbb{Q}}_p$-representation $\sigma(\mathrm{t})$ of $\mathrm{GL}_n(\mathcal{O}_K)$ associated to t. We choose a $\mathrm{GL}_n(\mathcal{O}_K)$-stable $\mathbb{Z}_p$-lattice L_t in $\sigma(\mathrm{t})$, form the tensor product $L_{\lambda,\mathrm{t}} = L_\mathrm{t} \otimes_{\mathbb{Z}_p} L_\lambda$ and write its semi-simplification modulo p as follows:

$$\left(L_{\lambda,\mathrm{t}} \otimes_{\mathbb{Z}_p} \overline{\mathbb{F}}_p\right)^{\mathrm{ss}} \simeq \bigoplus_{\sigma \in \mathcal{D}} \sigma^{\oplus n_{\lambda,\mathrm{t}}(\sigma)}$$

where the sums runs over the set $\mathcal{D}$ of Serre weights σ, that is the set of (isomorphism classes of) irreducible $\overline{\mathbb{F}}_p$-representations of $\mathrm{GL}_n(k_K)$.

Conjecture 1.1 (Breuil–Mézard) *There exists a family of integers* $(\mu_{\overline{\rho}}(\sigma))_{\overline{\rho},\sigma}$, *called* intrinsic multiplicities, *such that the following numerical equality holds:*

$$e\left(R_{\overline{\rho}}^{\lambda,\mathrm{t}} \otimes_{\mathbb{Z}_p} \overline{\mathbb{F}}_p\right) = \sum_{\sigma \in \mathcal{D}} n_{\lambda,\mathrm{t}}(\sigma)\, \mu_{\overline{\rho}}(\sigma) \tag{1}$$

for all triples $(\overline{\rho}, \lambda, \mathrm{t})$ *as above.*

The Breuil–Mézard conjecture was first formulated for 2-dimensional representations in [5]. Since then, it has attracted a lot of attention. Kisin [29] proved it when $K = \mathbb{Q}_p$ (and $n = 2$) by making use of the p-adic local Langlands correspondence for $\mathrm{GL}_2(\mathbb{Q}_p)$ and the (global) Taylor–Wiles–Kisin patching argument. Sander [42] and Paškūnas [39] gave a purely local alternative proof which has been extended later on by Hu and Tan to nonscalar split residual representations [27]. For a general

K, but still assuming $n = 2$, the conjecture was proved by Gee and Kisin [23] (see also [13, Appendix C]) when $\lambda = (0, 0)$ for each embedding (which corresponds to potentially Barsotti–Tate deformations).

The extension of the Breuil–Mézard conjecture to higher n came later. The formulation stated in dimension n in Conjecture 1.1 is due to Emerton and Gee [20] (see also [24]). The case of 3-dimensional representations was considered and partially solved by Herzig et al. [26] and Le et al. [33, 34].

1.2 Encoding Representations with Combinatorial Data

It turns out that, in some cases, all the objects that intervene in the statement of the Breuil–Mézard conjecture can be entirely described by combinatorial data. Besides, these explicit descriptions provide us with a new viewpoint on the conjecture, quite useful for attacking it.

1.2.1 The Case of 2-Dimensional Representations

We start with the case of GL_2, for which encodings are simpler and more is known. For simplicity, we assume further that $K/\mathbb{Q}_p$ is unramified, i.e. $K = \mathbb{Q}_{p^f}$ for some positive integer f. In dimension 2, the *irreducible* continuous representations $\overline{\rho} : G_K \to \mathrm{GL}_2(\overline{\mathbb{F}}_p)$ all take the form:

$$\overline{\rho} = \mathrm{Ind}_{G_{K'}}^{G_K} \left(\omega_{2f}^h \cdot \mathrm{nr}'(\theta) \right) \tag{2}$$

where K' is the unique unramified extension of degree 2 of K, ω_{2f} is the fundamental character of $G_{K'}$ of level $2f$ and $\mathrm{nr}'(\theta)$ denotes the unique unramified character of $G_{K'}$ sending the arithmetic Frobenius to θ. The parameters h and θ are an integer defined modulo $p^{2f} - 1$ and an element of $\overline{\mathbb{F}}_p$ respectively.

Similarly, we have a complete description of *tame* inertial types, that are, by definition, those inertial types $\mathrm{t} : I_K \to \mathrm{GL}_2(\overline{\mathbb{F}}_p)$ that factor through the tame inertia. Depending on whether they are reducible or not, they are of the form:

$$\mathrm{t} = \omega_f^\gamma \oplus \omega_f^{\gamma'} \quad \text{or} \quad \mathrm{t} = \mathrm{Ind}_{I_{K'}}^{I_K} \omega_{2f}^\gamma \tag{3}$$

with obvious notations. In the first case, we say that t has level f; otherwise, that it has level $2f$.

We now assume further that $\lambda = (0, 0)$ for each embedding. In this case, the integers $n_{\lambda,\mathrm{t}}(\sigma)$ are always 0 or 1 and we define $\mathcal{D}(\mathrm{t})$ as the set of Serre weights $\sigma \in \mathcal{D}$ for which $n_{\lambda,\mathrm{t}}(\sigma) = 1$. Similarly, it is conjectured that $\mu_{\overline{\rho}}(\sigma)$ is always 0 or 1 as well and we let $\mathcal{D}(\overline{\rho}) \subset \mathcal{D}$ be the locus over which $\mu_{\overline{\rho}}(\sigma)$ is strictly positive. Understanding the summation in the Breuil–Mézard conjecture (see Eq. (1)) then amounts to understanding the set $\mathcal{D}(\mathrm{t}, \overline{\rho}) = \mathcal{D}(\mathrm{t}) \cap \mathcal{D}(\overline{\rho})$.

It turns out that $\mathcal{D}(t)$ and $\mathcal{D}(\overline{\rho})$ admit very explicit combinatorial descriptions in terms of the parameters h, γ and γ' we introduced earlier. These descriptions first appeared in [2, 7] and were then simplified in [17]. Very roughly, once t (resp. $\overline{\rho}$) is fixed, the weights in $\mathcal{D}(t)$ (resp. in $\mathcal{D}(\overline{\rho})$) are parametrized by tuples $\underline{\varepsilon} = \{\varepsilon_0, \ldots, \varepsilon_{f-1}\} \in \{0, 1\}^f$. Each such tuple produces a Serre weight by a simple recipe and the set $\mathcal{D}(t)$ (resp. $\mathcal{D}(\overline{\rho})$) is finally obtained by putting together all such weights. We underline that $\mathcal{D}(t)$ and $\mathcal{D}(\overline{\rho})$ usually have cardinality strictly less than 2^f because some $\underline{\varepsilon}$ may actually fail to produce a weight and it also happens that two different $\underline{\varepsilon}$ lead to the same weight.

1.2.2 The Gene

Building on the previous results, we gave in [11] a purely combinatorial description of the intersection $\mathcal{D}(t, \overline{\rho}) = \mathcal{D}(t) \cap \mathcal{D}(\overline{\rho})$ in terms of the parameters h, γ and γ', assuming that the tame inertial type is reducible. More precisely, setting $q = p^f$ for simplicity, we considered the quantity $h - (q+1)\gamma' \bmod q^2 - 1$ and wrote its decomposition in radix p:

$$h - (q+1)\gamma' \equiv p^{2f-1}v_0 + p^{2f-2}v_1 + \cdots + pv_{2f-2} + v_{2f-1} \ (\mathrm{mod}\ q^2 - 1). \quad (4)$$

From the v_i's, we then formed a periodic sequence $\mathbb{X} = (X_i)_{i \in \mathbb{Z}}$ of period $2f$ assuming values in the finite set $\{\mathsf{A}, \mathsf{B}, \mathsf{AB}, \mathsf{O}\}$. The sequence $\mathbb{X}$ is called the *gene* of $(t, \overline{\rho})$ and it satisfies the following rules (see [11, Lemma B.1.3]):

- if $v_i = 0$ and $X_{i+1} = \mathsf{O}$, then $X_i = \mathsf{AB}$;
- if $v_i = 0$ and $X_{i+1} \neq \mathsf{O}$, then $X_i = \mathsf{A}$;
- if $v_i = 1$ and $X_{i+1} = \mathsf{O}$, then $X_i = \mathsf{O}$;
- if $v_i = 1$ and $X_{i+1} \neq \mathsf{O}$, then $X_i = \mathsf{B}$;
- if $v_i \geq 2$, then $X_i = \mathsf{O}$.

To each gene $\mathbb{X}$, we attached a set $\mathcal{W}(\mathbb{X})$ of *combinatorial weights*, which are sequences of length f with values in $\{0, 1\}$. We then proved (see [11, Theorem 3.1.2]) that, if $\mathbb{X}$ denotes the gene of $(t, \overline{\rho})$, there is a canonical bijection:

$$\mathcal{W}(\mathbb{X}) \overset{\sim}{\longrightarrow} \mathcal{D}(t, \overline{\rho}). \quad (5)$$

Beyond yielding an explicit description of $\mathcal{D}(t, \overline{\rho})$ and opening concrete and algorithmical perspectives on the Breuil–Mézard conjecture, the above result raises new questions because it somehow shows that the dependence of $\mathcal{D}(t, \overline{\rho})$ in t, $\overline{\rho}$ and even in the underlying prime number p itself, is very weak, given that the gene only retains little information about these data. In some sense, one can interpret the gene as the "skeleton" of the pair $(t, \overline{\rho})$ that captures its most fundamental combinatorial properties in view of the Breuil–Mézard conjecture. In this perspective, the construction $\mathbb{X} \mapsto \mathcal{W}(\mathbb{X})$ should be thought of as the core factory of Serre weights, while the bijections (5), for varying t, $\overline{\rho}$ and p, appear as many tangible incarnations of this manufacture.

1.2.3 Higher Dimension and Group-Theoretic Formulation

When we are moving to higher dimensions, the numerical descriptions we used previously cannot continue to be that simple but, interestingly, they have analogues which can be formulated in the language of group theory. In what follows, we continue to assume that $K = \mathbb{Q}_{p^f}$ for some positive integer f. In this setting, the relevant algebraic group is:

$$G = \left(\mathrm{Res}_{\mathcal{O}_K/\mathbb{Z}_p} \mathrm{GL}_{n/\mathcal{O}_K} \right) \times_{\mathbb{Z}_p} \overline{\mathbb{Z}}_p \simeq \prod_{\mathfrak{J}} \mathrm{GL}_{n/\overline{\mathbb{Z}}_p}$$

where $\mathfrak{J}$ is the set of embeddings $\mathcal{O}_K \hookrightarrow \overline{\mathbb{Z}}_p$ (or, equivalently, $k_K \hookrightarrow \overline{\mathbb{F}}_p$) and will be identified with $\mathbb{Z}/f\mathbb{Z}$ in what follows. We denote by T the diagonal maximal torus of G. We let R be the set of roots of (G, T) and W be the corresponding Weyl group. The Borel subgroup $B \subset G$ of upper triangular matrices determines a subset $R^+ \subset R$ of positive roots. The group of characters $X^\star(T)$ can be canonically identified with $(\mathbb{Z}^n)^{\mathfrak{J}} = (\mathbb{Z}^n)^f$ and the Weyl group W is isomorphic to $\mathfrak{S}_n^f$. We shall also need the extended Weyl group $\widetilde{W}$ of G, defined by $\widetilde{W} = W \ltimes X^\star(T) \simeq (\mathfrak{S}_n \ltimes \mathbb{Z}^n)^f$.

In this setting, a Serre weight is an (isomorphism class of) irreducible $\overline{\mathbb{F}}_p$-representation of $\mathrm{GL}_n(\mathbb{F}_{p^f})$. It follows from a somehow classical argument of the theory of representations of reductive groups that Serre weights can be parametrized by certain characters of G. Precisely, after [24, Lemma 9.2.4], we know that if we set:

$$X_0(T) = \left\{ \lambda \in X^\star(T) \text{ s.t. } \langle \lambda, \alpha^\vee \rangle = 0 \text{ for all } \alpha \in R^+ \right\}$$

$$X_1(T) = \left\{ \lambda \in X^\star(T) \text{ s.t. } 0 \leq \langle \lambda, \alpha^\vee \rangle \leq p - 1 \text{ for all } \alpha \in R^+ \right\}$$

and let π denote the shift $(x_i)_{i \in \mathbb{Z}/f\mathbb{Z}} \mapsto (x_{i+1})_{i \in \mathbb{Z}/f\mathbb{Z}}$ on $X^\star(T) \simeq (\mathbb{Z}^n)^f$, there is a bijection:

$$F : X_1(T)/(p-\pi)X_0(T) \xrightarrow{\sim} \mathcal{D} \tag{6}$$

taking a character λ to the restriction to $\mathrm{Res}_{\mathcal{O}_K/\mathbb{Z}_p} \mathrm{GL}_{n/\mathcal{O}_K}(\mathbb{F}_p) = \mathrm{GL}_n(\mathbb{F}_{p^f})$ of the representation of $G(\mathbb{F}_p)$ induced by the algebraic representation of G of highest weight λ.

We also have a description of tame inertial types in terms of elements of the group $\widetilde{W}$ [3, 33, 35]. Let $(s, \mu) \in \widetilde{W} = W \ltimes X^\star(T)$ and write $s = (s_0, \ldots, s_{f-1})$ with $s_i \in \mathfrak{S}_n$ for all i. Let r be the order of $s_0 s_{f-1} \cdots s_1 \in \mathfrak{S}_n$. We consider the unramified extension K' of K of degree r and let $k_{K'}$ denote its residue field. We let

$\tilde{\omega}_{rf} : I_K = I_{K'} \to \overline{\mathbb{Z}}_p^{\times}$ be the Teichmüller lift of the Serre fundamental character of level rf. For $j \in \mathfrak{J}$, we put:

$$\eta_j = \begin{cases} (n-1, \ldots, 1, 0) & j\text{-th coordinate} \\ (0, \ldots, 0) & \text{elsewhere} \end{cases}$$

and $\eta = \sum_{j \in \mathfrak{J}} \eta_j$. We define $\alpha'_{(s,\mu)} \in X^{\star}(T)^{\mathrm{Hom}_{k_K\text{-alg}}(k_{K'}, \overline{\mathbb{F}}_p)} \simeq X^{\star}(T)^r \simeq (\mathbb{Z}^n)^{rf}$ by:

$$\alpha'_{(s,\mu),j} = s_1^{-1} s_2^{-1} \cdots s_j^{-1} (\mu_j + \eta_j)$$

and finally set:

$$\tau(s, \mu + \eta) = \bigoplus_{1 \le i \le n} \tilde{\omega}_{rf}^{\sum_{j'=0}^{rf-1} \alpha'_{(s,\mu),j',i} \, p^{j'}}. \tag{7}$$

It is our tame inertial type.

Irreducible representations $\overline{\rho} : G_K \to \mathrm{GL}_n(\overline{\mathbb{F}}_p)$ can be encoded in a similar fashion. Moreover, when $n = 2$, it turns out that the explicit descriptions of $\mathcal{D}(t)$ and $\mathcal{D}(\overline{\rho})$ we have mentioned earlier can be rephrased in the language of group theory which was briefly sketched above (at least under sufficiently generic assumptions). We do not reproduce here the corresponding recipes (which involve the so-called p-dot product) but refer to Propositions 2.4.2 and 2.4.3 of [3] for more details. So far, we do not have any candidate for being a plausible replacement of the gene when $n > 2$. However, all the above constructions tend to show that, even though they now need to be formulated in the language of group theory, combinatorics is still here (and is maybe even more ubiquitous) in higher dimensions.

1.3 Explicit Computations of $R_{\overline{\rho}}^{\lambda,\mathrm{t}}$

The Breuil–Mézard conjecture is concerned with the special fibre of $R_{\overline{\rho}}^{\lambda,\mathrm{t}}$ but, of course, obtaining a complete description of the ring $R_{\overline{\rho}}^{\lambda,\mathrm{t}}$ is also of interest on its own. For example, explicit presentations of some $R_{\overline{\rho}}^{\lambda,\mathrm{t}}$ have been used by Emerton, Gee and Savitt [21] to prove important conjectures stated by Breuil in [4] about lattices in the cohomology of Shimura curves. In this subsection, we outline the standard strategy that is used to approach $R_{\overline{\rho}}^{\lambda,\mathrm{t}}$ and report on the results of some explicit computations.

1.3.1 Review on Kisin's Construction of $R_{\overline{\rho}}^{\lambda,\mathrm{t}}$

The main theoretical ingredient for studying deformations of $\overline{\rho}$ with prescribed Hodge type and inertial type is the theory of Breuil–Kisin, which provides a description of these deformations by means of semi-linear algebra. In our setting and assuming in addition that t is tame of level f and $K = \mathbb{Q}_{p^f}$ as we already did previously, a Breuil–Kisin module is a projective module $\mathfrak{M}$ over $\overline{\mathbb{Z}}_p \otimes_{\mathbb{Z}_p} \mathcal{O}_K[[u]] \simeq \overline{\mathbb{Z}}_p[[u]]^{\widetilde{}}$ equipped with two additional structures:

- a *Frobenius map* $\varphi_{\mathfrak{M}} : \mathfrak{M} \to \mathfrak{M}$ which is semi-linear (with respect to the endomorphism of $\overline{\mathbb{Z}}_p \otimes_{\mathbb{Z}_p} \mathcal{O}_K[[u]]$ acting by the identity on $\overline{\mathbb{Z}}_p$, by the Frobenius of $\mathcal{O}_K$ and taking u to u^p) and satisfies additional properties,
- a *descent data*, that is a linear action of the group $\mathrm{Gal}(K[\sqrt[e]{p}]/K) \simeq \mathbb{Z}/e\mathbb{Z}$ (with $e = p^f - 1$) that commutes with the Frobenius map.

A famous theorem of Kisin [30] indicates that these modules are in correspondence with $\overline{\mathbb{Z}}_p$-representations of G_K that become crystalline over the extension $K[\sqrt[e]{p}]$. Besides, the Hodge type (resp. the inertial type) of the latter can be easily read off from the form of the Frobenius map (resp. of the descent data) on the former. Even better, the reduction modulo p of the representation associated to a Breuil–Kisin module $\mathfrak{M}$ is uniquely and entirely described by the module $\mathfrak{M} \otimes_{\mathcal{O}_K[[u]]} k_K((u))$ equipped with its additional structures.

The Breuil–Kisin theory then looks particularly well suited for the study of the deformation rings $R_{\overline{\rho}}^{\lambda,\mathrm{t}}$ and it turns out that it indeed is. In [31], Kisin constructed a scheme $\mathcal{GR}_{\overline{\rho}}^{\lambda,\mathrm{t}}$ parametrizing the Breuil–Kisin modules $\mathfrak{M}$ of Hodge type λ, inertial type t and having the additional property that $\mathfrak{M} \otimes_{\mathcal{O}_K[[u]]} k_K((u))$ corresponds to the given representation $\overline{\rho}$. This scheme is moreover equipped with a morphism $\mathcal{GR}_{\overline{\rho}}^{\lambda,\mathrm{t}} \to \mathrm{Spec}\, R_{\overline{\rho}}$ whose schematic image has closure $\mathrm{Spec}\, R_{\overline{\rho}}^{\lambda,\mathrm{t}}$. One should be careful however that the morphism:

$$\kappa : \mathcal{GR}_{\overline{\rho}}^{\lambda,\mathrm{t}} \to \mathrm{Spec}\, R_{\overline{\rho}}^{\lambda,\mathrm{t}} \tag{8}$$

is *not* an isomorphism in general because two p-torsion Breuil–Kisin modules may correspond to the same Galois representation. It is however always an isomorphism on the generic fibre. The special fibre of $\mathcal{GR}_{\overline{\rho}}^{\lambda,\mathrm{t}}$ is denoted by $\overline{\mathcal{GR}}_{\overline{\rho}}^{\lambda,\mathrm{t}}$ and is called the *Kisin variety*;[1] in some sense, it measures the default for κ to be an isomorphism.

---

[1] The terminology "variety" is justified by the fact that the scheme $\overline{\mathcal{GR}}_{\overline{\rho}}^{\lambda,\mathrm{t}}$ is always of finite type over $\overline{\mathbb{F}}_p$.

1.3.2 Examples in Dimension 2: The Generic Case

The first examples of explicit calculations of certain rings $R_{\overline{\rho}}^{\lambda,\mathfrak{t}}$ have been carried out by Breuil and Mézard in [5] and [6]. They considered the case where $\overline{\rho}$ is 2-dimensional and absolutely irreducible, $\lambda = (0,0)$ for each embedding and $\mathfrak{t}$ is tame of level f. Under some additional assumptions of genericity on $\overline{\rho}$, they obtained, when $\mathcal{D}(\mathfrak{t}, \overline{\rho})$ is not empty:

$$R_{\overline{\rho}}^{\lambda,\mathfrak{t}} \simeq \frac{\overline{\mathbb{Z}}_p[[X_i, Y_i, i \in \mathfrak{I}_{\mathrm{II}}, Z_j, j \in \mathfrak{I} \setminus \mathfrak{I}_{\mathrm{II}}]]}{(X_i Y_i - p, i \in \mathfrak{I}_{\mathrm{II}})}. \tag{9}$$

for a certain subset $\mathfrak{I}_{\mathrm{II}}$ of $\mathfrak{I}$ (which depends on λ, $\mathfrak{t}$ and $\overline{\rho}$). The aforementioned genericity assumptions play a quite important role in Breuil and Mézard's argument. In fact, they imply that the underlying Kisin variety is reduced to one point, which itself ensures that the morphism κ of Eq. (8) is an isomorphism. The computation of $R_{\overline{\rho}}^{\lambda,\mathfrak{t}}$ then directly reduces to that of $\mathcal{GR}_{\overline{\rho}}^{\lambda,\mathfrak{t}}$.

Before moving to nongeneric cases, it is important to comment on the subset $\mathfrak{I}_{\mathrm{II}}$ which appeared in Eq. (9). The triviality of the Kisin variety indicates that the module over $\overline{\mathbb{F}}_p \otimes_{\mathbb{F}_p} k_K((u)) \simeq \overline{\mathbb{F}}_p((u))^{\mathfrak{I}}$ corresponding to $\overline{\rho}$ contains a unique lattice $\mathfrak{M}(\overline{\rho})$ that is a Breuil–Kisin module of type $(\lambda, \mathfrak{t})$. Breuil and Mézard then defined the *shape* of $\mathfrak{M}(\overline{\rho})$: it is a finite sequence $(g_0, \ldots, g_{f-1})$ assuming values in the finite set $\{\mathrm{I}, \mathrm{II}\}$ which, roughly speaking, is obtained by looking at the form of the matrix of $\varphi_{\mathfrak{M}(\overline{\rho})}$ in bases diagonalizing the action of the descent data. The set $\mathfrak{I}_{\mathrm{II}}$ is then formed by the indices i for which g_i is II.

The shape also plays a key role on the GL$_2$-side of the Breuil–Mézard conjecture: in our setting, the cardinality of $\mathcal{D}(\mathfrak{t}, \overline{\rho})$ is $2^{\mathrm{Card}(\mathfrak{I}_{\mathrm{II}})}$ (which is the Hilbert–Samuel multiplicity of the special fibre of the ring $R_{\overline{\rho}}^{\lambda,\mathfrak{t}}$ given by Eq. (9)) and, more precisely, we can explicitly parametrize the weights in $\mathcal{D}(\mathfrak{t}, \overline{\rho})$ by subsets of $\mathfrak{I}_{\mathrm{II}}$.

1.3.3 Examples in Dimension 2: The Nongeneric Case

Nongeneric cases are more complicated because they usually correspond to non-trivial Kisin varieties. In [10], we computed these Kisin varieties when, as above, $\overline{\rho}$ is absolutely irreducible, $\lambda = (0,0)$ for each embedding and $\mathfrak{t}$ is tame of level f. We recall that, in this setting, we have attached to the pair $(\mathfrak{t}, \overline{\rho})$ its gene $\mathbb{X}$ (see Sect. 1.2). Our results show that the Kisin variety is entirely determined by the gene. Being a little bit more precise, we showed that $\overline{\mathcal{GR}}_{\overline{\rho}}^{\lambda,\mathfrak{t}}$ is a closed subscheme of $\left(\mathbb{P}_{\overline{\mathbb{F}}_p}^1\right)^{\mathbb{Z}/f\mathbb{Z}}$ defined by equations of the form:

$$\lambda_i \, x_i \, y_{i+1} = \mu_i \, x_{i+1} \, y_i \tag{10}$$

where $[x_i : y_i]$ denotes the projective coordinates on the i-th copy of $\mathbb{P}^1_{\overline{\mathbb{F}}_p}$ and λ_i and μ_i are elements of $\{0, 1\}$ that can be read off from the gene $\mathbb{X}$.

It is important to observe that the notion of shape can be extended to the nongeneric case as well. Indeed, each $\overline{\mathbb{F}}_p$-point x of $\overline{\mathcal{GR}}_{\overline{\rho}}^{\lambda,\mathrm{t}}$ corresponds, by definition, to a Breuil–Kisin module and so has a well defined shape $g(x) = (g_0(x), \ldots, g_{f-1}(x)) \in \{\mathrm{I}, \mathrm{II}\}^f$ in the sense of Breuil and Mézard. In full generality, the shape is thus no longer a unique element in $\{\mathrm{I}, \mathrm{II}\}^f$ but a function on the $\overline{\mathbb{F}}_p$-points of the Kisin variety taking values in $\{\mathrm{I}, \mathrm{II}\}^f$. We proved moreover that this function is lower-continuous (for the partial ordering on the codomain defined by $\mathrm{I} < \mathrm{II}$) and thus defines a stratification on $\overline{\mathcal{GR}}_{\overline{\rho}}^{\lambda,\mathrm{t}}$ by locally closed subschemes. As well as the Kisin variety, the shape stratification is entirely determined by the gene.

We then proposed the following conjecture.

Conjecture 1.2

(i) *The generic fibre of $R_{\overline{\rho}}^{\lambda,\mathrm{t}}$ is determined by the Kisin variety equipped with its shape stratification.*
(ii) *The ring $R_{\overline{\rho}}^{\lambda,\mathrm{t}}$ is determined by the gene.*

Regarding the first item of the conjecture, we were actually much more precise and exhibited a candidate for being the generic fibre of $R_{\overline{\rho}}^{\lambda,\mathrm{t}}$. Besides, our candidate is rather explicit: it is defined as the formal neighborhood of the Kisin variety in a certain blow-up of $\left(\mathbb{P}^1_{\overline{\mathbb{Z}}_p}\right)^{\mathbb{Z}/f\mathbb{Z}}$. We refer to [10, §5.4] for a complete exposition of this construction.

Conjecture 1.2 is known is most cases where the Kisin variety is trivial. It has also been checked in [9] in one example where the Kisin variety is isomorphic to $\mathbb{P}^1_{\overline{\mathbb{F}}_p}$; in this case, the deformation ring we obtained is $\overline{\mathbb{Z}}_p[[X, Y, Z]]/(XY - p^2)$.

1.3.4 Higher Dimension and Group-Theoretic Formulation

In dimension 3, tamely potentially crystalline deformation rings for small Hodge–Tate weights and generic Galois representations have been studied in [34]. The authors also obtained explicit presentations of the deformation rings $R_{\overline{\rho}}^{\lambda,\mathrm{t}}$ in several cases. The equations they found are often quite similar to the one given in Eq. (9). The case of rank 2 unitary group has also been considered in [32]. In this setting, it turns out that the explicit computations of the corresponding $R_{\overline{\rho}}^{\lambda,\mathrm{t}}$'s boil down to the case of GL_2 (with an additional polarization structure on the Breuil–Kisin modules). The final equations they get are then again similar to Eq. (9).

In all these situations, although computations are certainly much more difficult, it is important to underline that the basic ingredients are the same: we continue to have a Kisin variety, together with a shape stratification and we hope that these two objects strongly govern the final form of the deformation space. In full generality (i.e. for any reductive group, even not necessarily GL_n), the Kisin variety and the shape stratification are defined using techniques coming from group theory: the Kisin variety is a subscheme of the affine Grassmannian defined by an explicit condition, which can be formulated in terms of the Cartan decomposition, while the shape function $x \mapsto g(x)$ is defined by means of the Iwahori–Cartan decomposition (and it now takes values in the Iwahori–Weyl group). After [12], we know moreover that Kisin varieties are related to Pappas–Zhu local models [41]; in this connection, the shape stratification corresponds to the canonical stratification by affine Schubert varieties.

2 The Field with One Element

The theory of the field with one element $\mathbb{F}_1$, starts with an observation of Tits [45] who notices some curious numerical coincidences in the theory of algebraic linear groups. The most fundamental example is given by the group GL_n itself. Indeed, its number of points over the finite field $\mathbb{F}_q$ is given by:

$$\mathrm{Card}\,\mathrm{GL}_n(\mathbb{F}_q) = (q^n - 1) \cdot (q^n - q) \cdots (q^n - q^{n-1})$$
$$= (q - 1)^n \cdot q^{n(n-1)/2} \cdot [n]_q \cdot [n-1]_q \cdots [1]_q$$

where $[i]_q = 1 + q + \cdots + q^{i-1}$ is the q-analogue of i and letting q tends to 1, we find:

$$\mathrm{Card}\,\mathrm{GL}_n(\mathbb{F}_q) \sim_{q \to 1} (q - 1)^n \cdot n!. \tag{11}$$

What is surprising is that $n!$ can be interpreted as the cardinality of the symmetric group $\mathfrak{S}_n$, which is nothing but the Weyl group of GL_n. Similar results hold more generally for a large family of groups, including the orthogonal groups, the symplectic groups and their scalar restrictions. After these observations, Tits asked if these numerical matchings could have deeper roots and proposed to build a geometry of the so-called *field with one element* with the objective to give a systematical and geometrical understanding of all the combinatorial structures and constructions which appear in the theory of Lie groups or algebraic groups.

Tits' vision was then popularized by Soulé who came up in [44] with a first tentative definition of affine varieties over $\mathbb{F}_1$. Later on, other constructions were proposed and the subject has attracted more and more attention over the last two decades [1, 8, 15, 16, 18, 36, 38, 46]; see [37] for a recent review on this topic. The theory of $\mathbb{F}_1$ is still not yet well established. However, several definitions of the category of schemes over $\mathbb{F}_1$ have been proposed over the years and significant progress towards Tits' initial dream have been realized. Besides, geometry over $\mathbb{F}_1$ has been extended to new contexts and has nowadays close interactions with Arakelov geometry and p-adic geometry. In particular, Bambozzi, Ben-Bassat and Kremnizer introduced in [1] analytic geometry over $\mathbb{F}_1$; notably they managed to construct a model of the Fargues–Fontaine curve [22] in their theory. To our knowledge, this was the first connection between the theory of Galois representations (incarnated here by p-adic Hodge theory) and the field with one element.

2.1 Clues in Favor of a 1-Adic Breuil–Mézard Conjecture

The main reason why we believe that a 1-adic version of the Breuil–Mézard conjecture is possible is that many objects involved in the formulation and/or the resolution of this conjecture have natural seeds in $\mathbb{F}_1$-geometry. In this subsection, we list the most important of them and comment on their $\mathbb{F}_1$-aspects.

First of all, we notice that the Weyl group W and its extended version $\widetilde{W}$, which both play a quite important role in the construction of Serre's weights and inertial types, have natural interpretations in characteristic one: the Weyl group is the set of $\mathbb{F}_1$-points of the underlying algebraic group (this property is Tits' dream, which is the main guide of the theory) while the extended Weyl group may be interpreted as its set of points over $\mathbb{F}_1(X)$ (see Eq. (12) in Section "Brief Review of Geometry over $\mathbb{F}_1$"). Moreover, the recipes used for constructing Serre's weights and inertial types from an element of $\widetilde{W}$ (see Eqs. (6) and (7)) are purely combinatorial and it is quite likely that they can be reformulated by means of $\mathbb{F}_1$-geometry.

The equations of the Kisin varieties we obtained in [10] (see Eq. (10)) show that all of them are defined over $\mathbb{F}_1$. Similarly, the deformation spaces computed in [5, 6] and [34] all appear as product of discs and annuli (see particularly Eq. (9)); as a consequence, they all come through scalar extensions from analytic spaces over $\mathbb{F}_1$ in the sense of [1]. Moreover, although the construction of blow-ups and formal neighborhoods was not addressed in [1], it looks quite plausible that the candidates for deformation spaces we introduced in [10] are defined over $\mathbb{F}_1$ as well.

All the above examples show furthermore a strong uniform behaviour with respect to p. In the language of $\mathbb{F}_1$-geometry, this uniformity means that a whole family of Kisin varieties (resp. of deformation spaces) parametrized by p comes by scalar extension to $\mathbb{F}_p$ (resp. to $\mathbb{Q}_p$) from a *unique* variety (resp. analytic variety) over $\mathbb{F}_1$. This result might suggest the existence of a common denominator of the theory of Kisin varieties (resp. deformation spaces) which is defined in characteristic one and underpins some of their features we observe over the p-adics. This expectation is strengthened by the fact that Kisin varieties have a deep group-theoretic interpretation (see last paragraph of Sect. 1.3). The same remark is valid for the shape stratifications as well: they have good chance to be visible in characteristic one, given that they are closely connected to affine Schubert varieties, which are themselves known to be defined over $\mathbb{F}_1$ [38].

The recipe of [11], giving a combinatorial description of the set of common Serre's weights in terms of the corresponding gene, also has a strong $\mathbb{F}_1$-flavor. Concretely, what we expect is that:

1. the gene is a sort of $\mathbb{F}_1$-encoding of the pair $(t, \overline{\rho})$,
2. the combinatorial weights of [11] are the mirror of a notion of Serre's weights in characteristic 1,
3. the association

$$\text{gene} \mapsto \text{set of combinatorial weights}$$

is the $\mathbb{F}_1$-incarnation of the construction $(t, \overline{\rho}) \mapsto \mathcal{D}(t, \overline{\rho})$.

Beyond the justifications coming from the constructions of [11], we underline that there is other evidence supporting that Serre's weights in characteristic 1 should have something to do with combinatorial weights (which are, we recall, sequences of length f assuming values in $\{0, 1\}$). Indeed, mimicking the usual definition in characteristic p, we expect Serre's weights in characteristic 1 to be interpreted as $\overline{\mathbb{F}}_1$-representations (whatever that means) of the group $\mathrm{GL}_2(\mathbb{F}_{1^f})$. But, following Tits' vision, we can write:

$$\mathrm{GL}_2(\mathbb{F}_{1^f}) = \big(\mathrm{Res}_{\mathbb{F}_{1^f}/\mathbb{F}_1} \mathrm{GL}_2\big)(\mathbb{F}_1) = \mathrm{Weyl}\big(\mathrm{Res}_{\mathbb{F}_{p^f}/\mathbb{F}_p} \mathrm{GL}_2\big) = (\mathbb{Z}/2\mathbb{Z})^f$$

and we already see the set $\{0, 1\}$ entering into the scene. More precisely, we can define $\mathrm{Sym}^k \mathbb{F}_1^2$ as the set $\{X^k, XY^{k-1}, \ldots, Y^k\}$ and let $\mathrm{GL}_2(\mathbb{F}_1) = \mathbb{Z}/2\mathbb{Z}$ act on it by letting its unique nontrivial element operate by swapping X and Y (see Appendix "Brief Review of Geometry over $\mathbb{F}_1$" for a justification of this definition). Similarly, if $\underline{k} = (k_0, \ldots, k_{f-1})$ is a tuple of integers, we let $\mathrm{Sym}^{\underline{k}} \mathbb{F}_1^2$ be the cartesian product of the $\mathrm{Sym}^{k_i} \mathbb{F}_1^2$ equipped with the induced action of $\mathrm{GL}_2(\mathbb{F}_{1^f}) = (\mathbb{Z}/2\mathbb{Z})^f$. It is then an easy exercise to check that $\mathrm{Sym}^{\underline{k}} \mathbb{F}_1^2$ is irreducible (i.e. the action is transitive) if and only if $k_i \in \{0, 1\}$ for all i.

We underline that Conjecture 1.2 is also in line with the above vision: roughly speaking, it stipulates that the mapping $(t, \overline{\rho}) \mapsto R_{\overline{\rho}}^{\lambda, t}$ descends over $\mathbb{F}_1$.

Combining the previous observations and being quite optimistic, one might hope that the Pappas–Rapoport spaces [40] and/or Emerton–Gee stacks [20] themselves have a model over $\mathbb{F}_1$ and that the irreducible components of the special fibre of the latter will be related to the set of Serre's weights in characteristic 1.

2.2 Major Challenges

We do not hide that, if possible, devising a 1-adic Langlands correspondence (or a 1-adic Breuil–Mézard conjecture) will definitely not be a simple task. Actually, although the geometry over $\mathbb{F}_1$ has already been developed quite a lot, many fundamental ingredients and objects of the Langlands programme are missing in characteristic 1.

To start with, we notice that extensions of $\mathbb{F}_1$, usually referred to as $\mathbb{F}_{1^n}$, have already been considered by several authors [15, 28, 44] but they have never been systematically studied. Moreover, in the above references, $\mathbb{F}_{1^n}$ is defined as the cyclotomic extension of $\mathbb{F}_1$ whose Galois group is isomorphic to $(\mathbb{Z}/n\mathbb{Z})^\times$, and not $\mathbb{Z}/n\mathbb{Z}$. This means in particular that we apparently do not have a nice analogue of the Frobenius endomorphism, which sounds annoying. An option for fixing this issue is to work with another version of $\mathbb{F}_{1^n}$ on which we impose by design the existence of a Frobenius of order n (see Appendix "Galois Theory over $\mathbb{F}_1$" for a first rough proposal).

Similarly, the field of 1-adic numbers $\mathbb{Q}_1$ and its extensions have not attracted much attention so far. We mention however that Connes introduced the ring of Witt vectors over $\mathbb{F}_1$ in [15] but we are afraid that Connes' treatment does not perfectly fit with our perspectives since it is eventually related to Banach algebras over the reals, and not over the p-adics. In Appendix "Galois Theory over $\mathbb{Q}_1$", using a (certainly too) naive definition of $\mathbb{Q}_1$, we start exploring its theory of finite extensions.

On a different note, it seems that the theory of representations of reductive groups over $\mathbb{F}_1$ has not yet been systematically studied. This could sound surprising given that representations of reductive groups over finite fields have been attracting a lot of attention for more than 50 years and the underlying theory includes a lot of combinatorial parts that have good chance to be "defined" over $\mathbb{F}_1$.

2.3 Conclusion

Although, clearly, many locks still need to be unlocked, we continue to believe that the 1-adic Langlands correspondence is possible. Besides, according to us, trying to develop it could be, at the same time, a wonderful motivation and guide for exploring the 1-adic world and for inspiring the p-adic Langlands correspondence by separating the universal combinatorial structures on the one hand and the more usual arithmetical properties on the other hand. We thus warmly encourage all contributions to this topic.

Appendix: Remarks on Galois Theory in Characteristic 1

In this appendix, we start exploring the Galois properties of the field with one element $\mathbb{F}_1$ and the field of 1-adic numbers $\mathbb{Q}_1$. Our aim is not at all to elaborate a complete and coherent theory but to share our intuition and point out some difficulties. In Section "Brief Review of Geometry over $\mathbb{F}_1$", we briefly review the usual constructions of geometries over $\mathbb{F}_1$. We then successively address the Galois theory of $\mathbb{F}_1$ and $\mathbb{Q}_1$ in Sections "Galois Theory over $\mathbb{F}_1$" and "Galois Theory over $\mathbb{Q}_1$" respectively.

Brief Review of Geometry over $\mathbb{F}_1$

Most modern theories of geometry over $\mathbb{F}_1$ start by defining the category $\mathbb{F}_1$ − **Vect** of $\mathbb{F}_1$-vector spaces. After Tits' observation that $\mathrm{GL}_d(\mathbb{F}_1)$ should be isomorphic to $\mathfrak{S}_d$, it is tempting to define a vector space over $\mathbb{F}_1$ simply as a set, its cardinality corresponding to its dimension over $\mathbb{F}_1$. By definition, an $\mathbb{F}_1$-linear morphism $V \to W$ is a set-theoretical *partially defined* function $f : V \to W$. The direct sum (resp. the tensor product) of two vector spaces is their disjoint union (resp. their cartesian product). Besides, from this point of view, the standard $\mathbb{F}_1$-vector space of dimension d, namely $\mathbb{F}_1^d$, is represented by the set $\{1, \ldots, d\}$; its group of automorphisms then coincides with $\mathfrak{S}_d$, i.e. $\mathrm{GL}_d(\mathbb{F}_1) = \mathfrak{S}^d$, as expected.

After vector spaces over $\mathbb{F}_1$, one introduces $\mathbb{F}_1$-algebras: they are, by definition, objects in commutative monoids in the category $\mathbb{F}_1$ − **Vect**, i.e. sets equipped with a partially defined law of commutative monoids which is usually denoted with the multiplicative convention. Examples of $\mathbb{F}_1$-algebras include $X^{\mathbb{N}} = \{1, X, X^2, \ldots\}$ and $X^{\mathbb{Z}}$, which should be thought of as $\mathbb{F}_1[X]$ and $\mathbb{F}_1(X)$ respectively. Similarly, the $\mathbb{F}_1$-algebra $\mathbb{F}_1[X_1, \ldots, X_d]$ is realized by the monoid $X_1^{\mathbb{N}} X_2^{\mathbb{N}} \cdots X_d^{\mathbb{N}}$ whose elements are monomials in $X_1, \ldots, X_d$.

If M is an $\mathbb{F}_1$-algebra, it makes sense to define M-modules: they are sets endowed with an action of M. The standard free module of rank d over M is $M^{\oplus d} = \{1, \ldots, d\} \times M$ where M acts by multiplication on the second coordinate. It is an easy exercise to check that the group of M-linear automorphisms of $M^{\oplus d}$ is the semi-direct product $\mathfrak{S}_d \ltimes (M^{\mathrm{gp}})^d$, where M^{gp} denotes the subgroup of invertible elements of M. In particular, we get:

$$\mathrm{GL}_d\left(\mathbb{F}_1(X)\right) = \mathfrak{S}_d \ltimes \mathbb{Z}^d \tag{12}$$

and thus obtain, at least in the case of GL_d, an $\mathbb{F}_1$-style interpretation of the extended Weyl group.

Up to this point, (almost) all theories agree on definitions but, when we come to $\mathbb{F}_1$-schemes, points of view start to diverge. Chronologically, the first approach is due to Deitmar [18] and it closely follows the classical theory of schemes: Deitmar introduced spectra of monoids, equipped them with a topology and a notion of sheaves and finally glued them to get $\mathbb{F}_1$-schemes. Soon after, Toen and Vaquié [46] developed the functorial point of view. Starting from the category $\mathbb{F}_1$ − **Vect** (or, more generally, with an abstract monoidal symmetric category $\mathcal{C}$), they defined the category $\mathbb{F}_1$ − **Alg** as we did before, introduced a notion of Zariski covering on it and finally defined $\mathbb{F}_1$-schemes as sheaves on $\mathbb{F}_1$ − **Alg** for this Grothendieck topology. In [47], Vezzani proved (in a slightly different context) that Deitmar's construction on the one hand and Toen–Vaquié's approach on the other hand are equivalent, in the sense that they give rise to the same category of $\mathbb{F}_1$-schemes. Besides, both viewpoints include a functor of scalar extension:

$$\mathbb{F}_1 - \mathbf{Sch} \to \mathbb{Z} - \mathbf{Sch}, \quad X \mapsto X_{\mathbb{Z}}$$

deriving from the construction $M \to \mathbb{Z}[M]$ at the level of monoids. Deitmar observed that toric varieties are defined over $\mathbb{F}_1$ but he also proved that these are essentially the only ones [19].

Another point of view on $\mathbb{F}_1$-schemes, which in some sense goes back to Soulé's original definition, was proposed in [16] by Connes and Consani. They suggested to define an $\mathbb{F}_1$-scheme as a triple $(\tilde{A}, X_{\mathbb{Z}}, e_X)$ where $\tilde{A}$ is an $\mathbb{F}_1$-algebra, $X_{\mathbb{Z}}$ is a classical scheme and $e_X : (\operatorname{Spec} \tilde{A})_{\mathbb{Z}} \to X_{\mathbb{Z}}$ is a morphism of schemes inducing a bijection on k-points for any field k. In some sense, this approach separates the purely combinatorial part, which is encoded by $\tilde{A}$, and the geometrical part, which is delegated to the classical theory through the scheme $X_{\mathbb{Z}}$. López Peña and Lorscheid [38] proved that this framework is more flexible in the sense that it allows for defining a much larger panel of varieties over $\mathbb{F}_1$; those include Grassmannians, split reductive groups, Schubert varieties, etc. Besides, in a subsequent paper, Lorscheid [36] concretized Tits' premonition by realizing the Weyl group of a split reductive group as its set of points over $\mathbb{F}_1$.

Beyond the field with one element, what we need for the purpose of this paper is the field of 1-adic numbers. Fortunately, this question has already been touched on in the literature by several authors. In [15], Connes came up with a definition of Witt vectors over $\mathbb{F}_1$. However, Connes' construction looks a bit disconnected to our needs as it is equipped with a scalar extension functor assuming values in Banach algebras over the reals, and not over the p-adics. In a different direction, Bambozzi, Ben-Bassat and Kremnizer [1] laid the foundations of the theory of analytic varieties over $\mathbb{F}_1$. Roughly speaking, their construction is similar to the ones we briefly sketched above except that, instead of starting with the category $\mathbb{F}_1 - \mathbf{Vect}$, they consider various categories of sets X equipped with a function $|\cdot| : X \to \mathbb{R}^+$ whose purpose is to model the norm map. They showed that balls and annuli of rigid geometry come from analytic varieties over $\mathbb{F}_1$; this is thus also the case for all the deformation spaces we encountered in Sect. 1.

Galois Theory over $\mathbb{F}_1$

It is a natural expectation that $\mathbb{F}_1$ should have a finite extension of degree n for all n with Galois group isomorphic to $\mathbb{Z}/n\mathbb{Z}$. In the literature [15, 28, 44], this extension $\mathbb{F}_{1^n}$ is usually defined as the cyclotomic extension of $\mathbb{F}_1$, that is the $\mathbb{F}_1$-algebra represented by the monoid $X^{\mathbb{Z}/n\mathbb{Z}}$. However, it appears that this point of view is not perfectly suited to our purpose for at least two reasons:

1. *The Galois theory is not the expected one.* Indeed, the group of automorphisms of the monoid $X^{\mathbb{Z}/n\mathbb{Z}}$ is $(\mathbb{Z}/n\mathbb{Z})^\times$ and not $\mathbb{Z}/n\mathbb{Z}$; in particular, we do not have a distinguished Frobenius endomorphism. This issue is maybe even more visible when we extend scalars to $\mathbb{Z}$ or $\mathbb{F}_p$. Indeed, with the above definition, one would get:

$$\mathbb{F}_{1^n} \otimes_{\mathbb{F}_1} \mathbb{F}_p = \mathbb{F}_p[X]/(X^n - 1)$$

which is certainly *not* $\mathbb{F}_{p^n}$ and which besides has a different Galois group.

2. *The formation of $\mathbb{F}_{1^n}$-points does not behave as desired.* If X is a scheme defined over $\mathbb{F}_{1^n}$, it seems reasonable to expect that the set of $\mathbb{F}_{1^n}$-points of X agrees with the set of $\mathbb{F}_1$-points of $\mathrm{Res}_{\mathbb{F}_{1^n}/\mathbb{F}_1} X$. For $X = \mathrm{GL}_d$, this results in:

$$\mathrm{GL}_d(\mathbb{F}_{1^n}) = \mathrm{Weyl}\big(\mathrm{Res}_{\mathbb{F}_{p^n}/\mathbb{F}_p} \mathrm{GL}_d\big) = (\mathfrak{S}_d)^n$$

where p, here, denotes any auxiliary prime number. This expectation is reinforced by the fact that the group $(\mathfrak{S}_d)^n$ plays a quite important role in the Breuil–Mézard conjecture as recalled in Sect. 1. However, if one lets $\mathbb{F}_{1^n}$ be the $\mathbb{F}_1$-algebra corresponding to the monoid $X^{\mathbb{Z}/n\mathbb{Z}}$, one would obtain $\mathrm{GL}_d(\mathbb{F}_{1^n}) = \mathfrak{S}_d \ltimes (\mathbb{Z}/n\mathbb{Z})^d$ (see the discussion before Eq. (12)) which is certainly *not* isomorphic to $(\mathfrak{S}_d)^n$ since even the cardinalities differ!

The conclusion of these observations is that, although the cyclotomic extension of $\mathbb{F}_1$ is undoubtedly an interesting object, it is probably not the $\mathbb{F}_{1^n}$ we need for the applications we have in mind. Moreover, one checks that there is unfortunately no $\mathbb{F}_1$-algebra meeting all our requirements. Instead, we propose to define from scratch a theory of $\mathbb{F}_{1^n}$-vector spaces as we did previously for $\mathbb{F}_1$, trying as much as possible to incorporate the Frobenius action and keep its desired properties.

Definition 2.1 An $\mathbb{F}_{1^n}$-*vector space* is a set. Given two $\mathbb{F}_{1^n}$-vector spaces V and W, an $\mathbb{F}_{1^n}$-*linear morphism* $f : V \to W$ is the datum of n partially defined set-theoretical functions $f_1, \ldots, f_n : V \to W$.

Again, the standard $\mathbb{F}_{1^n}$-vector space of dimension d is represented by the set $\{1, \ldots, d\}$. We denote it by $(\mathbb{F}_{1^n})^d$, or simply $\mathbb{F}_{1^n}$ when $d = 1$, in what follows. It is obvious from the definition that the automorphism group of $(\mathbb{F}_{1^n})^d$ is $(\mathfrak{S}_d)^n$, i.e. we have the expected equality $\mathrm{GL}_d(\mathbb{F}_{1^n}) = (\mathfrak{S}_d)^n$. Similarly, extending the definition of $\mathbb{F}_1(X)$ to our new setting, one checks that $\mathrm{GL}_d(\mathbb{F}_{1^n}(X)) = (\mathfrak{S}_d \ltimes \mathbb{Z}^d)^n$.

Moreover, we have an obvious scalar extension functor $\mathbb{F}_1 - \mathbf{Vect} \to \mathbb{F}_{1^n} - \mathbf{Vect}$ acting on objects by $V \mapsto V$ and on morphisms by $f \mapsto (f, f, \ldots, f)$. In what follows we shall use the notation $\mathbb{F}_{1^n} \otimes_{\mathbb{F}_1} V$ to denote the scalar extension of V from

$\mathbb{F}_1$ to $\mathbb{F}_{1^n}$. Regarding scalar restriction, there are two different options to define it, namely:

$$\mathrm{aRes}_{\mathbb{F}_{1^n}/\mathbb{F}_1} : \mathbb{F}_{1^n} - \mathbf{Vect} \longrightarrow \mathbb{F}_1 - \mathbf{Vect}$$
$$V \mapsto V^{\oplus n}$$
$$\mathrm{mRes}_{\mathbb{F}_{1^n}/\mathbb{F}_1} : \mathbb{F}_{1^n} - \mathbf{Vect} \longrightarrow \mathbb{F}_1 - \mathbf{Vect}$$
$$V \mapsto V^{\otimes n}.$$

(We recall that the direct sum and the tensor product over $\mathbb{F}_1$ are defined as the disjoint union and the cartesian product respectively.) The functor $\mathrm{aRes}_{\mathbb{F}_{1^n}/\mathbb{F}_1}$ (resp. $\mathrm{mRes}_{\mathbb{F}_{1^n}/\mathbb{F}_1}$) will be referred to as the *additive* (resp. the *multiplicative*) scalar restriction from $\mathbb{F}_{1^n}$ to $\mathbb{F}_1$; hence the notation. Both versions look interesting given than they both appear as adjoints of the scalar extensions. Precisely, for $V \in \mathbb{F}_{1^n} - \mathbf{Vect}$ and $W \in \mathbb{F}_1 - \mathbf{Vect}$, we have:

$$\mathrm{Hom}_{\mathbb{F}_{1^n} - \mathbf{Vect}}(V, \ \mathbb{F}_{1^n} \otimes_{\mathbb{F}_1} W) = \mathrm{Hom}_{\mathbb{F}_1 - \mathbf{Vect}}(\mathrm{aRes}_{\mathbb{F}_{1^n}/\mathbb{F}_1}(V), \ W),$$

$$\mathrm{Hom}_{\mathbb{F}_{1^n} - \mathbf{Vect}}(\mathbb{F}_{1^n} \otimes_{\mathbb{F}_1} W, \ V) = \mathrm{Hom}_{\mathbb{F}_1 - \mathbf{Vect}}(W, \ \mathrm{mRes}_{\mathbb{F}_{1^n}/\mathbb{F}_1}(V)).$$

Besides, $\mathrm{aRes}_{\mathbb{F}_{1^n}/\mathbb{F}_1}(V)$ and $\mathrm{mRes}_{\mathbb{F}_{1^n}/\mathbb{F}_1}(V)$ are both equipped with a Frobenius which acts by permuting cyclically the summands/factors. More concretely, the Frobenius action on $\mathrm{aRes}_{\mathbb{F}_{1^n}/\mathbb{F}_1}(V) \simeq \mathbb{Z}/n\mathbb{Z} \times V$ is given by $(i, x) \mapsto (i+1, x)$ and it is given on $\mathrm{mRes}_{\mathbb{F}_{1^n}/\mathbb{F}_1}(V) \simeq V^n$ by $(x_1, \ldots, x_n) \mapsto (x_2, \ldots, x_n, x_1)$. We observe in particular that $\mathrm{aRes}_{\mathbb{F}_{1^n}/\mathbb{F}_1}(\mathbb{F}_{1^n}) \simeq \mathbb{Z}/n\mathbb{Z}$. In this sense, our definition meets the most standard presentation of $\mathbb{F}_{1^n}$ [15, 28, 44]; the main difference is that we do not retain the group structure on $\mathbb{Z}/n\mathbb{Z}$ but replace it by a Frobenius structure given by the shift. This slight modification in the point of view is actually enough to retrieve a cyclic Galois group of order n.

Proposition 2.1 *The group of automorphisms of* $\mathrm{aRes}_{\mathbb{F}_{1^n}/\mathbb{F}_1}(\mathbb{F}_{1^n})$ *commuting with the Frobenius action is the cyclic group of order n generated by the Frobenius.*

Proof An automorphism of $\mathrm{aRes}_{\mathbb{F}_{1^n}/\mathbb{F}_1}(\mathbb{F}_{1^n})$ is, by definition, a bijection $f : \mathbb{Z}/n\mathbb{Z} \to \mathbb{Z}/n\mathbb{Z}$. Requiring that it commutes with the Frobenius amounts to saying that $f(x + 1) = f(x) + 1$ for all $x \in \mathbb{Z}/n\mathbb{Z}$. Clearly, a function satisfying this condition must be of the form $x \mapsto x + a$, i.e. f is a power of the Frobenius. $\square$

Proposition 2.1 does not extend *verbatim* if we replace $\mathrm{aRes}_{\mathbb{F}_{1^n}/\mathbb{F}_1}$ by $\mathrm{mRes}_{\mathbb{F}_{1^n}/\mathbb{F}_1}$; indeed, given that $\mathrm{mRes}_{\mathbb{F}_{1^n}/\mathbb{F}_1}(\mathbb{F}_{1^n})$ is reduced to one point, its group of automorphisms is trivial as well. However, we can recover the expected group if we consider all objects $V \in \mathbb{F}_{1^n} - \mathbf{Vect}$ at the same time: the group of Frobenius-preserving automorphisms of the *functor* $\mathrm{mRes}_{\mathbb{F}_{1^n}/\mathbb{F}_1}$ is cyclic of order n and generated by the Frobenius.

Passing to the limit, we can similarly set up a theory of vector spaces over $\bar{\mathbb{F}}_1 = \varinjlim_n \mathbb{F}_{1^n}$.

Definition 2.2 An $\bar{\mathbb{F}}_1$-*vector space* is a set. Given two $\bar{\mathbb{F}}_1$-vector spaces V and W, an $\bar{\mathbb{F}}_1$-*linear morphism* $f : V \to W$ is the datum of a sequence $(f_i)_{i \geq 0}$ of partially defined functions from V to W such that for all $x \in V$, the sequence $(f_i(x))_{i \geq 0}$ is periodic.

As before, we have a scalar extension functor $\mathbb{F}_1 - \mathbf{Vect} \to \bar{\mathbb{F}}_1 - \mathbf{Vect}$ which acts trivially on objects and takes a morphism f in $\mathbb{F}_1 - \mathbf{Vect}$ to the constant sequence $(f, f, \ldots)$. More generally, there is a functor $\mathbb{F}_{1^n} - \mathbf{Vect} \to \bar{\mathbb{F}}_1 - \mathbf{Vect}$ mapping an $\mathbb{F}_{1^n}$-linear morphism $(f_1, \ldots, f_n)$ to the sequence $(f_{i \bmod n})_{i \geq 0}$. If V is finite-dimensional over $\bar{\mathbb{F}}_1$ (i.e. if V is a finite set), any $\bar{\mathbb{F}}_1$-linear morphism with domain V comes from an $\mathbb{F}_{1^n}$-linear morphism for some n. Restrictions of scalars also exist in this context. Writing $\hat{\mathbb{Z}} = \varprojlim_n \mathbb{Z}/n\mathbb{Z}$, they are given by:

$$\mathrm{aRes}_{\bar{\mathbb{F}}_1/\mathbb{F}_1} : \quad \bar{\mathbb{F}}_1 - \mathbf{Vect} \longrightarrow \mathbb{F}_1 - \mathbf{Vect}$$
$$V \mapsto \hat{\mathbb{Z}} \times V$$
$$\mathrm{mRes}_{\bar{\mathbb{F}}_1/\mathbb{F}_1} : \quad \bar{\mathbb{F}}_1 - \mathbf{Vect} \longrightarrow \mathbb{F}_1 - \mathbf{Vect}$$
$$V \mapsto \{ \text{periodic sequences with values in } V \}.$$

Furthermore, $\mathrm{aRes}_{\bar{\mathbb{F}}_1/\mathbb{F}_1}(V)$ and $\mathrm{mRes}_{\bar{\mathbb{F}}_1/\mathbb{F}_1}(V)$ are both equipped with a Frobenius endomorphism: on $\mathrm{aRes}_{\bar{\mathbb{F}}_1/\mathbb{F}_1}(V)$, it is $(i, x) \mapsto (i+1, x)$ while it acts by shifting the sequence by 1 on $\mathrm{mRes}_{\bar{\mathbb{F}}_1/\mathbb{F}_1}(V)$. One checks that $\mathrm{Aut}_\varphi(\mathrm{aRes}_{\bar{\mathbb{F}}_1/\mathbb{F}_1}(\bar{\mathbb{F}}_1)) \simeq \mathrm{Aut}_\varphi(\mathrm{aRes}_{\bar{\mathbb{F}}_1/\mathbb{F}_1}) \simeq \mathrm{Aut}_\varphi(\mathrm{mRes}_{\bar{\mathbb{F}}_1/\mathbb{F}_1}) \simeq \hat{\mathbb{Z}}$ where Aut_φ means the Frobenius-preserving automorphisms.

Galois Theory over $\mathbb{Q}_1$

In what follows, we view $\mathbb{Z}_1$ (resp. $\mathbb{Q}_1$) as the $\mathbb{F}_1$-analytic algebra (in the sense of [1]) corresponding to the monoid $\varpi^{\mathbb{N}}$ (resp. $\varpi^{\mathbb{Z}}$) endowed with the norm $\|\varpi^v\| = r^v$ where r is a fixed real number in $(0, 1)$. Here ϖ is a formal notation for the uniformizer of $\mathbb{Z}_1$ and does not have further meaning. Our definition of $\mathbb{Z}_1$ might sound too naive as it seems to identify $\mathbb{Z}_1$ with $\mathbb{F}_1[[\varpi]]$; however, at least for the properties we want to illustrate in this appendix, making this confusion will not have undesirable consequences.

We now aim at defining several families of extensions of $\mathbb{Q}_1$ and studying their Galois properties. We start with unramified extensions: we set $\mathbb{Q}_{1^n} = \mathbb{F}_{1^n} \otimes_{\mathbb{F}_1} \mathbb{Q}_1$ for any positive integer n and $\mathbb{Q}_1^{\mathrm{ur}} = \bar{\mathbb{F}}_1 \otimes_{\mathbb{F}_1} \mathbb{Q}_1$. They are equipped with a Frobenius structure coming from the Frobenius on $\mathbb{F}_{1^n}$ (resp. $\bar{\mathbb{F}}_1$) and with a $\mathbb{Q}_1$-algebra structure materialized by the multiplication morphism by ϖ at the level of monoids. One checks that the group of automorphisms of $\mathrm{aRes}_{\mathbb{F}_{1^n}/\mathbb{F}_1}(\mathbb{Q}_{1^n})$ (resp.

of $\mathrm{aRes}_{\bar{\mathbb{F}}_1/\mathbb{F}_1}(\mathbb{Q}_1^{\mathrm{ur}}))$ commuting with both structures) is isomorphic to $\mathbb{Z}/n\mathbb{Z}$ (resp. to $\hat{\mathbb{Z}}$). We then get the expected Galois group.

We now come to the analogue of the tower of tamely ramified extensions. When p is an actual prime number, this tower is obtained by extracting e-th roots of the uniformizer for e coprime with p. There exists an obvious analogue of this construction over $\mathbb{Q}_1$: for any positive integer e (without any condition of coprimality), we consider the $\mathbb{F}_1$-analytic algebra $\mathbb{Q}_1[\sqrt[e]{\varpi}]$ defined by the underlying monoid $\varpi^{(1/e)\cdot\mathbb{Z}}$ equipped with the norm $\|\varpi^v\| = r^v$ ($v \in \frac{1}{e}\mathbb{Z}$). Clearly $\mathbb{Q}_1[\sqrt[e]{\varpi}]$ is an extension of $\mathbb{Q}_1$ and we can define generally $\mathbb{Q}_{1^n}[\sqrt[e]{\varpi}] = \mathbb{F}_{1^n} \otimes_{\mathbb{F}_1} \mathbb{Q}_1[\sqrt[e]{\varpi}]$ and $\mathbb{Q}_1^{\mathrm{ur}}[\sqrt[e]{\varpi}] = \bar{\mathbb{F}}_1 \otimes_{\mathbb{F}_1} \mathbb{Q}_1[\sqrt[e]{\varpi}]$. It is also possible to take the limit on e and define $\mathbb{Q}_1[\sqrt[\infty]{\varpi}]$ as the $\mathbb{F}_1$-analytic algebra associated to the monoid $\varpi^{\mathbb{Q}}$ with norm $\|\varpi^v\| = r^v$. We write $\mathbb{Q}_1^{\mathrm{tr}} = \bar{\mathbb{F}}_1 \otimes_{\mathbb{F}_1} \mathbb{Q}_1[\sqrt[\infty]{\varpi}]$; it is our candidate for being the maximal tamely ramified extension (or even an algebraic closure?) of $\mathbb{Q}_1$.

Devising a decent Galois theory for the extension $\mathbb{Q}_1^{\mathrm{tr}}/\mathbb{Q}_1^{\mathrm{ur}}$ looks more difficult. Indeed, given that a morphism of monoids $\mathbb{Q} \to \mathbb{Q}$ which acts by the identity on $\mathbb{Z}$ needs to be trivial, we conclude that there is no nontrivial morphism of $\mathbb{Q}_1^{\mathrm{ur}}$-algebras of $\mathbb{Q}_1^{\mathrm{tr}}$ in the sense of Definition 2.1. In order to explain how this issue can be tackled, it will be more convenient to work with finite extensions. For any positive integer n, we define $K_n = \mathbb{F}_1[\sqrt[n]{\varpi}]$; it is the $\mathbb{F}_{1^n}$-algebra represented by the monoid $\eta^{\mathbb{Z}}$ where we have set $\eta = \sqrt[n]{\varpi}$ for simplicity. As we said earlier, there are no nontrivial automorphisms of $\mathbb{Q}_{1^n}$-algebras of K_n. The subtlety is that such automorphisms do exist after restricting scalars to $\mathbb{F}_1$. An explicit example is given by the morphism

$$\sigma_n : \mathrm{aRes}_{\mathbb{F}_{1^n}/\mathbb{F}_1}(K_n) \to \mathrm{aRes}_{\mathbb{F}_{1^n}/\mathbb{F}_1}(K_n)$$

corresponding to the map:

$$\sigma_n^{\sharp} : \mathbb{Z}/n\mathbb{Z} \times \eta^{\mathbb{Z}} \longrightarrow \mathbb{Z}/n\mathbb{Z} \times \eta^{\mathbb{Z}}$$
$$(i, \eta^j) \mapsto (i+j, \eta^j).$$

One checks that $\sigma_n^{\sharp}$ is a morphism of monoids (where the factor $\mathbb{Z}/n\mathbb{Z}$ is endowed with its additive structure) acting by the identity on $\mathrm{aRes}_{\mathbb{F}_{1^n}/\mathbb{F}_1}(\mathbb{Q}_{1^n})$. Therefore, σ_n has all the virtues to be considered as an element of the Galois group $\mathrm{Gal}(K_n/\mathbb{Q}_{1^n})$, although it is not clear to us, here, why we need to retain the monoid structure on $\mathbb{Z}/n\mathbb{Z}$ whereas we argued earlier that it should be discarded in favor of the Frobenius structure. In any case, we have the following proposition.

Proposition 2.2 *The group of automorphisms of monoids of $\mathbb{Z}/n\mathbb{Z} \times \eta^{\mathbb{Z}}$ acting trivially on the submonoid $\mathbb{Z}/n\mathbb{Z} \times \varpi^{\mathbb{Z}}$ is cyclic of order n, generated by $\sigma_n^{\sharp}$.*

Proof Let f be an automorphism of $\mathbb{Z}/n\mathbb{Z} \times \eta^{\mathbb{Z}}$ satisfying the conditions of the proposition. By assumption f fixes $(1, 1)$ and $(0, \varpi)$. Write $f(0, \eta) = (a, \eta^b)$ with $a \in \mathbb{Z}/n\mathbb{Z}$ and $b \in \mathbb{Z}$. Since f is a morphism of monoids, we must have $(na, \eta^{nb}) =$

$(0, \varpi)$, showing that $b = 1$. Hence f takes the form $(i, \eta^j) \mapsto (i+aj, \eta^j)$ and the proposition follows. □

After what precedes, we are tempted to write:

$$\mathrm{Gal}(K_n/\mathbb{Q}_{1^n}) = \langle \sigma_n \rangle \simeq \mathbb{Z}/n\mathbb{Z}.$$

Moreover, noticing that σ_n commutes with the Frobenius, we conclude that:

$$\mathrm{Gal}(K_n/\mathbb{Q}_1) = \langle \varphi, \sigma_n \rangle \simeq (\mathbb{Z}/n\mathbb{Z})^2.$$

Passing to the limit, we would end up with $\mathrm{Gal}(\mathbb{Q}_1^{\mathrm{tr}}/\mathbb{Q}_1^{\mathrm{ur}}) \simeq \hat{\mathbb{Z}}$ and $\mathrm{Gal}(\mathbb{Q}_1^{\mathrm{tr}}/\mathbb{Q}_1) \simeq \hat{\mathbb{Z}}^2$.

In order to give more credit to this conclusion, we would like to make the comparison with the classical case of $\mathbb{Q}_p$ (where p is an actual prime number). For each positive integer n, we set $K_{p,n} = \mathbb{Q}_{p^n}[p^{1/(p^n-1)}]$; it is the maximal totally and tamely ramified Galois extension of $\mathbb{Q}_{p^n}$. As a consequence, the extensions $K_{p,n}$ are cofinal in the maximal tamely ramified extension of $\mathbb{Q}_p$. Besides, the Galois group of $K_{p,n}/\mathbb{Q}_p$ sits in the following exact sequence:

$$1 \longrightarrow \mathrm{Gal}(K_{p,n}/\mathbb{Q}_{p^n}) \longrightarrow \mathrm{Gal}(K_{p,n}/\mathbb{Q}_p) \longrightarrow \mathrm{Gal}(\mathbb{Q}_{p^n}/\mathbb{Q}_p) \longrightarrow 1$$

$$\uparrow \qquad\qquad\qquad\qquad\qquad\qquad \uparrow$$

cyclic of   cyclic of

order p^n-1 order n

which admits a section and provides a presentation of $\mathrm{Gal}(K_{p,n}/\mathbb{Q}_p)$ as a semi-direct product $\mathbb{Z}/n\mathbb{Z} \ltimes \mathbb{Z}/(p^n-1)\mathbb{Z}$ where $a \in \mathbb{Z}/n\mathbb{Z}$ acts on $\mathbb{Z}/(p^n-1)\mathbb{Z}$ by multiplication by p^a.

Naively setting $p = 1$ in the preceding, we find $p^n-1 = 0$, which does not really make sense since $\mathrm{Gal}(K_n/\mathbb{Q}_{1^n})$ cannot reasonably be of cardinality zero. Remembering what we did in Eq. (11), we instead write the factorization:

$$p^n - 1 = (p-1) \cdot (1 + p + \cdots + p^{n-1}) = (p-1) \cdot [n]_p.$$

At the level of groups, the above factorization reflects the fact that $\mathrm{Gal}(K_{p,n}/\mathbb{Q}_{p^n})$ admits a subgroup of order $[n]_p$, namely $\mathrm{Gal}(K_{p,n}/K_{p,1})$. When p goes to 1, the factor $[n]_p$ converges to n and then, passing to the limit, we expect the group $\mathrm{Gal}(K_n/K_1) = \mathrm{Gal}(K_n/\mathbb{Q}_{1^n})$ to be cyclic of order n, which is exactly what we have found earlier. Furthermore, when p tends to 1, the action of $\mathbb{Z}/n\mathbb{Z}$ on $\mathbb{Z}/(p^n-1)\mathbb{Z}$ (and consequently on all its subgroups) becomes trivial, confirming our prediction that $\mathrm{Gal}(K_n/\mathbb{Q}_1)$ should be a direct product of $\mathrm{Gal}(K_n/\mathbb{Q}_{1^n})$ by $\mathrm{Gal}(\mathbb{Q}_{1^n}/\mathbb{Q}_1)$.

Remark 2.1 There is however one small annoying point in what we have said: why is it legitimate to get rid of the factor $(p-1)$? If instead of discarding it without further discussion, we try to keep it, we come to the conclusion that there should be between $\mathbb{Q}_{1^n}$ and K_1 an extension of degree 0 or, say, of infinitesimal degree. This suggests that the extensions $\mathbb{Q}_{1^n}$ and K_1 need to be considered as different objects, which could be a way to explain why the prefactor $\mathbb{Z}/n\mathbb{Z}$ should be endowed with its Frobenius structure in the former case and with its monoid structure in the latter (see the discussion before Proposition 2.2).

Similar to how the factor $(q-1)^n$ in Eq. (11) corresponds to the n-dimensional torus of GL_n, it is tempting to interpret the Galois group of the ghost infinitesimal extension $K_1/\mathbb{Q}_{1^n}$ as the algebraic group $\mathbb{G}_m$ over $\mathbb{F}_1$. Similarly, it sounds plausible to interpret the cyclic group $\mathbb{Z}/n\mathbb{Z}$ (which is supposed to be the Galois group of K_n/K_1) as the group of $\mathbb{F}_1$-points of an algebraic group, maybe $\mathrm{aRes}_{\mathbb{F}_{1^n}/\mathbb{F}_1}(\mathbb{G}_a)$. All of this suggests that $\mathbf{Gal}(K_n/\mathbb{Q}_{1^n})$ could just be the pale reflection of an algebraic group $\mathbf{Gal}(K_n/\mathbb{Q}_{1^n})$ defined over $\mathbb{F}_1$ and sitting in an exact sequence of the form:

$$1 \to \mathrm{aRes}_{\mathbb{F}_{1^n}/\mathbb{F}_1}(\mathbb{G}_a) \to \mathbf{Gal}(K_n/\mathbb{Q}_{1^n}) \to \mathbb{G}_m \to 1.$$

And similarly, passing to the limit:

$$1 \to \mathrm{aRes}_{\bar{\mathbb{F}}_1/\mathbb{F}_1}(\mathbb{G}_a) \to \mathbf{Gal}(\mathbb{Q}_1^{\mathrm{tr}}/\mathbb{Q}_1^{\mathrm{ur}}) \to \mathbb{G}_m \to 1.$$

Beyond its own interest, this interpretation would provide us with a natural Frobenius structure (given by $i \mapsto i + 1$) on $\mathrm{Gal}(\mathbb{Q}_1^{\mathrm{tr}}/\mathbb{Q}_1^{\mathrm{ur}}) \simeq \hat{\mathbb{Z}}$ after taking $\mathbb{F}_1$-points.

References

1. F. Bambozzi, O. Ben-Bassat and K. Kremnizer. Analytic geometry over $\mathbb{F}_1$ and the Fargues–Fontaine curve. *Adv. in Math.* **356**, 106815 (2019).
2. K. Buzzard, F. Diamond and F. Jarvis. On Serre's conjecture for mod ℓ Galois representations over totally real fields. *Duke Math. J.* **155**, 105–161 (2010).
3. C. Breuil, F. Herzig, Y. Hu, S. Morra and B. Schraen. Gelfand–Kirillov dimension and mod p cohomology for GL_2. *Preprint*, 101pp. (2020).
4. C. Breuil. Sur un problème de compatibilité local-global modulo p pour GL_2. *J. Reine Angew. Math.* **692**, 1–76 (2014).
5. C. Breuil and A. Mézard. Multiplicités modulaires et représentations de $\mathrm{GL}_2(\mathbb{Z}_p)$ et de $\mathrm{Gal}(\overline{\mathbb{Q}}_p/\mathbb{Q}_p)$ en $l = p$. *Duke Math. J.* **115**, no. 2, 205–310 (2002).
6. C. Breuil and A. Mézard. Multiplicités modulaires raffinées. *Bull. Soc. Math. France* **142**, 127–175 (2014).
7. C. Breuil and V. Paškūnas. Towards a modulo p Langlands correspondence for GL_2. Memoirs of A.M.S. **216** (2012).
8. A. Connes, C. Consani and M. Marcolli. Fun with $\mathbb{F}_1$. *J. Number Theory* **129**, 1532–1561 (2009).

9. X. Caruso, A. David and A. Mézard. Calculs d'anneaux de déformations potentiellement Barsotti–Tate. *Transactions of the American Mathematical Society* **370**(9), 6041–6096 (2018).
10. X. Caruso, A. David and A. Mézard. Variétés de Kisin stratifiées et déformations potentiellement Barsotti–Tate. *Journal of the Institute of Mathematics of Jussieu* **15**(5), 1019–1064 (2018).
11. X. Caruso, A. David and A. Mézard. Combinatorics of Serre weights in the potentially Barsotti-Tate setting. *Preprint*, 56pp. (2021).
12. A. Caraiani and B. Levin. Kisin modules with descent data and parahoric local models. *Ann. Scient. ÉNS* **51**, 181–213 (2018).
13. A. Caraiani, M. Emerton, T. Gee and D. Savitt. Moduli stacks of two-dimensional Galois representations. *Preprint*, 123pp. (2019).
14. A. Caraiani, M. Emerton, T. Gee, D. Geraghty, V. Paskunas and S.W. Shin. Patching and the *p*-adic local Langlands correspondence. *Cambridge J. Math.* **4**, no. 2, 197–287 (2016).
15. A. Connes. The Witt construction in characteristic one and quantization. In: *Noncommutative geometry and global analysis* **546**, pp. 83–113, Amer. Math. Soc., Providence (2011).
16. A. Connes and C. Consani. Schemes over $\mathbb{F}_1$ and zeta functions. *Compos. Math.* **146**, 1383–1415 (2010).
17. A. David. Poids de Serre dans la conjecture de Breuil–Mézard. *Contemporary Math.* **691**, 133–156 (2017).
18. A. Deitmar. Schemes over $\mathbb{F}_1$. In: *Number fields and function fields two parallel worlds*, Progr. Math. **239**, pp. 87–100, Birkhäuser, Boston (2005).
19. A. Deitmar. $\mathbb{F}_1$-schemes and toric varieties. *Beiträge Algebra Geom.* **49**, 517–525 (2008).
20. M. Emerton and T. Gee. A geometric perspective on the Breuil–Mézard conjecture. *J. Inst. Math. Jussieu* **13**, no. 1, 183–223 (2014).
21. M. Emerton, T. Gee and D. Savitt. Lattices in the cohomology of Shimura curves. *Invent. Math.* **200**, no. 1, 1–96 (2015).
22. L. Fargues and J.-M. Fontaine. Courbes et fibrés vectoriels en théorie de Hodge *p*-adique. *Astérisque* **406**, 382pp. (2018).
23. T. Gee and M. Kisin. The Breuil–Mézard conjecture for potentially Barsotti–Tate representations. *Forum of Mathematics, Pi*, 56pp. (2014).
24. T. Gee, F. Herzig and D. Savitt. General Serre weights conjectures. *J. Eur. Math. Soc. (JEMS)* **20**, no. 12, 2859–2949 (2018).
25. G. Henniart. Sur l'unicité des types pour GL$_2$, Appendix of [5], *Duke Math J.* **115**, no. 2, 298–310 (2002).
26. F. Herzig, D. Le and S. Morra. On mod *p* local-global compatibility for GL$_3$ in the ordinary case. *Compo. Math.* **153**, 2215–2286 (2017).
27. Y. Hu and F. Tan. The Breuil–Mézard conjecture for non-scalar split residual representations. *Ann. Scient. ÉNS* **48**(6), 1383–1421 (2013).
28. M. Kapranov and A. Smirnov. Cohomology determinants and reciprocity laws: number field case. *Unpublished letter* (1995).
29. M. Kisin. The Fontaine–Mazur conjecture for GL$_2$. *J. Amer. Math. Soc.* **22**, 641–690 (2009).
30. M. Kisin. Moduli of finite flat group schemes and modularity. *Ann. of Math.* **107**, 1085–1180 (2009).
31. M. Kisin. Potentially semi-stable deformation rings. *J. Amer. Math. Soc.* **21**, 513–546 (2008).
32. K. Koziol and S. Morra. Serre weight conjectures for *p*-adic unitary groups of rank 2. To appear in *Algebra & Number Theory*, 79pp. (2020).
33. D. Le, B. Le Hung, B. Levin and S. Morra. Serre weights and Breuil's lattice conjecture in dimension three. *Forum of Math, Pi* **8**, e5, 135pp. (2020).
34. D. Le, B. Le Hung, B. Levin and S. Morra. Potentially crystalline deformation rings and Serre weight conjectures: shapes and shadows. *Inventiones Mathematica* **212**(1), 1–107 (2018).
35. D. Le, S. Morra and B. Schraen. Multiplicity one at full congruence level. *J. Inst. Math. Jussieu* **21**, no. 2, 637–658 (2022).
36. O. Lorscheid. Algebraic groups over the field with one element. *Math. Z.* **271**, 117–138 (2012).
37. O. Lorscheid. $\mathbb{F}_1$ for everyone. *Jahresber. Dtsch. Math.-Ver.* **120**, 83–116 (2018).

38. J. López Peña and O. Lorscheid. Torified varieties and their geometries over $\mathbb{F}_1$. *Math. Z.* **267**, 605–643 (2011).
39. V. Paškūnas. On the Breuil–Mézard conjecture. *Duke Math. J.* **164**(2), 297–359 (2015).
40. G. Pappas and M. Rapoport. φ-modules and coefficient spaces. *Mosc. Math. J.* **9**, 625–663 (2009).
41. G. Pappas and X. Zhu. Local models of Shimura varieties and a conjecture of Kottwitz. *Invent. Math.* **194**, 147–254 (2013).
42. F. Sander. A local proof of the Breuil–Mézard conjecture in the scalar semi-simplification case. *J. Lond. Math. Soc.* **94**(2), 447–461 (2016).
43. P. Schneider and E.-W. Zink. K-types for the tempered components of a p-adic general linear group. With an appendix by P. Schneider and U. Stuhler. *J. Reine Angew. Math.* **517**, 161–208 (1999).
44. C. Soulé. Les variétés sur le corps à un élément. *Mosc. Math. J.* **4**, 217–244 (2004).
45. J. Tits. Sur les analogues algébriques des groupes semi-simples complexes. In: *Colloque d'algèbre supérieure*, tenu à Bruxelles du 19 au 22 décembre 1956, Centre Belge de Recherches Mathématiques, Établissements Ceuterick, Louvain, pp. 261–289 (1957).
46. B. Toën and M. Vaquié. Au-dessous de Spec $\mathbb{Z}$. *J. K-Theory* **3**, 437–500 (2009).
47. A. Vezzani. Deitmar's Versus Toën–Vaquié's schemes over $\mathbb{F}_1$. *Math. Z.* **271**, 911–926 (2012).

Computational Number Theory, Past, Present, and Future

Henri Cohen

For Catriona Byrne, with thanks

1 Introduction

This paper is a *very personal* account of some computational aspects of number theory, especially in relation to the `Pari/GP` computer algebra system [104]. It is in no way exhaustive, but highlights significant advances that I have personally encountered.

Without going back too far in time, one of the pioneering figures in the subject was D.H. Lehmer (and to a lesser extent his father D.N. Lehmer) who introduced a number of methods some of which are still in use today. He is probably best known for the Lehmer conjecture dealing with finding a polynomial with smallest nonzero logarithmic Mahler measure [87] (see also [123] for recent work). I had the privilege of meeting him once in Berkeley when I was still in high school. Although not a number theorist per se, one can also mention A. Turing [133] who made extensive computations on the Riemann Hypothesis using a method that is still used and bears his name, see also [30].

More recently, in the 1960s and 1970s D. Shanks, who by training was not a professional mathematician, made a number of very significant contributions to computational number theory, see for instance [117]. To name a few, the baby-step giant step method, which allows to find a given element in a group of size N in time $O(N^{1/2})$, the infrastructure of real quadratic fields, which was in some sense a precursor to Arakelov theory, and an essential piece in the development of algorithms for computing class and unit groups, the Tonelli–Shanks algorithm for computing square roots modulo p, as well as less important but still interesting 6-letter methods (coming from FORTRAN) such as NUCOMP and SQUFOF, the

H. Cohen (✉)
Univ. Bordeaux, CNRS, INRIA, Bordeaux INP, IMB, UMR 5251, Talence, France
e-mail: cohen@math.u-bordeaux1.fr

© The Author(s), under exclusive license to Springer Nature Switzerland AG 2023
J.-M. Morel, B. Teissier (eds.), *Mathematics Going Forward*, Lecture Notes in Mathematics 2313, https://doi.org/10.1007/978-3-031-12244-6_38

latter allowing to factor 18-digit numbers on a pocket calculator using only 10 digits. I refer to [37] for details on all these methods.

In parallel and during the same period, more "serious" computational mathematics was being done: first and foremost, the computations of B. Birch and P. Swinnerton-Dyer in the 1960s on the Mordell–Weil group of rational elliptic curves, leading to the famous BSD conjecture [28], which is one of the most outstanding conjecture in number theory, on par with the Riemann Hypothesis; the work of D. Tingley [132] leading to tables of modular elliptic curves in Antwerp IV [29]; and later the work of J. Buhler proving the existence of an icosahedral weight 1 modular form in level 800 [35]; work of O. Atkin on many computational aspects of modular forms such as the Atkin–Lehner operators [8], non-congruence subgroups [10], congruences between modular forms, etc.; work of H. Stark on Stark units, leading to the Stark conjectures [124], as well as the first explicit (although at the time non-rigorous) computation of cuspidal Maass forms in level 1 [125] and also [31] for a modern and rigorous treatment; this was followed by extensive work of D. Hejhal on this subject [81]; work of A. Odlyzko on verifying the Riemann Hypothesis at very large heights and relations with the GUE hypothesis, and in particular the Odlyzko–Schönhage algorithm for computing $\zeta(s)$ for large $\Im(s)$ [103].

In the late 1970s and early 1980s, a flurry of activity took place around primality testing and factoring. This ultimately led to the two leading practical primality tests, the APRCL (Adleman–Pomerance–Rumely–Cohen–Lenstra) test using Jacobi sums and cyclotomic fields [48] and [49], and the ECPP (Elliptic Curve Primality Proving) algorithm using elliptic curves by O. Atkin and F. Morain [9]. Later, the AKS (Agrawal–Kayal–Saxena) test [1] proved that primality proving is polynomial-time, but this test is less practical than the previous ones.

For factoring, the decisive step was taken by J. Pollard (again a nonprofessional mathematician) which led to the NFS (Number Field Sieve) algorithm [90], other important algorithms being the MPQS (Multiple Polynomial Quadratic Sieve), in large part due to C. Pomerance [108], and the ECM (Elliptic Curve Method), due to H. W. Lenstra, Jr. [89]. All these algorithms are still in use today.

In parallel, R. Schoof used "ℓ-adic" techniques to create a polynomial time algorithm for counting points on elliptic curves over prime finite fields [116], which was later improved by O. Atkin and N. Elkies into the SEA (Schoof–Elkies–Atkin) algorithm [63]. For elliptic curves over fields of small characteristic, J.-F. Mestre [98] and T. Satoh [115] invented incredibly simple algorithms for this task (only a few lines of programming necessary, see for instance Algorithm 17.58 in [47] for Mestre's AGM based algorithm in characteristic 2). Later, this was largely generalized to all hyperelliptic curves (and other varieties) by K. Kedlaya [85]. These algorithms are "p-adic" in nature, as opposed to ℓ-adic above.

2 The Development of Pari/GP and Computational Number Theory Books

There are two ways of doing number theory on a computer: either program in a standard low-level computer language such as C or even directly in assembly, or use high-level software such as Maple or Mathematica. The first method is by far the most efficient, but is extremely cumbersome, since for instance multiprecision operations are not available, at least directly, in these languages. In the early 1980s, there existed a few Computer Algebra Systems, but first they were mostly tailored to perform computations in applied mathematics and numerical analysis and not number theory, and second they were very slow for the little number theory that they could do.

With the help of a few colleagues, first F. Dress, then C. Batut, D. Bernardi, and M. Olivier, in 1985 we embarked on the daunting task of writing a complete computer algebra system with two goals in mind: first speed and efficiency, second specifically tailored to number-theoretic computations, although we also included some numerical analysis tools. Once the basic functionality written (which required 2 years of hard work, including tens of thousands of lines of assembly language code), we included algorithms for working with common mathematical objects, and recent groundbreaking algorithms such as the LLL (Lenstra–Lenstra–Lovász) algorithm [91] (see [100] and [101] for later work), one of the most useful algorithms invented at the end of the twentieth century. In friendly competition with the KANT group led by M. Pohst, we also completed H. Zassenhaus' program consisting in computing rings of integers, unit groups, and class groups of algebraic number fields. The pioneering work of Hafner–McCurley [77], followed by that of J. Buchmann [33], led us to the first program able to compute class and unit groups of general number fields in reasonable time and at the simple press of a keystroke. I still recall our elation in seeing the class groups of several thousand cubic fields being computed in the blink of an eye, since previously, computing even a few could constitute a Masters thesis. Later, with the help of D. Bernardi and J.-F. Mestre, we also wrote a number of programs for working with elliptic curves over the rationals, not including rank computations.

In parallel with the writing of these programs, I decided to write down explicitly the algorithms used, since they were either scattered in the literature, or completely new such as Buchmann's algorithm. This resulted in quite a large manuscript (more than 500 pages), and on the occasion of the 1991 Arbeitstagung in Bonn I was introduced to Catriona Byrne, who after sending the book to a number of referees, accepted the manuscript, which was published with the title "A Course in Computational Number Theory" as Springer GTM 138 [37]. Since then we stayed in professional contact, at ICMs and on the occasion of the publication of my later books, and I was very pleased that she attended the conference given in Bordeaux for my 60th birthday in 2007.

I was of course very happy by the success of this book (I believe it has now reached a circulation of more than 10,000 copies, Springer can confirm this), but evidently almost 30 years later a large part is outdated since computational methods change much faster than mathematical theories do.

Note that previously almost all books on computational number theory were mostly conference proceedings such as [92] and [109], and/or devoted to a specific subject, with the exception of [140] in 1972, and [106] which appeared essentially at the same time as [37] in 1993.

Shortly after the book appeared, J. Martinet (who was both my thesis advisor and the chairman of our department) gave a course on class field theory, using the classical language of *moduli*, instead of the more modern language of adèles and idèles. That modern language had always frightened me, and I considered class field theory as a difficult subject. Martinet's course opened my eyes, and I soon realized that class field theory could, with some effort, be included in computer packages. For this purpose I had to develop some elementary but apparently new machinery for dealing with relative extensions, and also I rewrote Martinet's course (omitting many proofs) in a way which was more suited to computer implementation. Exactly as in the beginning of `Pari/GP`, this led both to an extensive `Pari` library for computing in relative extensions, computing ray class groups and ray class fields (helped by F. Diaz y Diaz and M. Olivier [43]), and to the writing of a new book, naturally called "Advanced topics" in computational algebraic number theory, published in 2000 as Springer GTM 193 [38]. This book contains in particular a very understandable description of class field theory in classical language, in large part based on Martinet's course.

For completeness (and self-advertising) I also mention my two other books [39] and [40] which are much less computationally oriented but contain a very large amount of material and exercises on modern number theory.

3   Arithmetic Statistics

A considerable number of additional algorithmic methods for number fields have since been found. In particular, in the domain of making *tables* of number fields, in addition to the brute force methods using theorems of Hunter and Martinet, class-field theoretic methods have been used to compute certain classes of number fields, in particular quartic fields [45] and [46], very elegant methods for computing cubic fields based on the Delone–Fadeev correspondence have been devised by K. Belabas [15], and much more recently the work of M. Bhargava [22] and [23] has led to much more efficient methods for computing S_4 quartic fields [138].

The rest of this section is more theoretical, but intimately connected to computational number theory.

The problem of *enumerating* number fields (usually, but not always, ordered by discriminant) has attracted a lot of attention. Denote by $N_n(G, X)$ the number of isomorphism classes of number fields of degree n, absolute discriminant less than X, and Galois group of their Galois closure isomorphic to G, and by $N_n(X)$ if G is not specified. The case $n = 2$ is trivial, The case $G = C_3$ is easy and first published by H. Cohn [52], $G = C_4$ and $G = C_2 \times C_2$ are due to A. Baily [11], although his formulas need to be corrected, and the general case where G is abelian has been treated by S. Mäki in her thesis (Helsinki, 1985), see also [96].

Non-abelian groups are much more difficult. The case of S_3 cubic fields was solved by Davenport and Heilbronn using the Delone–Fadeev correspondence [58], the case of D_4 quartic fields is due to F. Diaz y Diaz, M. Olivier, and the author [45], and S_4 quartic and S_5 quintic fields are due to the fundamental pioneering work of M. Bhargava [23, 24] using prehomogeneous vector spaces and a careful point counting inside multidimensional fundamental domains.

A folk conjecture predicts that the total number of number fields of absolute discriminant less than X should be asymptotic to $c \cdot X$ for a suitable positive constant c. A much more precise general conjecture due to G. Malle [97] predicts that $N_n(G, X) \sim c(G) \cdot X^{a(G)} \log(X)^{b(G)}$ for explicit constants $a(G)$ and $b(G)$ and a nonexplicit positive constant $c(G)$ (his initial conjectured value of $b(G)$ needs to be corrected in certain cases; note that this conjecture implies that $N_n(G, X) > 0$ for sufficiently large X, i.e., the truth of the inverse Galois problem, which is also conjectural), and in particular that $N_n(X) \sim c_n \cdot X$ for $n \geq 2$ for some $c_n > 0$.

Malle's conjecture is known to be true in a number of cases in addition to the ones already mentioned, but is far from being proved in general. For instance it predicts that $N_4(A_4, X) \sim c\, X^{1/2} \log(X)$, but the best known unconditional upper bound is $N_4(A_4, X) = O(X^{0.7785})$ [25] (even conditionally, the best known is $O(X^{2/3})$ [139]). For another example, it is not known whether $N_6(X) = O(X)$, the trivial bound being $O(X^2)$; it is only recently (April 2022) that several authors [6], [26] succeeded in improving on this trivial bound, the best result being $O(X^{61/32+\epsilon})$ for any $\epsilon > 0$, still far from the conjectured result. Even more frustrating, it is widely conjectured that the number of cubic fields with *given* discriminant X is $O(X^\epsilon)$ for any $\varepsilon > 0$, but the trivial class-field theoretic bound only gives $O(X^{1/2})$, and the current best bound is $O(X^{1/3})$ [65].

On the other hand, an important step towards the folk conjecture was made by J.-M. Couveignes [54], later slightly improved by other authors in [88], who show that $N_n(X) = O_n(X^{c \cdot \log(n)^2})$ for a suitable constant c.

Another aspect of arithmetic statistics, closely linked to the above-mentioned works via class field theory, is to study the distribution of class groups of number fields. The basic conjectures were proposed by Lenstra, Martinet, and the author in [50] and [51], although the latter should be modified in the presence of roots of unity, see for instance [134] and [13] for recent approaches and references.

These conjectures have been proved only in a very small number of cases: the theorems of Davenport–Heilbronn and the works of Bhargava et al. already mentioned, and in the special case $p = 2$ for quadratic fields in the remarkable work of A. Smith [122], following work of F. Gerth [73] and Fouvry–Klüners [72].

In a different direction, the so-called Odlyzko bounds, due in fact to many people (H. Stark, A. Odlyzko, G. Poitou, J.-P.-Serre, F. Diaz y Diaz, see the references in Odlyzko's survey [102]) led to considerable work towards finding number fields with smallest possible absolute discriminant for given degree and signature, see for instance [44].

A considerable amount of work has also been done on statistics for elliptic curves. Without going into too much detail, probably the most spectacular heuristic is due to B. Poonen and E. Rains [111], predicting in particular that there should exist only a finite number of (isomorphism classes of) elliptic curves defined over $\mathbb{Q}$ with Mordell–Weil rank strictly larger than 21 (the current record is due to N. Elkies with a curve of rank 28, see [64]), see [110] for additional references. Note that 21 would be very close to optimal since another result of Elkies shows that there are infinitely many elliptic curves defined over $\mathbb{Q}$ with rank greater or equal to 19, see again [64]. A different heuristic due to A. Granville and M. Watkins [136] also gives 21 as an upper bound.

4   Automorphic Forms, L-Functions, and `Pari/GP` Implementations

Notwithstanding all this work, in the past 25 years, the emphasis has turned away from number fields, and more toward more algebro-geometric objects and automorphic forms, in particular related to the Langlands program.

Already in the early 1990s, J. Cremona launched an extensive computation to tabulate elliptic curves defined over $\mathbb{Q}$ (he has reached conductor 500,000), and has written a very nice book giving all the details [56]. In addition, he provided the number-theoretic community with the `mwrank` program, which in many cases is able to compute the Mordell–Weil group of a rational elliptic curve.

Cremona's computation of elliptic curves defined over $\mathbb{Q}$ is based on the use of modular symbols for computing spaces of modular forms of weight 2 with trivial character. In collaboration with N. Skoruppa and D. Zagier, we developed algorithms for computing spaces of modular forms of any even weight with trivial character, later generalized by Skoruppa to forms with character (Nebentypus). The method was completely different since based on the Eichler–Selberg trace formula. We computed and even printed large tables, which were used by a small circle but were never published, but see below.

Since the 1990s a large number of papers appeared containing new or improved algorithms and programs in computational number theory. In particular, the ANTS (Algorithmic Number Theory Symposia) series held every 2 years since 1994 holds a wealth of information, see the 14 volumes of [3], and see also [47], which is more oriented towards cryptographic applications.

I will now focus on what I know best, without minimizing the importance of subjects that I do not mention.

The appearance of the Computer Algebra Systems `magma` [32] (headed by J. Cannon) and `Sage` [114] (headed by W. Stein) gave the (non-numerical) mathematical community powerful additional tools, although the strictly number-theoretic part of `Sage` mostly comes from the use of `Pari/GP`. One of the initial ingredients of `Sage` was a package written by W. Stein for computing spaces of modular forms using modular symbols, which is very nicely explained in his book

[126]. Many of the implementations that I will now describe are also available in magma (and of course also in Sage since it contains the Pari code).

In the Pari/GP system [104], a number of very important new implementations have been included, which can be roughly divided into five categories, although almost all these improvements are interwoven. This is the main strength of the Pari library: so many arithmetic functions are available and used internally in so many places that even localized improvements or better design concepts quickly have major impacts elsewhere.

4.1 Algebraic Number Fields

- A considerably more efficient computation of the class and unit groups of algebraic number fields due to the work of B. Allombert, K. Belabas and L. Grenié over 20 years [19] and [76].
- The systematic use of compact representations of elements and in particular of S-units in number fields by K. Belabas following H. Williams' original ideas, absolutely essential for many applications, see [34] and [120].
- A Thue equation solver, by G. Hanrot [27].
- A fast polynomial factorization engine over number fields by K. Belabas building on earlier work of X. Roblot and the revolutionary ideas introduced in the LLL method by M. van Hoeij [16] and [21].
- A large number of Galois-theoretic functions by B. Allombert [4].
- A complete rewrite of basic finite fields arithmetic (including polynomial factorization and many multimodular methods) by B. Allombert [5] and asymptotically fast linear algebra by P. Bruijn, including fast linear algebra over cyclotomic rings by B. Allombert [84].
- On the fly computation of number fields with given Galois group and local data by K. Belabas and the author, see in particular [46].
- A much more efficient program for Kummer extensions and computing class fields by K. Belabas, L. Grenié and A. Page using C. Fieker's ideas.

4.2 Elliptic and Hyperelliptic Curves

- Many new algorithms for elliptic curves over number fields and p-adic fields by B. Allombert, K. Belabas, and B. Perrin-Riou.
- Isogenies by H. Ivey-Law and B. Allombert.
- Modular and class polynomial computations by A. Enge, H. Ivey-Law, and A. Sutherland, see [66, 68, 129], and [128].
- Pairings by J. Milan and B. Allombert.
- ECPP implementation by J. Asuncion.

- Mordell–Weil group of elliptic curves by B. Allombert, K. Belabas and D. Simon, extending D. Simon's original GP scripts and considerably more efficient than J. Cremona's initial very useful mwrank program.
- Improvements of the Heegner point method using ideas of J. Cremona and M. Watkins by B. Allombert [135].
- Implementation of the SEA point-counting algorithm by B. Allombert, C. Doche, and S. Duquesne.
- Kedlaya's algorithm [85] to compute the characteristic polynomial of the Frobenius automorphism on a hyperelliptic curve by B. Allombert.
- Reduction of genus 2 curves by K. Belabas and Q. Liu [94].
- A port by B. Allombert of the ratpoints program written by M. Stoll [127] which searches for rational points of small height on hyperelliptic curves.

4.3 L-Functions and Automorphic Forms

- Numerical computation of arbitrary (motivic) L-functions, initially based on a paper and a GP script due to T. Dokchitser [61], but largely enhanced thanks to ideas of A. Booker, P. Molin, and the Pari group, see [41, 42], and [18].
- Computation of modular form spaces by K. Belabas and the author, again using the Eichler–Selberg trace formula, but enormously enhanced: in particular, it can compute modular forms of weight 1, of half-integral weight, expansions at arbitrary cusps, Petersson products, etc., see [17] for complete details.
- Isomorphisms of lattices by B. Allombert, porting B. Souvignier's implementation of the Plesken and Souvignier algorithm [105].
- Modular symbols by K. Belabas and B. Perrin-Riou [20] (after R. Pollack and G. Stevens [107]), analogous to W. Stein's initial one and more tailored towards the computation of p-adic L-functions attached to modular forms.
- Associative and central simple algebras due to A. Page, complementing similar work done by J. Voight.
- Hypergeometric motives and their L-functions by K. Belabas and the author, based on ideas of N. Katz, F. Rodriguez-Villegas and M. Watkins, see [14] and [113].
- Implementation of arbitrary Hecke Grössencharacters by P. Molin and A. Page [99].

4.4 Numerical Methods

A large number of arbitrary precision numerical methods, many of them new, have been implemented:

- A. Schönhage's polynomial root finding method, which guarantees to find all complex polynomial roots to a given accuracy, as implemented by X. Gourdon [75].
- A port by B. Allombert of the `fplll` software written by D. Stehlé implementing very efficient floating point versions of the LLL algorithm due to D. Stehlé and P. Nguyen [100].
- Numerous methods for numerical summation (in particular discrete Euler–McLaurin and Monien summation), numerical integration (in particular Gauss–Legendre integration and doubly-exponential methods), extrapolation (in particular Lagrange and Sidi extrapolation), asymptotic expansions, and efficient evaluation of continued fractions. All of these algorithms are explained in great detail (including GP code) in the recent book [18] of K. Belabas and the author.
- Computation of transcendental functions, both elementary, and higher transcendental functions, in particular hypergeometric functions, as well as p-adic transcendental functions.
- Multiple zeta values and multiple polylogarithms, based on work of P. Akhilesh [2] and the author.
- Computation of Dirichlet L-functions for large imaginary part of the argument, using either K. Fischer's `zetafast` algorithm [70] or the Riemann–Siegel formula, see [7] for $\zeta(s)$ and [119] for Dirichlet characters.

4.5 Software Enhancements

On the non-mathematical side, one can mention the following:

- The use of the highly optimized `gmp` multiprecision library to replace most of the integer arithmetic, in particular our own assembly code.
- The GP2C compiler written by B. Allombert which translates GP scripts into pure C code which can be 3 or 4 times faster and can be incorporated into standalone programs or be used to learn `libpari` programming.
- The possibility of using parallelism in `Pari/GP` programs (POSIX threads or MPI) with essentially no effort nor additional programming, also due to B. Allombert. The underlying mechanism is also heavily used internally in many algorithms, without user intervention, to benefit from the now ubiquitous multicore machines (pthreads) or launch massive jobs on computing clusters (MPI).

5 Additional Available Software and Algorithms

As already mentioned, both `magma` and `Sage` are very large systems containing much more than computational number theory. But in addition to the programs

provided by these systems, the most important additional resource is the LMFDB
(L-function and Modular Form Database) [93] and [57], which contains a huge
amount of interconnected tables related to computational number theory, which
can be trivially downloaded and used in Pari/GP, magma, or Sage. This is a
collaborative effort by almost a hundred people, and has become an essential tool
for working on the subject.

Unrelated but also very useful is the arb system [83] developed by F. Johansson
which guarantees the accuracy of numerical results by working in ball arithmetic,
and which in particular contains a very large number of transcendental functions.
Not only are the results *guaranteed*, but in addition the implementations are
considerably more efficient than in other systems. Note also the paritwine
package [67] which allows easy access to many arb functions inside a Pari/GP
session.

I would also like to mention the following additional works, again far from being
exhaustive:

*   Implementations of Hilbert modular forms by L. Dembelé and J. Voight [59], as
    well as later work.
*   Implementation of Bianchi modular forms by J. Cremona [55] as well as later
    work, see the LMFDB [93] and [57] for further references.
*   Implementation of certain types of Siegel modular forms by many people.
*   Implementation of p-adic L-functions by R. Pollack and C. Wuthrich.
*   Work of D. Farmer's group on creating L-functions "out of thin air", and in
    particular in making tables of GL_3 and GL_4 Maass forms [69].
*   Work of Poor and Yuen on paramodular forms, and in parallel of A. Brumer on
    abelian surfaces, in the direction of the paramodular conjecture [112] and [36].
*   The very efficient use of the p-adic Gross–Koblitz formula for counting points
    over finite fields by several people, see for instance [18].
*   Quasi-linear time computation of coefficients of motivic L-functions by D. Har-
    vey, K. Kedlaya, A. Sutherland et al., and application to classification of
    Sato–Tate groups, as well as other important contributions of these authors, see
    for instance [78, 79, 130, 131], and [71].
*   Chabauty–Coleman methods [95] to compute all rational points on curves and
    certain other varieties, and generalizations such as Kim's non-abelian Chabauty
    [53],
*   Generalizing Schoof's algorithm, the *polynomial time* (but not practical) al-
    gorithm of J.-M. Couveignes, B. Edixhoven[1] et al. [62], for computing the
    Ramanujan tau function and more generally Hecke eigenvalues.
*   The work of K. Khuri-Makdisi on efficient computations on Jacobians of curves
    [86].

---

[1] My friend and colleague Bas Edixhoven died suddenly and very prematurely on January 16,
2022.

- The recent proof by D. Harvey and J. van der Hoeven [80] of the existence of a $O(n\log(n))$ algorithm for multiplying n-bit numbers, which had been conjectured for more than 50 years.

6 The Future

In the same way that fundamental research is essential for practical applications, theoretical progress on mathematical conjectures is often essential for computational uses, but conversely, not only computational experiments often lead to important conjectures (I have already mentioned BSD and the Stark conjectures, but there are many other examples), but in conjunction with theoretical advances can lead to *proofs*. Let me specialize to number theory.

Many theorems in number theory are *non-effective*, in that one knows that some property is true for a "sufficiently large" (but unspecified) number, or that some quantity is larger than $C \cdot f(x)$ for a known function $f(x)$ but an unspecified constant C, or similar. It is then useless to do computations since we will never know when the "sufficiently large" is attained. Some other theorems are effective, but the implied constants are so large that the computations become unfeasible.

In these cases, computational methods can be applied *only* after some *theoretical* progress is made. Let me give a few examples.

- A special case of a well-known theorem of Brauer–Siegel implies that for any $\varepsilon > 0$ the class number $h(D)$ of an imaginary quadratic field of discriminant D is greater than $|D|^{1/2-\varepsilon}$ for $|D|$ sufficiently large, but *non-effectively*. In particular $h(D) \to \infty$ with $|D|$. The class number 1 problem ($h(D) > 1$ for $|D| > 163$) was famously solved by Heegner–Stark and Baker, and the class number 2 problem ($h(D) > 2$ for $|D| > 427$) by Stark and Baker using Baker's lower bounds for linear forms in logarithms. But it wasn't until the combined work of D. Goldfeld and B. Gross–D. Zagier that a very weak but explicit lower bound tending to infinity for $h(D)$ was found, using a very clever method, see [74] for an overview of all this. Computational methods could then be applied, and in this way M. Watkins [137] was able to solve the class number h problem for all $h \leq 100$ (one could go further if desired).
- A theorem of Siegel states that the number of *integral* points on any model of an elliptic curve defined over $\mathbb{Q}$ is finite, but non-effectively. Progress on this was made thanks to two advances, the main one being theoretical: thanks to Baker-type theorems on linear forms in *elliptic* logarithms, Siegel's theorem could be made effective, but with bounds of the type 10^{10^4}. The second advance was the crucial use of the LLL algorithm to reduce in 2 or 3 steps the bound to something manageable, so that finding all integral points on an elliptic curve is now routine (as long as the Mordell–Weil group is known), see [121] or Section 8.7 of [39].
- A theorem of Faltings (previously known as Mordell's conjecture) tells us that the number of rational points on a curve of genus $g \geq 2$ is finite, again non-

effectively. In this case, theoretical advances have been constant, but slow. When the rank r of the Mordell–Weil group of the Jacobian of the curve satisfies $r < g$, Faltings' result is in fact effective and due to C. Chabauty and R. Coleman. In the past few years, considerable progress has been made when $r = g$ and in some cases $r > g$, but we are still far from a satisfactory situation analogous to integral points on elliptic curves, see [53] for a survey.

• Several theorems in analytic number theory are effective, but with implied constants that would a priori seem inaccessible to computation. Once again, it is thanks to theoretical advances such as a very careful analysis of the so-called "minor arcs" or similar, that the theorems have been completed, usually after a very large computation. Two examples: Waring's problem for 4th powers (every integer $n > 13{,}792$ is a sum of 16 4th powers [60], and every integer is a sum of 19 such [12]) and Goldbach's conjecture for odd integers (every odd integer $n > 5$ is a sum of at most three primes) by H. Helfgott [82].

Therefore it seems reasonable to believe that future important progress in computational number theory will come from theoretical advances.

For instance, a very important problem (which may never be solved) is to find efficient algorithms for finding Euler factors and local conductors of motivic L-functions at *bad* primes. Apart from brute force searches (the search domain being finite), one of the most general methods consists in writing systems of linear or nonlinear equations and trying to solve them using the functional equation of the L-function, but these methods fail as soon as the problems get large.

For specific types of L-functions, one has specific theorems and/or algorithms, the most famous being Tate's algorithm for elliptic curves. One also has algorithms for curves of genus 2 and partial results in higher genus, for certain other varieties, for symmetric powers of modular forms, for Hecke L-functions, and for Artin L-functions.

A general algorithm would involve computing explicitly ℓ-adic cohomology groups, which for now seems out of reach except in special cases including those mentioned above.

Other future goals may be the generalization of the Riemann–Siegel formula to L-functions of degree larger than 1, for instance attached to classical modular forms, a rigorous understanding of Dokchitser's heuristic continued fraction method for computing inverse Mellin transforms, obtaining more efficient methods for finding L-functions knowing only their functional equation and arithmetic properties by building on the work of D. Farmer, algorithmic methods for more general automorphic forms, improvement in the computation of Mordell–Weil groups of abelian varieties including the use of n-descent for $n \geq 3$, and improvements of Chabauty-like methods for finding rational points.

One can also have even more inaccessible dreams: first, of finding a polynomial time algorithm for factoring, or at least faster than the number field sieve. But perhaps less inaccessible, it is quite frustrating that on the one hand, given a rational elliptic curve of rank 1, the Heegner point method can find a generator very efficiently, while if the curve has rank 2, say, and one rational point is known, there is no efficient general method for finding a second independent rational point. This is intimately linked to the fact that the BSD conjecture is totally open for curves of rank greater than or equal to 2.

One can also consider paradigm shifts in computational problems. Until rather recently, the Graal was to find polynomial time algorithms (possibly probabilistic or depending on unproved hypotheses such as GRH) for computational problems. Since then, the emphasis is sometimes more on finding quasi-linear algorithms (with respect to the input and output size). In view of the possible existence of quantum computers, the possibilities of quantum computability opens also a wide scope for research, the prototypical example being Shor's factoring algorithm [118], but many other applications of quantum computing have since been found.

References

1. M. Agrawal, N. Kayal, and N. Saxena. PRIMES is in P. *Ann. Math.* **160**, 781–793 (2004).
2. P. Akhilesh. Double tails of multiple zeta values. *J. Number Theory* **170**, 228–249 (2017).
3. Algorithmic Number Theory Symposia I to IX. Lecture Notes in Computer Science **877** (1994), **1122** (1996), **1423** (1998), **1838** (2000), **2369** (2002), **3076** (2004), **4076** (2006), **5011** (2008), **6197** (2010), Springer-Verlag; X, XIII, and XIV, Open Book Series **1** (2012), **2** (2018), **4** (2020), XI and XII, *J. of Computation and Math.* **17A** (2014), **19A** (2016), London Math. Soc.
4. B. Allombert. *An efficient algorithm for the computation of Galois automorphisms.* *Math. Comp.* **73**, 359–375 (2001).
5. B. Allombert. Explicit computation of isomorphisms between finite fields. *Finite Fields Appl.* **8**, 332–342 (2002).
6. T. Anderson et al., *Improved bounds on number fields of small degree*, arXiv:2204.01651 (2022).
7. J. Arias de Reyna. High precision computation of Riemann's zeta function by the Riemann–Siegel formula, I. *Math. Comp.* **80** (2011), 995–1009, and *II*, arXiv:2201.00342 (2022).
8. O. Atkin and J. Lehner. Hecke operators on $\Gamma_0(m)$. *Math. Annalen* **185**, 134–160 (1970).
9. O. Atkin and F. Morain. Elliptic curves and primality proving. *Math. Comp.* **61**, 29–68 (1993).
10. O. Atkin and P. Swinnerton-Dyer. *Modular forms on congruence subgroups.* Proc. Sympos. Pure Math. **19**, American Math. Soc. 1–25 (1971).
11. A. Baily. On the density of discriminants of quartic fields. *J. reine angew. Math.* **315**, 190–210 (1980).
12. R. Balasubramanian, J.-M. Deshouillers and F. Dress. Problème de Waring pour les bicarrés. I. *Comptes Rendus Acad. Sci Paris* **303** 85–88, and *II* 161–163 (1986).
13. A. Bartel, H. Johnston and H. W. Lenstra, Jr. Galois module structure of oriented Arakelov class groups. *arXiv:2005.11533* (2020).
14. F. Beukers, H. Cohen and A. Mellit. Finite hypergeometric functions. *Pure Appl. Math. Q.* **11**, 559–589 (2015).
15. K. Belabas. A fast algorithm to compute cubic fields. *Math. Comp.* **66**, 1213–1237 (1997).
16. K. Belabas. A relative van Hoeij algorithm over number fields. *J. Symbolic Computation* **37**, 641–668 (2004).
17. K. Belabas and H. Cohen. Modular forms in Pari/GP. In: *Research in the Math. Sciences* **5**, Springer, 46–64 (2018).
18. K. Belabas and H. Cohen. *Numerical Algorithms for Number Theory Using Pari/GP.* Math. surveys and monographs **254**, American Math. Soc. (2021).

19. K. Belabas, F. Diaz y Diaz and E. Friedman. Small generators of the ideal class group. *Math. Comp.* **77**, 1185–1197 (2008).
20. K. Belabas and B. Perrin-Riou. Overconvergent modular symbols and p-adic L-functions. *arXiv:2101.06960* (2021).
21. K. Belabas, M. Van Hoeij, J. Klüners and A. Steel. Factoring polynomials over global fields. *J. Théor. Nombres Bordeaux* **21**, 15–39 (2009).
22. M. Bhargava. Higher composition laws I, II, III, and IV. *Ann. Math.* **159**, 217–250, 865–886, 1329–1360 (2004) and **172**, 1559–1591 (2010).
23. M. Bhargava. The density of discriminants of quartic rings and fields. *Ann. Math.* **162**, 1031–1063 (2005).
24. M. Bhargava. The density of discriminants of quintic rings and fields. *Ann. Math.* **172**, 1559–1591 (2010).
25. M. Bhargava, A. Shankar, T. Taniguchi, F. Thorne, J. Tsimerman and Y. Zhao. Bounds on 2-torsion in class groups of number fields and integral points on elliptic curves. *J. Amer. Math. Soc.* **33**, 1087–1099 (2020).
26. M Bhargava, A. Shankar, and X. Wang, *An improvement on Schmidt's bound on the number of number fields of bounded discriminant and small degree*, arXiv:2204.01331 (2022).
27. Y. Bilu and G. Hanrot. Solving Thue equations of high degree. *J. Number Theory* **60**, 373–392 (1996).
28. B. Birch and P. Swinnerton-Dyer. Notes on elliptic curves I and II. *J. reine angew. Math.* **212**, 7–25 (1963), **218**, 79–108 (1965).
29. B. Birch and W. Kuyk (eds). *Modular forms in one variable IV*. Lecture Notes in Math. **476**, Springer-Verlag (1975).
30. A. Booker. Artin's conjecture, Turing's method, and the Riemann hypothesis. *Experiment. Math.* **15**, 385–407 (2006).
31. A. Booker, A. Strömbergsson and A. Venkatesh. Effective computation of Maass cusp forms. *Int. Math. Res. Not.*, Art. ID 71281, 34 (2006).
32. W. Bosma, J. Cannon and C. Playout. The Magma algebra system I. The user language. *J. Symbolic Comput.* **24**, 235–265 (1997).
33. J. Buchmann. On the computation of units and class numbers by a generalization of Lagrange's algorithm. *J. Number Theory* **26**, 8–30(1987).
34. J. Buchmann, C. Thiel and H. Williams. Short representations of quadratic integers. In: *Math. and its applications* **325**, Kluwer, 159–185 (1995).
35. J. Buhler. *Icosahedral Galois representations*. Lecture Notes in Math. **654**, Springer–Verlag (1978).
36. A. Brumer and K. Kramer. Paramodular abelian varieties of odd conductor. *Trans. Amer. Math. Soc.* **366**, 2463–2516 (2014).
37. H. Cohen. *A Course in Computational Algebraic Number Theory (fourth corrected printing)*. Graduate Texts in Math. **138**, Springer-Verlag (2000).
38. H. Cohen. *Advanced Topics in Computational Number Theory*. Graduate Texts in Math. **193**, Springer-Verlag (2000).
39. H. Cohen. *Number Theory I, Tools and Diophantine Equations*. Graduate Texts in Math. **239**, Springer-Verlag (2007).
40. H. Cohen. *Number Theory II, Analytic and Modern Tools*. Graduate Texts in Math. **240**, Springer-Verlag (2007).
41. H. Cohen. Computing L-functions: A survey. *J. Th. Nombres Bordeaux* **27**, 699–726 (2015).
42. H. Cohen. Computational number theory in relation with L-functions. In: I. Inam and E. Büyükaşik (eds.), *International Autumn School on Computational Number Theory*, 171–266, Birkhäuser (2019).
43. H. Cohen, F. Diaz y Diaz and M. Olivier. Computing ray class groups, conductors, and discriminants., *Math. Comp.* **67**, 773–795 (1998).
44. H. Cohen, F. Diaz y Diaz and M. Olivier. A table of totally complex number fields of small discriminant. In: *Proceedings ANTS XIII*, Lecture Notes in Computer Science **1423**, Springer-Verlag, 381–391 (1998).

45. H. Cohen, F. Diaz y Diaz and M. Olivier. Enumerating quartic dihedral extensions of $\mathbb{Q}$. *Compositio Math.* **133**, 65–93 (2002).
46. H. Cohen, F. Diaz y Diaz and M. Olivier. Constructing complete tables of quartic fields using Kummer theory. *Math. Comp.* **72**, 941–951 (2003).
47. H. Cohen and G. Frey (eds). *Handbook of elliptic and elliptic and hyperelliptic curve cryptography.* CRC Press (2006).
48. H. Cohen and H.W. Lenstra, Jr. Primality testing and Jacobi sums. *Math. Comp.* **42**, 297–330 (1984).
49. H. Cohen and A.K. Lenstra. Implementation of a new primality test. *Math. Comp.* **48**, 103–121 and S1–S4 (1987).
50. H. Cohen and H.W. Lenstra, Jr. *Heuristics on class groups of number fields.* Lecture Notes in Math. **1068**, 33–62, Springer-Verlag (1984).
51. H. Cohen and J. Martinet. Étude heuristique des groupes de classes des corps de nombres. *J. für die reine und angew. Math.* **404**, 39–76 (1990).
52. H. Cohn. The density of abelian cubic fields. *Proc. Amer. Math. Soc.* **5**, 476–477 (1954).
53. D. Corwin. *From Chabauty's method to Kim's non-abelian Chabauty's method.* Unpublished manuscript, 41 pp, (2021).
54. J.-M. Couveignes. Enumerating number fields. *Ann. Math.* **192**, 487–497 (2020) and *arXiv:1907.13617* (2019).
55. J. Cremona. Hyperbolic tessellations, modular symbols, and elliptic curves over complex quadratic fields. *Compositio Math.* **51**, 275–324 (1984).
56. J. Cremona. *Algorithms for modular elliptic curves (2nd ed.)* Cambridge Univ. Press (1997).
57. J. Cremona. The L-functions and modular forms database project. *Found. Comp. Math.* **16**, 1541–1553 (2016).
58. H. Davenport and H. Heilbronn. On the density of discriminants of cubic fields I. *Bull. London Math. Soc.* **1**, 345–348 (1969), and *II, Proc. Royal Soc. A* **322**, 405–420 (1971).
59. L. Dembélé and J. Voight. Explicit methods for Hilbert modular forms. In: *Elliptic curves, Hilbert modular forms, and Galois deformations*, 135–198, Birkhäuser (2013).
60. J.-M. Deshouillers, F. Hennecart and B. Landreau. Waring's problem for sixteen biquadrates - Numerical results. *J. Th. Nombres Bordeaux* **12**, 411–422 (2000).
61. T. Dokchitser. Computing special values of motivic L-functions. *Exp. Math.* **13**, 137–149 (2004).
62. B. Edixhoven and J.-M. Couveignes (eds.) *Computational aspects of modular forms and Galois representations.* Annals of Math. Studies **176**, Princeton Univ. Press (2011).
63. N. Elkies. Elliptic and modular curves over finite fields and related computational issues. In: *Computational perspectives in number theory* **7**, 21–76, American Math. Soc. (1998).
64. N. Elkies. Three lectures on elliptic surfaces and curves of high rank. *arXiv:0709.2908* (2007).
65. J. Ellenberg and A. Venkatesh. Reflection principles and bounds for class group torsion. *Int. Math. Res. Not.* **1** (2007).
66. A. Enge. Computing modular polynomials in quasi-linear time. *Math. Comp.* **78**, 1809–1824 (2009).
67. A. Enge and F. Johansson. `paritwine 0.1`. INRIA (2019). https://www.multiprecision.org/paritwine/.
68. A. Enge and A. Sutherland. Class invariants by the CRT method. In: *Proceedings ANTS IX*, Lecture Notes in Comp. Science **6197**, 142–156 (2010).
69. D. Farmer, S. Koutsioliotas and S. Lemurell. Maass forms on GL(3) and GL(4)., *Int. research notices* **22** and *arXiv:1212.4544* (2012).
70. K. Fischer. The Zetafast algorithm for computing zeta functions. *arXiv:1703.01414v7* (2017).
71. F. Fité, K. Kedlaya and A. Sutherland. Sato–Tate groups of abelian threefolds: a preview of the classification. *Contemp. Math.* **770**, 103–129 (2021).
72. E. Fouvry and J. Klüners. On the 4-rank of class groups of quadratic number fields. *Invent. Math.* **167**, 455–513 (2007).
73. F. Gerth. The 4-class ranks of quadratic fields. *Invent. Math.* **77**, 489–515 (1984).

74. D. Goldfeld. Gauss' class number problem for imaginary quadratic fields. *Bull. Amer. Math. Soc.* **13**, 23–37 (1985).
75. X. Gourdon. *Algorithmique du théorème fondamental de l'algèbre.* Rapport de recherche **1852** INRIA (1993).
76. L. Grenié and G. Molteni. Explicit versions of the prime ideal theorem for Dedekind zeta functions under GRH. *Math. Comp.* **85**, 889–906 (2016).
77. J. Hafner and K. McCurley. A rigorous subexponential algorithm for computation of class groups. *J. American Math. Soc.* **2**, 837–850 (1989).
78. D. Harvey. Counting points on hyperelliptic curves in average polynomial time. *Ann. Math.* **179**, 783–803 (2014).
79. D. Harvey. Computing zeta functions of arithmetic schemes. *Proc. London Math. Soc.* **111**, 1379–1401 (2015).
80. D. Harvey and J. van der Hoeven. Integer multiplication in time $O(n \log(n))$. *Annals of Math.* **193**, 563–617 (2021).
81. D. Hejhal. *The Selberg trace formula for* $\mathrm{PSL}_2(\mathbb{R})$. I and II, Lecture Notes in Math. **548** and **1001**, Springer-Verlag (1976 and 1983).
82. H. Helfgott. The ternary Goldbach problem. *arXiv:1501.05438v2* (2015).
83. F. Johansson. Arb: efficient arbitrary-precision midpoint-radius interval arithmetic. *IEEE Trans. on Computers* **66**, 1281–1292 (2017).
84. C. Jeannerod, C. Pernet and A. Storjohann. Rank-profile revealing Gaussian elimination and the CUP matrix decomposition. *J. Symbolic Comput.* **56**, 46–68 (2013).
85. K. Kedlaya. Counting points on hyperelliptic curves using Monsky–Washnitzer cohomology. *J. Ramanujan Math. Soc.* **16**, 318–330 (2001) and **18**, 417–418 (2003).
86. K. Khuri-Makdisi. Asymptotically fast group operations on Jacobians of general curves. *Math. Comp.* **76**, 2213–2239 (2007).
87. D.H. Lehmer. Factorization of certain cyclotomic functions. *Ann. Math. (2)* **34**, 461–479 (1933).
88. R. Lemke Oliver and F. Thorne. Upper bounds on number fields of given degree and bounded discriminant. *arXiv:2005.14110v2* (2020).
89. H.W. Lenstra, Jr. Factoring integers with elliptic curves. *Annals of Math.* **126**, 649–673 (1987).
90. A.K. Lenstra and H.W. Lenstra, Jr. (eds.) *The development of the number field sieve.* Lecture Notes in Math. **1554**, Springer-Verlag (1993).
91. A.K. Lenstra, H.W. Lenstra, Jr. and L. Lovász. Factoring polynomials with rational coefficients. *Math. Ann.* **261**, 515–534 (1982).
92. H.W. Lenstra and R. Tijdeman (eds). *Computational methods in number theory.* Math. Center Tracts **154/155**, Math. Centrum Amsterdam (1982).
93. The LMFDB collaboration. *The L-function and modular form database.* http://www.lmfdb. org and http://beta.lmfdb.org.
94. Q. Liu. *Modèles minimaux des courbes de genre deux. J. für die reine und angew. Math.* **453**, 137–164 (1994).
95. W. McCallum and B. Poonen. The method of Chabauty and Coleman. In: *Panoramas et synthèses* **36**, Soc. Math. de France, 99–117 (2012).
96. S. Mäki. The conductor density of abelian number fields. *J. London Math. Soc. (2)* **47**, 18–30 (1993).
97. G. Malle. On the distribution of Galois groups I., *J. Number Theory* **92**, 315–322 (2002), and II., *Exp. Math.* **13**, 129–135 (2004).
98. J.-F. Mestre. Unpublished, but see Section 17.3.2.b in [47].
99. P. Molin and A. Page. Computing groups of Hecke characters. arXiv:2210.02716.
100. P. Nguyen and D. Stehlé. Floating-point LLL revisited. In: *Proceedings Eurocrypt '2005*, Springer-Verlag (2005).
101. P. Nguyen and B. Vallée (eds). *The LLL algorithm: Survey and applications.* Information Security and Cryptography, Springer (2010).

102. A. Odlyzko. Bounds for discriminants and related estimates for class numbers, regulators, and zeros of zeta functions: a survey of recent results. *J. Th. Nombres Bordeaux* **2**, 114–141 (1990).
103. A. Odlyzko and A. Schönhage. Fast algorithms for multiple evaluations of the Riemann zeta function. *Trans. Amer. Math. Soc.* **309**, 797–809 (1988).
104. The PARI Group. `PARI/GP` *version 2.15.1*. Univ. Bordeaux (2022). http://pari.math.u-bordeaux.fr/.
105. W. Plesken and B. Souvignier. Computing isometries of lattices. *J. Symbolic Comp.* **24**, 327–334 (1997).
106. M. Pohst and H. Zassenhaus. *Algorithmic algebraic number theory (3rd ed.)* Cambridge Univ. Press (1993).
107. R. Pollack and G. Stevens. Overconvergent modular symbols and p-adic L-functions. *Ann. Sci. ENS* **44**, 1–42 (2011).
108. C. Pomerance. The quadratic sieve factoring algorithm. In: *EUROCRYPT*, Lecture Notes in Comp. Science **209**, 169–182, Springer-Verlag (1984).
109. A. Pethö, M. Pohst, H. Williams and H. Zimmer (eds). *Computational number theory*, Walter de Gruyter (1991).
110. B. Poonen. Heuristics for the arithmetic of elliptic curves. *arXiv:1711.10112v2* (2017).
111. B. Poonen and E. Rains. Random maximal isotropic subspaces and Selmer groups. *J. Amer. Math. Soc.* **25**, 245–269 (2012).
112. C. Poor and D. Yuen. Paramodular cusp forms. *Math. Comp.* **84**, 1401–1438 (2015).
113. D. Roberts and F. Rodriguez-Villegas. Hypergeometric motives. *arXiv:2109.00027* (2021).
114. The Sage Developers. *SageMath, the Sage Mathematics Software System version 9.7* (2022). https://www.sagemath.org.
115. T. Satoh. The canonical lift of an ordinary elliptic curve over a finite field and its point counting. *J. Ramanujan Math. Soc.* **15**, 247–270 (2000).
116. R. Schoof. Elliptic curves over finite fields and the computation of square roots mod p. *Math. Comp.* **44**, 483–494 (1985).
117. D. Shanks. Class number, a theory of factorization, and genera. In: *Proc. Sympos. Pure Math.* **20**, 415–440, American Math. Soc. (1971).
118. P. Shor. Polynomial-time algorithms for prime factorization and discrete logarithms on a quantum computer. *arXiv:quant-ph/9508027v2* (1996).
119. C.-L. Siegel. Contributions to the theory of the Dirichlet L-series and the Epstein zeta-functions. *Ann. of Math. (2)* **44**, 143–172 (1943).
120. A. Silvester, M. Jacobson and H. Williams. *Shorter compact representations in real quadratic fields.* In: *Number Theory and Cryptography*, Lecture Notes in Computer Science **8260**, 50–72 (2013).
121. N. Smart. *The algorithmic resolution of Diophantine equations.* London Math. Soc. student texts **41** (1998).
122. A. Smith. 2^∞-Selmer groups, 2^∞-class groups, and Goldfeld's conjecture. *arXiv:1702.02325v2* (2017).
123. C. Smyth. The Mahler measure of algebraic numbers: a survey. In: J. McKee and C. Smyth (eds.), *Number theory and polynomials*, 322–349, Cambridge Univ. Press (2008).
124. H. Stark. L-functions at $s = 1$ I, II, III, IV. *Adv. Math.* **7**, 301–343 (1971), **17**, 60–92 (1975), **22**, 64–84 (1976), **35**, 197–235 (1980).
125. H. Stark. Fourier coefficients of Maass waveforms. In: *Modular forms*, R. Rankin ed., 263–269, Ellis Horwood (1984).
126. W. Stein. *Modular forms, a computational approach.* Graduate Studies in Math. **79**, American Math. Soc. (2007).
127. M. Stoll, *Documentation for the ratpoints program*, arXiv:0803.3165v5 (2022).
128. A. Sutherland. Computing Hilbert class polynomials with the Chinese remainder theorem. *Math. Comp.* **80**, 501–538 (2011).
129. A. Sutherland. On the evaluation of modular polynomials. In: *Proceedings ANTS X*, 531–555, Open Book Series **1**, msp (2013).

130. A. Sutherland. Sato–Tate distributions. *Contemp. Math.* **740**, 197–248 (2019).
131. A. Sutherland. Counting points on superelliptic curves in average polynomial time. In: *Proceedings ANTS XIV*, 403–422, Open Book Series **4**, msp (2020), or *arXiv:2004.10189v4*.
132. D. Tingley. *Elliptic curves uniformized by modular functions*. Ph.D. thesis, Univ. of Oxford (1975).
133. A. Turing. Some calculations of the Riemann zeta-function. *Proc. London Math. Soc.* **3**, 99–117 (1953).
134. W. Wang and M. Matchett Wood. Moments and interpretations of the Cohen–Lenstra–Martinet heuristics. *arXiv:1907.11201v2* (2020).
135. M. Watkins. Some remarks on Heegner point computations. In: *Panoramas et synthèses* **36**, 81–97, Soc. Math. de France (2012).
136. M. Watkins. A discursus on 21 as a bound for ranks of elliptic curves over $\mathbb{Q}$, and sundry related topics. http://magma.maths.usyd.edu.au/~watkins/papers/DISCURSUS.pdf (2015).
137. M. Watkins. Class numbers of imaginary quadratic fields. *Math. Comp.* **73**, 907–938 (2004).
138. A.-E. Wilke. Thesis, in preparation.
139. S. Wong. Densities of quartic fields with even Galois groups. *Proc. Amer. Math. Soc.* **133**, 2873–2881 (2005).
140. H. Zimmer. *Computational problems, methods, and results in algebraic number theory.* Lecture Notes in Math. **262**, Springer–Verlag (1972).

The Four Exponentials Problem and Schanuel's Conjecture

Michel Waldschmidt

Personal Note
This Festschrift in honor of Springer's Editorial Director Dr. Catriona Byrne is a good opportunity for me to thank Catriona for her support in publishing a good part of my works with Springer Verlag. I had the pleasure of working with her also as an editor of this publisher. I am thankful also to the editors of this special volume of the Lectures Notes in Mathematics for their invitation to share my favorite open problems and the ones dear to my heart, with some background and context.

1 Introduction

I have worked on several conjectures. The one on which I spent much more time than the others is the so-called *four exponentials problem* (Lang [17, Chap. II § 1 p. 11]), which is also the first of the eight problems in Schneider's book on transcendental numbers [38]. This question is a very special case, arguably one of the easiest unsolved cases so far, of Schanuel's Conjecture (Lang [17, Chap. III, Historical note p. 30]). Over the years, I tried to prove Schanuel's conjecture; since very few results are known, in the process of trying to solve it, I added some hypotheses which might help, and, quite often, after some time, I came back trying to solve the four exponentials problem. Without success so far! It is hard to predict whether the special case of the four exponentials problem will be solved before the

M. Waldschmidt (✉)
Sorbonne Université, CNRS IMJ-PRG, Paris, France
e-mail: michel.waldschmidt@imj-prg.fr

© The Author(s), under exclusive license to Springer Nature Switzerland AG 2023
J.-M. Morel, B. Teissier (eds.), *Mathematics Going Forward*, Lecture Notes in Mathematics 2313, https://doi.org/10.1007/978-3-031-12244-6_39

very general case of Schanuel's conjecture. As an example, the following simple looking statement is open:

> Let t be a real number such that 2^t and 3^t are integers. Prove that t is a nonnegative integer.

While the four exponentials problem is still open, a weaker statement, the *six exponentials theorem* (Theorem 4.1), is known to be true; a special case is the following:

> Let t be a real number and p_1, p_2, p_3 be three distinct primes. Assume that the three numbers p_1^t, p_2^t and p_3^t are integers. Then t is a nonnegative integer.

For a complete proof of this result using interpolation determinants, see Waldschmidt [53].

The present paper deals only with Schanuel's conjecture and some of its consequences, including the four exponentials problem and the problem of algebraic independence of logarithms of algebraic numbers (of which Leopoldt's conjecture is a special case in the p-adic case). Further conjectures (including Grothendieck's conjecture on abelian periods, André's conjecture on motives, the conjecture of Kontsevich–Zagier on periods,...) also deserve to be discussed—see for instance Waldschmidt [52].

2 Leopoldt's Conjecture on the p-Adic Rank of the Group of Units of an Algebraic Number Field

When I started to do research in 1969 in Bordeaux, my thesis advisor, Jean Fresnel, suggested me to study Leopoldt's conjecture (Leopoldt [21]). At that time, Fresnel was interested in p-adic L-functions (Amice and Fresnel [2]) and Leopoldt's conjecture was comparatively recent. The goal is to prove that the p-adic rank of the group of units of an algebraic number field is the same as the usual rank given by Dirichlet's unit theorem. It amounts to saying that the p-adic regulator, which is defined as the usual regulator by replacing logarithms with p-adic logarithms, does not vanish. According to Fresnel, *since it amounts to proving that a determinant does not vanish, it should not be so difficult!*

For a subfield of an abelian extension of an imaginary quadratic field, the decomposition, due to Frobenius, of the *Gruppendeterminant* of the Galois group— see for instance Fresnel [10], Waldschmidt [41], Kanemitsu and Waldschmidt [16]—shows that the regulator splits into a product of linear forms with algebraic coefficients of logarithms of algebraic numbers. As a consequence, in this special case, as shown by J. Ax [3, Conjecture p. 587], Leopoldt's conjecture is a consequence of the p-adic version of a conjecture of A.O. Gel'fond on the linear independence, over the field $\overline{\mathbb{Q}}$ of algebraic numbers, of $\mathbb{Q}$-linearly independent logarithms of algebraic numbers. This linear independence result in the complex case has been achieved by the seminal work of A. Baker [4], by means of a far-reaching development of Gel'fond's method.

The p-adic analog of Baker's result was proved the year after by A. Brumer [6], who therefore solved Leopoldt's conjecture for these abelian extensions. As pointed out by Brumer, the translation to the p-adic case of transcendence methods had been worked out by J-P. Serre [39]. As a matter of fact, Serre was interested in an extension to several variables (in the p-adic case) of the six exponentials theorem for an application to ℓ-adic abelian representations (Serre [40], Henniart [14]).

In the general case, Leopoldt's conjecture is a special case of the p-adic version of the conjecture on algebraic independence of logarithms of algebraic numbers. This is why Fresnel suggested me to study the theory of transcendental numbers.

3 Conjecture on the Algebraic Independence of Logarithms of Algebraic Numbers

According to A.O. Gel'fond [11, Chap. III § 5 p. 177], *one may assume ... that the most pressing problem in the theory of transcendental numbers is the investigation of the measures of transcendence of finite sets of logarithms of algebraic numbers.* From a qualitative point of view, the statement is the following one (Lang [17, Chap. III, Historical note p. 31]), which, according to Lang, *has been conjectured for a long time (by anybody who has looked at the subject).*

Conjecture 3.1 (Algebraic Independence of Logarithms of Algebraic Numbers) *Let $\lambda_1, \ldots, \lambda_n$ be $\mathbb{Q}$-linearly independent complex numbers, such that the numbers $\alpha_i = e^{\lambda_i}$ $(i = 1, \ldots, n)$ are algebraic numbers. Then $\lambda_1, \ldots, \lambda_n$ are algebraically independent.*

In Calegari and Mazur [7, Conjecture 3.9], this conjecture is called *weak Schanuel.*

Under the assumptions of Conjecture 3.1, the conclusion of Baker's theorem [4] is that the numbers $1, \lambda_1, \ldots, \lambda_n$ are $\overline{\mathbb{Q}}$-linearly independent, while the conclusion of Conjecture 3.1 is that, for any nonzero polynomial P (with rational or algebraic coefficients) in n variables, the number $P(\lambda_1, \ldots, \lambda_n)$ does not vanish.

By abuse of language, we sometimes write $\lambda_i = \log \alpha_i$ $(i = 1, \ldots, n)$; the way Conjecture 3.1 is stated avoids the need to select a branch of the complex logarithm. For instance with $\lambda_1 = \log 2$ and $\lambda_2 = \log 2 + 2\pi i$, hence $\alpha_1 = \alpha_2 = 2$, Baker's theorem yields the linear independence of the numbers $1, \log 2, \pi$ over $\overline{\mathbb{Q}}$, while Conjecture 3.1 claims that $\log 2$ and π are algebraically independent (which is not yet proved).

Conjecture 3.1 is true for $n = 1$: a nonzero logarithm of an algebraic number is transcendental, according to the theorem of Hermite–Lindemann (Schneider [38, Chap. II § 4], Lang [17, Chap. III Corollary 1], Waldschmidt [43, Th. 3.1.1], [49, Th. 1.2]). This is essentially the only case where Conjecture 3.1 has been proved. Under the assumptions of Conjecture 3.1, the conclusion should be that the transcendence degree of the field $\mathbb{Q}(\lambda_1, \ldots, \lambda_n)$ is n. As a matter of fact, it is not yet known if the field generated by all logarithms of all nonzero algebraic numbers has a transcendence degree over $\mathbb{Q}$ of at least 2. However, for a conjecture which is equivalent to Conjecture 3.1, half of the result is proved (see inequalities (2) in Sect. 5 and (3) in Sect. 6). Hence, depending on the point of view, one may consider that we are half way toward proving Conjecture 3.1.

4 The Four Exponentials Problem and Six Exponentials Theorem

We would like to solve at least some special cases of Conjecture 3.1. For instance we would like to prove that there are no algebraic relations like

$$(\log \alpha_1)^2 = \log \alpha_2$$

involving nonzero logarithms of algebraic numbers $\log \alpha_1$ and $\log \alpha_2$. Very few results are known even for this very specific case. We will mainly work with homogeneous relations; among the simplest nonlinear ones is the following:

$$(\log \alpha_1)(\log \alpha_4) = (\log \alpha_2)(\log \alpha_3),$$

which amounts to considering the vanishing of the determinant

$$\begin{vmatrix} \log \alpha_1 & \log \alpha_2 \\ \log \alpha_3 & \log \alpha_4 \end{vmatrix}. \tag{1}$$

Here, $\log \alpha_i$ denote complex numbers such that $\alpha_i = e^{\log \alpha_i}$ are algebraic numbers. The four exponentials problem states that such a determinant can vanish if and only if either the two rows are linearly dependent over the rational number field $\mathbb{Q}$, or the two columns are linearly dependent over $\mathbb{Q}$. Since a 2×2 matrix has rank ≤ 1 if and only if it can be written as

$$\begin{pmatrix} x_1 y_1 & x_1 y_2 \\ x_2 y_1 & x_2 y_2 \end{pmatrix},$$

an equivalent form of the four exponentials problem is the following:

Conjecture 4.1 (Four Exponentials Problem) *Let x_1, x_2 be two complex numbers which are linearly independent over $\mathbb{Q}$ and let y_1, y_2 be two complex numbers which are linearly independent over $\mathbb{Q}$. Then at least one of the four numbers*

$$e^{x_1 y_1}, \ e^{x_1 y_2}, \ e^{x_2 y_1}, \ e^{x_2 y_2}$$

is transcendental.

Here is a sketch of proof of Conjecture 4.1 as a consequence of Conjecture 3.1 on algebraic independence of logarithms of algebraic numbers (Waldschmidt [49, Exercise 1.8]). As pointed out by D. Roy [30, p. 52], Conjecture 3.1 is equivalent to the following statement: *Let* $\lambda_1, \ldots, \lambda_n$ *be logarithms of algebraic numbers and let* $P \in \overline{\mathbb{Q}}[X_1, \ldots, X_n]$ *be a nonzero polynomial with algebraic coefficients such that* $P(\lambda_1, \ldots, \lambda_n) = 0$. *Then there is a vector subspace* $\mathcal{V}$ *of* $\mathbb{C}^n$, *rational over* $\mathbb{Q}$, *which is contained in the set of zeroes of* P *and contains the point* $(\lambda_1, \ldots, \lambda_n)$. To completes the proof, one uses the fact that if $\mathcal{V}$ a vector subspace of $\mathbb{C}^4$, which is rational over $\mathbb{Q}$ and is contained in the hypersurface $z_1 z_4 = z_2 z_3$, then there exists $(a : b) \in \mathbb{P}_1(\mathbb{Q})$ such that $\mathcal{V}$ is included either in the plane

$$\{(z_1, z_2, z_3, z_4) \in \mathbb{C}^4 ; a z_1 = b z_2, \ a z_3 = b z_4\}$$

or in the plane

$$\{(z_1, z_2, z_3, z_4) \in \mathbb{C}^4 ; a z_1 = b z_3, \ a z_2 = b z_4\}.$$

This four exponentials problem was proposed explicitly by S. Lang [17, Chap. II § 1 p. 11] and K. Ramachandra [25, p. 87–88]; it is also the first of the eight problems at the end of Schneider's book [38].

The following statement is weaker than Conjecture 4.1 but is proved:

Theorem 4.1 (Six Exponentials Theorem) *Let* x_1, x_2 *be two complex numbers which are linearly independent over* $\mathbb{Q}$, *and let* y_1, y_2, y_3 *be three complex numbers which are linearly independent over* $\mathbb{Q}$. *Then at least one of the six numbers*

$$e^{x_1 y_1}, \ e^{x_1 y_2}, \ e^{x_1 y_3}, \ e^{x_2 y_1}, \ e^{x_2 y_2}, \ e^{x_2 y_3}$$

is transcendental.

Equivalently, a 2×3 matrix with entries logarithms of algebraic numbers, having its two rows linearly independent over $\mathbb{Q}$ and its three columns linearly independent over $\mathbb{Q}$, has rank 2.

The six exponentials Theorem 4.1 was proved by Lang [17, Chap. II Th. 1] and Ramachandra [25]. The footnote on p. 67 of [25] reads:

> After writing this manuscript I came to know from professor C.L. Siegel that this is a result first due to Schneider and Siegel. The result is unpublished. This result may also be found in a recent paper by S. Lang, *Algebraic values of meromorphic functions*, Topology **5** (4), (1966), pp. 363–370. The results of this paper have something in common with Lang's results.

Indeed, one can infer from Alaoglu and Erdős [1, p. 455] that the six exponentials Theorem 4.1 and the four exponentials problem (Conjecture 4.1) were also known to C.L. Siegel. When I met A. Selberg at a conference organized by Kai-Man Tsang in Hong Kong in December 1993, he told me that he knew the proof of the six exponentials Theorem 4.1, but he did not publish it because it was too easy. He said he tried to solve the four exponentials problem (Conjecture 4.1), which was much more interesting, but he did not succeed.

A proof of Theorem 4.1 is given in Waldschmidt [43, Chapter 2 (Schneider's method)]. In his plenary lecture for the Journées Arithmétiques in Luminy in 1989, M. Laurent introduced a new idea for transcendence proofs, by means of interpolation determinants in place of an auxiliary function; the example he worked out was the six exponentials theorem (Laurent [20]). See also Waldschmidt [53] for the simplest case of rational integers.

The four exponentials problem (Conjecture 4.1) has been solved under the extra assumption that the field generated by the four numbers x_1, x_2, y_1, y_2 has transcendence degree ≤ 1 (Brownawell [5, Corollary 7], Waldschmidt [42, Corollary 4]). The proof uses a method of algebraic independence devised by A.O. Gel'fond. This result has been extended in Roy and Waldschmidt [37], where the determinant $X_1 X_4 - X_2 X_3$ is replaced by any homogeneous quadratic form; for the proof, Gel'fond's criterion is replaced by a Diophantine approximation result due to Wirsing.

For an explanation of the fact that the transcendence machinery has so far failed to solve the four exponentials problem, see Corollary 8.3 of Roy [33].

As mentioned above, the p-adic analog of the six exponentials theorem has been proved by J-P. Serre [39]. As shown in Roy [29, Corollary p. 450], a positive solution of the p-adic version of the four exponentials conjecture implies Leopoldt's conjecture for Galois extensions of $\mathbb{Q}$ with Galois group a dihedral group of order 6, 8 or 12 (hence, in particular, it implies Leopoldt's conjecture for number fields which are Galois extensions of $\mathbb{Q}$ of degree ≤ 7).

5 Rank of Matrices

We have seen that the four exponentials problem can be stated as the nonvanishing of the determinant (1). More generally, Conjecture 3.1 shows that a determinant, the entries of which are logarithms of algebraic numbers, can vanish only in *trivial* cases. A precise statement, with a definition of the meaning of *trivial*, is the following (Roy [30, p. 54] and Waldschmidt [49, Lemma 12.8]).

Definition 5.1 Let M be a matrix with entries in $\mathbb{C}$ and K a subfield of $\mathbb{C}$. Let $e_1, \ldots, e_t$ be a basis of the K-vector space spanned by the entries of M. Hence M can be written as

$$M = M_1 e_1 + \cdots + M_t e_t,$$

where the matrices $M_1, \ldots, M_t$ have entries in K. Let $X_1, \ldots, X_t$ be indeterminates. The rank of the matrix $M_1 X_1 + \cdots + M_t X_t$, with coefficients in the ring $K[X_1, \ldots, X_n]$ of polynomials in n variables, does not depend on the choice of the basis $e_1, \ldots, e_t$ and is denoted as $r_{\mathrm{str}, K}(M)$, which is called the *structural rank of* M with respect to K.

For any matrix M with complex coefficients and any field K, the upper bound $\mathrm{rk}(M) \leq r_{\mathrm{str},K}(M)$ is plain. Assume now that the entries of M are logarithms of algebraic numbers. Conjecture 3.1 implies $\mathrm{rk}(M) = r_{\mathrm{str},K}(M)$. From the six exponentials theorem, one deduces that when $r_{\mathrm{str}}(M) \geq 3$, then $\mathrm{rk}(M) \geq 2$. More generally, the lower bound

$$\mathrm{rk}(M) \geq \frac{1}{2} r_{\mathrm{str},\mathbb{Q}}(M) \qquad (2)$$

follows from Waldschmidt [44]. The lower bound (2) also holds in the p-adic case; it proves that the p-adic rank of the group of units of an algebraic number field is at least half of its usual rank (Waldschmidt [46]).

The proof of Waldschmidt [44] also yields an answer to the above mentioned question on ℓ-adic representations (Serre [40], Henniart [14]), while the complex version answers a question of A. Weil [54] on the characters of the idèle class group of an algebraic number field (Waldschmidt [45]).

6 The Strong Six Exponentials Theorem and the Strong Four Exponentials Problem

There is room between the four exponentials problem and the six exponentials theorem for a result involving five numbers (Waldschmidt [47]):

Theorem 6.1 (Five Exponentials Theorem) *Let γ be a nonzero algebraic number, x_1, x_2 be two complex numbers which are linearly independent over $\mathbb{Q}$, and y_1, y_2 be two complex numbers which are linearly independent over $\mathbb{Q}$. Then at least one of the five numbers*

$$e^{x_1 y_1}, \ e^{x_1 y_2}, \ e^{x_2 y_1}, \ e^{x_2 y_2}, \ e^{\gamma x_1/x_2}$$

is transcendental.

This result is weaker than the four exponentials problem (because of the assumption $\gamma \neq 0$) but does not imply the six exponentials theorem. A result which includes both the five and the six exponentials theorems (Theorems 6.1 and 4.1) is the next one (Waldschmidt [48, Corollary 2.3], [47, Corollary 2.1], [49, § 11.3.3, example 2, p. 386]).

Let x_1, x_2 be two complex numbers which are linearly independent over $\mathbb{Q}$, let y_1, y_2, y_3 be three complex numbers which are linearly independent over $\mathbb{Q}$ and let β_{ij} ($i = 1, 2$, $j = 1, 2, 3$) be six algebraic numbers. Then at least one of the six numbers

$$e^{x_1 y_1 - \beta_{11}}, \ e^{x_1 y_2 - \beta_{12}}, \ e^{x_1 y_3 - \beta_{13}}, \ e^{x_2 y_1 - \beta_{21}}, \ e^{x_2 y_2 - \beta_{22}}, \ e^{x_2 y_3 - \beta_{23}}$$

is transcendental.

A first generalization of this result has been achieved by D. Roy (Roy [26], [28, Corollary 2 p. 38], Waldschmidt [49, Corollary 11.16]), who considers matrices with entries which are linear combinations, with algebraic coefficients, of 1 and of logarithms of algebraic numbers. Denote by $\mathcal{L}$ the $\overline{\mathbb{Q}}$-vector space spanned by 1 and all logarithms of all nonzero algebraic numbers. A typical element of $\mathcal{L}$ is of the form

$$\beta_0 + \beta_1 \log \alpha_1 + \cdots + \beta_n \log \alpha_n$$

with algebraic numbers α_i and β_j.

Theorem 6.2 (D. Roy, Strong Six Exponentials Theorem) *Let x_1, x_2 be two complex numbers which are linearly independent over $\overline{\mathbb{Q}}$ and let y_1, y_2, y_3 be three complex numbers which are linearly independent over $\overline{\mathbb{Q}}$. Then at least one of the six numbers*

$$x_1 y_1, \quad x_1 y_2, \quad x_1 y_3, \quad x_2 y_1, \quad x_2 y_2, \quad x_2 y_3$$

does not belong to $\mathcal{L}$.

The *strong four exponentials problem* is the same statement as Theorem 6.2 with only two numbers y_1, y_2 instead of three.

Several consequences of the strong four exponentials problem are stated in Waldschmidt [50][1] and corollaries of the strong six exponentials theorem are derived in Waldschmidt [51].

A much more general statement than Theorem 6.2 is an extension by D. Roy [27] of the lower bound (2) to matrices having entries in $\mathcal{L}$: for such a matrix,

$$\mathrm{rk}(M) \geq \frac{1}{2} r_{\mathrm{str}, \overline{\mathbb{Q}}}(M) \tag{3}$$

see Waldschmidt [49, Th. 1.17 and Corollary 12.18].

As pointed out by D. Roy (Roy [27], [30, Conjecture 1.1] and Waldschmidt [49, Lemma 12.14]), Conjecture 3.1 on the algebraic independence of logarithms of algebraic numbers is equivalent to the statement that if the entries of M are in $\mathcal{L}$, then the rank $\mathrm{rk}(M)$ of the matrix M is always equal to its structural rank $r_{\mathrm{str}, \overline{\mathbb{Q}}}(M)$ with respect to $\overline{\mathbb{Q}}$. From this point of view, we can consider (3) as proving half of Conjecture 3.1.

As always, the situation is the same in the p-adic case, both for results and for conjectures. Applications to Leopoldt's Conjecture on the p-adic rank of the units of a number field have been derived in the following references: Emsalem, Kisilevsky

---

[1] Erratum: The right assumption in corollary 2.12, p. 346 of Waldschmidt [50] is that the three numbers 1, Λ_{11} and Λ_{21} are linearly independent over the field of algebraic numbers.

and Wales [9], Jaulent [15], Emsalem [8], Laurent [18, 19], Roy [29]. See also Calegari and Mazur [7, § 3 Remark p. 127] and Maksoud [22].

7 Schanuel's Conjecture

Conjecture 3.1 on the algebraic independence of logarithms of algebraic numbers is a special case of Schanuel's conjecture, which was proposed by Stephen Schanuel during a course given by Serge Lang at Columbia in the 1960s [17, Chap. III, Historical Note, p. 30–31].

Conjecture 7.1 (Schanuel's Conjecture) *Let* $x_1, \ldots, x_n$ *be* $\mathbb{Q}$*-linearly independent complex numbers. Then at least n of the 2n numbers*

$$x_1, \ldots, x_n, e^{x_1}, \ldots, e^{x_n}$$

are algebraically independent over $\mathbb{Q}$.

The conclusion is that the transcendence degree over $\mathbb{Q}$ of the field $\mathbb{Q}(x_1, \ldots, x_n, e^{x_1}, \ldots, e^{x_n})$ is at least n. This result is known when $x_1, \ldots, x_n$ are algebraic numbers: this is the Lindemann–Weierstrass Theorem. Conjecture 3.1 is the special case of Conjecture 7.1 where the n numbers $e^{x_1}, \ldots, e^{x_n}$ are assumed to be algebraic.

8 Roy's Conjecture

In his plenary talk at the Journées Arithmétiques in Rome in 1999 [31, 32], D. Roy proposed a new conjecture of his own and proved the remarkable and surprising result that it is equivalent to Schanuel's conjecture 7.1.
Denote by $\mathcal{D}$ the derivation

$$\mathcal{D} = \frac{\partial}{\partial X_0} + X_1 \frac{\partial}{\partial X_1}$$

on the field $\mathbb{C}(X_0, X_1)$.

Conjecture 8.1 (Conjecture of D. Roy) *Let* ℓ *be a positive integer,* $y_1, \ldots, y_\ell$ $\mathbb{Q}$*-linearly independent complex numbers,* $\alpha_1, \ldots, \alpha_\ell$ *nonzero complex numbers and* s_0, s_1, t_0, t_1, u *positive real numbers satisfying*

$$\max\{1, t_0, 2t_1\} < \min\{s_0, 2s_1\} < u \quad and \quad \max\{s_0, s_1 + t_1\} < u < \frac{1}{2}(1 + t_0 + t_1).$$

Assume that, for any sufficiently large positive integer N, there exists a nonzero polynomial $P_N \in \mathbb{Z}[X_0, X_1]$ *with partial degree* $\leq N^{t_0}$ *in* X_0, *partial degree* $\leq N^{t_1}$

in X_1 and height $H(P_N) \leq e^N$, *which satisfies*

$$\left| (\mathcal{D}^k P_N) \left(\sum_{j=1}^{\ell} m_j y_j, \prod_{j=1}^{\ell} \alpha_j^{m_j} \right) \right| \leq \exp(-N^u)$$

for any integers $k, m_1, \ldots, m_\ell$ in $\mathbb{N}$ with $k \leq N^{s_0}$ and $\max\{m_1, \ldots, m_\ell\} \leq N^{s_1}$. Then, we have the following lower bound for the transcendence degree:

$$\mathrm{trdeg}_{\mathbb{Q}} \mathbb{Q}(y_1, \ldots, y_\ell, \alpha_1, \ldots, \alpha_\ell) \geq \ell.$$

See also Waldschmidt [49, Conjecture 15.36]. Hence Schanuel's conjecture is equivalent to a purely algebraic statement, which bears some similarity to the available criteria of algebraic independence.

The proof of the equivalence between Schanuel's conjecture 7.1 and Roy's conjecture 8.1 involves a new interpolation formula for holomorphic functions of two complex variables (Roy [31, 32]). Refined interpolation formulae are proved in Roy [33] and Nguyen and Roy [24].

Several significant steps in the direction of Conjecture 8.1 were performed by D. Roy, first for the multiplicative group [34], next for the additive group [35] and then for the product of the additive group by the multiplicative group [36]. A refinement of Conjecture 8.1, again equivalent to Schanuel's conjecture, is devised by Nguyen Ngoc Ai Van in [23]. The statement which is proved in Nguyen and Roy [24] is similar to Conjecture 8.1 and is not restricted to the one-parameter subgroup $t \mapsto (t, \exp(t))$.

In [12], Luca Ghidelli refines the results of Roy [36] and Nguyen and Roy [24], replacing the total degree with multidegrees; his tool [13] is an extension of Roy's multiplicity lemma for the resultant, using the theory of multiprojective elimination initiated by P. Philippon and developed by G. Rémond.

This original point of view of D. Roy suggests a promising approach to proving Schanuel's conjecture: so far it is the only available strategy towards a proof of it.

Transcendence theory is going forward.

Acknowledgements Thanks to Damien Roy for his comments on a preliminary draft of this paper and to Claude Levesque for his support.

References

1. L. Alaoglu and P. Erdős. On highly composite and similar numbers. *Trans. Am. Math. Soc.* **56**, 448–469 (1944). https://doi.org/10.2307/1990319; http://msp.org/idx/zbl/03099757; http://msp.org/idx/mr/0011087

2. Y. Amice and J. Fresnel. Fonction zêta *p*-adique des corps de nombres abéliens réels. *Acta Arith.* **20**, 353–384 (1972). https://doi.org/10.4064/aa-20-4-353-384; http://msp.org/idx/zbl/03344662; http://msp.org/idx/mr/0337898

3. J. Ax. On the units of an algebraic number field. *Ill. J. Math.* **9**, 584–589 (1965). https://doi.org/10.1215/ijm/1256059299; http://msp.org/idx/zbl/03214676; http://msp.org/idx/mr/0181630

4. A. Baker. Linear forms in the logarithms of algebraic numbers. *Mathematika* **13**, 204–216 (1966). https://doi.org/10.1112/S0025579300003971; http://msp.org/idx/zbl/03257537; http://msp.org/idx/mr/0220680

5. W.D Brownawell. The algebraic independence of certain numbers related by the exponential function. *J. Number Theory* **6**, 22–31 (1974). https://doi.org/10.1016/0022-314X(74)90005-5; http://msp.org/idx/zbl/03432369; http://msp.org/idx/mr/0337804

6. A. Brumer. On the units of algebraic number fields. *Mathematika* **14**, 121–124 (1967). https://doi.org/10.1112/S0025579300003703; http://msp.org/idx/zbl/03272326; http://msp.org/idx/mr/0220694

7. F. Calegari and B. Mazur. Nearly ordinary Galois deformations over arbitrary number fields. *J. Inst. Math. Jussieu* **8**(1), 99–177 (2009). https://doi.org/10.1017/S1474748008000327; http://msp.org/idx/zbl/1211.11065; http://msp.org/idx/mr/2461903

8. M. Emsalem. Sur les idéaux dont l'image par l'application d'Artin dans une $\mathbb{Z}_p$-extension est triviale. *J. reine angew. Math.* **382**, 181–198 (1987). https://doi.org/10.1515/crll.1987.382.181; http://msp.org/idx/zbl/04006374; http://msp.org/idx/mr/0921171

9. M. Emsalem, H.H. Kisilevsky and D.B. Wales. Indépendance linéaire sur $\overline{\mathbb{Q}}$ de logarithmes p-adiques de nombres algébriques et rang p-adique du groupe des unités d'un corps de nombres. *J. Number Theory* **19**, 384–391 (1984). https://doi.org/10.1016/0022-314X(84)90079-9; http://msp.org/idx/zbl/03871495; http://msp.org/idx/mr/0769790

10. J. Fresnel. Rang p-adique du groupe des unités d'un corps de nombres. In: *Séminaire de théorie des nombres de Bordeaux (1968–1969)*, Exposé no. 9, pp. 1–18. http://eudml.org/doc/275157; http://www.numdam.org/item/STNB_1968-1969____A9_0/; http://msp.org/idx/zbl/03437276

11. A.O. Gel'fond. *Transcendental and algebraic numbers*. Moskva: Gosudarstv. Izdat. Tekhn.-Teor. Lit., 224 pages (1952). (English. Russian original) New York: Dover Publications (1960). http://msp.org/idx/zbl/03074959; http://msp.org/idx/zbl/0090.26103; http://msp.org/idx/mr/0057921; http://msp.org/idx/mr/0111736

12. L. Ghidelli. *Heights of multiprojective cycles and small value estimates in dimension two*. Tesi di Laurea Magistrale, Università di Pisa (2015). https://etd.adm.unipi.it/theses/available/etd-06262015-164322/

13. L. Ghidelli. Multigraded Koszul complexes, filter-regular sequences and lower bounds for the multiplicity of the resultant. *Preprint* (2019). https://doi.org/10.48550/arXiv.1912.04047

14. G. Henniart. *Représentations ℓ-adiques abéliennes*. Théorie des nombres, Sémin. Delange-Pisot-Poitou, Paris 1980–81, Prog. Math. **22**, 107–126, Birkhäuser, Boston, Mass (1982). http://msp.org/idx/zbl/03785011; http://msp.org/idx/mr/0693314

15. J.-F. Jaulent. Sur l'indépendance ℓ-adique de nombres algébriques. *J. Number Theory* **20**, 149–158 (1985). https://doi.org/10.1016/0022-314X(85)90035-6; http://msp.org/idx/zbl/03910509; http://msp.org/idx/mr/0790777

16. S. Kanemitsu and M. Waldschmidt. Matrices of finite abelian groups, finite Fourier transform and codes. In: *Number theory. Arithmetic in Shangri-La. Proceedings of the 6th China-Japan seminar held at Shanghai Jiao Tong University, Shanghai, China, August 15–17, 2011*, pp. 90–106, Hackensack, NJ: World Scientific (2013). https://doi.org/10.1142/9789814452458_0005; http://msp.org/idx/zbl/06296818; http://msp.org/idx/mr/3089011

17. S. Lang. *Introduction to transcendental numbers*. Addison-Wesley Series in Mathematics VI, 105 pages., Reading, Mass. Addison-Wesley Publishing Company (1966). Collected papers. Volume I: 1952–1970, New York, NY: Springer (2000). http://msp.org/idx/zbl/03232021; http://msp.org/idx/mr/0214547; http://msp.org/idx/zbl/01467752; http://msp.org/idx/mr/3087336

18. M. Laurent. Rang p-adique d'unités et action de groupes. *J. reine angew. Math.* **399**, 81–108 (1989). https://doi.org/10.1515/crll.1989.399.81; http://msp.org/idx/zbl/04089664; http://msp.org/idx/mr/1004134

19. M. Laurent. Rang p-adique d'unités: Un point de vue torique. In: *Sémin. Théor. Nombres, Paris/Fr. 1987–88*, Prog. Math. **81**, pp. 131–146 (1990), Birkhäuser, Boston, Mass. (1990). http://msp.org/idx/zbl/04212167; http://msp.org/idx/mr/MR1042768

20. M. Laurent. Sur quelques résultats récents de transcendance. Exposés présentés aux seizièmes journées arithmétiques, Luminy, France, 17–21 juillet 1989, pp. 209–230, Paris: Société Mathématique de France, Astérisque No. **198–200** (1991). https://smf.emath.fr/node/41992; http://msp.org/idx/zbl/00065929; http://msp.org/idx/mr/1144324

21. H.-W. Leopoldt. Zur Arithmetik in abelschen Zahlkörpern. *J. reine angew. Math.* **209**, 54–71 (1962). https://doi.org/10.1515/crll.1962.209.54; http://msp.org/idx/zbl/03323990; http://msp.org/idx/mr/0139602

22. A. Maksoud. On Leopoldt's and Gross's defects for Artin representations. Preprint (2022). https://doi.org/https://doi.org/10.48550/arXiv.2201.08203

23. Ngoc Ai Van Nguyen. A refined criterion for Schanuel's conjecture. *Chamchuri J. Math.* **1**(2), 25–29 (2009). http://www.math.sc.chula.ac.th/cjm/sites/default/files/02-2-CJM2009-013-GP.pdf; http://msp.org/idx/zbl/06149549; http://msp.org/idx/mr/3151096

24. Ngoc Ai Van Nguyen and D. Roy. A small value estimate in dimension two involving translations by rational points. *Int. J. Number Theory* **12**(5), 1273–1293 (2016). https://doi.org/10.1142/S1793042116500780; http://msp.org/idx/zbl/06589277; http://msp.org/idx/mr/3498626

25. K. Ramachandra. Contributions to the theory of transcendental numbers. I, II. *Acta Arith.* **14**, 65–72, 73–88 (1968). https://doi.org/10.4064/aa-14-1-65-72; http://msp.org/idx/zbl/03281860; http://msp.org/idx/mr/0224566

26. D. Roy. Matrices dont les coefficients sont des formes linéaires. In: *Sémin. Théor. Nombres, Paris/Fr. 1987–88*, Prog. Math. **81**, pp. 273–281, Birkhäuser, Boston, Mass. (1990). http://msp.org/idx/zbl/04171028; http://msp.org/idx/mr/1042774

27. D. Roy. Transcendance et questions de répartition dans les groupes algébriques. In: *Approximations diophantiennes et nombres transcendants. Comptes-rendus du colloque tenu au C.I.R.M. de Luminy, France, 18–22 Juin 1990* (pp. 249–274), Berlin: de Gruyter (1992). http://msp.org/idx/zbl/00124357; http://msp.org/idx/mr/1176534

28. D. Roy. Matrices whose coefficients are linear forms in logarithms. *J. Number Theory* **41**(1), 22–47 (1992). https://doi.org/10.1016/0022-314X(92)90081-Y; http://msp.org/idx/zbl/00039713; http://msp.org/idx/mr/1161143

29. D. Roy. On the v-adic independence of algebraic numbers. In: *Advances in number theory. The proceedings of the third conference of the Canadian Number Theory Association, held at Queen's University, Kingston, Canada, August 18–24, 1991*, pp. 441–451, Oxford: Clarendon Press (1993). http://msp.org/idx/zbl/00409845; http://msp.org/idx/mr/1368440

30. D. Roy. Points whose coordinates are logarithms of algebraic numbers on algebraic varieties. *Acta Math.* **175**(1), 49–73 (1995). https://doi.org/10.1007/BF02392486; http://msp.org/idx/zbl/00807537; http://msp.org/idx/mr/1353017

31. D. Roy. An arithmetic criterion for the values of the exponential function. *Acta Arith.* **97**(2), 183–194 (2001). https://doi.org/10.4064/aa97-2-6; http://msp.org/idx/zbl/01605193; http://msp.org/idx/mr/1824984

32. D. Roy. Une formule d'interpolation en deux variables. *J. Théor. Nombres Bordeaux.* **13**(1), 315–323 (2001). https://doi.org/10.5802/jtnb.324; http://msp.org/idx/zbl/02081368; http://msp.org/idx/mr/1838090

33. D. Roy. Interpolation formulas and auxiliary functions. *J. Number Theory* **94**(2), 248–285 (2002). https://doi.org/10.1006/jnth.2000.2595; http://msp.org/idx/zbl/01809761; http://msp.org/idx/mr/1916273

34. D. Roy. Small value estimates for the multiplicative group. *Acta Arith.* **135**(4), 357–393 (2008). https://doi.org/10.4064/aa135-4-5; http://msp.org/idx/zbl/05376856; http://msp.org/idx/mr/2465718

35. D. Roy. Small value estimates for the additive group. *Int. J. Number Theory* **6**(4), 919–956 (2010). https://doi.org/10.1142/S179304211000323X; http://msp.org/idx/zbl/05758252; http://msp.org/idx/mr/2661291

36. D. Roy. A small value estimate for $\mathbb{G}_a \times \mathbb{G}_m$. *Mathematika* **59**(2), 333–363 (2013). https:// doi.org/10.1112/S002557931200112X; http://msp.org/idx/zbl/06193344; http://msp.org/idx/ mr/3081775

37. D. Roy and M. Waldschmidt. Quadratic relations between logarithms of algebraic numbers. *Proc. Japan Acad. Ser. A* **71**(7), 151–153 (1995). https://doi.org/10.3792/pjaa.71.151; http:// msp.org/idx/zbl/00897674; http://msp.org/idx/mr/1363903

38. T. Schneider. *Einführung in die transzendenten Zahlen*. Die Grundlehren der Mathematischen Wissenschaften. Band 81. Berlin-Göttingen-Heidelberg: Springer-Verlag, 150 pp. (1957). *Introduction aux nombres transcendants*. Translated by P. Eymard Paris: Gauthier-Villars. viii, 151 pp. (1959). http://msp.org/idx/zbl/03160414; http://msp.org/idx/mr/0106890; http://msp. org/idx/zbl/0077.04703; http://msp.org/idx/mr/0086842

39. J.-P. Serre. Dépendance d'exponentielles p-adiques. *Théorie des Nombres, Sém. Delange-Pisot-Poitou* **7** (1965/66), No. 15, 1–14 (1967). https://eudml.org/doc/110667; http://www. numdam.org/item/?id=SDPP_1965-1966__7_2_A4_0; http://msp.org/idx/zbl/03401001

40. J.-P. Serre. *Abelian ℓ-adic representations and elliptic curves* (Vol. 7). Wellesley, MA: A K Peters (1968). (1997; reprint of the 1989 edition. First edition 1968.). https://doi.org/10.1201/ 9781439863862; http://msp.org/idx/zbl/01101830; http://msp.org/idx/mr/0263823; http://msp. org/idx/zbl/0186.25701; http://msp.org/idx/mr/1484415

41. M. Waldschmidt. Gruppendeterminant. In: *Séminaire de théorie des nombres de Bordeaux (1971)*, Exposé no. 1, pp. 1–10 (1971). https://eudml.org/doc/275172; http://www.numdam. org/item/STNB_1971____A1_0/; http://msp.org/idx/zbl/0293.20005

42. M. Waldschmidt. Solution du huitième problème de Schneider. *J. Number Theory* **5**, 191–202 (1973). https://doi.org/10.1016/0022-314X(73)90044-9; http://msp.org/idx/zbl/03412740; http://msp.org/idx/mr/0321884

43. M. Waldschmidt. *Nombres transcendants*. Lect. Notes Math. **402**, Springer (1974). https://doi. org/10.1007/BFb0065320; http://msp.org/idx/zbl/03472160; http://msp.org/idx/mr/0360483

44. M. Waldschmidt. Transcendance et exponentielles en plusieurs variables. *Invent. Math.* **63**, 97–127 (1981). https://doi.org/10.1007/BF01389195; http://msp.org/idx/zbl/03710265; http:// msp.org/idx/mr/0608530

45. M. Waldschmidt. Sur certains caractères du groupe des classes d'idèles d'un corps du nombres. In: *Théorie des nombres, Sémin. Delange-Pisot-Poitou, Paris 1980–81*, Prog. Math. **22**, 323–335, Birkhäuser, Boston, Mass. (1982). http://msp.org/idx/zbl/03777623; http://msp.org/idx/ mr/693328

46. M. Waldschmidt. A lower bound for the p-adic rank of the units of an algebraic number field. In: *Topics in classical number theory, Colloq. Budapest 1981*, Vol. II, Colloq. Math. Soc. János Bolyai **34**, 1617–1650 (1984). http://msp.org/idx/zbl/03859250; http://msp.org/idx/mr/ 0781200

47. M. Waldschmidt. On the transcendence methods of Gel'fond and Schneider in several variables. In: *New advances in transcendence theory, Proc. Symp., Durham/UK 1986*, pp. 375–398, Cambridge University Press (1988). http://msp.org/idx/zbl/04077344; http://msp.org/idx/ mr/0972013

48. M. Waldschmidt. Dependence of logarithms of algebraic points. In: *Number theory. Vol. II. Diophantine and algebraic, Proc. Conf., Budapest/Hung. 1987*, Colloq. Math. Soc. János Bolyai **51**, pp. 1013–1035, North-Holland Publishing Company, Budapest (1990). http://msp. org/idx/zbl/04175044; http://msp.org/idx/mr/1058258

49. M. Waldschmidt. *Diophantine approximation on linear algebraic groups. Transcendence properties of the exponential function in several variables*. Grundlehren der Mathematischen Wissenschaften (Vol. 326), Berlin: Springer (2000). https://doi.org/10.1007/978-3-662-11569-5; http://msp.org/idx/zbl/01467843; http://msp.org/idx/mr/1756786

50. M. Waldschmidt. Variations on the six exponentials theorem. In: *Algebra and number theory. Proceedings of the silver jubilee conference, Hyderabad, India, December 11–16, 2003*, pp. 338–355, New Delhi: Hindustan Book Agency (2005). http://msp.org/idx/zbl/02226145; http://msp.org/idx/mr/2193363

51. M. Waldschmidt. Further variations on the six exponentials theorem. *Hardy-Ramanujan J.* **28**, 1–9 (2005). https://doi.org/10.46298/hrj.2005.86; http://msp.org/idx/zbl/05054470; http://msp. org/idx/mr/2192074

52. M. Waldschmidt. Questions de transcendance: grandes conjectures, petits progrès. In: *Transcendance et irrationalité*, pp. 49–67, Société Mathématique de France, Journées Annuelles 2012, Paris (2012). http://msp.org/idx/zbl/06616539; http://msp.org/idx/mr/3467286

53. M. Waldschmidt. Six Exponentials Theorem – Irrationality. *Resonance: Journal of Science Education* **27**(4), 599–606 (2022). Translation of: Le théorème des six exponentielles restreint à l'irrationalité. *Revue de la filière mathématiques RMS, avril 2021, 131ème année, N.3, 2020–2021*, 26–33 (2021). https://doi.org/10.1007/s12045-022-1351-0

54. A. Weil. On a certain type of characters of the idèle-class group of an algebraic number-field. In: *Proc. internat. Sympos. algebraic number theory, Tokyo & Nikko Sept. 1955*, pp. 1–7. *Œuvres scientifiques. Collected papers. Vol. II (1951–1964)*. Reprint of the 2009/1979 edition. Springer Collected Works in Mathematics, pp. 255–261 (1955). http://msp.org/idx/zbl/03122464; http://msp.org/idx/mr/0083523; http://msp.org/idx/zbl/1317.01045; http://msp.org/idx/mr/3309921

Part XII
Probability and Applications

In their short notes *Research going forward?* and *The future of probability*, leading experts Varadhan and Protter make independent, quick and brilliant summaries of probability and its relation to mathematics, economics, finance and society. Varadhan notes that "Although Kolmogorov provided the axiomatic foundations of probability, making it a part of formal mathematics, at heart, it remains applied with its growth derived from tackling problems coming from physics, engineering statistics as well as computer science." Philip Protter expects "the research in Math Finance to migrate from the analysis of option pricing and hedging to the larger problems of the stability of the entire financial system. There is much interesting mathematics to be done in this regard."

Geoffrey Grimmett's celebratory article *Selected problems in probability theory* contains a personal and idiosyncratic selection of a few open problems in discrete probability theory. These include certain well-known questions concerning Lorentz scatterers and self-avoiding walks, and also some problems of percolation type.

Michel Ledoux's article on stochastic optimization *Optimal matching of random samples and rates of convergence of empirical measures* points out that "whereas the conceptual problems of the past are now largely resolved, contemporary questions arise frequently where the intuitive apparatus of sub-fields collide" and that "many prominent problems are to be found at the conjunction of probability and discrete geometry". The article is a survey of some recent developments and challenging open questions on the random optimal matching problem.

In his essay *Space-time stochastic calculus and white noise*, Bernt Øksendal proposes that white noise can be a powerful tool in the study of multi-parameter stochastic calculus in general. He concludes that "the multi-parameter stochastic calculus is a mostly unexplored area of research. It is clearly important, not only for the study of stochastic partial differential equations driven by space-time Brownian motion but also for many other applications."

Research Going Forward?

S. R. S. Varadhan

It requires extraordinary skill at clairvoyance to make predictions about anything, particularly so about mathematics going forward. Since I have no such ability I will limit myself to a few observations. Probability theory is the study of random objects. The subject evolved as the random objects that were studied changed and became more complex. It started with numbers and proceeded to sequences, functions, generalized functions, surfaces, metric spaces, various combinatorial structures like graphs, trees, triangulations, tilings etc, and required new tools.

It is hard to tell what set of random objects we will be investigating in the future. There is so much randomness around us. Understanding the random nature of the objects requires learning from the data that is available about them. We are now able to collect vast amounts of data. Organizing it and inferring some thing from it has itself become an important area of study, one in which rapid progress is being made. Although Kolmogorov provided the axiomatic foundations of probability, making it a part of formal mathematics, at heart, it remains applied with its growth derived from tackling problems coming from physics, engineering statistics as well as computer science. This means the direction in which probability theory develops will depend very much on from where the push comes.

Developments come in two types. Most of the time it is steady progress in incremental steps culminating in achieving the planned goal. Here one knows the target. But one does not really know if and when they will reach the goal. But every so often there are breakthroughs made possible by radical shift in the point of view that produces a beautiful solution to the problem, perhaps deserving a place in 'The Book'. They are not predictable.

A more basic and important advance has to do with discovering relations or connections between different areas. I can think of a paper by Kakutani [1] that

S. R. S. Varadhan (✉)
Courant Institute of Mathematical Sciences, New York University, New York, NY, USA
e-mail: varadhan@cims.nyu.edu

© The Author(s), under exclusive license to Springer Nature Switzerland AG 2023   595
J.-M. Morel, B. Teissier (eds.), *Mathematics Going Forward*, Lecture Notes
in Mathematics 2313, https://doi.org/10.1007/978-3-031-12244-6_40

proved that the distribution $\pi(x, dy)$ of the random exit point $y \in \partial G$ of Brownian particle starting from x inside a smooth domain G in the plane is the same as the Harmonic measure on the boundary, i.e.

$$u(x) = \int_{\partial G} f(y)\pi(x, dy)$$

is the Harmonic function in G with boundary value $f(y)$ on ∂G. While every advanced student in probability is aware of it now, it was no doubt a surprise finding at that time.

Kolmogorov's forward and backward differential equations connect PDEs with Markov processes. Schramm–Loewner theory uses conformal invariance to identify the scaling limits of various random structures in the plane. Conformal Field Theory of Mathematical Physics is connected to the theory of Gaussian Free Fields in probability. These are just some examples that indicate surprising connections between various branches of mathematics and mathematical physics. It makes going forward unpredictable but rewarding.

I am sure many more such connections will be made in the future but there is no way to predict what they will be or when they will be made.

Reference

1. S. Kakutani. Two-dimensional Brownian motion and harmonic functions. *Proc. Imp. Acad. Tokyo* **20**, 706–714 (1944). (Reprinted in *Shizuo Kakutani: Selcted Papers*, Vol. 2, Birkhäuser (1986))

The Future of Probability

Philip Protter

Probability has always been a bit troubled within the mathematics paradigm. Mathematicians used to view it with contempt: In the pecking order of mathematical sub-disciplines, it was viewed as just below point-set topology. It wasn't even considered mathematics *per se* until Kolmogorov's fundamental and groundbreaking works of the late 1920s/early 1930s gave it a rigorous foundation. Before Kolmogorov probabilists used E for expectation instead of an integral sign as in the work of Riemann, Borel, and Lebesgue. Actually, we still do of course, only now we know that it's equivalent to integration.

Some of the great Russian and French mathematicians of the eighteenth and nineteenth centuries explored aspects of probability theory. Notable among them were d'Alembert, De Moivre, and—especially—Pierre Simon de Laplace. Laplace even wrote an early treatise on Probability Theory [8]. In the mid eighteenth century, according to Hans Fischer, "The value of probabilistic research was determined less by internal mathematical criteria, but rather by the quality of its application to "real" situations. Laplace's CLT met the latter point in an excellent manner. The results of all applications of this theorem matched with "good sense" and thus confirmed Laplace's well-known saying:

Basically, probability is only good sense reduced to a calculus." [6]

Even such a luminary as the Baron Louis Augustin Cauchy, whose name is given to the Cauchy Distribution in probability theory, had contempt for probability theory, because it lacked the mathematical rigor that he championed. Indeed, he got into a fight with Bienaymé over the claims of the universality of the central limit theorem, using the distribution that bears his name as a counterexample. The fight

P. Protter (✉)
Statistics Department, Columbia University, New York, NY, USA
e-mail: pep2117@columbia.edu

between Bienaymé and Cauchy was played out across the pages of the prestigious *Comptes Rendus* from July to September of 1853 [9].

The modern era of research, however, began with the work of Doob, who gave a rigorous foundation for the study of stochastic processes, and practically single handedly developed the theory of martingales. His goal was to study a probabilistic interpretation of potential theory, involving Markov processes, martingales, supermartingales, excessive functions, the whole shebang. See [4] for his *magnum opus*. Doob was especially proud of the martingale convergence theorem, which he once explained to me that, together with the ergodic theorem, were the only convergence theorems in analysis that concluded convergence simply from structure, and were not of the type that $X_n \to X$ implies $\int X_n dP \to \int X dP$. The work of Doob alone made such an impact on mathematics that suddenly it became (mildly) fashionable to hire probabilists in mathematics departments.

In my professional lifetime several things happened to make mathematicians continue to sit up and notice probability from time to time (and therefore hire probabilists). The ones I observed began with Fefferman and Stein [5] showing the dual of H^1 is BMO. This had been an open problem posed by G.H. Hardy. The key insight for us probabilists is that Fefferman used Burkholder's theory of martingale transforms in a key way in his proof. Most purely mathematical proofs before had not taken an excursion into Probability Theory to prove a purely mathematical result. I was just beginning graduate school when this ground breaking result was published. I saw it as a job creation program for young probabilists, who could join math departments and explain the theory to their curious new colleagues.

A half decade later I was lucky enough to hear Hörmander give a series of lectures at the Institute for Advanced Study, in the Fall of 1977. Hörmander had made a breakthrough in hypoelliptic PDEs, and his work and proofs were hard to understand, especially on the intuitive level. I remember one of my friends commented after one of Hörmander's lectures that she "needed an instant replay." I think said friend was Dina Taiani.

The French mathematician P. Malliavin came up with a highly original use of Brownian motion to explain, in an intuitive way, Hörmander's results. In a way, Malliavin was continuing the program of Doob, that of using Brownian motion and probabilistic ideas to give insights into hard analysis. In the process, *en passant*, Malliavin created a new calculus of variations theory, which is now known as the Malliavin calculus. It has proved to be a highly useful tool for a large variety of areas. 'Recently' M. Hairer [7] has given a concise and insightful treatment of the Malliavin calculus along with a simple proof of Hörmander's Theorem, using the probabilistic tools Malliavin created.

Also in the 1970s, a truly great decade for innovation in probability theory, came the seminal papers of P. Samuelson, F. Black, M. Scholes, and R. Merton, using the Itô calculus to correctly price financial options, and as a consequence creating a revolution in Mathematical Finance, which finally culminated in the two papers of F. Delbaen and W. Schachermayer, in the 1990s that tied together an absence of arbitrage opportunities with the theory of martingales, and local martingales, even sigma martingales! These results have been nicely presented in a book by

Delbaen and Schachermayer [3]. This created, once again, a job program for young probabilists, although this time not in the academy, but rather in the big banks and investment houses who suddenly needed "quants" to study and understand the new advances in the fair pricing of financial options. This theory appealed to me, because it used the Itô calculus in new, and creative ways. I'm not sure Doob, or for that matter K. Itô, ever envisaged abstract stochastic calculus to study the tiny subarea of financial mathematics. P.A. Meyer, the great French probabilist who did fundamental work developing the theory of stochastic integration for semimartingales, was known to have contempt for its possible use in high finance.

Student demand to learn the new 'Financial Mathematics' once again led Mathematics Departments and other departments, too (Statistics, Operations Research, mathematically oriented Business Schools) to hire probabilists, this time to teach their students financial mathematics. So, because of the new job opportunities in the banking sector of the economy, probabilists who knew what is loosely known as the Black–Scholes theory, found jobs in academia too.

More recently, we should mention what is known as the Stochastic Loewner Equation (SLE for short). L. Bieberbach made a conjecture early in the twentieth century (1916) that a certain class of holomorphic functions had a series expansion where the nth term had the property $|a_n| \leq n$ [1]. This conjecture remained open for a long time, and the Czech mathematician Charles Loewner proved it was true for the third coefficient $|a_3| \leq 3$. To do this, he related the problem to partial differential equations, and in doing so created what is now known as the Loewner equation. My former colleague at Purdue University, L. de Branges, finally proved the Bieberbach conjecture in 1985.

The Loewner equation relates to probability via a stunning result of O. Schramm in 2000. Revolving around work of G. Lawler and W. Werner, they gave scaling limits of a range of stochastic processes, relating to critical percolation, the critical Ising model, the double-dimer model, self-avoiding walks, and other critical statistical mechanics models that exhibit conformal invariance.

In contemporary probability theory much attention has gone to the Kardar–Parisi–Zhang equation (and its variants) of statistical physics. Work in this area, especially by my Columbia colleague I. Corwin [2], has led to rigorous proofs of various predictions from physics. This is an exciting area which is still developing.

I have given a sketch of some of what has gone before, but now I am asked to forecast what is to come. Really?

Well, I cut my teeth on strong Markov processes, and then martingales, and since most of my recent work has been in the Statistics of Stochastic Processes, especially as they relate to Mathematical Finance, let me restrict my pseudo-insights to that area.

In economics, the fear caused by the near total collapse of the world's economic system in 2008, related to the bubble in housing prices, is still with us. I expect the research in Math Finance to migrate from the analysis of option pricing and hedging to the larger problems of the stability of the entire financial system. There is much interesting mathematics to be done in this regard.

In the good old days, banks going bankrupt was an important issue. During the depression of the 1930s, many banks did exactly that, and families lost their life savings. To give people a renewed confidence in the banking sector, the Federal Government (of the United States), proposed insurance backed by the full strength of the government, up to $100,000 per account. This was later raised to $250,000 per account, which seems like a lot, until you consider housing prices in New York City, for example. For businesses, the $250k guarantee is small, if not tiny.

Nevertheless, the guarantees worked and people began to trust their money to banks again. Controls had been instituted to prevent the rise of mega banks. For example, a given bank could operate in only one state. This favored the big states with financial centers, such as California, Illinois, and of course New York.

In New York State, banks were limited to one county only, a regulation to protect and foster smaller, local, community banks. New York City alone has five counties; otherwise Citibank, Chase and JP Morgan would have overwhelmed the other banks. Everything worked well, so of course the government, in its wisdom, decided to deregulate, and relax the controls on the big banks. This is analogous to having an illness well controlled by medication, and then saying you feel so good, that you're stopping the medicine. Guess what happens?

Academics are mostly powerless to change the stupidity of our governments, but we can, at the least, show them where their actions will take us, and why. The community has already had limited success in this direction as regards climate change, so hopefully we can also wake up the politicians to the dangers of their rash actions as regards finance. Mathematics is important in this regard. One example is when I read two articles in the same issue of a magazine, both by economists. One article argued that the price of gold was in a bubble, and it convinced me. The second article explained why the price of gold was not in a bubble, and I found it equally convincing! That illustrated, at least to me, of how desperate was the need for mathematical models, rather than just the stories economists tell so well. History tells us it will not be easy. Good luck, everyone!

Acknowledgements This project was supported in part by NSF grant DMS-2106433.

References

1. L. Bieberbach. Über die Koeffizienten derjenigen Potenzreihen, welche eine schlichte Abbildung des Einheitskreises vermitteln. *S. B. Preuss Akad. Wiss.* **138**, 940–955 (1916).
2. I. Corwin. Macdonald processes, quantum integrable systems and the Kardar–Parisi–Zhang universality class. In: *Proceedings of the International Congress of Mathematicians 2014*, Vol III, pp. 1007–1034 (2014).
3. F. Delbaen and W. Schachermayer. *The Mathematics of Arbitrage.* Springer (2005).
4. J. Doob. *Classical Potential Theory and Its Probabilistic Counterpart.* Springer (1984).
5. C. Fefferman and C. Stein. H^p-spaces of several variables. *Acta Math.* **129**, 137–193 (1972).
6. H. Fischer. *A History of the Central Limit Theorem.* Springer, 2010.

7. M. Hairer. On Malliavin's Proof of Hörmander's Theorem. *Bulletin des Sciences Mathématiques* **135**(6), 2011.
8. P.S. de Laplace. *Théorie Analytique des Probabilités*. Paris, V. Courcier (1812).
9. S. Stigler. Cauchy and the Witch of Agnesi: An Historical Note on the Cauchy Distribution. *Biometrika*, Vol. **61**, 375–380 August (1974).

Selected Problems in Probability Theory

Geoffrey R. Grimmett

Dedicated in friendship to Catriona Byrne

Personal Remarks

The editorial team of Springer Mathematics has become almost family for many of us worldwide, with Catriona Byrne at its heart. She has come to know us better than we know ourselves, always with sympathy, and with an honest and constructive approach to occasionally challenging areas of professional debate. Through our numerous collaborations, she and I have kindled a warm friendship that will persist into the next phase of our adventures. We wish her many happy years free from the woes of authors, editors, readers, and publishers.

1 Introduction

Probability has been a source of many tantalising problems over the centuries, of which the St Petersburg paradox, Fermat's problem of the points, and Bertrand's random triangles feature still in introductory courses. Whereas the conceptual problems of the past are now largely resolved, contemporary questions arise frequently where the intuitive apparatus of sub-fields collide. Many prominent problems are to be found at the conjunction of probability and discrete geometry. This short and idiosyncratic article summarises some of these. This account is personal and incomplete, and is to be viewed as a complete review of nothing. The bibliography is not intended to be complete, and apologies are extended to those whose work has been omitted.

G. R. Grimmett (✉)
Statistical Laboratory, Centre for Mathematical Sciences, Cambridge University, Cambridge, UK
e-mail: grg@statslab.cam.ac.uk

© The Author(s), under exclusive license to Springer Nature Switzerland AG 2023   603
J.-M. Morel, B. Teissier (eds.), *Mathematics Going Forward*, Lecture Notes in Mathematics 2313, https://doi.org/10.1007/978-3-031-12244-6_42

The questions highlighted here vary from the intriguing to the profound. Whereas some may seem like puzzles with limited consequence, others will require new machinery and may have far-reaching implications.

The problem of counting self-avoiding walks is introduced in Sect. 2, with emphasis on the existence of critical exponents and the scaling limit in two dimensions, followed by a fundamental counting problem on a random percolation cluster. Section 3 is devoted to Lorentz scatterers and the Ehrenfest wind/tree model, followed by Poissonian mirrors in two dimensions, and finally Manhattan pinball. Two well-known conjectures concerning product measures are presented in Sect. 4, namely the bunkbed conjecture and the negative association of a uniform spanning forest. Section 5 is concerned with the identification of criticality for the randomly oriented square lattice. In the final Sect. 6, we present two basic problems associated with a model for a dynamic spatial epidemic, provoked in part by COVID-19.

2 Self-Avoiding Walks

2.1 Origins

Self-avoiding walks were first introduced in the chemical theory of polymerisation (see [10, 32]), and their properties have received much attention since from mathematicians and physicists (see, for example, [5, 14, 30]).

A path in an infinite graph $G = (V, E)$ is called *self-avoiding* if no vertex is visited more than once. Fix a vertex $v \in V$, and let $\Sigma_n(v)$ be the set of n-step self-avoiding walks (SAWs) starting at v. The principal combinatorial problem is to determine how the cardinality $\sigma_n(v) := |\Sigma_n(v)|$ grows as $n \to \infty$, and the complementary probability problem is to establish properties of the shape of a randomly selected member of $\Sigma_n(v)$. Progress has been striking but limited.

2.2 Asymptotics

It is now regarded as elementary that the so-called *connective constant* $\kappa = \kappa(G)$, given by

$$\log \kappa = \lim_{n \to \infty} \frac{1}{n} \log \sigma_n(v),$$

exists when G is quasi-transitive, and is independent of the choice of v. Thus, in this case

$$\sigma_n(v) = \kappa^{(1+o(1))n}.$$

The correction term is much harder to understand. We shall not make precise the concept of a d-dimensional lattice, but for definiteness the reader may concentrate on the hypercubic lattice $\mathbb{Z}^d$.

Conjecture 2.1 *For $d \geq 2$ there exists a* critical exponent *$\gamma = \gamma_d$ such that the following holds. Let G be a d-dimensional lattice. There exists a constant $A > 0$ such that*[1]

$$\sigma_n \sim A n^{\gamma - 1} \kappa(G)^n \qquad as\ n \to \infty. \tag{1}$$

Furthermore,

$$\gamma = \begin{cases} \frac{43}{32} & when\ d = 2, \\ 1 & when\ d \geq 4. \end{cases}$$

See [5, 30] for further discussion and results so far, and the papers [19, 20] of Hara and Slade when $G = \mathbb{Z}^d$ with $d \geq 5$, for which case they prove that $\gamma = 1$. Of particular interest is the case when $G = \mathbb{H}$, the hexagonal lattice. By a beautiful exact calculation that verifies an earlier conjecture of Nienhuis [31] based in conformal field theory, Duminil-Copin and Smirnov [8] proved that

$$\kappa(\mathbb{H}) = \sqrt{1 + \sqrt{2}}.$$

The proof reveals a discrete holomorphic function that is highly suggestive of a connection to a Schramm–Loewner evolution (see [23]), namely the following.

Question 2.1 Does a uniformly distributed n-step SAW from the origin of $\mathbb{H}$ converge weakly, when suitably rescaled, to the Schramm–Loewner random curve $SLE_{8/3}$?

Progress on this question should come hand-in-hand with a calculation of the associated critical exponent $\gamma = \frac{43}{32}$. Gwynne and Miller [17] have proved the corresponding weak limit in the universe of Liouville quantum gravity.

2.3 Self-Avoiding Walks in a Random Environment

How does the sequence (σ_n) behave when the underlying graph G is random? For concreteness, we consider here the infinite cluster I of bond percolation on $\mathbb{Z}^2$ with edge-density $p > \frac{1}{2}$ (see [11]).

---

[1] A logarithmic correction is in fact expected in (1) when $d = 4$.

Question 2.2 Does the limit $\mu(v) := \lim_{n \to \infty} \sigma_n(v)^{1/n}$ exist a.s., and satisfy $\mu(v) = \mu(w)$ a.s. on the event $\{v, w \in I\}$?

Related discussion, including of the issue of deciding when $\mu(v) = p\mu(\mathbb{Z}^2)$ a.s. on the event $\{v \in I\}$, may be found in papers of Lacoin [24, 25]. The easier SAW problem on (deterministic) weighted graphs is considered in [16].

3 Lorentz Scatterers

3.1 Background

The scattering problem of Lorentz [27] gives rise to the following general question. Scatterers are distributed randomly about $\mathbb{R}^d$. Light is shone from the origin in a given direction, and is subjected to reflection at the scatterers. Under what circumstances is the light ray: (i) bounded, (ii) unbounded, (iii) diffusive? While certain special cases are understood, the general question remains open. The problems mentioned here are concerned with *aperiodic* distributions of scatterers; the periodic case is rather easier.

Fig. 1 A NW and a NE mirror. Each is reflective on both sides

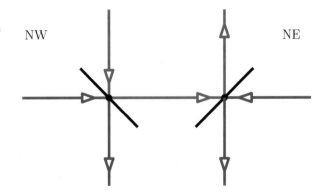

3.2 Ehrenfest Wind/Tree Model

The following notorious problem on the square lattice $\mathbb{Z}^2$ has resisted solution for many years. Let $p \in [0, 1]$. At each vertex of $\mathbb{Z}^2$ is placed a mirror with probability p, or alternatively nothing. Mirrors are plane and two-sided. Each mirror is designated a *north-east* (NE) mirror with probability $\frac{1}{2}$, or alternatively a *north-west* (NW) mirror. The states of different vertices are independent. The meanings of the mirrors are illustrated in Fig. 1.

Light is shone from the origin in a given compass direction, say north, and it is reflected off the surface of any mirror encountered. The problem is to decide whether or not the light ray is unbounded.

Question 3.1 Let $\theta(p)$ be the probability that the light ray is unbounded. For what values of p is it the case that $\theta(p) > 0$?

It is trivial that $\theta(0) = 1$. By considering bond percolation (with density $p/2$) on the diagonal lattice of Fig. 2, and using the fact that there is no percolation when $p = 1$, one obtains the less trivial fact that $\theta(1) = 0$ (see [11]). Very little more is known rigorously about the answer to Question 3.1.

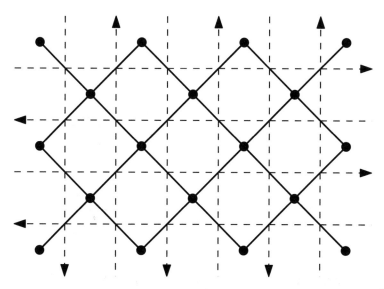

Fig. 2 From the original (dashed) square lattice $\mathbb{Z}^2$ one may construct a diagonal lattice $\widehat{\mathbb{Z}}^2$. In fact there are two such diagonal lattices, and this fact may be used to obtain some information about the power-law behaviour of the light ray when $p = 1$. The Manhattan orientations are not relevant to the usual Ehrenfest model, but are provided to facilitate the discussion of Manhattan pinball in Sect. 3.4

3.3 Poisson Mirrors

Here is version of the wind/tree model in the two-dimensional continuum $\mathbb{R}^2$. Let Π be a rate-1 Poisson process in $\mathbb{R}^2$. Let $\epsilon > 0$, and let μ be a probability measure on $[0, \pi)$. We possess an infinity of two-sided, plane mirrors of length ϵ, and we centre one at each point in Π; the inclination to the horizontal of each mirror is random with law μ, and different mirrors have independent inclinations. Think of a mirror as being a randomly positioned, closed line segment of length ϵ, and let M denote the union of these segments. We call μ *degenerate* if it is concentrated on a single atom, and shall assume μ is non-degenerate. See Fig. 3.

Light is shone from the origin at an angle α to the horizontal. Let I_α be the indicator function of the event that the light ray is unbounded. Some convention is adopted for the zero-probability event that the light strikes an intersection of two or more mirrors.

Fig. 3 Light from the origin is reflected off the mirrors

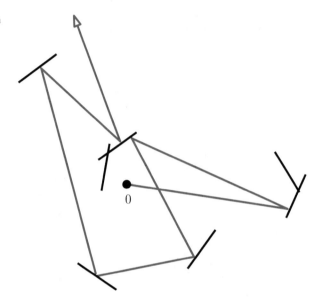

We may assume that the origin 0 does not lie in M. Let C be the component of $\mathbb{R}^2 \setminus M$ containing 0, and let $\{0 \leftrightarrow \infty\}$ be the event that C is unbounded. It is a standard result of so-called needle percolation (see [36]) that there exists $\epsilon_c = \epsilon_c(\mu) \in (0, \infty)$ such that

$$\mathbb{P}_\mu(0 \leftrightarrow \infty) \begin{cases} > 0 & \text{when } \epsilon < \epsilon_c, \\ = 0 & \text{when } \epsilon > \epsilon_c. \end{cases}$$

(Here and later, the subscript μ keeps track of the choice of μ.) Obviously, on the event that $0 \leftrightarrow \infty$, we have that $I_\alpha = 0$ for all α. Therefore,

$$\mathbb{P}_\mu(I_\alpha = 1 \text{ for some } \alpha) = 0, \qquad \epsilon > \epsilon_c.$$

The converse issue is much harder and largely open.

Question 3.2 Suppose μ is non-degenerate.

(a) Does there exist $\epsilon'_c = \epsilon'_c(\mu) > 0$ such that $\theta_\mu(\epsilon) > 0$ for $\epsilon < \epsilon'_c$?
(b) Could it be that $\epsilon'_c(\mu) = \epsilon_c(\mu)$?
(c) In particular, what happens when μ is the uniform measure on $[0, \pi)$?

Let $\mathcal{Q}$ be the set of probability measures μ that are non-degenerate and have support in the rational angles $\pi\mathbb{Q}$. Suppose $\mu \in \mathcal{Q}$ and $0 < \epsilon < \epsilon_c(\mu)$. Harris [21] has shown the striking fact that, $\mathbb{P}_\mu$-a.s. on the event that $0 \leftrightarrow \infty$, we have that $I_\alpha = 1$ for (Lebesgue) almost every α.

This leads to a deterministic question. Let K be the set of mirror configurations for which $0 \leftrightarrow \infty$ but $I_\alpha = 0$ for all α.

Question 3.3 Is K non-empty?

Harris's theorem implies in effect that $\mathbb{P}_\mu(K) = 0$ when $\mu \in \mathcal{Q}$ and $\epsilon \neq \epsilon_c(\mu)$. Question 3.2(c) hints at the possibility that $\mathbb{P}_\mu(K) = 0$ when μ is the uniform measure on $[0, \pi)$ and $\epsilon < \epsilon(\mu)$.

Here is a final question concerning diffusivity. Let $\mu \in \mathcal{Q}$, and denote by $X_\alpha(t)$ the position at time t of the light ray that leaves the origin at angle α.

Question 3.4 Is it the case that, on the event I_α, $X_\alpha(\cdot)$ is diffusive? That is, the limit $\sigma^2 := \lim_{t \to \infty} t^{-1} \mathrm{var}(X_\alpha(t))$ exists in $(0, \infty)$, and, when normalized, $X_\alpha(t)$ is asymptotically normally distributed.

Related work on Lorentz models in the so-called Boltzmann–Grad limit may be found in [28, 29].

3.4 Manhattan Pinball

Here is a variant of the Ehrenfest model motivated by a problem of quantum localization, [7, Sec. 4.2] and [37, p. 238]. Draw $\mathbb{Z}^2$ and the diagonal lattice $\widehat{\mathbb{Z}}^2$ as in Fig. 2; each edge of $\mathbb{Z}^2$ receives its Manhattan orientation as indicated in the figure. Consider bond percolation with density q on the diagonal lattice. Along each open edge of $\widehat{\mathbb{Z}}^2$ we place a two-sided plane mirror. Light is shone from the origin along a given one of the two admissible directions, and it is reflected by any mirror that it encounters (such reflections are automatically consistent with the Manhattan orientations). Let $\theta(q)$ be the probability that the light ray is unbounded.

Question 3.5 Could it be that $\theta(q) = 0$ for all $q > 0$?

It follows as in Sect. 3.2 that $\theta(q) = 0$ for $q \geq \frac{1}{2}$, and it has been proved by Li in [26] that there exists $\epsilon > 0$ such that $\theta(q) = 0$ when $q > \frac{1}{2} - \epsilon$. The proof uses the method of enhancements; see [1] and [11, Sect. 3.3].

4 Two Stochastic Inequalities

4.1 Bunkbed Inequality

The mysterious 'bunkbed' inequality was posed by Kasteleyn around 1985 (see [6, Rem. 5], and also [18]). Of its various flavours, we select the following. Let $G = (V, E)$ be a finite simple graph. From G we construct two copies denoted $G_1 = (V_1, E_1)$ and $G_2 = (V_2, E_2)$. For $v \in V$ we write v_i for the copy of v lying in V_i. We now attach G_1 and G_2 by adding edges $\langle v_1, v_2 \rangle$ for each $v \in V$. This new graph is denoted $\widetilde{G}$, and it may be considered as the product graph $G \times K_2$ where K_2 is the complete graph on two vertices (that is, an edge). We may think of the G_i as being 'horizontal' and the extra edges as being 'vertical'.

Each edge of $\widetilde{G}$ is declared *open* with probability p, independently of the states of other edges. Write $\mathbb{P}_p$ for the appropriate product measure. For two vertices u_i, v_j of $\widetilde{G}$, we write $\{u_i \leftrightarrow v_j\}$ for the event that there exists a u_i/v_j path using only open edges.

Conjecture 4.1 *For $u, v \in V$, we have*

$$\mathbb{P}_p(u_1 \leftrightarrow v_1) \geq \mathbb{P}_p(u_1 \leftrightarrow v_2).$$

There is uncertainty over whether this was the exact conjecture of Kasteleyn. For example, it is suggested in [22] (and perhaps elsewhere) that Kasteleyn may have made the stronger conjecture that the inequality holds even after conditioning on the set T of open vertical edges.

Some special cases of the bunkbed conjecture have been proved (see the references in [22], and more recently [35]), but the general question remains open.

4.2 Negative Correlation

Our next problem is quite longstanding (see [33]) and remains mysterious. In a nutshell it is to prove that the uniform random forest measure (USF) has a property of negative association.

Let $G = (V, E)$ be a finite graph which, for simplicity, we assume has neither loops nor multiple edges. A subset $F \subseteq E$ is called a *forest* if (V, F) has no cycles. Let $\mathcal{F}$ be the set of all forests in G and let Φ be a random forest chosen uniformly from $\mathcal{F}$. We call Φ *edge-negatively associated* if

$$\mathbb{P}(e, f \in \Phi) \leq \mathbb{P}(e \in \Phi)\mathbb{P}(f \in \Phi), \qquad e, f \in E, \ e \neq f. \tag{2}$$

Conjecture 4.2 *For all graphs G, the random forest Φ is edge-negatively associated.*

One may formulate various forms of negative dependence, amongst which the edge-negative association of (2) is quite weak. One may conjecture that Φ has a stronger variety of such dependence. Further discussion may be found in [13, Sec. 3.9] and [33].

Experimental evidence for Conjecture 4.2 is quite strong. A similar conjecture may be made for uniform measure on the set of $F \subseteq E$ such that (V, F) is *connected* (abbreviated to UCS). In contrast, uniform spanning tree (UST) is well understood via the Kirchhoff theory of electrical networks, and further by Feder and Mihail [9]. USF, UCS, and UST are special cases of the so-called random-cluster measure with cluster-weighting factor q satisfying $q < 1$ (see [13, Sects 1.5, 3.9]).

In recent work, [3, 4], the percolative properties of the weighted random forest (or 'arboreal gas') on $\mathbb{Z}^d$ have been explored. It turns out that there is a phase transition if and only if $d \geq 3$.

5 Randomly Oriented Square Lattice

The following percolation-type problem remains open. Consider the square lattice $\mathbb{Z}^2$ and let $p \in [0, 1]$. Each horizontal edge is oriented rightward with probability p, and otherwise leftward. Each vertical edge is oriented upward with probability p, and otherwise downward. Write $\vec{\mathbb{Z}}^2$ for the ensuing randomly oriented network.

Let $\theta(p)$ denote the probability that the origin 0 is the endpoint of an infinite path of $\vec{\mathbb{Z}}^2$ that is oriented away from 0. The challenge is to determine for which p it is the case that $\theta(p) > 0$. It is elementary that $\theta(0) = 1$, and that $\theta(p) = \theta(1 - p)$. It is less obvious that $\theta(\frac{1}{2}) = 0$ (see [11, 1st edn]), which is proved via a coupling with bond percolation. By a comparison with oriented percolation, we have that $\theta(p) > 0$ if $p > \vec{p}_c$, where $\vec{p}_c$ is the critical point of oriented percolation on $\mathbb{Z}^2$; it is not difficult to deduce by the enhancement method (see [1] and [11, Sect. 3.3]) that there exists $p' \in (\frac{1}{2}, \vec{p}_c)$ such that $\theta(p) > 0$ when $p > p'$. It is believed that $\vec{p}_c \sim 0.64$, and proved that $\vec{p}_c < 0.6735$.

Question 5.1 Is it the case that $\theta(p) > 0$ for $p \neq \frac{1}{2}$.

It is shown in [12] that, for all p, $\vec{\mathbb{Z}}^2$ is either critical or supercritical in the following sense: if any small positive density of oriented edges is added at random, then there is a strictly positive probability that the origin is the endpoint of an infinite self-avoiding oriented path in the resulting graph.

6 Dynamic Stochastic Epidemics

The recent pandemic has inspired a number of mathematical problems, including the following stochastic model (see [15]). Particles are placed at time 0 at the points of a rate-1 Poisson process in $\mathbb{R}^d$, where $d \geq 1$. Each particle diffuses around $\mathbb{R}^d$ according to a Brownian motion, independently of other particles. At any given time, each particle is in one of the states S (susceptible), I (infected), R (removed/dead).

At time 0 there exists a unique particle in state I, and all others are in state S. The infection/removal rules are as follows.

(a) If an infected particle comes within distance 1 of a susceptible particle, the latter particle is infected.
(b) An infected particle remains infected for a period of time having the exponential distribution with parameter α, and is then removed.

We call this the 'diffusion model'.

Survival is said to occur if, with a strictly positive probability, infinitely many particle are ultimately infected. It is proved in [15] that, when $d \geq 1$ and α is sufficiently large, survival does not occur. The following two questions (amongst others) are left open.

Question 6.1

(i) When $d = 1$, could it be that there is no survival for any $\alpha > 0$?
(ii) When $d \geq 2$, does survival occur for sufficiently small $\alpha > 0$?

In a variant of this problem termed the 'delayed diffusion model', a much fuller picture is known. Suppose, instead of the above, a particle moves only when it is infected; susceptible particles are stationary. The answers to Question 6.1(i, ii) are then no and yes, respectively , and indeed (when $d \geq 2$) there exists a critical value $\alpha_c(d) \in (0, \infty)$ of α marking the onset of survival. The key difference between the two systems is that the latter model has a property of monotonicity that is lacking in the former.

The delayed diffusion model is a continuous space/time cousin of the discrete-time 'frog model' of [2] (see also [34]), with the addition of removal. Further relevant references may be found in [15].

References

1. M. Aizenman and G.R. Grimmett. Strict monotonicity of critical points in percolation and ferromagnetic models. *J. Statist. Phys.* **63**, 817–835 (1991).
2. O.S.M. Alves, F.P. Machado and S.Yu. Popov. Phase transition for the frog model. *Electron. J. Probab.* **7**, paper no. 16, 21 pp. (2002).
3. R. Bauerschmidt, N. Crawford and T. Helmuth. Percolation transition for random forests in $d \geq 3$. https://arxiv.org/abs/2107.01878 (2021).
4. R. Bauerschmidt, N. Crawford, T. Helmuth and A. Swan. Random spanning forests and hyperbolic symmetry. *Commun. Math. Phys.* **381**, 1223–1261 (2021).
5. R. Bauerschmidt, H. Duminil-Copin, J. Goodman and G. Slade. Lectures on self-avoiding walks. In: *Probability and Statistical Physics in Two and More Dimensions*, D. Ellwood, C.M. Newman, V. Sidoravicius and W. Werner (eds.), pp. 395–476, CMI/AMS publication, Clay Mathematics Institute Proceedings **15** (2012).

6. J. van den Berg and J. Kahn. A correlation inequality for connection events in percolation. *Ann. Probab.* **29**, 123–126 (2001).
7. J. Cardy. Quantum network models and classical localization problems. *Int. J. Mod. Phys. B* **24**, 1989–2014 (2010).
8. H. Duminil-Copin and S. Smirnov, The connective constant of the honeycomb lattice equals $\sqrt{2 + \sqrt{2}}$. *Ann. Math.* **175**, 1653–1665 (2012).
9. T. Feder and M. Mihail. Balanced matroids. In: *Proceedings of the 24th ACM Symposium on the Theory of Computing*, pp. 26–38 (1992).
10. P. Flory. *Principles of Polymer Chemistry.* Cornell University Press 1953.
11. G.R. Grimmett. *Percolation.* Second edition. Springer, Berlin (1999).
12. G.R. Grimmett. Infinite paths in randomly oriented lattices. *Rand. Struct. Algor.* **18**, 257–266 (2001).
13. G.R. Grimmett. *The Random-Cluster Model.* Springer, Berlin (2006).
14. G.R. Grimmett and Z. Li. Self-avoiding walks and connective constants. In: *Sojourns in Probability Theory and Statistical Physics, III*, V. Sidoravicius (ed), pp. 215–241, Proceedings in Mathematics & Statistics **300**, Springer (2019).
15. G.R. Grimmett and Z. Li. Brownian snails with removal: epidemics in diffusing populations. https://arxiv.org/abs/2009.02495 (2020).
16. G.R. Grimmett and Z. Li. Weighted self-avoiding walks. *J. Alg. Comb.* **52**, 77–102 (2020).
17. E. Gwynne and J. Miller. Convergence of the self-avoiding walk on random quadrangulations to $SLE_{8/3}$ on $\sqrt{8/3}$-Liouville quantum gravity. *Ann. Sci. de l'Ecole Norm. Sup.* **54**, 305–405 (2021).
18. O. Häggström. Probability on bunkbed graphs. In: *Proceedings of FPSAC'03, Formal Power Series and Algebraic Combinatorics, Linköping, Sweden* (2003).
19. T. Hara and G. Slade. The lace expansion for self-avoiding walk in five or more dimensions. *Rev. Math. Phys.* **4**, 235–327 (1992).
20. T. Hara and G. Slade. Self-avoiding walk in five or more dimensions. I. The critical behaviour. *Commun. Math. Phys.* **147**, 101–136 (1992).
21. M. Harris. Nontrivial phase transition in a continuum mirror model. J. Theoret. Probab. **14**, 299–317 (2001).
22. T. Hutchcroft, A. Kent and P. Nizić-Nikolac. The bunkbed conjecture holds in the $p \uparrow 1$ limit. https://arxiv.org/abs/2110.00282 (2021).
23. A. Kemppainen. *Schramm–Loewner Evolution.* SpringerBriefs in Mathematical Physics **24**, Springer, Cham (2017).
24. H. Lacoin. Existence of a non-averaging regime for the self-avoiding walk on a high-dimensional infinite percolation cluster. *J. Statist. Phys.* **154**, 1461–1482 (2014).
25. H. Lacoin. Non-coincidence of quenched and annealed connective constants on the supercritical planar percolation cluster. *Probab. Th. Rel. Fields* **159**, 777–808 (2014).
26. L. Li. On the Manhattan pinball problem. *Electron. Commun. Probab.* **26**, 1–11 (2021).
27. H.A. Lorentz. The motion of electrons in metallic bodies, I, II, III. *Koninklijke Akademie van Wetenschappen te Amsterdam, Section of Sciences* **7**, 438–453, 585–593, 684–691 (1905).
28. C. Lutsko and B. Tóth. Invariance principle for the random Lorentz gas—beyond the Boltzmann–Grad limit. *Commun. Math. Phys.* **379**, 589–632 (2020).
29. C. Lutsko and B. Tóth. Invariance principle for the random wind-tree process. *Ann. Henri Poincaré* **22**, 3357–3389 (2021).
30. N. Madras and G. Slade. *Self-Avoiding Walks.* Birkhäuser, Boston (1993).
31. B. Nienhuis. Exact critical points and critical exponents of $O(n)$ models in two dimensions. *Phys. Rev. Lett.* **49**, 1062–1065 (1982).
32. W.J.C. Orr. Statistical treatment of polymer solutions at infinite dilution. *Trans. Faraday Soc.* **43**, 12–27 (1947).
33. R. Pemantle. Towards a theory of negative dependence. *J. Math. Phys.* **41**, 1371–1390 (2000).
34. A.F. Ramírez and V. Sidoravicius. Asymptotic behavior of a stochastic growth process associated with a system of interacting branching random walks. *C. R. Math. Acad. Sci. Paris* **335**, 821–826 (2002).

35. T. Richthammer. Bunkbed conjecture for complete bipartite graphs and related classes of graphs. https://arxiv.org/abs/2204.12931 (2022).
36. R. Roy. Percolation of Poisson sticks on the plane. *Probab. Th. Rel. Fields* **89**, 503–517 (1991).
37. T. Spencer. Duality, statistical mechanics and random matrices. In: *Current Developments in Mathematics*, pp. 229–260, International Press, Somerville (2012).

Optimal Matching of Random Samples and Rates of Convergence of Empirical Measures

Michel Ledoux

Personal Note

It is a pleasure to dedicate this contribution to Catriona Byrne, in recognition of her many years at the service of the scientific community, with dedication, professionalism, deep scientific knowledge and expertise, and cordiality. With special thoughts to the many rewarding and friendly exchanges over the years, as author and editor.

Optimal matching problems have been investigated from various viewpoints in computer science, algorithmic analysis and physics, while rates of convergence of empirical measures to their common distribution is a central topic in probability and mathematical statistics.

Perfect matching problems (on bipartite graphs), also called assignment problems, are combinatorial optimization problems classically studied within operation research and algorithmic, combinatorics, graph theory and mathematical physics (cf. e.g. [32, 33]). Classical applications to planning, allocation of resources, traveling salesman problems, expand nowadays to networks and complex systems. Linear programming relaxation within assignment and optimal transport problems also provide useful tools in machine learning and data science [34].

The random version of the matching problems addresses optimization of Euclidean additive functionals in geometric probability [37, 41] and rates of convergence of empirical measures. It opened recently fascinating challenges, which are active parts of current research. The close relationship with mass transportation in particular favored the novel use of tools from convex analysis, probability theory and partial differential equations (pde). This note describes a few of these stimulating

---

M. Ledoux (✉)

Institut de Mathématiques de Toulouse, Université de Toulouse – Paul-Sabatier, Toulouse, France

e-mail: ledoux@math.univ-toulouse.fr

J.-M. Morel, B. Teissier (eds.), *Mathematics Going Forward*, Lecture Notes in Mathematics 2313, https://doi.org/10.1007/978-3-031-12244-6_43

questions for the Euclidean random optimal matching problem and associated rates
of convergence of empirical measures.

1 Euclidean Random Optimal Matching and Rates of Convergence of Empirical Measures

Given points $x_1, \ldots, x_n$ and $y_1, \ldots, y_n$ in $\mathbb{R}^d$, and $p \geq 1$, the optimal matching
problem raises the question of estimating

$$\inf_\sigma \frac{1}{n} \sum_{i=1}^n |x_i - y_{\sigma(i)}|^p$$

where the infimum runs over all permutations σ of $\{1, \ldots, n\}$ (and $| \cdot |$ is, for
example at this stage, the Euclidean distance on $\mathbb{R}^d$). That is, the task is to match
the points of a sample $(x_1, \ldots, x_n)$ with the ones of another sample $(y_1, \ldots, y_n)$
minimizing a given cost function. The typical values of p are $p = 1$ and $p = 2$,
also $p = \infty$ (with then $\inf_\sigma \max_{1 \leq i \leq n} |x_i - y_{\sigma(i)}|$).

The question may be formulated equivalently in the closely related mass
transportation framework. Given $p \geq 1$, the Kantorovich distance (cf. [40] e.g.)
between two probability measures ν and μ on the Borel sets of $\mathbb{R}^d$ with a finite p-th
moment is defined by

$$W_p(\nu, \mu) = \inf_\pi \left(\int_{\mathbb{R}^d \times \mathbb{R}^d} |x - y|^p \mathrm{d}\pi(x, y) \right)^{1/p} \tag{1}$$

where the infimum is taken over all couplings π on $\mathbb{R}^d \times \mathbb{R}^d$ with respective
marginals ν and μ. The W_p metrics are monotone increasing with p. In the limit
$p \to \infty$, $W_\infty(\nu, \mu)$ may be understood as the infimum over all couplings π of

$$\operatorname{esssup}_\pi \left\{ |x - y| \, ; \, (x, y) \in \mathbb{R}^d \times \mathbb{R}^d \right\}$$

(for measures ν and μ with bounded support).

Given samples $(x_1, \ldots, x_n)$ and $(y_1, \ldots, y_n)$ of points in $\mathbb{R}^d$, if $\nu = \frac{1}{n} \sum_{i=1}^n \delta_{x_i}$
and $\mu = \frac{1}{n} \sum_{i=1}^n \delta_{y_i}$ are the empirical measures on the respective samples, the
right-hand side of (1) to the power p takes the form

$$\inf_\pi \sum_{i,j=1}^n |x_i - y_j|^p \pi_{ij}$$

where $\pi_{ij} = \pi(\{x_i, y_j\})$, $i, j = 1, \ldots, n$. Since π has marginals ν and μ, for
every i or j, $\sum_{i=1}^n \pi_{ij} = \sum_{j=1}^n \pi_{ij} = \frac{1}{n}$, and the set of those matrices π is convex

and compact in $\mathbb{R}^{n^2}$, so that by the Birkhoff–von Neumann theorem the infimum is achieved on a permutation matrix $\pi_{ij} = \frac{1}{n} 1_{\{j=\sigma(i)\}}$. As a consequence

$$\inf_\sigma \frac{1}{n} \sum_{i=1}^n |x_i - y_{\sigma(i)}|^p = W_p^p(\nu, \mu) = W_p^p\left(\frac{1}{n} \sum_{i=1}^n \delta_{x_i}, \frac{1}{n} \sum_{i=1}^n \delta_{y_i}\right).$$

The matching problem is thus translated equivalently as a discrepancy problem between empirical measures in Kantorovich distances.

The random optimal matching problem deals with samples $(X_1, \ldots, X_n)$ and $(Y_1, \ldots, Y_n)$ of independent and identically distributed random variables in $\mathbb{R}^d$ (with a finite p-th moment), and a first order analysis aims at studying the order of growth in n of the averages

$$\mathbb{E}\left(\inf_\sigma \frac{1}{n} \sum_{i=1}^n |X_i - Y_{\sigma(i)}|^p\right). \tag{2}$$

If $X_1, \ldots, X_n$ are independent random variables in $\mathbb{R}^d$ with common distribution μ, and if $\mu_n = \frac{1}{n} \sum_{i=1}^n \delta_{X_i}$, $n \geq 1$, is the empirical measure on the sample, simple arguments from the triangle and Jensen's inequalities compare (2) to the average $\mathbb{E}(W_p^p(\mu_n, \mu))$. The latter is then sometimes referred to as a semi-discrete matching as opposed to bipartite matching for the former. Almost surely, the sequence μ_n, $n \geq 1$, of empirical measures converges weakly to the common distribution μ, a central question of interest and study in probability and statistics. The strength of the approximation of μ by the empirical μ_n is indeed of basic importance in statistical applications, and orders of decay in various probabilistic distances have been considered. One of them is thus the Kantorovich distance that attracted a lot of attention (the convergence of $W_p(\mu_n, \mu)$ to 0 is equivalent to the weak convergence of μ_n towards μ plus convergence of p-moments). By standard concentration tools, not developed here, rates on $\mathbb{E}(W_p^p(\mu_n, \mu))$ may often be turned into bounds on $W_p^p(\mu_n, \mu)$ with high probability.

More general probabilistic dependences in the random sample $(X_1, \ldots, X_n)$ may be considered. Spectral measures of random matrices is one such instance, that gave rise to numerous recent contributions.

The exposition here is devoted to the Euclidean random optimal matching problem in the semi-discrete form and to the rate of convergence of the empirical measure to the reference measure in Kantorovich distance. As such, the discussion will be mostly focused on (lower and upper) bounds on the sequence of expectations

$$\mathbb{E}(W_p^p(\mu_n, \mu)) \tag{3}$$

where $\mu_n = \frac{1}{n} \sum_{i=1}^n \delta_{X_i}$, $n \geq 1$, and $X_1, \ldots, X_n$ are independent identically distributed in $\mathbb{R}^d$ with common distribution μ with a finite p-th moment, as well as on possible exact (renormalized) limits (although the study of the limits for the bipartite matching problem does in general require further details). The note surveys some basic results in the area, and features challenging open questions of the current research on the asymptotic rates of (3). The basic parameters entering the discussion are $p \geq 1$, the distribution μ and the dimension d of the state space. Throughout the text, the notation $A \approx B$ expresses that $C^{-1}B \leq A \leq CB$ for some constant $C > 0$ independent of n, possibly depending on the dimension d and the parameters of the underlying distribution μ. Similarly $A \lesssim B$ and $A \gtrsim B$ indicate that $A \leq CB$ and $A \geq C^{-1}B$ respectively. The bibliography is not extensive, and often concentrated only on reference texts or articles.

2   The One-Dimensional Case

The one-dimensional case is of particular nature due to explicit representations of the Kantorovich metrics $W_p(\nu, \mu)$ by monotone transport map of the distributions ν and μ on the Borel sets of $\mathbb{R}$. For example,

$$W_1(\nu, \mu) = \int_{\mathbb{R}} |G(x) - F(x)| dx$$

where $G(x) = \nu(]-\infty, x])$, $F(x) = \mu(]-\infty, x])$, $x \in \mathbb{R}$, are the distribution functions of ν and μ respectively. There are similar representation formulas for $W_p(\nu, \mu)$, $p \geq 1$, in terms of the inverse distribution functions, quantiles or order statistics of empirical measures (cf. e.g. [11]).

   On the basis of these explicit representations, rather precise descriptions of the rates of convergence of empirical measures in Kantorovich distances are available (cf. [11]). For example, $\mathbb{E}(W_1(\mu_n, \mu))$ is typically of the order of $1/\sqrt{n}$ for large families of distributions μ, and a precise statement is that

$$\mathbb{E}\big(W_1(\mu_n, \mu)\big) \lesssim \frac{1}{\sqrt{n}}$$

if and only if $\int_{\mathbb{R}} \sqrt{F(x)(1 - F(x))}\, dx < \infty$ (which is the case for instance if $\int_{\mathbb{R}} |x|^q d\mu < \infty$ for some $q > 2$). The lower bound $\mathbb{E}(W_1(\mu_n, \mu)) \gtrsim 1/\sqrt{n}$ holds true for any μ (with a first moment).

   However, when $p > 1$ some differences occur already on basic examples emphasizing the size of the support of μ as influencing the rate. The standard rate $1/n^{p/2}$ is the rule for compactly supported laws, but in general it cannot be obtained under moment conditions only. For example, by comparison with W_1,

$$\mathbb{E}\big(W_2^2(\mu_n, \mu)\big) \lesssim \frac{1}{n}$$

if and only if $\int_{\mathbb{R}}[F(x)(1 - F(x))/f(x)]\,dx < \infty$, where f is the density of the absolutely continuous component of μ. Such a characterization, which admits a version for any $p > 1$, is of particular interest for log-concave measures for which two-sided comparison inequalities may be achieved. As an illustration, while $\mathbb{E}(W_p^p(\mu_n, \mu))$ is of order $1/n^{p/2}$ for any $p \geq 1$ if μ is uniform on a compact interval, for μ the (standard) Gaussian distribution on $\mathbb{R}$,

$$\mathbb{E}\big(W_p^p(\mu_n, \mu)\big) \approx \begin{cases} \frac{1}{n^{p/2}} & \text{if } 1 \leq p < 2, \\[2mm] \frac{\log\log n}{n} & \text{if } p = 2, \\[2mm] \frac{1}{n(\log n)^{p/2}} & \text{if } p > 2. \end{cases} \tag{4}$$

While the rate is therefore the same as in the uniform case for $1 \leq p < 2$, two changes occur as $p = 2$ and $p > 2$. Further models are of interest, such as for instance the exponential distribution (see [11]).

Theoretical studies have been completed by various numerical simulations in the physics literature, covering related random assignment problems and their sharp asymptotic behaviors [15], such as for example the exact renormalized limit $\lim_{n\to\infty}(n/\log\log n)\,\mathbb{E}(W_2^2(\mu_n, \mu)) = 1$ in the Gaussian case obtained in [10].

3 The Ajtai–Komlós–Tusnády Theorem in Dimension 2 and the Ultimate Matching Conjecture

In the bipartite formulation (2) and for $p = 1$, the famous Ajtai–Komlós–Tusnády theorem in dimension 2 expresses that

$$\mathbb{E}\left(\inf_\sigma \frac{1}{n}\sum_{i=1}^n |X_i - Y_{\sigma(i)}|\right) \approx \sqrt{\frac{\log n}{n}} \tag{5}$$

for samples of independent random variables uniformly distributed on the unit square $[0, 1]^2$. It has been established in [1] by the transportation method on dyadic decompositions and combinatorial arguments, then reproved and deepened by P. Shor [35] and M. Talagrand (cf. [39] and the references therein) via generic chaining tools. The point is that, from the Kantorovich dual representation (see [40]),

$$W_1(\nu, \mu) = \sup\left[\int_{\mathbb{R}^d} \varphi\,d\nu - \int_{\mathbb{R}^d} \varphi\,d\mu\right]$$

where the supremum is taken over 1-Lipschitz maps $\varphi : \mathbb{R}^d \to \mathbb{R}$, and as such the study enters the framework of bounds on stochastic processes. Together with a Fourier representation of the parameter set as an ellipsoid, it allows indeed for the powerful use of majorizing measures and generic chaining methods (for which a complete account is the monograph [39]). The methodology covers similarly the values of $p \geq 1$, with order of growth $(\log n / n)^{p/2}$, but the limiting case $p = \infty$ shows an interesting logarithmic correction described by the Leighton–Shor grid-matching theorem [31]

$$\mathbb{E}\left(\inf_{\sigma} \max_{1 \leq i \leq n} |X_i - Y_{\sigma(i)}| \right) \approx \frac{(\log n)^{3/4}}{\sqrt{n}} . \tag{6}$$

This type of analysis furthermore led P. Shor [35] to a striking statement, improving upon the upper bound in the Ajtai–Komlós–Tusnády theorem, in which the coordinates of the variables (in $\mathbb{R}^2$, indicated by the superscripts 1 and 2) do not play the same role, namely

$$\mathbb{E}\left(\inf_{\sigma} \max \left(\frac{1}{n} \sum_{i=1}^{n} |X_i^1 - Y_{\sigma(i)}^1|, \max_{1 \leq i \leq n} |X_i^2 - Y_{\sigma(i)}^2| \right) \right) \lesssim \sqrt{\frac{\log n}{n}} . \tag{7}$$

In this framework, the "ultimate matching conjecture" promoted by M. Talagrand [39] would be that, for every $\alpha_1, \alpha_2 > 0$ with $\frac{1}{\alpha_1} + \frac{1}{\alpha_2} = \frac{1}{2}$, there is a constant $C > 0$ such that

$$\mathbb{E}\left(\inf_{\sigma} \max_{j=1,2} \left(\sum_{i=1}^{n} \exp\left(\frac{1}{C}\sqrt{\frac{n}{\log n}} |X_i^j - Y_{\sigma(i)}^j| \right)^{\alpha_j} \right) \right) \leq Cn.$$

Using on the one hand that $e^{a^4} \geq a^4$, and on the other hand that $\sum_{i=1}^{n} e^{a_i^4} \geq \exp(\max_{1 \leq i \leq n} a_i^4)$ together with Jensen's inequality, the case $\alpha_1 = \alpha_2 = 4$ would provide a neat common generalization of (the upper bounds in) the Ajtai–Komlós–Tusnády and Leighton–Shor theorems, and at the same time improve upon (7) corresponding to $\alpha_1 = 2$ and $\alpha_2 = \infty$ (with $\max_{1 \leq i \leq n} |X_i^2 - Y_{\sigma(i)}^2|$ in the $j = 2$ coordinate). A partial version of the conjecture as well as a suitable formulation in dimension $d \geq 3$ are discussed in [39].

Turning back to rates of empirical measures (or semi-discrete matching), for μ the uniform distribution on the unit cube $[0, 1]^d$ of $\mathbb{R}^d$, for any $d \geq 1$ and $p \geq 1$,

$$\mathbb{E}\big(W_p^p(\mu_n, \mu) \big) \approx \begin{cases} \frac{1}{n^{p/2}} & \text{if } d = 1, \\[2mm] \left(\frac{\log n}{n} \right)^{p/2} & \text{if } d = 2, \\[2mm] \frac{1}{n^{p/d}} & \text{if } d \geq 3. \end{cases} \tag{8}$$

The case $d = 2$ is thus the Ajtai–Komlós–Tusnády theorem, which, as for $d = 1$, develops an unusual rate with respect to the uniform spacing $1/n^{1/d}$ of n points in $[0, 1]^d$. However, this natural spacing is fully reflected in dimension $d \geq 3$, which makes this case easier than $d = 2$. Indeed, in dimension 2, there are irregularities at all scales in the distribution of a random sample $(X_1, \ldots, X_n)$ which combine to create the $\log n$ factor, while in higher dimension, there are still irregularities at many different scales but they cannot combine (see [39]). The complete range of parameters $p \geq 1$ and $d \geq 1$ in (8) is implicit in the paper [1] and the study [39]. See [27] for an independent proof relying on the mass transportation and pde methodology exposed in the subsequent Sect. 5. In the same vein, a simple Fourier analytic proof of the Ajtai–Komlós–Tusnády theorem is provided in [12] (see also [39]). The articles [18, 21] consider distributions with compact support and densities with respect to the Lebesgue measure uniformly bounded from below and above.

When $p = \infty$, the rates are respectively $1/\sqrt{n}$ in dimension 1, $(\log n)^{3/4}/\sqrt{n}$ in dimension 2 (the Leighton–Shor theorem (6)), and $(\log n/n)^{1/d}$ in dimension $d \geq 3$ [36] (and its extension [21]).

4 General Distributions and Higher Dimension

Beyond the uniform distribution, the corresponding results for more general distributions μ, in particular with unbounded support, gave rise to a number of contributions and open questions. The one-dimensional case is extensively discussed in [11], and already develops unusual phenomena as mentioned in Sect. 2.

The Ajtai–Komlós–Tusnády theorem (5) in dimension 2 for $p = 1$ extends to large families of distributions (see [39]). For example,

$$\mathbb{E}\big(W_1(\mu_n, \mu)\big) \lesssim \sqrt{\frac{\log n}{n}}$$

as soon as $\int_{\mathbb{R}^2} |x|^q d\mu < \infty$ for some $q > 2$.

A non-trivial aspect of the Ajtai–Komlós–Tusnády theorem consists also in the lower bound $\mathbb{E}(W_1(\mu_n, \mu)) \gtrsim \sqrt{\log n/n}$ (besides the proofs in [1] and [39], see [5] for a new proof by mass transportation-pde arguments). Lower bounds are usually not covered by general tools and for general distributions. Actually, for irregular laws, the decay can be faster, see among others [7, 8, 17].

When $p > 1$, and in higher dimension $d \geq 1$, the picture is more diversified. The general investigations of [13, 17, 20], based on couplings on dyadic decompositions together with a randomization argument by a Poisson variable, typically yield that, if for example $\int_{\mathbb{R}^d} |x|^q d\mu < \infty$ for some $q > \frac{p}{1-\min(p/d,1/2)}$, then

$$\mathbb{E}\big(W_p^p(\mu_n,\mu)\big) \lesssim \frac{1}{n^{\min(p/d,1/2)}} . \tag{9}$$

(The case $p = d/2$ actually involves some extra logarithmic factor.) At this level of generality, these results are essentially optimal, and suitably extend the uniform example when $p < d/2$. With respect to the Ajtai–Komlós–Tusnády theorem, one structural aspect of the proof of the general bounds (9) (due in particular to the randomization step) is however that, for $d = 1$ or 2, these will never yield anything better than a rate of the order of $1/\sqrt{n}$, and are essentially restricted to $p < d/2$ in higher dimension.

The Gaussian model is a good test example to appreciate the potential range of decay. Let thus, in the following, μ be the standard Gaussian distribution on $\mathbb{R}^d$ with density $(2\pi)^{-d/2} e^{-\frac{1}{2}|x|^2}$ with respect to the Lebesgue measure, and in particular moments of all orders. By a contraction argument, the Gaussian rates are always larger than the uniform ones from (8). The one-dimensional case is pictured in (4). In dimension $d = 2$, with respect to (9), it holds true that

$$\mathbb{E}\big(W_p^p(\mu_n,\mu)\big) \approx \begin{cases} \left(\frac{\log n}{n}\right)^{p/2} & \text{if } 1 \le p < 2, \\ \frac{(\log n)^2}{n} & \text{if } p = 2, \end{cases} \tag{10}$$

which extends the uniform model for $1 \le p < 2$, while a specific new feature appears at $p = 2$ as a consequence of the unbounded support. The proof of the case $1 \le p < 2$ and of the upper bound for $p = 2$ in [27] is based on the mass transportation and pde approach presented in the next section together with a localization step, while the lower bound for $p = 2$ in [38] relies on the generic chaining ideas of [39] together with a scaling argument (an alternate proof using the transportation-pde method is presented in [28]).

In higher dimension $d \ge 3$, the general bounds (9) yield that $\mathbb{E}(W_p^p(\mu_n,\mu)) \lesssim 1/n^{p/d}$ whenever $1 \le p < d/2$. This has been extended to $1 \le p < d$ in [30] by a specific Gaussian analysis of the associated Mehler kernel. In this range $1 \le p < d$, $d \ge 3$, the rates for the Gaussian are therefore the same as the ones for the compact uniform model.

As identified by (10) when $p = d = 2$, the case $p = d$ might be of special interest. A possible conjecture for $d \ge 3$ might be that

$$\mathbb{E}\big(W_d^d(\mu_n,\mu)\big) \approx \frac{(\log n)^{d/2}}{n} .$$

This is suggested as a lower bound in the note [38] (upper bounds with extra logarithmic factors are obtained in [30]). It is certainly possible that the tools developed in [38] could lead to more conclusions, also for $p > d \ge 2$ (and for more general distributions than the Gaussian with exponential tail decay), but this is essentially open at this point. Actually, there is no clear conjecture for $p > d \geq 2$ (including $p = \infty$) at this point.

5 Mass Transportation, PDE, and Exact Limits and Asymptotic Expansions

On the basis of the Ajtai–Komlós–Tusnády theorem (8) for the uniform distribution μ on the unit cube $[0, 1]^d$ of $\mathbb{R}^d$, the question of the exact asymptotic behavior of $\mathbb{E}(W_p^p(\mu_n, \mu))$ as $n \to \infty$ becomes natural. Again the one-dimensional case may be addressed rather simply, for example $\mathbb{E}(W_2^2(\mu_n, \mu)) = 1/6n$ (for any n).

Things are much more challenging in higher dimension, and actually some deep structural issues are underlying the picture, in particular motivated by conjectures raised by S. Caracciolo et al. [16] in the physics literature. As a major recent development in this regard, answering one of these conjectures, the landmark contribution [5] by L. Ambrosio, F. Stra and D. Trevisan achieved the exact (renormalized) limit for the uniform measure μ on $[0, 1]^2$ for $p = 2$,

$$\lim_{n \to \infty} \frac{n}{\log n} \, \mathbb{E}\big(W_2^2(\mu_n, \mu)\big) \; = \; \frac{1}{4\pi} \, . \tag{11}$$

The result actually applies to the uniform measure on a two-dimensional compact Riemannian manifold M of volume one (the results are invariant under rescaling of the measure), with the Euclidean distance in the definition of W_2 being replaced by the Riemannian distance. The factor $1/4\pi$ captures the common small time behavior of the trace of the Laplace operator Δ in the form of

$$\lim_{t \to 0} 4\pi t \int_M p_t(x, x) \mathrm{d}\mu(x) \; = \; 1$$

where $p_t(x, y)$, $t > 0$, $x, y \in M$, is the associated heat kernel (generating the heat semigroup P_t, $t > 0$). The result has been extended in [4] to measures on a bounded connected domain in $\mathbb{R}^2$ with Lipschitz boundary, with Hölder continuous density uniformly strictly positive and bounded from above. The method of proof is based on a deep analysis combining mass transportation and pde tools following an Ansatz put forward in [16]. If $T = \nabla \psi$ is the optimal transport map between two probability densities ρ_0 and ρ_1 (on M), the associated Monge–Ampère equation $\rho_1(\nabla \psi) \det(\nabla^2 \psi) = \rho_0$ is turned, via the linearization $\rho_j \approx 1$, into $\psi \approx \frac{1}{2}|x|^2 + f$ where f solves the Poisson equation $-\Delta f = \rho_1 - \rho_0$. In this way, the Kantorovich metric W_2 is approximated by an energy functional represented by a dual Sobolev norm through the observation that, whenever $g : M \to \mathbb{R}$ is smooth with $\int_M g d\mu = 0$, by a Taylor expansion on $d\nu_\varepsilon = (1 + \varepsilon g) d\mu$ as $\varepsilon \to 0$ (cf. [40])

$$\lim_{\varepsilon \to 0} \frac{1}{\varepsilon^2} W_2^2(\nu_\varepsilon, \mu) = \|g\|_{H^{-1,2}(\mu)}^2$$

where $H^{-1,2}(\mu)$ is the dual Sobolev norm described by the trace

$$\|g\|_{H^{-1,2}(\mu)}^2 = \int_M |\nabla(-\Delta)^{-1} g|^2 d\mu = \int_M g(-\Delta)^{-1} g \, d\mu$$

with $(-\Delta)^{-1} = \int_0^\infty P_t dt$. On the basis of this Ansatz, the proof of the limit (11) proceeds by regularization by the heat kernel and approximation by the energy functional, the leading term in (11), as well as the full rates in (8), reflecting the behaviour of the Green function (of the associated heat kernel) depending in particular on the dimension. More precise descriptions of the optimal map, rather than only the transport cost, are developed in [3], and in [23, 24] in connection with the behavior of the optimal transport map in the Lebesgue-to-Poisson problem together with a refined large-scale regularity theory for the Monge–Ampère equation. In case of the 2-dimensional sphere, a proof of the optimal matching rate is provided in [26] via gravitational allocation (the paper also describes related algorithmic questions of interest).

Still in dimension $d = 2$, the case $p \neq 2$ is completely open (and the value $p = 1$ should be of particular interest). For $p = 2$, the paper [16] (see also [9]) actually suggests moreover that for some value $\xi \in \mathbb{R}$,

$$\mathbb{E}\big(W_2^2(\mu_n, \mu)\big) = \frac{1}{4\pi} \frac{\log n}{n} + \frac{\xi}{n} + o\Big(\frac{1}{n}\Big). \tag{12}$$

Towards this conjecture, but still far from the answer, it is shown in [2] that

$$\left| \mathbb{E}\big(W_2^2(\mu_n, \mu)\big) - \frac{1}{4\pi} \frac{\log n}{n} \right| \lesssim \frac{\sqrt{\log n \, \log \log n}}{n}.$$

A further conjecture in this framework would be that

$$n\big[W_2^2(\mu_n, \mu) - \mathbb{E}\big(W_2^2(\mu_n, \mu)\big)\big] \to \chi$$

in distribution where χ is some centered random variable with an explicit distribution as a quadratic form of a Gaussian free field (see [22, 29]). Under the conjecture (12), it would hold that

$$n\Big[W_2^2(\mu_n, \mu) - \frac{1}{4\pi} \log n\Big] \to \xi + \chi$$

in distribution, which would be the ultimate description of the limiting behaviour of $W_2^2(\mu_n, \mu)$ (provided the limiting value ξ is identified). For the matter of comparison, in dimension $d = 1$ for μ the Lebesgue measure on $[0, 1]$, $\mathbb{E}(W_2^2(\mu_n, \mu)) = 1/6n$ while $n\,W_2^2(\mu_n, \mu)$ converges in law to $\int_0^1 B(t)^2 dt$ with B a Brownian bridge on $[0, 1]$ (in particular $\mathbb{E}\left(\int_0^1 B(t)^2 dt\right) = 1/6$) [6].

The identification of the limits in dimension $d \geq 3$ seems to raise even higher difficulties. Let still μ denote the uniform measure on $[0, 1]^d$. In [25], M. Goldman and D. Trevisan showed that, for every $d \geq 3$ and $p \geq 1$, the limit

$$\lim_{n \to \infty} n^{p/d}\, \mathbb{E}\big(W_p^p(\mu_n, \mu)\big)$$

exists and is strictly positive. The result actually extends previous works, basically covering $p < d/2$, in [7, 14, 17, 19] making use of subadditivity arguments on dyadic and combinatorial partitionings. The new ingredient in [25] is the coupling of subadditivity with the optimal transport and pde approach which has been successful in dimension 2. However, since the error in the smoothing by the heat kernel and the energy functional are of the same order in higher dimension $d \geq 3$, a delicate feature is that the leading term in the asymptotics of the Kantorovich rate might not be given anymore by the dual Sobolev norm (while higher orders are), so that identification of the limit is a serious task. Actually the prediction in [9, 16], for $p = 2 < d$, would be that

$$\mathbb{E}\big(W_2^2(\mu_n, \mu)\big) = \frac{c_d}{n^{2/d}} + \frac{\xi}{4\pi^2}\frac{1}{n} + o\Big(\frac{1}{n}\Big),$$

but c_d is not clearly conjectured, while ξ should be explicitly given in terms of the Epstein function.

References

1. M. Ajtai, J. Komlós, G. Tusnády. On optimal matchings. *Combinatorica* **4**, 259–264 (1984).
2. L. Ambrosio, F. Glaudo. Finer estimates on the 2-dimensional matching problem. *J. Éc. polytech. Math.* **6**, 737–765 (2019).
3. L. Ambrosio, F. Glaudo, D. Trevisan. On the optimal map in the 2-dimensional random matching problem. *Discrete Contin. Dyn. Syst.* **39**, 7291–7308 (2019).
4. L. Ambrosio, M. Goldman, D. Trevisan. On the quadratic random matching problem in two-dimensional domains. *Preprint* (2021).
5. L. Ambrosio, F. Stra, D. Trevisan. A PDE approach to a 2-dimensional matching problem. *Probab. Theory Related Fields* **173**, 433–478 (2019).
6. E. del Barrio, E. Giné, F. Utzet. Asymptotics for L_2 functionals of the empirical quantile process, with applications to tests of fit based on weighted Wasserstein distances. *Bernoulli* **1**, 131–189 (2005).
7. F. Barthe, C. Bordenave. Combinatorial optimization over two random point sets. *Séminaire de Probabilités XLV, Lecture Notes in Math.* **2078**, 483–535 (2013).

8. D. Benedetto, E. Caglioti. Euclidean random matching in $2d$ for non-constant densities. *J. Stat. Phys.* **181**, 854–869 (2020).
9. D. Benedetto, E. Caglioti, S. Caracciolo, M. D'Achille, G. Sicuro, A. Sportiello. Random assignment problems on $2d$ manifolds. *J. Stat. Physics* **183**, 1–40 (2021).
10. P. Berthet and J. Fort. Exact rate of convergence of the expected W_2 distance between the empirical and true Gaussian distribution. *Electron. J. Probab.* **25**, paper no. 12 (2020).
11. S. Bobkov and M. Ledoux. *One-dimensional empirical measures, order statistics, and Kantorovich transport distances.* Mem. Amer. Math. Soc. **261**, no. 1259 (2019).
12. S. Bobkov and M. Ledoux. A simple Fourier analytic proof of the AKT optimal matching theorem. *Ann. Appl. Probab.* **31**, 2567–2584 (2021).
13. E. Boissard, T. Le Gouic. On the mean speed of convergence of empirical and occupation measures in Wasserstein distance. *Ann. Inst. Henri Poincaré Probab. Stat.* **50**, 539–563 (2014).
14. J. Boutet de Monvel and O. Martin. Almost sure convergence of the minimum bipartite matching functional in Euclidean space. *Combinatorica* **22**, 523–530 (2002).
15. S. Caracciolo, M. D'Achille and G. Sicuro. Anomalous scaling of the optimal cost in the one-dimensional random assignment problem. *J. Stat. Phys.* **174**, 846–864 (2019).
16. S. Caracciolo, C. Lucibello, G. Parisi and G. Sicuro. Scaling hypothesis for the Euclidean bipartite matching problem. *Phys. Rev. E* **90**, 012118 (2014).
17. S. Dereich, M. Scheutzow and R. Schottstedt. Constructive quantization: approximation by empirical measures. *Ann. Inst. Henri Poincaré Probab. Stat.* **49**, 1183–1203 (2013).
18. V. Divol. A short proof on the rate of convergence of the empirical measure for the Wasserstein distance. *Preprint* (2021).
19. V. Dobrić and J. Yukich. Asymptotics for transportation cost in high dimensions. *J. Theoret. Probab.* **8**, 97–118 (1995).
20. N. Fournier and A. Guillin. On the rate of convergence in Wasserstein distance of the empirical measure. *Probab. Theory Related Fields* **162**, 707–738 (2015).
21. N. García Trillos and D. Slepčev. On the rate of convergence of empirical measures in ∞-transportation distance. *Canadian J. Math.* **67**, 1358–1383 (2015).
22. M. Goldman and M. Huesmann. A fluctuation result for the displacement in the optimal matching problem. *Ann. Probab.* **50**, 1446–1477 (2022).
23. M. Goldman, M. Huesmann and F. Otto. A large-scale regularity theory for the Monge–Ampère equation with rough data and application to the optimal matching problem. *Preprint* (2018).
24. M. Goldman, M. Huesmann and F. Otto. Quantitative linearization results for the Monge–Ampère equation. *Commun. Pure Appl. Math.* **74**, 2483–2560 (2021).
25. M. Goldman and D. Trevisan. Convergence of asymptotic costs for random Euclidean matching problems. *Probab. Math. Phys.* **2**, 341–362 (2021).
26. N. Holden, Y. Peres and A. Zhai. Gravitational allocation for uniform points on the sphere. *Ann. Probab.* **49**, 287–321 (2021).
27. M. Ledoux. On optimal matching of Gaussian samples. *Zap. Nauchn. Sem. S.-Peterburg. Otdel. Mat. Inst. Steklov. (POMI) 457, Veroyatnost' i Statistika.* **25**, 226–264 (2017).
28. M. Ledoux. On optimal matching of Gaussian samples II. *Preprint* (2018).
29. M. Ledoux. A fluctuation result in dual Sobolev norm for the optimal matching problem. *Preprint* (2019).
30. M. Ledoux and J.-X. Zhu. On optimal matching of Gaussian samples III. *Probab. Math. Statist.* **41**, 237–265 (2021).
31. T. Leighton and P. Shor. Tight bounds for minimax grid matching with applications to the average case analysis of algorithms. *Combinatorica* **9**, 161–187 (1989).
32. L. Lovász and M.D. Plummer. *Matching theory.* American Mathematical Society, Chelsea Publishing (2009).
33. M. Mézard and A. Montanari. *Information, physics, and computation.* Oxford University Press (2009).
34. G. Peyré and M. Cuturi. Computational optimal transport: With applications to data science. *Foundations and Trends in Machine Learning* **11**, 355–607 (2019).

35. P. Shor. How to pack better than Best Fit: Tight bounds for average-case on-line bin packing. In: *Proc. 32nd Annual Symposium on Foundations of Computer Sciences*, pp. 752–759 (1991).
36. P. Shor and J. Yukich. Minimax grid matching and empirical measures. *Ann. Probab.* **19**, 1338–1348 (1991).
37. J.M. Steele. *Probability theory and combinatorial optimization.* SIAM (1997).
38. M. Talagrand. Scaling and non-standard matching theorems. *Comptes Rendus Acad. Sciences Paris, Mathématique* **356**, 692–695 (2018).
39. M. Talagrand. *Upper and lower bounds for stochastic processes. Decomposition theorems.* Second edition. Ergebnisse der Mathematik und ihrer Grenzgebiete **60**, Springer (2021).
40. C. Villani. *Topics in optimal transportation.* Graduate Studies in Mathematics **58**, American Mathematical Society (2003).
41. J. Yukich. *Probability theory of classical Euclidean optimization problems.* Lecture Notes in Mathematics **1675**, Springer (1998).

Space-Time Stochastic Calculus and White Noise

Bernt Øksendal

*Dedicated to Catriona Byrne, in gratitude for her support and
encouragement through 40 years*

1 Introduction

My interest in the interplay between mathematical analysis and probability theory
goes back to the beginning of my studies at the University of Oslo in the mid
1960s. But it really gained momentum in the late 1970s when I started studying
the beautiful little book *Stochastic Integrals* by Henry McKean [13]. My colleague
at the time, Knut Aase, and I ran a little seminar on the book at Agder College (now
the University of Agder) and we were both fascinated by this new calculus that the
book presented, namely the *Itô calculus!*

Then in 1982 Sandy Davie and Alan Sinclair offered me a Research Fellowship
and invited me to spend one semester at the Department of Mathematics, University
of Edinburgh. There they asked me to give a course on stochastic differential
equations (SDEs). I knew nothing about SDEs, but started immediately to study
the subject intensively, in an effort to at least stay ahead of my (very advanced)
audience. It was a rewarding experience, which opened up a new world, consisting
of both interesting new mathematics and a number of important applications, e.g.
to modelling of dynamical systems with noise, filtering theory, optimal stopping,
stochastic control and (subsequently) mathematical finance. I was enthusiastic about
this new field of mathematics and started lecturing about it, almost like preaching a
gospel, at a number of places, including Agder College, Eötvös Loránd University,
the University of Tromsø, and California Institute of Technology (Caltech). Every
time I used the opportunity to polish my lecture notes, based on useful feedback
from the audience.

B. Øksendal (✉)
Department of Mathematics, University of Oslo, Oslo, Norway
e-mail: oksendal@math.uio.no

In 1984 I submitted my lecture notes to Springer, and asked them to consider the manuscript for publication, for example in the Springer Lecture Notes in Mathematics (LNM). At a conference in Lancaster the same year I was fortunate to meet the recently appointed Springer Mathematical Editor, namely *Catriona Byrne*, who was handling my manuscript. She saw the potential of the manuscript and recommended that it be considered for publication in the new book series Universitext. The reviews were all very positive, and in 1985 the first edition of my book *Stochastic Differential Equations* [18] appeared, and it became a great success. I think one reason for the success of the book was that it filled a gap in the literature. There were already several excellent books available, written by top experts, but my book was written from the point of view of an enthusiastic beginner, who was not trying to humiliate the reader but to work with the reader to understand the topic. Some years later a professor once told me that although more recent and more polished editions of my book were available, he gave his students the first edition to start with, because it was somehow more "raw" and direct, without all the technical subtleties (that should be taken seriously later) and therefore easier as a first encounter with the field.

I will always be grateful to Catriona for her encouragement and support, not just for the first edition of this book, but also for later editions [17] and for the other books I wrote later with coauthors. I think she is an important reason for the success of Springer among the mathematical community.

2   Equations with Noise and White Noise Theory

One of the fascinations with stochastic analysis is the interplay between mathematical analysis and probability theory. And perhaps the most spectacular example of such interplay is the topic of dynamical systems subject to *noise*. Noise in some form is everywhere in our society. For example, it can be in the form of mechanical noise from machines, noise due to lack of information in the system, or environmental noise due to random fluctuations in weather. A classic example is the model for population growth:
Let $Y(t)$ denote the density of a given population at time t. Then the most basic model for the growth of $Y(t)$ is the differential equation

$$\frac{\mathrm{d}Y(t)}{\mathrm{d}t} = \alpha_0(t)Y(t); \quad Y(0) = y_0 \text{ (constant)} \tag{1}$$

where $\alpha_0(t)$ is a given function, representing the relative growth rate. The solution of this differential equation is

$$Y(t) = y_0 \exp\left(\int_0^t \alpha_0(s)\mathrm{d}s\right).$$

A weakness of this model is that it does not take into account that there might be unpredictable, random changes, or "noise" in the environment. If we add "noise" in the relative growth rate, the equation gets the form

$$\frac{dY(t)}{dt} = \alpha(t) + \sigma(t)W(t); \quad Y(0) = y_0, \tag{2}$$

where $W(t)$ represents "noise" at time t and σ is a given "noise" coefficient. The question is what properties such a "noise" process $W(t)$ should have. If we try to represent $W(t)$ by a stochastic process defined on a filtered probability space $(\Omega, \mathcal{F}, \{\mathcal{F}_t\}_{t \geq 0}, \mathbb{P})$, then ideally, i.e. if we think of noise as "white" in some sense, we could require that $W(t_1)$ and $W(t_2)$ are independent if $t_1 \neq t_2$ and that $W(t)$ is normalised, in the sense that $\mathbb{E}[W(t)] = 0$ (where $\mathbb{E}$ denotes expectation with respect to $\mathbb{P}$) and $\mathbb{E}[W^2(t)] < \infty$ for all t. But it turns out that no such measurable stochastic process exist. However, since the Brownian motion process $B(t) = B(t, \omega); t \geq 0, \omega \in \Omega$ is continuous and has stationary, independent increments, one could try to put

$$W(t) = \frac{dB(t)}{dt}. \tag{3}$$

This derivative does not exist in the ordinary (strong) sense, but since $t \mapsto B(t, \omega)$ is continuous for a.a. ω, it is weakly differentiable for a.a. ω, and we could try to interpret it weakly, in the sense that we regard the equation (2) as an integral equation, that is

$$Y(t) = y_0 + \int_0^t Y(s)\Big(\alpha(s) + \sigma(s)\frac{dB(s)}{ds}\Big)ds$$

$$= y_0 + \int_0^t Y(s)\alpha(s)ds + \int_0^t \sigma(s)Y(s)dB(s).$$

This last integral can be made rigorous as an *Itô integral*. It is sometimes written in the following short-hand, differential form

$$dY(t) = \alpha(t)Y(t)dt + \sigma(t)Y(t)dB(t); \quad Y(0) = y_0.$$

Applying the Itô rules of stochastic calculus (the Itô formula), we find the following well-known solution

$$Y(t) = y_0 \exp\Big(\int_0^t \sigma(s)dB(s) + \int_0^t \{\alpha(s) - \tfrac{1}{2}\sigma^2(s)\}ds\Big). \tag{4}$$

2.1 A More Elaborate Model

In the model above we are only considering the population at one given point x in space and the noise is in time only. In an attempt to get a more realistic model, let us consider the population density $Y(t, x)$ at the time $t \in \mathbb{R}$ and at the point $x \in \mathbb{R}^n$, where n is the dimension of the space. For simplicity of the notation, let us assume that $n = 1$ in the following. In this case, it is natural to assume that the "noise" depends on both time and space also, i.e. that it is represented by a 2-parameter process $W(t, x)$. This is also relevant in many other situations, for example in temperature modelling or, more generally, weather modelling. Assuming this, and arguing as in (2) and allowing the coefficients to depend on both t and x, we arrive at the following 2-parameter stochastic differential equation for $Y(t, x)$:

$$\frac{\partial^2 Y(t, x)}{\partial t \partial x} = \alpha(t, x) Y(t, x) dt dx + \sigma(t, x) W(t, x) dt dx.$$

Now suppose we proceed as above, and try to represent $W(t, x)$ weakly as

$$W(t, x) = \frac{\partial^2 B(t, x)}{\partial t \partial x}, \tag{5}$$

where $B(t, x); t \geq 0, x \in \mathbb{R}$ is a 2-parameter Brownian motion, also called a *Brownian sheet*. (See the next section for an explanation.) Then we arrive at the following stochastic integral equation

$$Y(t, x) = y_0 + \int_0^t \int_0^x \alpha(s, z) Y(s, z) ds dz + \int_0^t \int_0^x \sigma(s, z) Y(s, z) B(ds, dz),$$

where the last integral is the *space-time (2-parameter) Itô integral with respect to* $B(\cdot, \cdot)$, as constructed in [5, 23, 24]. See also [14, 20] and [22]. As in the 1-parameter case, we will use the following short-hand differential notation

$$\frac{\partial^2 Y(t, x)}{\partial t \partial x} = \alpha(t, x) Y(t, x) dt dx + \sigma(t, x) Y(t, x) B(dt, dx); \quad t, x \geq 0 \tag{6}$$

$$Y(t, 0) = Y(0, x) = y_0; \quad \text{for all } t, x \geq 0.$$

It follows from Theorem 2.4.1 in [14] that this stochastic partial differential equation (SPDE) has a unique adapted solution. The question is:

Problem 2.1 *Can we find a formula for the solution of* (6), *like we did in the 1-parameter case* (4)?

In view of the many applications of white noise theory to 1-parameter stochastic calculus (see e.g. [1–4, 6, 7, 9–11, 19] and also the spectacular white noise solution of general SDEs in [12]), it is natural to ask if such an interplay can be useful also in the multi-parameter case. To support this idea we will show that white noise theory can be used to find the solution of (6). We will explain how after a short review of *multi-parameter white nose calculus*. It is mainly based on the presentation in [9], and we refer to that book for proofs and more details.

3 Multi-Parameter White Noise Calculus

The basic idea of white noise analysis, due to Hida [8], is to consider white noise W rather than Brownian motion B as the fundamental object. Within this framework, we will see that Brownian motion can indeed be expressed as the integral of white noise.

3.1 The d-Parameter White Noise Probability Space

In the following d will denote a fixed positive integer, interpreted as either the time, space or time-space dimension of the system we consider. More generally, we will call d the parameter dimension. Let $\mathcal{S}(\mathbb{R}^d)$ be the Schwartz space of rapidly decreasing smooth (C^∞) real-valued functions on $\mathbb{R}^d$. Recall that $\mathcal{S}(\mathbb{R}^d)$ is a Fréchet space under the family of seminorms

$$\|f\|_{k,\alpha} := \sup_{x \in \mathbb{R}^d} \{(1 + |x|^k)|\partial^\alpha f(x)|\},$$

where k is a nonnegative integer, $\alpha = (\alpha_1, \cdots, \alpha_d)$ is a multi-index of nonnegative integers $\alpha_1, \cdots, \alpha_d$ and

$$\partial^\alpha f = \frac{\partial^{|\alpha|} f}{\partial x_1^{\alpha_1} \cdots \partial x_d^{\alpha_d}} \quad \text{where} \quad |\alpha| := \alpha_1 + \cdots + \alpha_d.$$

The dual $\Omega = \mathcal{S}'(\mathbb{R}^d)$ of $\mathcal{S}(\mathbb{R}^d)$, equipped with the weak-star topology, is the space of *tempered distributions*. This space will be the base of our basic probability space, which we explain in the following:

As events we will use the family $\mathcal{F} = \mathcal{B}(\mathcal{S}'(\mathbb{R}^d))$ of Borel subsets of $\mathcal{S}'(\mathbb{R}^d)$, and our probability measure $\mathbb{P}$ is defined by the following result:

Theorem 3.1 (The Bochner–Minlos Theorem) *There exists a unique probability measure $\mathbb{P}$ on $\mathcal{B}(\mathcal{S}'(\mathbb{R}^d))$ with the following property:*

$$\mathbb{E}[e^{i\langle \cdot, \varphi \rangle}] := \int_{\mathcal{S}'} e^{i\langle \omega, \varphi \rangle} d\mu(\omega) = e^{-\frac{1}{2}\|\varphi\|^2}; \quad i = \sqrt{-1}$$

for all $\varphi \in \mathcal{S}(\mathbb{R}^d)$, *where* $\|\varphi\|^2 = \|\varphi\|_{L^2(\mathbb{R}^d)}^2$, $\langle \omega, \varphi \rangle = \omega(\varphi)$ *is the action of* $\omega \in \mathcal{S}'(\mathbb{R}^d)$ *on* $\varphi \in \mathcal{S}(\mathbb{R}^d)$ *and* $\mathbb{E} = \mathbb{E}_{\mathbb{P}}$ *denotes the expectation with respect to* $\mathbb{P}$.

We will call the triplet $(\mathcal{S}'(\mathbb{R}^d), \mathcal{B}(\mathcal{S}'(\mathbb{R}^d)), \mathbb{P})$ the *white noise probability space*, and $\mathbb{P}$ is called the *white noise probability measure*.

The measure $\mathbb{P}$ is also often called the (normalized) *Gaussian measure* on $\mathcal{S}'(\mathbb{R}^d)$. It is not difficult to prove that if $\varphi \in L^2(\mathbb{R}^d)$ and we choose $\varphi_n \in \mathcal{S}(\mathbb{R}^d)$ such that $\varphi_n \to \varphi$ in $L^2(\mathbb{R}^d)$, then

$$\langle \omega, \varphi \rangle := \lim_{n\to\infty} \langle \omega, \varphi_n \rangle \quad \text{exists in} \quad L^2(\mathbb{P})$$

and is independent of the choice of $\{\varphi_n\}$. In particular, if we define

$$\widetilde{B}(x) := \widetilde{B}(x_1, \cdots, x_d, \omega) = \langle \omega, \chi_{[0,x_1] \times \cdots \times [0,x_d]} \rangle; \quad x = (x_1, \cdots, x_d) \in \mathbb{R}^d,$$

where $[0, x_i]$ is interpreted as $[x_i, 0]$ if $x_i < 0$, then $\widetilde{B}(x, \omega)$ has an x-continuous version $B(x, \omega)$, which becomes a *d-parameter Brownian motion*, in the following sense:

By a *d-parameter Brownian motion* we mean a family $\{B(x, \cdot)\}_{x \in \mathbb{R}^d}$ of random variables on a probability space $(\Omega, \mathcal{F}, \mathbb{P})$ such that

- $B(0, \cdot) = 0$ almost surely with respect to $\mathbb{P}$,
- $\{B(x, \omega)\}$ is a continuous and Gaussian stochastic process, and, further,
- for all $x = (x_1, \cdots, x_d)$, $y = (y_1, \cdots, y_d) \in \mathbb{R}_+^d$, $B(x, \cdot)$, $B(y, \cdot)$ have the covariance $\prod_{i=1}^d x_i \wedge y_i$. For general $x, y \in \mathbb{R}^d$ the covariance is $\prod_{i=1}^d \int_{\mathbb{R}} \theta_{x_i}(s) \theta_{y_i}(s) \mathrm{d}s$, where $\theta_x(t_1, \ldots, t_d) = \theta_{x_1}(t_1) \cdots \theta_{x_d}(t_d)$, with

$$\theta_{x_j}(s) = \begin{cases} 1 & \text{if } 0 < s \le x_j \\ -1 & \text{if } x_j < s \le 0 \\ 0 & \text{otherwise.} \end{cases}$$

It can be proved that the process $\widetilde{B}(x, \omega)$ defined above has a modification $B(x, \omega)$ which satisfies all these properties. This process $B(x, \omega)$ then becomes a *d-parameter Brownian motion*.

We remark that for $d = 1$ we get the classical (1-parameter) Brownian motion $B(t)$ if we restrict ourselves to $t \ge 0$. For $d \ge 2$ we get what is often called the *Brownian sheet*.

With this definition of Brownian motion it is natural to define the d-parameter Wiener–Itô integral of $\varphi \in L^2(\mathbb{R}^d)$ by

$$\int_{\mathbb{R}^d} \varphi(x) \mathrm{d}B(x, \omega) := \langle \omega, \varphi \rangle; \quad \omega \in \mathcal{S}'(\mathbb{R}^d).$$

We see that by using the Bochner–Minlos theorem we have obtained an easy construction of d-parameter Brownian motion that works for any parameter dimension d. Moreover, we get a representation of the space Ω as the Fréchet space $\mathcal{S}'(\mathbb{R}^d)$. This is an advantage in many situations, for example in the construction of the Hida–Malliavin derivative, which can be regarded as a stochastic gradient on Ω. See Sect. 4 and e.g. [7] and the references therein.

3.2 Chaos Expansion in Terms of Hermite Polynomials

The *Hermite polynomials* $h_n(x)$ are defined by

$$h_n(x) = (-1)^n e^{1/2x^2} \frac{d^n}{dx^n} (e^{-1/2x^2}); \quad n = 0, 1, 2, \cdots .$$

We see that the first Hermite polynomials are

$$h_0(x) = 1, \ h_1(x) = x, \ h_2(x) = x^2 - 1, \ h_3(x) = x^3 - 3x,$$

$$h_4(x) = x^4 - 6x^2 + 3, \ h_5(x) = x^5 - 10x^3 + 15x, \cdots .$$

The *Hermite functions* $\xi_n(x)$ are defined by

$$\xi_n(x) = \pi^{-1/4}((n-1)!)^{-1/2} e^{-1/2x^2} h_{n-1}(\sqrt{2}x); \quad n = 1, 2, \cdots .$$

Some important properties of the Hermite functions are the following:

- $\xi_n \in \mathcal{S}(\mathbb{R})$ for all n
- The collection $\{\xi_n\}_{n=1}^{\infty}$ constitutes an orthonormal basis for $L^2(\mathbb{R})$.
- $\sup_{x \in \mathbb{R}} |\xi_n(x)| = O(n^{-1/12})$.

We now use these functions to define an orthogonal basis for $L^2(\mathbb{P})$:

In the following, we let $\delta = (\delta_1, \cdots, \delta_d)$ denote d-dimensional multi-indices with $\delta_1, \cdots, \delta_d \in \mathbb{N}$. It follows that the family of tensor products

$$\xi_\delta := \xi_{(\delta_1, \cdots, \delta_d)} := \xi_{\delta_1} \otimes \cdots \otimes \xi_{\delta_d} \ ; \ \delta \in \mathbb{N}^d$$

forms an orthogonal basis for $L^2(\mathbb{R}^d)$. Let $\delta^{(j)} = (\delta_1^{(j)}, \delta_2^{(j)}, \cdots, \delta_d^{(j)})$ be the jth multi-index number in some fixed ordering of all d-dimensional multi-indices $\delta = (\delta_1, \cdots, \delta_d) \in \mathbb{N}^d$. We may assume that this ordering has the property that

$$i < j \Rightarrow \delta_1^{(i)} + \delta_2^{(i)} + \cdots + \delta_d^{(i)} \le \delta_1^{(j)} + \delta_2^{(j)} + \cdots + \delta_d^{(j)},$$

i.e., that the $\{\delta^{(j)}\}_{j=1}^{\infty}$ occur in increasing order.

Now define

$$\eta_j := \xi_{\delta^{(j)}} = \xi_{\delta_1^{(j)}} \otimes \cdots \otimes \xi_{\delta_d^{(j)}}; \quad j = 1, 2, \cdots.$$

We regard multi-indices as elements of the space $(\mathbb{N}_0^{\mathbb{N}})_c$ of all sequences $\alpha = (\alpha_1, \alpha_2, \cdots)$ with elements $\alpha_i \in \mathbb{N}_0$ and with compact support, i.e., with only finitely many $\alpha_i \neq 0$. Put

$$\mathcal{J} = (\mathbb{N}_0^{\mathbb{N}})_c.$$

Definition 3.2 Let $\alpha = (\alpha_1, \alpha_2, \cdots) \in \mathcal{J}$. Then we define

$$H_\alpha(\omega) := \prod_{i=1}^{\infty} h_{\alpha_i}(\langle \omega, \eta_i \rangle); \quad \omega \in \mathcal{S}'(\mathbb{R}^d).$$

Theorem 3.3 (Wiener–Itô Chaos Expansion Theorem) *Every $f \in L^2(\mathbb{P})$ has a unique representation*

$$f(\omega) = \sum_{\alpha \in \mathcal{J}} c_\alpha H_\alpha(\omega),$$

where $c_\alpha \in \mathbb{R}$ for all α.

Moreover, the following isometry holds:

$$\|f\|_{L^2(\mathbb{P})}^2 = \sum_{\alpha \in \mathcal{J}} \alpha! c_\alpha^2.$$

Example 3.1 The d-parameter Brownian motion $B(x, \omega)$ is defined by:

$$B(x, \omega) = \langle \omega, \psi \rangle,$$

where

$$\psi(y) = \chi_{[0,x_1] \times \cdots \times [0,x_d]}(y).$$

Proceeding as above, we write

$$\psi(y) = \sum_{j=1}^{\infty} (\psi, \eta_j) \eta_j(y) = \sum_{j=1}^{\infty} \left(\int_0^x \eta_j(u) du \right) \eta_j(y),$$

where we have used the multi-index notation

$$\int_0^x \eta_j(u)du = \int_0^{x_d} \cdots \int_0^{x_1} \eta_j(u_1, \cdots, u_d)du_1 \cdots du_d = \prod_{k=1}^d \int_0^{x_k} \xi_{\beta_k^{(j)}}(t_k)dt_k$$

when $x = (x_1, \cdots, x_d)$. Therefore,

$$B(x, \omega) = \langle \omega, \sum_{j=1}^\infty \int_0^x \eta_j(u)du\eta_j \rangle = \sum_{j=1}^\infty \left(\int_0^x \eta_j(u)du \right) \langle \omega, \eta_j \rangle.$$

We conclude that $B(x, \omega) = B(x_1, x_2, ..., x_d, \omega)$ has the expansion

$$B(x, \omega) = \sum_{j=1}^\infty \int_0^x \eta_j(u)du\, H_{\epsilon^{(j)}}(\omega)$$

$$= \sum_{j=1}^\infty \left(\int_0^{x_1} \int_0^{x_2} \cdots \int_0^{x_d} \eta_j(u)du_1 du_2...du_d \right) H_{\epsilon^{(j)}}(\omega). \tag{7}$$

3.3 The Stochastic Test Function Spaces $(\mathcal{S})$ and the Stochastic Distribution Space $(\mathcal{S})^*$

We have seen that the growth condition

$$\sum_\alpha \alpha! c_\alpha^2 < \infty \tag{8}$$

assures that

$$f(\omega) := \sum_\alpha c_\alpha H_\alpha(\omega) \in L^2(\mathbb{P}).$$

By replacing condition (8) by various other conditions we will obtain a family of stochastic test functions and stochastic generalized functions that relates to $L^2(\mathbb{P})$ in a way that is analogous to the spaces $\mathcal{S}(\mathbb{R}^d) \subset L^2(\mathbb{R}^d) \subset \mathcal{S}'(\mathbb{R}^d)$. These spaces provide a favourable setting for the study of stochastic (ordinary and partial) differential equations.

As an analogue, recall the characterizations of $\mathcal{S}(\mathbb{R}^d)$ and $\mathcal{S}'(\mathbb{R}^d)$ in terms of Fourier coefficients: As above we let $\{\delta^{(j)}\}_{j=1}^\infty = \{(\delta_1^{(j)}, \cdots, \delta_d^{(j)})\}_{j=1}^\infty$ be a fixed ordering of all d-dimensional multi-indices $\delta = (\delta_1, \cdots, \delta_d) \in \mathbb{N}^d$. In general, if $\alpha = (\alpha_1, \cdots, \alpha_j, \cdots) \in \mathcal{J}$, $\beta = (\beta_1, \cdots, \beta_j, \cdots) \in (\mathbb{R}^\mathbb{N})_c$ are two finite sequences, we will use the notation

$$\alpha^\beta = \alpha_1^{\beta_1} \alpha_2^{\beta_2} \cdots \alpha_j^{\beta_j} \cdots \quad \text{where} \quad \alpha_j^0 = 1.$$

Theorem 3.4 (Reed and Simon [21], Theorem V. 13–14)

(a) *Let $\varphi \in L^2(\mathbb{R}^d)$, so that*

$$\varphi = \sum_{j=1}^\infty a_j \eta_j, \tag{9}$$

where $a_j = (\varphi, \eta_j)$; $j = 1, 2, \cdots$, are the Fourier coefficients of φ with respect to $\{\eta_j\}_{j=1}^\infty$. Then $\varphi \in \mathcal{S}(\mathbb{R}^d)$ if and only if

$$\sum_{j=1}^\infty a_j^2 (\delta^{(j)})^\gamma < \infty,$$

for all d-dimensional multi-indices $\gamma = (\gamma_1, \cdots, \gamma_d)$.

(b) *The space $\mathcal{S}'(\mathbb{R}^d)$ can be identified with the space of all formal expansions*

$$T = \sum_{j=1}^\infty b_j \eta_j \tag{10}$$

such that

$$\sum_{j=1}^\infty b_j^2 (\delta^{(j)})^{-\theta} < \infty$$

for some d-dimensional multi-index $\theta = (\theta_1, \cdots, \theta_d)$.

If this condition holds, then the action of $T \in \mathcal{S}'(\mathbb{R}^d)$ given by (10) on $\varphi \in \mathcal{S}(\mathbb{R}^d)$ given by (9) is

$$\langle T, \varphi \rangle = \sum_{j=1}^\infty a_j b_j.$$

We now formulate a stochastic analogue of this result. The following quantity is crucial: If $\gamma = (\gamma_1, \cdots, \gamma_j, \cdots) \in (\mathbb{R}^\mathbb{N})_c$ (i.e., only finitely many of the real numbers γ_j are nonzero), we use the short-hand notation

$$(2\mathbb{N})^\gamma := \prod_{j=1}^{\infty} (2j)^{\gamma_j}.$$

Definition 3.5 (The Hida Spaces of Stochastic Test Functions and Stochastic Distributions)

(a) **The stochastic test function space** $(\mathcal{S})$

Let $(\mathcal{S})$ consist of the functions

$$f = \sum_{\alpha \in \mathcal{J}} c_\alpha H_\alpha \in L^2(\mathbb{P}) \quad \text{with } c_\alpha \in \mathbb{R}$$

such that

$$\|f\|_k^2 := \sum_{\alpha \in \mathcal{J}} c_\alpha^2 (\alpha!)^2 (2\mathbb{N})^{k\alpha} < \infty \quad \text{for all } k \in \mathbb{N}$$

equipped with the projective topology.

(b) **The stochastic distribution spaces** $(\mathcal{S})^*$

Let $q \in \mathbb{R}$. We say that the formal sum $F = \sum_{\alpha \in \mathcal{J}} b_\alpha H_\alpha$ belongs to the *Hida distribution Hilbert space* $(\mathcal{S})_{-q}$ if

$$\|F\|_{-q}^2 := \sum_{\alpha \in \mathcal{J}} \alpha! c_\alpha^2 (2\mathbb{N})^{-\alpha q} < \infty. \tag{11}$$

We define the *Hida stochastic distribution space* $(\mathcal{S})^*$ as the union $(\mathcal{S})^* = \bigcup_{q \in \mathbb{R}} (\mathcal{S})_{-q}$ equipped with the inductive topology.

Note that $(\mathcal{S})^*$ can be regarded as the dual of $(\mathcal{S})$ as follows: The action of $F = \sum_\alpha b_\alpha H_\alpha \in (\mathcal{S})^*$ on $f = \sum_\alpha a_\alpha H_\alpha \in (\mathcal{S})$, where $b_\alpha, a_\alpha \in \mathbb{R}$, is given by

$$\langle F, f \rangle = \sum_\alpha \alpha! a_\alpha b_\alpha.$$

We have the inclusions

$$(\mathcal{S}) \subset L^2(\mathbb{P}) \subset (\mathcal{S})^*.$$

Example 3.2 (Singular White Noise) One of the most useful properties of $(\mathcal{S})^*$ is that it contains the *singular* or *pointwise white noise*:

The d-parameter singular white noise process is defined by the formal expansion

$$W(x) = W(x, \omega) = \sum_{k=1}^{\infty} \eta_k(x) H_{\epsilon^{(k)}}(\omega); \quad x \in \mathbb{R}^d. \tag{12}$$

From the definition one van verify that

$$W(x, \omega) \in (\mathcal{S})^*.$$

By comparing the expansion (12) for singular white noise $W(x)$ and the expansion (7) for Brownian motion $B(x)$, we see that

$$W(x) = \frac{\partial^d}{\partial x_1 \cdots \partial x_d} B(x) \quad \text{in } (\mathcal{S})^*. \tag{13}$$

In particular, for $d = 1$, we put $x_1 = t$ and get the familiar identity

$$W(t) = \frac{\mathrm{d}}{\mathrm{d}t} B(t) \text{ in } (\mathcal{S})^*.$$

3.4 The Wick Product

Since $x \mapsto B(x, \omega)$ is continuous a.s., it is weakly differentiable a.s., and we see that the identity (13) also holds in the weak distribution sense (in $\mathcal{S}'(\mathbb{R}^d)$), a.s. For that matter, one could argue that we might as well work on $\mathcal{S}'(\mathbb{R}^d)$ (a.s.) However, there is no tractable product operator on $\mathcal{S}'(\mathbb{R}^d)$. One of the (many) advantages of working on $(\mathcal{S})^*$ is that it has a natural multiplication called the *Wick product*, which is a binary operation on both $(\mathcal{S})$ and $(\mathcal{S})^*$, and is fundamental in the study of stochastic (ordinary and partial) differential equations. In general, one can say that the use of this product corresponds to—and extends naturally—the use of Itô integrals. We now explain this in more detail.

Definition 3.6 The *Wick product* $F \diamond G$ of two elements

$$F = \sum_{\alpha} a_{\alpha} H_{\alpha}, \ G = \sum_{\alpha} b_{\alpha} H_{\alpha} \in (\mathcal{S})^* \quad \text{with} \quad a_{\alpha}, b_{\alpha} \in \mathbb{R}$$

is defined by

$$F \diamond G = \sum_{\alpha, \beta} (a_\alpha, b_\beta) H_{\alpha+\beta}.$$

An important property of the spaces $(S)^*$, (S) is that they are closed under Wick products:

Lemma 3.7

(a) $F, G \in (S)^* \Rightarrow F \diamond G \in (S)^*$;
(b) $f, g \in (S) \Rightarrow f \diamond g \in (S)$.

It is easy to see directly from the definition that the Wick product is commutative, associative and distributive over addition.

The *Wick powers* $F^{\diamond k}$; $k = 0, 1, 2, \cdots$ of $F \in (S)^*$ are defined inductively as follows:

$$F^{\diamond 0} = 1.$$

$$F^{\diamond k} = F \diamond F^{\diamond(k-1)} \text{ for } k = 1, 2, \cdots.$$

The *Wick exponential* of $F \in (S)^*$ is defined by

$$\exp^\diamond F = \sum_{n=0}^\infty \frac{1}{n!} F^{\diamond n}; \text{ if convergent in } (S)^*.$$

3.5   Wick Product and Hermite Polynomials

There is a striking connection between Wick powers and Hermite polynomials h_n; $n = 0, 1, 2, ...$:

Theorem 3.8 *Choose $\varphi \in L^2(\mathbb{R}^d)$ and define the random variable w_φ by*

$$w_\varphi(\omega) = \langle \omega, \varphi \rangle.$$

Then

$$w_\varphi^{\diamond n} = \|\varphi\|^n h_n\left(\frac{w_\varphi}{\|\varphi\|}\right) \tag{14}$$

where $\|\varphi\| = \|\varphi\|_{L^2(\mathbb{R}^d)}$.

This result can be used to get an explicit formula for the Wick exponential:

Theorem 3.9 (The Wick Exponential) *Let* $\varphi \in L^2(\mathbb{R}^d)$. *Then*

$$\exp^\diamond \left(\int_{\mathbb{R}^d} \varphi(x) B(dx) \right) = \exp \left(\int_{\mathbb{R}^d} \varphi(x) B(dx) - \tfrac{1}{2} \int_{\mathbb{R}^d} \varphi^2(x) dx \right).$$

Proof We may assume that $\varphi = c\eta_1$, in which case we get, with $w(\varphi) = \int_{\mathbb{R}^d} \varphi(x) B(dx)$,

$$\exp^\diamond[w(\varphi)] = \sum_{n=0}^\infty \frac{1}{n!} w(\varphi)^{\diamond n} = \sum_{n=0}^\infty \frac{1}{n!} c^n \langle \omega, \eta_1 \rangle^{\diamond n}$$

$$= \sum_{n=0}^\infty \frac{c^n}{n!} H_{\epsilon_1}^{\diamond n}(\omega) = \sum_{n=0}^\infty \frac{c^n}{n!} H_{n\epsilon_1}(\omega)$$

$$= \sum_{n=0}^\infty \frac{c^n}{n!} h_n(\langle \omega, \eta_1 \rangle) = \exp \left[c \langle \omega, \eta_1 \rangle - \tfrac{1}{2} c^2 \right]$$

$$= \exp \left[w(\varphi) - \tfrac{1}{2} \|\varphi\|^2 \right],$$

where we have used the generating property of the Hermite polynomials.   □

3.6 Wick Multiplication and Itô Integration

One of the most striking features of the Wick product is its relation to Itô integration. In short, this relation can be expressed as follows:

Theorem 3.10 *Let* $Y(x)$ *be an Itô integrable process. Then*

$$\int_{\mathbb{R}^d} Y(x) B(dx) = \int_{\mathbb{R}^d} Y(x) \diamond W(x) dx. \tag{15}$$

Here the left-hand side denotes the Itô integral of the stochastic process $Y(x) = Y(x_1, x_2, ..., x_d, \omega)$ with respect to $B(dx) = B(dx_1 dx_2 ... dx_d)$, while the right-hand side is to be interpreted as an $(\mathcal{S})^*$-valued (Pettis) Lebesgue integral of $Y \diamond W$ in $(\mathcal{S})^*$.

This relation explains why the Wick product is so natural and important in stochastic calculus. It is also the key to the fact that Itô calculus (with Itô's formula, etc.) with ordinary multiplication is equivalent to ordinary calculus with Wick multiplication. To illustrate this, consider the example with $d = 1$, $x = t$ and

$Y(t) = B(t) \cdot \chi_{[0,T]}(t)$. Then by the Itô formula the left-hand side of (15) becomes

$$\int\limits_0^T B(t) \mathrm{d}B(t) = \frac{1}{2}B^2(T) - \frac{1}{2}T, \tag{16}$$

while (formal) Wick calculation makes the right-hand side equal to

$$\int\limits_0^T B(t) \diamond W(t)\mathrm{d}t = \int\limits_0^T B(t) \diamond B'(t)\mathrm{d}t = \frac{1}{2}B(T)^{\diamond 2},$$

which is equal to (16) by virtue of (14).

4 The Space-Time Hida–Malliavin Calculus

It is natural to ask if also the Hida–Malliavin derivative can be extended to the space-time case. As in previous sections we assume that the Brownian motion $B(x)$; $x \in \mathbb{R}^d$, is constructed on the space $(\Omega, \mathcal{B}, \mathbb{P})$ with $\Omega = \mathcal{S}'(\mathbb{R}^d)$. Note that any $\gamma \in L^2(\mathbb{R}^d)$ can be regarded as an element of $\Omega = \mathcal{S}'(\mathbb{R}^d)$ by the action

$$\langle \gamma, \varphi \rangle = \int_{\mathbb{R}^d} \gamma(x)\varphi(x)\mathrm{d}x; \quad \varphi \in \mathcal{S}(\mathbb{R}^d).$$

Following the approach in [7] we define the Hida–Malliavin derivative as follows:

Definition 4.1

(i) Let $F \in L^2(\mathbb{P})$ and let $\gamma \in L^2(\mathbb{R}^d)$ be deterministic. Then the *directional derivative of F in $(\mathcal{S})^*$ in the direction γ* is defined by

$$\mathcal{D}_\gamma F(\omega) = \lim_{\varepsilon \to 0} \frac{1}{\varepsilon}[F(\omega + \varepsilon\gamma) - F(\omega)] \tag{17}$$

whenever the limit exists in $(\mathcal{S})^*$.

(ii) Suppose there exists a function $\psi : \mathbb{R}^d \mapsto (\mathcal{S})^*$ such that

$$\int_{\mathbb{R}^d} \psi(x)\gamma(x)\mathrm{d}x \quad \text{exists in } (\mathcal{S})^* \text{ and}$$

$$\mathcal{D}_\gamma F = \int_{\mathbb{R}^d} \psi(x)\gamma(x)\mathrm{d}x, \quad \text{for all } \gamma \in L^2(\mathbb{R}^d). \tag{18}$$

Then we say that F is *Hida–Malliavin differentiable* in $(\mathcal{S})^*$ and we write

$$\psi(x) = \mathrm{D}_x F, \quad x \in \mathbb{R}^d.$$

We call $\mathrm{D}_x F \in (\mathcal{S})^*$ the *Hida–Malliavin derivative of F at x*.

Example 4.2 Suppose $F(\omega) = \langle \omega, f \rangle = \int_{\mathbb{R}^d} f(x) B(\mathrm{d}x)$, $f \in L^2(\mathbb{R}^d)$. Then

$$\mathcal{D}_\gamma F = \frac{1}{\varepsilon} \big[\langle \omega + \varepsilon\gamma, f \rangle - \langle \omega, f \rangle \big] = \langle \gamma, f \rangle = \int_{\mathbb{R}^d} f(x)\gamma(x)\mathrm{d}x.$$

Therefore F is Hida–Malliavin differentiable and

$$\mathrm{D}_x \left(\int_{\mathbb{R}^d} f(u) B(\mathrm{d}u) \right) = f(x) \text{ for a.a. } x \in \mathbb{R}^d.$$

As in the 1-parameter case one can prove the following:

Lemma 4.3 (Chain Rule) *Let $F \in L^2(\mathbb{P})$ be Hida–Malliavin differentiable, with $\mathrm{D}_x F \in L^2(\lambda \times \mathbb{P})$, where λ denotes Lebesgue measure. Suppose that $\varphi \in C^1(\mathbb{R})$ and $\varphi'(F)\mathrm{D}_x F \in L^2(\lambda \times \mathbb{P})$. Then $\varphi(F)$ is also Hida–Malliavin differentiable and*

$$\mathrm{D}_x\big(\varphi(F)\big) = \varphi'(F)\mathrm{D}_x F \quad \text{for a.a. } x \in \mathbb{R}^d. \tag{19}$$

4.1 The General Hida–Malliavin Derivative

The Hida–Malliavin derivative can be expressed in terms of the Wiener–Itô chaos expansion as follows: Recall that

$$H_\alpha(\omega) := \prod_{i=1}^{\infty} h_{\alpha_i}(\theta_i); \quad \text{where } \theta_i = \langle \omega, \eta_i \rangle$$

Then by the chain rule (19) we have,

$$\mathrm{D}_x H_\alpha = \sum_{k=1}^{m} \prod_{j \neq k} h_{\alpha_j}(\theta_j) \alpha_k h_{\alpha_k - 1}(\theta_k)\eta_k(x) = \sum_{k=1}^{m} \alpha_k \eta_k(x) H_{\alpha - \epsilon^{(k)}}. \tag{20}$$

In view of this, the following definition is natural:

Definition 4.4 (The General Hida–Malliavin Derivative) If $F = \sum_{\alpha \in \mathcal{J}} c_\alpha H_\alpha \in (\mathcal{S})^*$ we define the *Hida–Malliavin derivative* $\mathrm{D}_x F$ of F at x in $(\mathcal{S})^*$ by the following expansion:

$$D_x F = \sum_{\alpha \in \mathcal{J}} \sum_{k=1}^{\infty} c_\alpha \alpha_k \eta_k(x) H_{\alpha - \epsilon(k)}, \tag{21}$$

whenever this sum converges in $(\mathcal{S})^*$.

This extension of the Hida–Malliavin derivative makes it possible to deal with more general cases. In short, taking the Hida–Malliavin derivative $D_x F$ may take you from $L^2(\mathbb{P})$ into $(\mathcal{S})^*$, but conditioning with respect to $\mathcal{F}_x$ brings you back to $L^2(\lambda \times \mathbb{P})$. For example, in the case $d = 1$ the following extension of the Clark–Ocone representation theorem [17] was proved in [1]:

Theorem 4.5 (Generalised Clark–Ocone Theorem (d=1)) *Let* $F \in L^2(\mathcal{F}_T, \mathbb{P})$. *Then* $D_t F \in (\mathcal{S})^*$ *for all* t, $\mathbb{E}[D_t F | \mathcal{F}_t] \in L^2(\lambda \times \mathbb{P})$ *and*

$$F = \mathbb{E}[F] + \int_0^T \mathbb{E}[D_t F | \mathcal{F}_t] dB(t).$$

It is natural to ask if a similar result can be obtained in the general parameter case:

Problem 4.6 Is there a d-parameter version of Theorem 4.5?

5 Solving the Population Growth Equation

We now have the machinery from white noise theory we need to solve the space-time stochastic partial differential equation (6):

Theorem 5.1 *Suppose* $\alpha(t, x)$ *and* $\sigma(t, x)$ *are deterministic and in* $L^2(\mathbb{R}^2)$. *Then the solution* $Y(t, x)$ *of the equation*

$$\frac{\partial^2 Y(t, x)}{\partial t \partial x} = \alpha(t, x) Y(t, x) dt dx + \sigma(t, x) Y(t, x) B(dt, dx); \quad t, x \geq 0,$$

$$Y(t, 0) = Y(0, x) = y_0 \text{ (constant) for all } t, x \geq 0,$$

can be written as

$$Y(t, x) =$$

$$y_0 \sum_{n=0}^{\infty} \frac{1}{n!} \sum_{k=0}^{n} \frac{\|\sigma\|^k}{k!(n-k)!} \left[\left(\int_0^t \int_0^x \alpha(s, z) ds dz \right)^{n-k} h_k \left(\int_0^t \int_0^x \frac{\sigma(s,z)}{\|\sigma\|} B(ds, dz) \right) \right].$$

Proof First note that the equation can be written

$$Y(t, x) = y_0 + \int_0^t \int_0^x K(s, z) \diamond Y(s, z) ds dz, \tag{22}$$

where

$$K(s, z) = \alpha(s, z) + \sigma(s, z) W(t, x), \text{ with } W(t, x) = \frac{\partial^2}{\partial t \partial x} B(t, x). \tag{23}$$

Substituting for $Y(s, z)$ in (22) we get

$$Y(u) = y_0 + \int_0^u K(u_1) \diamond \left(y_0 + \int_0^{u_1} K(u_2) \diamond Y(u_2) du_2 \right) du_1,$$

where we put $u = (t, x), u_j = (s_j, z_j); j = 1, 2, \ldots.$
Repeating this we obtain

$$Y(u) = y_0 \left(1 + \int_0^u K(u_1) du_1 + \int_0^u \left(\int_0^{u_1} K(u_1) \diamond K(u_2) du_2 \right) du_1 \right.$$
$$\left. + \cdots + \int_0^u \left(\int_0^{u_1} \cdots \int_0^{u_n} K(u_1) \diamond K(u_2) \diamond \cdots \diamond K(u_{n+1}) du_1 du_2 \ldots du_n du_{n+1} \right) + R_n \right.$$
$$\tag{24}$$

where

$$R_n =$$

$$y_0 \int_0^u \left(\int_0^{u_1} \cdots \int_0^{u_n} K(u_1) \diamond K(u_2) \diamond \cdots \diamond K(u_{n+1}) \diamond Y(u_{n+1}) du_1 du_2 \ldots du_n \right) du_{n+1}$$

It follows from the Våge inequality [9], Theorem 3.3.1, that $R_n \to 0$ in $(\mathcal{S})^*$ as $n \to \infty$. Therefore we get, by letting $n \to \infty$ in (24),

$$Y(t, x) = y_0 \sum_{n=0}^{\infty} \int_0^u \left(\int_0^{u_1} \cdots \int_0^{u_n} K(u_1) \diamond K(u_2) \diamond \cdots \diamond K(u_{n+1}) du_1 du_2 \ldots du_n \right) du_{n+1}$$

$$= y_0 \sum_{n=0}^{\infty} \frac{1}{n! n!} \int_{[0, u]^n} K(u_1) \diamond K(u_2) \diamond \cdots \diamond K(u_n)(v) du_1 \ldots du_n$$

$$= y_0 \sum_{n=0}^{\infty} \frac{1}{n! n!} \left(\int_{[0, u]} K(v) dv \right)^{\diamond n}$$

$$= y_0 \sum_{n=0}^{\infty} \frac{1}{n! n!} \sum_{k=0}^{n} \frac{n!}{k!(n-k)!} \left(\int_0^t \int_0^x \alpha(s, z) ds dz \right)^{\diamond (n-k)} \left(\int_0^t \int_0^x \sigma(s, x) B(ds, dz) \right)^{\diamond k}$$

$$= y_0 \sum_{n=0}^{\infty} \frac{1}{n!n!} \sum_{k=0}^{n} \frac{n!}{k!(n-k)!} \left(\int_0^t \int_0^x \alpha(s,z)\mathrm{d}s\mathrm{d}z \right)^{n-k} \|\sigma\|^k h_k \left(\int_0^t \int_0^x \frac{\sigma(s,z)}{\|\sigma\|} B(\mathrm{d}s,\mathrm{d}z) \right)$$

$$= y_0 \sum_{n=0}^{\infty} \frac{1}{n!} \sum_{k=0}^{n} \frac{\|\sigma\|^k}{k!(n-k)!} \left(\int_0^t \int_0^x \alpha(s,z)\mathrm{d}s\mathrm{d}z \right)^{n-k} h_k \left(\int_0^t \int_0^x \frac{\sigma(s,z)}{\|\sigma\|} B(\mathrm{d}s,\mathrm{d}z) \right).$$

$\square$

Remark 5.1 A direct consequence of this result is that

$$\mathbb{E}[Y(t,x)] = y_0 \sum_{n=0}^{\infty} \frac{\|\sigma\|^n}{n!n!} \left(\int_0^t \int_0^x \alpha(s,z)\mathrm{d}s\mathrm{d}z \right)^n,$$

and if we assume that $\alpha = 0$, we get

$$Y(t,x) = y_0 \sum_{n=0}^{\infty} \frac{\|\sigma\|^n}{n!n!} h_n \left(\int_0^t \int_0^x \frac{\sigma(s,z)}{\|\sigma\|} B(\mathrm{d}s,\mathrm{d}z) \right).$$

Remark 5.2 It follows from the proof that this result holds for any random $y_0 \in L^2(\mathbb{P})$, if we replace the product by a Wick product and interpret the equation in the Wick–Itô–Skorohod sense, that is

$$\frac{\partial^2 Y(t,x)}{\partial t \partial x} = \alpha(t,x)Y(t,x)\mathrm{d}t\mathrm{d}x + \sigma(t,x)Y(t,x) \diamond W(t,x)\mathrm{d}t\mathrm{d}x; \quad t,x \geq 0,$$

$$Y(t,0) = Y(0,x) = y_0 \in L^2(\mathbb{P}) \text{ for all } t,x \geq 0.$$

Then the solution is

$$Y(t,x) =$$

$$y_0 \diamond \sum_{n=0}^{\infty} \frac{1}{n!} \sum_{k=0}^{n} \frac{\|\sigma\|^k}{k!(n-k)!} \left(\int_0^t \int_0^x \alpha(s,z)\mathrm{d}s\mathrm{d}z \right)^{n-k} h_k \left(\int_0^t \int_0^x \frac{\sigma(s,z)}{\|\sigma\|} B(\mathrm{d}s,\mathrm{d}z) \right),$$

which has expectation

$$\mathbb{E}[Y(t,x)] = \mathbb{E}[y_0] \sum_{n=0}^{\infty} \frac{\|\sigma\|^n}{n!n!} \left(\int_0^t \int_0^x \alpha(s,z)\mathrm{d}s\mathrm{d}z \right)^n.$$

Note that no adaptedness conditions are needed.

6 Concluding Remarks

The multi-parameter stochastic calculus is a mostly unexplored area of research. It is clearly crucial in the study of stochastic partial differential equations driven by space-time Brownian motion. See e.g. [14–16, 22] and also [9], which includes SPDEs driven by space-time Poisson random measure white noise as well. But it is also important for many other applications. In this informal note I have tried to illustrate that white noise calculus can be a powerful tool in the study of multi-parameter stochastic calculus in general.

Acknowledgements I am grateful to Nacira Agram and Yaozhong Hu for valuable comments.

References

1. K. Aase, B. Øksendal, N. Privault and J. Ubøe. White noise generalizations of the Clark–Haussmann–Ocone theorem with application to mathematical finance. *Finance Stoch.* **4**(4), 465–496 (2000).
2. K. Aase, B. Øksendal and J. Ubøe. Using the Donsker delta function to compute hedging strategies. *Potential Analysis* **14**, 351–374 (2001).
3. N. Agram and B. Øksendal. A Hida–Malliavin white noise calculus approach to optimal control. *Inf. Dim. Anal. Quant. Probab. Rel. Top.* **21**(3), 1850014 (2018).
4. N. Agram and B. Øksendal. A financial market with singular drift and no arbitrage. *Math. and Finan. Econ.* **15**, 477–500 (2020). https://doi.org/10.1007/s11579-020-00284-9
5. R. Cairoli. Sur une équation differentielle stochastique. *C. R. Math. Acad. Sc. Paris* **274** Ser. A, 1739–1742 (1972).
6. O. Draouil and B. Øksendal. A Donsker delta functional approach to optimal insider control and applications to finance. *Comm. Math. Stat. (CIMS)* **3**, 365–421 (2015). Erratum: *Comm. Math. Stat. (CIMS)* **3**, 535–540 (2015).
7. G. Di Nunno, B. Øksendal and F. Proske. *Malliavin Calculus for Lévy Processes with Applications to Finance.* Second Edition. Springer (2009).
8. T. Hida. *Brownian Motion.* Springer (1980).
9. H. Holden, B. Øksendal, J. Ubøe and T. Zhang. *Stochastic Partial Differential Equations.* Second Edition. Springer (2010).
10. Y. Hu and B. Øksendal. Wick approximation of quasilinear stochastic differential equations. In: *Stochastic Analysis and Related Topics*, Vol 5, Körezlioglu et al (editors), pp. 203–231, Birkhäuser (1996).
11. T. Hida and L. Streit. *Let Us Use White Noise.* World Scientific (2017)
12. A. Lanconelli and F. Proske. On explicit strong solutions of SDE's and the Donsker delta function of a diffusion. *Inf. Dim. Anal. Quant. Probab. Rel. Top.* **7**, 437–447 (2004).
13. H.P. McKean. *Stochastic Integrals.* Academic Press (1969).
14. D. Nualart. *The Malliavin Calculus and Related Topics.* Second Edition. Springer (2006).
15. D. Nualart and B. Rozovskii. Weighted Sobolev spaces and bilinear SPDEs driven by space-time white noise. *J. Funct. Anal.* **149**, 200–225 (1997).
16. D. Nualart and M. Zakai. Generalised Brownian functionals and the solution to a stochastic partial differential equation. *J. Funct. Anal.* **84**, 279–296 (1989).
17. D. Ocone. Malliavin calculus and stochastic integral representations of functionals of diffusion processes. *Stochastics* **12**(3-4), 161–185 (1984).
18. B. Øksendal. *Stochastic Differential Equations.* Sixth edition. Springer (2013).

19. B. Øksendal and E.E. Røse. A white noise approach to insider trading. In: *Let Us Use White Noise*, Hida & Streit (editors), pp. 191–203, World Scientific (2017).
20. E. Pardoux. *Partial Differential Equations*. Lecture notes, Fudan University, Shanghai (2007).
21. M. Reed and B. Simon. *Methods of Modern Mathematical Physics I*. Academic Press (1980).
22. J.B. Walsh. *An introduction to stochastic partial differential equations*. Lecture Notes in Mathematics **1180**, Springer (1986).
23. E. Wong. The dynamics of random fields. *Proc. U.S.-Japan Joint Seminar on Stochastic Methods in Dynamical Problems*, Kyoto (1971).
24. G.J. Zimmermann. Some sample function properties of the two-parameter Gaussian process. *Ann. Math. Statist.* **43**, 1235–1246 (1972).

LECTURE NOTES IN MATHEMATICS

Editors in Chief: J.-M. Morel, B. Teissier;

Editorial Policy

1. Lecture Notes aim to report new developments in all areas of mathematics and their applications – quickly, informally and at a high level. Mathematical texts analysing new developments in modelling and numerical simulation are welcome.

   Manuscripts should be reasonably self-contained and rounded off. Thus they may, and often will, present not only results of the author but also related work by other people. They may be based on specialised lecture courses. Furthermore, the manuscripts should provide sufficient motivation, examples and applications. This clearly distinguishes Lecture Notes from journal articles or technical reports which normally are very concise. Articles intended for a journal but too long to be accepted by most journals, usually do not have this "lecture notes" character. For similar reasons it is unusual for doctoral theses to be accepted for the Lecture Notes series, though habilitation theses may be appropriate.

2. Besides monographs, multi-author manuscripts resulting from SUMMER SCHOOLS or similar INTENSIVE COURSES are welcome, provided their objective was held to present an active mathematical topic to an audience at the beginning or intermediate graduate level (a list of participants should be provided).

   The resulting manuscript should not be just a collection of course notes, but should require advance planning and coordination among the main lecturers. The subject matter should dictate the structure of the book. This structure should be motivated and explained in a scientific introduction, and the notation, references, index and formulation of results should be, if possible, unified by the editors. Each contribution should have an abstract and an introduction referring to the other contributions. In other words, more preparatory work must go into a multi-authored volume than simply assembling a disparate collection of papers, communicated at the event.

3. Manuscripts should be submitted either online at www.editorialmanager.com/lnm to Springer's mathematics editorial in Heidelberg, or electronically to one of the series editors. Authors should be aware that incomplete or insufficiently close-to-final manuscripts almost always result in longer refereeing times and nevertheless unclear referees' recommendations, making further refereeing of a final draft necessary. The strict minimum amount of material that will be considered should include a detailed outline describing the planned contents of each chapter, a bibliography and several sample chapters. Parallel submission of a manuscript to another publisher while under consideration for LNM is not acceptable and can lead to rejection.

4. In general, **monographs** will be sent out to at least 2 external referees for evaluation.

   A final decision to publish can be made only on the basis of the complete manuscript, however a refereeing process leading to a preliminary decision can be based on a pre-final or incomplete manuscript.

   Volume Editors of **multi-author works** are expected to arrange for the refereeing, to the usual scientific standards, of the individual contributions. If the resulting reports can be forwarded to the LNM Editorial Board, this is very helpful. If no reports are forwarded or if other questions remain unclear in respect of homogeneity etc, the series editors may wish to consult external referees for an overall evaluation of the volume.

5. Manuscripts should in general be submitted in English. Final manuscripts should contain at least 100 pages of mathematical text and should always include

   - a table of contents;
   - an informative introduction, with adequate motivation and perhaps some historical remarks: it should be accessible to a reader not intimately familiar with the topic treated;
   - a subject index: as a rule this is genuinely helpful for the reader.
   - For evaluation purposes, manuscripts should be submitted as pdf files.

6. Careful preparation of the manuscripts will help keep production time short besides ensuring satisfactory appearance of the finished book in print and online. After acceptance of the manuscript authors will be asked to prepare the final LaTeX source files (see LaTeX templates online: https://www.springer.com/gb/authors-editors/book-authors-editors/manuscriptpreparation/5636) plus the corresponding pdf- or zipped ps-file. The LaTeX source files are essential for producing the full-text online version of the book, see http://link.springer.com/bookseries/304 for the existing online volumes of LNM). The technical production of a Lecture Notes volume takes approximately 12 weeks. Additional instructions, if necessary, are available on request from lnm@springer.com.

7. Authors receive a total of 30 free copies of their volume and free access to their book on SpringerLink, but no royalties. They are entitled to a discount of 33.3 % on the price of Springer books purchased for their personal use, if ordering directly from Springer.

8. Commitment to publish is made by a *Publishing Agreement*; contributing authors of multiauthor books are requested to sign a *Consent to Publish form*. Springer-Verlag registers the copyright for each volume. Authors are free to reuse material contained in their LNM volumes in later publications: a brief written (or e-mail) request for formal permission is sufficient.

Addresses:
Professor Jean-Michel Morel, Ecole Normale Supérieure Paris-Saclay, Paris, France
E-mail: moreljeanmichel@gmail.com

Professor Bernard Teissier, Equipe Géométrie et Dynamique,
Institut de Mathématiques de Jussieu – Paris Rive Gauche, Paris, France
E-mail: bernard.teissier@imj-prg.fr

Springer: Ute McCrory, Mathematics, Heidelberg, Germany,
E-mail: lnm@springer.com

Printed in the United States
by Baker & Taylor Publisher Services